The Applied Dynamics
of Ocean Surface Waves

THE APPLIED DYNAMICS OF OCEAN SURFACE WAVES

CHIANG C. MEI
Department of Civil Engineering
Massachusetts Institute of Technology

A Wiley-Interscience Publication
JOHN WILEY & SONS
New York   Chichester   Brisbane   Toronto   Singapore

PHYSICS

66666-8864

Library of Congress Cataloging in Publication Data:
Mei, Chiang C.
    The applied dynamics of ocean surface waves.

    "A Wiley-Interscience publication."
    Bibliography: p.
    Includes index.
    1. Ocean waves.   I. Title.

GC211.2.M44   1982   551.47'022   82-8639
ISBN 0-471-06407-6   AACR2

Printed in the United States of America

10 9 8 7 6 5 4 3 2 1

CORRECTIONS TO THE TEXT

The Applied Dynamics of Ocean Surface Waves by Chiang C. Mei

page 12, second of Eq. (4.12): *For* $(gk)^{1/2}$ *read* $(g/k)^{1/2}$

Page 36, line 4 after Eq. (2.2): *For* $x = y - 0$ *read* $x = y = 0$

Page 41, Eq. (2.32): *For* $\int_0^\infty d\psi \cdots$ *read* $\int_0^\pi d\psi \cdots$

Page 66, Eq. (3.1): *For dy' read ky'*

Page 374, Eqs. (11.14) and (11.15): *For* $\iiint\limits_{S_B}$ *read* $\iint\limits_{S_B}$

Page 678, Eq. (2.17): *For* $\mathbf{T}_{\bar{\alpha}}\mathbf{n}, \mathbf{T}_{\alpha}\mathbf{n}$ *read* $\mathbf{T}_{\bar{\alpha}} \cdot \mathbf{n}, \mathbf{T}_{\alpha} \cdot \mathbf{n}$

Page 680, Eq. (2.27): *For* $(\mathbf{T})_{ij} = T_{ij}$ *read* $(\mathbf{T})_{ij} = \tau_{ij}$

Equation (2.28): *For* $\dfrac{\partial V_i}{\partial x_i}$ *read* $\dfrac{\partial V_j}{\partial x_i}$

Page 680, three lines above Eq. (2.30): Replace everything from "as follows.† The pore fluid..." to the end of the footnote, by: as follows (Raats, 1968). Consider a mixture volume $\mathcal{V}$ surrounded by $\mathcal{S}$. Let $\mathcal{S}_{ws}$ be the union of all contact surfaces between the two phases throughout $\mathcal{V}$ and $\mathcal{S}_{ww}$ be the part of $\mathcal{S}$ occupied by the pore fluid. The total pressure force on the solid by the fluid can be expressed in two ways

$$\iint\limits_{\mathcal{S}_{ws}} p\mathbf{n}\, d\mathcal{S} = \iint\limits_{\mathcal{S}_w} p\mathbf{n}\, d\mathcal{S} - \iint\limits_{\mathcal{S}_{ww}} p\mathbf{n}\, d\mathcal{S} \qquad (2.30)$$

where $\mathcal{S}_w = \mathcal{S}_{ws} \cup \mathcal{S}_{ww}$ is closed, and $\mathbf{n}$ points out of the fluid. Using the fact that $(d\mathcal{S})_{ww} = n_w\, d\mathcal{S}$ and Gauss' theorem, we transform the surface integrals to

Page 681: Replace the first three lines by

$$\iint\limits_{\mathcal{S}_w} p\mathbf{n}\, d\mathcal{S} - \iint\limits_{\mathcal{S}} pn_w\mathbf{n}\, d\mathcal{S} = \iiint\limits_{\mathcal{V}_w} \nabla p\, d\mathcal{V} - \iiint\limits_{\mathcal{V}} \nabla(pn_w)\, d\mathcal{V}$$

Since $(d\mathcal{V})_w = n_w\, d\mathcal{V}$ the two volume integrals may be combined to give the total pressure force on the solid phase by the fluid phase

Page 681, Eq. (2.31): *For* $-\dfrac{n_w^2}{K}(\mathbf{v}_w - \mathbf{v}_s)$ *read* $\dfrac{n_w^2}{K}(\mathbf{v}_w - \mathbf{v}_s)$

Page 681, line below Eq. (2.33): *For* $\mathbf{v}_s = \nabla n_w = 0$ *read* $\mathbf{v}_s = 0$

Page 683, Eq. (2.44): *For* $(\mathbf{v}_s' - \mathbf{v}_w')$ *read* $(\mathbf{v}_w' - \mathbf{v}_s')$

Page 683, Equation above Eq. (2.45): *For P_0 read p^0*

Page 684, Eq. (2.47): *For $-\nabla p_t'$ read* $-\nabla \dfrac{\partial p'}{\partial t}$

Page 687, Eq. (3.14): *For $\nabla \cdot \mathbf{v}$ read $\nabla \cdot \tilde{\mathbf{v}}$*

Page 688, Eq. (3.19): *For* $\dfrac{\partial}{\partial t}(\tilde{\sigma}_{ij} - \tilde{p}\delta_{ij})$ *read* $\dfrac{\partial}{\partial \tau}(\tilde{\sigma}_{ij} - \tilde{p}\delta_{ij})$

Page 688, third line from bottom: *For ∂y^2 read ∂z^2*

Page 692, sixth line from the bottom: *For $\tilde{p}$ read $\hat{p}$*

Addendum to References: Raats, P. A. C. (1968) Forces acting upon the solid phase of a porous medium, *J. Appl. Math. Phys.* **19**: 606–613.

To My Parents
and
My Wife

Preface

A substantial growth of knowledge in the dynamics of ocean surface waves has been witnessed over the past 20 years. While many advances have been stimulated by purely scientific inquiry in geophysics, the pace of progress has also been quickened by the increase in large engineering projects both offshore and along the coast. A major construction project now demands not only careful estimates of wave conditions near the site but also reliable predictions of the effects on and of the construction itself. With a view to bringing together scientific and engineering aspects of ocean waves, educational and research programs have naturally been established in a number of universities and industries.

This book is the outgrowth of my lecture notes for a two-semester course taught at M.I.T. since 1974 to graduate students in civil and ocean engineering, with occasional participants from physical oceanography. The aim of the book is to present selected theoretical topics on ocean-wave dynamics, including basic principles and applications in coastal and offshore engineering, all from the deterministic point of view. The bulk of the material deals with the linearized theory which has been well developed in the research literature. The inviscid linearized theory is covered in Chapters One to Five and again in Seven. Frictional effects caused directly or indirectly by viscosity are treated in Chapters Six, Eight, and Nine. A special effect of breaking waves on beaches is examined in Chapter Ten. Chapters Nine and Ten focus on the secondary effects of nonlinearity. The cases where nonlinearity is of primary importance are the subjects of Chapters Eleven and Twelve, for shallow and deep waters, respectively. The last chapter (Thirteen) is on wave-induced stresses in a porous but deformable seabed, which is a problem vital to offshore engineering. In the construction of a gravity platform, the cost of the foundation alone can be as high as 40% of the total. Under the influence of waves, the strength of a porous seabed is affected to varying degrees by fluid in the pores. Hence hydrodynamics is an essential part of the problem. In this chapter a well-known fluid-dynamic reasoning is applied to a soil model which includes fluid and solid phases. I hope the material will stimulate further interaction among researchers in different disciplines.

Most parts of this book have been used either for my own lectures or for self-paced reading by the students. Since contributions by mathematical scientists have always been prominent in this field, the use of certain analytical techniques which may be less familiar to many potential readers cannot be avoided. Therefore, considerable space is devoted to the informal explanation and demonstration of those techniques not customarily discussed in a course on advanced calculus. The derivations of most of the results are given in detail in order to reduce possible frustrations to those who are still acquiring the requisite skills. A few exercises are included; nearly all of them demand some effort. For additional exercises, I have usually suggested term papers based on the student's own survey of literature.

Studies on waves in general, and on water waves in particular, have always been enriched by cross fertilization among diverse fields of science and engineering, including physics, mathematics, oceanography, electrical engineering, and others. A conscientious effort has been made in this book to reflect this fact which I hope will induce more engineers and scientists to join their talents for further challenges of the sea.

Several important areas which are either beyond my own experience or have been treated in other books are not included here. The mechanisms of wave generation by wind and many aspects of resonant interactions have been admirably surveyed by Phillips (1977) and by LeBlond and Mysak (1978). On the statistical *description* of random sea waves, a detailed discussion of the basic aspects may be found in Price and Bishop (1974). For the statistical *mechanics* of sea waves one should consult Phillips (1977) and West (1981). The rapid advance on steep waves, spearheaded by M. S. Longuet-Higgins, is of obvious interest to engineers and oceanographers alike; the numerous papers by him and his associates on the subject cannot be matched for clarity and thoroughness. Waves due to advancing bodies belong to the realm of ship hydrodynamics; the definitive treatises by Stoker (1957), Wehausen and Laitone (1960), and Newman (1977), and all the past proceedings of the Naval Hydrodynamics Symposium should be consulted. Wave-induced separation around small bodies is at the heart of force prediction for offshore structures; it is a subject where experiments play the leading role and has been expertly covered in a recent book by Sarpkaya and Issacson (1981). Storm surges are also omitted.

In a book containing many mathematical expressions, freedom from error can be strived for but is hard to achieve. I shall be grateful to readers who wish to inform me of any oversights that remain.

CHIANG C. MEI

Cambridge, Massachusetts
July 1982

Acknowledgments

All students who participated in my lectures have contributed in bringing this book to its present form. During the final stages of preparation, invaluable criticism and suggestions have been received from Professors P. L.-F. Liu, Cornell University, O. S. Madsen, M.I.T., J. N. Newman, M.I.T., D. D. Serman, Universidad Nacional del Sur, Argentina, Dr. M. A. Foda, Kuwait Institute for Scientific Research, D. F. McTigue, Sandia National Laboratories, D. K.-P. Yue, Science Applications Inc., S. C. Dai and X. C. Zhou, both of Academia Sinica, China, and H. C. Graber, M.I.T. To each of them my heart-felt thanks are due. I have benefited from 15 years of close collaboration with Professor Newman; his personal influence and contributions should be particularly apparent from the contents of Chapter Seven. Chapter Thirteen is based largely on joint works with Dr. M. A. Foda. In all other chapters, materials have been freely drawn from my joint learning experience with many past and present collaborators.

I gratefully acknowledge the kindness of the following colleagues for sending me photographs of certain figures: Dr. J. R. Houston of Waterways Experiment Station, Vicksburg (Figs. 11.3–11.5, Chapter Four, and Fig. 10.5, Chapter Five), Dr. J. Lau, formerly of Florida State University (Plate 1, Chapter Nine), and Dr. H. C. Yuen of TRW, Inc. (Figs. 6.4 and 6.5, Chapter Twelve).

I also take this opportunity to express my gratitude and admiration for Professor T. Y. T. Wu of Cal Tech whose approach to engineering science I try to emulate in this book. By his open-mindedness and exemplary dedication to both basics and practice, the late Professor A. T. Ippen of the Civil Engineering Department at M.I.T. inspired all those around him to contribute to engineering, in whatever style. In addition, my colleagues at the same department, Professors D. R. F. Harleman, P. S. Eagleson, and J. F. Kennedy, who is now at the University of Iowa, have rendered timely help which either exposed me to new challenges or enabled me to do what I enjoyed.

I am also indebted to two patrons of research, the Office of Naval Research (Fluid Dynamics Program) and the National Science Foundation (Civil and

Mechanical Engineering, and Earthquake Engineering Programs) whose sustained support has contributed greatly to my learning of what is reported in this book.

The enormous amount of typing and retyping could not have been accomplished without the professionalism of Mrs. Beth Quivey and Mrs. Laureen Luszcz. All the drawings are by Ms. Donna Hall. Aside from contributing countless hours of typing, editorial counseling, and proofreading, my wife, Caroline, has always been a fountain of support, encouragement, and constructive criticism. To her, *vielen dank*!

C. C. M.

Contents

ONE. INTRODUCTION   1

  1.1   Review of Basic Formulation for an Incompressible Fluid of
        Constant Density 2
        1.1.1   Governing Equations 2
        1.1.2   Boundary Conditions for an Inviscid Irrotational
                 Flow 4
  1.2   Linearized Approximation for Small-Amplitude Waves 6
  1.3   Elementary Notions of a Propagating Wave 9
  1.4   Progressive Water Waves on Constant Depth 11
  1.5   Group Velocity 14
        1.5.1   A Kinematic View 14
        1.5.2   A Dynamic View: Energy Flux 16

**TWO. PROPAGATION OF TRANSIENT WAVES IN OPEN WATER
OF ESSENTIALLY CONSTANT DEPTH**   20

  2.1   Two-Dimensional Transient Problems 20
        2.1.1   Transient Disturbance Due to an Initial Displacement
                 on the Free Surface 23
        2.1.2   Energy Propagation, Group Velocity 28
        2.1.3   Leading Waves Due to a Transient Disturbance 30
        2.1.4   Tsunami Due to Tilting of the Bottom 32
  2.2   Three-Dimensional Transient Response to Bottom
        Disturbances 35
        2.2.1   Two-Dimensional Propagation of Tsunami Due to
                 Impulsive Bottom Displacement 39
        2.2.2   Leading Waves of a Two-Dimensional Tsunami 44
  2.3   The Propagation of a Dispersive Wave Packet 47
  2.4   Slowly Varying Wave Train by Multiple-Scale Analysis 50

THREE.   REFRACTION BY SLOWLY VARYING DEPTH
OR CURRENT   59

3.1   Geometrical Optics Approximation for Progressive Waves over a Gradually Varying Bottom 59

3.2   Ray Theory for Sinusoidal Waves, Fermat's Principle 63

3.3   Straight and Parallel Depth Contours 66
  3.3.1   Geometry of Rays 66
  3.3.2   Amplitude Variation 70
  3.3.3   The Neighborhood of a Straight Caustic 72

3.4   Circular Depth Contours 74
  3.4.1   Geometry of Rays 74
  3.4.2   Amplitude Variation 81
  3.4.3   A Circular Island 83

3.5   An Approximate Equation Combining Diffraction and Refraction on a Slowly Varying Bottom—The Mild-Slope Equation 86

3.6   Geometrical Optics Approximation for Refraction by Slowly Varying Currents and Depth 89

3.7   Physical Effects of Simple Steady Currents on Waves 98
  3.7.1   Uniform Current on Constant Depth 99
  3.7.2   Oblique Incidence on a Shear Current over Constant Depth 100
  3.7.3   Colinear Waves and Current 105

FOUR.   LONG WAVES OF INFINITESIMAL AMPLITUDE OVER
BOTTOM WITH APPRECIABLE VARIATIONS   109

4.1   Formulation of Linearized Long Wave Theory 109
  4.1.1   Governing Equations 109
  4.1.2   Quasi-One-Dimensional Waves in a Long Channel of Slowly Varying Cross Section 112
  4.1.3   Further Remarks on the Radiation Condition 113

4.2   Straight Depth Discontinuity—Normal Incidence 116
  4.2.1   The Solution 116
  4.2.2   Justification of the Matching Conditions at the Junction 120
  4.2.3   The Near Field for a Rectangular Step 124

4.3   Straight Depth Discontinuity—Oblique Incidence 127

4.4   Scattering by a Shelf or Trough of Finite Width 130

4.5   Transmission and Reflection by a Slowly Varying Depth 135

4.6   Trapped Waves on a Stepped Ridge 140

4.7   Some General Features of One-Dimensional Problems—Trapped Modes and Scattering Matrix 146
  4.7.1   A Qualitative Discussion of Trapped Waves 146

4.7.2   The Scattering Matrix $[S(\alpha)]$ 147
4.7.3   Trapped Modes as Imaginary Poles of $[S(\alpha)]$ 150
4.7.4   Properties of $[S(\alpha)]$ for Real α 152
4.8   Edge Waves on a Constant Slope 154
4.9   Circular Bottom Contours 156
4.9.1   General Aspects 156
4.9.2   Scattering of Plane Incident Waves by a Circular Sill 159
4.10   Head-Sea Incidence on a Slender Submarine Topography—The Parabolic Approximation 162
4.11   A Numerical Method Based on Finite Elements 168
4.11.1   Introduction 168
4.11.2   The Variational Principle 170
4.11.3   Finite-Element Approximation 173
Appendix 4.A   Partial Wave Expansion of the Plane Wave 181

FIVE.   HARBOR OSCILLATIONS EXCITED BY INCIDENT LONG WAVES   183

5.1   Introduction 183
5.2   Formulation for Harbor Oscillation Problems 185
5.3   Natural Modes in a Closed Basin of Simple Form and Constant Depth 187
5.3.1   A Rectangular Basin 187
5.3.2   A Circular Basin 188
5.4   Concept of Radiation Damping: A Model Example 190
5.5   Diffraction through a Narrow Gap 193
5.6   Scattering by a Long and Narrow Canal or a Bay 199
5.6.1   General Solution 199
5.6.2   An Open Narrow Bay 202
5.7   A Rectangular Harbor with a Narrow Entrance 206
5.7.1   Solution by Matched Asymptotic Expansions 209
5.7.2   Resonant Spectrum and Response for Non-Helmholtz Modes 212
5.7.3   The Helmholtz Mode 216
5.7.4   Numerical Results and Experiments 217
5.7.5   Effects of Finite Entry Channel 217
5.8   The Effects of Protruding Breakwater 220
5.8.1   Representation of Solution 221
5.8.2   Reduction to an Integral Equation 223
5.8.3   Approximate Solution by Variational Method 225
5.8.4   Numerical Results 228
5.9   A Harbor with Coupled Basins 231
5.10   A Numerical Method for Harbors of Complex Geometry 234
5.11   Harbor Response to Transient Incident Wave 238

Appendix 5.A   The Source Function for a Rectangular Basin 247
Appendix 5.B   Summation of the $\tilde{G}$ Series 249
Appendix 5.C   Proof of a Variational Principle 250
Appendix 5.D   Evaluation of an Integral 251

SIX.   EFFECTS OF HEAD LOSS AT A CONSTRICTION ON
THE SCATTERING OF LONG WAVES: HYDRAULIC THEORY   253

6.1   One-Dimensional Scattering by a Slotted or Perforated
       Breakwater 254
       6.1.1   The Field Equations 254
       6.1.2   The Matching Conditions and the Near Field 255
       6.1.3   The Coefficients f and L 258
       6.1.4   Equivalent Linearization 261
       6.1.5   Approximate and Exact Solutions 262
6.2   Effect of Entrance Loss on Harbor Oscillations 268
       6.2.1   The Boundary-Value Problem 268
       6.2.2   Local and Mean Square Response in Harbor 271
       6.2.3   Approximations for Narrow Entrance 272
       6.2.4   Small Radiation and Friction Damping 273
       6.2.5   Large Friction Damping 276
       6.2.6   Numerical Results for General W 276
Appendix 6.A   Approximation of an Integral for $ka \ll 1$ 278

SEVEN.   FLOATING BODY DYNAMICS: DIFFRACTION AND
RADIATION BY LARGE BODIES   282

7.1   Introduction 282
7.2   Linearized Equations of Motion for a Constrained
       Floating Body 285
       7.2.1   The Kinematic Condition 285
       7.2.2   Conservation of Linear Momentum 287
       7.2.3   Conservation of Angular Momentum 290
       7.2.4   Summary of Dynamic Equations for a
                Floating Body 296
7.3   Simple Harmonic Motion 300
       7.3.1   Decomposition into Diffraction and Radiation
                Problems 300
       7.3.2   Exciting and Restoring Forces; Added Mass and
                Radiation Damping for Bodies of Arbitrary Shape 302
7.4   Formal Representations of Velocity Potential When
       $h = $ Constant 304
       7.4.1   Away from the Body 304
       7.4.2   The Entire Fluid Domain 308
7.5   Scattering by a Vertical Cylinder with Circular Cross
       Section 312

7.6   General Identities for the Diffraction and Radiation of
      Simple Harmonic Waves 318
      7.6.1   Relations between Two Radiation Problems and Their
              Consequences 319
      7.6.2   Relations between Two Diffraction Problems 321
      7.6.3   One Diffraction Problem and One Radiation
              Problem 326
7.7   Numerical Solution by Hybrid Element Method 330
      7.7.1   The Variational Formulation 331
      7.7.2   The Approximate Solution 333
      7.7.3   A Theoretical Property of the Hybrid Element
              Method 335
      7.7.4   A Numerical Example 336
7.8   Remarks on the Numerical Methods by Integral Equations 340
      7.8.1   The Integral Equations 340
      7.8.2   Irregular Frequencies 341
7.9   Power Absorption by Floating Bodies 344
      7.9.1   Introduction 344
      7.9.2   A Two-Dimensional Beam–Sea Absorber—
              Salter's Cam 347
      7.9.3   Optimum Efficiency of Three-Dimensional
              Absorbers 351
7.10 Drift Forces 365
7.11 Principles of Calculating the Transient Motion of a
      Floating Body 371
      7.11.1 Radiated Waves Caused by Impulsive Motion
              of a Floating Body 372
      7.11.2 Relation to the Frequency Response 375
      7.11.3 Exciting Force Caused by Scattering of Transient
              Incident Waves 376
      7.11.4 Linearized Equations of Transient Motion of a
              Floating Body 378
      Appendix 7.A   Derivation of Green's Functions 379

EIGHT. VISCOUS DAMPING IN SMALL-AMPLITUDE WAVES   384

8.1   Introduction 384
8.2   Linearized Equations of Viscous Flows and the Laminar
      Boundary Layer 384
8.3   Damping Rate and the Process of Energy Transfer 388
      8.3.1   The Entire Fluid 391
      8.3.2   Meniscus Boundary Layer 392
      8.3.3   Wall Boundary Layer 393
      8.3.4   Interior Core 393
      8.3.5   The Damping Rate 394
8.4   Damping Rate by a Perturbation Analysis 395

8.5   Details for Standing Waves in a Circular Basin 401
8.6   The Effect of Air on the Damping of Deep Water Waves 406
8.7   The Turbulent Boundary Layer Near a Rough Bottom 411
       8.7.1   The Boundary-Layer Structure 411
       8.7.2   The Friction Coefficient 414
       8.7.3   Bottom Friction on the Damping of Shallow Water
                Waves in a Basin 415
       Appendix 8.A   An Equipartition Theorem 417

NINE.   MASS TRANSPORT DUE TO VISCOSITY   419

9.1   Introduction 419
9.2   Mass Transport Near the Sea Bottom—General Theory 420
9.3   Bottom Mass Transport Under Long Crest Waves 427
9.4   Bottom Mass Transport Near a Small Structure 434
9.5   Remarks on Induced Streaming Outside the Stokes
       Boundary Layer 439
9.6   Creeping Flow Theory of Mass Transport in a Channel of
       Finite Depth 443
9.7   Further References 450

TEN.   CURRENTS INDUCED BY BREAKING WAVES   451

10.1   Introduction 451
10.2   Depth and Time-Averaged Equations for the Mean Motion 453
       10.2.1   Averaged Equation of Mass Conservation 454
       10.2.2   Averaged Equation of Momentum Conservation 455
       10.2.3   Some Preliminary Simplifications 458
       10.2.4   Summary of Approximate Averaged Equations 463
10.3   Radiation Stresses in the Shoaling Zone—Small-Amplitude
       Waves on Constant or Nearly Constant Depth 464
10.4   Empirical Knowledge of Breaking Waves 467
       10.4.1   Breaking of Standing Waves on a Slope 467
       10.4.2   Types of Breakers on Mild Beaches 468
       10.4.3   Maximum Wave Height 469
10.5   The Structure of a Uniform Longshore Current on a
       Straight Beach 471
       10.5.1   The Shoaling Zone: $x > x_b$ 471
       10.5.2   The Surf Zone: $x < x_b$ 474
10.6   Other Empirical Hypotheses or Improvements 479
       10.6.1   Bottom Friction 479
       10.6.2   Lateral Turbulent Diffusion S''_{xy} 484
10.7   Currents Behind an Offshore Breakwater 485
       10.7.1   The Wave Field 486
       10.7.2   The Mean Motion 491
10.8   Current Around a Conical Island 496
       10.8.1   The Wave Field 497

10.8.2   The Mean Motion 497

10.9   Related Works on Nearshore Currents 502

ELEVEN.   NONLINEAR LONG WAVES IN SHALLOW WATER   504

11.1   Derivation and Classification of Approximate Equations 504

11.2   Nondispersive Waves in Water of Constant Depth 512

    11.2.1   Analogy to Gas Dynamics 512

    11.2.2   Method of Characteristics for One-Dimensional Problems 513

    11.2.3   Simple Waves and Constant States 517

    11.2.4   Expansion and Compression Waves—Tendency of Breaking 518

11.3   Nonbreaking Waves on a Slope 521

    11.3.1   Standing Waves of Finite Amplitude 524

    11.3.2   Matching with Deep Water 527

    11.3.3   Transient Response to Initial Inputs 530

11.4   Subharmonic Resonance of Edge Waves 532

11.5   Dispersive Waves of Permanent Form and the Korteweg–de Vries (KdV) Equation 540

    11.5.1   Solitary Waves 541

    11.5.2   Cnoidal Waves 543

    11.5.3   The Korteweg–de Vries (KdV) Equation 549

11.6   Nonlinear Dispersive Standing Waves on a Horizontal Bottom 550

11.7   Evolution of an Initial Pulse 554

11.8   Fission of Solitons by Decreasing Depth 560

11.9   Viscous Damping of Solitary Waves 564

11.10   Remarks on Modeling Large-Scale Tsunamis 572

11.11   Evolution of Periodic Waves over Constant Depth— Harmonic Generation 578

    11.11.1 The Initial Stage of Near-Resonant Interaction 580

    11.11.2 Governing Equations for Coupled Harmonics 584

    11.11.3 Exact Solution of the Two-Harmonics Problem 587

11.12   Nonlinear Resonance in a Narrow Bay 593

Appendix 11.A   Evaluation of Certain Integrals in Section 11.4 600

Appendix 11.B   Reduction of an Integral in Section 11.9 602

Appendix 11.C   The Square of a Fourier Series 602

TWELVE.   SOME ASPECTS OF NONLINEAR WAVES IN WATER OF INTERMEDIATE OR GREAT DEPTH   605

12.1   Introduction 605

12.2   Evolution Equations for Slowly Modulated Weakly Nonlinear Waves 607

    12.2.1   Finite and Constant Depth 607

    12.2.2   Infinite Depth 616

12.3   Uniform Stokes' Waves 618

12.4   Side-Band Instability of Stokes' Waves 620

12.5   Permanent Envelope in Deep Water: Nonlinear Solutions
        of the Evolution Equation 628

12.6   Transient Evolution of One-Dimensional Wave Envelope
        on Deep Water 632
        12.6.1   Evolution of a Single Pulse 636
        12.6.2   Evolution of the Front of a Uniform Wave Train 641
        12.6.3   Periodic Modulation of a Uniform Wave Train—
                  Evolution Beyond the Initial Stage of Instability 641

12.7   Remarks on Variable Depth 645
        12.7.1   Very Mild Slope: $h = h(x_2) = h(\varepsilon^2 x)$ 646
        12.7.2   Mild Slope: $h = h(x_1) = h(\varepsilon x)$ 647

12.8   Diffraction of Steady Stokes' Waves by a Thin Wedge or
        a Slightly Slanted Breakwater 650

12.9   Second-Order Wave Forces on a Fixed Body 657

12.10   Numerical Solution for Steep Waves 665

THIRTEEN.   WAVE-INDUCED STRESSES IN A
POROELASTIC SEABED   673

13.1   Introduction 673

13.2   Governing Equations 675
        13.2.1   Conservation of Mass and Momentum 675
        13.2.2   Empirical Assumptions 679
        13.2.3   Static Equilibrium 681
        13.2.4   Linearized Equations for Water and Solid
                  Skeleton 682

13.3   The Boundary-Layer Approximation 684
        13.3.1   The Outer Problem 684
        13.3.2   Boundary-Layer Correction Near the Mud Line 688
        13.3.3   Outline of Solving Poroelastic Boundary-
                  Value Problems 692

13.4   Progressive Sea Waves over a Porous Seabed 693
        13.4.1   Infinitely Thick Seabed 693
        13.4.2   Seabed of Finite Thickness 695

13.5   Response to Localized Oscillating Pressure   699
        13.5.1   General Solution 699
        13.5.2   Disturbance Due to a Horizontal Circular Cylinder
                  Resting on a Seabed of Infinite Thickness 700

13.6   Concluding Remarks 704

Appendix 13.A.   Love's Relation between Displacement and Airy's
                 Stress Function 705

REFERENCES   706

AUTHOR INDEX   729

SUBJECT INDEX   735

The Applied Dynamics
of Ocean Surface Waves

ONE

Introduction

Many types of waves involving different physical factors exist in the ocean. As in the elementary problem of a spring–mass system, all waves must be associated with some kind of restoring force. It is therefore convenient to make a crude classification of ocean waves according to the restoring force, as shown in Table 1.1.

Wind waves and swell, generated by local and distant storms are the most directly experienced by mankind. Occurring less frequently but with occasionally disastrous consequences are the tsunamis which usually refer to long-period $[\sim O(1 \text{ h})]$ oscillations caused by large submarine earthquakes or landslides. Within the same broad range of time scales, waves can also exist as a result of human activities (ship motion, explosion, and so on). Since these waves are the most prominent on the water surface and their main restoring force is gravity, they are called the *surface gravity waves*. The shorter term, *surface waves*, is often used if the exclusion of surface capillary waves is understood.

Important in the science of oceanography are the internal gravity waves along the thermoclines which are horizontal layers of sharp density stratification beneath the sea surface. The associated wave motion is generally not pronounced on the surface except for some indirect signs of its presence. These waves contribute to the process of mixing and affect the eddy viscosity of ocean currents. Storm surges are the immediate consequence of local weather and can inflict severe damages to human life and properties by innundating the coast.

In nature, several restoring forces can be in effect at the same time, hence the distinction between various waves listed in Table 1.1 is not always very sharp.

This book will be limited to wave motions having time scales such that compressibility and surface tension at one extreme and earth rotation at the other are of little direct importance. Furthermore, the vertical stratification of sea water is assumed to be small enough within the depth of interest. Therefore, we shall only be concerned with the surface gravity waves, that is, wind waves, swell, and tsunamis. Discussions of all other waves listed in Table 1.1

1

Table 1.1 Wave Type, Physical Mechanism, Activity Region.

Wave Type	Physical Mechanism	Typical Period[a]	Region of Activity
Sound	Compressibility	10^{-2}–10^{-5} s	Ocean interior
Capillary ripples	Surface tension	$< 10^{-1}$ s	Air–water interface
Wind waves and swell	Gravity	1–25 s	Air–water interface
Tsunami	Gravity	10 min–2 h	Air–water interface
Internal waves	Gravity and density stratification	2 min–10 h	Layer of sharp density change
Storm surges	Gravity and Earth rotation	1–10 h	Near coastline
Tides	Gravity and Earth rotation	12–24 h	Entire ocean layer
Planetary waves	Gravity, Earth rotation and variation of latitude or ocean depth	O(100 days)	Entire ocean layer

[a] In seconds (s), minutes (min), hours (h), and days.

can be found in the oceanographic treatises by Hill (1962) and LeBlond and Mysak (1978).

In this chapter we first review the basic equations of fluid motion and some general deductions for inviscid, irrotational flows. Linearized equations for infinitesimal waves are then derived. After introducing the general notions of propagating waves, we examine the properties of simple harmonic progressive waves on constant depth. An elementary discussion of group velocity will be given from both kinematic and dynamic points of view.

1.1 REVIEW OF BASIC FORMULATION FOR AN INCOMPRESSIBLE FLUID OF CONSTANT DENSITY

1.1.1 Governing Equations

In a wide variety of gravity wave problems, the variation of water density is insignificant over the temporal and spatial scales of engineering interest. The fundamental conservation laws are adequately described by the following Navier–Stokes equations:

$$\text{mass:} \qquad\qquad \nabla \cdot \mathbf{u} = 0, \qquad\qquad (1.1)$$

$$\text{momentum:} \quad \left(\frac{\partial}{\partial t} + \mathbf{u} \cdot \nabla\right)\mathbf{u} = -\nabla\left(\frac{P}{\rho} + gz\right) + \nu\nabla^2\mathbf{u}, \qquad (1.2)$$

where $\mathbf{u}(\mathbf{x}, t)$ is the velocity vector (u, v, w), $P(\mathbf{x}, t)$ the pressure, ρ the density, g the gravitational acceleration, ν the constant kinematic viscosity, and $\mathbf{x} = (x, y, z)$ with the z axis pointing vertically upward.

One of the most important deductions from these equations is concerned with the vorticity vector $\mathbf{\Omega}(\mathbf{x}, t)$ defined by

$$\mathbf{\Omega} = \nabla \times \mathbf{u} \tag{1.3}$$

which is twice the rate of local rotation. By taking the curl of Eq. (1.2) and using Eq. (1.1), we can show that

$$\left(\frac{\partial}{\partial t} + \mathbf{u} \cdot \nabla\right)\mathbf{\Omega} = \mathbf{\Omega} \cdot \nabla \mathbf{u} + \nu \nabla^2 \mathbf{\Omega}. \tag{1.4}$$

Physically, the preceding equation means that following the moving fluid, the rate of change of vorticity is due to stretching and twisting of vortex lines and to viscous diffusion (see, e.g., Batchelor, 1967). In water where ν is small ($\cong 10^{-2}$ cm^2/s) the last term in Eq. (1.4) is negligible except in regions of large velocity gradient and strong vorticity. A good approximation applicable in nearly all of the fluid is

$$\left(\frac{\partial}{\partial t} + \mathbf{u} \cdot \nabla\right)\mathbf{\Omega} = \mathbf{\Omega} \cdot \nabla \mathbf{u}. \tag{1.5}$$

An important class of problems is one where $\mathbf{\Omega} \equiv 0$ and is called the *irrotational flow*. Taking the scalar product of Eq. (1.5) and $\mathbf{\Omega}$, we have

$$\left(\frac{\partial}{\partial t} + \mathbf{u} \cdot \nabla\right)\frac{\Omega^2}{2} = \Omega^2[\mathbf{e}_\Omega \cdot (\mathbf{e}_\Omega \cdot \nabla \mathbf{u})]$$

where $\mathbf{e}_\Omega$ is the unit vector along $\mathbf{\Omega}$. Since the velocity gradient is finite in any physically realizable situation, the maximum of $\mathbf{e}_\Omega \cdot (\mathbf{e}_\Omega \cdot \nabla \mathbf{u})$ must be a finite value, $M/2$, say. The magnitude $\Omega^2(\mathbf{x}, t)$ following a fluid particle cannot exceed $\Omega^2(\mathbf{x}, 0)e^{Mt}$. Consequently, if there is no vorticity anywhere at $t = 0$, the flow will remain irrotational for all time.

For an inviscid irrotational flow, the velocity $\mathbf{u}$ can be expressed as the gradient of a scalar potential Φ

$$\mathbf{u} = \nabla \Phi. \tag{1.6}$$

Conservation of mass requires that the potential satisfies Laplace's equation

$$\nabla^2 \Phi = 0. \tag{1.7}$$

If the velocity potential is known, then the pressure field can be found from the momentum equation (1.2). By using the vector identity

$$\mathbf{u} \cdot \nabla \mathbf{u} = \nabla \frac{\mathbf{u}^2}{2} - \mathbf{u} \times (\nabla \times \mathbf{u})$$

and irrotationality, we may rewrite Eq. (1.2), with $\nu = 0$, as

$$\nabla \left[\frac{\partial \Phi}{\partial t} + \frac{1}{2} |\nabla \Phi|^2 \right] = - \nabla \left(\frac{P}{\rho} + gz \right).$$

Upon integration with respect to the space variables, we obtain

$$-\frac{P}{\rho} = gz + \frac{\partial \Phi}{\partial t} + \tfrac{1}{2} |\nabla \Phi|^2 + C(t) \tag{1.8}$$

where $C(t)$ is an arbitrary function of t and can usually be omitted by redefining Φ without affecting the velocity field. Equation (1.8) is called the Bernoulli equation. The first term, gz, on the right-hand side of Eq. (1.8) is the hydrostatic contribution, whereas the rest is the hydrodynamic contribution to the total pressure P.

1.1.2   Boundary Conditions for an Inviscid Irrotational Flow

Two types of boundaries interest us: the air–water interface which will also be called the free surface, and the wetted surface of an impenetrable solid. Along these two boundaries the fluid is assumed to move only tangentially. Let the instantaneous equation of the boundary be

$$F(\mathbf{x}, t) = z - \zeta(x, y, t) = 0, \tag{1.9}$$

where ζ is the height measured from $z = 0$, and let the velocity of a geometrical point $\mathbf{x}$ on the moving free surface be $\mathbf{q}$. After a short time dt, the free surface is described by

$$F(\mathbf{x} + \mathbf{q}\, dt, t + dt) = 0 = F(\mathbf{x}, t) + \left(\frac{\partial F}{\partial t} + \mathbf{q} \cdot \nabla F \right) dt + O(dt)^2.$$

In view of Eq. (1.9), it follows that

$$\frac{\partial F}{\partial t} + \mathbf{q} \cdot \nabla F = 0$$

for small but arbitrary dt. The assumption of tangential motion requires $\mathbf{u} \cdot \nabla F = \mathbf{q} \cdot \nabla F$ which, in turn, implies that

$$\frac{\partial F}{\partial t} + \mathbf{u} \cdot \nabla F = 0 \qquad \text{on } z = \zeta, \tag{1.10}$$

or, equivalently,

$$\frac{\partial \zeta}{\partial t} + \frac{\partial \Phi}{\partial x}\frac{\partial \zeta}{\partial x} + \frac{\partial \Phi}{\partial y}\frac{\partial \zeta}{\partial y} = \frac{\partial \Phi}{\partial z} \qquad \text{on } z = \zeta. \qquad (1.11)$$

Equation (1.10) or (1.11) is referred to as the *kinematic* boundary condition. In the special case where the boundary is the wetted surface of stationary solid S_B, $\partial \zeta / \partial t = 0$ and Eq. (1.10) reduces to

$$\frac{\partial \Phi}{\partial n} = 0 \qquad \text{on } S_B. \qquad (1.12)$$

On the sea bottom B_0 at the depth $h(x, y)$, Eq. (1.9) becomes $z + h(x, y) = 0$ and Eq. (1.12) may be written

$$\frac{\partial \Phi}{\partial z} = \frac{\partial \Phi}{\partial x}\frac{\partial h}{\partial x} + \frac{\partial \Phi}{\partial y}\frac{\partial h}{\partial y} \qquad \text{on } B_0. \qquad (1.13)$$

On the air–water interface, both ζ and Φ are unknown and it is necessary to add a *dynamical* boundary condition concerning forces.

For most of the topics of interest in this book, the wavelength is so long that surface tension is unimportant; the pressure just beneath the free surface must equal the atmospheric pressure P_a above. Applying Eq. (1.8) on the free surface, we have

$$-\frac{P_a}{\rho} = g\zeta + \frac{\partial \Phi}{\partial t} + \tfrac{1}{2}|\nabla \Phi|^2 \qquad \text{on } z = \zeta. \qquad (1.14)$$

The two conditions, (1.11) and (1.14), may be combined into one in terms of Φ by taking the total derivative of Eq. (1.14):

$$\left(\frac{\partial}{\partial t} + \mathbf{u} \cdot \nabla\right)\frac{P_a}{\rho} + \left(\frac{\partial}{\partial t} + \mathbf{u} \cdot \nabla\right)\left(\frac{\partial \Phi}{\partial t} + \frac{u^2}{2} + g\zeta\right) = 0, \qquad z = \zeta.$$

$$(1.15)$$

Using Eq. (1.11) and

$$\mathbf{u} \cdot \nabla \frac{\partial \Phi}{\partial t} = \frac{\partial}{\partial t}\frac{1}{2}u^2,$$

we have from Eq. (1.15)

$$\frac{D}{Dt}\frac{P_a}{\rho} + \left[\frac{\partial^2 \Phi}{\partial t^2} + g\frac{\partial \Phi}{\partial z} + \frac{\partial u^2}{\partial t} + \frac{1}{2}\mathbf{u} \cdot \nabla u^2\right] = 0, \qquad z = \zeta. \qquad (1.16)$$

Furthermore, if P_a = constant, the above condition becomes

$$\frac{\partial^2 \Phi}{\partial t^2} + g\frac{\partial \Phi}{\partial z} + \frac{\partial}{\partial t}(\mathbf{u})^2 + \frac{1}{2}\mathbf{u}\cdot\nabla\mathbf{u}^2 = 0, \qquad z = \zeta \qquad (1.17)$$

which is essentially a condition for Φ. Not only do nonlinear terms appear in these boundary conditions, but the position of the free surface is also an unknown quantity. An exact analytical theory for water-wave problems is therefore almost impossible.

When the motion of the air above is significant, the atmospheric pressure cannot always be prescribed *a priori*; the motions of air and water are, in general, coupled. Indeed, the exchange of momentum and energy between air and sea is at the heart of the theory of surface-wave generation by wind. However, we shall limit our attention to sufficiently localized regions in the absence of direct wind action. Air can then be ignored for most purposes because of its comparatively small density.

1.2 LINEARIZED APPROXIMATION FOR SMALL-AMPLITUDE WAVES

Let us assume that certain physical scales of motion can be anticipated *a priori*. In particular, let

$$\begin{pmatrix} \lambda/2\pi \\ \omega^{-1} \\ A \\ A\omega\lambda/2\pi \end{pmatrix} \quad \text{characterize} \quad \begin{pmatrix} x, y, z, h \\ t \\ \zeta \\ \Phi \end{pmatrix}, \qquad (2.1)$$

where λ, ω, and A are the typical values of wavelength, frequency, and free-surface amplitude respectively. We have assigned the scale for Φ to be $A\omega\lambda/2\pi$ so that the velocity has the scale $A\omega$ which is expected near the free surface. We now introduce dimensionless variables and denote them by primes as follows:

$$\begin{pmatrix} \Phi \\ x, y, z, h \\ t \\ \zeta \end{pmatrix} = \begin{pmatrix} A\omega\lambda\Phi'/2\pi \\ \lambda(x', y', z', h')/2\pi \\ t'/\omega \\ A\zeta' \end{pmatrix}. \qquad (2.2)$$

When these variables are substituted into Eqs. (1.7), (1.11), (1.12) and (1.14), a set of dimensionless equations are obtained:

$$\nabla^2 \Phi' = \left(\frac{\partial^2}{\partial x'^2} + \frac{\partial^2}{\partial y'^2} + \frac{\partial^2}{\partial z'^2} \right) \Phi' = 0, \qquad -h' < z' < \varepsilon\zeta', \quad (2.3)$$

$$\frac{\partial \Phi'}{\partial n'} = 0, \qquad z' = -h', \tag{2.4}$$

$$\frac{\partial \zeta'}{\partial t'} + \varepsilon \left(\frac{\partial \phi}{\partial x'} \frac{\partial \zeta'}{\partial x'} + \frac{\partial \phi'}{\partial y'} \frac{\partial \zeta'}{\partial y'} \right) = \frac{\partial \phi'}{\partial z'} \tag{2.5}$$

$$\text{on } z' = \varepsilon\zeta',$$

$$\frac{\partial \Phi'}{\partial t'} + \left(\frac{2\pi g}{\omega^2 \lambda} \right) \zeta' + \frac{\varepsilon}{2} (\nabla' \Phi')^2 = -P_a' = -\frac{2\pi P_a}{\rho A \omega^2 \lambda} \tag{2.6}$$

where $\varepsilon = 2\pi A/\lambda = 2\pi \times$ amplitude/wavelength = wave slope. Since the scales are supposed to reflect the physics properly, the dimensionless variables must all be of order unity; the importance of each term above is measured solely by the coefficients in front.[†]

Let us now consider small-amplitude waves in the sense that the wave slope is small: $\varepsilon \ll 1$. The free-surface boundary conditions can be simplified by noting that the unknown free surface differs by an amount of $O(\varepsilon)$ from the horizontal plane $z' = 0$. Thus, we can expand Φ' and its derivatives in a Taylor series:

$$f'(x', y', \varepsilon\zeta', t') = f'\Big|_0 + \varepsilon\zeta' \frac{\partial f'}{\partial z'}\Big|_0 + \frac{(\varepsilon\zeta')^2}{2!} \frac{\partial^2 f'}{\partial z'^2}\Big|_0 + O(\varepsilon^3)$$

where $f|_0$ means $f(x, y, 0, t)$, etc. To the leading order of $O(1)$, the free-surface conditions become approximately

$$\frac{\partial \zeta'}{\partial t'} = \Phi_{z'}'$$

$$\frac{\partial \Phi'}{\partial t'} + \frac{2\pi g}{\omega^2 \lambda} \zeta' = -P_a' \qquad z' = 0.$$

[†]If the scales have been chosen properly, the normalized variables and their derivatives should indeed be of order unity. The relative importance of each term in an equation is entirely indicated by the dimensionless coefficient multiplying the term. If under certain conditions the solution of the approximate problem exhibits behavior which violates the original assumptions on the order of magnitude, then the scales initially chosen are no longer valid. New scales, hence new approximations, must be found to reflect the physics. It is not an exaggeration to say that estimating the scales is the first step toward the approximate solution of a physical problem.

As a procedural point, when the choices of the scales are limited so that only one dimensionless parameter appears, the formalism of nondimensionalization can often be omitted for brevity, although its essence must always be clearly understood.

Only linear terms remain in these conditions which are now applied at a known plane $z' = 0$. Together with Eqs. (2.3) and (2.4) the approximate problem is completely linearized. Returning to physical variables, we have

$$\nabla^2 \Phi = 0, \qquad -h < z < 0, \tag{2.7}$$

$$\frac{\partial \Phi}{\partial n} = 0, \qquad z = -h, \tag{2.8}$$

$$\frac{\partial \zeta}{\partial t} = \frac{\partial \Phi}{\partial z}, \tag{2.9}$$

$$z = 0.$$

$$\frac{\partial \Phi}{\partial t} + g\zeta = -\frac{P_a}{\rho} \tag{2.10}$$

Furthermore, Eqs. (2.9) and (2.10) may be combined to give

$$\frac{\partial^2 \Phi}{\partial t^2} + g\frac{\partial \Phi}{\partial z} = -\frac{1}{\rho}\frac{\partial P_a}{\partial t}, \qquad z = 0 \tag{2.11}$$

which can also be obtained by linearizing Eq. (1.16).

The total pressure inside the fluid can be related to Φ by linearizing the Bernoulli equation:

$$P = -\rho g z + p, \qquad \text{where } p = -\rho \frac{\partial \Phi}{\partial t} = \text{dynamic pressure.} \tag{2.12}$$

These conditions must be supplemented by initial conditions and the boundary conditions at the body and at infinity, if appropriate.

It is worthwhile to remark further on the assumption of zero viscosity in the context of linear approximation. Near a solid boundary, the potential theory allows a finite slip in the tangential direction, but in reality all velocity components must vanish. There must be a thin boundary layer to smooth the transition from zero to a finite value. Thus,

$$\frac{\partial}{\partial x_N} \gg \frac{\partial}{\partial x_T'}, \frac{\partial}{\partial x_T''}$$

where x_N, x_T', and x_T'' form a locally orthogonal coordinate system with x_N normal to the solid surface and x_T' and x_T'' tangential to it. It follows from the linearized momentum equation that the tangential velocity $\mathbf{u}_T$ satisfies

$$\frac{\partial \mathbf{u}_T}{\partial t} \cong \nu \frac{\partial^2 \mathbf{u}_T}{\partial x_N^2} - \frac{1}{\rho}\nabla_T p$$

inside the boundary layer. With the wave period as the time scale, the boundary-layer thickness δ must be of the order

$$\delta \sim \left(\frac{2\nu}{\omega}\right)^{1/2}.$$

For water, $\nu \cong 0.01 \text{ cm}^2/\text{s}$; in model experiments the typical period is 1 s so that $\delta \sim 0.056$ cm, which is far too small compared with the wavelength of usual interest. In the ocean, swells of 10-s periods are common; $\delta \sim 0.17$ cm. But the boundary layer near the natural sea bottom is usually turbulent for most of the period. As will be discussed later, a typical experimental value of eddy viscosity is about 100ν; the thickness of the turbulent boundary layer for a 10-s period is then about $\leq O(10)$ cm, which is still quite small. Thus, the boundary-layer region is but a tiny fraction of a fluid volume whose dimensions are comparable to a wavelength, and the global influence on the wave motion is small over distances of several wavelengths or time of several periods.

1.3 ELEMENTARY NOTIONS OF A PROPAGATING WAVE

Let us consider a special form of the free surface

$$\zeta(x, y, t) = \text{Re } A e^{i(\mathbf{k}\cdot\mathbf{x}-\omega t)} = A \cos(\mathbf{k} \cdot \mathbf{x} - \omega t) \tag{3.1}$$

where i is the imaginary unit $(-1)^{1/2}$ and

$$\mathbf{k} = (k_1, k_2), \qquad \mathbf{x} \equiv (x, y). \tag{3.2}$$

For the convenience of mathematical manipulation, the exponential form is often preferred, and for brevity the sign Re (the real part of) is often omitted, that is,

$$\zeta(x, y, t) = A e^{i(\mathbf{k}\cdot\mathbf{x}-\omega t)} \tag{3.3}$$

is used to mean the same as Eq. (3.1). What sort of free surface does this expression describe?

To a stationary observer, ζ oscillates in time with the period $T = 2\pi/\omega$ between the two extremes A and $-A$. If we take a three-dimensional snapshot at a fixed t with ζ as the vertical ordinate and (x, y) as the horizontal coordinates, the variation of ζ in (x, y) describes a periodic topography. In a plane $y = \text{const}$, ζ is seen to vary periodically in the x direction between A and $-A$ with the spatial period $2\pi/k_1$. Similarly in a plane $x = \text{const}$, ζ varies periodically in the y direction between A and $-A$ with the spatial period $2\pi/k_2$. Thus, along the x direction the number of crests per unit length is $k_1/2\pi$, and along the y direction the number of crests per unit length is $k_2/2\pi$.

Let us define the *phase function* S by

$$S(x, y, t) = k_1 x + k_2 y - \omega t = \mathbf{k} \cdot \mathbf{x} - \omega t. \qquad (3.4)$$

For a fixed time, $S(x, y, t) = \text{const} = S_0$ describes a straight line with the normal vector

$$\mathbf{e}_k = \left(\frac{k_1}{k}, \frac{k_2}{k} \right), \qquad \text{where } k = \left(k_1^2 + k_2^2 \right)^{1/2} = |\mathbf{k}|. \qquad (3.5)$$

Along this straight line, the surface height is the same everywhere. In particular, the waves are the highest (crests) when $S_0 = 2n\pi$ and the lowest (troughs) when $S_0 = (2n + 1)\pi$. When S_0 increases by 2π, the surface height is repeated. Lines of different S_0 are parallel to each other if k_1 and k_2 are constant. We call these lines the *phase lines*. If we take a snapshot and cut a cross section along the direction $\mathbf{e}_k$, the profile of ζ is sinusoidal with the wavelength $\lambda = 2\pi/k$. Alternatively, we may say that the number of waves per unit length along the $\mathbf{k}$ direction is $k/2\pi$. Hence k is called the *wavenumber*, and $\mathbf{k}$ the *wavenumber vector* with k_1 and k_2 as its components. The maximum deviation A from the mean $z = 0$ is called the *amplitude*.

Let us follow a particular phase line $S = S_0$. As time t progresses, the position of the phase line changes. What is the velocity of the phase line? Evidently, if the observer travels with the same velocity $d\mathbf{x}/dt$, the phase line appears stationary, that is,

$$dS = \nabla S \cdot d\mathbf{x} + \frac{\partial S}{\partial t} dt = 0.$$

From Eq. (3.4) it follows that

$$\mathbf{k} = \nabla S = \mathbf{e}_k |\nabla S|, \qquad (3.6a)$$

$$-\omega = \frac{\partial S}{\partial t}, \qquad (3.6b)$$

and that

$$\mathbf{e}_k \cdot \frac{d\mathbf{x}}{dt} = \frac{-\partial S/\partial t}{|\nabla S|} = \frac{\omega}{k} \equiv C. \qquad (3.7)$$

Thus, the speed at which the phase line advances normal to itself is ω/k, which is called the *phase speed* C. Equations (3.6a) and (3.6b) can be regarded as the definitions of ω and $\mathbf{k}$, that is, the frequency is the time rate, and the wavenumber is the spatial rate, of phase change.

1.4 PROGRESSIVE WATER WAVES ON CONSTANT DEPTH

For simple harmonic motion with frequency ω, linearity of the problem allows separation of the time factor $e^{-i\omega t}$ as follows:

$$\left.\begin{array}{l} \zeta(x, y, t) = \eta(x, y) \\ \Phi(x, y, z, t) = \phi(x, y, z) \\ \mathbf{u}(x, y, z, t) \to \mathbf{u}(x, y, z) \\ P(x, y, z, t) + \rho g z = p(x, y, z) \end{array}\right\} e^{-i\omega t}. \tag{4.1}$$

Note that for the fluid velocity the same symbols are used for the original variables and for the spatial factor. The linearized governing equations (2.7) to (2.10) can be reduced to

$$\nabla^2 \phi = 0, \qquad -h < z < 0, \tag{4.2}$$

$$\frac{\partial \phi}{\partial z} = 0, \qquad z = -h, \tag{4.3}$$

$$\frac{\partial \phi}{\partial z} + i\omega\eta = 0, \tag{4.4}$$

$$z = 0,$$

$$g\eta - i\omega\phi = \frac{-p_a}{\rho}, \tag{4.5}$$

where Eqs. (4.4) and (4.5) may be combined as

$$g\frac{\partial \phi}{\partial z} - \omega^2\phi = \frac{i\omega}{\rho}p_a, \qquad z = 0. \tag{4.6}$$

Let us seek a two-dimensional solution which represents a progressive wave without direct atmospheric forcing, that is, $p_a = 0$ and

$$\eta = Ae^{ikx}. \tag{4.7}$$

The potential that satisfies Eqs. (4.2) and (4.3) is easily seen to be

$$\phi = B \cosh k(z + h)e^{ikx}.$$

To satisfy the free-surface conditions with $p_a = 0$, we require

$$B = -\frac{igA}{\omega}\frac{1}{\cosh kh},$$

and

$$\omega^2 = gk \tanh kh \tag{4.8}$$

so that

$$\phi = -\frac{igA}{\omega}\frac{\cosh k(z+h)}{\cosh kh}e^{ikx}. \qquad (4.9)$$

Thus, for a given frequency ω the progressive wave must have the proper wavenumber given by Eq. (4.8). In dimensionless form

$$\omega\left(\frac{h}{g}\right)^{1/2} = (kh\tanh kh)^{1/2}.$$

The dimensionless frequency $\omega(h/g)^{1/2}$ and the dimensionless wavenumber kh vary as shown in Fig. 4.1. In particular, the limiting approximations are

$$\omega \simeq (gh)^{1/2}k, \qquad kh \ll 1,$$
$$\omega \simeq (gk)^{1/2}, \qquad kh \gg 1. \qquad (4.10)$$

Since $kh = 2\pi h/\lambda$ is essentially the depth-to-wavelength ratio, the terms *long waves* and *shallow-water waves* refer to $kh \ll 1$, while *short waves* and *deep-water waves* refer to $kh \gg 1$. For fixed h, shorter waves have higher frequencies. In shallow water, waves of a fixed frequency have shorter length in smaller depth since $k \simeq \omega/(gh)^{1/2}$.

The phase speed C is given by

$$C = \frac{\omega}{k} = \left(\frac{g}{k}\tanh kh\right)^{1/2} \qquad (4.11)$$

which is plotted in dimensionless form in Fig. 4.1. For long and short waves the limiting relations are

$$C = (gh)^{1/2}, \qquad kh \ll 1,$$
$$C = (gk)^{1/2}, \qquad kh \gg 1. \qquad (4.12)$$

$$g/k$$

In general, for the same depth, longer waves have faster speeds. It will be shown in Chapter Two that a localized initial disturbance can be thought of as a Fourier superposition of periodic disturbances with wavelengths ranging over a continuous spectrum. As time passes the longer waves lead the shorter waves. As the disturbances propagate outward the longest and shortest waves become further and further apart with intermediate waves marching in between. The phenomenon that waves of different frequencies travel at different speeds is called *dispersion*. Clearly, if the relation between ω and k for a sinusoidal wave is nonlinear, the medium is dispersive. Equation (4.8) or its equivalent, Eq. (4.11), is therefore called the *dispersion relation*.

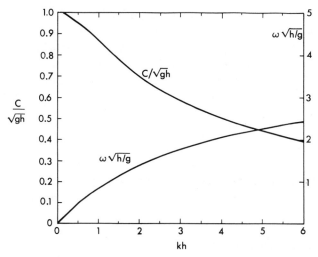

Figure 4.1 Dispersion curves for a progressive wave.

From the linearized Bernoulli equation the dynamic pressure (without $-\rho gz$) is

$$\frac{p}{\rho} = i\omega\phi = gA\frac{\cosh k(z+h)}{\cosh kh}e^{ikx} = g\eta\frac{\cosh k(z+h)}{\cosh kh}. \qquad (4.13)$$

The velocity field is

$$u = \frac{gkA}{\omega}\frac{\cosh k(z+h)}{\cosh kh}e^{ikx}, \qquad (4.14)$$

$$v = 0, \qquad (4.15)$$

$$w = -\frac{igkA}{\omega}\frac{\sinh k(z+h)}{\cosh kh}e^{ikx}. \qquad (4.16)$$

For very deep water, $kh \gg 1$,

$$(\phi, u, v, w, p) = \left(-\frac{ig}{\omega}, \frac{gk}{\omega}, 0, -\frac{igk}{\omega}, \rho g\right)Ae^{kz}e^{ikx}, \qquad (4.17)$$

and for very shallow water, $kh \ll 1$,

$$(\phi, u, v, w, p) = \left(-\frac{ig}{\omega}, \frac{gk}{\omega}, 0, 0, \rho g\right)Ae^{ikx}. \qquad (4.18)$$

Several distinctive features of the shallow water results deserve mentioning:
(i) The dependence on z disappears; (ii) the vertical velocity is negligible; and

(iii) the dynamic pressure is $\rho g \eta$ and the total pressure $P = \rho g(\zeta - z)$, which is hydrostatic in terms of depth below the free surface.

Finally, we know from Section 1.2 that when the spatial scale is $1/k$, the condition for linearization is $kA \ll 1$. Let us check the linearizing assumption again by comparing a nonlinear term with a linear term, both evaluated at the free surface $z = 0$. For arbitrary kh, we have from Eqs. (4.11) and (4.14)

$$\left(\frac{u \, \partial u / \partial x}{\partial u / \partial t} \right)_{z=0} \sim \left(\frac{uk}{\omega} \right)_{z=0} \sim \left(\frac{u}{C} \right)_{z=0} = \frac{kA}{\tanh kh} \qquad \text{for all } kh.$$

Note that for $kh \ll 1$, the above ratio becomes A/h. Therefore, in shallow water the linearized theory is indeed a very restricted approximation.

1.5   GROUP VELOCITY

One of the most important concepts in dispersive waves is the group velocity, for which two views may be examined to understand its significance.

1.5.1   A Kinematic View

Suppose that there is a group of sinusoidal waves with a continuous but narrow range of wavelengths centered around $k = k_0$. The free-surface displacement may be represented by

$$\zeta = \int_{k_0 - \Delta k}^{k_0 + \Delta k} A(k) e^{i[kx - \omega(k)t]} \, dk, \qquad \frac{\Delta k}{k_0} \ll 1 \qquad (5.1)$$

where $A(k)$ is the wavenumber spectrum with ω and k satisfying the dispersion relation

$$\omega = \omega(k). \qquad (5.2)$$

By Taylor expansion we write

$$\omega = \omega[k_0 + (k - k_0)] = \omega(k_0) + (k - k_0)\left(\frac{d\omega}{dk} \right)_{k_0} + O(k - k_0)^2.$$

Denoting

$$\frac{k - k_0}{k_0} = \xi, \quad \omega_0 = \omega(k_0), \quad \text{and} \quad \left(\frac{d\omega}{dk} \right)_0 = \left(\frac{d\omega}{dk} \right)_{k_0} \equiv C_g, \qquad (5.3)$$

we have for sufficiently smooth $A(k)$ and to the crudest approximation

$$\zeta \simeq A(k_0) e^{i(k_0 x - \omega_0 t)} \int_{-\Delta k/k_0}^{\Delta k/k_0} \left\{ \exp\left[ik_0 \xi (x - C_g t)\right] \right\} k_0 \, d\xi$$

$$= 2 A(k_0) \frac{\sin \Delta k (x - C_g t)}{(x - C_g t)} e^{i(k_0 x - \omega_0 t)} = \tilde{A} e^{i(k_0 x - \omega_0 t)}, \qquad (5.4)$$

where

$$\tilde{A} = 2 A(k_0) \frac{\sin \Delta k (x - C_g t)}{(x - C_g t)}. \qquad (5.5)$$

Because of the factor $\exp[i(k_0 x - \omega_0 t)]$ in Eq. (5.4), ζ may be viewed as a locally sinusoidal wave train with a slowly modulated amplitude $\tilde{A}$. In particular, the envelope defined by $\tilde{A}$ is in the form of wave groups as shown in Fig. 5.1, and advances at the speed C_g. Therefore, C_g is called the *group velocity*. The distance between two adjacent nodes of an envelope, hence the modulation length scale of the amplitude, is roughly $\pi/\Delta k$ and is much greater than the length of the constituent waves $2\pi/k_0$.

For water waves on a constant depth, it follows by differentiating the dispersion relation (4.8) that

$$C_g = \frac{d\omega}{dk} = \frac{1}{2} \frac{\omega}{k} \left(1 + \frac{2kh}{\sinh 2kh}\right) = \frac{C}{2} \left(1 + \frac{2kh}{\sinh 2kh}\right). \qquad (5.6)$$

For deep water $kh \gg 1$,

$$C_g \simeq \frac{1}{2} C \simeq \frac{1}{2} \left(\frac{g}{k}\right)^{1/2}, \qquad (5.7)$$

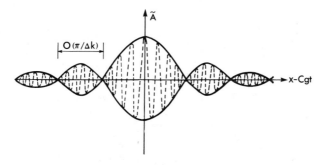

Figure 5.1 A group of waves within a narrow frequency band.

while for shallow water, $kh \ll 1$,

$$C_g \simeq C \simeq (gh)^{1/2}. \tag{5.8}$$

Since the phase velocity exceeds the group velocity for general depths, the individual wave crests travel from the tail toward the front of a group.

It will be shown more generally in Section 2.4 that C_g is the velocity of any slowly varying envelope, Eq. (5.5) being a special case.

1.5.2 A Dynamic View: Energy Flux

Let us first compute the average energy in a uniform progressive wavetrain beneath a unit square of the free surface. Denoting the time average over a period by an overhead bar, we have the kinetic energy in the whole fluid column

$$K.E. = \frac{\rho}{2} \overline{\int_{-h}^{\zeta} [\mathbf{u}(\mathbf{x}, t)]^2 \, dz} \cong \frac{\rho}{2} \int_{-h}^{0} \overline{\left\{ [\operatorname{Re} u(\mathbf{x}) e^{-i\omega t}]^2 + [\operatorname{Re} w(\mathbf{x}) e^{-i\omega t}]^2 \right\}} \, dz,$$

$$\tag{5.9}$$

where for second-order $O(kA)^2$ accuracy the upper limit has been replaced by $z = 0$, and $\mathbf{u}$ may be approximated by the first-order result, Eqs. (4.14) and (4.16). Note that for any two sinusoidal functions,

$$a = Ae^{-i\omega t} \quad \text{and} \quad b = Be^{-i\omega t},$$

the following formula is true:

$$\overline{ab} = \frac{1}{T} \int_0^T dt \, ab = \tfrac{1}{2} \operatorname{Re}(AB^*) = \tfrac{1}{2} \operatorname{Re}(A^*B), \tag{5.10}$$

where ( )* denotes the complex conjugate. The proof is left as an exercise. With Eqs. (4.14), (4.16), and (5.10), Eq. (5.9) becomes

$$K.E. = \frac{\rho}{4} \left(\frac{gkA}{\omega} \right)^2 \frac{1}{\cosh^2 kh} \int_{-h}^{0} [\cosh^2 k(z+h) + \sinh^2 k(z+h)] \, dz$$

$$= \frac{\rho}{4} \left(\frac{gkA}{\omega} \right)^2 \frac{\sinh 2kh}{2k \cosh^2 kh} = \tfrac{1}{4} \rho g A^2, \tag{5.11}$$

where use is made of the following formula:

$$\int_0^{kh} \cosh^2 \xi \, d\xi = \tfrac{1}{4}(\sinh 2kh + 2kh), \tag{5.12}$$

and the dispersion relation. On the other hand, the potential energy in the fluid column due to wave motion is

$$\text{P.E.} = \overline{\int_0^\zeta \rho g z \, dz} = \tfrac{1}{2}\rho \, \overline{g\zeta^2} = \tfrac{1}{4}\rho g A^2 \tag{5.13}$$

since $\rho g \, dz$ is the weight of a thin horizontal slice whose height above the mean free surface is z. The total energy is

$$E = \text{K.E.} + \text{P.E.} = \tfrac{1}{2}\rho g A^2. \tag{5.14}$$

Note that the kinetic and potential energies are equal; this property is called the *equipartition of energy*. Let us consider a vertical cross section of unit width along the crest. The rate of energy flux across this section is equal to the mean rate of work done by the dynamic pressure, that is,

$$\begin{array}{c}\text{Rate of} \\ \text{energy flux}\end{array} = \begin{array}{c}\text{Rate of} \\ \text{pressure working}\end{array} = \overline{\int_{-h}^\zeta p(\mathbf{x}, t) u(\mathbf{x}, t) \, dz} \cong -\rho \int_{-h}^0 \overline{\Phi_t \Phi_x} \, dz$$

$$\tag{5.15}$$

which can be calculated to be

$$\text{Rate of energy flux} = \frac{1}{2}\rho g A^2 \left[\frac{1}{2} \frac{\omega}{k} \left(1 + \frac{2kh}{\sinh 2kh} \right) \right] = E C_g. \tag{5.16}$$

Hence the group velocity has the dynamical meaning of the velocity of energy transport. In contrast, the phase speed is merely a kinematic quantity and is not always identifiable with the transport of any dynamical substance.

As an immediate application, consider a long wave tank of unit width with sinusoidal waves generated at one end. Many periods after the start of the wavemaker, the envelope is uniform almost everywhere except near the wave front which may look like Fig. 5.2. Since the rate of energy input by the

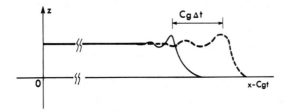

Figure 5.2   The envelope front of a sinusoidal wave train.

wavemaker at the left (say at $x = 0$) is EC_g, the rate of lengthening of the wave region must be C_g. Thus the wave front must propagate at the group velocity. Details of the wave-front evolution will be discussed in Section 2.4.

Exercise 5.1

Consider a two-layered fluid system over a horizontal bottom. The lighter fluid above has the density of ρ, while the heavier fluid below has the density of ρ'. Let the free surface be at $z = 0$, the interface at $z = -h$, and the bottom at $z = -h'$. Show that a sinusoidal progressive wave must satisfy the dispersion relation:

$$\left(\frac{\omega^2}{gk}\right)^2 \{\rho'\text{coth } kh \text{ coth } k(h' - h) + \rho\}$$

$$-\frac{\omega^2}{gk}\rho'\{\text{coth } kh + \text{coth } k(h' - h)\} + \rho' - \rho = 0.$$

Study the two possible modes corresponding to the two solutions ω_1^2 and ω_2^2 for the same k.

In particular, when $h' \sim \infty$ show that

$$\omega_1^2 = gk \quad \text{and} \quad \omega_2^2 = gk\frac{\rho' - \rho}{\rho'\text{coth } kh + \rho} < \omega_1^2$$

and that the amplitude ratio of interface to free surface is

$$e^{-kh} \quad \text{and} \quad -\frac{\rho'}{\rho' - \rho}e^{kh}$$

for the first and second mode, respectively. Plot the group velocity as a function of k for each mode.

Exercise 5.2   Capillary Waves

Surface tension on the free surface introduces a pressure difference between the atmospheric pressure P_a above and the water pressure P below. The difference is given by the Laplace formula (see, e.g., Landau and Lifshitz, 1959, p. 237 ff)

$$P - P_a \cong -T(\zeta_{xx} + \zeta_{yy}), \qquad \text{on } z \cong 0, \tag{1}$$

where the right-hand side is proportional to the surface curvature and T is the surface tension coefficient. For the water–air interface at 20°C, $T = 74$ dyn/cm in cgs units. Reformulate the boundary conditions on the free surface and study a plane progressive wave on deep water: $\Phi \propto e^{kz}e^{i(kx-\omega t)}$. Show that

$$\omega^2 = gk + \frac{Tk^3}{\rho}.$$

Show further that the phase velocity has a minimum value C_m which satisfies

$$\frac{C^2}{C_m^2} = \frac{1}{2}\left(\frac{\lambda}{\lambda_m} + \frac{\lambda_m}{\lambda}\right) = \frac{1}{2}\left(\frac{k_m}{k} + \frac{k}{k_m}\right),$$

where

$$\lambda_m = \frac{2\pi}{k_m} = 2\pi\left(\frac{T}{g\rho}\right)^{1/2}.$$

What are the numerical values of λ_m and C_m for water and air?
  Discuss the variations of ω, C and C_g versus k or λ.

Propagation of Transient Waves in Open Water of Essentially Constant Depth

Disturbances generated by excitations of finite duration such as submarine earthquakes, landslides, explosions, and so on, produce transient waves. Because of dispersion, the propagation of transient water waves is considerably more complex than many other kinds of waves in nature. To facilitate a clear understanding of the physical consequences of dispersion, we will study in this chapter simple models of source mechanism and ocean depth for which detailed analyses are possible. In Sections 2.1 and 2.2 we will study the so-called Cauchy-Poisson problem of waves due to an impulsive source of some kind, with particular emphasis given to the behavior away from the source. Many practical issues concerning earthquake-induced waves in nature are omitted for simplicity. In Sections 2.3 and 2.4 the role of dispersion in slowly modulated wave groups will be investigated. The mathematical technique of multiple-scale expansions is introduced to arrive very directly at the asymptotic result for large time and to prepare the ground for later inclusion of nonlinearity.

2.1 TWO-DIMENSIONAL TRANSIENT PROBLEMS

Let us consider an ocean of constant depth without other rigid boundaries. Assume that the disturbances on the free surface and at the bottom are independent of y. The problem can be formulated in the x, z plane. Thus, the velocity potential $\Phi(x, z, t)$ satisfies

$$\nabla^2 \Phi = \frac{\partial^2 \Phi}{\partial x^2} + \frac{\partial^2 \Phi}{\partial z^2} = 0. \qquad (1.1)$$

On the free surface the following conditions hold:

$$\frac{\partial \zeta}{\partial t} = \frac{\partial \Phi}{\partial z}, \qquad z = 0 \tag{1.2a}$$

$$\frac{\partial \Phi}{\partial t} + g\zeta = -\frac{P_a(x, t)}{\rho}, \qquad z = 0, \tag{1.2b}$$

where $P_a(x, t)$ is prescribed. Let the seafloor be denoted by $z = -h + H(x, t)$. If the ground motion is known, continuity of normal velocity gives

$$\frac{\partial \Phi}{\partial z} = \frac{\partial H}{\partial t} + \frac{\partial \Phi}{\partial x} \frac{\partial H}{\partial x} \qquad \text{on } z = -h + H(x, t). \tag{1.3}$$

Within the framework of linearization we assume that the amplitudes of H, $\partial H/\partial t$ and $\partial H/\partial x$ are small so that the quadratic term is negligible; hence

$$\frac{\partial \Phi}{\partial z} = \frac{\partial H}{\partial t} \equiv W(x, t) \qquad \text{on } z \cong -h. \tag{1.4}$$

Initial conditions must be further prescribed. To see what initial data are needed, let us employ the method of Laplace transform defined by

$$\bar{f}(s) = \int_0^\infty e^{-st} f(t) \, dt, \tag{1.5a}$$

$$f(t) = \frac{1}{2\pi i} \int_\Gamma e^{st} \bar{f}(s) \, ds, \tag{1.5b}$$

where Γ is a vertical line to the right of all singularities of $\bar{f}(s)$ in the complex s plane. The transforms of Eqs. (1.1) and (1.4) give

$$\nabla^2 \bar{\Phi}(x, z, s) = 0, \qquad -h < z < 0, \tag{1.6}$$

$$\frac{\partial \bar{\Phi}}{\partial z} = \bar{W}(x, s), \qquad z = 0. \tag{1.7}$$

From Laplace transforms of conditions (1.2a, b) we obtain

$$-\zeta(x, 0) + s\bar{\zeta}(x, s) = \frac{\partial \bar{\Phi}(x, 0, s)}{\partial z}, \tag{1.8}$$

$$-\Phi(x, 0, 0) + s\bar{\Phi}(x, 0, s) + g\bar{\zeta}(x, s) = -\frac{\bar{P}_a(x, s)}{\rho}, \tag{1.9}$$

which can be combined to give

$$\frac{\partial \overline{\Phi}}{\partial z} + \frac{s^2}{g}\overline{\Phi} = -\frac{s\overline{P}_a}{\rho g} - \zeta(x,0) + \frac{s}{g}\Phi(x,0,0), \qquad z = 0. \qquad (1.10)$$

From the above equation it is immediately clear that we only need to prescribe the initial data $\Phi(x,0,0)$ and $\zeta(x,0)$ on the free surface, but nowhere else, because time derivatives appear only in the free-surface conditions. A more mathematical study on the uniqueness of the initial-value problem has been made by Finkelstein (1957).

What is the physical significance of $\Phi(x,0,0)$? Assume that, before $t = 0$, all is calm, but at $t = 0$ an impulsive pressure $P_a(x,t) = I\delta(t)$ is applied on the free surface. Integrating Bernoulli's equation from $t = 0 -$ to $t = 0 +$, we obtain

$$\Phi(x,0,0+) - \Phi(x,0,0-) + \int_{0-}^{0+} g\zeta \, dt = \frac{I}{\rho}\int_{0-}^{0+} \delta(t) \, dt = \frac{I}{\rho}.$$

Since $\Phi(x,0,0-) = 0$ and ζ must be finite, we obtain $\Phi(x,0,0+) = I/\rho$. Thus, the initial value of Φ represents physically an impulsive pressure acting on the free surface at an instant slightly earlier than $t = 0 +$.

Equations (1.6), (1.7), and (1.10) now define a boundary-value problem which formally resembles that of a simple harmonic case. For any finite t it is expected that no motion is felt at a great distance from the initial disturbance so that $\Phi(x,t) \to 0$ as $|x| \to \infty$, which implies that $\overline{\Phi} \to 0$ as $|x| \to \infty$. Since the region does not involve any finite bodies, the problem can be readily solved by applying the exponential Fourier transform with respect to x, defined by

$$\tilde{f}(k) = \int_{-\infty}^{\infty} e^{-ikx}f(x) \, dx, \qquad f(x) = \frac{1}{2\pi}\int_{-\infty}^{\infty} e^{ikx}\tilde{f}(k) \, dk. \qquad (1.11)$$

The Fourier–Laplace transform of Φ satisfies

$$\frac{d^2\tilde{\overline{\Phi}}}{dz^2} - k^2\tilde{\overline{\Phi}} = 0, \qquad -h < z < 0, \qquad (1.12)$$

$$\frac{d\tilde{\overline{\Phi}}}{dz} + \frac{s^2}{g}\tilde{\overline{\Phi}} = F(k,s), \qquad z = 0, \qquad (1.13)$$

$$\frac{d\tilde{\overline{\Phi}}}{dz} = \tilde{\overline{W}}, \qquad z = -h, \qquad (1.14)$$

where

$$F(k,s) \equiv -\frac{s\tilde{\overline{P}}_a(k,s)}{\rho g} - \tilde{\zeta}(k,0) + \frac{s}{g}\tilde{\Phi}(k,0,0). \qquad (1.15)$$

The general solution of Eq. (1.12) is

$$\tilde{\bar{\Phi}} = A \cosh k(z + h) + B \sinh k(z + h).$$

The coefficients A and B are found from the boundary conditions (1.13) and (1.14) with the following result:

$$\tilde{\bar{\Phi}} = \frac{1/\cosh kh}{s^2 + gk \tanh kh} \left[gF \cosh k(z + h) + \frac{\tilde{\bar{W}}}{k} (s^2 \sinh kz - gk \cosh kz) \right].$$

$$(1.16)$$

Clearly, the first and second addends in the brackets represent, respectively, all the disturbances on the free surface and on the bottom. Taking the inverse Fourier and Laplace transforms, we have, formally,

$$\Phi(x, z, t) = \frac{1}{2\pi} \int_{-\infty}^{\infty} dk \, e^{ikx} \frac{1}{2\pi i} \int_{\Gamma} ds \, e^{st} \tilde{\bar{\Phi}}(k, z, s). \quad (1.17)$$

To obtain the free-surface height we use Eq. (1.2b),

$$\zeta(x, t) = \frac{-P_a}{\rho g} - \frac{1}{g} \frac{\partial \Phi}{\partial t} (x, 0, t)$$

$$= \frac{-P_a}{\rho g} + \frac{1}{2\pi} \int_{-\infty}^{\infty} dk \, e^{ikx} \int_{\Gamma} \frac{ds}{2\pi i} e^{st} \frac{-s}{g} \tilde{\bar{\Phi}}(k, 0, s), \quad (1.18)$$

where $\tilde{\bar{\Phi}}$ is given by Eq. (1.16). The task is now to extract information from Eqs. (1.17) and (1.18). Two special cases will be studied in the following subsections.

2.1.1 Transient Disturbance Due to an Initial Displacement on the Free Surface

Here we let

$$P_a(x, t) = W(x, t) = \Phi(x, 0, 0) = 0 \quad \text{and} \quad \zeta(x, 0) \equiv \zeta_0(x) \neq 0,$$

$$(1.19)$$

hence

$$\tilde{\bar{W}} = 0, \quad F = -\tilde{\zeta}_0(k). \quad (1.20)$$

Equation (1.18) gives the free-surface height

$$\zeta = \frac{1}{4\pi i} \int_{-\infty}^{\infty} dk \; e^{ikx} \tilde{\zeta}_0(k) \int_{\Gamma} \frac{s \, e^{st} \, ds}{s^2 + gk \tanh kh}. \tag{1.21}$$

The s integral can be easily evaluated. The integrand has two real poles at

$$s = \pm i\omega, \quad \text{with } \omega = \left(gk \tanh kh \right)^{1/2}. \tag{1.22}$$

For $t < 0$ we introduce a closed semicircular contour in the right half s plane as shown in Fig. 1.1. Since the factor multiplying e^{st} in the integrand vanishes uniformly as $s \to \infty$, the line integral along the great semicircular arc is zero by Jordan's lemma. By Cauchy's residue theorem, the original s integral is zero, there being no singular points in the semicircle. Not suprisingly, then,

$$\zeta = 0, \quad t < 0. \tag{1.23}$$

For $t > 0$ we must choose a semicircle in the left half of the s plane. By Jordan's lemma again, the line integral along the semicircle vanishes, leaving only the residues for the two poles at $\pm i\omega$.

$$\frac{1}{2\pi i} \int_{\Gamma} \frac{s \, e^{st} \, ds}{s^2 + \omega^2} = \frac{1}{2\pi i} \int_{\Gamma} \frac{s \, e^{st} \, ds}{(s + i\omega)(s - i\omega)} = \cos \omega t, \quad t > 0.$$

Substituting into Eq. (1.21), we get

$$\zeta(x, t) = \frac{1}{2\pi} \int_{-\infty}^{\infty} dk \; e^{ikx} \cos \omega t \, \tilde{\zeta}_0(k). \tag{1.24}$$

Clearly, $\cos \omega t$ is even in k. In general, we can split $\zeta_0(x)$ into even and odd parts with respect to x: ζ_0^e and ζ_0^o. It follows from the definition of Fourier

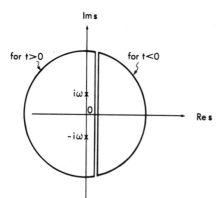

Figure 1.1 Contours for inverse Laplace transform.

transform that

$$\tilde{\zeta}_0(k) = 2\int_0^\infty dx \cos kx\, \zeta_0^e(x) - 2i\int_0^\infty dx \sin kx\, \zeta_0^o(x),$$

$$\equiv \tilde{\zeta}_0^e(k) + \tilde{\zeta}_0^o(k)$$

where $\tilde{\zeta}_0^e$ is real and even in k and $\tilde{\zeta}_0^o$ is imaginary and odd in k.

For simplicity, let ζ_0 be even in x; Eq. (1.24) may now be written

$$\zeta(x, t) = \frac{1}{\pi}\int_0^\infty dk\, \tilde{\zeta}_0^e \cos kx \cos \omega t$$

$$= \frac{1}{2\pi}\operatorname{Re}\int_0^\infty dk\, \tilde{\zeta}_0^e \big[e^{i(kx - \omega t)} + e^{i(kx + \omega t)} \big]. \qquad (1.25)$$

The first and second terms in the brackets above represent right- and left-going waves, respectively.

For a better physical understanding, approximations are necessary. At large t we can employ the *method of stationary phase* devised by Kelvin. Heuristically, the idea is as follows.

Consider the integral

$$I(t) = \int_a^b f e^{itg}\, dk \qquad (1.26)$$

where f and g are smooth functions of k. When t is large, the phase tg of the sinusoidal part oscillates rapidly as k varies. If one plots the integrand versus k, there is very little net area under the curve due to cancellation unless there is a point at which the phase is stationary, that is,

$$g'(k) = 0, \qquad k = k_0. \qquad (1.27)$$

In the neighborhood of this stationary point the oscillating factor of the integrand of Eq. (1.26) may be written

$$e^{itg(k_0)}\exp\{it[g(k) - g(k_0)]\}.$$

The real part of $\exp\{it[g(k) - g(k_0)]\}$ varies slowly, as sketched in Fig. 1.2, while the imaginary part slowly crosses the k axis at $k = k_0$. Therefore, a significant contribution to the integral can be expected from this neighborhood. If we approximate $g(k)$ by the first two terms of the Taylor expansion;

$$g(k) \cong g(k_0) + \tfrac{1}{2}(k - k_0)^2 g''(k_0),$$

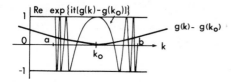

Figure 1.2   The real part of $\exp\{it[g(k) - g(k_0)]\}$.

then the integral may be written

$$I \cong e^{itg(k_0)}f(k_0)\int_{-\infty}^{\infty} dk \exp\left[\tfrac{1}{2}i(k - k_0)^2 t g''(k_0)\right],$$

where the limits (a, b) have been approximated by $(-\infty, \infty)$. Using the fact that

$$\int_{-\infty}^{\infty} e^{\pm itk^2}\, dk = \left(\frac{\pi}{t}\right)^{1/2} e^{\pm i\pi/4},$$

we finally have

$$I \cong e^{itg(k_0)}f(k_0)\left(\frac{2\pi}{t\,|g''(k_0)|}\right)^{1/2} e^{\pm i\pi/4}, \tag{1.28}$$

where the $\pm$ sign is to be taken if $g''(k_0) \gtrless 0$. It can be shown by a more elaborate analysis that the error is of order $O(t^{-1})$. Also if there is no stationary point in the range (a, b), the integral is at most of order $O(t^{-1})$. This and other information can be found in Stoker (1957) or Carrier, Krook, and Pearson (1966).

   Returning to Eq. (1.25), we need certain properties of the dispersion curve as sketched in Fig. 1.3. Consider $x > 0$. For the first integral

$$g(k) = k\frac{x}{t} - \omega,$$

it may be seen from Fig. 1.3b that there is a stationary point at

$$\frac{x}{t} = \omega'(k_0) = C_g(k_0) \qquad \text{if } \frac{x}{t} < (gh)^{1/2}. \tag{1.29}$$

In the same interval $(0, \infty)$ of k, there is no stationary point for the second integral. It follows from Eq. (1.28) that

$$\zeta \cong \frac{1}{2\pi}\tilde{\zeta}_0^e(k_0)\left[\frac{2\pi}{t\,|\omega''(k_0)|}\right]^{1/2}\cos\left[k_0 x - \omega(k_0)t + \frac{\pi}{4}\right] + O(t^{-1}),$$

$$x < (gh)^{1/2}t, \tag{1.30}$$

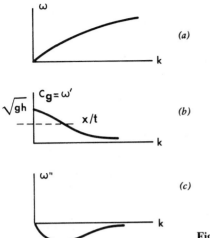

Figure 1.3 Variations of ω, ω', and ω'' with k.

where use is made of the fact that $\omega''(k) < 0$ (Fig. 1.3c), and

$$\zeta \cong O(t^{-1}), \qquad x > (gh)^{1/2}t. \tag{1.31}$$

Now let us examine the physics represented by Eq. (1.30). An observer moving at a certain speed x/t lower than $(gh)^{1/2}$ sees a train of sinusoidal waves of wavenumber k_0 [and frequency $\omega(k_0)$] whose group velocity equals x/t. The amplitude of the wavetrain decays as $O(t^{-1/2})$. For large x/t we see from Fig. 1.3(a) that k_0 is small, hence, a faster moving observer sees longer waves which are also of larger amplitude since $(|\omega''(k_0)|)^{1/2}$ is less. The precise shape of $\zeta_0(x)$ affects $\tilde{\zeta}_0(k)$ and the amplitude of the dispersed waves. For example, if

$$\zeta_0(x) = \frac{Sb}{\pi(x^2 + b^2)}$$

which is a symmetrical bump of area S and characteristic spread b, we find that

$$\tilde{\zeta}_0(k) = \tilde{\zeta}_0^e(k) = Se^{-|k|b}.$$

If the spread b is large, then $\tilde{\zeta}_0^e$ is not appreciable except for small k_0 or long leading waves. As b increases, the amplitude of a given k_0 decreases.

Summing up the views of many observers for the same t, we obtain a snapshot of the free surface (see Fig. 1.4). Thus, at a constant t, long waves are found toward the front, and short waves toward the rear. Now consider the snapshot at a later time $t_2 > t_1$. Both observers have now moved to the right. The spatial separation, however, has increased. In particular, let $\xi_1 \approx \xi_2$ so that

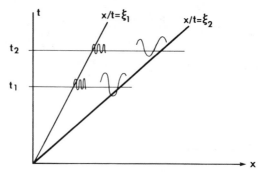

Figure 1.4   Space–time plot of dispersive waves between two moving observers.

between them $k, \omega \approx$ const. The total extent of a monochromatic wavetrain with k, ω now stretches with increasing t, implying that wave crests are created in the course of propagation.

To follow a particular wave crest at its phase speed, an observer must travel at a varying speed since k_0 and $C(k_0)$ do not remain constant as the crest moves into new territory. However, if one moves at the group velocity of the waves of length $2\pi/k_0$, one only sees sine waves of this length catching up from behind and then running away toward the front, since their phase velocity exceeds the group velocity.

A similar picture exists for the left-going disturbance.

2.1.2 Energy Propagation, Group Velocity

Consider a rightward traveling disturbance only. Equation (1.30) holds for large t and represents a progressive wave with amplitude

$$A = \frac{\tilde{\zeta}_0^e(k_0)}{2\pi} \left(\frac{2\pi}{t\,|\,\omega''(k_0)\,|} \right)^{1/2}, \tag{1.32}$$

which decays slowly as $t^{-1/2}$.

The wave energy density of this progressive wave is approximately

$$E \approx \rho \frac{gA^2}{2} = \tfrac{1}{2}\rho g\,|\,\zeta\,|^2 = \frac{\rho g}{2} \left| \frac{\tilde{\zeta}_0^e(k_0)}{2\pi} \left(\frac{2\pi}{t\,|\,\omega''(k_0)\,|} \right)^{1/2} \right|^2 = \frac{\rho g\,|\,\tilde{\zeta}_0^e(k_0)\,|^2}{4\omega t\,|\,\omega''(k_0)\,|}. $$

$$\tag{1.33}$$

At any given t, consider the waves between two observers moving at $C_{g_1} \equiv C_g(k_1)$ and $C_{g_2} \equiv C_g(k_2)$, that is, between the two rays

$$\frac{x_1}{t} = C_{g_1} \quad \text{and} \quad \frac{x_2}{t} = C_{g_2}$$

in the space–time diagram. The total wave energy between them is

$$\int_{x_2}^{x_2} E \, dx \cong \int_{x_1}^{x_2} \frac{\rho g |\tilde{\zeta}_0^e(k_0)|^2}{4\pi t |\omega''(k_0)|} \, dx. \tag{1.34}$$

Since $x = \omega'(k_0)t$ for fixed t and $\omega''(k_0) < 0$, we have

$$\frac{dx}{t} = \omega''(k_0) \, dk_0 = -|\omega''(k_0)| \, dk_0. \tag{1.35}$$

Now for $x_2 > x_1$, $k_2 < k_1$ (see Figs. 1.3a, b), Eq. (1.34) becomes

$$\int_{x_1}^{x_2} E \, dx \cong \int_{k_2}^{k_1} \frac{|\tilde{\zeta}_0^e(k_0)|^2}{4\pi} \, dk_0 = \text{const}, \tag{1.36}$$

which is constant in time. Therefore, the total energy of the waves between two observers moving at the local group velocities is conserved. This interpretation, due to Jeffreys and Jeffreys (1953), further strengthens the significance of the group velocity as discussed in Chapter One.

Whitham (1965) has shown that the asymptotic result by stationary phase for large x and t agrees with the so-called *geometrical optics theory*. From Eq. (1.29) we obtain, by differentiations with respect to x and t,

$$1 = \omega''(k)k_x t \quad \text{and} \quad 0 = \omega''(k)k_t t + \omega'$$

so that

$$k_x = \frac{1}{t\omega''(k)}, \qquad k_t = -\frac{\omega'}{t\omega''(k)}. \tag{1.37}$$

It follows that

$$\frac{\partial k}{\partial t} + \omega' \frac{\partial k}{\partial x} = 0,$$

which may also be written

$$\frac{\partial k}{\partial t} + \frac{\partial \omega}{\partial x} = 0. \tag{1.38}$$

Since

$$dk = \frac{\partial k}{\partial t} \, dt + \frac{\partial k}{\partial x} \, dx,$$

we see from Eq. (1.37) that along the curve $dx/dt = C_g = \omega'$, $dk = 0$; hence k

remains constant. Furthermore, by multiplying Eq. (1.33) with ω^{-1} and differentiating with respect to t and x, we obtain straightforwardly

$$\frac{\partial}{\partial t}\left(\frac{E}{\omega}\right) + \frac{\partial}{\partial x}\left(C_g \frac{E}{\omega}\right) = 0. \tag{1.39}$$

Both Eqs. (1.38) and (1.39) are basic results of the geometrical optics approximation and are of general validity for slowly varying, nearly periodic wavetrains as will be elaborated in Chapter Three.

2.1.3 Leading Waves Due to a Transient Disturbance

The fastest waves correspond to $k \simeq 0$ and move at the speed near $(gh)^{1/2}$. In the neighborhood of the wave front, $g'(k) \simeq x/t - (gh)^{1/2}$ is small, and the phase is nearly stationary. Furthermore $\omega''(k) \simeq -(gh)^{1/2}h^2 k$ is also very small and the approximation of Eq. (1.30) is not valid. A better approximation is needed (Kajiura, 1963).

Since $k \simeq 0$, we expand the phase function for small k as follows:

$$g(k) = k\frac{x}{t} - \omega(k) \simeq k\left(\frac{x}{t}\right) - (gh)^{1/2}\left(k - \frac{k^3 h^2}{6} + \cdots\right)$$

$$= k\left[\frac{x}{t} - (gh)^{1/2}\right] + \frac{(gh)^{1/2}}{6}h^2 k^3 + \cdots. \tag{1.40}$$

Near the leading wave, $x/t - (gh)^{1/2}$ can be zero; we must retain the term proportional to k^3. Again, only the first integral in Eq. (1.25) matters so that

$$\zeta = \frac{1}{2\pi}\int_0^\infty dk\, \tilde{\zeta}_0^e(k)\cos(kx - \omega t) + O\left(\frac{1}{t}\right)$$

$$\simeq \frac{1}{2\pi}\tilde{\zeta}_0^e(0)\int_0^\infty \cos\left\{k\left[x - (gh)^{1/2}t\right] + \left[\frac{(gh)^{1/2}h^2 t}{6}\right]k^3\right\} dk$$

where use is made of the fact that $\tilde{\zeta}_0^e$ is real. With the change of variables

$$Z^3 = \frac{2\left[x - (gh)^{1/2}t\right]^3}{(gh)^{1/2}h^2 t} \quad \text{and} \quad k\left[x - (gh)^{1/2}t\right] = Z\alpha,$$

the integral above becomes

$$\zeta \sim \frac{(2)^{1/3}\tilde{\zeta}_0^e(0)}{\pi\left((gh)^{1/2}h^2 t\right)^{1/3}}\int_0^\infty d\alpha \cos\left(Z\alpha + \frac{\alpha^3}{3}\right),$$

which can be expressed in terms of Airy's function of Z:

$$\text{Ai}(Z) \equiv \frac{1}{\pi} \int_0^\infty d\alpha \cos\left(Z\alpha + \frac{\alpha^3}{3}\right). \qquad (1.41)$$

Thus, we have

$$\zeta \sim \left[\frac{2}{(gh)^{1/2}h^2t}\right]^{1/3} \tilde{\zeta}_0^e(0)\text{Ai}\left\{\left[\frac{2}{(gh)^{1/2}h^2t}\right]^{1/3}\left[x - (gh)^{1/2}t\right]\right\}.$$

$$(1.42)$$

$\text{Ai}(Z)$ is oscillatory for $Z < 0$ and decays exponentially for $Z > 0$. Its variation is shown in Fig. 1.5.

The physical picture is as follows: For a fixed t, Z is proportional to $x - (gh)^{1/2}t$ which is the distance from the wave front $x = (gh)^{1/2}t$. At a fixed instant the amplitude is small ahead of the front, and the highest peak is at some distance behind. Toward the rear, the amplitude and the wavelength decrease. Since Z is proportional to $t^{-1/3}$, the snapshots at different times are of the same form except that the spatial scale is proportional to the factor $t^{1/3}$, meaning that the same wave form is being stretched out in time. During the evolution the amplitude decays as $t^{-1/3}$ while the rest of the wavetrain decays as $t^{-1/2}$. Thus, the head lives longer than the rest of the body. Note that the amplitude of the leading wave is proportional to $\tilde{\zeta}_0^e(0)$ which is equal to the total area of the initial displacement $\zeta_0^e(x)$.

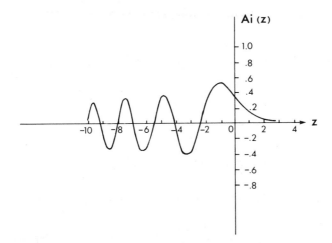

Figure 1.5 Leading wave due to a symmetrical surface hump or trough. The ordinate is $\zeta((gh)^{1/2}h^2t/2)^{1/3}[\tilde{\zeta}_0^e(0)]^{-1}$, see Eq. (1.42).

2.1.4   Tsunami Due to Tilting of the Bottom

Tsunamis are the water waves generated by earthquakes. If the seafloor displacement is known in the area of the earthquake, the water-wave problem is a purely hydrodynamic one. Unfortunately, direct measurements near the epicenter are too difficult to make, and a good deal of effort has been centered on using water-wave records measured at larger distances from the epicenter to infer roughly the nature of tectonic movement. Hence, there has been considerable theoretical studies on water waves due to a variety of ground movement.

Among the many features of tsunamis as recorded near a coast, two have been frequently (but not always) reported (Shepard, 1963). One feature is that the arrival of a tsunami is often preceded by the withdrawal of water from the beaches, and the other is that the first crest may not be the largest. In this section we shall show an idealized model which reproduces these features qualitatively.

We shall assume that there is no disturbance on the free surface

$$\zeta(x,0) = \Phi(x,0) = P_a(x,0,t) = 0. \tag{1.43}$$

On the seafloor $z = -h$, the ground displacement $H(x, t)$ is prescribed. Thus, $W = \partial H / \partial t$ is known and the transformed solution follows from Eq. (1.16).

$$\tilde{\Phi} = \frac{\tilde{\tilde{W}}}{k \cosh kh} \frac{s^2 \sinh kz - gk \cosh kz}{s^2 + gk \tanh kh}. \tag{1.44}$$

The free-surface displacement is

$$\zeta = \frac{1}{2\pi} \int_{-\infty}^{\infty} dk \frac{e^{ikx}}{\cosh kh} \frac{1}{2\pi i} \int_{\Gamma} ds \frac{s\tilde{\tilde{W}} e^{st}}{s^2 + \omega^2}, \tag{1.45}$$

where $\omega = (gk \tanh kh)^{1/2}$, as before. We further restrict the ground motion to be a sudden displacement accomplished in an infinitesimal interval of time,

$$H(x,0-) = 0 \quad \text{but} \quad H(x,0+) = H_0(x).$$

The ground velocity can be represented by a δ function

$$\frac{\partial \Phi}{\partial z} = W(x,t) = H_0(x)\,\delta(t)$$

so that $\tilde{\tilde{W}} = \tilde{H}_0(k)$. The s integral can be immediately evaluated to give

$$\zeta = \frac{1}{2\pi} \int_{-\infty}^{\infty} dk \frac{\tilde{H}_0(k)}{\cosh kh} \frac{1}{2} \left[e^{i(kx+\omega t)} + e^{i(kx-\omega t)} \right]. \tag{1.46}$$

Again any $H_0(x)$ can be thought of as the sum of $H_0^o(x)$ and $H_0^e(x)$ which are odd and even in x, respectively. By linearity, the two parts may be treated separately first and their results superimposed later. It is easily shown that the even part $H_0^e(x)$ has effects very similar to the previous example of symmetrical initial displacement on the free surface, the only difference being the factor $(\cosh kh)^{-1}$ which cuts down the influence of the short waves. We shall, therefore, only focus our attention to the odd part.

Let us introduce

$$H_0^o(x) = \frac{dB}{dx} \tag{1.47}$$

so that $\tilde{H}_0^o(k) = ik\tilde{B}(k)$. Since $\tilde{H}_0^o(k)$ is odd, $\tilde{B}$ must be real and even in k; hence,

$$\zeta = \frac{1}{2\pi} \int_{-\infty}^{\infty} dk \frac{e^{ikx}}{\cosh kh} ik\tilde{B}(k) \frac{1}{2}(e^{i\omega t} + e^{-i\omega t})$$

$$= \frac{1}{2\pi} \frac{d}{dx} \int_{-\infty}^{\infty} dk \frac{e^{ikx}}{\cosh kh} \tilde{B}(k) \frac{1}{2}(e^{i\omega t} + e^{-i\omega t})$$

$$= \frac{1}{2\pi} \frac{d}{dx} \operatorname{Re} \int_{0}^{\infty} dk \frac{e^{ikx}}{\cosh kh} \tilde{B}(k)(e^{i\omega t} + e^{-i\omega t}). \tag{1.48}$$

For large t and away from the leading waves, the integrals can be dealt with by the stationary phase method just as before, and many of the same qualitative features should be expected with the important difference that $\zeta \sim t^{-2/3}$ for $x/t = \text{const}$. Let us only look at the neighborhood of the *leading waves* propagating to $x > 0$. Again, the second integral dominates and the important contribution comes from the neighborhood of $k \simeq 0$. Hence

$$\operatorname{Re} \int_{0}^{\infty} dk \frac{e^{i(kx - \omega t)}}{\cosh kh} \tilde{B}(k)$$

$$\cong \operatorname{Re} \tilde{B}(0) \int_{0}^{\infty} dk \, e^{ikx} e^{-i\omega t}$$

$$\cong \operatorname{Re} \tilde{B}(0) \int_{0}^{\infty} dk \exp\left(i\left\{k\left[x - (gh)^{1/2}t\right] - \tfrac{1}{6}(gh)^{1/2}h^2k^3t\right\}\right)$$

$$= 2\tilde{B}(0)\left[\frac{2}{(gh)^{1/2}h^2t}\right]^{1/3} \operatorname{Ai}\left\{\left[\frac{2}{(gh)^{1/2}h^2t}\right]^{1/3}\left[x - (gh)^{1/2}t\right]\right\},$$

as discussed earlier. Differentiating with respect to x, we have

$$
\zeta \simeq \tilde{B}(0) \left[\frac{2}{(gh)^{1/2}h^2t} \right]^{1/3} \frac{d}{dx} \text{Ai} \left\{ \left[\frac{2}{(gh)^{1/2}h^2t} \right]^{1/3} \left[x - (gh)^{1/2}t \right] \right\}
$$

$$
= \tilde{B}(0) \left(\frac{2}{(gh)^{1/2}h^2t} \right)^{2/3} \text{Ai}' \left\{ \left[\frac{2}{(gh)^{1/2}h^2t} \right]^{1/3} \left[x - (gh)^{1/2}t \right] \right\},
$$

$$(1.49)$$

where

$$
\text{Ai}'(Z) \equiv \frac{d}{dZ} \text{Ai}(Z).
$$

The leading wave attenuates with time as $t^{-2/3}$ which is much faster than the case of a pure rise or fall (where $\zeta \sim t^{-1/3}$). This result is due to the fact that the ground movement is half positive and half negative, thereby reducing the net effect. The function $\text{Ai}'(Z)$ behaves as shown in Fig. 1.6. Note that

$$
\tilde{B}(0) = \int_{-\infty}^{\infty} B(x)\, dx = \int_{-\infty}^{\infty} dx \int_{-\infty}^{x} H_0^o(x')\, dx' = \int_{-\infty}^{\infty} x H_0^o(x)\, dx.
$$

Thus, if the ground tilts down on the right and up on the left, $\tilde{B}(0) > 0$ and the wave front propagating to the right is led by depression of water surface (hence withdrawal from a beach). The subsequent crests increase in amplitude. On the left side, $x < 0$, the wave front has the opposite phase and is led by a crest. If, however, the ground tilt is opposite in direction, that is, down on the left and up on the right, then the right-going wave front should be led by an elevation.

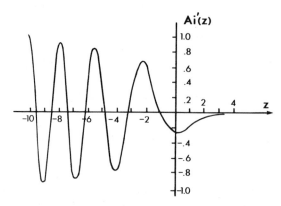

Figure 1.6   Leading wave due to antisymmetric ground tilt $\zeta [\tilde{B}(0)]^{-1}((gh)^{1/2}h^2t/2)^{2/3}$, see Eq. (1.49).

Kajiura pointed out that retaining the terms gk^3h^2 in $\omega(k)$ implies keeping dispersion to the lowest order, and the same results, Eqs. (1.42) and (1.49), may be obtained alternatively by invoking the long-wave approximation at the outset, which is clearly appropriate far away from the source. It will be shown in Chapter Eleven that such an approximation is given by the linearized Boussinesq equations which are, in one dimension, equivalent to

$$\frac{\partial^2 \zeta}{\partial t^2} = gh\left(\frac{\partial^2 \zeta}{\partial x^2} + \frac{h^2}{3}\frac{\partial^4 \zeta}{\partial x^4}\right). \tag{1.50}$$

The vicinity of the tsunamic source has been examined by Kajiura (1963) and Momoi (1964a, b; 1965a, b).

Exercise 2.1

Show that by solving Eq. (1.50) exactly with the initial conditions

$$\zeta(x,0) = \tilde{\zeta}_0(0)\delta(x), \qquad \zeta_t(x,0) = 0,$$

the answer is given by Eq. (1.42).

Exercise 2.2: Cauchy–Poisson Problem for Capillary–Gravity Waves

Consider a free surface with capillarity (refer to Exercise 5.2, Section 1.5). Solve for the two-dimensional free-surface response to a localized initial elevation: $\zeta(x,0) = (b/\pi)(x^2 + b^2)^{-1}$. Deduce the asymptotic result for large t and fixed x/t and describe the physical picture. Examine in particular the case where the stationary point is a zero of $\omega''(k)$.

Exercise 2.3: Waves on a Running Stream

Consider a river of constant depth h and uniform streaming velocity U. Formulate the linearized initial-boundary-value problem for the potential Φ of the disturbed flow defined by

$$\text{total velocity} = U\mathbf{i} + \nabla\Phi, \qquad \Phi = \Phi(x, z, t),$$

where x, y, z refer to the coordinates fixed in space. Examine the effect of U on the dispersion relation $\omega = \omega(k; U)$ for a progressive wave.

If at $t = 0$, a localized impulsive pressure $P = P_0\delta(x)\delta(t)$ is applied externally on the free surface, find the asymptotic behavior of $\zeta(x, t)$ for large t and x including the wave front. Describe the physics and the effects of U.

2.2 THREE-DIMENSIONAL TRANSIENT RESPONSE TO BOTTOM DISTURBANCES

If the source of disturbance is confined to a limited horizontal area, waves will propagate in all horizontal directions and the fluid motion will be three dimensional. We illustrate only the case of a tsunami due to a sudden motion of the seafloor (Kajiura, 1963).

The governing equation for the velocity potential $\Phi(x, y, z, t)$ is the three-dimensional Laplace equation. Let there be no disturbance on the free surface at any time. On the bottom, the ground motion is two dimensional:

$$\frac{\partial \Phi}{\partial z} = W(x, y, t), \qquad z = -h, \tag{2.1}$$

where W differs from zero only in a finite area. Furthermore,

$$\Phi, \nabla\Phi \to 0 \qquad \text{as } r = (x^2 + y^2)^{1/2} \to \infty. \tag{2.2}$$

One can solve the initial-boundary-value problem by Laplace transform with respect to t and two-dimensional Fourier transform with respect to x and y. Here a method by superposition of sources will prove to be quite efficient. Consider an impulsive disturbance concentrated at the origin $x = y = 0$, $z = -h$ at time $t = 0 +$. Denote the potential response by $G(x, y, z, t)$; then

$$\frac{\partial G}{\partial z} = \delta(x)\delta(y)\delta(t - 0 +), \qquad z = -h \tag{2.3}$$

instead of Eq. (2.1). Otherwise, G satisfies the same conditions as Φ, that is,

$$\nabla^2 G = 0, \tag{2.4}$$

$$G_{tt} + gG_z = 0, \qquad z = 0, \tag{2.5}$$

$$G = G_t = 0, \qquad t = 0, \quad z = 0, \tag{2.6}$$

$$G, \nabla G, \to 0, \qquad r \to \infty, \qquad t \text{ finite.} \tag{2.7}$$

Once $G(x, y, z, t)$ is found, Φ can be expressed immediately by

$$\Phi(x, y, z, t) = \int_0^t d\tau \int_{-\infty}^{\infty} dx' \int_{-\infty}^{\infty} dy' \, W(x', y', \tau) G(x - x', y - y', z, t - \tau). \tag{2.8}$$

Physically, Eq. (2.8) represents the superposition of elemental impulsive sources whose intensity at $x = x'$, $y = y'$, $z = -h$, and $t = \tau$ is $W(x', y', \tau) \, dx' \, dy' \, d\tau$. We record here for later use that the Laplace transforms of Φ, W, and G are related by the convolution theorem as follows:

$$\overline{\Phi}(x, y, s) = \iint_{-\infty}^{\infty} dx' \, dy' \, \overline{W}(x', y', s) \overline{G}(x - x', y - y', z, s). \tag{2.9}$$

$G(x, y, z, t)$ is considerably easier to construct because a point source has axial symmetry. Let us define $\delta(r)$ by

$$\delta(x)\delta(y) = \frac{\delta(r)}{2\pi r}$$

in the sense that the area integrals of both sides are equal, that is,

$$\int_0^{2\pi} d\theta \int_0^\infty \frac{r\,dr\,\delta(r)}{2\pi r} = 1 = \iint_{-\infty}^\infty dx\,dy\,\delta(x)\delta(y).$$

Equation (2.3) may be rewritten

$$\frac{\partial G}{\partial z} = \frac{1}{2\pi r}\delta(r)\delta(t - 0+); \tag{2.10}$$

now the problem of G does not involve θ.

Because of axial symmetry, a Hankel transform with the Bessel function $J_0(kr)$ as the weighting function can be applied. Define the Hankel transform (see Sneddon, 1951) by

$$\hat{f}(k) = \int_0^\infty rJ_0(kr)f(r)\,dr, \tag{2.11a}$$

then the inverse transform is

$$f(r) = \int_0^\infty kJ_0(kr)\hat{f}(k)\,dk. \tag{2.11b}$$

Let us define the compound Laplace–Hankel transform of G by $\hat{\bar{G}}$

$$\hat{\bar{G}} = \int_0^\infty e^{-st}\,dt\int_0^\infty rJ_0(kr)G\,dr.$$

In polar coordinates,

$$\frac{1}{r}\frac{\partial}{\partial r}\left(r\frac{\partial G}{\partial r}\right) + \frac{\partial^2 G}{\partial z^2} = 0, \qquad -h < z < 0, 0 \le r < \infty. \tag{2.12}$$

Applying Hankel transform to the first term, integrating by parts, using the boundary conditions at $r = 0$ and ∞, and invoking the differential equation satisfied by J_0, we can show that

$$\int_0^\infty dr\,rJ_0(kr)\left(\frac{1}{r}\frac{\partial}{\partial r}r\frac{\partial G}{\partial r}\right) = -k^2\hat{G}.$$

Thus, the Laplace–Hankel transform of Eq. (2.12) is

$$\frac{d^2}{dz^2}\hat{\overline{G}} - k^2\hat{\overline{G}} = 0.$$ (2.13)

The transform of the free-surface condition is,

$$\hat{\overline{G}}_z + \frac{s^2}{g}\hat{\overline{G}} = 0,$$ (2.14)

and the transform of the bottom boundary condition (2.10) is

$$\hat{\overline{G}}_z = \frac{1}{2\pi}.$$ (2.15)

The solution to Eq. (2.13), subject to the boundary conditions (2.14) and (2.15), is

$$\hat{\overline{G}} = \frac{1}{2\pi}\frac{1}{s^2 + \omega^2}\frac{s^2\sinh kz - gk\cosh kz}{k\cosh kh}$$ (2.16)

with $\omega^2 = gk\tanh kh$. Inverting the Hankel transform, we have

$$\overline{G}(r, z, s) = \int_0^\infty kJ_0(kr)\hat{\overline{G}}(k, z, s)\,dk, \qquad r = (x^2 + y^2)^{1/2}.$$ (2.17)

If the point disturbance is not at the origin but at some other point $\mathbf{r}'$, we must replace $|\mathbf{r}|$ by $|\mathbf{r} - \mathbf{r}'|$ so that

$$\overline{G}(|\mathbf{r} - \mathbf{r}'|, z, s) = \int_0^\infty kJ_0(k|\mathbf{r} - \mathbf{r}'|)\hat{\overline{G}}(k, z, s)\,dk,$$ (2.18)

where

$$x = r\cos\theta, \qquad y = r\sin\theta,$$

$$x' = r'\cos\theta', \qquad y' = r'\sin\theta',$$ (2.19)

$$|\mathbf{r} - \mathbf{r}'| \equiv [(x - x')^2 + (y - y')^2]^{1/2} = [r^2 + r'^2 - 2rr'\cos(\theta - \theta')]^{1/2}.$$

When Eq. (2.18) is substituted into Eq. (2.9), it follows that

$$\overline{\Phi}(r, \theta, z, s) = \int_0^\infty r'\,dr'\int_0^{2\pi} d\theta'\,\overline{W}(r', \theta', s)\int_0^\infty kJ_0(k|\mathbf{r} - \mathbf{r}'|)\hat{\overline{G}}(k, z, s)\,dk$$

(2.20)

The potential Φ can then be obtained by Laplace inversion.

The Laplace transform of the free-surface displacement is

$$\bar{\zeta} = -\frac{s}{g}\bar{\phi}\Big|_{z=0} = \frac{1}{2\pi}\int_0^\infty r'\,dr'\int_0^\pi d\theta'\,\overline{W}(r',\theta',s)$$

$$\times \int_0^\infty kJ_0(k\,|\mathbf{r}-\mathbf{r}'|)\frac{1}{\cosh kh}\frac{s}{s^2+\omega^2}\,dk. \qquad (2.21)$$

We now study a few specific cases.

2.2.1 Two-Dimensional Tsunami Due to Impulsive Bottom Displacement

In the special case of impulsive displacement

$$W(r,\theta,t) = \mathcal{W}(r,\theta)\delta(t-0+), \qquad (2.22)$$

the Laplace transform is $\overline{W} = \mathcal{W}(r,\theta)$. Inversion of the Laplace transform of Eq. (2.21) is immediate:

$$\zeta = \frac{1}{2\pi}\int_0^\infty r'\,dr'\int_0^{2\pi}d\theta'\,\mathcal{W}(r',\theta')\int_0^\infty\frac{kJ_0(k\,|\mathbf{r}-\mathbf{r}'|)}{\cosh kh}\cos\omega t\,dk. \quad (2.23)$$

Further progress can be made by expressing $J_0(k\,|r-r'|)$ as a series with the aid of the famous addition theorem (Watson, 1958, pp. 358–359)

$$J_0\left(k[r^2+r'^2-2rr'\cos(\theta-\theta')]^{1/2}\right) = \sum_{n=0}^\infty \varepsilon_n J_n(kr)J_n(kr')\cos n(\theta-\theta')$$

$$(2.24)$$

where ε_n is the Jacobi symbol ($\varepsilon_0 = 1$, $\varepsilon_n = 2$, $n = 1,2,3,\dots$). Substituting (2.24) into (2.23) and denoting

$$\frac{1}{2\pi}\int_0^\infty r'\,dr\int_0^{2\pi}d\theta'\,\mathcal{W}(r',\theta')J_n(kr')\binom{\cos n\theta'}{\sin n\theta'} = \binom{W_n^c(k)}{W_n^s(k)}, \quad (2.25)$$

we have

$$\zeta(r,\theta,t) = \sum_{n=0}^\infty \varepsilon_n\int_0^\infty kJ_n(kr)\frac{\cos\omega t}{\cosh kh}(W_n^c\cos n\theta + W_n^s\sin n\theta)\,dk.$$

$$(2.26)$$

In principle, given $\mathcal{W}(r,\theta)$ we may perform the integration in Eq. (2.25) and obtain $W_n^s(k)$ and $W_n^c(k)$ so that the final solution can be obtained by numerical integration and summation.

To get some physical ideas we consider the following two simple examples:
(i) Axially symmetric displacement:

$$\mathcal{W}(r, \theta) = W_0(r). \tag{2.27}$$

Due to the orthogonality of $\{\cos n\theta\}$ and $\{\sin n\theta\}$, it follows that

$$W_0^c = \int_0^\infty r W_0(r') J_0(kr')\,dr' = \hat{W}_0(k), \qquad n = 0,$$

$$W_n^c = 0, \quad W_n^s = 0 \qquad \text{all } n \neq 0.$$

We have, therefore,

$$\zeta(r, \theta, t) = \int_0^\infty k J_0(kr) \frac{\cos \omega t}{\cosh kh} \hat{W}_0(k)\,dk$$

$$= \int_0^\infty r' J_0(kr') W_0(r')\,dr' \int_0^\infty k J_0(kr) \frac{\cos \omega t}{\cosh kh}\,dk, \tag{2.28}$$

which can be deduced directly by Hankel transform with $J_0(kr)$, without recourse to the source function G.

(ii) Displacement which is antisymmetric about the y axis:

$$\mathcal{W}(r, \theta) = W_1(r) \cos \theta. \tag{2.29}$$

It is easy to show that

$$W_1^c = \frac{1}{2} \int_0^\infty r W_1 J_1(kr')\,dr', \quad W_n^c = 0 \qquad \text{all } n \neq 1, \qquad W_n^s = 0 \qquad \text{all } n.$$

The integral is just the Hankel transform of W_1 with J_1 as the weighting function. Since $\varepsilon_1 = 2$, we have

$$\zeta(r, \theta, t) = \cos \theta \int_0^\infty k J_1(kr) \frac{\cos \omega t}{\cosh kh}\,dk \int_0^\infty r' W_1(r') J_1(kr')\,dr'. \tag{2.30}$$

The above result can also be directly obtained by Hankel transform with the weighting function J_1.

In general, one may need many terms in the series of Eq. (2.26) to model a more general disturbance.

Let us examine the asymptotic behavior for large r and t for antisymmetric impulsive displacement only, leaving the symmetrical case as an exercise. Writing

$$F(k) = k \frac{\hat{W}_1}{\cosh kh} \quad \text{and} \quad \hat{W}_1(k) = \int_0^\infty r J_1(kr) W_1(r)\,dr,$$

and using the identity

$$J_1(kr) = \frac{1}{2\pi} \int_0^{2\pi} d\psi \exp[-i(\psi - kr \sin \psi)] = \frac{1}{\pi} \int_0^{\pi} d\psi \cos(\psi - kr \sin \psi)$$

(2.31)

which may be proved readily from the partial wave expansion, Appendix 4.A, Eq. (A.5), we may rewrite Eq. (2.30) as

$$\zeta(r, \theta, t) = \cos \theta \, \mathrm{Re} \, \frac{1}{\pi} \int_0^{\infty \pi} d\psi \int_0^{\infty} dk \, F(k) \cos(\psi - kr \sin \psi) e^{-i\omega t}$$

$$= \cos \theta \, \mathrm{Re} \, \frac{1}{2\pi} \int_0^{\pi} d\psi \left\{ e^{-i\psi} \int_0^{\infty} dk \, F(k) e^{ikr \sin \psi - i\omega t} \right.$$

$$\left. + e^{i\psi} \int_0^{\infty} dk \, F(k) e^{-ikr \sin \psi - i\omega t} \right\}.$$

(2.32)

Now consider the first double integral above

$$I_1 = \int_0^{\pi} d\psi \, e^{-i\psi} \int_0^{\infty} dk \, F(k) e^{it[k(r/t)\sin \psi - \omega(k)]}.$$

(2.33)

The phase function depends on two variables, k and ψ, and a stationary phase point can be sought in the strip $k \geq 0$, $0 \leq \psi \leq \pi$ by equating to zero partial derivatives with respect to k and ψ simultaneously. A comprehensive account for higher-dimensional stationary phase methods may be found in Papoulis (1968). Let us take the obvious route of first keeping ψ fixed and finding the stationary phase contribution along k, and then repeating the process for ψ. Thus, for large t, fixed r/t and $\sin \psi$, one may apply the method of stationary phase

$$g(k) = k \frac{r}{t} \sin \psi - \omega(k),$$

(2.34a)

$$g'(k) = \frac{r}{t} \sin \psi - \omega'(k),$$

(2.34b)

$$g'' = -\omega''(k) > 0.$$

(2.34c)

There is a stationary point at the zero of $g'(k)$ since $\sin \psi > 0$ in the range $0 < \psi < \pi$. The approximate value for the k integral is

$$\left(\frac{2\pi}{t|\omega''(k)|} \right)^{1/2} F(k) \exp\left\{ it\left[k \frac{r}{t} \sin \psi - \omega(k) \right] + i\frac{\pi}{4} \right\},$$

where the stationary point k depends on ψ through Eq. (2.34b).

By a similar analysis, the remaining integral in Eq. (2.32) has no stationary point, hence is of the order $O(1/t)$. The integral I_1 becomes

$$I_1 = \int_0^\pi d\psi\, e^{i(-\psi+\pi/4)}\left[\frac{2\pi}{t\,|\omega''(k)|}\right]^{1/2} F(k)e^{it[k(r/t)\sin\psi - \omega(k)]} + O\left(\frac{1}{t}\right).$$

(2.35)

The ψ integral can be approximated once more by the method of stationary phase for large t and fixed r/t. The phase function and its first two derivatives are

$$f(\psi) = k\frac{r}{t}\sin\psi - \omega(k),$$

(2.36a)

$$\frac{df}{d\psi} = k\frac{r}{t}\cos\psi + \frac{dk}{d\psi}\left[\frac{r}{t}\sin\psi - \omega'(k)\right],$$

(2.36b)

$$\frac{d^2f}{d\psi^2} = -k\frac{r}{t}\sin\psi + \frac{r}{t}\cos\psi\frac{dk}{d\psi} + \frac{d^2k}{d\psi^2}\left[\frac{r}{t}\sin\psi - \omega'(k)\right].$$

(2.36c)

By the use of Eq. (2.34b), the point of stationary phase is clearly at $\psi = \pi/2$. Incorporating this result into Eqs. (2.36b) and (2.36c), we obtain

$$\frac{r}{t} - \omega'(k_0) = 0,$$

(2.37)

for the stationary point, now denoted by k_0, and

$$\left.\frac{d^2f}{d\psi^2}\right|_{\psi=\pi/2} = -k_0\frac{r}{t} < 0.$$

(2.38)

Equation (1.28) may be applied to Eq. (2.35), yielding

$$I_1 = -i\pi\left[\frac{2\pi}{t\,|\omega''(k_0)|}\right]^{1/2} F(k_0)\left(\frac{2}{\pi k_0 r}\right)^{1/2} e^{i[k_0 r - \omega(k_0)t]} + O\left(\frac{1}{t}\right).$$

The second double integral in Eq. (2.32) does not have a stationary point in

$k = [0, \infty]$, hence it is of $O(1/t)$. Finally, the total displacement is

$$
\zeta(r, \theta, t) = \cos \theta \, \mathrm{Re} \, \frac{-i}{2} \left(\frac{2\pi}{t \, | \, \omega''(k_0) \, |} \right)^{1/2}
$$

$$
\times F(k_0) \left(\frac{2}{\pi k_0 r} \right)^{1/2} e^{i[k_0 r - \omega(k_0) t]} + O\left(\frac{1}{t} \right)
$$

$$
= \tfrac{1}{2} \cos \theta \left[\frac{2\pi}{t \, | \, \omega''(k_0) \, |} \right]^{1/2} F(k_0) \left(\frac{2}{\pi k_0 r} \right)^{1/2}
$$

$$
\times \sin[k_0 r - \omega(k_0) t] + O\left(\frac{1}{t} \right). \tag{2.39}
$$

The preceding result can also be obtained by using the asymptotic formula of $J_1(kr)$ for large kr,

$$
J_1(kr) \simeq \left(\frac{2}{\pi kr} \right)^{1/2} \cos \left(kr - \frac{3\pi}{4} \right),
$$

and then applying the method of stationary phase only once. However, the legitimacy of assuming large kr when k ranges from 0 to ∞ needs confirmation and we have taken a more cautious route here.

The physical features of dispersion are almost the same as in the one-dimensional case and need not be elaborated. It is only necessary to point out that the amplitude decay rate is different since for $r/t = $ const

$$
\zeta \simeq \frac{1}{2} \frac{1}{t} \cos \theta F(k_0) \left[\frac{2}{| \, \omega''(k_0) \, |} \right]^{1/2} \left(\frac{2}{k_0(r/t)} \right)^{1/2} \sin(k_0 r - \omega_0 t),
$$

$$
\tag{2.40}
$$

where k_0 depends on r/t according to Eq. (2.38). Thus, the individual waves found near $r/t = $ const decay at the rate of $O(1/t)$ which is due to the radial spreading of two-dimensional waves. The antisymmetric nature of the source is exactly carried over to the propagating waves by the factor $\cos \theta$; the wave is the greatest along the x direction $\theta = 0$, and insignificant along the axis of antisymmetry $\theta = \pi/2, 3\pi/2$.

For a more explicit result it is necessary to prescribe $W_1(r)$. For example, one may assume that

$$W_1(r) = \frac{A}{a}(a^2 - r^2)^{1/2}, \quad r < a,$$

$$= 0, \quad r > a. \tag{2.41}$$

It can be inferred from a formula in Erdelyi (1954, II, p. 24, No. 25) that

$$\hat{W}_1(k) = \frac{A}{a}\int_0^\infty rJ_1(kr)(a^2 - r^2)^{1/2}\,dr = \frac{A}{a}\left(\frac{\pi}{2}\right)^{1/2}\frac{a^2}{k}J_1^2\left(\frac{ka}{2}\right).$$

Hence

$$F(k) = \frac{k}{\cosh kh}\hat{W}_1 = \left(\frac{\pi}{2}\right)^{1/2}\frac{Aa}{\cosh kh}J_1^2\left(\frac{ka}{2}\right) \tag{2.42}$$

which shows the effect of the size (a) of the source area. Through the Bessel function, $F(k)$ oscillates in ka, which is a manifestation of interference of waves from different parts of the source area.

2.2.2 Leading Waves of a Two-Dimensional Tsunami

Let us continue the antisymmetric example with the specific W_1 given by Eq. (2.41). In the zone of leading waves $kh \ll 1$, but for sufficiently large r, the leading wave must have some finite wavelength so that

$$kr \gg 1.$$

We may either express $J_1(kr)$ as an integral, Eq. (2.31), and carry out the stationary phase approximation for the ψ integral first, or take the asymptotic approximation of $J_1(kr)$ for large kr. Either way the result is

$$\zeta \simeq \cos\theta\,\mathrm{Re}\int_0^\infty dk\,F(k)\left(\frac{2}{\pi kr}\right)^{1/2}\frac{1}{2}\left[e^{ikr - i\omega t - i3\pi/4} + e^{-ikr - i\omega t + i3\pi/4}\right]. \tag{2.43}$$

For leading waves $kh \ll 1$, only the first integrand matters and we can expand

$$\omega \cong (gh)^{1/2}\left(k - \frac{k^3 h^2}{6}\right),$$

$$F(k) \cong Aa\left(\frac{\pi}{2}\right)^{1/2}\left(\frac{ka}{4}\right)^2 = \frac{Aa^3}{16}\left(\frac{\pi}{2}\right)k^2.$$

It follows that

$$\zeta \cong \frac{\cos\theta}{2}\frac{Aa^3}{16}\frac{1}{r^{1/2}}\operatorname{Re}e^{-i3\pi/4}\int_0^\infty dk\,k^{3/2}$$

$$\times\exp\left(i\left\{k[r-(gh)^{1/2}t]+\frac{(gh)^{1/2}h^2k^3t}{6}\right\}\right).\qquad(2.44)$$

This integral cannot be expressed in terms of known functions. Let us first rewrite it as follows:

$$\int_0^\infty = h^{-5/2}\int_0^\infty d(kh)(kh)^{3/2}\exp\left(i\left\{kh\left[\frac{r}{h}-\left(\frac{g}{h}\right)^{1/2}t\right]+\frac{(kh)^3}{6}\left(\frac{g}{h}\right)^{1/2}t\right\}\right)$$

$$(2.45)$$

and introduce the new variables (Kajiura, 1963, p. 549),

$$\frac{(kh)^3}{g}\left(\frac{g}{h}\right)^{1/2}t=u^6\quad\text{or}\quad kh=u^2\left[\left(\frac{g}{h}\right)^{1/2}\frac{t}{6}\right]^{-1/3}.$$

Then the integral of Eq. (2.45) becomes

$$h^{-5/2}\left[\left(\frac{g}{h}\right)^{1/2}\frac{t}{6}\right]^{-5/6}\int_0^\infty du\,2u^4e^{i(u^2p+u^6)},\qquad(2.46)$$

with

$$p=\frac{r/h-(g/h)^{1/2}t}{[(g/h)t/6]^{1/3}}.\qquad(2.47)$$

Equation (2.46) can be rewritten

$$-2h^{-5/2}\left[\left(\frac{g}{h}\right)^{1/2}\frac{t}{6}\right]^{-5/6}\frac{d^2}{dp^2}\int_0^\infty du\,e^{i(u^2p+u^6)},$$

whereupon Eq. (2.44) becomes

$$\zeta\cong\frac{\cos\theta}{2^{1/2}}\frac{Aa^3}{16r^{1/2}}h^{-5/2}\left(\left(\frac{g}{h}\right)^{1/2}\frac{t}{6}\right)^{-5/6}\frac{d^2}{dp^2}\operatorname{Re}(1+i)\int_0^\infty du\,e^{i(u^2p+u^6)}.$$

$$(2.48)$$

For $p = 0$, that is, if the observer is exactly at $r = (gh)^{1/2}t$, the integral in Eq. (2.46) can be evaluated by letting $u^6 = \tau$,

$$\int_0^\infty du\, u^4 e^{iu^6} = \int_0^\infty d\tau\, \tau^{-1/6} e^{i\tau} = \Gamma\!\left(\frac{5}{6}\right) e^{i5\pi/12}.$$

For general p we follow Kajiura and define

$$T(p) = \text{Re}(1 + i)\int_0^\infty du\, e^{i(u^2 p + u^6)}, \tag{2.49}$$

then

$$\zeta = \cos\theta \frac{Aa^3}{16(2r)^{1/2}} \frac{T_{pp}}{h^{5/2}\big((g/h)^{1/2}t/6\big)^{5/6}}. \tag{2.50}$$

The variations of T, $-T_p$, and T_{pp} are plotted in Fig. 2.1. Since the coefficient

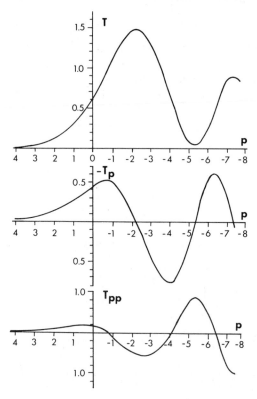

Figure 2.1 T, $-T_p$, and T_{pp} as functions of p. (From Kajiura, 1963, *Bull. Earthquake Res. Inst.* Univ. Tokyo.)

of T_{pp} in Eq. (2.48) is proportional to

$$r^{-1/2}t^{-5/6} = \left(\frac{r}{t}\right)^{-1/2}t^{-4/3} \sim \left(\frac{r}{t}\right)^{5/6}r^{-4/3},$$

we conclude that near the wave front $r/t \cong (gh)^{1/2}$ the wave amplitude decays as $t^{-4/3}$ or $r^{-4/3}$. If $A < 0$, the ground tilts down on the right $-\pi/2 < \theta < \pi/2$ and up on the left. To an observer on the right the leading waves are a low trough followed by a high crest, similar to the two-dimensional case.

2.3   THE PROPAGATION OF A DISPERSIVE WAVE PACKET

Let us leave the idealized tsunamis and study the evolution of a slowly modulated wave group in order to achieve further understanding of dispersion. Consider a disturbance traveling to the right, which may be represented as a superposition of sinusoidal waves with a continuum of wave lengths:

$$\zeta(x, t) = \text{Re} \int_{-\infty}^{\infty} \mathcal{C}(k)e^{i(kx-\omega(k)t)} \, dk. \tag{3.1}$$

The result may be generated by a wavemaker at $x \sim -\infty$ in a long tank from $t \sim -\infty$. The amplitude spectrum $\mathcal{C}(k)$ is determined by the initial disturbance (see Section 2.1). We shall leave $\omega(k)$ to be general so that the water wave is just a special case. Consider, in particular, a wave packet with a Gaussian envelope:

$$\zeta(x,0) = \text{Re} \, A_0 e^{ik_0 x}e^{-x^2/4\sigma^2}. \tag{3.2}$$

The amplitude spectrum is obtained by inverse Fourier transform

$$\mathcal{C}(k) = \frac{1}{2\pi}\int_{-\infty}^{\infty} \zeta(x,0)e^{-ikx} \, dx$$

$$= \frac{A_0}{2\pi}\int_{-\infty}^{\infty} e^{-i(k-k_0)x-x^2/4\sigma^2} \, dx$$

$$= \frac{A_0}{2\pi}\int_{-\infty}^{\infty} e^{-[x/2\sigma+i(k-k_0)\sigma]^2-(k-k_0)^2\sigma^2} \, dx \tag{3.3}$$

after completing the square. Letting $u = x/2\sigma + i(k - k_0)\sigma$, we have further,

$$\mathcal{C} = \frac{A_0}{2\pi}2\sigma e^{-(k-k_0)^2\sigma^2}\int_{\Gamma} e^{-u^2} \, du,$$

where the contour Γ is a straight line from $-\infty + i(k - k_0)\sigma$ to $\infty + i(k -$

$k_0)\sigma$ in the complex u plane. Since e^{-u^2} is analytic in the strip between Γ and the real u axis, by Cauchy's theorem the contour can be replaced by the real axis. Using the well-known result

$$\int_{-\infty}^{\infty} e^{-u^2} \, du = \pi^{1/2}, \tag{3.4}$$

we have

$$\mathcal{C}(k) = \frac{A_0\sigma}{\pi^{1/2}} e^{-(k-k_0)^2\sigma^2}. \tag{3.5}$$

Thus, the wave profile at any t is

$$\zeta = \mathrm{Re}\frac{A_0\sigma}{\pi^{1/2}} \int_{-\infty}^{\infty} e^{-(k-k_0)^2\sigma^2+i(kx-\omega t)} \, dk. \tag{3.6}$$

Let us study the behavior of the above integral when σk_0 is very large, that is, when the original envelope is very flat or the amplitude spectrum is very sharp near $k = k_0$. The integrand attenuates quickly away from $k = k_0$ so that one may approximate $\omega(k)$ by a few terms in the Taylor expansion:

$$\omega(k) = \omega_0 + (k - k_0)\omega_0' + \tfrac{1}{2}(k - k_0)^2\omega_0'' + \cdots,$$

where $\omega_0 \equiv \omega(k_0)$, $\omega_0' \equiv \omega'(k_0)$, and so on. After we let $u = k - k_0$, ζ becomes

$$\zeta \cong \mathrm{Re}\frac{'A_0\sigma}{\pi^{1/2}} e^{i(k_0x-\omega_0t)} \cdot \int_{-\infty}^{\infty} e^{-(\sigma^2+i\omega_0''t/2)u^2+i(x-\omega_0't)u} \, du. \tag{3.7}$$

Completing the squares and using Eq. (3.4), we get

$$\zeta \cong \mathrm{Re}\, A_0 \frac{e^{i(k_0x-\omega_0t)}}{\left[1 + (i/2)\omega_0''t\right]^{1/2}} \exp\left\{\frac{-(x - \omega_0't)^2}{4\sigma^2\left[1 + (i/2)\omega_0''t\right]}\right\}. \tag{3.8}$$

Clearly, the envelope moves at the group velocity $C_g = \omega_0'$; its maximum is at $x = C_g t$ and attenuates as $t^{1/2}$ for large t. In addition, the length of the envelope, which is measured by

$$2\sigma\left(1 + \frac{i}{2}\omega_0''t\right)^{1/2},$$

increases as $t^{1/2}$ for large t. Therefore, the whole wave group flattens during propagation.

Compared with Eq. (5.4), Chapter One, Eq. (3.8) is an improved approximation. Several other observations are useful here. Let us regard Eq. (3.8) as slowly modulated sinusoidal waves. The energy contained in the entire wave group can be approximated by

$$\frac{1}{4}\rho g A_0^2 \int_{-\infty}^{\infty} dx \left| \frac{\exp\{-(x - \omega_0' t)^2 / 4\sigma^2 [1 + (i/2)\omega_0'' t]\}}{[1 + (i/2)\omega_0'' t]^{1/2}} \right|^2$$

$$= \frac{\rho g A_0^2}{4} \int_{-\infty}^{\infty} dx \frac{\exp\{-(x - \omega_0' t)^2 / 2\sigma^2 [1 + (\omega_0'' t)^2/4]\}}{(1 + (\omega_0'' t)^2/4)^{1/2}}$$

$$= \frac{\rho g A_0^2}{4} 2^{1/2} \sigma \int_{-\infty}^{\infty} e^{-u^2} du = \frac{\rho g A_0^2}{22^{1/2}} \pi^{1/2} \sigma, \tag{3.9}$$

after an obvious change of variables. The total energy in the initial profile, Eq. (3.2), is

$$\frac{\rho g}{4} A_0^2 \int_{-\infty}^{\infty} e^{-x^2/2\sigma^2} dx = \frac{\rho g}{2(2)^{1/2}} A_0^2 \pi^{1/2} \sigma. \tag{3.10}$$

Thus energy is conserved, as it must be.

It may be noted that to the order σ^{-2} where σ^{-1} designates the slow rate of envelope modulation, the curvature of the dispersion curve ω'' is involved. To see that this is no accident, let us examine the elementary example of two wavetrains with slightly different wavelengths $k_+ = k + dk$ and $k_- = k - dk$ with $dk/k \ll 1$,

$$\zeta = A_0\{\exp[i(k_+ x - \omega_+ t)] + \exp[i(k_- x - \omega_- t)]\}, \tag{3.11}$$

where $\omega_\alpha = \omega(k_\alpha)$, $\alpha = 1, 2$. Expanding ω to $O(dk)^2$, we have

$$\omega_\pm = \left[\omega \pm \omega' dk + \tfrac{1}{2}\omega''(dk)^2 + \cdots\right]_k, \tag{3.12}$$

so that

$$\zeta \cong 2A_0\cos\left[dk(x - C_g t)\right]\exp\left\{i\left(kx - \left[\omega + \tfrac{1}{2}\omega''(dk)^2\right]t\right)\right\}. \tag{3.13}$$

Over the space and time scale $O(dk)^{-1}$, the envelope modulates and moves at C_g; however, over the time scale $O(dk)^{-2}$, the phase, in particular the frequency, changes. This example suggests that there is a cascade of time scales $O(1), O(dk)^{-1}, O(dk)^{-2}, \ldots$, and so on.

Finally, for any amplitude spectrum sharply peaked around k_0 (the Gaussian spectrum (3.5) being a special case), Eq. (3.1) may be approximated by

$$\zeta(x, t) \cong \text{Re}\{A(x, t)e^{i(k_0 x - \omega_0 t)}\}, \tag{3.14}$$

where

$$A(x, t) = \int_{-\infty}^{\infty} dk \, \mathcal{Q}(k) \exp\left\{i\left[(k - k_0)x - \left[(k - k_0)\omega_0'\right.\right.\right.$$
$$\left.\left.\left. + \tfrac{1}{2}(k - k_0)^2 \omega_0''\right]t\right]\right\}. \tag{3.15}$$

It is readily verified that A satisfies the following differential equation:

$$\frac{\partial A}{\partial t} + \omega_0' \frac{\partial A}{\partial x} = \frac{i}{2}\omega_0'' \frac{\partial^2 A}{\partial x^2}. \tag{3.16}$$

Such a simple equation obviously makes subsequent analysis easy, as will be demonstrated in the next section. To prepare the ground for further extension to nonlinear problems where formally exact solutions are often not possible, we shall also rederive Eq. (3.16) by an alternative method directly from the governing equations, and not from the integral solution.

2.4 SLOWLY VARYING WAVE TRAIN BY MULTIPLE-SCALES ANALYSIS

Let us incorporate the difference in scales as suggested at the end of Section 2.3. by formally introducing slow variables

$$\begin{aligned} x_1 &= \mu x, & x_2 &= \mu^2 x, \dots, \\ t_1 &= \mu t, & t_2 &= \mu^2 t, \dots, \end{aligned} \tag{4.1}$$

where $\mu \ll 1$ measures the ratio of two time scales, and then treating these variables in a perturbation analysis as if they were independent. This device may appear very artificial to the newcomer, but it has been shown to agree with other methods in old problems and to be an especially powerful tool for problems with weak nonlinearity. For a systematic exposition, see Cole (1968) and Nayfeh (1973).

Assume that

$$\Phi(x, z, t) = \Phi(x, x_1, x_2, \dots; z; t, t_1, t_2, \dots),$$
$$\zeta(x, t) = \zeta(x, x_1, x_2, \dots; t, t_1, t_2, \dots). \tag{4.2}$$

Derivatives with respect to x and t must now be replaced by

$$\frac{\partial}{\partial x} \to \frac{\partial}{\partial x} + \mu \frac{\partial}{\partial x_1} + \mu^2 \frac{\partial}{\partial x_2} + \cdots, \tag{4.3}$$

so that

$$\frac{\partial^2}{\partial x^2} \to \frac{\partial^2}{\partial x^2} + 2\mu \frac{\partial^2}{\partial x \, \partial x_1} + \mu^2 \left(\frac{\partial^2}{\partial x_1^2} + 2 \frac{\partial^2}{\partial x \, \partial x_2} \right) + \cdots. \tag{4.4}$$

Similar replacements must be made for time derivatives, while z derivatives remain the same. We now restrict ourselves to slowly varying sinusoidal waves and assume a perturbation series as follows:

$$\Phi = \left(\psi_0 + \mu \psi_1 + \mu^2 \psi_2 + \cdots \right) e^{i(kx - \omega t)}, \tag{4.5}$$

where

$$\psi_\alpha = \psi_\alpha(x_1, x_2, \ldots; z; t_1, t_2, \ldots), \qquad \alpha = 0, 1, 2, \ldots. \tag{4.6}$$

Substituting Eq. (4.4) to Eq. (4.6) into Laplace's equation and separating by powers of μ, we obtain

$$O(\mu^0): \qquad -k^2 \psi_0 + \frac{\partial^2 \psi_0}{\partial z^2} = 0, \tag{4.7a}$$

$$O(\mu): \qquad -k^2 \psi_1 + \frac{\partial^2 \psi_1}{\partial z^2} = -2ik \frac{\partial \psi_0}{\partial x_1}, \tag{4.7b}$$

$$O(\mu^2): \qquad -k^2 \psi_2 + \frac{\partial^2 \psi_2}{\partial z^2} = -\left(2ik \frac{\partial \psi_1}{\partial x_1} + \frac{\partial^2 \psi_0}{\partial x_1^2} + 2ik \frac{\partial \psi_0}{\partial x_2} \right). \tag{4.7c}$$

Similarly, the free-surface boundary condition gives

$$O(\mu^0): \qquad g \frac{\partial \psi_0}{\partial z} - \omega^2 \psi_0 = 0, \tag{4.8a}$$

$$O(\mu^1): \qquad g \frac{\partial \psi_1}{\partial z} - \omega^2 \psi_1 = 2i\omega \frac{\partial \psi_0}{\partial t_1}, \tag{4.8b}$$

$$O(\mu^2): \qquad g \frac{\partial \psi_2}{\partial z} - \omega^2 \psi_2 = 2i\omega \frac{\partial \psi_1}{\partial t_1} - \left(\frac{\partial^2 \psi_0}{\partial t_1^2} - 2i\omega \frac{\partial \psi_0}{\partial t_2} \right). \tag{4.8c}$$

On the bottom we have

$$\frac{\partial \psi_0}{\partial z} = \frac{\partial \psi_1}{\partial z} = \frac{\partial \psi_2}{\partial z} = 0, \qquad z = -h. \qquad (4.9a, b, c)$$

It is obvious that the solution for ψ_0 governed by Eqs. (4.7a), (4.8a), and (4.9a), is simply

$$\psi_0 = -\frac{igA}{\omega} \frac{\cosh k(z+h)}{\cosh kh}, \qquad A = A(x_1, x_2, \ldots; t_1, t_2, \ldots) \quad (4.10)$$

with $\omega^2 = gk \tanh kh$. The amplitude A is thus far undetermined. Now ψ_1 is governed by the inhomogeneous boundary-value problem (4.7b), (4.8b), and (4.9b). Since the homogeneous version of the boundary-value problem has ψ_0 as a nontrivial solution, the inhomogeneous problem must satisfy a solvability condition,[†] which follows by applying Green's theorem to ψ_0 and ψ_1:

$$\int_{-h}^{0} dz \left[\psi_0 \left(\frac{\partial^2 \psi_1}{\partial z^2} - k^2 \psi_1 \right) - \psi_1 \left(\frac{\partial^2 \psi_0}{\partial z^2} - k^2 \psi_0 \right) \right] = \left[\psi_0 \frac{\partial \psi_1}{\partial z} - \psi_1 \frac{\partial \psi_0}{\partial z} \right]_{-h}^{0}.$$

$$(4.11)$$

If Eqs. (4.7a, b) are used on the left, and Eqs. (4.8a, b) and (4.9a, b) are used on the right, we get, from the preceding theorem.

$$-\frac{\partial A}{\partial x_1} \left[\frac{gk}{\omega} \frac{1}{\cosh^2 kh} \int_{-h}^{0} \cosh^2 k(z+h)\, dz \right] = \frac{\partial A}{\partial t_1}.$$

In view of Eq. (5.12), Chapter One, it follows that

$$\frac{\partial A}{\partial t_1} + C_g \frac{\partial A}{\partial x_1} = 0. \qquad (4.12)$$

The solution can be easily verified to be $A(x_1 - C_g t_1)$, implying that the envelope propagates at the group velocity without change of form. This general result includes Eq. (5.5), Chapter One, as a special case. In addition, it also applies to the front of a gradually started and steadily maintained wavetrain as anticipated in Section 1.5.2.

[†] This condition is related to a very general mathematical theorem called *Fredholm alternative*. In the context of boundary-value problems the theorem can be stated as follows (see, e.g., Garabedian, 1964): "Either the inhomogeneous boundary-value problem is solvable whatever the forcing terms may be, or the corresponding homogeneous problem has one or more eigen functions (non-trivial solutions). In the first case the inhomogeneous solution is unique. In the second case the inhomogeneous problem is solvable if and only if the forcing terms are orthogonal to all the eigen functions of the homogeneous problem."

In our problem, ψ_0 is the eigenfunction to the homogeneous problem and ψ_1 is the solution to the inhomogeneous problem; Eq. (4.12) is the "orthogonality" condition.

Before proceeding to the next order, let us digress for an immediate application. When sinusoidal disturbances are generated in a localized region, there is a mathematical ambiguity in the boundary condition at infinity in the steady-state formulation. For example, merely stating that the disturbance should remain finite (in one dimension) or die out (in two or three dimensions) at infinity does not guarantee the uniqueness of the solution; a much stronger condition must be imposed. Within the strict limits of a steady-state formulation this condition is usually stated as follows: "A locally generated sinusoidal disturbance must propagate outward to infinity." This important statement is called the *radiation condition*. A justification, even for a special case, is certainly desirable.

Let us regard the steady state as the limit of $t \to \infty$ of an initial-value problem. In particular, consider the one-dimensional waves in the domain $x > 0$ due to the sinusoidal oscillation of a wavemaker at $x = 0$. Let the amplitude of the sinusoidal wave near $x = 0$ vary slowly from 0 at $t \sim -\infty$ to a constant A_0 at $t \sim +\infty$ according to some law $A(t) = \overline{A}(t_1)$. (See Fig. 4.1.) We then expect the solution at any $x_1 > 0$ to be given by Eq. (4.10) with $A(x, t) = \overline{A}(t_1 - x_1/C_g)$. By causality, the wave amplitude vanishes at sufficiently large x_1 for any finite t; hence $A(x, t) \cong \overline{A}(-x_1/C_g) \downarrow 0$ as $x_1 \uparrow +\infty$. In view of Fig. 4.1 this is possible only if $C_g > 0$. Since C_g and k are of the same sign, we must have $k > 0$, which implies from Eq. (4.5) that $\psi_0 e^{i(kx - \omega t)}$ propagates to the right, that is, *outgoing*.

In principle, the radiation condition can be deduced from an initial-value problem without the assumption of a slow start. However, the required analysis is long (see Stoker, 1948, 1957).

The permanence of form implied by Eq. (4.12) is only true on the scale $O(\mu^{-1})$, that is, with respect to x_1 and t_1. Let us pursue the next order to observe changes over a longer distance or time $O(\mu^{-2})$, that is, with respect to x_2 and t_2.

First, we leave it to the reader to show that the inhomogeneous solution for ψ_1 is

$$\psi_1 = -\frac{g}{\omega k}\frac{Q \operatorname{sh} Q}{\omega k \operatorname{ch} kh}\frac{\partial A}{\partial x_1}, \qquad Q \equiv k(z + h), \qquad \operatorname{sh} \equiv \sinh, \qquad \operatorname{ch} \equiv \cosh$$

$$(4.13)$$

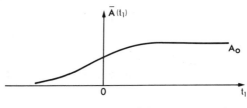

Figure 4.1

which satisfies the boundary conditions because of Eq. (4.12). A homogeneous solution is omitted because it can be considered as contained in ψ_0. Substituting Eqs. (4.10) and (4.13) into the right-hand sides of Eqs. (4.7c) and (4.8c), we obtain

$$\frac{\partial^2 \psi_2}{\partial z^2} - k^2 \psi_2 = \frac{2ig}{\omega} \frac{\partial^2 A}{\partial x_1^2} \frac{Q \, \mathrm{sh} \, Q}{\mathrm{ch} \, kh} + \frac{ig}{\omega} \left(\frac{\partial^2 A}{\partial x_1^2} + 2ik \frac{\partial A}{\partial x_2} \right) \frac{\mathrm{ch} \, Q}{\mathrm{ch} \, kh},$$

(4.14a)

$$\frac{\partial \psi_2}{\partial z} - \frac{\omega^2}{g} \psi_2 = i \left(\frac{2h \, \mathrm{sh} \, kh}{\mathrm{ch} \, kh} C_g + \frac{C_g^2}{\omega} \right) \frac{\partial^2 A}{\partial x_1^2} + 2 \frac{\partial A}{\partial t_2}, \qquad z = 0, \quad (4.14b)$$

$$\frac{\partial \psi_2}{\partial z} = 0, \qquad z = -h.$$

(4.14c)

In arriving at Eq. (4.14b), use is made of Eq. (4.12). Now the argument of solvability can be repeated which, after lengthy but straightforward algebra, gives a surprisingly simple result:

$$\frac{\partial A}{\partial t_2} + C_g \frac{\partial A}{\partial x_2} = \frac{i}{2} \omega'' \frac{\partial^2 A}{\partial x_1^2},$$

(4.15)

where

$$\omega'' = \frac{d^2 \omega}{dk^2} = \frac{C_g}{k} (1 - 2kh \, \mathrm{th} \, kh) - \frac{C_g^2}{\omega} + \frac{C}{2k} (2kh \, \mathrm{cth} \, 2kh - 1)$$

$$= \frac{-g}{4\omega k \sigma} \left\{ T^2 - 2khT(1 - T^2) + (kh)^2 (1 - T^2)^2 + 4(kh)^2 T^2 (1 - T^2) \right\},$$

with th $\equiv$ tanh, cth $\equiv$ cotanh and $T \equiv$ tanh kh for brevity.

The two orders (4.12) and (4.15) may be combined, and the small parameter μ may be dropped to give:

$$\frac{\partial A}{\partial t} + C_g \frac{\partial A}{\partial x} = \frac{i}{2} \omega'' \frac{\partial^2 A}{\partial x^2},$$

(4.16)

which governs the slow modulation of the envelope and is exactly Eq. (3.16). In terms of the coordinate system moving at C_g, that is, with the group

$$\xi = x - C_g t,$$

Equation (4.16) becomes the Schrödinger equation in quantum mechanics,

$$\frac{\partial A}{\partial t} = \frac{i\omega''}{2} \frac{\partial^2 A}{\partial \xi^2}. \tag{4.17}$$

This equation involves only one space coordinate and is, therefore, easier to deal with than the boundary-value problem involving x and z. Let us apply it to a new example.

Evolution of the Front of a Sinusoidal Wavetrain

Consider a sinusoidal disturbance originated from a wavemaker which is switched on at some earlier time and is maintained steadily afterward. Ultimately, at a fixed station, a steady sinusoidal motion is attained. It is interesting to examine the development of the wave front. This problem was first studied by Wu (1957) for an oscillating point pressure suddenly started at $t = 0$ on the surface of an infinitely deep water with surface tension. The same problem, without surface tension, was later studied by Miles (1962). The transient wave front caused by a vertical plate rolling in the free surface was also examined by Mei (1966a). The approach of all these authors was to start from an exact solution and then to perform an asymptotic approximation of the integral representation for large x and t. The mathematics contained some subtleties in that the major contribution came from near the pole of a principal-valued integral.

Let us use Eq. (4.17) which should be applicable for the wave front once it is far enough from the source of disturbance. The precise history of the starting process is not important. We now seek the solution to satisfy the boundary conditions that

$$A \to 0 \qquad \text{as } \xi \to \infty, \tag{4.18a}$$
$$A \to A_0 \qquad \text{as } \xi \to -\infty, \tag{4.18b}$$

that is, the envelope changes from the constant A_0 behind the wave front to zero ahead of the front. No restriction is made on kh.

The boundary-value problem defined by Eqs. (4.17) and (4.18a, b) can be solved by the similarity method familiar in boundary-layer theory or heat conduction. Since Eq. (4.17) resembles the heat equation, we anticipate a similarity solution of the form

$$A = A_0 f(\gamma) \qquad \text{where } \gamma = -\frac{\xi}{t^{1/2}}. \tag{4.19}$$

It follows from Eq. (4.17) that

$$f'' - \frac{i\gamma}{|\omega''|} f' = 0, \tag{4.20}$$

with the boundary conditions

$$f \to 1, \qquad \gamma \sim \infty, \tag{4.21a}$$

$$f \to 0, \qquad \gamma \sim -\infty. \tag{4.21b}$$

Equation (4.20) may be integrated to give

$$f = C \int_{-\infty}^{\gamma} \exp\left(\frac{iu^2}{2|\omega''|}\right) du,$$

which satisfies Eq. (4.21b). To satisfy Eq. (4.21a), we insist that

$$1 = C \int_{-\infty}^{\infty} \exp\left(\frac{iu^2}{2|\omega''|}\right) du.$$

Since,

$$\int_0^{\infty} e^{i\tau^2} d\tau = \frac{\pi^{1/2}}{2} e^{i\pi/4}, \tag{4.22}$$

we obtain

$$C = e^{-i\pi/4} (2\pi |\omega''|)^{-1/2},$$

and the solution,

$$\frac{A}{A_0} = e^{-i\pi/4} (2\pi |\omega''|)^{-1/2} \int_{-\infty}^{\gamma} du \exp\left(\frac{iu^2}{2|\omega''|}\right). \tag{4.23}$$

The above result may be expressed alternatively

$$\frac{A}{A_0} = e^{-i\pi/4} (2\pi |\omega''|)^{-1/2} \left[\int_{-\infty}^{0} + \int_{0}^{\gamma} \right] du \exp\left(\frac{iu^2}{2|\omega''|}\right)$$

$$= \frac{1}{2} + e^{-i\pi/4} (2\pi |\omega''|)^{-1/2} \int_{0}^{\gamma} du \exp\left(\frac{iu^2}{2|\omega''|}\right) \tag{4.24}$$

after the use of Eq. (4.22). Introducing $u^2/|\omega''| = \pi v^2$, we get

$$\frac{A}{A_0} = \frac{1}{2} + \frac{e^{-i\pi/4}}{2^{1/2}} \int_{0}^{\beta} e^{i\pi v^2/2} \, dv$$

$$= \frac{e^{-i\pi/4}}{2^{1/2}} \left[\frac{1+i}{2} + \int_{0}^{\beta} dv \left(\cos\frac{\pi v^2}{2} + i \sin\frac{\pi v^2}{2} \right) \right], \tag{4.25}$$

where

$$\beta = -\xi(\pi|\omega''|t)^{-1/2}. \tag{4.26}$$

Since

$$C(\beta) = \int_0^\beta \cos\frac{\pi v^2}{2}\, dv \tag{4.27a}$$

and

$$S(\beta) = \int_0^\beta \sin\frac{\pi v^2}{2}\, dv \tag{4.27b}$$

are the Fresnel cosine and sine integrals, Eq. (4.25) may be written

$$\frac{A}{A_0} = \frac{e^{-i\pi/4}}{2^{1/2}}\left\{\left[\frac{1}{2} + C(\beta)\right] + i\left[\frac{1}{2} + S(\beta)\right]\right\}. \tag{4.28}$$

The magnitude $|A/A_0|$ is given by

$$\left|\frac{A}{A_0}\right| = \frac{1}{2^{1/2}}\left\{\left[\frac{1}{2} + C(\beta)\right]^2 + \left[\frac{1}{2} + S(\beta)\right]^2\right\}^{1/2} \tag{4.29}$$

which is plotted in Fig. 4.2. It is interesting that Eq. (4.29) also describes the

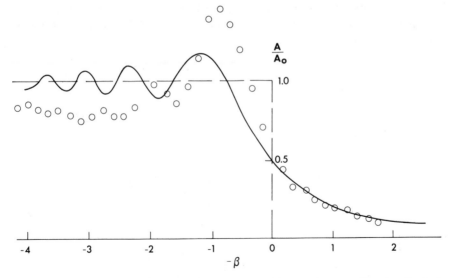

Figure 4.2 Comparison of theory (Miles 1962) with experiments. The amplitude is measured at 160 ft from the wave maker; frequency is $\omega = 5.52$ rad/s. (From Longuet-Higgins, 1964, *Proceedings, Tenth Syposium on Naval Hydrodynamics*. Reproduced by permission of U.S. Office of Naval Research.)

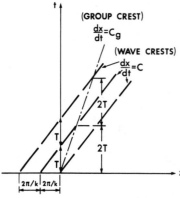

Figure 4.3 Effect of group velocity versus phase velocity in deep water $C_g = C/2$; the time interval is $2T$ between two successive instants when a wave crest coincides with the envelope crest.

variation across a shadow boundary in a diffraction problem which will be examined later. To an observer fixed at x which is far from the wavemaker, the envelope first grows monotonically to $A_0/2$ when $t = x/C_g$, then approaches the steady-state limit A_0 in an oscillatory manner. The transition region stretches out with time as $t^{1/2}$. Longuet-Higgins (1974) has performed experiments which support this theory qualitatively. Some quantitative discrepancy exists which is likely related to nonlinear effects. In particular, for sufficiently high amplitude the observed crests break at the first peak of the envelope. Since the envelope velocity in deep water is one-half of the crest velocity, the time interval between two successive crests arriving at the peak of the envelope is two wave periods (see Fig. 4.3). If the waves at the peak are high enough to break, one finds the time interval between two successive breakers to be $2T$. This phenomenon can be observed in white caps (Donelan, Longuet-Higgins, and Turner, 1972).

THREE

Refraction by Slowly Varying Depth or Current

When a train of plane monochromatic waves enters a zone of slowly varying depth, the wavenumber can be expected to change with depth in accordance with Eq. (4.8), Chapter One, resulting in a gradual change in the phase velocity. In general, the spacings between equal phase lines and the amplitude of crests and troughs will vary from place to place. Similar changes can also occur for waves riding on a current whose intensity varies in horizontal directions. These phenomena, which are related mainly to the variation in phase velocity, are of course well known in optics and acoustics and are called *refraction*. In this chapter, we develop an approximate *ray theory* (or *geometrical optics theory*) for the effects of varying depth (Sections 3.1–3.4) and of varying current (Sections 3.6 and 3.7) on the propagation of infinitesimal waves. The evolution equations will be deduced by the so-called WKB method which is a special version of the multiple-scales method. While these equations are normally solved numerically for practical problems, we shall extract from them physical insights through a variety of analytical examples. A brief discussion is also made, for varying depth only, on the local remedy needed when the ray approximation fails. In a numerical treatment of natural topography, this failure can be overcome more straightforwardly by incorporating diffraction in the so-called *mild-slope equation*, which we derive in Section 3.5. Other more mathematical aspects not treated here can be found in the excellent surveys of Meyer (1979a) for varying depth and Peregrine (1976) for varying current.

3.1 GEOMETRICAL OPTICS APPROXIMATION FOR PROGRESSIVE WAVES OVER A GRADUALLY VARYING BOTTOM

We assume that the typical wavelength is much less than the horizontal length scale of depth variation. A small parameter can be introduced as follows:

$$\mu = O\left(\frac{\nabla h}{kh}\right) \ll 1. \tag{1.1}$$

To be more general, let us allow slow modulation in time too, which may be the result of initial conditions. Following Keller (1958), we introduce the slow coordinates

$$\bar{x} = \mu x, \qquad \bar{y} = \mu y, \qquad \bar{t} = \mu t. \tag{1.2}$$

The linearized governing equations become

$$\mu^2\left(\Phi_{\bar{x}\bar{x}} + \Phi_{\bar{y}\bar{y}}\right) + \Phi_{zz} = 0, \qquad -h(\bar{x}, \bar{y}) < z < 0, \tag{1.3}$$

$$\mu^2\phi_{\bar{t}\bar{t}} + g\Phi_z = 0, \qquad z = 0, \tag{1.4}$$

$$\Phi_z = -\mu^2\left(\Phi_{\bar{x}}h_{\bar{x}} + \Phi_{\bar{y}}h_{\bar{y}}\right), \qquad z = -h(\bar{x}, \bar{y}). \tag{1.5}$$

The second key step, typical of WKB, is to introduce the following expansion in anticipation that waves are progressive:

$$\Phi = \left[\phi_0 + (-i\mu)\phi_1 + (-i\mu)^2\phi_2 + \cdots\right]e^{iS/\mu}, \tag{1.6}$$

where

$$\phi_j = \phi_j(\bar{x}, \bar{y}, z, \bar{t}) \qquad \text{for } j = 0, 1, 2, \ldots; \quad \text{and} \quad S = S(\bar{x}, \bar{y}, \bar{t}).$$

The intuitive basis for this assumption is that while wave amplitude varies with the slow coordinates $\bar{x}, \bar{y}, \bar{t}$, the phase varies with the fast coordinates $(\bar{x}, \bar{y}, \bar{t})\mu^{-1}$. By straightforward differentiation we get

$$\mu^2\phi_{\bar{t}\bar{t}} = -(-i\mu)^2\phi_{\bar{t}\bar{t}}$$

$$= -\left\{S_{\bar{t}}^2\left(\phi_0 + (-i\mu)\phi_1 + (-i\mu)^2\phi_2 + \cdots\right)\right.$$

$$+ (-i\mu)\left[S_{\bar{t}\bar{t}}\left(\phi_0 + (-i\mu)\phi_1 + \cdots\right)\right.$$

$$\left.+ 2S_{\bar{t}}\left(\phi_{0\bar{t}} + (-i\mu)\phi_{1\bar{t}} + \cdots\right)\right] + (-i\mu)^2\left(\phi_{0\bar{t}\bar{t}} + \cdots\right)\Big\}e^{iS/\mu},$$

$$\overline{\nabla}\Phi = \left\{\left[\overline{\nabla}\phi_0 + (-i\mu)\overline{\nabla}\phi_1 + \cdots\right] + \frac{i\overline{\nabla}S}{\mu}\left[\phi_0 + (-i\mu)\phi_1 + \cdots\right]\right\}e^{iS/\mu},$$

$$\mu^2\overline{\nabla}\cdot\overline{\nabla}\phi = -(-i\mu)^2\overline{\nabla}\cdot\overline{\nabla}\phi$$

$$= -(-i\mu)^2\left\{\left[\overline{\nabla}^2\phi_0 + (-i\mu)\overline{\nabla}^2\phi_1 + \cdots\right]\right.$$

$$+ \left[\overline{\nabla}\phi_0 + (-i\mu)\overline{\nabla}\phi_1 + \cdots\right]\cdot\frac{i\overline{\nabla}S}{\mu}$$

$$+ \frac{1}{-i\mu}\left[\overline{\nabla}\cdot\phi_0\overline{\nabla}S + (-i\mu)\overline{\nabla}\cdot(\phi_1\overline{\nabla}S) + \cdots\right]$$

$$\left.+ \left(\frac{i\overline{\nabla}S}{\mu}\right)^2\left[\phi_0 + (-i\mu)\phi_1 + \cdots\right]\right\}e^{iS/\mu}.$$

Let us define

$$\mathbf{k} = \overline{\nabla} S \tag{1.7a}$$

and

$$\omega = -S_{\bar{t}}, \tag{1.7b}$$

which represent the local wavenumber vector and frequency, respectively. Substituting Eqs. (1.7) into Eqs. (1.3)–(1.5) and separating the orders, we obtain at $O(-i\mu^0)$

$$\phi_{0zz} - k^2\phi_0 = 0, \qquad -h < z < 0, \tag{1.8}$$

$$\phi_{0z} - \frac{\omega^2}{g}\phi_0 = 0, \qquad z = 0, \tag{1.9}$$

$$\phi_{0z} = 0, \qquad z = -h; \tag{1.10}$$

and at $O(-i\mu)$

$$\phi_{1zz} - k^2\phi_1 = \mathbf{k} \cdot \overline{\nabla}\phi_0 + \overline{\nabla} \cdot (\mathbf{k}\phi_0), \qquad -h < z < 0, \tag{1.11}$$

$$\phi_{1z} - \frac{\omega^2}{g}\phi_1 = -\frac{[\omega\phi_{0\bar{t}} + (\omega\phi_0)_{\bar{t}}]}{g}, \qquad z = 0, \tag{1.12}$$

$$\phi_{1z} = \phi_0 \mathbf{k} \cdot \overline{\nabla} h, \qquad z = -h. \tag{1.13}$$

Equations (1.8)–(1.10) and (1.11)–(1.13) define two boundary-value problems governed by ordinary differential equations. The solution to Eqs. (1.8)–(1.10) is, formally,

$$\phi_0 = -\frac{igA}{\omega}\frac{\cosh k(z+h)}{\cosh kh}, \tag{1.14}$$

with

$$\omega^2 = gk \tanh kh. \tag{1.15}$$

Thus, $\omega(\bar{x}, \bar{y}, \bar{t})$ and $k(\bar{x}, \bar{y}, \bar{t})$ are related to the local depth $h(\bar{x}, \bar{y})$ by the same dispersion relation as if h were constant. The amplitude $A(\bar{x}, \bar{y}, \bar{t})$ is still arbitrary.

To get a condition on A we examine the solvability of ϕ_1 by applying Green's formula [Eq. (4.11), Chapter Two] to ϕ_0 and ϕ_1. Making use of all conditions (1.8)–(1.10) and (1.11)–(1.13), we get

$$\int_{-h}^{0} dz\, \phi_0\big[(\mathbf{k} \cdot \overline{\nabla}\phi_0) + \overline{\nabla} \cdot (\mathbf{k}\phi_0)\big] = -\frac{1}{g}\big\{\phi_0[\omega\phi_{0\bar{t}} + (\omega\phi_0)_{\bar{t}}]\big\}_{z=0}$$

$$-\{\phi_0^2\}_{z=-h}\mathbf{k} \cdot \overline{\nabla} h.$$

By Leibniz's rule

$$D \int_b^a f \, dz = \int_b^a Df \, dz + (Da)(f)_{z=a} - (Db)(f)_{z=b}, \qquad (1.16)$$

where D is either $\partial/\partial \bar{t}$, $\partial/\partial \bar{x}$, or $\partial/\partial \bar{y}$; the integral on the left and the last term on the right can be combined to give

$$\overline{\nabla} \cdot \int_{-h}^0 dz \, \mathbf{k} \phi_0^2 + \frac{1}{g} \frac{\partial}{\partial \bar{t}} \left[\omega \phi_0^2 \right]_{z=0} = 0.$$

By the use of Eqs. (1.14) and (1.15), and the definitions of E and C_g [Eqs. (5.14) and (5.6), Chapter One] which are still valid here except that they are functions of ($\bar{x}$, $\bar{y}$, and $\bar{t}$), it is easy to verify that

$$\overline{\nabla} \cdot \left(\frac{E}{\omega} \mathbf{C}_g \right) + \frac{\partial}{\partial \bar{t}} \left(\frac{E}{\omega} \right) = 0. \qquad (1.17)$$

In the classical mechanics of oscillators, a similar ratio of energy to frequency is called *action* and is also found to be invariant when properties of the oscillator change slowly (adiabatically). Hence E/ω has come to be known as the *wave action* and Eq. (1.17) states its conservation while being transported by the group velocity.

In summary, the phase function of slowly varying water waves is governed by Eq. (1.15) with k and ω given by Eq. (1.7). S is thus governed by a highly nonlinear first-order partial differential equation which is called the *eikonal equation* in optics. Once the phase is found, the amplitude is solved from the wave action equation (1.17).

Let us also note that definition (1.7) implies

$$\overline{\nabla} \times \mathbf{k} = 0 \qquad (1.18)$$

$$\frac{\partial \mathbf{k}}{\partial \bar{t}} + \overline{\nabla} \omega = 0 \qquad (1.19)$$

The one-dimensional version of Eq. (1.19)

$$\frac{\partial k}{\partial \bar{t}} + \frac{\partial \omega}{\partial \bar{x}} = 0 \qquad (1.20)$$

is most easily interpreted physically. By definition, k is the number of equal phase lines per unit distance, hence the *density* of equal phase lines. By definition also, ω is the number of equal phase lines passing a fixed station, hence the *flux* of equal phase lines. Between the two stations $\bar{x}$ and $\bar{x} + d\bar{x}$, the net rate of out-flux of phase lines is $(\partial \omega / \partial \bar{x}) \, d\bar{x}$, while the rate of decrease of phase lines in the control volume is $-(\partial k/\partial t)\,d\bar{x}$. It is clear that Eq. (1.20) is simply a law of wave crest conservation.

In the next few sections we shall confine ourselves to strictly sinusoidal waves and study several examples which have their analogs in optics (Luneberg, 1964). Since the purpose of deducing approximate equations is already achieved, there is no need to distinguish the slow variables from the physical variables. All bars will now be removed.

Exercise 1.1

A two-layer ocean with densities ρ and ρ' has a slowly varying bottom $z = -h(x, y)$. The interface is at $z = 0$, while the mean free surface is at $x = h'$. Invoke the rigid-lid approximation and analyze a progressive train of internal waves by WKB approximation. Show that at the leading order $O(\mu^0)$ energy is $E = \frac{1}{2}\Delta\rho g A^2$ with $\Delta\rho = \rho - \rho'$, while the dispersion relation and group velocity are, respectively,

$$\omega^2 = \frac{\Delta\rho g k}{\rho'\cot kh' + \rho\coth kh},$$

$$C_g = \frac{C}{2}\left[1 + \frac{\omega^2}{g\,\Delta\rho}\left(\rho'h'\operatorname{csch}^2 kh' + \rho h\operatorname{csch}^2 kh\right)\right].$$

From the solvability condition at $O(\mu)$, show that Eq. (1.17) is also true.

3.2 RAY THEORY FOR SINUSOIDAL WAVES, FERMAT'S PRINCIPLE

If the waves are steady, $\partial/\partial t = 0$, then Eq. (1.19) implies that $\omega = $ const. The problem involves waves purely sinusoidal in time. From Eq. (1.17) the amplitude variation is governed by

$$\nabla \cdot \left(E\mathbf{C}_g\right) = 0. \tag{2.1}$$

Imagine the x-y plane to be filled with $\mathbf{k}$ vectors which vary with position in magnitude and direction. Starting from a given point, let us draw a curve which is tangent to the local $\mathbf{k}$ vector at every point along the curve. Such a curve is called a wave *ray* and is always orthogonal to the local crests or phase lines $S = $ const. From different starting points different rays can be drawn. Two adjacent rays form a ray channel. Consider a segment of a ray channel whose widths at the two ends are $d\sigma_0$ and $d\sigma$ (Fig. 2.1). Integrate Eq. (2.1) along the closed contour formed by the boundaries of the ray segment. From the Gauss divergence theorem and the fact that $\mathbf{C}_g$ is tangent to the ray, it follows that the energy fluxes through both ends are the same

$$EC_g\,d\sigma = \left(EC_g\,d\sigma\right)_0 = \text{const.} \tag{2.2}$$

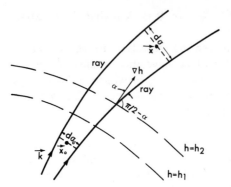

Figure 2.1 A ray channel and depth contours.

The variation of amplitude along a ray then follows the law

$$\frac{A}{A_0} = \left[\frac{(C_g)_0}{C_g} \frac{d\sigma_0}{d\sigma} \right]^{1/2},\tag{2.3}$$

where the ratio $d\sigma/d\sigma_0$ is called the *ray separation factor*.

The problem now is to find the rays, or their orthogonals, which are just the phase lines $S(x, y) = $ const. Once the rays are located and the amplitude at station 0 is known, the amplitude at any other station along the ray is found immediately.

Taking the square of Eq. (1.7a), we obtain a nonlinear partial differential equation for S:

$$|\nabla S|^2 = k^2 \quad \text{or} \quad \left(\frac{\partial S}{\partial x} \right)^2 + \left(\frac{\partial S}{\partial y} \right)^2 = k^2,\tag{2.4}$$

the right-hand side being known from the dispersion relation. Equation (2.4) is called the *eikonal* equation which can be treated in a most general way by the method of characteristics. We present below a more elementary approach.

Let $y(x)$ represent a particular ray; its slope must be given by

$$y' = \frac{dy}{dx} = \frac{\partial S}{\partial y} \bigg/ \frac{\partial S}{\partial x}.$$

It follows from Eq. (2.4) that

$$\left(1 + y'^2 \right)^{1/2} = \frac{k}{\partial S/\partial x} \quad \text{and} \quad \frac{ky'}{\left(1 + y'^2 \right)^{1/2}} = \frac{\partial S}{\partial y}.$$

The derivative of the second equation above gives

$$\frac{d}{dx}\frac{ky'}{(1+y'^2)^{1/2}} = \frac{\partial^2 S}{\partial y\,\partial x} + \frac{\partial^2 S}{\partial y^2}y' = \left(\frac{\partial^2 S}{\partial y\,\partial x}\frac{\partial S}{\partial x} + \frac{\partial^2 S}{\partial y^2}\frac{\partial S}{\partial y}\right)\Big/\frac{\partial S}{\partial x}$$

$$= \left[\frac{1}{2}\frac{\partial}{\partial y}(\nabla S)^2\right]\Big/\frac{\partial S}{\partial x}$$

$$= \left(\frac{\partial k}{\partial y}\right)(1+y'^2)^{1/2},$$

or

$$\frac{d}{dx}\left[\frac{ky'}{(1+y'^2)^{1/2}}\right] = (1+y'^2)^{1/2}\frac{\partial k}{\partial y} \qquad \text{with } k = k(x, y(x)). \quad (2.5)$$

Equation (2.5) is a nonlinear ordinary differential equation for the ray $y(x)$. Once the initial point is known, the ray path can be solved numerically.

Before discussing specific examples, it is interesting to establish the correspondence between Eq. (2.5) and the celebrated *Fermat's principle* which states: "If P_0 and P_1 are two points on a ray and

$$L = \int_{P_0}^{P_1} k\,ds \qquad (2.6)$$

is an integral along a certain path joining P_0 and P_1, then L is an extremum if and only if the path coincides with the ray." It is well known in the calculus of variations (see Hildebrand, 1964 p. 355ff) that the functional

$$L = \int_{P_0}^{P_1} F[x, y(x), y'(x)]\,dx \qquad (2.7)$$

is extremum if and only if F satisfies the following Euler's equation:

$$\frac{d}{dx}\left(\frac{\partial F}{\partial y'}\right) = \frac{\partial F}{\partial y}. \qquad (2.8)$$

If we let

$$L = \int_{P_0}^{P_1} k(1+y'^2)^{1/2}\,dx$$

and identify

$$F = k\left(1 + \frac{y'^2}{k(1 + y'^2)^{1/2}}\right)^{1/2},$$

then Eq. (2.5) is precisely Euler's equation for Fermat's principle.

   We have now seen that the eikonal equation and Fermat's principle are but two ways of expressing the same thing. Let us consider a few cases where the ray geometry can be easily found. Indeed, all the cases have their counterparts in optics (Luneberg, 1964).

3.3   STRAIGHT AND PARALLEL DEPTH CONTOURS

3.3.1   Geometry of Rays

Let all the contours be parallel to the y axis so that $h = h(x)$ and $k = k(x)$. The Euler equation (2.5) gives

$$\frac{d}{dx} \frac{k\,dy'}{(1 + y'^2)^{1/2}} = 0, \tag{3.1}$$

implying that

$$\frac{ky'}{(1 + y'^2)^{1/2}} = K = \text{const.} \tag{3.2}$$

Since

$$\frac{y'}{(1 + y'^2)^{1/2}} = \frac{dy}{ds} = \sin \alpha \tag{3.3}$$

where α is the angle between the ray and the positive x axis, Eq. (3.2) is easily recognized as the well-known Snell's law:

$$k \sin \alpha = K = k_0 \sin \alpha_0 \quad \text{or} \quad \frac{\sin \alpha}{C} = \frac{\sin \alpha_0}{C_0} \tag{3.4}$$

where k_0 and α_0 refer to a known point (x_0, y_0) on the ray. Solving for y' from Eq. (3.2), we have

$$\frac{dy}{dx} = \frac{\pm K}{(k^2 - K^2)^{1/2}}. \tag{3.5}$$

the above result can also be arrived at more simply. Indeed, Eq. (3.4) is just the consequence of Eq. (1.18) with $\partial/\partial y = 0$, while Eq. (3.5) follows from the geometrical definition of a ray:

$$\frac{dy}{dx} = \frac{k \sin \alpha}{k \cos \alpha}.$$

The equation of the ray is, upon integration,

$$y - y_0 = \pm \int_{x_0}^{x} \frac{K\, dx}{\left[k^2(x) - K^2\right]^{1/2}}. \tag{3.6}$$

Clearly, a ray can exist only where $k^2 > K^2$.

On the other hand, since a wave phase line is orthogonal to the rays, its slope must be given by

$$\frac{dy}{dx} = \mp \frac{1}{K}(k^2 - K^2)^{1/2}.$$

The equation of the phase line is therefore

$$\mp Ky = \int^{x} dx (k^2 - K^2)^{1/2} + \text{const.}$$

A good deal be learned from Eqs. (3.5) and (3.6) without restricting $k(x)$ to any explicit form. The following cases give some idea of the possible varieties.

Case 1: Plane Wave Incident on a Ridge or a Beach

Let a plane incident wave approach from the left, $x \sim -\infty$. The incident rays are all parallel and enter the ridge at $x = x_0 < 0$ at the angle α_0. Since $k_0 \sin \alpha_0 = K < k$ everywhere, the square root $(k^2 - K^2)^{1/2}$ is always real, and since $dy/dx > 0$, the positive sign is to be taken in Eqs. (3.5) and (3.6). As h decreases, k increases, and dy/dx decreases; thus, when a ray passes over the ridge, it first becomes increasingly normal to the depth contours. After the peak is passed, the ray then turns away from the normal. The ray path is sketched in Fig. 3.1.

As a limiting case, let the summit of the ridge rise above the mean water level, resulting in a beach on either side. Consider a ray with $k = k_0$ at $x = x_0$ approaching from the left at the incidence angle α_0. The ray turns toward the depth contours, and finally strikes the shorelines perpendicularly since $k \uparrow \infty$ for $h \downarrow 0$.

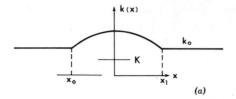

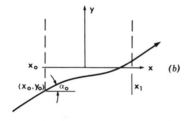

Figure 3.1   An incident ray passing over a submarine ridge. (a) Variation of $k(x)$ for a ridge; (b) an incident ray with $K < k_0 = k_{min}$.

The following choice of k, due to Pocinki (1950), is a special model for a beach which begins at $x = a$ and ends at the shoreline $x = b$.

$$\frac{k}{k_0} = 1, \qquad x > a,$$

$$\frac{k}{k_0} = \frac{1 - a/b}{1 - x/b}, \qquad a > x > b.$$

Substituting into Eq. (3.5), we get

$$\frac{dy}{dx} = \frac{(1 - x/b)\left[(\sin\alpha_0)/(1 - a/b)\right]}{\left\{1 - \left[(\sin\alpha_0)/(1 - a/b)\right]^2 (1 - x/b)^2\right\}^{1/2}}, \qquad a > x > b.$$

Let

$$\beta = \frac{\sin\alpha_0}{1 - a/b}, \qquad \xi = 1 - \frac{x}{b}, \qquad \eta = \frac{y}{b},$$

then the ray differential equation becomes

$$d\eta = \frac{-\beta\xi\, d\xi}{(1 - \beta^2\xi^2)^{1/2}} = \frac{1}{\beta} d(1 - \beta^2\xi^2)^{1/2}$$

which is easily integrated to give

$$\xi^2 + (\eta - \eta_c)^2 = \frac{1}{\beta^2},$$

or

$$(x - b)^2 + (y - y_c)^2 = \frac{(b - a)^2}{\sin^2\alpha_0}.$$

Hence the rays are a family of circular arcs centered at $x = b$, and $y = y_c$. The parameter y_c is related to the coordinate y_0 where the ray intersects the contour $x = a$. By letting $x = a$ and $y = y_0$ in the last formula, we find

$$y_c = y_0 - (b - a)\cos\alpha_0.$$

Case 2: Wave Trapping on a Ridge

If $k_{max} > K = k_0\sin\alpha_0 > k_{min}$ (Fig. 3.2a), then rays can exist only in the region $b < x < a$, where $k > K$. Let such a ray originate from x_0 at the angle α_0 with $0 < \alpha_0 < \frac{1}{2}\pi$. From x_0 to a, $dy/dx > 0$ and y is given by Eq. (3.6) with the positive sign. The ray approaches the point $x = a$ and $y = y_a$ where

$$y_a = y_0 + \int_{x_0}^{a} \frac{K\,dx}{(k^2 - K^2)^{1/2}}.$$

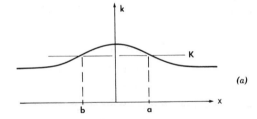

(a)

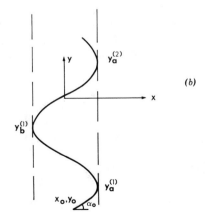

(b)

Figure 3.2 Wave trapping on a ridge: (a) variation of k over a ridge; (b) a trapped ray with $k_{max} > K > k_{min}$.

For a sufficiently smooth bottom where k can be expanded as a Taylor series near $x = a$,

$$k^2 = K^2 + (x - a)(k^2)'_a + \cdots \qquad \text{if } (k^2)'_a \equiv (k^2)'|_{x=a} \neq 0, \quad (3.7)$$

the integral is finite. The slope dy/dx is, however, infinite; hence the line $x = a$ is the envelope of all the rays and is called a *caustic*. Because of the crossing of the adjacent rays, the equation for the amplitude variation (2.3) ceases to be valid. A more refined treatment in the neighborhood of a caustic will be described in Section 3.3.3. After the point (a, y_a), $dy/dx < 0$; the ray turns back and is described by Eq. (3.6) with the negative sign until it reaches the line $x = b$, which is another caustic enveloping all rays. Thus, the ray bounces back and forth within the two caustics while advancing in the positive y direction (Fig. 3.2b). No simple harmonic waves with the stated K are possible outside the range $b < x < a$. This phenomenon is called *wave trapping*.

External excitation of the trapped waves is possible by meteorological forcing on the free surface (atmospheric pressure or wind). For such high values of K ($> k_{min}$) there is no simple harmonic wave away from the ridge. It is not possible to excite the ridge wave by a simple harmonic incident wave from either side of the ridge, according to the linearized mechanism. This does not, however, rule out the excitation of transient waves from outside sources.

Case 3 Submarine Trough

For a trough connecting two sides of equal depth, $k(x)$ varies as shown in Fig. 3.3a. If an incident wave is such that $K = k_0 \sin \alpha_0 = K_2 < k_{min}$, it will simply bend first toward, then away from the axis of the trough and pass the trough to the right side as shown in Fig. 3.3b. However, if $K = K_1$ is sufficiently large, then no rays can exist in the region where $k < K$ and the line $x = x_1$ where $k(x_1) = K_1$ is a caustic. The ray must then turn around to the same side where it started. For fixed $k_0 > k_{min}$, a sufficiently large value of K may be achieved if the incidence angle α_0 is sufficiently close to $\frac{1}{2}\pi$. The incident ray then makes a small acute angle to the contours; this is called *glancing* incidence. At the critical value $k_0 \sin \alpha_0 = k_{min}$, the incident ray becomes asymptotically parallel to the depth contours.

3.3.2 Amplitude Variation

In this simple case, $\partial/\partial y = 0$ and Eq. (2.1) can be integrated to give

$$EC_g \cos \alpha = \frac{1}{2}\rho g A^2 C_g \cos \alpha = \text{const.} \qquad (3.8)$$

Let the subscript $(\ )_0$ denote values at the reference depth h_0, then the

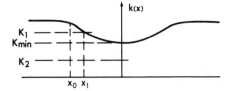

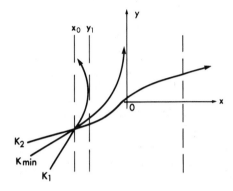

Figure 3.3 Wave rays over a submarine trough.

amplitude ratio is

$$\frac{A}{A_0} = \left[\frac{(C_g)_0 \cos \alpha_0}{C_g \cos \alpha}\right]^{1/2}$$

$$= \left[\frac{k \cos \alpha_0}{k_0 \cos \alpha} \frac{(1 + 2kh/\sinh 2kh)_0}{1 + 2kh/\sinh 2kh}\right]^{1/2}. \tag{3.9}$$

In very shallow water, $\cos \alpha \to 1$, $C_g \simeq C \simeq (gh)^{1/2}$ and

$$\frac{A}{A_0} \simeq (C_{g_0} \cos \alpha_0)^{1/2} (gh)^{-1/4}, \tag{3.10}$$

hence the amplitude increases with decreasing depth. The $1/4$ power dependence is often called Green's law. Combined with the shortening of the wavelength $[k \simeq \omega(gh)^{-1/2}]$, the local wave slope increases with decreasing depth as $kA \propto h^{-3/4}$. For sufficiently small depth, the small-amplitude assumption underlying the linearized theory ceases to be valid and nonlinear effects become important. For a beach of constant slope the assumption that $(dh/dx)kh^{-1} \ll 1$ inherent in the WKB method also breaks down ultimately. Under certain conditions to be elaborated in Chapter Ten, incident progressive waves may break in very shallow water. For normal incidence on a plane beach, experiments by Eagleson (1956) confirm Eq. (3.9) almost up to the first line of breaking.

3.3.3 The Neighborhood of a Straight Caustic

The inadequacy of the ray approximation may be remedied near a straight caustic with reasonable ease. In terms of the stretched variables defined by Eq. (1.2) we let the $\bar{y}$ axis be the caustic, and the incident and reflected rays be to the left. Then near $\bar{x} = 0$ we may approximate

$$k^2 \cong K^2 - \gamma\bar{x} \quad \text{with } \gamma > 0 \tag{3.11}$$

provided that dk/dx does not vanish at $\bar{x} = 0$. It follows that

$$k_1 = (-\gamma\bar{x})^{1/2} \quad \text{and} \quad \int k_1 \, d\bar{x} = \tfrac{2}{3}\gamma^{1/2}(-\bar{x})^{3/2} \tag{3.12}$$

where k_1 is the x component of $\mathbf{k}$. According to the ray approximation (3.9), we have

$$A = A_0 \left(\frac{C_g k_1}{k}\right)_0^{1/2} \left(\frac{K}{C_g}\right)_{\bar{x}=0}^{1/2} (-\gamma\bar{x})^{-1/4} \equiv \tau(-\gamma\bar{x})^{-1/4}. \tag{3.13}$$

The free surface to the left of the caustic is

$$\eta = \tau(-\gamma\bar{x})^{-1/4} e^{iK\bar{y}/\mu} \left\{ \exp\left[i\frac{\gamma^{1/2}}{\mu}\frac{2}{3}(-\bar{x})^{3/2}\right] \right.$$

$$\left. + R \exp\left[-i\frac{\gamma^{1/2}}{\mu}\frac{2}{3}(-\bar{x})^{3/2}\right] \right\} \tag{3.14}$$

where the first term in { } is the incident wave and the second term is the reflected wave whose complex amplitude R is yet unknown.

From the above result the amplitude increases without bound as $\bar{x} \to 0$. An improved local theory near the caustic must therefore retain the highest derivative of the amplitude with respect to $\bar{x}$. Substituting

$$\Phi = \frac{-igX(\bar{x})}{\omega} \frac{\cosh k(z + h)}{\cosh kh} \exp\left(\frac{iK\bar{y}}{\mu} - \frac{i\omega\bar{t}}{\mu}\right) \tag{3.15}$$

into Eq. (1.3) and keeping the leading terms and the highest derivatives with respect to $\bar{x}$, we get

$$\mu^2 X_{\bar{x}\bar{x}} + (k^2 - K^2)X \cong 0. \tag{3.16}$$

Now $k^2 - K^2$ changes sign at $\bar{x} = 0$, being positive for $\bar{x} < 0$ and negative for $\bar{x} > 0$. The solution is oscillatory for $\bar{x} < 0$ and monotonic for $\bar{x} > 0$. The point $\bar{x} = 0$ is called a *turning point* in mathematical physics. Invoking Eq.

(3.11), we have from Eq. (3.16)

$$\mu^2 X_{\bar{x}\bar{x}} - \gamma \bar{x} X \cong 0,$$ (3.17)

which is a good approximation in the region $\bar{x} = O(\mu)^{2/3}$, that is, $x = O(\mu^{-1/3})$. With the new variable

$$\sigma = \gamma^{1/3} \bar{x} \mu^{-2/3},$$ (3.18)

Eq. (3.17) becomes the Airy equation

$$X_{\sigma\sigma} - \sigma X = 0$$ (3.19)

whose general solution is

$$X = a \,\mathrm{Ai}(\sigma) + b \,\mathrm{Bi}(\sigma).$$ (3.20)

The Airy function Ai has already been plotted in Fig. 1.5, Chapter Two. It is also known that for large $|\sigma|$:

$$\mathrm{Ai}(\sigma) \sim \frac{1}{2\pi^{1/2}} \sigma^{-1/4} \exp\left(-\frac{2}{3}\sigma^{3/2}\right), \qquad \sigma \sim \infty,$$ (3.21a)

$$\sim \frac{1}{\pi^{1/2}} (-\sigma)^{1/4} \sin\left[\frac{2}{3}(-\sigma)^{3/2} + \frac{\pi}{4}\right], \qquad \sigma \sim -\infty,$$ (3.21b)

and

$$\mathrm{Bi}(\sigma) \sim \frac{1}{\pi^{1/2}} \sigma^{-1/4} \exp\left[\frac{2}{3}(\sigma^{3/2})\right], \qquad \sigma \sim \infty,$$ (3.22a)

$$\sim \frac{1}{\pi^{1/2}} \sigma^{-1/4} \cos\left[\frac{2}{3}(-\sigma)^{3/2} + \frac{\pi}{4}\right], \qquad \sigma \sim -\infty.$$ (3.22b)

If there are no other caustics or solid boundaries within the region $\bar{x} = O(\mu^{2/3}) > 0$, the solution $\mathrm{Bi}(\sigma)$ must be discarded; hence

$$\eta = a \,\mathrm{Ai}(\sigma) e^{iK\bar{y}/\mu}.$$ (3.23)

The coefficient a and the amplitude R of the reflected wave must be found by matching Eq. (3.23) with Eq. (3.14) for $-\sigma \gg 1$. With Eq. (3.21b) we rewrite Eq. (3.23)

$$\eta \cong \frac{a}{2i\pi^{1/2}} \left(-\frac{\gamma^{1/3}}{\mu^{2/3}}\bar{x}\right)^{-1/4} \left\{\exp\left[i\frac{2}{3}\frac{\gamma^{1/2}}{\mu}(-\bar{x})^{3/2} + \frac{i\pi}{4}\right]\right.$$

$$\left. - \exp\left[-i\frac{2}{3}\frac{\gamma^{1/2}}{\mu}(-\bar{x})^{3/2} - \frac{i\pi}{4}\right]\right\}$$ (3.24)

for $\sigma \sim -\infty$. Equations (3.14) and (3.24) are now required to match, hence

$$a = 2\pi^{1/2} i e^{i\pi/4} \tau (\gamma\mu)^{-1/6}, \tag{3.25a}$$

and

$$R = e^{i\pi/2}. \tag{3.25b}$$

For a given incident wave at $\bar{x} = \bar{x}_0$, τ is known. The coefficient a can be found at once. It is interesting that the largest amplitude is now finite and occurs before the caustic is reached. The reflected wave has the same amplitude as the incident wave but differs from the latter in phase by $\frac{1}{2}\pi$.

For a submarine trough there can be two parallel caustics. If the distance between them is not too great, the residual effect of Ai(σ) from the left caustic may penetrate the right caustic, inducing a transmitted wave. The details involve a similar treatment of the right caustic with both Ai and Bi used. Another case where $dk^2/d\bar{x} = 0$ but $d^2k^2/d\bar{x}^2 \neq 0$ is more complicated but may be analyzed in principle by modifying Eq. (3.11).

The approximate treatment of the caustic region presented in this subsection is of the boundary-layer type well known in viscous flow theory. An alternative procedure is to seek a single representation which is uniformly valid everywhere. Such theories have been developed in other physical contexts for curved caustics (see Ludwig, 1966; Nayfeh, 1973) and have been applied to water waves (Chao, 1971). Experimental confirmation of this kind of theory is also available for simple topography and sufficiently small amplitudes (Chao and Pierson, 1972).

3.4   CIRCULAR DEPTH CONTOURS

This class of problems was first studied by Arthur (1946) for water waves; analogous examples are also known in optics (Luneberg, 1964).

3.4.1   Geometry of Rays

In cylindrical polar coordinates (r, θ), the water depth, and hence the magnitude of the wavenumber vector, depend only on r, that is, $h = h(r)$, $k = k(r)$. To get Euler's equation for the ray we start from Fermat's principle and extremize the integral

$$L = \int k(r)(1 + r^2\theta'^2)^{1/2} \, dr \tag{4.1}$$

where $\theta' \equiv d\theta/dr$. Euler's equation is then

$$\frac{d}{dr} \left\{ \frac{\partial}{\partial \theta'} \left[k(1 + r^2\theta'^2)^{1/2} \right] \right\} = 0,$$

or

$$\frac{kr^2\theta'}{(1 + r^2\theta'^2)^{1/2}} = \text{const} = \kappa \qquad (4.2)$$

along a ray where κ is a constant characterizing the ray. Solving for θ', we obtain

$$\frac{d\theta}{dr} = \frac{\pm|\kappa|}{r(k^2r^2 - \kappa^2)^{1/2}}. \qquad (4.3)$$

This differential equation can be integrated formally to give

$$\theta - \theta_0 = \pm|\kappa| \int_{r_0}^{r} \frac{dr}{r(k^2r^2 - \kappa^2)^{1/2}}, \qquad (4.4)$$

where r_0 and θ_0 refer to a known point passed by the ray.

What is the significance of the constant κ? With the help of Fig. 4.1, Eq. (4.2) may be rewritten

$$\kappa = kr\frac{r\,d\theta}{(dr^2 + r^2\,d\theta^2)^{1/2}} = kr\frac{r\,d\theta}{ds} = kr\sin\alpha \qquad (4.5)$$

where α is the angle between the ray and the normal (radius) vector to the depth contour intersected by the ray. If at the point r_0, θ_0 the incidence angle is $\alpha = \alpha_0$, then

$$\kappa = k_0 r_0 \sin\alpha_0. \qquad (4.6)$$

Thus, the constant κ is determined by the initial position and direction of the ray.

In contrast to the case of straight and parallel contours, r appears on the right-hand side of Snell's law (4.5) as an additional factor. To help understand

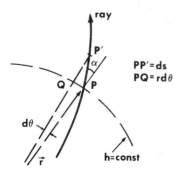

ray

P'

α

PP'=ds
PQ=rdθ

Q P

dθ

$\bar{r}$

h=const

Figure 4.1 A smooth bottom with circular coutours.

this difference, let us examine in an elementary way a bottom having a stepwise depth variation with radial symmetry, that is,

$$k = k_i = \text{const}, \qquad r_{i-1} \leq r \leq r_i, \qquad i = 1, 2, 3.$$

Here the subscripts 1, 2, 3 designate the regions and not the vector components. Consider a ray passing through regions 1, 2, and 3 (see Fig. 4.2). The ray in region i leaves the discontinuity at $r = r_{i-1}$ at the angle α_{i-1} and is incident on $r = r_i$ at the angle α'_i, being a straight line segment in between. Applying Snell's law at the junction $r = r_1$, we obtain

$$k_1 \sin \alpha'_1 = k_2 \sin \alpha_1. \tag{4.7}$$

It is important to note that $\alpha_1 \neq \alpha'_2$; in fact, it is evident from Fig. 4.2 that

$$\sin \alpha_1 = \frac{CD}{AC} = \frac{r_2 \Delta\theta}{AC}, \qquad \sin \alpha'_2 = \frac{AB}{AC} = \frac{r_1 \Delta\theta}{AC},$$

hence

$$r_1 \sin \alpha_1 = r_2 \sin \alpha'_2. \tag{4.8}$$

Combining Eq. (4.8) with Eq. (4.7), we have

$$k_1 r_1 \sin \alpha'_1 (= k_2 r_1 \sin \alpha_1) = k_2 r_2 \sin \alpha'_2.$$

Clearly, the same argument can be extended for many successive rings so that $k_n r_n \sin \alpha'_n = \text{const}$ which is the discontinuous version of Eq. (4.5). Thus, the appearance of r is due to the *curvature* of the depth contours.

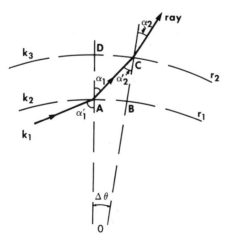

Figure 4.2 Circular step bottom.

From Eq. (4.4), it is clear that rays only exist in the regions where $k^2r^2 > \kappa^2$. The critical radius at which

$$k^2r^2 = \kappa^2, \qquad r = r_*\tag{4.9}$$

will be denoted by r_* and the corresponding θ by θ_* with

$$\theta_* - \theta_0 = \pm \int_{r_0}^{r_*} \frac{|\kappa|\, dr}{r(k^2r^2 - \kappa^2)^{1/2}}.\tag{4.10}$$

From Eq. (4.3), $dr/d\theta = 0$ at (r_*, θ_*); the ray is either the closest to, or the farthest from, the origin. The proper choice of sign in the preceding formula can be made by considering the sign of $dr/d\theta$ as will be illustrated in later examples.

Let us also deduce the equation for the constant-phase lines. Denoting the ray by $r = f(\theta)$ and the constant-phase line by $r = g(\theta)$, we use the fact that the two are orthogonal to obtain

$$\nabla[r - f(\theta)] \cdot \nabla[r - g(\theta)] = 0,$$

or

$$g' = -\frac{r^2}{f'}.$$

Since

$$f'(\theta) = \pm \frac{r}{\kappa}(k^2r^2 - \kappa^2)^{1/2}$$

the differential equation for a phase line is

$$\frac{dr}{d\theta} = \mp \frac{\kappa r}{(k^2r^2 - \kappa^2)^{1/2}},\tag{4.11}$$

which may be integrated to give

$$\kappa\theta \pm \int \frac{dr}{r}(k^2r^2 - \kappa^2)^{1/2} = \text{const.}\tag{4.12}$$

Several types of k are now examined for the physical implications:

Case 1: $0 < kr < \infty$ and kr is monotonic in r

In very shallow water, $k \sim h^{-1/2}$; we have $kr \to 0$ when $r \to 0$ as long as $rh^{-1/2} \to 0$. A submerged circular shoal falls into this category. Let the point

$P_0(r_0, \theta_0)$ be the initial point. Then

$$\theta - \theta_0 = -|\kappa| \int_{r_0}^{r} \frac{dr}{r(k^2 r^2 - \kappa^2)^{1/2}}, \qquad (4.13)$$

where the negative sign is chosen since $d\theta/dr < 0$. This equation is valid until the point P_* is reached where $r = r_*$ is minimum. Beyond this point the ray is given by

$$\theta - \theta_* = \int_{r_*}^{r} \frac{|\kappa|\, dr}{r(k^2 r^2 - \kappa^2)^{1/2}}, \qquad (4.14)$$

where the positive sign is chosen. Since the ray is obviously symmetrical about the radius vector $\theta = \theta_*$, we may incorporate both branches of the ray into one equation,

$$|\theta - \theta_*| = \int_{r_*}^{r} \frac{|\kappa|\, dr}{r(k^2 r^2 - \kappa^2)^{1/2}}. \qquad (4.15)$$

The geometry is shown in Fig. 4.3.

Suppose that there is a plane wave incident from $x \sim -\infty$ toward a circular shoal. Beyond $r = r_0$, the bottom is assumed to be horizontal so that $k = k_0$ for $r > r_0$. The incident rays are originally parallel to the x axis. Among them, those rays which are initially outside the strip $|y| \leq r_0$ do not intersect the circle $r = r_0$ and advance without deflection. Consider a ray initially within the strip $-r_0 < y < 0$ entering the shoal at the angle α_0 with respect to the radius vector; it first bends toward, and then away from, the center after a minimum r_*. Since the ray must be symmetrical about the minimum radius vector $\theta = \theta_*$, the angle between the outgoing ray and the radius vector at the point of departure must be $\pi - \alpha_0$ (see Fig. 4.3). Let the total angle by which the ray is deflected be β. It is clear that

$$\beta = \pi - \alpha_0 + \theta_0'$$

where θ_0' is the direction of the point at which the ray leaves the shoal:

$$\theta_0' - \theta_* = \int_{r_*}^{r_0} \frac{|\kappa|\, dr}{r(k^2 r^2 - \kappa^2)^{1/2}}.$$

Similarly, rays entering the shoal from $0 < y < r_0$ bend first toward, and then away from, the center of the shoal. Thus, on the lee side of the shoal, rays from opposite sides of the x axis intersect, and the progressive waves associated with these rays interfere. In particular, at any point on the positive x axis, the resulting amplitude is twice that associated with a single ray because of

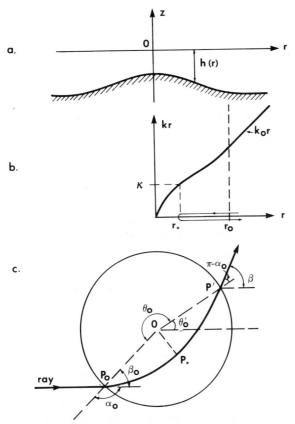

Figure 4.3 Submerged shoal: (a) side view of topography; (b) variation of kr versus r; (c) ray geometry.

symmetry. At a point not on the x axis, the intersecting rays may interfere either destructively or constructively, depending on the wave phases.

Consider rays from the same side of the x axis, say, $-r_0 < y < 0$. Since there is no deflection and $\beta = 0$ for the two extreme values of α_0: π (normal incidence) and $\pi/2$ (glancing incidence), and since $\beta > 0$ for intermediate values of α_0, there must be a positive maximum for β. Similarly, a ray entering the upper half of the shoal must have a negative maximum for β. It follows that the bundle of rays from the same side of the x axis must intersect one another in addition to intersecting those from the other side. A cusp-like caustic will develop on the lee side of the shoal, and a local remedy more complicated than that in Section 3.3.3 can be constructed to yield finite amplitude.

Case 2: kr First Decreases to a Minimum, then Increases

Here is a case of a circular island with the shoreline at $r = b$ (see Fig. 4.4). Let $h \downarrow 0$ as $r - b \downarrow 0$, then $k \downarrow h^{-1/2}$ from the dispersion relation, and $kr \uparrow bh^{-1/2}$.

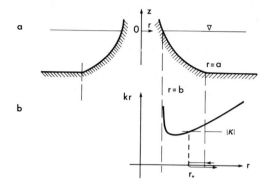

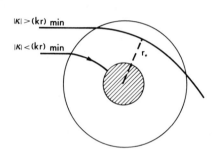

Figure 4.4 Circular island: (a) side view of topography; (b) kr versus r; (c) ray geometry.

At large r, $kr \to k_0 r$. All the incident rays intersecting the outer circle $r = a$ have $|\kappa|$ less than $k_0 a$. Those sufficiently close to the island axis satisfy $\kappa^2 < (kr)^2_{\min}$ so that they reach the shore eventually. However, those farther from the axis satisfy $(kr)_{\min} < \kappa^2 < (k_0 a)^2$ and will be repelled by the island without reaching the shore. The critical ray is one which has such an entry angle that $\alpha_0 = \alpha_0^C$, where $|\sin \alpha_0^C| = (kr)_{\min}/k_0 r_0$. A more explicit example will be given shortly.

Case 3: Wave Trapping on a Ring-Shaped Ridge

If the depth variation is as shown in Fig. 4.5a, then a local maximum kr is possible at some finite r. A ray originating at r_0, θ_0 with an inclination $\alpha_0 > \pi$, is given at first by

$$\theta - \theta_0 = \int_{r_0}^r \frac{|\kappa| \, dr}{r(k^2 r^2 - \kappa^2)^{1/2}}$$

so that $dr/d\theta > 0$ until $r = r_1$, $\theta = \theta_1'$. The ray then bends back to larger r with

$$\theta - \theta_1' = \int_{r_1}^r \frac{-|\kappa| \, dr}{r(k^2 r^2 - \kappa^2)^{1/2}},$$

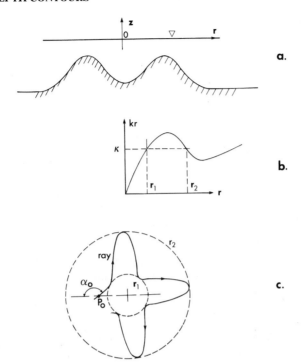

Figure 4.5 Trapped waves on a ring-shaped ridge: (*a*) side view of topography; (*b*) *kr* versus *r*; (*c*) ray geometry.

and advances clockwise, undulating between two caustic circles $r = r_1$ and $r = r_2$. Previous arguments show that the ray is symmetrical about the radius vector $\theta = \theta_1'$ and θ_2', and so on. Clearly, the configuration of the ray is repeated after every angular period:

$$\Delta\theta = 2|\kappa| \int_{r_1}^{r_2} \frac{dr}{r(k^2 r^2 - \kappa^2)^{1/2}}.$$

Furthermore, if $\Delta\theta$ is a rational multiple of 2π, the ray will return to its original point and form a closed curve. Thus, the condition

$$\frac{\Delta\theta}{2\pi} = \frac{n}{m}, \qquad m, n = 1, 2, 3, \ldots$$

determines the "eigenvalues" of free oscillation trapped on the ridge. Direct meteorological or transient incident waves may excite these modes, which may pose potential hazards to ocean structures built on the ridge.

3.4.2 Amplitude Variation

Consider the ray separation factor for a plane wave incident on a circular zone of refraction $r \leq r_0$. Let the incident rays be parallel to the negative x axis as before. We note from Fig. 4.3 that

$$\theta_0 = \pi + \beta_0$$

where $\beta_0 = \pi - \alpha_0$. It follows from Eq. (4.4) that

$$\theta = \pi + \beta_0 \pm \int_{r_0}^{r} \frac{|\kappa| \, dr}{r(k^2 r^2 - \kappa^2)^{1/2}}, \tag{4.16}$$

where

$$|\kappa| = k_0 r_0 |\sin \alpha_0| = k_0 r_0 |\sin \beta_0|$$

and a refracted ray is characterized by its point of entry r_0, θ_0, or β_0. Consider two neighboring rays of slightly different incident angles β_0 and $\beta_0 + d\beta_0$. From Fig. 4.6 we have, at any circle $r < r_0$,

$$d\sigma = \overline{AB} \cos \beta = r \Delta\theta \cos \beta = r \cos \beta \, d\beta_0 \left(\frac{\partial \theta}{\partial \beta_0} \right)_{r=\text{const}}$$

because $\Delta\theta$ is measured along the circle of radius r. Now at the initial circle $r = r_0$, $\Delta\theta = \Delta\beta_0$ and

$$d\sigma_0 = \overline{A_0 B_0} \cos \beta_0 = r_0 \, d\beta_0 \cos \beta_0.$$

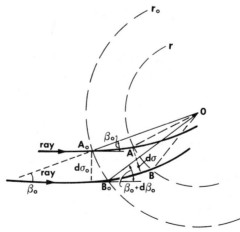

Figure 4.6   Geometry of a ray channel for circular depth contours.

Hence the separation factor is

$$\frac{d\sigma}{d\sigma_0} = \frac{r\cos\beta}{r_0\cos\beta_0}\left(\frac{\partial\theta}{\partial\beta_0}\right)_{r=\text{const}}.$$

Since $kr\sin\beta = \kappa = \text{const}$ and

$$\cos\beta = \left(1 - \left(\frac{\kappa}{kr}\right)^2\right)^{1/2},$$

we have

$$\frac{d\sigma}{d\sigma_0} = \left\{r\left[1 - \left(\frac{\kappa}{kr}\right)^2\right]^{1/2}\left(\frac{\partial\theta}{\partial\beta_0}\right)_{r=\text{const}}\right\}(r_0\cos\beta_0)^{-1} \qquad (4.17)$$

where $\partial\theta/\partial\beta_0$ may be obtained from Eq. (4.16).

Example: *A Circular Island (Pocinki, 1950)*

Take

$$kr = k_0 a \cdot \frac{\ln(a/b)}{\ln(r/b)}, \qquad b < r < a,$$
$$= k_0 r, \qquad a < r, \qquad\qquad (4.18)$$

so that the shoreline is at $r = b$ and the toe of the island is at $r = a$. The variation is shown in Fig. 4.7a. Near the shore $r = b$ where the depth is small, $h \sim k^{-2} \sim r^2 \ln^2 r/b$ and $dh/dr \sim 0$ as $r \to b$. The beach is therefore very flat. It is evident that in this case all the rays entering the island toe eventually intersect the shoreline at right angles.

With this choice of k, the ray equation is easily integrated:

$$\pm(\theta - \theta_0) = \int_{r=a}^{r=r} \frac{\ln(r/b)d[\ln(r/b)]k_0a|\sin\alpha_0|}{[(k_0a)^2\ln^2(a/b) - (k_0a)^2\sin^2\alpha_0\ln^2(r/b)]^{1/2}} \qquad (4.19)$$

where the $+(-)$ sign is chosen for those rays entering the shoal in the second (third) quadrant. Letting

$$D = \frac{\ln(a/b)}{\sin\beta_0}, \qquad \rho = \ln\frac{r}{b}, \qquad\qquad (4.20)$$

we rewrite and integrate Eq. (4.19):

$$\pm (\theta - \theta_0) = \int_{\ln(a/b)}^{\ln(r/b)} \frac{\rho \, d\rho}{\left(D^2 - \rho^2\right)^{1/2}} = \left(D^2 - \ln^2\frac{a}{b}\right)^{1/2} - \left(D^2 - \ln^2\frac{r}{b}\right)^{1/2},$$

$$(4.21)$$

or, equivalently,

$$\ln\frac{r}{b} = \left\{ D^2 - \left[\pm(\theta - \theta_0) - \left(D^2 - \ln^2\frac{a}{b}\right)^{1/2} \right]^2 \right\}^{1/2}.$$

Dividing throughout by $\ln(a/b)$, we have, finally,

$$\frac{\ln(r/b)}{\ln(a/b)} = \left\{ \csc^2\beta_0 - \left[\pm\frac{\theta - \theta_0}{\ln(a/b)} - \cot\beta_0 \right]^2 \right\}^{1/2}, \qquad \theta_0 = \pi + \beta_0 \quad (4.22)$$

which has been plotted by Pocinki (see Fig. 4.7c).

Let (r_b, θ_b) be the point where the outermost ray entering the shoal in the third quadrant ($\beta_0 = \frac{1}{2}\pi$, i.e., $\theta_0 = \frac{3}{2}\pi$) intersects the shore. Since $\sin \beta_0 = 1$ and $\cot \beta_0 = 0$, we have from Eq. (4.21)

$$\theta_b - \frac{3\pi}{2} = \ln\frac{a}{B}.$$

Since $\theta_b > 3\pi/2$ the plus sign has been chosen. Similarly, the outermost ray which enters the shoal in the second quadrant ($\beta_0 = -\pi/2$, $\theta_0 = \pi/2$) intersects the shore at the point $r = b$ and

$$\theta_b' = \frac{\pi}{2} - \ln\frac{a}{b}.$$

If $\ln(a/b) < \pi/2$, there is a portion of the shoreline within the range $-[\pi/2 - \ln(a/b)] < \theta < \pi/2 - \ln(a/b)$ which is shielded from the incident waves; this range has been called the *lee shore* by Arthur (1946).

If $\ln(a/b) > \pi/2$, the rays from one side of the axis cross those from the other side in the lee of the island. The resulting wave amplitude can be computed by superposition and proper account of the phases. At the value $\ln(a/b) = \pi/2$ or $a/b \cong 4.81$, the outermost rays from both sides meet the shore at $\theta = 0$; the lee shore then disappears.

To find the separation factor, consider again the rays entering from the third quadrant. Differentiating Eq. (4.21) in accordance with Eq. (4.17), we get

$$\frac{\partial}{\partial\beta_0}\left(D^2 - \ln^2\frac{r}{b}\right)^{1/2} = D\frac{\partial D}{\partial\beta_0}\left(D^2 - \ln^2\frac{r}{b}\right)^{-1/2},$$

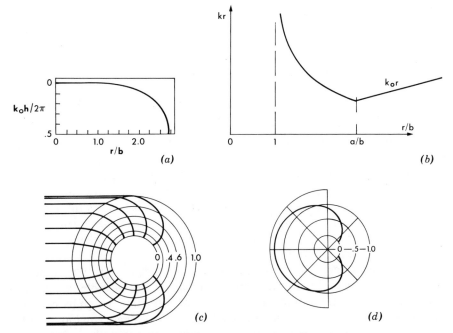

Figure 4.7 A circular island (Pocinki): (*a*) bottom profile; (*b*) kr versus r; (*c*) ray geometry; (*d*) polar plot of $(d\sigma_0/d\sigma)^{1/2}$ at the shore $r = b$.

so that

$$\frac{\partial \theta}{\partial \beta_0} = 1 - D\frac{\partial D}{\partial \beta_0}\left[\left(D^2 - \ln^2\frac{a}{b}\right)^{-1/2} - \left(D^2 - \ln^2\frac{r}{b}\right)^{-1/2}\right].$$

Now from the definition equation for D we find

$$\frac{\partial D}{\partial \beta_0} = -\cos \beta_0 \ln \frac{a}{b}\, \sin^{-2}\beta_0.$$

It follows after a little algebra that

$$\frac{\partial \theta}{\partial \beta_0} = \left\{1 + \frac{\ln(a/b)}{\sin^2\beta_0}\left[1 - \frac{\cos \beta_0}{\left(1 - R^2\sin^2\beta_0\right)^{1/2}}\right]\right\}, \qquad (4.23)$$

where

$$R \equiv \frac{\ln(r/b)}{\ln(a/b)}.$$

The separation factor is therefore

$$\frac{d\sigma}{d\sigma_0} = \frac{r\cos\beta}{a\cos\beta_0}\frac{\partial\theta}{\partial\beta_0} = \frac{r\left[1 - (k_0 a)^2\sin^2\beta_0/(kr)^2\right]^{1/2}}{a\cos\beta_0}\left(\frac{\partial\theta}{\partial\beta_0}\right) \quad (4.24)$$

with $\partial\theta/\partial\beta_0$ given by Eq. (4.23). The inverse square root of Eq. (4.24) is plotted for $r = b$ in Fig. 4.7d.

The approach presented in Sections 3.3 and 3.4 is a semi-inverse one in that some convenient form of k is assumed and the depth variation must be found from the dispersion relation. Thus, for different frequencies the same k corresponds to different depths. The more direct problem of prescribing ω and $h(\mathbf{x})$ must usually be solved by numerical methods. This is not a difficult task for straight or circular contours. For general contours computer methods have been developed by Skovgaard, Jonsson, and Bertelsen (1976) who also incorporated the additional effects of bottom friction.

Over a general topography, caustics of various kinds are possible. Although a local remedy or uniformly valid approximation is still possible in principle (Ludwig, 1966), incorporation in a numerical refraction program becomes cumbersome in practice. In the next section an approximate equation will be deduced for slowly varying depth but without the assumption of rays, as implied by Eq. (1.6). Since the new equation can be effectively solved by modern numerical methods, the intricate theory of caustics can be circumvented altogether for general topography and will not be pursued here.

3.5 AN APPROXIMATE EQUATION COMBINING DIFFRACTION AND REFRACTION ON A SLOWLY VARYING BOTTOM—THE MILD-SLOPE EQUATION

A merit of the ray approximation is to reduce a three-dimensional problem to many one-dimensional problems along ray channels. Yet, near a caustic, additional consideration must be given to variations transverse to a ray. Hence the problem is, at least locally, two dimensional in the horizontal plane. There are also situations where the problem is basically two dimensional, as for instance the obstruction of incident waves by a vertical cylinder on a slowly varying seabed. These two-dimensional effects are related to *diffraction* which will be studied more extensively later in this book. It is therefore desirable to obtain an approximation which accounts for slow depth variations and allows rapid horizontal variations associated with diffraction.

In the case of constant depth, the velocity potential may be written

$$\phi = -\frac{ig\eta}{\omega}f, \quad (5.1)$$

where

$$f = \frac{\cosh k(z + h)}{\cosh kh} \quad \text{and} \quad \omega^2 = gk \tanh kh. \tag{5.2}$$

From Laplace's equation, $\eta(\mathbf{x}, y)$ is found to satisfy the two-dimensional Helmholtz equation

$$\nabla^2\eta + k^2\eta = 0 \tag{5.3}$$

which describes diffraction. It is plausible to expect that for slowly varying depth, Eqs. (5.1) and (5.2) still apply with k and h referring to their local values. Based on this idea, Berkhoff (1972) deduced a convenient equation for $\eta(x, y)$. Various derivations of the same result have been reported by Schonfeld (1972), Jonsson and Brinkjaer (1973), Smith and Sprinks (1975), and Lozano and Meyer (1976). Here we present the arguments of Smith and Sprinks.

The exact governing equations for ϕ and f may be written

$$\frac{\partial^2\phi}{\partial z^2} + \nabla^2\phi = 0, \qquad -h \le z \le 0; \qquad \nabla = \left(\frac{\partial}{\partial x}, \frac{\partial}{\partial y}\right), \tag{5.4a}$$

$$\frac{\partial\phi}{\partial z} - \frac{\omega^2}{g}\phi = 0, \qquad z = 0, \tag{5.4b}$$

$$\frac{\partial\phi}{\partial z} = -\nabla h \cdot \nabla\phi, \qquad z = -h, \tag{5.4c}$$

while f satisfies Eqs. (1.8)–(1.10). Considering Eq. (5.4a) as an ordinary differential equation in z, and applying Green's formula for ϕ and f, we get, upon using Eqs. (5.4a)–(5.4c) and (1.8)–(1.10),

$$\int_{-h}^{0} \left(k^2\phi f + f\nabla^2\phi\right) dz = \left(f\nabla h \cdot \nabla\phi\right)_{-h}. \tag{5.5}$$

Now we invoke Eqs. (5.1) and (5.2) and note that

$$\nabla\phi = -\frac{ig}{\omega}\left(f\nabla\eta + \eta\frac{\partial f}{\partial h}\nabla h\right),$$

$$\nabla^2\phi = -\frac{ig}{\omega}\left(f\nabla^2\eta + 2\frac{\partial f}{\partial h}\nabla\eta\cdot\nabla h + \eta\frac{\partial^2 f}{\partial h^2}(\nabla h)^2 + \eta\frac{\partial f}{\partial h}\nabla^2 h\right).$$

Equation (5.5) may be written

$$\int_{-h}^{0} \left\{ f^2 \nabla^2 \eta + 2f \frac{\partial f}{\partial h} \nabla \eta \cdot \nabla h + \eta f \frac{\partial^2 f}{\partial h^2} (\nabla h)^2 + \eta f \frac{\partial f}{\partial h} \nabla^2 h + k^2 \eta f^2 \right\} dz$$

$$= -\nabla h \cdot \nabla \eta f^2 \bigg|_{-h} - \eta (\nabla h)^2 f \frac{\partial f}{\partial h} \bigg|_{-h}. \qquad (5.6)$$

By Leibniz' rule, the first two terms on the left of Eq. (5.6) may be combined with the first term on the right, yielding

$$\nabla \cdot \int_{-h}^{0} f^2 \nabla \eta \, dz + \int_{-h}^{0} k^2 f^2 \eta \, dz$$

$$= -f \frac{\partial f}{\partial h} \bigg|_{-h} \eta (\nabla h)^2 - \int_{-h}^{0} \eta f \frac{\partial^2 f}{\partial h^2} (\nabla h)^2 \, dz - \int_{-h}^{0} \eta f \frac{\partial f}{\partial h} \nabla^2 h \, dz.$$

Since $\nabla h / kh = O(\mu) \ll 1$ and $\nabla \eta / k\eta = O(1)$, every term on the right-hand side of Eq. (5.6) is of $O(\mu^2)$ relative to the left-hand side and may be omitted. Upon integration and using Eq. (5.12), Chapter One, we obtain, finally,

$$\nabla \cdot (b \nabla \eta) + \omega^2 c \eta = 0, \qquad (5.7)$$

where

$$b = gh \frac{\tanh kh}{kh} \frac{1}{2} \left(1 + \frac{2kh}{\sinh 2kh} \right) = CC_g, \qquad (5.8a)$$

$$c = \frac{1}{2} \left(1 + \frac{2kh}{\sinh skh} \right) = \frac{C_g}{C}. \qquad (5.8b)$$

Smith and Sprinks (1975) further estimated that the so-called *evanescent modes* (see Section 7.4.1), which represent localized effects, are of $O(\mu^2)$ and hence negligible. They pointed out that while Eq. (5.7) is accurate with an error of $O(\mu^2)$, the solution may be valid only up to $O(\mu)$ because the accumulated error in the phase can reach $O(\mu)$ after a propagation distance of $O(1/\mu)$.

In the special case of arbitrary constant kh, Eq. (5.7) reduces to the Helmholtz equation (5.3). On the other hand, for small but variable depth $kh \ll 1$, Eq. (5.7) reduces to

$$\nabla \cdot (h \nabla \eta) + \frac{\omega^2}{g} \eta = 0 \qquad (5.9)$$

which will be shown in Chapter Four to be valid even if $\nabla h / kh = O(1)$. Therefore, Eq. (5.7) provides an interpolation for the whole range of wavelengths as long as the seabed slope is small, and is now known as the *mild-slope* equation (Jonsson and Skovgaard, 1979).

By the simple transformation

$$\eta = b^{-1/2}\xi, \tag{5.10}$$

Eq. (5.7) may be rewritten

$$\nabla^2\xi + \kappa^2\xi = 0, \tag{5.11}$$

where

$$\kappa^2(x, y) = \frac{\omega^2 c}{b} - \frac{\nabla^2 b}{2b} + \frac{|\nabla b|^2}{4b^2}. \tag{5.12}$$

Equation (5.11) is well known in the acoustics of inhomogeneous media, with κ being the index of refraction. Many approximate analytical techniques exist in classical physics for special classes of κ^2. For coastal problems, numerical techniques are required. In Chapter Four one such technique will be described for Eq. (5.9); the necessary modification for Eq. (5.7) or (5.11) is straightforward and has been carried out by Houston (1981).

3.6   GEOMETRICAL OPTICS APPROXIMATION FOR REFRACTION BY SLOWLY VARYING CURRENT AND DEPTH

In addition to depth variation, the presence of current in the ocean affects the propagation of waves. Of practical interest in coastal problems are the tidal currents near a river inlet or a harbor entrance. During flood tides the current and the waves are in the same direction, resulting in the lengthening of waves and reduction of wave heights. However, during ebb tides the waves are shortened and steepened by the opposing current, often to the extent of inducing breaking (See aerial photograph of Hambolt Bay, California, Johnson, 1947). If there are submarine bars in the entrance channel of a harbor, the combined effect of shoaling and current over the bars can create significant choppiness on the sea surface and therefore present hazards to navigation. Entrance channels to many harbors on the northern Pacific coast of the United States are rarely passable for small fishing boats during ebb tides in the winter (Issacs, 1948). The best time for passage is at the end of the flood tide when the water depth is the greatest and current speed is the least.

In this section we shall lay the theoretical foundation for the combined effects of currents and depth on small-amplitude waves. In particular, attention will be focused on strong currents which affect waves but are not affected by waves. We shall also assume, as is frequently the case in nature, that the characteristic time and distance of the current are very much greater than those of the waves. A systematic theory of this class of problems began with

Longuet-Higgins and Stewart (1961) and Whitham (1962), while important extensions have been made by Bretherton and Garrett (1969) and Phillips (1977). A different derivation of the basic equations will be presented below by extending the WKB formalism of Section 3.2.

To describe the magnitude of a quantity in this section, reference to the characteristic wavelength $2\pi/k$ and wave period $2\pi/\omega$ (which are related by the dispersion relation) is always implied. It will be assumed that the depth h varies slowly in horizontal coordinates x_i $(i = 1, 2)$, while the current varies slowly both in x_i and t. The long length and time scales are L and T such that

$$(\omega T)^{-1} \sim (kL)^{-1} \sim \frac{h}{L} = O(\mu) \ll 1. \tag{6.1}$$

While the horizontal velocity components of the strong current U_i are $O(gh)^{1/2}$, and the small wave velocities u_i and w are $O[(kA)(gh)^{1/2}]$, we shall for brevity speak of U_i, h, and so on, as being $O(1)$, and u_i and w as being $O(kA)$. Similarly, the operators $\partial/\partial t$ and $\partial/\partial x_i$ $(i = 1, 2)$ are $O(\mu)$ when acting on a current-related quantity and $O(1)$ when acting on a wave-related quantity, for example,

$$\frac{\partial U_i}{\partial t}, \frac{\partial U_j}{\partial x_j}, \frac{\partial h}{\partial x_i} \cdots \sim O(\mu);$$

$$\frac{\partial u_i}{\partial x_j}, \frac{\partial w}{\partial x_i}, \ldots \sim O(kA);$$

and

$$\frac{\partial}{\partial z} = O(1)$$

on all quantities.

Consider first the current $\mathbf{U} = (U_i, W)$ without waves. If dissipation is ignored, then the governing equations are

$$\frac{\partial U_i}{\partial x_i} + \frac{\partial W}{\partial z} = 0, \tag{6.2}$$

$$\frac{\partial U_i}{\partial t} + U_j \frac{\partial U_i}{\partial x_j} + W \frac{\partial U_i}{\partial z} = -\frac{1}{\rho} \frac{\partial P}{\partial x_i}, \quad i = 1, 2, \tag{6.3}$$

$$\frac{\partial W}{\partial t} + U_j \frac{\partial W}{\partial x_j} + W \frac{\partial W}{\partial z} = -\frac{1}{\rho} \frac{\partial P}{\partial z} - g. \tag{6.4}$$

Because of Eq. (6.1) and continuity, the vertical current velocity is small,

$W = O(\mu)$. It follows from Eq. (6.4) that the pressure is nearly hydrostatic:

$$P = \rho g(\bar{\zeta} - z) + O(\mu^2) \tag{6.5}$$

where $\bar{\zeta}$ is the associated free-surface displacement. On the free surface and at the sea bottom, the kinematic boundary conditions are

$$W = \frac{\partial \bar{\zeta}}{\partial t} + U_j \frac{\partial \bar{\zeta}}{\partial x_j}, \qquad\qquad z = \bar{\zeta}(x_i, t), \tag{6.6}$$

$$W = -U_j \frac{\partial h}{\partial x_j} \qquad\qquad z = -h(x_i). \tag{6.7}$$

Now the vorticity vector has the following horizontal components:

$$\Omega_j = \frac{\partial U_j}{\partial z} - \frac{\partial W}{\partial x_j}. \tag{6.8}$$

The term $\partial W/\partial x_j$ above is $O(\mu^2)$; thus, as long as

$$\Omega_j \leqq O(\mu^2), \tag{6.9}$$

U_j is independent of z to the order $O(\mu^2)$, that is,

$$\frac{\partial U_j}{\partial z} = O(\mu^2). \tag{6.10}$$

Equation (6.9) will be assumed. Recall, however, that the vertical vorticity component is allowed to be of the order $O(\mu)$. With Eq. (6.10) the horizontal momentum equations may be approximated by

$$\frac{\partial U_i}{\partial t} + U_j \frac{\partial U_i}{\partial x_j} = -g \frac{\partial \bar{\zeta}}{\partial x_i} + O(\mu^2 U_i). \tag{6.11}$$

Integrating the continuity equation vertically from $z = -h$ to $z = \bar{\zeta}$ and using Eqs. (6.6) and (6.7), we obtain

$$\frac{\partial \bar{\zeta}}{\partial t} + \frac{\partial}{\partial x_i}\left[U_i(\bar{\zeta} + h)\right] = 0. \tag{6.12}$$

Equations (6.5), (6.11), and (6.12) constitute the so-called Airy's theory of finite-amplitude long waves which will be discussed more extensively in Chapter Eleven. For present purposes, U_i and $\bar{\zeta}$ will be assumed to be known. We merely point out that the free-surface displacement $\bar{\zeta}$ is of the order $O(h)$.

In the special case of steady current, $\partial/\partial t = 0$, Eq. (6.11) may be rewritten

$$U_j \frac{\partial}{\partial x_j} \left(\frac{U_i U_i}{2} + g\bar{\zeta} \right) = 0,$$

which implies

$$\bar{\zeta} = -\frac{U_i U_i}{2g} + \text{const} \tag{6.13}$$

along a streamline.

Consider next the wave fluctuations that are superposed on the current. The velocity components (u_i, w) and the pressure p of the wave field are less than their counterparts in the current by the factor $O(kA)$. Continuity requires that

$$\frac{\partial u_i}{\partial x_i} + w = 0. \tag{6.14}$$

The momentum equations may be linearized by omitting $O(kA)^2$. We further discard linearized terms of $O(\mu^2 kA)$ or less, that is, $w\,\partial U_i/\partial z$ and $u_i\,\partial W/\partial x_i$, and obtain

$$\frac{\partial u_i}{\partial t} + U_j \frac{\partial u_i}{\partial x_j} + W \frac{\partial u_i}{\partial z} + u_j \frac{\partial U_i}{\partial x_j} = -\frac{1}{\rho} \frac{\partial p}{\partial x_i}, \tag{6.15}$$

$$\frac{\partial w}{\partial t} + U_j \frac{\partial w}{\partial x_j} + W \frac{\partial w}{\partial z} + w \frac{\partial W}{\partial z} = -\frac{1}{\rho} \frac{\partial p}{\partial z}. \tag{6.16}$$

The remaining terms contain $O(kA)$ and $O(\mu kA)$. By differentiating Eqs. (6.15) and (6.16) and adding the results, we have

$$\frac{\partial^2 p}{\partial x_i \partial x_i} + \frac{\partial^2 p}{\partial z^2} = -2\rho \left(\frac{\partial u_i}{\partial x_j} \frac{\partial U_j}{\partial x_i} + \frac{\partial w}{\partial z} \frac{\partial W}{\partial z} \right) \tag{6.17}$$

after omitting again terms of $O(\mu^2 kA)$. Equation (6.17) will be treated as the governing equation for p.

At the sea bottom the wave field also has no normal velocity

$$w = -u_j \frac{\partial h}{\partial x_j}, \qquad z = -h(x_i). \tag{6.18}$$

It is desirable to infer from this condition a boundary condition for p. Differentiating Eq. (6.18) with respect to x_j and noting that w and u_i are already evaluated on $z = -h(x_i)$, we get

$$\frac{\partial w}{\partial x_j} - \frac{\partial h}{\partial x_j}\frac{\partial w}{\partial z} = -\frac{\partial u_i}{\partial x_j}\frac{\partial h}{\partial x_i} + O(\mu^2 kA).$$

With this result, Eq. (6.16) may be rewritten

$$-\frac{1}{\rho}\frac{\partial p}{\partial z} = -\frac{\partial u_j}{\partial t}\frac{\partial h}{\partial x_j} + \left(U_j\frac{\partial h}{\partial x_j} + W\right)\frac{\partial w}{\partial z} - U_j\frac{\partial u_i}{\partial x_j}\frac{\partial h}{\partial x_i} - w\frac{\partial U_j}{\partial x_j}$$

$$= -\frac{\partial h}{\partial x_i}\left(\frac{\partial u_i}{\partial t} + U_j\frac{\partial u_i}{\partial x_j}\right) + O(\mu^2 kA)$$

$$= \frac{\partial h}{\partial x_i}\frac{1}{\rho}\frac{\partial p}{\partial x_i} + O(\mu^2 kA) \tag{6.19}$$

after invoking Eqs. (6.7), (6.15), and (6.18).

On the free surface, the exact kinematic boundary condition states that

$$\frac{\partial}{\partial t}(\bar{\zeta} + \zeta) + (U_j + u_j)\frac{\partial}{\partial x_j}(\bar{\zeta} + \zeta) = W + w \qquad \text{on } z = \bar{\zeta} + \zeta.$$

Upon linearizing and using Eq. (6.6), we get

$$\frac{\partial\zeta}{\partial t} + U_j\frac{\partial\zeta}{\partial x_j} + u_j\frac{\partial\bar{\zeta}}{\partial x_j} = w + \zeta\frac{\partial W}{\partial z} \qquad \text{on } z = \bar{\zeta}. \tag{6.20}$$

The dynamic condition is that the total pressure does not vary as one follows the fluid motion along the free surface,

$$\frac{\partial p}{\partial t} + (U_j + u_j)\frac{\partial p}{\partial x_j} + (W + w)\frac{\partial p}{\partial z}$$

$$+ \left[\frac{\partial}{\partial t} + (U_j + u_j)\frac{\partial}{\partial x_j} + (W + w)\frac{\partial}{\partial z}\right]\rho g(\bar{\zeta} - z) = 0, \qquad z = \bar{\zeta} + \zeta.$$

With the help of Eq. (6.6) the above equation can be linearized to give

$$\frac{\partial p}{\partial t} + U_j\frac{\partial p}{\partial x_j} + W\frac{\partial p}{\partial z} + \rho g\left(u_j\frac{\partial\bar{\zeta}}{\partial x_j} - w - \zeta\frac{\partial W}{\partial z}\right) = O(kA)^2, \qquad z = \bar{\zeta}.$$

$$\tag{6.21}$$

In order to express this boundary condition for p only, we shall differentiate

Eq. (6.21) and make use of Eqs. (6.16) and (6.20). Again, it must be realized that

$$\frac{\partial}{\partial x_i} W(x_i, \bar{\zeta}, t) = \left[\frac{\partial}{\partial x_i} W(x_j, z, t) + \frac{\partial \bar{\zeta}}{\partial x_i} \frac{\partial}{\partial z} W(x_j, z, t) \right]_{z=\bar{\zeta}} \quad (6.22a)$$

and

$$\frac{\partial}{\partial t} W(x_i, \bar{\zeta}, t) = \left[\frac{\partial}{\partial t} W(x_i, z, t) + \frac{\partial \bar{\zeta}}{\partial t} \frac{\partial}{\partial z} W(x_i, z, t) \right]_{z=\bar{\zeta}}. \quad (6.22b)$$

With this kind of care the following boundary condition is obtained on the free surface:

$$\left(\frac{\partial}{\partial t} + U_j \frac{\partial}{\partial x_j} \right)^2 p + 2W \frac{\partial}{\partial z} \left(\frac{\partial p}{\partial t} + U_j \frac{\partial P}{\partial x_j} \right) - g \frac{\partial \bar{\zeta}}{\partial x_j} \frac{\partial p}{\partial x_j} + g \frac{\partial p}{\partial z} = 0,$$

$$z = \bar{\zeta} \quad (6.23)$$

where derivatives with respect to t and x_i are taken before letting $z = \bar{\zeta}$.

We now introduce the slow variables $\bar{x}_i = \mu x_i$ and $\bar{t} = \mu t$, so that $U_j = U_j(\bar{x}_i, \bar{t})$, $\bar{\zeta} = \bar{\zeta}(\bar{x}_i, \bar{t})$, and $h = h(\bar{x}_i)$, and assume the WKB expansions:

$$p(x_i, z, t) = \left[p_0 + (-i\mu) p_1 + \cdots \right] e^{iS/\mu}, \quad (6.24a)$$

$$u_i(x_i, z, t) = \left[u_{0_i} + (-i\mu) u_{1_i} + \cdots \right] e^{iS/\mu}, \quad (6.24b)$$

and so on, where

$$p_0 = p_0(\bar{x}_i, z, \bar{t}), \qquad u_{0_i} = u_{0_i}(\bar{x}_i, z, \bar{t})$$

$$S = S(\bar{x}_i, \bar{t}), \qquad \text{and so·on.} \quad (6.25)$$

When Eqs. (6.24a) and (6.24b) are substituted into Eqs. (6.17), (6.19), and (6.23), a sequence of perturbation equations results. At the order $O(\mu^0)$ we have simply

$$\frac{\partial^2 p_0}{\partial z^2} - k^2 p_0 = 0, \qquad -h < z < \bar{\zeta}, \quad (6.26)$$

$$\frac{\partial p_0}{\partial z} - \frac{\sigma^2}{g} p_0 = 0, \qquad z = \bar{\zeta}, \quad (6.27)$$

$$\frac{\partial p_0}{\partial z} = 0, \qquad z = -h, \quad (6.28)$$

where

$$\sigma = \omega - U_j k_j, \tag{6.29}$$

with

$$k_i \equiv \frac{\partial S}{\partial \bar{x}_i}, \qquad \omega \equiv -\frac{\partial S}{\partial \bar{t}}, \qquad k_i k_i = k^2.$$

We shall call ω the *absolute* and σ the *intrinsic* frequency, respectively. From these definitions, it is clear that

$$\frac{\partial k_i}{\partial \bar{t}} + \frac{\partial \omega}{\partial \bar{x}_i} = 0, \tag{6.30}$$

that is, wave crests are conserved, and

$$\frac{\partial k_i}{\partial \bar{x}_j} = \frac{\partial k_j}{\partial \bar{x}_i}, \tag{6.31}$$

that is, **k** is irrotational. The solution to Eqs. (6.26)–(6.28) is

$$p_0 = \rho g A \frac{\cosh k(z + h)}{\cosh k\bar{h}} \tag{6.32}$$

where $A = A(\bar{x}_i, \bar{t})$, $\bar{h} = \bar{\zeta} + h$ is the total mean depth, and

$$\sigma^2 = gk \tanh k\bar{h}. \tag{6.33}$$

It may be shown from the horizontal momentum equation (6.15) that

$$u_{0_i} = \frac{k_i p_0}{\rho \sigma}. \tag{6.34}$$

Let us denote the group velocity relative to the current by

$$C_g = \frac{\partial \sigma}{\partial k}\bigg|_{\bar{h}} = \frac{1}{2} \frac{\sigma}{k} \left(1 + \frac{2k\bar{h}}{\sinh 2k\bar{h}} \right), \tag{6.35}$$

and

$$C_{g_i} = C_g \frac{k_i}{k}.$$

Two useful results on wave kinematics may be deduced at this stage. Differentiating Eq. (6.29) with respect to time and using Eqs. (6.30) and (6.31), we find

$$\frac{\partial \omega}{\partial \bar{t}} + \left(U_i + C_{g_i} \right) \frac{\partial \omega}{\partial \bar{x}_i} = k_i \frac{\partial U_i}{\partial \bar{t}} + \frac{\partial \sigma}{\partial \bar{h}} \bigg|_k \frac{\partial \bar{h}}{\partial \bar{t}}. \tag{6.36}$$

When the current is steady, the absolute frequency ω does not change for an observer traveling at the absolute group velocity $U + C_g$. Alternatively, one may start from Eq. (6.30) and get

$$\frac{\partial k_i}{\partial \bar{t}} + \left(U_j + C_{g_j} \right) \frac{\partial k_i}{\partial \bar{x}_j} = -k_j \frac{\partial U_j}{\partial \bar{x}_i} - \frac{\partial \sigma}{\partial \bar{h}} \bigg|_k \frac{\partial \bar{h}}{\partial \bar{x}_i} \tag{6.37}$$

after using the irrotationality of **k** [cf. Eq. (6.31)]. In principle, Eq. (6.37) may be solved numerically for the rays which are everywhere tangent to the local **k** vector.

We must now seek information regarding the variation of the wave amplitude $A(\bar{x}_i, \bar{t})$. To this end the order $O(\mu)$ problem is needed,

$$\frac{\partial^2 p_1}{\partial z^2} - k^2 p_1 = \frac{\partial}{\partial \bar{x}_j} (k_j p_0) + k_j \frac{\partial p_0}{\partial \bar{x}_j} + 2\rho \left(k_j u_{0_i} \frac{\partial U_j}{\partial \bar{x}_i} + k_i u_{0_i} \frac{\partial U_j}{\partial \bar{x}_j} \right),$$

$$-h < z < \bar{\zeta} \quad (6.38)$$

$$g \frac{\partial p_1}{\partial z} - \sigma^2 p_1 = -\frac{\partial}{\partial \bar{t}} (\omega p_0) - \omega \frac{\partial p_0}{\partial \bar{t}} - 2U_j \left[\omega \frac{\partial p_0}{\partial \bar{x}_j} - \frac{\partial}{\partial \bar{t}} (k_j p_0) \right]$$

$$+ U_i U_j \left[\frac{\partial}{\partial \bar{x}_j} (k_i p_0) + k_i \frac{\partial p_0}{\partial \bar{x}_j} \right]$$

$$+ \left(\frac{\partial U_i}{\partial \bar{t}} + U_j \frac{\partial U_i}{\partial \bar{x}_j} \right) k_i p_0 - 2 \left(\frac{\partial \bar{\zeta}}{\partial \bar{t}} + U_j \frac{\partial \bar{\zeta}}{\partial \bar{x}_j} \right) (\omega - U_i k_i) \frac{\partial p_0}{\partial z}$$

$$- g \frac{\partial \bar{\zeta}}{\partial \bar{x}_j} k_j p_0, \qquad z = \bar{\zeta}, \quad (6.39)$$

and

$$\frac{\partial p_1}{\partial z} = k_j \frac{\partial h}{\partial \bar{x}_j} p_0, \qquad z = -h. \tag{6.40}$$

In Eq. (6.39), the derivatives of p_0 with respect to $\bar{t}$ and $\bar{x}_i$ are taken before letting $z = \bar{\zeta}(\bar{x}_i, \bar{t})$. Making use of Eq. (6.22), we may condense the right-hand side of Eq. (6.39) to

$$-\frac{\partial}{\partial \bar{t}}(\omega p_0) - \omega \frac{\partial p_0}{\partial \bar{t}} - 2U_j \left[\omega \frac{\partial p_0}{\partial \bar{x}_j} - \frac{\partial}{\partial \bar{t}}(k_j p_0) \right]$$

$$+ U_i U_j \left[\frac{\partial}{\partial \bar{x}_j}(k_i p_0) + k_i \frac{\partial p_0}{\partial \bar{x}_j} \right] + p_0 \left[k_j \frac{\partial U_j}{\partial \bar{t}} + k_i U_j \frac{\partial U_i}{\partial \bar{x}_j} \right] - g k_j p_0 \frac{\partial \bar{\zeta}}{\partial \bar{x}_j}$$

where p_0 is now evaluated at $z = \bar{\zeta}$ first before derivatives are taken with respect to $\bar{t}$ and $\bar{x}_i$. Applying Green's formula to p_0 and p_1, we then get, after a little rearrangement,

$$\frac{\partial}{\partial \bar{t}}\left(\frac{E}{\sigma} \right) + \frac{\partial}{\partial \bar{x}_i}\left[(U_i + C_{g_i})\frac{E}{\sigma} \right] + \frac{2E}{\sigma^2}\left[\frac{\partial \sigma}{\partial \bar{t}} + (U_i + C_{g_i})\frac{\partial \sigma}{\partial \bar{x}_i} \right]$$

$$+ \frac{2E}{\sigma} \frac{\partial U_i}{\partial \bar{x}_j}\left[\frac{C_g}{C} \frac{k_i k_j}{k^2} + \left(\frac{C_g}{C} - \frac{1}{2} \right)\delta_{ij} \right] = 0 \qquad (6.41)$$

where $E = \frac{1}{2}\rho g A^2$. As will be discussed more extensively in Chapter Ten, the quantity

$$S_{ij} = E\left[\frac{C_g}{C} \frac{k_i k_j}{k^2} + \left(\frac{C_g}{C} - \frac{1}{2} \right)\delta_{ij} \right] \qquad (6.42)$$

is a component of the *radiation stress tensor* associated with averaged momentum fluxes in a sinusoidal wavetrain. Equation (6.41) may thus be rewritten

$$\left\{ \frac{\partial}{\partial \bar{t}}\left(\frac{E}{\sigma} \right) + \frac{\partial}{\partial \bar{x}_i}\left[(U_i + C_{g_i})\frac{E}{\sigma} \right] \right\} +$$

$$\left\{ \frac{2E}{\sigma^2}\left[\frac{\partial \sigma}{\partial \bar{t}} + (U_i + C_{g_i})\frac{\partial \sigma}{\partial \bar{x}_i} \right] + \frac{2}{\sigma}S_{ij}\frac{\partial U_i}{\partial \bar{x}_j} \right\} = 0. \qquad (6.43)$$

It can be shown (Bretherton and Garrett, 1968) by differentiating Eq. (6.33), and using Eqs. (6.36), (6.37) and (6.12), that quantities in the second pair of curly brackets in Eq. (6.43) vanish identically, therefore,

$$\frac{\partial}{\partial \bar{t}}\left(\frac{E}{\sigma} \right) + \frac{\partial}{\partial \bar{x}_i}\left[(U_i + C_{g_i})\frac{E}{\sigma} \right] = 0. \qquad (6.44)$$

The *wave action* defined with the intrinsic frequency σ is again conserved!

Using the same identity, we may also rewrite Eq. (6.43)

$$\frac{\partial E}{\partial \bar{t}} + \frac{\partial}{\partial \bar{x}_i}\left[(U_i + C_{g_i})E\right] + \frac{1}{2}S_{ij}\left(\frac{\partial U_i}{\partial \bar{x}_j} + \frac{\partial U_j}{\partial \bar{x}_i}\right) = 0 \qquad (6.45)$$

after noting that $S_{ij} = S_{ji}$. This result was first deduced by Longuet-Higgins and Stewart (1961) and means physically that work done to the current strain by the radiation stress tends to reduce wave energy.

As in Section 3.1 the ordering parameter μ may be dropped from the final results and the original coordinates restored.

Equation (6.44) forms the starting point for further analysis. In principle, one computes **k** first; Eq. (6.44) is then integrated along the path of the convected ray for the wave amplitude. There are some numerical techniques for this purpose (see Dingemans, 1978 for a survey). We shall only examine a few analytical examples in subsequent sections.

The law of wave-action conservation is a very general result valid in many different physical contexts involving slowly varying media. In fact Eq. (6.44) has been deduced for general non-dissipative dynamical systems by Bretherton and Garrett (1969) by applying WKB expansions to a variational principle involving a Lagrangian. The same result can be obtain even more rapidly by first averaging the Lagrangian and then taking the variations. These approaches require that the problem be first formulated as a variational principle; this is not always a trivial task. For a clear, authoritative and comprehensive treatment of the averaged Lagrangian method, the more advanced reader should consult Whitham (1974).

3.7 PHYSICAL EFFECTS OF SIMPLE STEADY CURRENTS ON WAVES

When current and waves are both steady, $\partial/\partial t = 0$, we get from Eqs. (6.29) and (6.30)

$$\omega = \text{const} = \sigma + \mathbf{U} \cdot \mathbf{k}. \qquad (7.1)$$

Let $y = y(x)$ be the equation for a ray. With Eq. (7.1), Eq. (2.5) may be solved numerically by first noting that

$$\mathbf{U} \cdot \mathbf{k} = Uky'\left(1 + y'^2\right)^{-1/2}. \qquad (7.2)$$

The wave amplitude may be obtained from

$$\nabla \cdot \left[(\mathbf{U} + \mathbf{C}_g)\frac{E}{\sigma}\right] = 0. \qquad (7.3)$$

Clearly, in addition to rays which follow $\mathbf{k}$, one must construct curves that are everywhere tangential to the local absolute group velocity $\mathbf{U} + \mathbf{C}_g$.

The implication of Eq. (7.1) is best seen through the simplest example of a uniform current.

3.7.1 Uniform Current on Constant Depth

Introducing Galilean transformation

$$\mathbf{x}' = \mathbf{x} - \mathbf{U}t, \tag{7.4}$$

a progressive wave in the rest frame may be written

$$e^{i(\mathbf{k}\cdot\mathbf{x} - \omega t)} = e^{i[\mathbf{k}\cdot(\mathbf{x}' + \mathbf{U}t) - \omega t]} = e^{i[\mathbf{k}\cdot\mathbf{x} - (\omega - \mathbf{k}\cdot\mathbf{U})t]}. \tag{7.5}$$

Thus in the moving frame the effective frequency is

$$\sigma = \omega - \mathbf{k}\cdot\mathbf{U} = \omega - Uk\cos\alpha \tag{7.6}$$

where α denotes the angle between $\mathbf{k}$ and $\mathbf{U}$. We observe immediately that $\sigma \gtreqless \omega$ if $|\alpha| \gtreqless \pi/2$; this result is the well-known Doppler's effect.

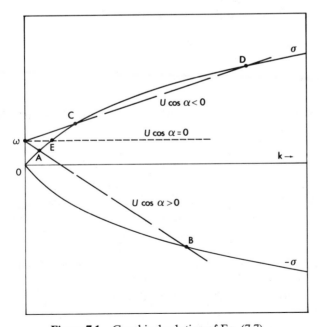

Figure 7.1 Graphical solution of Eq. (7.7).

Let us orient our coordinate system such that $\mathbf{k} = (k, 0)$ with $k > 0$. Equation (6.33) may be rewritten

$$\omega - kU\cos\alpha = \pm\sigma(k) = \pm\left(gk\tanh k\bar{h}\right)^{1/2}. \tag{7.7}$$

This dispersion relation can be solved graphically from the intersection of the straight line $y = \omega - kU\cos\alpha$ and the curves $y = \pm\sigma(k)$ as shown in Fig. 7.1. When $\alpha = \pm\pi/2$, the current has no effect on waves (see point E).

For $U\cos\alpha > 0$, there are two solutions corresponding to the points A and B. In comparison with the waves without current (point E), the waves represented by A are lengthened, and their intrinsic phase and group velocities are increased. The opposite is true for the waves represented by B. In particular, the intrinsic phase and group velocities are negative, but they are both less than $U\cos\alpha$. Hence the crests and wave energy are swept along by the fast current.

For $U\cos\alpha < 0$, the current has a component opposing the waves. If $-U\cos\alpha > (g\bar{h})^{1/2}$, no waves of any length are possible. For a smaller $-U\cos\alpha$ there is a threshold such that only one solution exists for any given ω. At this threshold, the net velocity of energy transport vanishes,

$$C_g + U\cos\alpha = 0. \tag{7.8}$$

Thus, although the crests appear to propagate upstream ($\sigma/k > 0$), energy is held stationary in space.

For still smaller $-U\cos\alpha$, two solutions exist. The waves represented by point C satisfy

$$\frac{\sigma}{k} > C_g > -U\cos\alpha,$$

hence both the crests and wave energy move upstream. The waves represented by D satisfy

$$\frac{\sigma}{k} > -U\cos\alpha > C_g.$$

Now wave energy is swept downstream while the crests move upstream. This type of wave owes its existence to the finite curvature of the σ-curve, hence to dispersion.

3.7.2 Oblique Incidence on a Shear Current over Constant Depth

Let the shear current be in the positive y direction,

$$U_1 = 0, \qquad U_2 = V(x) > 0. \tag{7.9}$$

The associated $\bar{\zeta}$ may be taken to be zero. Since $\partial/\partial y = 0$, $k_2 = k \sin \alpha = \text{const}$ from Eq. (6.31). The direction of the ray is determined from

$$k_1 = \left[k^2(x) - k_2^2\right]^{1/2}, \qquad (7.10)$$

where

$$\omega - V(x)k_2 = (gk \tanh kh)^{1/2}. \qquad (7.11)$$

The equation of the ray is similar to Eq. (3.6)

$$y - y_0 = \pm \int_{x_0}^{x} \frac{dx \, k_2}{\left(k^2(x) - k_2^2\right)^{1/2}}. \qquad (7.12)$$

Assume that V increases from zero for $x < 0$ to a peak at $x = 0$, and decreases to zero for $x > 0$ as shown in Fig. 7.2. Consider first waves propagating with the current, that is, $k_2 > 0$. The variation of $k(x)$ implied by Eq. (7.11) is

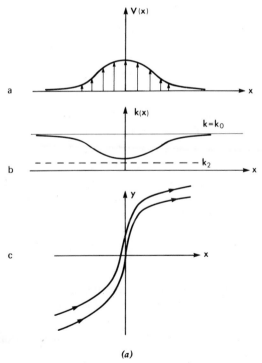

(a)

Figure 7.2 (a) $0 < k_2 < k_{\min}$: rays pass through a shear current after deflection. (b) $k_0 > k_2 > k_{\min}$: rays are bent back by a shear current. (c) $k_2 < 0$ and $k_0 < |k_2| < k_{\max}$: rays are trapped in a shear current.

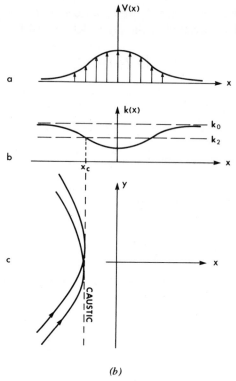

(b)

Figure 7.2   (Continued)

qualitatively as shown in Figs. 7.2a and 7.2b. If $0 < k_2 < k_{min}$, then a ray can penetrate the current (see Fig. 7.2a). If, however, $0 < k_{min} < k_2$, the square root in Eq. (7.12) is real only for $x < x_c$. Rays incident from outside must bend backward after touching the caustic at $x = x_c$, where $k_1 = 0$. These geometrical varieties resemble those over the submarine trough in Section 2.3. If the waves propagate against the same current, $k_2 < 0$, the variation of $k(x)$ is shown in Fig. 7.2c in accordance with

$$\omega + V|k_2| = (gk \tanh kh)^{1/2}. \tag{7.13}$$

Now rays can exist only within the current when $k_0 < k_2 < k_{max}$, that is, they can be trapped near the current peak. This situation resembles the submarine ridge in Section 2.3.

From the wave-action equation (7.3) the amplitude variation can be found

$$\frac{k_1 EC_g}{k(\omega - Vk_2)} = \left[\frac{k_1 EC_g}{k(\omega - Vk_2)}\right]_0 = \text{const}.$$

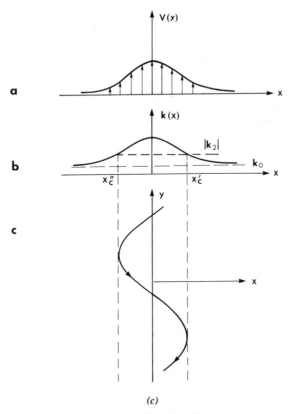

(c)

Figure 7.2 (Continued)

It follows that

$$\frac{A^2}{A_0^2} = \frac{C_{g0}\cos\alpha_0}{C_g\cos\alpha}\frac{\omega - Vk_2}{\omega - V_0k_2}. \tag{7.14}$$

In the special case of deep water, $kh \gg 1$, more explicit results may be obtained. In particular, Eq. (7.11) gives

$$\omega - Vk_2 = (gk)^{1/2}. \tag{7.15}$$

With the further simplification that $V_0 = 0$, Eq. (7.14) becomes

$$\frac{A^2}{A_0^2} = \frac{\cos\alpha_0\left[1 - (V/C_0)\sin\alpha_0\right]^2}{\left\{1 - \left(\sin^2\alpha_0/\left[1 - (V/C_0)\sin\alpha_0\right]^4\right)\right\}^{1/2}}. \tag{7.16}$$

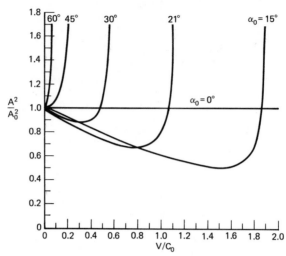

Figure 7.3 Wave amplitude versus current velocity for various angles of incidence α.

Figure 7.3 gives the amplitude ratio for various values of incidence angle α_0 and local current strength V/C_0.

Along a caustic, $\alpha = \pm\pi/2$ in Eq. (7.14) and the amplitude cannot be correctly predicted by the ray approximation. Local improvement can be effected in the manner of Section 2.3.3. McKee (1974) and Peregrine and Smith (1975) have given a detailed treatment of various types of caustics in a current in the context of linearized theory. Smith (1976) further included nonlinearity in order to accommodate fairly large amplitudes. These highly theoretical investigations have been stimulated by the *giant waves* in Agulhas Current, as vividly described in the introduction of Smith's paper:

> During the closure of the Suez Canal a number of ships, particularly oil tankers, have reported extensive damage caused by giant waves off the south–east coast of South Africa (Mallory, 1974). Two particularly unfortunate vessels were the *World Glory*, which broke in two and sank in June 1968, and the *Neptune Sapphire*, which lost 60 m of its bow section in August 1973. We can only speculate that giant waves may account for many ships which have been lost without trace off this coast. When returning from the Persian Gulf the tankers take advantage of the rapid Agulhas Current, and all except one of the eleven incidents listed by Captain Mallory (1974) involved vessels riding on the current. By examining weather charts, Mallory showed that when the incidents occurred the dominant wind-produced waves were opposed by the current.

Smith's account suggests that the situation depicted in Fig. 7.2c is likely encountered in nature.

3.7.3 Colinear Waves and Current

Assume that both waves and current are parallel to the x axis and the current speed changes in x:

$$U_1 = U(x), \qquad U_2 = 0, \qquad \mathbf{k} = (k, 0). \tag{7.17}$$

This kind of current, which must be accompanied by a vertical velocity component to satisfy continuity, Eq. (7.2), is called *upwelling* (or *downwelling*). Again $\omega = $ const so that

$$\omega = kU + \sigma = k_0 U_0 + \sigma_0 = \text{const} \tag{7.18}$$

where $(\ )_0$ signifies the value at a reference point. Defining C and C_0, which are the phase velocities with respect to the moving fluid, by

$$C^2 = \left(\frac{\sigma}{k}\right)^2 = \frac{g}{k} \tanh k\bar{h}, \qquad C_0^2 = \left(\frac{\sigma_0}{k_0}\right)^2 = \frac{g}{k_0} \tanh k_0 \bar{h}_0, \tag{7.19}$$

we have from Eq. (7.18)

$$\frac{k}{k_0}\left(\frac{U}{C_0} + \frac{C}{C_0}\right) = \frac{U_0}{C_0} + 1. \tag{7.20}$$

On the other hand, from Eq. (7.19)

$$\frac{C^2}{C_0^2} = \frac{k_0}{k} \frac{\tanh k\bar{h}}{\tanh k_0 \bar{h}} \tag{7.21}$$

which is an implicit equation for k/k_0 in terms of C/C_0. Equations (7.20) and (7.21) can be combined to give

$$\frac{C^2}{C_0^2} = \frac{(U/C_0 + C/C_0)}{(U_0/C_0 + 1)} \frac{\tanh k\bar{h}}{\tanh k_0 \bar{h}_0}. \tag{7.22}$$

Solving C/C_0 formally from Eq. (7.22) by pretending that k is constant, we get

$$\frac{C}{C_0} = \frac{1}{2T}\left[1 + \left(1 + 4T\frac{U}{C_0}\right)^{1/2}\right] \tag{7.23}$$

where

$$T = \left(1 + \frac{U_0}{C_0}\right)\frac{\tank k_0 \bar{h}_0}{\tank k\bar{h}}. \tag{7.24}$$

Clearly, there is a critical current velocity at which the square root in Eq. (7.23) vanishes, that is,

$$U = -\frac{C_0}{4T} \qquad (7.25a)$$

which implies that

$$\frac{C}{C_0} = \frac{1}{2T} \qquad (7.25b)$$

and

$$U = -\frac{C}{2}. \qquad (7.25c)$$

When $-U$ is greater than $C/2$, C/C_0 is complex and no propagation is possible.

Consider the limit of deep water $k\bar{h}, k_0\bar{h}_0 \gg 1$, and $U_0 = 0$. Equation (7.23) reduces to

$$\frac{C}{C_0} = \frac{1}{2}\left[1 + \left(1 + 4\frac{U}{C_0}\right)^{1/2}\right] \qquad (7.26)$$

Thus, a current following the waves ($U > 0$) increases the phase velocity and lengthens the waves. On the other hand, an opposing current ($U < 0$) reduces the phase velocity and shortens the waves. The critical speed is at $U = -C/2 = -C_0/4$.

For arbitrary depth, the wavelength has been calculated for an extensive range of current and depth by Jonsson, Skovgaard, and Wang (1971). Figure 7.4 shows a few sample results computed by Brevik and Aas (1980) who also conducted some experiments which confirmed the present theory.

As for the amplitude, Eq. (7.3) gives

$$\left(C_g + U\right)\frac{E}{\sigma} = \text{const}$$

so that

$$\frac{A^2}{A_0^2} = \frac{\sigma}{\sigma_0}\frac{C_{g0} + U_0}{C_g + U}. \qquad (7.27)$$

which becomes unbounded when $U = -C_g$. For deep water, $C_g \to \frac{1}{2}C$, $\sigma \to$

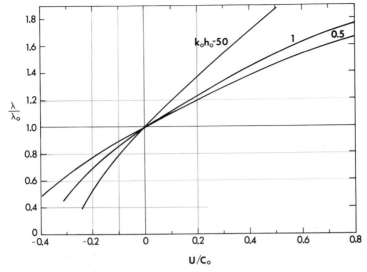

Figure 7.4 Change of wavelength due to colinear current. (From Brevik and Aas, 1980, *Coastal Engineering*. Reproduced by permission of Elsevier Scientific Publishing Co.)

g/C; Eq. (7.27) may be written

$$\frac{A^2}{A_0^2} = \frac{C_0}{C}\frac{C_0/2 + U_0}{C/2 + U} = \frac{C_0}{C}\frac{\tfrac{1}{2} + U_0/C_0}{\tfrac{1}{2}C/C_0 + U/C_0}, \qquad k\bar{h} \gg 1. \quad (7.28)$$

By using Eq. (7.26), C/C_0 may be eliminated to give

$$\frac{A^2}{A_0^2} = \frac{\tfrac{1}{2} + U_0/C_0}{U/C_0 + \tfrac{1}{2}\left[\tfrac{1}{2} + \tfrac{1}{2}(1 + 4U/C_0)^{1/2}\right]}\frac{1}{\left[\tfrac{1}{2} + \tfrac{1}{2}(1 + 4U/C_0)^{1/2}\right]},$$

$$k\bar{h} \gg 1. \quad (7.29)$$

Again, when the critical current speed $U = -\tfrac{1}{2}C = -\tfrac{1}{4}C_0$ is reached, the wave amplitude becomes infinite. This prediction is often used to infer breaking and is the idea behind the so-called *hydraulic breakwater*. In fact, the ray approximation ceases to be valid here. A more refined linear theory accounting for reflection near the critical speed can be found in Peregrine (1976). Nonlinear effects, however, can also be important and have been, to some extent, explored by Crapper (1972).

Numerical results for A/A_0 can be obtained straightforwardly for arbitrary $k\bar{h}$, by obtaining the mean surface height from Eq. (6.13), then solving for k from Eq. (7.22). Finally, Eq. (7.27) gives A/A_0. Figure 7.5 shows some sample results for $U_0 = 0$ by Brevik and Aas (1980), who also performed some

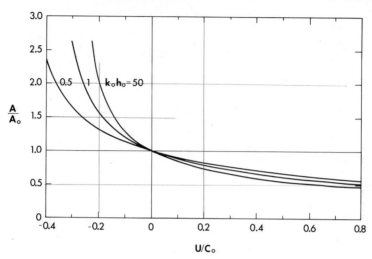

Figure 7.5 Change of wave amplitude due to colinear current. (From Brevik and Aas, 1980, *Coastal Engineering*. Reproduced by permission of Elsevier Scientific Publishing Co.)

experiments which supported the theory here. More extensive computations have been reported by Jonsson, Skovgaard and Wang (1970).

For the purpose of computing practical cases where the slow variations of depth and current can be quite arbitrary, Eq. (5.7) has been generalized by Booij (1981) using the elegant theory of Lagrangian. Both refraction and diffraction are included. Actual computations can, however, be expensive and further approximation must be added.

FOUR

Long Waves of Infinitesimal Amplitude over Bottom with Appreciable Variations

When a propagating wave enters a region where the fractional change of depth within a wavelength is significant, the phenomenon of scattering occurs in which reflection becomes appreciable. The simple ray theory which ignores reflection is no longer adequate. Before discussing the scattering of dispersive waves, let us examine similar problems for long waves in shallow water where dispersion is not important. For mathematical simplicity we shall deal largely with discontinuous depth variation. An interesting aspect of varying depth is wave trapping, that is, the phenomenon whereby waves are effectively confined in a part of the ocean. This topic has been discussed for short waves in Chapter Three. Long-wave trapping problems associated with sloping beaches, continental shelves, and oceanic ridges will be discussed here by way of several simple models such as a rectangular sill or shelf, and a uniformly sloping beach, and so on. In addition, several general aspects of the scattering matrix will be studied. Since only a few continuous variations can be solved analytically with a good deal of algebra, approximate or numerical methods are needed and will be discussed at the end of this chapter.

4.1 FORMULATION OF LINEARIZED LONG-WAVE THEORY

4.1.1 Governing Equations

In Section 1.4 we have observed that for infinitesimal waves on constant depth, water motion in long waves is essentially horizontal, implying that the vertical variation is weak and the pressure is hydrostatic. This observation has been reaffirmed in Section 3.6 in the derivation of the nonlinear equations for large-scale currents which are just long waves of finite amplitudes. Thus, long-wave motions are approximately two dimensional.

Linearizing Eqs. (6.11) and (6.12), Chapter Three, for infinitesimal amplitudes

$$\frac{|\zeta|}{h} \ll 1, \tag{1.1}$$

and changing the notations from $\mathbf{U}$ to $\mathbf{u}$ and from $\bar{\zeta}$ to ζ, we have for mass conservation

$$\frac{\partial \zeta}{\partial t} + \nabla \cdot (h\mathbf{u}) = 0, \tag{1.2}$$

and for momentum conservation

$$\frac{\partial \mathbf{u}}{\partial t} = -g\nabla \zeta. \tag{1.3}$$

The total pressure is still hydrostatic:

$$P = \rho g(\zeta - z). \tag{1.4}$$

Eliminating $\mathbf{u}$ from Eqs. (1.2) and (1.3), we obtain

$$g\nabla \cdot (h\nabla \zeta) = \frac{\partial^2 \zeta}{\partial t^2} \tag{1.5}$$

which is a hyperbolic partial differential equation with variable coefficients.

If the waves are sinusoidal in time with radian frequency ω, we may separate time and space dependences by

$$\zeta = \eta(x, y)e^{-i\omega t},$$

$$\mathbf{u}(x, y, t) \rightarrow \mathbf{u}(x, y)e^{-i\omega t}. \tag{1.6}$$

From Eqs. (1.2) and (1.3) the spatial factors are related by

$$i\omega\eta = \nabla \cdot (h\mathbf{u}) \tag{1.7}$$

$$\mathbf{u} = -\frac{ig}{\omega}\nabla\eta, \tag{1.8}$$

and

$$\nabla \cdot (h\nabla\eta) + \frac{\omega^2}{g}\eta = 0. \tag{1.9}$$

For constant depth ($h = $ const), Eq. (1.5) becomes the classical wave equation:

$$\nabla^2 \zeta = \frac{1}{gh} \frac{\partial^2 \zeta}{\partial t^2},$$ (1.10)

while Eq. (1.9) becomes the Helmholtz equation:

$$\nabla^2 \eta + k^2 \eta = 0, \qquad k = \frac{\omega}{(gh)^{1/2}}.$$ (1.11)

If the lateral boundary is a rigid vertical wall, the boundary condition must be such that the normal flux vanishes. It follows from Eq. (1.8) that

$$\frac{\partial \zeta}{\partial n} = 0 \quad \text{or} \quad \frac{\partial \eta}{\partial n} = 0$$ (1.12)

which implies that the free-surface height is either a maximum or a minimum. If the boundary is a relatively steep beach and if the waves are not steep enough to break (see Section 10.5), Eq. (1.12) may be modified to

$$\lim_{h \to 0} h \mathbf{u} \cdot \mathbf{n} = 0 \quad \text{or} \quad \lim_{h \to 0} h \frac{\partial \zeta}{\partial n} = 0.$$ (1.13)

On the other hand, the boundary condition along a rubble-mound breakwater or a mild beach with breaking waves cannot be easily stated because dissipation on these boundaries is a nonlinear process which is difficult to describe mathematically.

Finally, an appropriate condition must be specified at infinity.

To reinforce the above heuristic argument, let us give a more formal derivation for Eqs. (1.1) and (1.4) by taking the long-wave limit of the general linearized theory. The reasoning follows that of Friedrichs (1948) for nonlinear long waves and is partially similar to that of Section 3.1. Let us normalize all variables according to the scales anticipated on physical grounds:

$$(x', y') = k(x, y), \quad z' = \frac{z}{h_0}, \quad h' = \frac{h}{h_0},$$
$$t' = \left[k(gh_0)^{1/2}\right]t, \quad \zeta' = \frac{\zeta}{A_0}, \quad \Phi = \frac{1}{k}\frac{A}{h_0}(gh_0)^{1/2}\phi,$$ (1.14)

where $\omega \sim (gh_0)^{1/2}k$. The normalization for t and Φ is suggested by Eq. (2.2), Chapter One. The dimensionless governing equations are precisely Eqs. (1.3)–(1.5), Chapter Three, if ($^-$) is replaced by ( )' everywhere. For brevity the primes will be omitted from here on.

Let us introduce the series

$$\phi = \phi_0 + \mu^2\phi_2 + \mu^4\phi_4 + \cdots . \tag{1.15}$$

At the order $O(\mu^0)$ we have

$$\frac{\partial^2\phi_0}{\partial z^2} = 0, \qquad -h < z < 0, \tag{1.16}$$

$$\frac{\partial\phi_0}{\partial z} = 0, \qquad z = 0, -h, \tag{1.17}$$

so that $\phi_0 = \phi_0(x, y, t)$. At the order $O(\mu^2)$ we have

$$\frac{\partial^2\phi_2}{\partial z^2} = -\nabla^2\phi_0, \qquad -h < z < 0, \tag{1.18}$$

$$\frac{\partial\phi_2}{\partial z} = -\frac{\partial^2\phi_0}{\partial t^2}, \qquad z = 0, \tag{1.19}$$

$$\frac{\partial\phi_2}{\partial z} = -\nabla h \cdot \nabla\phi_0, \qquad z = -h. \tag{1.20}$$

Application of Green's formula [Eq. (4.11), Chapter Two] to ϕ_0 and ϕ_2 gives a solvability condition for ϕ_2,

$$\nabla \cdot (h\nabla\phi_0) = \frac{\partial^2\phi_0}{\partial t^2}. \tag{1.21}$$

If the physical variables are restored and the linearized Bernoulli equation $g\zeta = \Phi_t$ is used, then Eq. (1.21) leads to Eq. (1.5).

4.1.2 Quasi-One-Dimensional Waves in a Long Channel of Slowly Varying Cross Section

For a long channel of rectangular cross section with width much less than the length longitudinal scale, the lateral variation is expected to be much less significant than that of the longitudinal variation. Intuitively this is true because the boundary conditions of zero normal flux on the banks of a narrow channel imply that the transverse variation of ζ is negligible everywhere. The motion should be describable by a one-dimensional equation; a heuristic derivation is given below.

Let x be the longitudinal axis, y the transverse axis, $b(x)$ the width, and $h(x, y)$ the depth. Let $y = a_1(x)$ and $a_2(x)$ be the banks, then

$$b = a_2 - a_1, \qquad \text{Area } A = \int_{a_1}^{a_2} h \, dy. \tag{1.22}$$

Integrating the continuity equation (1.2) from one bank to the other, we have

$$\int_{a_1}^{a_2} \frac{\partial \zeta}{\partial t} \, dy + \frac{\partial}{\partial x} \int_{a_1}^{a_2} hu \, dy + \left[h \left(v - u \frac{da}{dx} \right) \right]_{y=a_1}^{y=a_2} = 0,$$

where use is made of the Leibniz rule. The integrated terms vanish along both banks, hence,

$$\frac{\partial \zeta}{\partial t} + \frac{1}{b} \frac{\partial Au}{\partial x} = 0. \tag{1.23}$$

When the transverse variations of ζ and u are ignored, the momentum equation reads approximately

$$\frac{\partial u}{\partial t} = -g \frac{\partial \zeta}{\partial x}. \tag{1.24}$$

Combining Eqs. (1.23) and (1.24), we have

$$\frac{\partial}{\partial x} \left(A \frac{\partial \zeta}{\partial x} \right) - \frac{b}{g} \frac{\partial^2 \zeta}{\partial t^2} = 0. \tag{1.25}$$

For sinusoidal waves $\zeta = \eta e^{-i\omega t}$, the governing equation is

$$\frac{\partial}{\partial x} \left(A \frac{\partial \eta}{\partial x} \right) + \frac{\omega^2 b}{g} \eta = 0, \tag{1.26}$$

which is of the well-known Sturm–Liouville type. It should be emphasized that so far the wavelength and the longitudinal scale of channel geometry are considered to be of the same order.

Exercise 1.1

Use Friedrichs' method to deduce Eq. (1.25) by a perturbation analysis.

4.1.3 Further Remarks on the Radiation Condition

Recall from Section 2.4 that for steady sinusoidal problems, it is necessary to impose the radiation condition that waves due to localized disturbances propagate outward. An equivalent approach is to begin with an initial-value problem and to regard the steady state as the limit of $t \to \infty$. Another alternative is to insist on a steady-state formulation but to include a small damping, which may be either real or fictitious, and then to require that the simple harmonic solution vanish at infinity. When damping is allowed to diminish at the end, the ultimate result satisfies the radiation condition also. The artifice of fictitious damping is due to Rayleigh.

In shallow water, one may imagine bottom friction to be the physical source of damping. Let the friction force by modelled by $2\varepsilon\mathbf{u}$, where ε is a small positive coefficient. The momentum equation reads

$$\frac{\partial \mathbf{u}}{\partial t} = -g\nabla\zeta - 2\varepsilon\mathbf{u}. \tag{1.27}$$

which can be combined with the continuity equation (1.2), to give

$$\nabla \cdot gh\nabla\zeta - 2\varepsilon\frac{\partial\zeta}{\partial t} = \frac{\partial\zeta^2}{\partial t^2}. \tag{1.28}$$

For simple harmonic motion Eq. (1.28) becomes

$$\nabla \cdot gh\nabla\eta + (\omega^2 + 2i\varepsilon\omega)\eta = 0, \tag{1.29}$$

which may be written

$$\nabla \cdot gh\nabla\eta + (\omega + i\varepsilon)^2\eta = 0 \tag{1.30}$$

for small ε. The boundary condition at infinity is that η must be bounded. The ultimate steady state is then the limit of the solution when $\varepsilon\downarrow0$.

Instead of the physical or pseudophysical approach of introducing damping, a mathematically equivalent way is to say that η satisfies

$$\nabla \cdot gh\nabla\eta + \omega'^2\eta = 0, \tag{1.31}$$

where ω' is complex with a small but positive imaginary part.

To see the implication of "damping" or complex ω', consider one-dimensional scattering by a localized obstacle. In the zones of constant h the scattered wave is

$$e^{ik'|x|} \quad \text{or} \quad e^{-ik'|x|},$$

where

$$k' = k + i\varepsilon(gh)^{-1/2} \tag{1.32}$$

with $k = \omega(gh)^{-1/2}$. For the solution to be bounded as $|x| \to \infty$, $e^{-ik'|x|}$ must be discarded. In the limit of $\varepsilon\downarrow0$, the disturbance becomes

$$\eta^S \sim e^{ik|x|}, \qquad k|x| \gg 1, \tag{1.33}$$

which implies outgoing waves. Hence the complex ω' implies the radiation condition. As is easily verified, the radiation condition can be expressed

$$\left(\frac{\partial}{\partial x} \mp ik\right)\eta^S \to 0, \qquad |kx| \to \infty. \tag{1.34}$$

In two-dimensional scattering by localized objects, the solution in a sea of constant depth can be constructed by superposition of the following terms:

$$\begin{Bmatrix} H_n^{(1)}(k'r) \\ H_n^{(2)}(k'r) \end{Bmatrix} \begin{Bmatrix} \sin n\theta \\ \cos n\theta \end{Bmatrix}.$$

Because of the asymptotic behavior of the Hankel functions

$$\begin{Bmatrix} H_n^{(1)}(k'r) \\ H_n^{(2)}(k'r) \end{Bmatrix} \sim \left(\frac{2}{\pi kr} \right)^{1/2} \exp\left[\pm i\left(k'r - \frac{\pi}{4} - \frac{n\pi}{2} \right) \right], \qquad (1.35)$$

$H_n^{(2)}$ must be discarded when k' is complex with a positive real part. At the limit of $\varepsilon \downarrow 0$, the general solution for the scattered waves may be written

$$\eta^S = \sum_{n=0}^{\infty} (\alpha_n \cos n\theta + \beta_n \sin n\theta) H_n^{(1)}(kr). \qquad (1.36)$$

For $kr \gg 1$ η^S behaves as

$$\eta^S \simeq \left[\sum (\alpha_n \cos n\theta + \beta_n \sin n\theta) e^{-in\pi/2} \right] \left(\frac{2}{\pi kr} \right)^{1/2} e^{ikr - i\pi/4}$$

$$\equiv A(\theta)\left(\frac{2}{\pi kr} \right)^{1/2} e^{ikr - i\pi/4} \qquad (1.37)$$

which is again an outgoing wave. Equation (1.37) is therefore the radiation condition for two-dimensional scattering by finite objects. Alternatively, this condition may be expressed

$$(kr)^{1/2}\left(\frac{\partial}{\partial r} - ik \right)\eta^S \to 0, \qquad kr \gg 1. \qquad (1.38)$$

We hasten to emphasize that Eq. (1.38) is much stronger than the mere requirement that $\eta^S \downarrow 0$ at infinity.

The above remarks suggest a simple routine for constructing transient solutions from simple harmonic solutions. If the disturbance begins at some finite time, then $\zeta \to 0$ as $t \to -\infty$. With damping we expect $\zeta \to 0$ also as $t \to +\infty$. Thus, the Fourier transform may be applied

$$\eta = \frac{1}{2\pi} \int_{-\infty}^{\infty} \zeta(\mathbf{x}, t) e^{i\omega t}\, dt. \qquad (1.39)$$

The transform of Eq. (1.28) is just Eq. (1.29) and the boundary-value problem is precisely the one for damped harmonic disturbance $\eta(\mathbf{x}, \omega')$. Consequently, the transient disturbance may be obtained by inversion:

$$\zeta(\mathbf{x}, t) = \int_{-\infty}^{\infty} \eta(\mathbf{x}, \omega') e^{-i\omega t} \, d\omega \tag{1.40}$$

which amounts to linear superposition of damped simple harmonic solutions. Since $\omega' = \omega + i\varepsilon$, Eq. (1.40) may be rewritten

$$\zeta(\mathbf{x}, t) = e^{\varepsilon t} \int_{-\infty}^{\infty} \eta(\mathbf{x}, \omega') e^{-i\omega' t} \, d\omega$$

$$= e^{\varepsilon t} \int_{-\infty + i\varepsilon}^{\infty + i\varepsilon} \eta(\mathbf{x}, \omega') e^{-i\omega' t} \, d\omega'. \tag{1.41}$$

The inviscid solution is simply the limit of $\varepsilon \downarrow 0$, with the understanding that the Fourier integral in Eq. (1.41) is along a path slightly above the real axis in the complex ω' plane. Now that the final goal has been achieved, one can forget the artifice of damping and simply say that $\zeta(\mathbf{x}, t)$ is the Fourier integral of the simple harmonic solution which satisfies the radiation condition

$$\zeta(\mathbf{x}, t) = \int_{-\infty}^{\infty} \eta(\mathbf{x}, \omega) e^{-i\omega t} \, d\omega \tag{1.42}$$

where the path of integration must be slightly above the real ω axis.

The ideas in this subsection can be generalized to the three-dimensional case of arbitrary kh.

4.2   STRAIGHT DEPTH DISCONTINUITY—NORMAL INCIDENCE

4.2.1   The Solution

Consider a simple ocean where the depth changes discontinuously at $x = 0$, and $h = h_1$ for $x < 0$ and $h = h_2$ for $x > 0$, where h_1 and h_2 are unequal constants. Two incident waves of frequency ω arrive from $x \sim \pm\infty$. From each wavetrain some energy is transmitted beyond the step and some is reflected backward, creating scattered waves propagating away from the step. The problem is to find the transmitted and the reflected waves.

The wave on either side of $x = 0$ satisfies

$$\frac{\partial \zeta}{\partial t} + \frac{\partial}{\partial x}(hu) = 0, \tag{2.1}$$

and

$$\frac{\partial u}{\partial t} + g\frac{\partial \zeta}{\partial x} = 0, \tag{2.2}$$

where $\zeta = (\zeta_1, \zeta_2)$, $u = (u_1, u_2)$, and $h = (h_1, h_2)$ for $x < 0$, $x < 0$, respectively. We must now find matching conditions at $x = 0$. Subject to further scrutiny let us assume for the moment that Eqs. (2.1) and (2.2) are valid even across the depth discontinuity and can be integrated with respect to x from $x = 0_-$ to $x = 0_+$. Since the interval of integration is infinitesimal and $\partial \zeta / \partial t$ and $\partial u / \partial t$ are finite, the first terms of Eqs. (2.1) and (2.2) do not contribute to the result, so that

$$\lim_{x \to 0_-} h_1 u_1 = \lim_{x \to 0_+} h_2 u_2, \tag{2.3}$$

$$\lim_{x \to 0_-} \zeta_1 = \lim_{x \to 0_+} \zeta_2. \tag{2.4}$$

These conditions (Lamb, 1932) relate ζ and the flux uh across the discontinuity.

For simple harmonic motion we use Eq. (1.9) so that the spatial factors satisfy

$$\frac{d^2 \eta_m}{dx^2} + k_m^2 \eta_m = 0, \qquad m = 1, 2, \tag{2.5}$$

with

$$k_m = \frac{\omega}{(gh_m)^{1/2}}. \tag{2.6}$$

The spatial part of the velocity is given by

$$u_m = -\frac{ig}{\omega} \frac{d\eta_m}{dx}. \tag{2.7}$$

The matching conditions at the junction are

$$\eta_1 = \eta_2, \tag{2.8a}$$

$$h_1 \frac{\partial \eta_1}{\partial x} = h_2 \frac{\partial \eta_2}{\partial x}. \tag{2.8b}$$

To complete the formulation we must add the radiation condition that the disturbance caused by the incident wave can only be outgoing. Thus, if there is only one incident wave from the left (or right) the waves on the right (left) side must be only right- (or left-) going. More generally, we assume that there are incident waves coming from both infinities $A_- e^{ik_1 x}$ and $B_+ e^{-ik_2 x}$. The general solution should be of the following form:

$$\eta_1 = A_- e^{ik_1 x} + B_- e^{-ik_1 x} \qquad \text{for } x < 0, \tag{2.9}$$

and

$$\eta_2 = B_+ e^{-ik_2 x} + A_+ e^{ik_2 x} \qquad \text{for } x > 0. \tag{2.10}$$

The amplitudes of the incident waves A_- and B_+ are known and those of the scattered waves A_+ and B_- are to be found. Applying the matching conditions (2.8a) and (2.8b), we obtain

$$A_+ + B_+ = A_- + B_- ,$$

$$k_1 h_1 (A_- - B_-) = k_2 h_2 (-B_+ + A_+),$$

which are readily solved to give

$$B_- = \frac{(k_1 h_1 - k_2 h_2) A_- + 2 k_2 h_2 B_+}{k_1 h_1 + k_2 h_2}, \tag{2.11}$$

$$A_+ = \frac{2 k_1 h_1 A_- - (k_1 h_1 - k_2 h_2) B_+}{k_1 h_1 + k_2 h_2}. \tag{2.12}$$

The results may be written more compactly in matrix form as

$$\{A^S\} = [S]\{A^I\} \tag{2.13}$$

with

$$\{A^I\} = \begin{Bmatrix} A_- \\ B_+ \end{Bmatrix}, \qquad \{A^S\} = \begin{Bmatrix} A_+ \\ B_- \end{Bmatrix}, \tag{2.14}$$

and

$$[S] = (k_1 h_1 + k_2 h_2)^{-1} \begin{bmatrix} 2 k_1 h_1 & -(k_1 h_1 - k_2 h_2) \\ k_1 h_1 - k_2 h_2 & 2 k_2 h_2 \end{bmatrix}$$

$$= \begin{bmatrix} T_1 & R_2 \\ R_1 & T_2 \end{bmatrix}. \tag{2.15}$$

The matrix $[S]$ is called the *scattering matrix*.

To understand the meaning of T_1, T_2, R_1, and R_2, let there be an incident wave from the left only so that $A_- \neq 0$ and $B_+ = 0$. It is clear that

$$\frac{A_+}{A_-} = T_1 = \frac{2 k_1 h_1}{k_1 h_1 + k_2 h_2}, \tag{2.16a}$$

$$\frac{B_-}{A_-} = R_1 = \frac{k_1 h_1 - k_2 h_2}{k_1 h_1 + k_2 h_2}. \tag{2.16b}$$

Thus, T_1 and R_1 can be defined respectively as the *transmission* and *reflection* coefficients when the incident wave originates from the side of $h = h_1$. T_2 and R_2 are similarly defined for an incident wave from h_2. Since $k_m h_m = \omega(h_m/g)^{1/2}$, we have

$$T_1 = \frac{2(h_1)^{1/2}}{(h_1)^{1/2} + (h_2)^{1/2}} = \frac{2}{1 + (h_2/h_1)^{1/2}}, \tag{2.17a}$$

$$R_1 = \frac{(h_1)^{1/2} - (h_2)^{1/2}}{(h_1)^{1/2} + (h_2)^{1/2}} = \frac{1 - (h_2/h_1)^{1/2}}{1 + (h_2/h_1)^{1/2}}. \tag{2.17b}$$

The variation of T_1 and R_1 with the depth ratio is shown in Fig. 2.1. Note that the phase of the reflected wave does not change when the incident wave is from the deeper side, but changes by π when it is from the shallower side. We leave it as an exercise to prove that the energy transported by the scattered waves (reflected and transmitted) equals the energy transported by the incident wave. For very shallow shelf, $h_2/h_1 \ll 1$,

$$T_1 = 2\left(1 - \left(\frac{h_2}{h_1}\right)^{1/2}\right) \qquad R_1 = 1 - 2\left(\frac{h_2}{h_1}\right)^{1/2}. \tag{2.18}$$

The reflection coefficient $R_1 \cong 1$, so that the sum of the incident and reflected waves represents essentially a standing wave with an antinode of amplitude $2A^I$ at $x = 0$. It must be cautioned that nonlinear effects so far ignored here can be very important for small enough h_2. Although the transmitted wave amplitude increases to twice the incident wave amplitude due to the decrease in depth h_2, very little energy gets through because the rate of energy flux is $T_1^2 C_g \propto (h_2)^{1/2}$. In the extreme case of $h_2/h_1 \gg 1$ the reflection coefficient $R_1 = -1$, so that the total wave system in $x < 0$ is also a standing wave but with a node at $x = 0$.

Exercise 2.1

Consider a shelf of depth h_1 in the region $x < x_1$ connected to an ocean of greater depth h_2 in the region $x > x_2$. In the transition $x_1 < x < x_2$, the depth

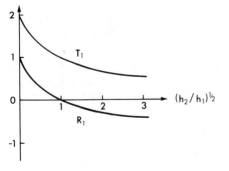

Figure 2.1 Transmission (T_1) and reflection (R_1) coefficients for a step shelf with normally incident waves.

is given by $h = ax^2$, with $h_1 = ax_1^2$, $h_2 = ax_2^2$ and $x_2 - x_1 \gg h_1$ or h_2. Let a train of long-period waves be incident normally from the ocean. Show that the scattering coefficients are

$$T = \frac{ib}{\mu^{1/2}\Delta},$$

and

$$R = i\sinh\left(\frac{b}{2}\ln\frac{1}{\mu}\right)\frac{\exp\left[-2i(\omega^2/ga)^{1/2}\right]}{\Delta},$$

where

$$b = \left(1 - 4\frac{\omega^2}{ga}\right)^{1/2}, \qquad \mu = \frac{x_1}{x_2},$$

and

$$\Delta = 2\left(\frac{\omega^2}{ga}\right)^{1/2}\sinh\left(\frac{b}{2}\ln\frac{1}{\mu}\right) + ib\cosh\left(\frac{b}{2}\ln\frac{1}{\mu}\right).$$

Plot the results and examine the effects of ω^2/ga and μ (Kajiura, 1961).

4.2.2   Justification of the Matching Conditions at the Junction

Although the conditions of matching Eqs. (2.8a) and (2.8b) are intuitively reasonable, they are deduced on the basis of Eqs. (2.1) and (2.2) which are valid only when the vertical motion is negligible compared with the horizontal and when $\partial/\partial x$ is small. However, these assumptions are no longer valid in the neighborhood of the step. Is our theory in Section 4.2.1 still valid? This question was the subject of a paper by Bartholomeuz (1958) who started from a formulation for arbitrary kh and proved rigorously that the result of the foregoing section is the correct asymptotic limit of $k_m h_m \to 0$. His argument was very lengthy and involved some very difficult mathematics. We present below a simpler argument via the *matched asymptotics method*, which is a slightly elaborated version of the boundary-layer approximation in Section 3.3.3, and has been used effectively in many long-wave problems by Ogilvie (1960), Tuck (1975), and others.

As a start we divide the physical region into a *near field* and a *far field* according to the dominant scales in each region. For example, the length scale on the incident side far away from the junction is the wavelength $1/k_1$, hence

$$\eta_1 = A\left(e^{ik_1x} + Re^{-ik_1x}\right) \tag{2.19}$$

properly describes the waves. This region of $O(k_1^{-1})$ is a far field. In the eyes of a far-field observer, the neighborhood of the step appears so small that the first few terms of the Taylor expansion of Eq. (2.19) sufficiently approximate the free surface there; hence

$$\eta_1 \simeq A[1 + R + (1 - R)ik_1x] + O(k_1x)^2 \qquad \text{as } k_1x \to 0. \qquad (2.20)$$

To another similar observer on the transmission side of the far field, the wave is described by

$$\eta_2 = ATe^{ik_2x}, \qquad (2.21)$$

which tends to

$$\eta_2 \simeq AT(1 + ik_2x) + O(k_2x)^2 \qquad (2.22)$$

in the neighborhood of the step. For shallow water the Bernoulli equation gives

$$\phi = \frac{-ig}{\omega}\eta,$$

so that

$$\phi_1^{\text{far}} \to \frac{-ig}{\omega}A[(1 + R) + (1 - R)ik_1x], \qquad k_1x \to 0, \qquad (2.23)$$

$$\phi_2^{\text{far}} \to \frac{-ig}{\omega}AT(1 + ik_2x), \qquad k_2x \to 0. \qquad (2.24)$$

Now the neighborhood of the step constitutes a near field where the motion is two dimensional and the characteristic scale is the local depth $h(h_1$ or $h_2)$. The equation of motion and the boundary condition on the step are

$$\frac{\partial^2\phi}{\partial x^2} + \frac{\partial^2\phi}{\partial z^2} = 0, \qquad (2.25)$$

$$\frac{\partial\phi}{\partial n} = 0. \qquad (2.26)$$

Although the exact linearized boundary condition on the free surface is

$$\frac{\partial\phi}{\partial z} - \frac{\omega^2}{g}\phi = 0, \qquad (2.27)$$

the two terms above are in the ratio

$$\frac{\omega^2\phi/g}{(\partial\phi/\partial z)} = O\left(\frac{\omega^2 h}{g}\right) = O(k^2 h^2).$$

Hence condition (2.27) is approximately

$$\phi_z \cong 0 \qquad\qquad (2.28)$$

with an error of $(kh)^2$. Physically, Eq. (2.28) implies that the near-field observer is oblivious of the long-scale waves and sees, at any instant, a current passing a confined channel with a step as in Fig. 2.2. The formal solution to this simplified potential flow problem can, in principle, be obtained by conformal mapping or other means.

Thus far, the near- and far-field solutions contain coefficients yet undetermined. The next step of the matched asymptotics method is to require that these solutions be joined smoothly in intermediate regions, which appear to be very near the junction to the far-field observer but very far from the junction to the near-field observer; in other words,

$$\phi^{far}\big|_{|kx|\ll 1} = \phi^{near}\big|_{|x/h|\gg 1} + O(kh)^2. \qquad (2.29)$$

Before carrying out the matching, let us write down the far-field (outer) approximation of ϕ^{near}:

$$\phi^{near} = C - DUh_1 + Ux, \qquad \frac{x}{h_1} \sim -\infty$$

$$= C + DUh_1 + U\frac{h_1}{h_2}x \qquad \frac{x}{h_2} \sim +\infty. \qquad (2.30)$$

Note in particular that the additive constants at $x \sim \pm\infty$ are different by $2DUh_1$; in fact, D is related to the unknown constant U as follows. By continuity, we have at any x,

$$Uh_1 = \int_{-h}^0 \frac{\partial\phi}{\partial x}\,dz = \frac{\partial}{\partial x}\int_{-h}^0 \phi\,dz + \frac{\partial h}{\partial x}\phi(x, -h)$$

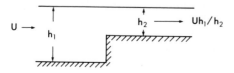

Figure 2.2   Near field of a step shelf.

which gives, after integration from x_1 to x_2 where $-x_1/h_1$ and $x_2/h_2 \gg 1$,

$$\left[\int_{-h}^{0} \phi \, dz\right]_{x_1}^{x_2} = Uh_1(x_2 - x_1) - \int_{x_1}^{x_2} \frac{\partial h}{\partial x} \phi(x, -h) \, dx. \tag{2.31}$$

Since Eq. (2.30) applies at x_1 and x_2, the left-hand side of Eq. (2.31) may be written

$$C(h_2 - h_1) + DUh_1(h_1 + h_2) + Uh_1(x_2 - x_1),$$

whereas the right-hand side of Eq. (2.31) may be written

$$Uh_1(x_2 - x_1) + C(h_2 - h_1) - \int_{-\infty}^{\infty} \frac{\partial h}{\partial x} [\phi(x, -h) + C] \, dx.$$

Substituting these into Eq. (2.31), we get

$$D = \frac{-1}{h_1 + h_2} \int_{-\infty}^{\infty} dx \frac{\partial h}{\partial x} \frac{\phi(x, -h) + C}{Uh_1}. \tag{2.32}$$

Since $\phi + C$ must be of the order of Uh_1, D is a dimensionless number of order unity and depends only on the geometry of the near field. The explicit value of D can be worked out for a rectangular step as in Section 4.2.3.

Supposing that the near field and hence D are already known in terms of C and U, let us match Eqs. (2.23) and (2.24) with Eq. (2.30). By equating the coefficients of like powers of x, we get

$$C - Uh_1 D = -\frac{igA}{\omega}(1 + R),$$

$$U = -\frac{igA}{\omega}(1 - R)ik_1,$$

$$C + Uh_1 D = -\frac{igA}{\omega} T,$$

$$U\frac{h_1}{h_2} = -\frac{igA}{\omega} Tik_2.$$

These equations can be solved for R, T, U, and C; the results are

$$R = -\frac{1 - s + 2iDk_1h_1}{1 + s - 2iDk_1h_1}, \tag{2.33}$$

$$T = \frac{2s}{1 + s - 2iDk_1h_1}, \tag{2.34}$$

$$h_1 U = -\frac{igA}{\omega} ik_2 h_2 \frac{2s}{1 + s - 2iDk_1h_1}, \tag{2.35}$$

and

$$C = -\frac{igA}{\omega}\frac{2s - iDk_1h_1}{1 + s - 2iDk_1h_1},\qquad(2.36)$$

where

$$s \equiv \frac{k_1h_1}{k_2h_2}.\qquad(2.37)$$

Since D is real and of order unity [see Eq. (2.32)], it only affects the phases of R, T, U, and C but may be ignored for their magnitudes, with an error of $O(kh)^2$. This conclusion is consistent with Bartholomeuz (1958) and was deduced in this manner by Tuck (1976). Thus, the simple requirements of Eq. (2.8) are justified.

Exercise 2.2

If there is a ship with a draft $z = -H(x)$, show that Eq. (2.32) is generalized to

$$Uh_1D = \frac{1}{h_1 + h_2}\int_{-\infty}^{\infty} dx\left\{-\frac{\partial h}{\partial x}[\phi(x_1, -h) + C] + \frac{\partial H}{\partial x}[\phi(x_1, -H) + C]\right\}.$$

4.2.3 The Near Field for a Rectangular Step

In general, the near field of an abrupt transition must be solved numerically as a classical problem of steady potential flow. For a rectangular step, the solution can be achieved analytically by the theory of complex functions (see Milne-Thomson, 1967). Let us introduce the complex variable $z = x + jy$ and the complex velocity potential $W(z)$ with $\phi(x, y) = \mathrm{Re}_j W(z)$. Note that the imaginary unit here is denoted by j in order to be distinguished from the unit i used for time variation. Although both i and j are $(-1)^{1/2}$, each is to be regarded as real with respect to the other when they appear together. In particular the real velocity potential is to be interpreted by

$$\Phi(x, y, t) = \mathrm{Re}_i \mathrm{Re}_j W(z)e^{-i\omega t}$$

$$= \mathrm{Re}_i e^{-i\omega t}\mathrm{Re}_j(\phi + j\psi)$$

$$= \mathrm{Re}_i e^{-i\omega t}\phi = \mathrm{Re}_i e^{-i\omega t}(\phi_1 + i\phi_2)$$

$$= \phi_1\cos \omega t + \phi_2\sin \omega t,$$

where ϕ_1 and ϕ_2 are real with respect to both i and j.

The physical strip in the z plane can be mapped to the upper half of the ζ plane, as shown in Fig. 2.3, by the Schwarz–Christoffel formula

$$\frac{dz}{d\zeta} = \frac{K}{\zeta} \left(\frac{\zeta - 1}{\zeta - c^2} \right)^{1/2}. \tag{2.38}$$

Clearly, the complex potential $W = \phi + j\psi$ is a source of strength Uh_1 at the origin of the ζ plane,

$$W = \frac{Uh_1}{\pi} \ln \zeta + \text{const.} \tag{2.39}$$

To fix K and c^2 we note that the complex velocity is

$$\frac{dW}{dz} = \frac{dW}{d\zeta} \bigg/ \frac{dz}{d\zeta} = \frac{Uh_1}{\pi K} \left(\frac{\zeta - c^2}{\zeta - 1} \right)^{1/2}.$$

Since $\zeta \sim \infty$ near A, $dW/dz \cong Uh_1/\pi K = Uh_1/h_2$; hence

$$K = \frac{h_2}{\pi}.$$

Near B, $\zeta \sim 0$ and $dW/dz \cong Uh_1 c/\pi K = U$; hence

$$c = \frac{h_2}{h_1}.$$

To integrate Eq. (2.38) we introduce a t plane by

$$\zeta = \frac{t^2 - c^2}{t^2 - 1} \tag{2.40a}$$

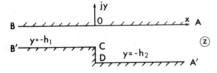

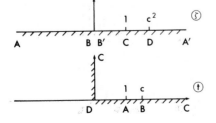

Figure 2.3 Mapping of the physical strip in the z plane to the upper half of the ζ plane.

or

$$t = \left(\frac{\zeta - c^2}{\zeta - 1} \right)^{1/2}, \tag{2.40b}$$

which maps the upper half ζ plane to the first quadrant of t as shown in Fig. 2.3. Taking the logarithmic derivative of Eq. (2.40) and combining with Eq. (2.38), we can integrate z in terms of t, with the result

$$z + jh_1 = \frac{h_2}{\pi} \left(\frac{1}{c} \ln \frac{t - c}{t + c} - \ln \frac{t - 1}{t + 1} \right), \tag{2.41}$$

where the constant jh_1 is chosen so that the images of point C are as shown in the z and t planes.

Now putting Eq. (2.40a) into Eq. (2.39), we get

$$W = \frac{Uh_1}{\pi} \ln \frac{t^2 - c^2}{t^2 - 1}. \tag{2.42}$$

For a given t in the first quadrant, we can find z from Eq. (2.41) and the corresponding W from Eq. (2.42). Now the near-field solution is complete.

The asymptotic approximations in the neighborhoods of A and B are needed. Let t approach the point B from the left, $t \to c - 0$, then

$$z + jh_1 \cong \frac{h_2}{\pi} \left[\frac{1}{c} \ln (t - c) - \frac{1}{c} \ln 2c - \ln \frac{c - 1}{c + 1} \right],$$

and

$$W \cong \frac{Uh_1}{\pi} \left[\ln(t - c) + \ln \frac{2c}{c^2 - 1} \right].$$

After $\ln(t - c)$ is eliminated, it follows that

$$W \cong Uz + \frac{Uh_1}{\pi} \left[j\pi + \ln 2c + c \ln \frac{c - 1}{c + 1} + \ln \frac{2c}{c^2 - 1} \right]. \tag{2.43}$$

Let t approach A from the right, then

$$z + jh_1 \cong \frac{h_2}{\pi} \left[\frac{1}{c} \left(\ln \frac{c - 1}{c + 1} + j\pi \right) - \ln(t - 1) + \ln 2 \right],$$

and

$$W \cong \frac{Uh_1}{\pi} \left[\ln(c^2 - 1) + j\pi - \ln 2 - \ln(t - 1) \right].$$

Eliminating $\ln(1-t)$, we get

$$W \cong \frac{Uh_1 z}{h_2} + \frac{Uh_1}{\pi}\left[\ln(1-c^2) - 2\ln 2 + j\pi - \frac{1}{c}\ln\frac{c-1}{c+1}\right]. \quad (2.44)$$

We may now subtract the additive constants of Eqs. (2.43) and (2.44) to give

$$2D = \frac{Uh_1}{\pi}\left[\frac{c^2+1}{c}\ln\frac{c+1}{c-1} - 2\ln\frac{4c}{c^2-1}\right] \quad (2.45)$$

which was given by Tuck (1976) and confirms the order estimate of D in Section 4.2.2.

4.3   STRAIGHT DEPTH DISCONTINUITY—OBLIQUE INCIDENCE

Consider a plane wavetrain arriving at an angle θ_1 with respect to the depth discontinuity (Fig. 3.1). Let the y axis be the discontinuity and x be normal to it. The depths on two sides are h_1, $x < 0$, and h_2, $x > 0$, where $h_1 \neq h_2$ in general.

Let the incident wave come from $x \to -\infty$,

$$\eta_I = Ae^{i(\alpha_i x + \beta y)} \qquad \text{so that } \alpha_1^2 + \beta^2 = k_1^2. \quad (3.1)$$

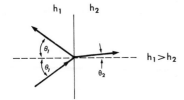

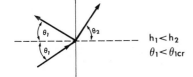

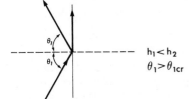

Figure 3.1   Directions of wave vectors at a step.

The incident wavenumber vector is inclined at the angle

$$\theta_1 = \tan^{-1}(\beta/\alpha_1) \tag{3.2}$$

with respect to the x axis. The solutions should be of the following form:

$$\eta_1 = A(e^{i\alpha_1 x} + Re^{-i\alpha_1 x})e^{i\beta y}, \quad \alpha_1^2 + \beta^2 = k_1^2, \quad x < 0, \tag{3.3}$$

$$\eta_2 = ATe^{i(\alpha_2 x + \beta y)}, \quad\quad\quad\quad \alpha_2^2 + \beta^2 = k_2^2, \quad x > 0, \tag{3.4}$$

so that on the left side there is a reflected wave toward the left, and on the right side a transmitted wave toward the right. The reflection and transmission coefficients R and T must be determined by matching the surface height and the volume flux at $x = 0$. Substituting Eqs. (3.3) and (3.4) into Eqs. (2.8a) and (2.8b), we obtain

$$1 + R = T, \tag{3.5a}$$

$$h_1(i\alpha_1 - Ri\alpha_1) = h_2 i\alpha_2 T. \tag{3.5b}$$

The solutions for R and T are formally the same as for normal incidence if we replace k_1 and k_2 by α_1 and α_2 in Eqs. (2.16) and (2.17), that is,

$$T = \frac{2\alpha_1 h_1}{\alpha_1 h_1 + \alpha_2 h_2} \tag{3.6a}$$

$$R = \frac{\alpha_1 h_1 - \alpha_2 h_2}{\alpha_1 h_1 + \alpha_2 h_2}. \tag{3.6b}$$

Certain features of the solution deserve attention. The directions of the incident and transmitted waves are given by

$$\tan \theta_1 = \frac{\beta}{\left(k_1^2 - \beta^2\right)^{1/2}}, \tag{3.7}$$

and

$$\tan \theta_2 = \frac{\beta}{\left(k_2^2 - \beta^2\right)^{1/2}}. \tag{3.8}$$

For $h_1 > h_2$, $k_1 < k_2$; hence $\theta_1 > \theta_2$. If the transmission side is the shallower of the two, the wavenumber vector of the transmitted wave is directed more closely to the x axis than to the incident wave vector. On the other hand, if $h_1 < h_2$ so that the incident side is the shallower, $\theta_1 < \theta_2$ and the transmitted wave turns away from the normal x axis. This result is just the phenomenon of refraction discussed in Chapter Two for slowly varying depth, and the transmitted wave may be called the refracted wave. For a fixed frequency, k_1 and k_2 are fixed by h_1 and h_2. If we increase β toward k_1 (i.e., increase the angle of incidence), a stage will be reached such that $k_2 = \beta$ because $k_2 < k_1$. At this stage, $\theta_2 = \pi/2$ and the transmitted wave propagates along the discontinuity ($\alpha_2 = 0$). The critical angle of incidence is

$$(\theta_1)_{cr} = \tan^{-1}\frac{k_2}{\alpha_1} = \tan^{-1}\frac{k_2}{(k_1^2 - k_2^2)^{1/2}} \tag{3.9}$$

Since $\alpha_2 = 0$, the reflection coefficient is $R = 1$; hence there is total reflection. The transmitted wave has crests parallel to the x axis with equal amplitude along the crests.

What happens when β increases further? $\alpha_2 = (k_2^2 - \beta^2)^{1/2}$ becomes imaginary, and $\tan\theta_2$ loses meaning. Let us go back to the original solution and rewrite $\alpha_2 = i\gamma_2$, $\gamma_2 = (\beta^2 - k_2^2)^{1/2}$ so that γ_2 is real and positive:

$$\eta_2 = a_1 T e^{-\gamma_2 x} e^{i\beta y}. \tag{3.10}$$

The general solution actually contains $e^{\gamma_2 x}$ and $e^{-\gamma_2 x}$; we take only $e^{-\gamma_2 x}$ for the bounded solution at $x \sim \infty$. Thus,

$$T = \frac{2\alpha_1 h_1}{\alpha_1 h_1 + i\gamma_2 h_2} \tag{3.11a}$$

$$R = \frac{\alpha_1 h_1 - i\gamma_2 h_2}{\alpha_1 h_1 + i\gamma_2 h_2}. \tag{3.11b}$$

Clearly $|R| = 1$, so that reflection is perfect. With Eqs. (3.11) the solution may be renormalized to give

$$\eta_1 = A\cos(\alpha_1 x + \delta)e^{i\beta y}, \tag{3.12}$$

$$\eta_2 = A\frac{\alpha_1 h_1}{(\alpha_1^2 h_1^2 + \gamma_2^2 h_2^2)^{1/2}}e^{-\gamma_2 x}e^{i\beta y}, \tag{3.13}$$

where δ is a phase angle

$$\tan\delta = \frac{\gamma_2 h_2}{\alpha_1 h_1}.$$

This solution requires a new interpretation. On the deep side, $x > 0$, the transmitted wave propagates along the y axis with the amplitude exponentially attenuating away from the maximum at $x = 0$. The larger the angle β, the faster the attenuation.

4.4   SCATTERING BY A SHELF OR TROUGH OF FINITE WIDTH[†]

Consider an ocean bottom with a stepwise variation of depth as shown in Fig. 4.1. An obliquely incident wave of unit amplitude arrives from $x \sim -\infty$. What are the effects of the finite size of the step?

The general solution over each flat region can be written

$$\eta_1 = e^{i\beta y}(e^{i\alpha_1(x+a)} + R'e^{-i\alpha_1(x+a)}), \qquad x < -a, \qquad (4.1)$$

$$\eta_2 = e^{i\beta y}(Ae^{i\alpha_2 x} + Be^{-i\alpha_2 x}), \qquad -a < x < a, \qquad (4.2)$$

$$\eta_3 = T'e^{i\beta y}e^{i\alpha_3(x-a)}, \qquad x > a. \qquad (4.3)$$

We may define

$$R = R'e^{-2i\alpha_1 a} \qquad (4.4)$$

as the reflection coefficient, and

$$T = T'e^{-i(\alpha_1 + \alpha_3)a} \qquad (4.5)$$

as the transmission coefficient. The coefficients A, B, R', and T' must be found by matching η and $h\,\partial\eta/\partial x$ at the two edges.

Matching at $x = -a$, we get

$$1 + R' = Ae^{-i\alpha_2 a} + Be^{i\alpha_2 a}, \qquad (4.6)$$

and

$$\alpha_1 h_1(1 - R') = \alpha_2 h_2(Ae^{-i\alpha_2 a} - Be^{i\alpha_2 a}), \qquad (4.7)$$

whereas matching at $x = a$, we get

$$Ae^{i\alpha_2 a} + Be^{-i\alpha_2 a} = T', \qquad (4.8)$$

and

$$\alpha_2 h_2(Ae^{i\alpha_2 a} - Be^{-i\alpha_2 a}) = \alpha_3 h_3 T'. \qquad (4.9)$$

The simultaneous equations (4.6)–(4.9) remain to be solved. The bookkeeping

---

[†] This problem is analogous to the problem of a square-well potential in quantum mechanics (see Bohm, 1951, p. 242ff).

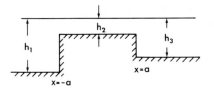

Figure 4.1 A submerged ridge.

may be simplified with the following new substitutions:

$$s_{\mu\nu} = \frac{\alpha_\mu h_\mu}{\alpha_\nu h_\nu} \quad \text{with } \mu, \nu = 1, 2, 3 \text{ (no summation).} \tag{4.10}$$

Equations (4.6)–(4.9) become

$$Ae^{-i\alpha_2 a} + Be^{i\alpha_2 a} = 1 + R', \tag{4.11}$$

$$Ae^{-i\alpha_2 a} - Be^{i\alpha_2 a} = s_{12}(1 - R'), \tag{4.12}$$

$$Ae^{i\alpha_2 a} + Be^{-i\alpha_2 a} = T', \tag{4.13}$$

$$Ae^{i\alpha_2 a} - Be^{-i\alpha_2 e} = s_{32}T'. \tag{4.14}$$

From Eqs. (4.13) and (4.14) we can express A and B in terms of T' or T:

$$A = \tfrac{1}{2}T'e^{-i\alpha_2 a}(1 + s_{32}), \tag{4.15}$$

$$B = \tfrac{1}{2}T'e^{i\alpha_2 a}(1 - s_{32}). \tag{4.16}$$

Eliminating A and B from Eqs. (4.11) and (4.12), we can solve for R' and T':

$$R' = \frac{e^{-2i\alpha_2 a}\left[-(1 - s_{12})(1 + s_{32}) + (1 + s_{12})(1 - s_{32})e^{2i\alpha_2 a}\right]}{\Delta}, \tag{4.17}$$

$$T' = \frac{4s_{12}}{\Delta}, \tag{4.18}$$

where

$$\Delta = (1 + s_{12})(1 + s_{32})e^{-2i\alpha_2 a} - (1 - s_{12})(1 - s_{32})e^{2i\alpha_2 a}. \tag{4.19}$$

Finally, A and B can be obtained from Eqs. (4.15) and (4.16) with the help of Eq. (4.18).

For physical implications let us examine a special case where the depths on both sides of the step are equal, $h_1 = h_3$. We now have $\alpha_1 = \alpha_3$, so that

$$s_{12} = s_{32} = \frac{\alpha_1 h_1}{\alpha_2 h_2} \equiv s = \frac{h_1}{h_2}\left(\frac{\omega^2/gh_1 - \beta^2}{\omega^2/gh_2 - \beta^2}\right)^{1/2}. \tag{4.20}$$

Note that $s > 1$ if the center region is a shelf and $s < 1$ if it is a trough. The transmission and reflection coefficients are

$$T' = \frac{4s}{(1+s)^2 e^{-2i\alpha_2 a} - (1-s)^2 e^{2i\alpha_2 a}}, \qquad (4.21)$$

$$R' = \frac{-(1-s^2)(e^{-i\alpha_2 a} - e^{i\alpha_2 a})}{(1+s)^2 e^{-i\alpha_2 a} - (1-s)^2 e^{2i\alpha_2 a}}. \qquad (4.22)$$

The energy of the transmitted and reflected waves are proportional to

$$\begin{pmatrix} |T'|^2 \\ |R'|^2 \end{pmatrix} = \begin{pmatrix} |T|^2 \\ |R|^2 \end{pmatrix}$$

$$= \begin{pmatrix} \dfrac{4s^2}{(1-s^2)^2 \sin^2 2\alpha_2 a} \end{pmatrix} \left[4s^2 + (1-s^2)^2 \sin^2 2\alpha_2 a \right]^{-1}. \qquad (4.23)$$

It is straightforward to show that $|R|^2 + |T|^2 = 1$ which means that the energy of the scattered waves is the same as the energy of the incident waves. An important physical feature is that $|R|^2$ and $|T|^2$ vary periodically with $2\alpha_2 a$. In particular, for $2\alpha_2 a = n\pi$, $n = 0, 1, 2, 3, \ldots$, that is, $4a/\lambda_2 = 0, 1, 2, 3, \ldots$, where $\lambda_2 = 2\pi/\alpha_2$, $|R|^2 = 0$, and $|T|^2 = 1$ so that there is total transmission and the shelf is transparent to the incident wave. Minimum transmission and maximum reflection occur when $\sin^2 2\alpha_2 a = 1$ or

$$2\alpha_2 a = \left(n - \tfrac{1}{2}\right)\pi, \qquad n = 1, 2, 3, \ldots$$

that is,

$$\frac{4a}{\lambda_2} = \frac{1}{2}, \frac{3}{2}, \frac{5}{2}, \ldots \ldots$$

The corresponding values are

$$\min |T|^2 = \frac{4s^2}{(1+s^2)^2}, \qquad (4.24)$$

$$\max |R|^2 = \frac{(1-s^2)^2}{(1+s^2)^2} \qquad (4.25)$$

whose dependence on s^2 is shown in Fig. 4.2a. The dependence of $|T|$ and $|R|$ on $2\alpha_2 a$ is oscillatory as shown in Fig. 4.2b.

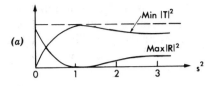

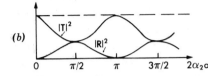

Figure 4.2 Properties of $|T|^2$ and $|R|^2$. (a) Effects of varying $s = \alpha_1 h_1/\alpha_2 h_2$; (b) effects of varying $2\alpha_2 a$.

The wave on the shelf is obtained by substituting Eq. (4.21) into Eqs. (4.15) and (4.16) with $s_{12} = s_{32} = s$, that is,

$$A = \tfrac{1}{2}T'(1 + s)e^{-i\alpha_2 a}, \qquad B = \tfrac{1}{2}T'(1 - s)e^{i\alpha_2 a} \qquad (4.26)$$

and then into Eq. (4.2). Omitting the intermediate steps, we give the final result:

$$\eta_2 = \frac{2s\left[(1 + s)e^{i\alpha_2(x-a)} + (1 - s)e^{-i\alpha_2(x-a)}\right]}{(1 + s)^2 e^{-2i\alpha_2 a} - (1 - s)^2 e^{2i\alpha_2 a}} \qquad (4.27)$$

so that the square of the envelope is

$$|\eta_2|^2 = \frac{4s^2\left[\cos^2\alpha_2(x - a) + s^2\sin^2\alpha_2(x - a)\right]}{4s^2 + (1 - s^2)^2\sin^2 2\alpha_2 a}. \qquad (4.28)$$

The free surface within $-a < x < a$ is composed of two wavetrains traveling in opposite directions, resulting in interference and a partially standing wave with amplitude varying along x. In particular, at the edge $x = a$,

$$|\eta_2|^2 = \frac{4s^2}{4s^2 + (1 - s^2)^2\sin^2 2\alpha_2 a}, \qquad x = a \qquad (4.29)$$

so that the interference is destructive, that is, $|\eta_2|^2$ is the smallest, when $2\alpha_2 a = (n - \tfrac{1}{2})\pi$, and constructive, that is, $|\eta_2|^2$ is the largest, when $2\alpha_2 a = n\pi$. Since the fluid at $x = a$ acts as a piston for the motion in the region of $x > a$, its amplitude of motion determines the amplitude of the transmitted wave.

The features of interference just deduced mathematically can also be explained physically. When a wave first strikes the edge at $x = -a$, part of the wave is transmitted into $-a < x < a$ and part is reflected. Upon reaching the edge at $x = a$, the transmitted wave undergoes the same scattering process, whereupon part of the wave is transmitted to $x > a$ and part is reflected toward the edge $x = 0$. This back-and-forth process of transmission and reflection is repeated indefinitely for all the waves of the periodic train. The total left-going wave in $x < -a$ is the sum of the reflected waves from $x = -a$ and all the transmitted waves from $-a < x < a$ to $x < -a$, whereas the total right-going wave in $x > a$ is the sum of all the transmitted waves from $-a < x < a$ to $x > a$. Now if $4a$ is an integral multiple of wavelength λ_2, each time a typical wave crest completes a round trip, being reflected from $x = a$ to $x = -a$ and back to $x = a$, its phase is changed by π twice. Therefore, all the crests which arrive at $x = a$ at the same time after different numbers of round trips $\ldots, -2, -1, 0, 1, 2, \ldots$ have the same phase; they interfere *constructively* and the total amplitude at $x = a$ is consequently the larger. On the other hand, if $4a$ is an odd multiple of half-wavelength $\lambda_2/2$, then after a round trip a typical wave crest is opposite in phase with the other crests which lag or lead by an odd number of round trips. This interference is *destructive*, resulting in the smallest net amplitude at $x = a$.

Furthermore, taking the x derivative of $|\eta_2|^2$, we find $\partial |\eta_2|^2/\partial x \propto \sin 2\alpha_2(x - \alpha)$ so that the intensity $|\eta_2|^2$ is extremum at $2\alpha_2(x - \alpha) = n\pi$, that is, $x - a = \frac{1}{2}n\lambda_2$. It follows from Eq. (4.28) that the extremum values are

$$\text{Extr}\,|\eta_2|^2 = \frac{4s^2}{4s^2 + (1 - s^2)^2\sin^2 2\alpha_2 a} \qquad \text{if } n = \text{even,}$$

and

$$\text{Extr}\,|\eta_2|^2 = \frac{4s^4}{4s^2 + (1 - s^2)\sin^2 2\alpha_2 a} \qquad \text{if } n = \text{odd.}$$

In both cases the extremum values are the largest when $2\alpha_2 a$ are integer multiples of π. Thus, the peaks of T versus $2\alpha_2 a$ coincide with the peak response in the region $0 < x < a$.

Finally, we consider the limit of a short step $\alpha_2 a \downarrow 0$. By Taylor's expansion of Eq. (4.23), it follows that

$$|T|^2 = 1 - \frac{(1 - s^2)^2}{s^2}(\alpha_2 a)^2 + O(\alpha_2 a)^4,$$

$$|R|^2 = \frac{(1 - s^2)^2}{s^2}(\alpha_2 a)^2 + O(\alpha_2 a)^4.$$

Thus, a barrier much shorter than the wavelength is quite transparent to incident waves. In nature, real fluid effects introduce flow separation, hence dissipation, and change the above conclusion in important ways.

4.5 TRANSMISSION AND REFLECTION BY A SLOWLY VARYING DEPTH

There are certain special depth profiles (linear, parabolic) for which analytical solutions are possible (Kajiura, 1961). The mathematics involved is straightforward but can be tedious. For slowly varying bottoms, it is possible to derive some general but approximate results which are physically revealing.

For a depth varying with the scale which is much longer than the local wavelength, the classical WKB approximation is a natural starting point. Assuming one-dimensional topography, that is, $h = h(x)$, the equation of motion is, from Eq. (1.26),

$$\frac{d}{dx}\left(gh \frac{d\eta}{dx} \right) + \omega^2 \eta = 0.$$ (5.1)

Consider a wave propagating in the positive x direction:

$$\eta = A(\bar{x}) e^{iS(\bar{x})/\mu},$$ (5.2)

where $\bar{x} = \mu x$ with μ being a small parameter characterizing the bottom slope. As in Section 2.1, we denote

$$k(\bar{x}) = \frac{1}{\mu} \frac{dS}{dx} = \frac{dS}{d\bar{x}}.$$ (5.3)

Substituting the derivatives of η into Eq. (5.1), we get

$$(-ghk^2 + \omega^2)A + \mu \left\{ g \frac{dh}{d\bar{x}} ikA + gh \left[ik \frac{dA'}{d\bar{x}} + ih \frac{d(kA)}{d\bar{x}} \right] \right\}$$

$$+ \mu^2 \left(g \frac{dh}{d\bar{x}} \frac{dA}{d\bar{x}} + gh \frac{d^2A}{d\bar{x}^2} \right) = 0.$$

From $O(\mu^0)$ the dispersion relation follows

$$\frac{\omega^2}{g} = k^2 h,$$ (5.4)

while from $O(\mu)$ we have, after some simple manipulations,

$$\frac{d}{dx}(khA^2) = 0$$

so that

$$(khA^2) = \text{const} \equiv E_0^2 \equiv (khA^2)_{x \sim -\infty}.$$ (5.5)

In terms of E_0 the leading order solution is

$$\eta = \frac{E_0}{(kh)^{1/2}} e^{iS/\mu} = \frac{E_0}{(kh)^{1/2}} \exp\left[\frac{i}{\mu} \int^{\bar{x}} k(\bar{x})\, d\bar{x}\right], \qquad (5.6)$$

where E_0^2 is proportional to the energy flux of the wave incident from $x \sim -\infty$.

It is possible to superimpose waves traveling in both directions so that the general solution is

$$\eta \cong \frac{1}{(kh)^{1/2}} \left(E_0 e^{iS/\mu} + F_0 e^{-iS/\mu}\right), \qquad (5.7)$$

where F_0^2 is proportional to the incident wave energy from the right,

$$F_0^2 = (khA^2)_{x \sim +\infty}.$$

The solution (5.6) or (5.7) gives no account of reflection. Extensions of the preceding analysis to weak reflection were developed in other physical contexts by Bremmer (1951) and others and applied to shallow-water waves by Ogawa and Yoshida (1959). For a very good summary see Kajiura (1961) or Wait (1962); their reasoning is followed here.

Beginning from the mass and momentum equations, it is convenient to define $uh = Q$ so that

$$i\omega\eta = \frac{dQ}{dx}, \qquad (5.8)$$

$$i\omega Q = gh\frac{d\eta}{dx}. \qquad (5.9)$$

From Eq. (5.7) the discharge to the leading order is given by

$$i\omega Q \cong ig(kh)^{1/2}\left(E_0 e^{iS/\mu} - F_0 e^{-iS/\mu}\right), \qquad (5.10)$$

the omitted term being $O(\mu)$. We now follow Bremmer and replace E_0 and F_0 in Eqs. (5.7) and (5.10) with two unknown functions E and F, that is,

$$\eta = \frac{1}{(kh)^{1/2}}\left(Ee^{iS/\mu} + Fe^{-iS/\mu}\right), \qquad (5.11)$$

$$i\omega Q = ig(kh)^{1/2}\left(Ee^{iS/\mu} - Fe^{-iS/\mu}\right), \qquad (5.12)$$

which are now taken as the exact solution to Eqs. (5.8) and (5.9). Upon substitution we obtain a pair of equations governing E and F:

$$\frac{dE}{dx}e^{iS/\mu} - \frac{dF}{dx}e^{-iS/\mu} = -\frac{\mu}{(kh)^{1/2}}\frac{d(kh)^{1/2}}{d\bar{x}}(Ee^{iS/\mu} - Fe^{-iS/\mu}),$$

and

$$\frac{dE}{dx}e^{iS/\mu} - \frac{dF}{dx}e^{-iS/\mu} = \frac{\mu}{(kh)^{1/2}}\frac{d(kh)^{1/2}}{d\bar{x}}(Ee^{iS/\mu} + Fe^{-iS/\mu}).$$

The derivatives dE/dx and dF/dx can be solved to give

$$\frac{dE}{dx} = \frac{\mu}{(kh)^{1/2}}\frac{d(kh)^{1/2}}{d\bar{x}}Fe^{-2iS/\mu}, \tag{5.13a}$$

$$\frac{dF}{dx} = \frac{\mu}{(kh)^{1/2}}\frac{d(kh)^{1/2}}{d\bar{x}}Ee^{2iS/\mu}, \tag{5.13b}$$

which are still exact. We now introduce the perturbation expansions:

$$E = E_0 + \mu E_1 + \mu^2 E_2 + \cdots,$$

$$F = F_0 + \mu F_1 + \mu^2 F_2 + \cdots.$$

Direct substitution yields

$$\frac{dE_0}{dx} = \frac{dF_0}{dx} = 0,$$

$$\frac{dF_{n+1}}{dx} = \left[\frac{d}{d\bar{x}}\ln(kh)^{1/2}\right]E_n e^{2iS/\mu},$$

and

$$\frac{dF_{n+1}}{dx} = \left[\frac{d}{d\bar{x}}\ln(kh)^{1/2}\right]E_n e^{2iS/\mu},$$

which can be integrated to give

$$E_0 = \text{const}, \qquad F_0 = \text{const}, \tag{5.14a}$$

$$E_{n+1} = \int_{-\infty}^{x}\left[\frac{d}{d\bar{x}}\ln(kh)^{1/2}\right]F_n e^{-2iS/\mu}\,dx, \tag{5.14b}$$

$$F_{n+1} = \int_{\infty}^{x}\left[\frac{d}{d\bar{x}}\ln(kh)^{1/2}\right]E_n e^{2iS/\mu}\,dx. \tag{5.14c}$$

The lower limits of integration are chosen such that

$$E_n(-\infty) = 0, \qquad F_n(\infty) = 0, \qquad n = 1, 2, 3, \ldots .$$

From now on the parameter μ may be set to unity and $\bar{x}$ restored to x. The solution is complete.

As a special case, let $F_0 = 0$ so that the incident wave is from left to right. Then,

$$E = E_0 + \mu E_1 + \mu^2 E_2 + \mu^3 E_3 + \cdots, \qquad x \sim +\infty \qquad (5.15)$$

represents the transmitted wave, while

$$F = F_1 + {}^2 F_2 + \cdots, \qquad x \sim -\infty \qquad (5.16)$$

represents the reflected wave. To $O(\mu)$ the reflection coefficient is

$$R_1 = \left(\frac{F_1}{E_0} \right)_{x \sim -\infty} = -\int_{-\infty}^{\infty} dx \left[\frac{d}{dx} \ln(kh)^{1/2} \right] e^{2iS/\mu}. \qquad (5.17)$$

The integrals can be carried out by quadrature once ω and $h(x)$ are prescribed.

To gain some physical understanding let us assume that h differs slightly from a constant h_0,

$$h = h_0[1 + q(x)], \qquad q \ll 1, \qquad (5.18)$$

then

$$\frac{\omega^2}{g} = k^2 h_0(1 + q),$$

and

$$k = \frac{\omega}{(gh_0)^{1/2}} (1 + q)^{-1/2} \cong k_0 \left(1 - \frac{q}{2} \right)$$

so that

$$kh \cong k_0 h_0 \left(1 + \frac{q}{2} \right).$$

Expanding the logarithm, we get

$$\frac{d}{dx} \ln(kh)^{1/2} \cong \frac{1}{4} \frac{dq}{dx},$$

and

$$\int_0^x k\,dx \cong k_0 x.$$

Thus, for small perturbations the reflection is approximately

$$R_1 = -\int_{-\infty}^{\infty} \frac{1}{4}\frac{dq}{dx}e^{2ik_0 x}\,dx. \tag{5.19}$$

Let us consider several special cases where the depth changes from one constant to another. If the depth changes discontinuously by an amount Δh_0, that is,

$$q = \Delta H(x), \qquad H(x) = \text{Heaviside function} \tag{5.20}$$

where $\delta \ll 1$, we have

$$R_1 = -\tfrac{1}{4}\Delta \tag{5.21}$$

which is constant. The above result can also be deduced as a limit of Eq. (2.17), even though a discontinuity is not consistent with the assumption of slow variation.

If the depth changes linearly from $x = -\tfrac{1}{2}L$ to $x = +\tfrac{1}{2}L$ by an amount Δ, then

$$q = \left(\frac{\Delta}{L}\right)x, \tag{5.22}$$

so that

$$R_1 = -\frac{\Delta}{4}\int_{-L/2}^{L/2}\frac{e^{ik_0 x}\,dx}{L} = -\frac{\Delta}{4}\frac{\sin k_0 L}{k_0 L} \tag{5.23}$$

which oscillates with $k_0 L$; the envelope diminishes as $k_0 L \to \infty$.

Finally, if the transition is infinitely smooth and can be represented by an error function of x, then

$$\frac{d}{dx}q = \frac{\Delta}{\pi^{1/2}L}e^{-x^2/L^2} \tag{5.24}$$

is Gaussian so that

$$R_1 = -\frac{\Delta}{4\pi^{1/2}L}\int_{-\infty}^{\infty}e^{-(x/L)^2}e^{2ik_0 x}\,dx = -\frac{\Delta}{4}e^{-k_0^2 L^2}. \tag{5.25}$$

Note that for this case R_1 diminishes exponentially in $(k_0 L)^2$.

The preceding examples differ from one another significantly in their rates of attenuation with respect to $k_0 L$; the smoother profile attenuates faster with increasing $k_0 L$. This fact can be proven more generally from Eq. (5.17) (Felsen and Marcuvitz, 1973). Let h, hence k also, be different from constant only within the range $x_1 < x < x_2$ with $x_2 - x_1 = L$. We rewrite Eq. (5.17)

$$R_1 = -\int_{x_1}^{x_2} e^{2iS} \frac{d}{dx} \ln(kh)^{1/2} \, dx.$$

If dh/dx is finite at the end points $x = x_1, x_2$ but d^2h/dx^2 is not, we can integrate by parts once to get

$$R_1 = -\left\{ e^{2iS} \frac{1}{2ik} \left[\frac{d}{dx} \ln(kh)^{1/2} \right] \right\}_{x_1}^{x_2}$$

$$+ \int_{x_1}^{x_2} dx \, e^{2iS} \frac{d}{dx} \left\{ \frac{1}{2ik} \left[\frac{d}{dx} \ln(kh)^{1/2} \right] \right\}.$$

From the integrated term above it is clear that $R_1 = O(k_0 L)^{-1}$, which agrees with Eq. (5.23). If d^2h/dx^2 is finite at the ends but d^3h/dx^3 is not, then the last integral above can be partially integrated once more to give a term that is of the order $O(k_0 L)^{-2}$. More generally, if $d^n h/dx^n$ is finite at both ends, then $R_1 = O(k_0 L)^{n-1}$. If the profile is infinitely smooth which implies that $x_1 \to -\infty$ and $x_2 \to \infty$, then R_1 decays faster than any algebraic power of $k_0 L_0$.

The result that reflection depends strongly on the smoothness at two points invites mathematical curiosity, since such a local property can hardly be so influential from the physical standpoint. Indeed Eq. (5.25) implies that the reflection coefficient is transcendentally small for infinitely smooth topographies. In a highly mathematical paper, Meyer (1979b) abandoned the WKB approximation and showed the reflection coefficient to be of the form $\exp\left[-\alpha(k_0 L)^{1/2}\right]$ instead for both a Gaussian ridge and a shelf of hyperbolic-tangent profile. Since the improvement is concerned only with a small quantity, we do not pursue the matter further here. The interested reader may consult Meyer's paper for details and references.

4.6 TRAPPED WAVES ON A STEPPED RIDGE

As shown in Section 4.3, certain sinusoidal waves may exist at a depth discontinuity, but are unable to propagate from shallow to deep water. Let us study what can happen on a shelf with two edges at a finite distance $2a$ apart. It will be found that *eigenfrequencies* exist which correspond to modes trapped over the shelf. These modes are analogous to the so-called *bound states* in a square-well potential in quantum mechanics and to *Love waves* in a layered elastic half-space. Indeed the analyses of trapped long waves (Snodgrass,

Munk, and Miller, 1962; Longuet-Higgins, 1967) can be borrowed from any standard treatise on quantum mechanics (e.g., Bohm, 1951).

Consider the geometry of Fig. 4.1 with $h = h_2$ over the ridge $-a < x < a$; the general solution is

$$\eta_2 = (Be^{i\alpha_2 x} + Ce^{-i\alpha_2 x})e^{i\beta y} \qquad (6.1)$$

where $\alpha_2 = (k_2^2 - \beta^2)^{1/2}$. We are only interested in the solution which attenuates to zero at infinites on either side of the ridge; hence

$$\eta_1 = Ae^{\gamma_1(x+a)}e^{i\beta y}, \qquad x < -a, \qquad (6.2)$$

and

$$\eta_3 = De^{-\gamma_1(x-a)}e^{i\beta y}, \qquad x > a \qquad (6.3)$$

where $\gamma_1 = (\beta^2 - k_1^2)^{1/2}$. It is assumed that

$$k_2 > \beta > k_1 \qquad \text{or} \qquad \beta(gh_1)^{1/2} > \omega > \beta(gh_2)^{1/2}. \qquad (6.4)$$

The coefficients A, B, C, and D are as yet arbitrary. Continuity of η and $h\,\partial\eta/\partial x$ at $x = \pm a$ gives four conditions

$$A = Be^{-i\alpha_2 a} + Ce^{i\alpha_2 a}, \qquad (6.5)$$

$$\gamma_1 h_1 A = i\alpha_2 h_2 (Be^{-i\alpha_2 a} - Ce^{i\alpha_2 a}), \qquad (6.6)$$

$$D = Be^{i\alpha_2 a} + Ce^{-i\alpha_2 a}, \qquad (6.7)$$

$$-\gamma_1 h_1 D = i\alpha_2 h_2 (Be^{i\alpha_2 a} - Ce^{-i\alpha_2 a}). \qquad (6.8)$$

For a nontrivial solution the coefficient determinant of the simultaneous Eqs. (6.5)–(6.8) must vanish:

$$(\gamma_1 h_1 - i\alpha_2 h_2)^2 e^{-2i\alpha_2 a} - (\gamma_1 h_1 + i\alpha_2 h_2)^2 e^{2i\alpha_2 a} = 0 \qquad (6.9)$$

which can be manipulated to give

$$\tan 2\alpha_2 a = \frac{2\gamma_1 h_1 \alpha_2 h_2}{(\alpha_2 h_2)^2 - (\gamma_1 h_1)^2}. \qquad (6.10)$$

Let

$$\frac{\gamma_1 h_1}{\alpha_2 h_2} = \tan\delta, \qquad (6.11)$$

then Eq. (6.10) becomes

$$\tan 2\alpha_2 a = \frac{2\tan\delta}{1 - \tan^2\delta} = \tan 2\delta,$$

hence

$$\delta = \tfrac{1}{2}n\pi + \alpha_2 a. \tag{6.12}$$

Taking the tangent of both sides of Eq. (6.12), we get

$$\tan\delta = \frac{\gamma_1 h_1}{\alpha_2 h_2} = \left(\begin{array}{c} \tan\alpha_2 a \\ -\cot\alpha_2 a \end{array}\right), \qquad \left(\begin{array}{c} n = \text{even} \\ n = \text{odd} \end{array}\right), \tag{6.13}$$

or

$$\frac{h_1}{h_2}\frac{\left(\beta^2 - k_1^2\right)^{1/2}}{\left(k_2^2 - \beta^2\right)^{1/2}} = \left(\begin{array}{c} \tan\left(k_2^2 - \beta^2\right)^{1/2}a \\ -\cot\left(k_2^2 - \beta^2\right)^{1/2}a \end{array}\right), \qquad \left(\begin{array}{c} n = \text{even} \\ n = \text{odd} \end{array}\right).$$

Since

$$k_2^2 = \frac{\omega^2}{gh_2} \quad \text{and} \quad k_1^2 = \frac{\omega^2}{gh_1} = \frac{h_2}{h_1}k_2^2,$$

Eq. (6.13) may be expressed in terms of k_2 alone,

$$\frac{h_1}{h_2}\frac{\left[\beta^2 - (h_2/h_1)k_2^2\right]^{1/2}}{\left(k_2^2 - \beta^2\right)^{1/2}} = \left(\begin{array}{c} \tan\left(k_2^2 - \beta^2\right)^{1/2}a \\ -\cot\left(k_2^2 - \beta^2\right)^{1/2}a \end{array}\right), \qquad \left(\begin{array}{c} n = \text{even} \\ n = \text{odd} \end{array}\right).$$

$$\tag{6.14}$$

By the change of variables

$$\xi = a\left(k_2^2 - \beta^2\right)^{1/2} = \alpha_2 a, \tag{6.15}$$

we obtain

$$a\left(\beta^2 - \frac{h_2}{h_1}k_2^2\right)^{1/2} = \left[k_2^2 a^2\left(1 - \frac{h_2}{h_1}\right) - \xi^2\right]^{1/2}$$

so that Eq. (6.14) becomes

$$\frac{(h_2/h_1)\xi}{\left(\xi_*^2 - \xi^2\right)^{1/2}} = \left(\begin{array}{c} \cot\xi \\ -\tan\xi \end{array}\right), \qquad \left(\begin{array}{c} n = \text{even} \\ n = \text{odd} \end{array}\right), \tag{6.16}$$

with

$$\xi_*^2 = k_2^2 a^2 \left(1 - \frac{h_2}{h_1}\right) = \frac{(\omega a)^2}{g h_2}\left(1 - \frac{h_2}{h_1}\right). \qquad (6.17)$$

For a given ω and geometry, ξ_* is fixed; ξ is solved from Eq. (6.16), and the eigenwavenumber α_2 follows from Eq. (6.15). Let us consider odd and even n separately.

n odd: The eigenvalues can be found graphically and correspond to intersections of the curve $y_1 = \tan \xi$ with the curve

$$y_2 = -\frac{h_2}{h_1} \frac{\xi}{\left(\xi_*^2 - \xi^2\right)^{1/2}}$$

as shown in Fig. 6.1a. The $y_2(\xi)$ curve is odd in ξ, passes though the origin, and approaches $\pm\infty$ as ξ approaches $\pm\xi_*$.

From the same figure it is clear that the roots appear in pairs $\pm\xi_n$ and only $+\xi_n$ needs to be considered. For $\frac{1}{2}\pi < \xi_* < \frac{3}{2}\pi$ there is one mode with

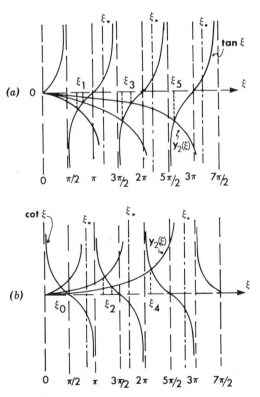

Figure 6.1 Graphic solution of eigenvalues: (*a*) odd *n*, (*b*) even *n*.

$\frac{1}{2}\pi < \xi_1 < \pi$. For $\frac{3}{2}\pi < \xi_* < \frac{5}{2}\pi$ there are two modes ξ_1 and ξ_3 with $\frac{3}{2}\pi < \xi_3 < 2\pi$. In general if $(n - \frac{1}{2})\pi < \xi_* < (n + \frac{1}{2})\pi$, there are n modes $\xi_1, \xi_3, \ldots, \xi_{2n-1}$ (n odd) with the mth root in the range

$$\left(m - \tfrac{1}{2}\right)\pi < \xi_m < m\pi.$$

Thus, there is a new trapped mode for every increase of ξ_* by π, which can be achieved by increasing ωa or decreasing shelf depth h_2 for a fixed h_2/h_1.

n even: We need to examine the intersections of $y_1 = \cot \xi$ with

$$y_2 = \frac{h_2}{h_1} \frac{\xi}{\left(\xi_*^2 - \xi^2\right)^{1/2}}.$$

The intersections are shown in Fig. 6.1b. For $0 < \xi_* < \pi$ there is one trapped mode, $\pm\xi_0$ with $0 < \xi_0 < \frac{1}{2}\pi$; for $\pi < \xi_* < 2\pi$ there are two modes $\pm\xi_0$ and $\pm\xi_2$, with $\pi < \xi_2 < \frac{3}{2}\pi$. In general, for $(n + 1)\pi > \xi_* > n\pi$ there are $n + 1$ modes: $\xi_0, \xi_2, \ldots, \xi_{2n}$.

Summarizing, the roots ξ_n form a sequence $\xi_0 < \xi_1 < \xi_2 < \cdots$; the corresponding eigenwavenumbers also form an increasing sequence $\alpha_0 < \alpha_1 < \alpha_2 < \alpha_3 < \cdots$.

What are the free surfaces of these modes like? From Eqs. (6.5) and (6.6), we have

$$\frac{B}{C} = -e^{2i\alpha_2 a}\left(\frac{\gamma_1 h_1 + i\alpha_2 h_2}{\gamma_1 h_1 - i\alpha_2 h_2}\right)$$

$$= -e^{2i\alpha_2 a}\frac{\tan \delta + i}{\tan \delta - i}$$

$$= e^{2i\alpha_2 a}\frac{\cos \delta - i\sin \delta}{\cos \delta + i\sin \delta} = e^{2i(\alpha_2 a - \delta)} = e^{in\pi}. \tag{6.18}$$

The last equality follows from Eq. (6.12). For $n = $ even, $B = C$; from Eqs. (6.5) and (6.7), $A = D$, and the displacement is proportional to $\cos \alpha_2 x$, see Eqs. (6.1)–(6.3). Hence $n = $ even corresponds to an even mode. Similarly, for $n = $ odd, $B = -C$ and $A = -D$; the displacement is proportional to $\sin \alpha_2 x$ and odd in x. The first few modes are sketched in Fig. 6.2.

The even modes can be regarded alternatively as the trapped modes on an idealized continental shelf of width a with $x = 0$ being the coastline. As an approximate model for the California Shelf, Miles (1972) has taken $h_2 = 600$ m, $h_1 = 3600$ m, and $\xi_* = 2.19\pi$ which for $a = 70$ km corresponds to $T = 2\pi/\omega = 27.78$ min $= \frac{1}{6} \times 10^4$ s. The three trapped modes are at $\xi_0 \cong \frac{1}{2}\pi$, $\xi_2 \cong 1.45\pi$, and $\xi_4 \cong 2.15\pi$, so that $2\pi/\alpha_0 = 280$ km, $2\pi/\alpha_2 = 96$ km, and $2\pi/\alpha_4 = 65$ km, which can be important to resonate the so-called Helmholtz mode in a harbor, a subject to be discussed in the next chapter. For the same ω

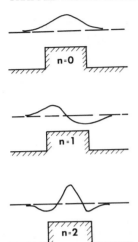

Figure 6.2 Trapped modes on a ridge.

and β there is no solution which corresponds to imaginary γ_1; no propagation in the x direction is possible. It follows that trapped waves over an infinitely long ridge or shelf of finite width cannot be excited by sinusoidal incident waves according to linearized theory. However, excitation is possible by transient waves, wind stress, or on a ridge (or shelf) of finite length. Furthermore, nonlinear mechanisms of excitation cannot be ruled out.

Exercise 6.1   Waves Trapped in a Jet-Like Stream

In Chapter Three the presence of a variable current is seen to be similar to a variable depth in influencing waves. Examine shallow-water long waves in a strong current $U = 0$, $V = V(x)$, on a sea of variable depth $z = -h(x)$. Show that the equations for the wave perturbations are

$$\frac{\partial \zeta}{\partial t} + V\frac{\partial \zeta}{\partial y} + \frac{\partial uh}{\partial x} + h\frac{\partial v}{\partial y} = 0, \tag{6.19}$$

$$\frac{\partial \mathbf{u}}{\partial t} + V\frac{\partial \mathbf{u}}{\partial y} + u\frac{\partial V}{\partial x}\mathbf{e}_y = -g\nabla\zeta \tag{6.20}$$

where the current set-down is zero.

Assume

$$\begin{pmatrix} \zeta(x, y, t) \\ \mathbf{u}(x, y, t) \end{pmatrix} = \begin{pmatrix} \hat{\zeta}(x) \\ \hat{\mathbf{u}}(x) \end{pmatrix} e^{i(\beta y - \omega t)}, \tag{6.21}$$

and show that

$$\left(h\hat{\zeta}'\right)' + \frac{2\beta V'}{\omega - \beta V}\hat{\zeta}' + \left[\left(\frac{\omega - \beta V}{g}\right)^2 - \beta^2 h\right]\hat{\zeta} = 0. \tag{6.22}$$

Thus, if $V(x) > 0$ within a finite strip $|x| < a$ and vanishes outside, and if $\beta < 0$ (waves oppose the current), waves satisfying the condition

$$\frac{\omega^2}{gh} < \beta^2 < \frac{(\omega - \beta V)^2}{gh} \tag{6.23}$$

will be trapped in the current.

For the special case where $h = $ const for all x, $V = $ const if $|x| < a$ and $V = 0$ if $|x| > a$, study the eigenvalues β of the trapped waves. Analyze also a scattering problem where $\beta^2 < \omega^2/gh$.

4.7   SOME GENERAL FEATURES OF ONE-DIMENSIONAL PROBLEMS—TRAPPED MODES AND THE SCATTERING MATRIX

4.7.1   A Qualitative Discussion of Trapped Waves

Let us discuss qualitatively the existence of trapped modes for a continuous one-dimensional topography $h = h(x)$. Substituting $\zeta = X(x)\exp[i(\beta y - \omega t)]$ into Eq. (1.5), we obtain

$$(hX')' + \left(\frac{\omega^2}{g} - \beta^2 h\right) X = 0. \tag{7.1}$$

The qualitative features may be studied in the so-called *phase plane* of X and Y, where Y is defined by

$$Y = hX' \quad \text{or} \quad X' = \frac{1}{h} Y. \tag{7.2}$$

Equation (7.1) may be written

$$Y' + \left(\frac{\omega^2}{g} - \beta^2 h\right) X = 0. \tag{7.3}$$

Division of Eq. (7.3) by Eq. (7.2) gives

$$\frac{dY}{dX} = \frac{-(\omega^2/g - \beta^2 h) X}{(1/h) Y} \tag{7.4}$$

which is a first-order equation in X and Y with x as the parameter. In the phase plane the solution to Eq. (7.4) is represented by a trajectory. Assume that $h(x)$ approaches a finite constant at infinity, that is,

$$h(x) \to h_\infty, \quad |x| \to \infty,$$

and assume further that

$$\beta^2 h_0 < \frac{\omega^2}{g} < \beta^2 h_\infty \tag{7.5}$$

so that there are two points (x_1, x_2) at which

$$\frac{\omega^2}{g} = \beta^2 h(x_l), \qquad l = 1, 2$$

(see Fig. 7.1). In the theory of differential equations, these two points are called the *turning points* on opposite sides of which the solution behaves differently. In the ranges $x > x_2$ and $x < x_1$, the factor $h(\omega^2/g - \beta^2 h)$ is negative, and the solution $X(x)$ is monotonic in x. In particular, for large x

$$h\left(\beta^2 h - \frac{\omega^2}{g}\right) \to h_\infty\left(\beta^2 h_\infty - \frac{\omega^2}{g}\right) \equiv H^2 > 0$$

so that $X \propto e^{-H|x|}$. In the phase plane the solution point (X, Y) approaches the origin as $|x| \to \infty$ along the straight lines $Y = \pm H h_\infty^{1/2} X$ which are integrals of the limiting form of Eq. (7.4). Within the range $x_1 < x < x_2$, the coefficient $h(\omega^2/g - \beta^2 h)$ is positive and the solution $X(x)$ is in general oscillatory in x. The trajectory in the phase plane can wind around the origin and cross the X, Y axes. Several possible solutions are sketched in Fig. 7.2 for both the phase plane and the physical plane. Thus, a wavy surface exists only in the range (x_1, x_2) and decays exponentially outside it; this feature is precisely the characteristics of trapped waves. If $\omega^2/g > \beta^2 h_\infty$, there is no monotonic region and the entire fluid can have wave-like motion; waves are no longer trapped. If $\omega^2/g < \beta^2 h_0$, there is no periodic wave anywhere.

4.7.2 The Scattering Matrix $[S(\alpha)]$

Consider oblique incidence on a submarine ridge with $h(x) \to h_\infty$ as $|x| \to \infty$. We rewrite Eq. (7.1)

$$(aX')' + \left(\frac{\omega^2}{gh_\infty} - \beta^2 a\right) X = 0 \qquad \text{with } a(x) = \frac{h}{h_\infty}, \tag{7.6a}$$

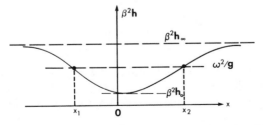

Figure 7.1 Variation of $\beta^2 h$ with x for a submarine ridge.

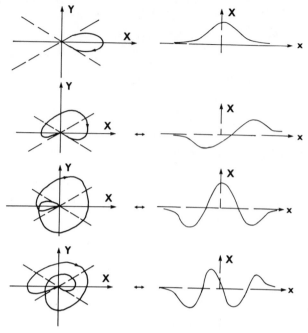

Figure 7.2 The solution trajectories in the phase plane (left) and the corresponding free surface (right) for the first few trapped modes. Dash lines in the phase planes are $Y = \pm H h_\infty^{1/2} x$. Arrows along the phase plane curves show the direction of increasing x.

or, since $\omega^2/gh_\infty = \alpha^2 + \beta^2$,

$$(aX')' + \left[\alpha^2 + \beta^2(1 - a)\right] X = 0 \tag{7.6b}$$

which is a Sturm–Liouville equation. By the transformation $X = a^{-1/2}\xi$, Eq. (7.6b) becomes the time-independent Schrödinger equation in quantum mechanics:

$$\xi'' + \left[\lambda - U(x)\right]\xi = 0$$

where

$$\lambda = -\beta^2, \quad -U = h^{-1}\left(\alpha^2 + \beta^2 + \frac{3}{4}\frac{h^{-2}}{h} - \frac{1}{2}\frac{h'}{h} - \frac{h''}{2}\right).$$

Many general properties of Schrödinger equations are known in quantum scattering theory. We shall only discuss a few of these properties with respect to Eq. (7.6). For a more extensive study reference should be made to Roseau (1952) and Sitenko (1971).

Let us consider a general scattering problem whose asymptotic behavior is

$$X \sim A_- e^{i\alpha x} + B_- e^{-i\alpha x}, \qquad x \sim -\infty, \qquad (7.7a)$$

$$\sim A_+ e^{i\alpha x} + B_+ e^{-i\alpha x}, \qquad x \sim +\infty, \qquad (7.7b)$$

Thus, A_- and B_+ correspond to the incoming waves from the left and the right, respectively, and A_+ and B_- correspond to the scattered waves toward the right and the left, respectively.

Let us also define $f_1(x, \alpha)$ to be the solution to the left-scattering problem

$$f_1(x, \alpha) \sim \frac{1}{T_1} e^{i\alpha x} + \frac{R_1}{T_1} e^{-i\alpha x}, \qquad x \sim -\infty, \qquad (7.8a)$$

$$\sim e^{i\alpha x}, \qquad x \sim +\infty, \qquad (7.8b)$$

and $f_2(x, \alpha)$ to be the solution to the right-scattering problem

$$f_2 \sim e^{-i\alpha x}, \qquad x \sim -\infty, \qquad (7.9a)$$

$$\sim \frac{1}{T_2} e^{-i\alpha x} + \frac{R_2}{T_2} e^{i\alpha x}, \qquad x \sim +\infty. \qquad (7.9b)$$

In quantum mechanics f_1 and f_2 are called *Jost functions*. Now f_1 and f_2 are linearly independent solutions since their Wronksian

$$W(f_1, f_2) \equiv f_1 f_2' - f_2 f_1' = -\frac{2i\alpha}{T_1} \qquad \text{(from } x \sim -\infty\text{)}$$

$$= -\frac{2i\alpha}{T_2} \qquad \text{(from } x \sim +\infty\text{)}$$

does not vanish in general. As a by-product the preceding equation implies at once

$$T_1 = T_2, \qquad (7.10)$$

that is, the right and left transmission coefficients are equal even if $h(x)$ is not symmetrical. Also, due to the linear independence of f_1 and f_2 we may express X in Eqs. (7.7) as a linear combination,

$$X = Cf_1 + Df_2. \qquad (7.11)$$

Comparing the asymptotic values of Eqs. (7.11) and (7.7), we get

$$\frac{C}{T_1} = A_- , \qquad C\frac{R_1}{T_1} + D = B_- ,$$

$$C + D\frac{R_2}{T_2} = A_+ , \qquad D\frac{1}{T_2} = B_+ .$$

A_+ and B_- may be solved by eliminating C and D,

$$A_+ = T_1 A_- + R_2 B_+ , \qquad B_- = R_1 A_- + T_2 B_+$$

which may be expressed in matrix form:

$$\begin{pmatrix} A_+ \\ B_- \end{pmatrix} = [S]\begin{pmatrix} A_- \\ B_+ \end{pmatrix} \tag{7.12a}$$

with

$$[S] = \begin{bmatrix} T_1 & R_2 \\ R_1 & T_2 \end{bmatrix}. \tag{7.12b}$$

As before, $[S]$ is the *scattering matrix*, or S matrix, which is the generalization of Eqs. (2.13) and (2.15) for a very special depth profile.

4.7.3 Trapped Modes as Imaginary Poles of $[S(\alpha)]$

In the result for a rectangular shelf let us replace $\gamma_1 = i\alpha_1$, so that the eigenvalue condition (6.9) becomes

$$(1 + s)^2 e^{-2i\alpha_2 a} - (1 - s)^2 e^{2i\alpha_2 a} = 0 \qquad \text{with } s = \frac{\alpha_1 h_1}{\alpha_2 h_2}.$$

In view of Eqs. (4.21) and (4.22) the above equation amounts to the vanishing of the denominators of R and T of the scattering problem, that is, the trapped modes correspond to the positive imaginary poles of R and T in the complex α plane. That the two physically different problems should be so connected mathematically arouses curiosity. We now give a general theory for arbitrary $h(x)$ as long as $h \to h_\infty$ when $|h| \to \infty$.

The Jost solutions $f_1(x, \alpha)$ and $f_1(x, -\alpha)$ are linearly independent since their Wronskian

$$W[f_1(x, \alpha), f_1(x, -\alpha)] = f_1(x, \alpha)f_1'(x, -\alpha) - f_1(x, -\alpha)f_1'(x, \alpha)$$

$$= -2i\alpha$$

does not vanish after using the asymptotic values at $x \sim +\infty$. Any solution such as $f_2(x, \alpha)$ can be expressed as a linear superposition of $f_1(x, \alpha)$ and $f_1(x, -\alpha)$. From the behavior at $x \sim \infty$, it is easy to see that

$$f_2(x, \alpha) = \frac{R_2}{T_2} f_1(x, \alpha) + \frac{1}{T_2} f_1(x, -\alpha),$$

or

$$T_2 f_2(x, \alpha) = R_2 f_1(x, \alpha) + f_1(x, -\alpha). \tag{7.13}$$

Differentiating the preceding equation with respect to x, we have

$$T_2 f_2'(\alpha, x) = R_2 f_1'(x, \alpha) + f_1'(x, -\alpha). \tag{7.14}$$

Let us solve T_2 and R_2 from Eqs. (7.13) and (7.14):

$$T_2 = -\frac{2i\alpha}{W\{f_1(x, \alpha), f_2(x, \alpha)\}},$$

$$R_2 = \frac{W\{f_1(x, -\alpha), f_2(x, \alpha)\}}{W\{f_1(x, \alpha), f_2(x, \alpha)\}}.$$

If there are poles for T_2, they must correspond to the zeroes of

$$W\{f_1(x, \alpha), f_2(x, \alpha)\} = 0.$$

Let the poles be denoted by α_n. First, they must also be the poles of R_2, hence of $[S(\alpha)]$. Second, at these poles $1/T_2 = 0$ and $R_2/T_2 = $ finite so that f_2 behaves asymptotically as

$$f_2 \sim e^{-i\alpha_n x}, \qquad x \sim -\infty,$$

$$\sim \left(\frac{R_2}{T_2}\right)_{\alpha_n} e^{i\alpha_n x}, \qquad x \sim +\infty. \tag{7.15}$$

Assume that these poles are complex with positive imaginary parts so that f_2 decays exponentially to zero as $|x| \to \infty$, that is,

$$\alpha_n = \delta_n + i\gamma_n, \qquad \gamma_n > 0.$$

From Eq. (7.6) it is simple to derive

$$[a(XX^{*\prime} - X^*X')]' = 2i(\text{Im }\alpha^2) |X|^2. \tag{7.16}$$

Now letting $X = f_2$, integrating both sides of Eq. (7.16) from $-\infty$ to ∞ and using the exponential behavior at $|x| \to \infty$, we get

$$\operatorname{Im} \alpha_n^2 \int_{-\infty}^{\infty} |f_2|^2 \, dx = 0$$

which implies

$$\operatorname{Im} \alpha_n^2 = 0 \qquad \text{or} \qquad \delta_n = 0. \tag{7.17}$$

Thus, the poles are purely imaginary and the trapped modes are *monotonically decaying* at large x.

For such an eigenvalue $\alpha_n = i\gamma_n$, the eigenfunction X may be taken to be real. Multiplying Eq. (7.6a) by X and integrating by parts, we get

$$\int_{-\infty}^{\infty} a(X')^2 \, dx + \int_{-\infty}^{\infty} \left(\beta^2 a - \frac{\omega^2}{gh_\infty} \right) X^2 \, dx = 0.$$

Since $(\beta^2 a - \omega^2/gh_\infty) \to (\beta^2 - \omega^2/gh_\infty) > 0$ as $x \to \pm\infty$, the above equality implies that $\beta^2 a - \omega^2/gh_\infty < 0$ for some range of x; otherwise X is trivially zero. The condition (7.5) for the existence of trapped modes is again confirmed.

4.7.4   Properties of $[S(\alpha)]$ for Real α

Returning to Eqs. (7.7) for the scattering problem, we now examine some other properties of the S matrix. Consider α to be real, then from Eqs. (7.6) and its complex conjugate, it can be shown that

$$a(XX^{*\prime} - X^*X') = \text{const.} \tag{7.18}$$

Equating the asymptotic values of the left-hand side at $x \sim -\infty$ and $x \sim +\infty$, we get

$$|A_+|^2 + |B_-|^2 = |A_-|^2 + |B_+|^2 \tag{7.19}$$

which states that the energy of the incoming waves equals the energy of the outgoing waves. In view of Eq. (7.12a), Eq. (7.19) may be written

$$|A_+|^2 + |B_-|^2 = \{A_+, B_-\} \begin{Bmatrix} A_+^* \\ B_-^* \end{Bmatrix} = \{A_-, B_+\}[S]^T[S^*] \begin{Bmatrix} A_-^* \\ B_+^* \end{Bmatrix}$$

where $[S]^T$ is the transpose of $[S]$. It follows that

$$[S]^T[S^*] = I = \begin{bmatrix} 1 & 0 \\ 0 & 1 \end{bmatrix} \tag{7.20}$$

which is called the *unitary* property of the S matrix. By the definition of $[S]$, Eq. (7.20) implies

$$\begin{bmatrix} T_1 & R_1 \\ R_2 & T_2 \end{bmatrix} \begin{bmatrix} T_1^* & R_2^* \\ R_1^* & T_2^* \end{bmatrix} = \begin{bmatrix} 1 & 0 \\ 0 & 1 \end{bmatrix}, \tag{7.21}$$

which yields three independent relations:

$$|T_1|^2 + |R_1|^2 = 1 \tag{7.22a}$$

$$|T_2|^2 + |R_2|^2 = 1 \tag{7.22b}$$

$$T_1 R_2^* + R_1 T_2^* = 0 \tag{7.22c}$$

Equations (7.22a) and (7.22b) again represent energy conservation. Because of Eq. (7.10), Eq. (7.22c) implies that

$$|R_1| = |R_2| . \tag{7.23}$$

From Eqs. (7.7), the asymptotic behavior of the complex conjugate of X is

$$X^* \sim A_-^* e^{-i\alpha x} + B_-^* e^{i\alpha x}, \qquad x \sim -\infty, \tag{7.24a}$$

$$\sim A_+^* e^{-i\alpha x} + B_+^* e^{i\alpha x} \qquad x \sim +\infty. \tag{7.24b}$$

By comparison with Eqs. (7.7) it is clear that A_-^*, B_-^*, A_+^*, and B_+^* may be substituted for B_-, A_-, B_+, and A_+, respectively, so that Eq. (7.12a) may be written

$$\begin{Bmatrix} B_+^* \\ A_-^* \end{Bmatrix} = [S] \begin{Bmatrix} B_-^* \\ A_+^* \end{Bmatrix} \tag{7.25a}$$

or

$$\begin{Bmatrix} B_+ \\ A_- \end{Bmatrix} = [S^*] \begin{Bmatrix} B_- \\ A_+ \end{Bmatrix}. \tag{7.25b}$$

On the other hand, we rewrite the solution X [Eqs. (7.7)] as

$$X \sim B_- e^{i(-\alpha)x} + A_- e^{-i(-\alpha)x}, \qquad x \sim -\infty, \tag{7.26a}$$

$$B_+ e^{i(-\alpha)x} + A_+ e^{-i(-\alpha)x}, \qquad x \sim +\infty, \tag{7.26b}$$

which can be regarded as a problem with ω replaced by $(-\omega)$, and α replaced by $-\alpha$. Now B_- and A_+ are the incoming waves and A_- and B_+ are the outgoing waves. By analogy to Eq. (7.12) we get

$$\begin{Bmatrix} B_+ \\ A_- \end{Bmatrix} = [S(-\alpha)] \begin{Bmatrix} B_- \\ A_+ \end{Bmatrix}. \tag{7.27}$$

Upon comparing Eqs. (7.25b) and (7.27), we conclude that

$$[S^*(\alpha)] = [S(-\alpha)]. \tag{7.28}$$

In summary, Eq. (7.18) which is a consequence of Green's formula, has led to considerable information regarding the far fields. This approach will be further explored in Chapter Seven.

Exercise 7.1

Consider a channel which has a varying cross section only in the finite part of x and approaches constant width and depth at infinities: $(b, h) \to (b_1, h_1)$ as $x \sim -\infty \to (b_2, h_2)$ as $x \to +\infty$. Let (R_1, T_1) and (R_2, T_2) be the left- and right-scattering coefficients, respectively. Show that

$$k_1 A_1 \left(1 - |R_1|^2\right) = k_2 A_2 |T_1|^2, \qquad \text{where } A_1 = b_1 h_1 \quad \text{and} \quad A_2 = b_2 h_2 \tag{7.29a}$$

$$\frac{T_1}{T_2} = \frac{k_1 A_1}{k_2 A_2}. \tag{7.29b}$$

and that

$$|R_1| = |R_2| \tag{7.30a}$$

$$|T_1 T_2| = 1 - |R_2|^2. \tag{7.30b}$$

4.8 EDGE WAVES ON A CONSTANT SLOPE

As a special case of continuous depth variation we consider a straight and long beach with constant slope (Eckart, 1951). Let the mean shoreline coincide with the y axis, and let water be in the region $x > 0$. The bottom is described by

$$z = -h = -sx, \qquad x > 0, s = \text{const.} \tag{8.1}$$

Because the coefficients are constant in y and t, we try the solution

$$\zeta = \eta(x) e^{i(\beta y - \omega t)}. \tag{8.2}$$

Equation (1.9) then gives

$$x\eta'' + \eta' + \left(\frac{\omega^2}{sg} - \beta^2 x\right)\eta = 0. \tag{8.3}$$

By the following transformation

$$\xi = 2\beta x, \qquad \eta = e^{-\xi/2}f(\xi). \tag{8.4}$$

Eq. (8.3) may be rewritten

$$\xi f'' + (1 - \xi)f' + \left[\frac{\omega^2}{2\beta sg} - \frac{1}{2}\right]f = 0 \tag{8.5}$$

which belongs to the class of confluent hypergeometric equations (more specifically Kummer's equation, see Abramowitz and Stegun, 1972). In general, there are two homogeneous solutions, one of which is singular at the shoreline $\xi = 0$ and must be discarded. Nontrivial solutions which render η finite at $\xi = 0$ and zero as $\xi \to \infty$ exist when ω corresponds to the following discrete values:

$$\frac{\omega^2}{2\beta sg} = n + \frac{1}{2}, \qquad n = 0, 1, 2, 3, \ldots . \tag{8.6}$$

The associated eigenfunctions are proportional to Laguerre polynomials

$$L_n(\xi) = \frac{(-)^n}{n!}\left[\xi^n - \frac{n^2}{1!}\xi^{n-1} + \frac{n^2(n-1)^2}{2!}\xi^{n-2}\right.$$

$$\left. - \frac{n^2(n-1)^2(n-2)^2}{3!}\xi^{n-3} + \cdots + (-)^n n!\right]; \tag{8.7}$$

for example, $L_0 = 1$, $L_1 = 1 - \xi$, $L_2 = 1 - 2\xi + \frac{1}{2}\xi^2$, and so on. The first few modes are plotted in Fig. 8.1 with the higher mode attenuating faster in the offshore direction. Because these eigenfunctions correspond to modes which are appreciable only near the shore, they are called *edge waves*. These eigenfunctions are orthonormal in the following sense:

$$\int_0^\infty e^{-\xi}L_n L_m \, d\xi = \delta_{nm}. \tag{8.8}$$

For $m = n$ Eq. (8.8) implies that each mode has a finite energy.

Edge waves are of interest in coastal oceanography because the largest amplitude, hence the largest run-up, occurs at the shore. They are also believed to be responsible for rip currents near the shore when the shorter breaking

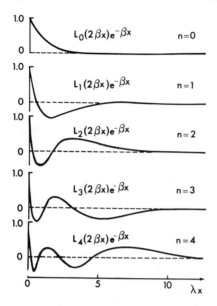

Figure 8.1 Profiles of some edge wave modes.

waves are present. For large beach angles, similar edge waves were discovered by Stokes (1847); the complete edge wave spectrum which includes both discrete and continuous parts was discovered by Ursell (1952).

Several mechanisms for generating edge waves are possible in nature. On a large scale (typical wavelength 200 miles, period 6 h, amplitude 3 ft) edge waves can be excited by wind stress directly above the water. Munk, Snodgrass, and Carrier (1956) and Greenspan (1958) have studied the effect of pressure deviation in a storm traveling parallel to the coast; their results are relevant to storm surges. Smaller-scale edge waves can be excited by a nonlinear mechanism of subharmonic resonance (Guza and Bowen, 1976; Minzoni and Whitham, 1977) as will be discussed in Chapter Eleven. Medium-scale edge waves of periods 1–5 min can also be excited by a long group of short swells through a nonlinear mechanism, which will be further commented on in Chapter Twelve.

4.9   CIRCULAR BOTTOM CONTOURS

4.9.1   General Aspects

Next in complexity to straight and parallel contours is the topography with concentric circular contours. In polar coordinates (r, θ), $h = h(r)$; the long-wave equation becomes

$$\nabla^2 \eta + \frac{h'}{h} \frac{\partial \eta}{\partial r} + \frac{\omega^2}{gh} \eta = 0, \tag{9.1}$$

with

$$\nabla^2 = \frac{\partial^2}{\partial r^2} + \frac{1}{r}\frac{\partial}{\partial r} + \frac{1}{r^2}\frac{\partial^2}{\partial\theta^2}. \tag{9.2}$$

Let us consider, for integer n,

$$\eta = R(r)e^{i(n\theta - \omega t)} \tag{9.3}$$

so that R satisfies

$$R'' + \left(\frac{1}{r} + \frac{h'}{h}\right)R' + \left(\frac{\omega^2}{gh} - \frac{n^2}{r^2}\right)R = 0, \tag{9.4}$$

or

$$(hrR')' + \left(\frac{\omega^2}{g} - \frac{n^2 h}{r^2}\right)rR = 0. \tag{9.4'}$$

Now the behavior of R will be exponential or oscillatory according to

$$\frac{\omega^2}{gn^2} \gtrless \frac{h}{r^2}: \quad \begin{array}{l}\text{oscillatory}\\\text{exponential}\end{array}. \tag{9.5}$$

Consider a submerged island with monotonically increasing h, $0 < h(0) < h(r) < h(\infty)$. For a fixed n, h/r^2 behaves as shown in Fig. 9.1. The solution is oscillatory outside a critical circle $r = r_0$ where

$$\frac{\omega^2}{g} = \left(\frac{n^2 h}{r^2}\right), \quad r = r_0. \tag{9.6}$$

In comparison with one-dimensional topography, the factor $1/r^2$ drastically changes the situation and perfect trapping is no longer possible.

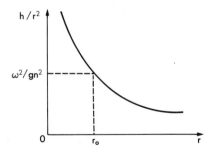

Figure 9.1 h/r^2 versus r for a submerged island.

Consider an island with a shore at $r = a$ so that $h(r) = 0$ for $0 < r < a$ and increases monotonically for $r > a$. Then $n^2 h/r^2$ behaves as shown in Fig. 9.2. For $\omega^2/gn^2 < \max(h/r^2)$, the free surface is oscillatory near the shore $a < r < r_1$, exponentially attenuating in the region $r_1 < r < r_2$, and oscillatory again outside ($r > r_2$). Therefore, the island slope acts as a barrier of finite thickness to trap waves of sufficiently low frequency, or waves of fixed ω but sufficiently large n. For a large n, the barrier appears thicker and is more effective in trapping energy, which means less energy leakage. These facts have important consequences on the resonance of trapped waves over a submarine seamount by externally incident waves.

Longuet–Higgins (1967) considered a special example of a stepped circular sill where

$$h = h_1, \qquad 0 < r < a$$

$$= h_2, \qquad r > a \quad \text{with} \quad \frac{h_1}{h_2} < 1 \qquad (9.7)$$

as shown in Fig. 9.3a. The variation of h/r^2 is shown in Fig. 9.3b. For sufficiently low ω or high n there is also an annular region $r_1 < r < a$ over the sill where oscillations occur. This annulus of oscillations surrounds a central core of monotonic motion $(0 < r < r_1)$ and is separated from the oscillating ocean ($r > r_2$) by a barrier $a < r < r_2$. If h_1/h_2 is very small, the barrier is high so that trapping is very effective, though still imperfect, and resonance of higher modes near the edge over the sill can be severe.

This mechanism of energy trapping is of practical interest in offshore engineering where geological conditions may dictate the site of construction to be on a seamount. The inherent danger of such a site is not always evident to the designer. During a tour of a Texas tower on Brown's Bank, off the east coast of the United States, a party of inspectors, caught by a Nor'easter, saw the storm waves moving the iron collars up and down the piles by as much as 100 ft and threatening to break up the structure. These collars weighed several tons each and were installed to protect the feet of ·the piles. This phenomenon was believed to be an evidence of the resonant mechanism discussed here (Meyer, 1970).

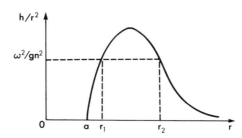

Figure 9.2 h/r^2 versus r for an island with shoreline radius a.

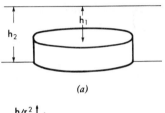

(a)

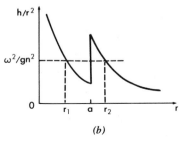

(b)

Figure 9.3 A submerged circular sill: (a) geometry; (b) h/r^2 versus r.

In the following subsection, we describe further details of Longuet–Higgins' example for which the analytical solution is relatively simple. For other smooth circular topographies, a WKB approximation is possible. However, for still more general two-dimensional topography, numerical methods are inevitable.

4.9.2 Scattering of Plane Incident Waves by a Circular Sill

Consider the circular sill as shown in Fig. 9.3. The top of the sill is at the depth $z = -h_2$. The adjacent water is assumed to be of constant depth h_1. For simple harmonic waves with frequency ω, the wavenumber is $k_2 = \omega/(gh_2)^{1/2}$ in the region above the sill $r < a$ and is $k_1 = \omega/(gh_1)^{1/2}$ around the sill $r > a$. Let the incident wave approach from $x \sim -\infty$ with unit amplitude so that

$$\eta^I = e^{ik_1 x}. \tag{9.8}$$

In the region $r > a$, there must be radiated (scattered) waves which propagate away to $r \sim \infty$. Hence if

$$\eta_1 = \eta^I + \eta^R, \tag{9.9}$$

η^R must satisfy, in polar coordinates,

$$\nabla^2 \eta^R + k_1^2 \eta^R = \frac{1}{r} \frac{\partial}{\partial r} \left(r \frac{\partial \eta^R}{\partial r} \right) + \frac{1}{r^2} \frac{\partial^2 \eta^R}{\partial \theta^2} + k_1^2 \eta^R = 0, \tag{9.10}$$

and must be outgoing as $r \to \infty$. Above the sill the displacement satisfies

$$\nabla^2 \eta_2 + k_2^2 \eta_2 = \frac{1}{r} \frac{\partial}{\partial r} \left(r \frac{\partial \eta_2}{\partial r} \right) + \frac{1}{r^2} \frac{\partial^2 \eta_2}{\partial \theta^2} + k_2^2 \eta_2 = 0. \tag{9.11}$$

At the edge of the sill $r = a$, the pressure and the flux must be matched:

$$\eta_1 = \eta_2, \qquad\qquad\qquad\qquad (9.12)$$
$$r = a.$$
$$h_1 \frac{\partial \eta_1}{\partial r} = h_2 \frac{\partial \eta_2}{\partial r}, \qquad\qquad\qquad (9.13)$$

By separation of variables, it is easy to show that the general solution to Eqs. (9.10) and (9.11) must consist of linear combinations of Bessel functions, that is,

$$\eta_R \sim \cos n\theta \big[J_n(k_1 r), \, Y_n(k_1 r) \big],$$

$$\eta_2 \sim \cos n\theta \big[J_n(k_2 r), \, Y_n(k_2 r) \big].$$

In Appendix 4.A it will be shown that the incident wave can be expanded as a series of *partial waves*, each with an angular dependence of $\cos n\theta$, $n = 0, 1, 2, \ldots$:

$$e^{ikx} = e^{ikr\cos\theta} = \sum_0^\infty \varepsilon_n (i)^n \cos n\theta J_n(kr) \qquad (9.14)$$

where ε_n are the Jacobi symbols defined by $\varepsilon_0 = 1$, $\varepsilon_n = 2$, for $n = 1, 2, \ldots$.
We propose the following solution for η:

$$\eta_1 = \sum_0^\infty \varepsilon_n (i)^n \cos n\theta \big[J_n(k_1 r) + B_n H_n^{(1)}(k_1 r) \big], \qquad r > a, \qquad (9.15)$$

$$\eta_2 = \sum_0^\infty \varepsilon_n (i)^n \cos n\theta \big[A_n J_n(k_2 r) \big], \qquad r < a, \qquad (9.16)$$

where A_n and B_n are to be determined. In Eq. (9.16) only J_n's have been kept to ensure boundedness at the center $r = 0$. In Eq. (9.15), only $H_n^{(1)}$'s have been kept so that the scattered waves are outgoing. Since $H_n^{(2)}$ is never used here, we shall omit the superscript on the Hankel functions and write simply

$$H_n(k_2 r) \equiv H_n^{(1)}(k_2 r). \qquad\qquad (9.17)$$

The coefficients A_n and B_n must be chosen so that the matching conditions at $r = a$, Eqs. (9.12) and (9.13), are satisfied; thus

$$A_n J_n(k_2 a) = J_n(k_1 a) + B_n H_n(k_1 a),$$

$$k_2 h_2 A_n J_n'(k_2 a) = k_1 h_1 \big[J_n'(k_1 a) + B_n H_n'(k_1 a) \big],$$

where primes denote derivations with respect to the argument.

In terms of

$$s = \frac{k_2 h_2}{k_1 h_1} \quad \left(= \left(\frac{h_2}{h_1} \right)^{1/2} = \frac{k_1}{k_2} \right), \quad v = k_2 a, \qquad (9.18)$$

the solutions for A_n and B_n are:

$$A_n = \frac{-[J_n(sv)H_n'(sv) - J_n'(sv)H_n(sv)]}{\Delta_n} = \frac{-2i}{\pi sv \Delta_n}, \qquad (9.19a)$$

and

$$B_n = \frac{J_n(v)J_n'(sv) - sJ_n'(v)J_n(sv)}{\Delta_n}, \qquad (9.19b)$$

where

$$\Delta_n \equiv -J_n(v)H_n'(sv) + sJ_n'(v)H_n(sv). \qquad (9.19c)$$

Use has been made of the Wronskian identity

$$J_n(\zeta)H_n'(\zeta) - J_n'(\zeta)H_n(\zeta) = \frac{2i}{\pi\zeta} \qquad (9.20)$$

which may be verified by writing the Bessel equation in Sturm–Liouville form and by using the asymptotic behavior of J_n and H_n. When Eqs. (9.19a)–(9.19c) are substituted into Eqs. (9.15) and (9.16), the solution for η is fully determined.

The modal responses over the sill have been calculated by Longuet-Higgins (1967), as shown in Fig. 9.4. Note that for the lowest mode $n = 0$, the resonant amplification ratio is nearly 8, while increasing n leads to much higher and sharper resonant peaks. Of course, the narrowness of a peak implies that the corresponding mode is difficult to excite unless the incident wavetrain is precisely tuned. If the tuning is good, then a weak but persistent incident wave can cause a large response. This feature can also be anticipated from Fig. 9.3 where a larger n leads to a thicker outer barrier which makes it hard to acquire energy from the incident waves. On the other hand, once the energy is trapped within the outer barrier, it does not escape easily. These features have further ramifications in the transient excitation by short-lived incident waves,[†] as will be discussed in Chapter Five for the similar problem of harbor resonance. The large numerical values of some of the amplification factors suggest that nonlinear and/or frictional effects may also be important near these resonant peaks, when the incident waves are steady.

[†]Longuet-Higgins (1967) studied the special case of an impulsive incident wave proportional to $\delta(x - (gh_1)^{1/2}t)$.

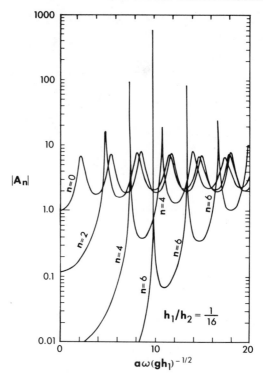

Figure 9.4 Graph of A_n for $h_1/h_2 = 1/16$ and $n = 0, 2, 4,$ and 6, giving the amplitude of the response as a function of the frequency of the incident wave. (From Longuet-Higgins, 1967, *J. Fluid Mech*. Reproduced by permission of Cambridge University Press.)

4.10 HEAD-SEA INCIDENCE ON A SLENDER TOPOGRAPHY—THE PARABOLIC APPROXIMATION

Before embarking on a general numerical method, let us describe the approximate analysis of a slender island or a sill attacked by waves incident along the longitudinal axis of the topography (head-sea incidence). A crucial aspect of this method was originated in electromagnetics and has been further developed in acoustics (see Tappert, 1977, for a review). The present adaptation is based on the work of Mei and Tuck (1980), and the critical appraisal by Bigg (1982). Similar ideas have also been extended to deep-water waves (Haren and Mei, 1981) and to weakly nonlinear waves (Yue and Mei, 1980, see Section 12.8).

Consider first an island with vertical sides in a shallow sea of constant depth h. The length L of the island is assumed to be much greater than the half-breadth B and the incident wavelength $2\pi/k$, namely,

$$\frac{B}{L} = \mu \ll 1 \quad \text{and} \quad kL = \omega L(gh)^{-1/2} \gg 1. \tag{10.1}$$

For head-sea incidence on a slender obstacle, the waves should roughly remain propagating forward with the amplitude modulated slowly in both horizontal directions, that is,

$$\eta(x, y) = A(x, y)e^{ikx} \tag{10.2}$$

where A varies slowly in x and y. Substituting Eq. (10.2) into (1.11), we get

$$2ikA_x + A_{yy} + A_{xx} = 0 \tag{10.3}$$

Now the length scale of A along x is L, $2ikA_x/A_{xx} = O(kL)$, hence A_{xx} is of secondary importance. In order to get something nontrivial, we retain A_{yy}; the implied length scale along y is then $O[L(kL)^{-1/2}]$. Let us introduce the *outer* length scales as follows:

$$X = \frac{x}{L}, \quad Y = y(\mu^{\alpha/2}L), \tag{10.4}$$

and express

$$\eta = A(X, Y)e^{iKX\mu^{-\alpha}} \tag{10.5}$$

with $kL = K\mu^{-\alpha}$ and $K = O(1)$; the leading order approximation to Eq. (10.3) is

$$2iKA_X + A_{YY} = 0 \tag{10.6}$$

with a relative error $O(\mu^\alpha)$. This is called the *parabolic approximation* and the region defined by Eq. (10.4) will be referred to as the parabolic region. Because of the resemblance to the heat equation, the following initial and boundary conditions

$$A = 1 \quad x = 0, |y| < \infty, \tag{10.7}$$

$$A \to 1 \quad x > 0, |y| \uparrow \infty \tag{10.8}$$

are natural, where the incident wave amplitude has been taken to be unity. The usual radiation condition is relevant to the region where $O(x, y) = L$ which lies outside the parabolic region, however.

Assuming symmetry about the x axis, we need be concerned only with the side $y > 0$. The no-flux condition on the island wall then requires

$$\frac{\partial \eta}{\partial y} = \frac{dW}{dx} \frac{\partial \eta}{\partial x} \qquad \text{on } y = W(x). \qquad (10.9)$$

Letting,

$$W(x) = Bb(x), \qquad 0 < b < 1 \qquad (10.10)$$

and using Eq. (10.12), we get, in terms of the normalized variables,

$$\frac{\partial A}{\partial Y} = iK\mu^{1-\alpha/2}b'A, \qquad \text{on } Y = \mu^{1-\alpha/2}b(x) \qquad (10.11)$$

with a relative error of $O(\mu^{\alpha})$. With $\alpha = 2$, both sides of Eq. (10.11) are balanced

$$\frac{\partial A}{\partial Y} = iKA \qquad \text{on } Y = b(x). \qquad (10.12)$$

The initial-boundary-value problem as defined by Eqs. (10.6), (10.7), (10.8) and (10.12) can be solved in general by numerical methods for heat conduction in one dimension (Y). Here X plays the role of time, and the computation marches forward in X over discrete steps much greater than a wavelength, hence is much more economical than the direct numerical solution of the Helmholtz equation. For the special case of a parabolic half-body, $b = \sqrt{X}$, the approximate problem can be solved quickly by the method of similarity as in Section 2.4; the result is:

$$A = 1 - \left(\int_{\infty}^{Y/\sqrt{X}} e^{iK\xi^2/2} \, d\xi \right) \left(e^{iK/2} - ik \int_{1}^{\infty} e^{iK\xi^2/2} \, d\xi \right)^{-1}. \qquad (10.13)$$

Thus, the amplitude along the island remains constant. Without the slenderness assumption the paraboilc cylinder can be solved exactly in terms of parabolic coordinates (see Jones, 1964, p. 467).

We now turn to a sill whose top is submerged at the depth $h_0 < h$. Let the half-breadth B be much smaller than $\mu^{\alpha/2}$ (i.e., $\alpha < 2$); the sill appears practically as a thin line to the *outer* observer in the parabolic region, causing a distributed flux $\partial A/\partial Y = V(X)$ along the X axis with $V \neq 0$, for $0 < X < 1$, and $V = 0$, for $X > 1$. This problem resembles the heat conduction in a semi-infinite rod with the variation of heat flux prescribed at one end. The solution is formally

$$A(X, Y) = 1 - \frac{1+i}{2(\pi K)^{1/2}} \int_0^X \frac{d\xi\, V(\xi)}{(X-\xi)^{1/2}} \exp \frac{iKY^2}{2(X-\xi)}, \qquad 0 < X < 1.$$

$$(10.14)$$

Near the sill the appropriate *inner* variables are

$$X = \frac{x}{L}, \qquad \overline{Y} = \frac{y}{\mu L}. \qquad (10.15)$$

Assume the inner solution to be of the form

$$\eta = A(X, \overline{Y})e^{iKX\mu^{-\alpha}}, \qquad (10.16)$$

then outside the sill this solution must satisfy Eq. (1.11) which yields

$$\frac{\partial^2 A}{\partial \overline{Y}^2} + \mu^{2-\alpha}2iK\frac{\partial A}{\partial X} + \mu^2\frac{\partial^2 A}{\partial X^2} = 0, \qquad \overline{Y} > b(X). \qquad (10.17)$$

Omitting terms of order $O(\mu^{2-\alpha})$, we get

$$A = B + C(\overline{Y} - b). \qquad (10.18)$$

Over the sill, the governing Helmholtz equation is of the same form as Eq. (1.11) but k^2 must be replaced by $k_0^2 = \omega^2/gh_0$. Substitution of Eq. (10.16) then gives

$$\frac{\partial^2 A}{\partial \overline{Y}^2} + K^2\left(\frac{h}{h_0} - 1\right)\mu^{2(1-\alpha)}A = O(\mu^{2-\alpha}A). \qquad (10.19)$$

The solution which is symmetric about the X axis is

$$A = \overline{A}\cos\left[K\left(\frac{h}{h_0} - 1\right)^{1/2}\mu^{2(1-\alpha)}\overline{Y}\right], \qquad 0 < \overline{Y} < b, \qquad (10.20)$$

where $\overline{A}(X)$ is the amplitude along the axis. Requiring the inner solutions (10.18) and (10.20) to be continuous and to have equal normal flux across $\overline{Y} = b$, we get, to the leading order,

$$B = \overline{A}\cos\left[\mu^{1-\alpha}K\left(\frac{h}{h_0} - 1\right)^{1/2}b\right], \qquad (10.21)$$

and

$$-h_0 \bar{A} \mu^{1-\alpha} K \left(\frac{h}{h_0} - 1 \right)^{1/2} \sin \left(\mu^{1-\alpha} K \left(\frac{h}{h_0} - 1 \right)^{1/2} b \right) = hC. \quad (10.22)$$

We now match the inner approximation of Eq. (10.14) for small Y,

$$A(X, Y) \cong 1 - \frac{1+i}{2(\pi K)^{1/2}} \int_0^X \frac{d\xi \, V(\xi)}{(X - \xi)^{1/2}} + VY + \cdots, \qquad Y \gg 1,$$

$$(10.23)$$

with the outer approximation of Eq. (10.20) for large $\bar{Y} \gg 1$,

$$A \cong B + C\bar{Y}, \qquad (10.24)$$

yielding

$$B = 1 + \frac{1+i}{2(\pi K)^{1/2}} \int_0^X \frac{d\xi \, V(\xi)}{(X - \xi)^{1/2}}, \qquad (10.25)$$

and

$$C = V\mu^{1-\alpha/2}. \qquad (10.26)$$

From Eqs. (10.21), (10.22), (10.25), and (10.26), B, C, and $\bar{A}$ may be eliminated to give

$$\frac{1+i}{2(\pi K)^{1/2}} \int_0^X \frac{d\xi \, V(\xi)}{(X - \xi)^{1/2}} = 1 + Z(X)V(X), \qquad (10.27)$$

where

$$Z(X) = \frac{h\mu^{\alpha/2}}{h_0 K (h/h_0 - 1)^{1/2}} \cot \left[\mu^{1-\alpha} K \left(\frac{h}{h_0} - 1 \right)^{1/2} b \right]. \quad (10.28)$$

We insist that $Z = O(1)$ to make all terms in Eq. (10.27) of equal order of magnitude, thus,

$$\alpha = 2/3 \quad \text{and} \quad \frac{h}{h_0} = \mu^{-2/3}, \qquad (10.29)$$

that is, the sill must be much shallower than the surrounding sea bottom.

The flux V can be solved numerically from the integral equation (see Mei and Tuck, 1980); afterward the amplitude $\bar{A}$ along the sill axis can be calculated.

Bigg (1982) has evaluated the present theory against numerical solutions of the full Helmholtz equation. He cautions that Eq. (10.27) can be used only when Eq. (10.29) is met and $\bar{A}$ and V do not change rapidly over the sill. Numerically, when K increases to the value such that

$$\cot\left[\mu^{1-\alpha}K\left(\frac{h}{h_0} - 1\right)^{1/2} b_{\max}\right] = 0,$$

V and $\bar{A}$ become unbounded at the station where the sill is the widest. Although this suggests resonance over the sill, the results are grossly invalid. Several numerical examples reported by Mei and Tuck (1980) are unacceptable for these reasons. We present in Fig. 10.1 the center-line amplitude along a parabolic sill with pointed ends for a range of parameters where Eq. (10.27) is reliable. The wave amplitude outside the sill follows from Eq. (10.14) and is omitted.

The parabolic approximation has been applied to short waves propagating over a slowly varying bottom. For example, Radder (1979) has examined a

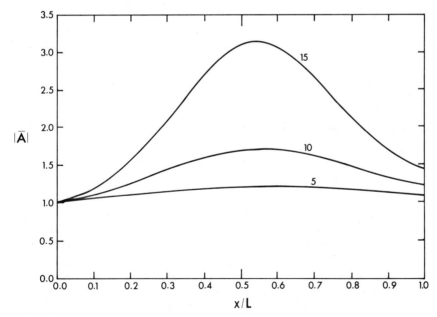

Figure 10.1 Variation of wave amplitude along the centerline of a submerged sill in head seas. The sides of the sill are a pair of parabolas $y/L = \pm\mu x(L - x)/L^2$. Numbers next to each curve indicate the value of $kL = \omega L(gh)^{-1/2}$ of the incident wave.

submerged circular shoal, neglecting the bending of rays. Diffraction around curved rays has been studied by Liu and Mei (1976a) in the shadow boundary created by the tip of a long breakwater (see Section 10.7 for the limiting case of straight rays). Extensions by Lozano and Liu (1980) can now deal with the local curvature of rays and the neighborhood of a focus. Booij (1981) has treated the combined effects of varying current and depth. As this approximation expedites numerical computations, it deserves to be further developed for practical problems.

4.11 A NUMERICAL METHOD BASED ON FINITE ELEMENTS

4.11.1 Introduction

The topography of the coastal region is seldom simple. To predict the water response to wave excitations, it is desirable to have a method which is capable of treating arbitrary geometry and depth. The *method of finite elements* is particularly suitable to fill this need because the element sizes and shapes can be freely varied to discretize the region of irregularity. After scoring spectacular successes in structural mechanics, this method has been extended to a variety of water-wave problems. While further refinements are still being made, recent advances have already found applications in actual design and planning on a large scale. We therefore devote this section to describing some essentials in the context of long waves, emphasizing those aspects which are unique to exterior problems. For the standard aspects of the theory of finite elements, the reader is referred to many excellent texts, for example, Zienkiewicz (1971) and Tong and Rossetti (1976).

A common procedure of the finite-element method for a linear boundary-value problem involves the following key steps:

1 Express the boundary-value problem as a variational principle where a certain functional is stationary.

2 Discretize the region into finite elements (such as triangles, quadrilaterals, and so on, for two-dimensional problems).

3 Select interpolating functions which approximate the solution inside the finite elements (linear, quadratic, and so on). The interpolating functions involve unknown coefficients.

4 For each element, perform the necessary differentiation and integration called for in the functional. For linear problems the functional is quadratic and can be expressed as a bilinear form in the unknown coefficients.

5 Assemble the element bilinear forms so that the total functional is expressed as a global bilinear form.

6 Extremize the functional with respect to each unknown coefficient and obtain a set of linear algebraic equations for the coefficients.

7 Solve the equations for the coefficients.

8 Compute quantities of physical interest.

Variational principles are not always easy to derive nor do they exist for all physical problems. A more general and straightforward approach is the so-called Galerkin weak formulation which will be touched upon in Chapter Seven for two- and three-dimensional problems with vertical variation. For problems of our interest, it can be shown that the variational formulation is equivalent to the weak formulation. Both formulations involve derivatives of order lower than those present in the original differential equation; therefore, they allow the use of a class of interpolating functions which are less differentiable (hence more general).

Usually the problems in structural mechanics involve a region of finite extent (a plate, a shell, a machine part, an airplane fuselage, and so on), whereas those in fluid mechanics often involve a theoretically infinite region. One may, of course, attempt to use a large but finite exterior boundary on which the condition at infinity is applied. After a numerical solution is completed for this finite region, a new and more remote boundary is introduced and the calculation is repeated. This procedure is continued until further expansion gives rise to negligible corrections to the solution. For wave problems the exterior boundary must be at least several wavelengths away from the bodies to give any degree of accuracy, while within each wavelength there must be sufficient grid points for good resolution. Thus, in order to study a wide range of wavelengths, one must use either a different grid for each narrow range of wavelengths or a single but enormous total region dictated by the longest waves, with very fine grids dictated by the shortest waves. Clearly, neither alternative is economical and many artifical devices such as fictitious damping have been proposed.

To study diffraction by an island where the water depth is constant, except in the vicinity of the island, Berkhoff (1972) divided the fluid into two regions by a circle which surrounds the island but is in the domain of constant depth. Only the interior of the circle is discretized into finite elements, while the solution in the exterior is represented by a continuous distribution of sources along the circle. The source solution is analytic and satisfies exactly the governing equation and the boundary condition at infinity. The source strength along the circle is unknown, however, and must be solved along with the interior of the circle by requiring the continuity of pressure and normal velocity across the circle. Different types of interpolating functions are used in different regions; hence the method may be called a *hybrid-element method* (HEM). The element with an analytical interpolating function is called the *superelement,* after Tong, Pian, and Lasry (1973).

Considerable freedom remains in the manner of enforcing the continuity of pressure and normal velocity between the finite elements and the superelement. An optimal choice is to cast the two matching conditions as *natural* boundary conditions in a variational principle, so that they are automatically satisfied in the numerical procedure. This variational approach has been successfully used for two-dimensional scattering and radiation problems (Chen and Mei 1974a, b; Bai and Yeung, 1974), two-dimensional ship waves (Mei and Chen, 1976), three-dimensional scattering problems (Yue, Chen, and Mei, 1976, 1978), as well as for wave problems involving fluid-structure interactions (Mei, Foda, and Tong, 1979). The procedure of Chen and Mei is described below.

4.11.2 The Variational Principle

We assume that complicated topography, such as large structures, curved coastlines, varying depths, and so on, are localized within a closed contour C, as shown in Fig. 11.1. Beyond C, the depth is every where constant, and the fluid region is denoted by $\overline{\Omega}$. The incident plane wave is in the direction θ_I:

$$\eta^I = Ae^{ikr\cos(\theta-\theta_I)} = A \sum_{n=0}^{\infty} \varepsilon_n i^n J_n(kr)\cos n(\theta - \theta_I). \qquad (11.1)$$

Within $\overline{\Omega}$, the scattered wave, denoted by $\bar{\eta}^S$, must satisfy the Helmholtz equation and the radiation condition and may be represented exactly by the Fourier–Bessel expansion:

$$\bar{\eta}^S = \alpha_0 H_0(kr) + \sum_{n=1}^{\infty} H_n(kr)(\alpha_n\cos n\theta + \beta_n\sin n\theta) \qquad (11.2)$$

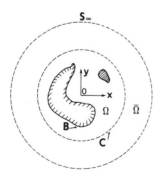

Figure 11.1   Definitions.

where H_n's are Hankel functions of the first kind. The expansion coefficients α_n, $n = 0, 1, 2, \ldots$, and β_n, $n = 1, 2, 3, \ldots$, are to be found. The total displacement in $\bar{\Omega}$ will be denoted by $\bar{\eta}$ $(= \eta^I + \bar{\eta}^S)$. The region $\bar{\Omega}$ is the *superelement*.

Let the fluid region within C be denoted by Ω and the corresponding displacement by η. Then η must satisfy Eq. (1.9) in Ω and Eq. (1.13) on B. Across C the pressure and normal velocity must be continuous:

$$\eta = \bar{\eta}, \qquad \frac{\partial \eta}{\partial n} = \frac{\partial \bar{\eta}}{\partial n} \qquad \text{on } C. \tag{11.3}$$

We shall now show that the stationarity of the functional

$$J(\eta, \bar{\eta}) = \iint_{\Omega} \frac{1}{2}\left[h(\nabla \eta)^2 - \frac{\omega^2}{g} \eta^2 \right] dA + \int_C h\left[\left(\tfrac{1}{2}\bar{\eta}^S - \eta^S \right) \frac{\partial \bar{\eta}}{\partial n} - \tfrac{1}{2}\bar{\eta}^S \frac{\partial \eta^I}{\partial n} \right] ds$$

$$\tag{11.4}$$

where $\eta^S \equiv \eta - \eta^I$, is equivalent to the boundary-value problem defining η in Ω and $\bar{\eta}^S$ in $\bar{\Omega}$. To prove the equivalence, we take the first variation of J:

$$\delta J = \iint_{\Omega}\left[h\nabla \eta \cdot \nabla \delta\eta - \frac{\omega^2}{g} \eta \delta\eta \right] dA$$

$$+ \int_C h\left[\left(\tfrac{1}{2}\delta\bar{\eta}^S - \delta\eta^S \right) \frac{\partial \bar{\eta}}{\partial n} + \left(\tfrac{1}{2}\bar{\eta}^S - \eta^S \right) \frac{\partial \delta\bar{\eta}}{\partial n} - \tfrac{1}{2}\delta\bar{\eta}^S \frac{\partial \eta^I}{\partial n} \right] ds.$$

By partial integration and Gauss' theorem, the first term above may be written

$$\iint_{\Omega} h\nabla \eta \cdot \nabla \delta\eta \, dA = \iint_{\Omega}\left[\nabla \cdot (h \, \delta\eta \, \nabla \eta) - \delta\eta \, \nabla \cdot (h\nabla \eta) \right] dA$$

$$= -\int_{\Omega} \delta\eta \, \nabla \cdot (h\nabla \eta) \, dA + \iint_C h\delta\eta \frac{\partial \eta}{\partial n} \, ds + \int_B h\delta\eta \frac{\partial \eta}{\partial n} \, ds.$$

Since η^I is known, $\delta\eta^I = 0$ and

$$\delta\bar{\eta}^S = \delta\bar{\eta} \quad \text{and} \quad \delta\eta^S = \delta\eta.$$

It follows that

$$\delta J = -\iint_\Omega \delta\eta \left[\nabla \cdot (h\nabla\eta) + \frac{\omega^2}{g}\eta \right] dA$$

$$+ \int_C h \left[(\tfrac{1}{2}\delta\bar\eta - \delta\eta)\frac{\partial\bar\eta}{\partial n} + (\tfrac{1}{2}\bar\eta^S - \eta^S)\frac{\partial\delta\bar\eta}{\partial n} \right.$$

$$\left. - \tfrac{1}{2}\delta\bar\eta \frac{\partial\eta^I}{\partial n} + \delta\eta \frac{\partial\eta}{\partial n} \right] ds + \int_B h\,\delta\eta \frac{\partial\eta}{\partial n}\,ds$$

$$= -\iint_\Omega \delta\eta \left[\nabla \cdot (h\nabla\eta) + \frac{\omega^2}{g}\eta \right] dA$$

$$+ \int_C h \left[\delta\eta \left(\frac{\partial\eta}{\partial n} - \frac{\partial\bar\eta}{\partial n} \right) - \frac{\partial\delta\bar\eta}{\partial n}(\eta^S - \bar\eta^S) \right] ds$$

$$+ \frac{1}{2}\int_C h \left[\delta\bar\eta \frac{\partial\bar\eta^S}{\partial n} - \bar\eta^S \frac{\partial\delta\bar\eta}{\partial n} \right] ds + \int_B h\,\delta\eta \frac{\partial\eta}{\partial n}\,ds. \qquad (11.5)$$

The next to the last integral in the preceding equation is equal to

$$I(C) = \frac{h}{2} \int_C \left[\delta\bar\eta^S \frac{\partial\bar\eta^S}{\partial n} - \bar\eta^S \frac{\partial\delta\bar\eta^S}{\partial n} \right] ds.$$

Now apply Green's formula to $\bar\eta^S$ and $\delta\bar\eta^S$ for the region $\bar\Omega$. Since both $\bar\eta^S$ and $\delta\bar\eta^S$ satisfy the Helmholtz equation exactly, the above integral I is unchanged if C is replaced by a circle C_∞ of great radius. Along C_∞, both $\bar\eta^S$ and $\delta\bar\eta^S$ satisfy the radiation condition (1.38); hence

$$I(C_\infty) = 0, \qquad \text{implying } I(C) = 0.$$

With I removed from Eq. (11.5), it is clear that if Eqs. (1.9), (1.13b), (11.3a), and (11.3b) are satisfied by η, then

$$\delta J = 0. \qquad (11.6)$$

Conversely, if $\delta J = 0$, it is necessary that η satisfies Eq. (1.9) in Ω as the Euler equation, Eq. (1.13b) on B, and Eqs. (11.3a) and (11.3b) on C as the *natural* boundary conditions. The equivalence is thus proved.

    The stationary functional involves integrals in and on the boundaries of Ω and may be used as the basis for obtaining an approximate solution in Ω. The method of finite elements is only one of the possibilities.

4.11.3 Finite-Element Approximation

Let us discretize the integrals in Eq. (11.4) by dividing the water area in Ω into a network of finite elements whose sizes should be much less than both the typical wavelength and the scale of local topographical variation. Since the integrands in Eq. (11.4) contain only first-order derivatives of η, it is only necessary to require the continuity of η in the finite-element domain. For simplicity we choose the three-node triangular elements, within each of which η is approximated by linear interpolating functions

$$\eta^2(x, y) = \sum_{i=1}^{3} N_i^2 \eta_i^e = \underset{1 \times 3}{\{N^e\}^T} \underset{3 \times 1}{\{\eta^e\}} \tag{11.7}$$

where superscripts e represent quantities associated with an element. Here $\{\ \}$ denotes a column vector and the superscript T denotes the transpose; hence $\{\ \}^T$ is a row vector. The underlying products indicate the dimensions of a matrix, for example, 1×3 means that the matrix has 1 row and 3 columns. More explicitly,

$$\{\eta^e\}^T = (\eta_1^e, \eta_2^e, \eta_3^e) \tag{11.8a}$$

$$\{N^e\}^T = (N_1^e, N_2^e, N_3^e) \tag{11.8b}$$

where

$$N_i^e(x, y) = \frac{a_i + b_i x + c_i y}{2\Delta^e}, \qquad i = 1, 2, 3 \tag{11.9a}$$

$$a_1 = x_2^e y_3^e - x_3^e y_2^e, \quad b_1 = y_2^e - y_3^e, \quad c_1 = x_3^e - x_2^e \tag{11.9b}$$

Δ^e = area of element e

$$= \frac{1}{2} \begin{vmatrix} 1 & x_1^e & y_1^e \\ 1 & x_2^e & y_2^e \\ 1 & x_3^e & y_3^e \end{vmatrix}. \tag{11.9c}$$

Other coefficients $a_{2,3}$, $b_{2,3}$, and $c_{2,3}$ may be obtained by permutation while x_i, y_i, and the coefficient η_i^e represent, respectively, the x, y coordinates and the value of η^e at node i. Let the depth within an element be approximated by a plane,

$$h^e(x, y) = \sum_{i=1}^{3} N_i^e h_i^e = \underset{1 \times 3}{\{N^e\}^T} \underset{3 \times 1}{\{h^e\}}, \tag{11.10}$$

h_i^e being the water depth at node i. When integration is performed for the area integral, we obtain

$$I_1 = \iint_\Omega \frac{1}{2}\left[h(\nabla\eta)^2 - \frac{\omega^2}{g}\eta^2 \right] dA = \frac{1}{2} \sum_{e\in\Omega} \underset{1\times3}{\{n^e\}^T} \underset{3\times3}{[K_1^e]} \underset{3\times1}{\{\eta^e\}} \quad (11.11)$$

where the element stiffness matrix $[K_1^e]$ has the components

$$K_{1_{ij}}^e = h_\beta^e \iint_e N_\beta^e \frac{\partial N_i^e}{\partial x_\alpha}\frac{\partial N_j^e}{\partial x_\alpha} dx\, dy - \frac{\omega^2}{g} \iint_e N_i^e N_j^e \, dx\, dy. \quad (11.12)$$

When Eq. (11.9) is substituted into (11.12), the integrals over any triangle e can be evaluated explicitly as

$$\iint_e N_i^e N_j^e \, dx\, dy = \tfrac{1}{6}\Delta^e, \quad i = j$$

$$\qquad\qquad\qquad = \tfrac{1}{2}\Delta^e, \quad i \neq j, \tag{11.13a}$$

$$\iint_e \frac{\partial N_i^e}{\partial x_\alpha}\frac{\partial N_j^e}{\partial x_\alpha} dx\, dy = \frac{1}{4\Delta^e}(b_i b_j + c_i c_j), \tag{11.13b}$$

$$h_\beta^e \iint_e N_\beta^e \frac{\partial N_i^e}{\partial x_\alpha}\frac{\partial N_j^e}{\partial x_\alpha} dx\, dy = \frac{h_1^e + h_2^e + h_3^e}{12\Delta^e}(b_i b_j + c_i c_j). \tag{11.13c}$$

The element stiffness matrix $[K_1^e]$ is symmetric. Now assemble all the element nodal displacements and define a total nodal displacement vector $\{\eta\}$ with E components where E is the total number of nodes in and on the boundaries of Ω. In so doing, the element nodal points must be reindexed to take into account the fact that the same node can belong to several adjacent elements. The element stiffness matrices must be assembled accordingly to give a total stiffness matrix $[K_1]$. Finally, I_1 is written

$$I_1 = \tfrac{1}{2} \underset{1\times E}{\{\eta\}^T} \underset{E\times E}{[K_1]} \underset{E\times1}{\{\eta\}} . \tag{11.14}$$

Because $[K_1^e]$ is symmetric for all elements, $[K_1]$ is also symmetric.

The line integral in Eq. (11.4) may be written as follows:

$$\int_C h\left[\left(\tfrac{1}{2}\bar{\eta}^S - \eta^S\right)\left(\frac{\partial \bar{\eta}^S}{\partial n} + \frac{\partial \eta^I}{\partial n}\right) - \tfrac{1}{2}\bar{\eta}^S \frac{\partial \eta^I}{\partial n}\right] ds$$

$$= \int_C h \tfrac{1}{2}\bar{\eta}^S \frac{\partial \bar{\eta}^S}{\partial n} ds - \int_C h(\eta - \eta^I)\frac{\partial \bar{\eta}^S}{\partial n} ds - \int_C h(\eta - \eta^I)\frac{\partial \eta^I}{\partial n} ds$$

$$= \tfrac{1}{2}\int_C h\bar{\eta}^S \frac{\partial \bar{\eta}^S}{\partial n} ds \cdots I_2$$

$$- \int_C h\eta \frac{\partial \bar{\eta}^S}{\partial n} ds \cdots I_3$$

$$- \int_C h\eta \frac{\partial \eta^I}{\partial n} ds \cdots I_4 \qquad\qquad (11.15)$$

$$+ \int_C h\eta^I \frac{\partial \bar{\eta}^S}{\partial n} ds \cdots I_5$$

$$+ \int_C h\eta^I \frac{\partial \eta^I}{\partial n} ds \cdots I_6.$$

For simplicity, C is assumed to be a circle of radius R. For computational purposes we truncate the series (11.1) and (11.2) at the term m. With Eq. (11.2), the line integral I_2 can be evaluated analytically by invoking the orthogonality of sines and cosines:

$$I_2 = \frac{\pi}{2} kRh\left[2\alpha_0^2 H_0 H_0' + \sum_{n=1}^{m}\left(\alpha_n^2 + \beta_n^2\right)H_n H_n'\right],$$

where

$$H_n = H_n^{(1)}(kR), \qquad H_n' \equiv \left[\frac{d}{d(kr)}H_n^{(1)}(kr)\right]_{r=R}$$

and R is the radius of C. Note that h on and outside C has been assumed to be constant. Define the column vector

$$\{\mu\}_{1 \times M}$$

such that

$$\{\mu\}_{1 \times M}^{T} = \{\alpha_0, \alpha_1, \beta_1, \alpha_2, \beta_2, \ldots, \alpha_m, \beta_m\}$$

with $M = 2m + 1$, and $[K_2]$ the diagonal matrix

$$[K_2]_{M \times M} = \pi k R h \operatorname{diag}\{2H_0' H_0, H_1' H_1, H_1' H_1, \ldots, H_m' H_m, H_m' H_m\}.$$

The integral I_2 can now be written

$$I_2 = \tfrac{1}{2} \{\mu\}^T \underset{1 \times M \; M \times M \; M \times 1}{[K_2]} \{\mu\} . \tag{11.16}$$

For integrals I_3 and I_4, we define for convenience the subset $\{\hat{\eta}\}$ of $\{\eta\}$ which lies on C (see Fig. 11.2)

$$\{\hat{\eta}\}^T_{1 \times P} = \{\hat{\eta}_1, \hat{\eta}_2, \ldots, \hat{\eta}_P\}.$$

Approximating the line integral over C by the sum of the integrals along the straight elements $\hat{\eta}_P \hat{\eta}_1, \hat{\eta}_1 \hat{\eta}_2, \ldots, \hat{\eta}_{P-1} \hat{\eta}_P$, η by its linear interpolations, and $\bar{\eta}^S$ by its value at the center of each arc element on C, namely, at $\theta = \theta_j$ (see Fig. 11.2), we obtain

$$I_3 \cong -\tfrac{1}{2} k h \sum_{j=1}^{P} L_j (\hat{\eta}_{j-1} + \hat{\eta}_j) \left[\alpha_0 H_0' + \sum_{n=0}^{} H_n' (\alpha_n \cos n\theta_j + \beta_n \sin n\theta_j) \right]$$

where L_j^e is the length of the element segment j. This may be written in matrix form

$$I_3 = \{\hat{\eta}\}^T \underset{1 \times P \; P \times M \; M \times 1}{[K_3]} \{\mu\} \tag{11.17}$$

where $[K_3]$ is a fully populated $P \times M$ matrix

$$[K_3]_{P \times M} = -\frac{kh}{2} \times$$

$$\begin{bmatrix} 2H_0' L_1 & \cdots & H_n'(\cos n\theta_P + \cos n\theta_1)L_1 & H_n'(\sin n\theta_P + \sin n\theta_1)L_1 & \cdots \\ 2H_0' L_2 & \cdots & H_n'(\cos n\theta_1 + \cos n\theta_2)L_2 & H_n'(\sin n\theta_1 + \sin n\theta_2)L_2 & \cdots \\ \vdots & \vdots & & & \\ 2H_0' L_P & \cdots & H_n'(\cos n\theta_{P-1} + \cos n\theta_P)L_P & H_n'(\sin n\theta_{P-1} + \sin n\theta_P)L_P \ldots \end{bmatrix}$$

$$\tag{11.18}$$

in which $n = 1, 2, \ldots, m$.

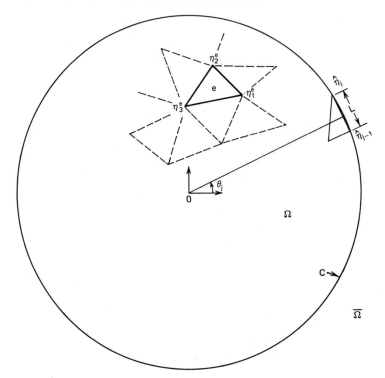

Figure 11.2 Typical boundary and interior elements.

Similarly, the integral I_4 is obtained:

$$I_4 = -\tfrac{1}{2}kh \sum_{j=1}^{P} L_j\big[i\cos(\theta_j - \theta_I)\big]\exp\big[ikR\cos(\theta_j - \theta_I)\big]$$

$$\cdot(\hat{\eta}_{j-1} + \hat{\eta}_j) \equiv -\{Q_4\}^T\{\hat{\eta}\} \tag{11.19}$$

where

$$\{Q_4\}^T = \tfrac{1}{2}kh\big\{(q_P - q_1)L_1, (q_1 + q_2)L_2,\ldots, (q_{P-1} + q_P)L_p\big\},$$

$$q_j = i\cos(\theta_j - \theta_1)\exp\big[ikR\cos(\theta_j - \theta_I)\big], \qquad j = 1, 2,\ldots, P.$$

Finally, the integral I_5 can be evaluated analytically to give

$$I_5 = -\underset{1 \times M}{\{Q_5\}^T} \underset{M \times 1}{\{\mu\}} \tag{11.20}$$

where

$$\underset{1 \times M}{\{Q_5\}^T} = 2\pi R k h \{J_0 H_0', \dots, i^m J_m H_m' \cos m\theta_I, i^m J_m H_m' \sin m\theta_I\}.$$

I_6 is a known constant and drops out upon extremization, hence it is of no interest.

Now we summarize the functional

$$J = \tfrac{1}{2}\{\eta\}^T[K_1]\{\eta\} + \tfrac{1}{2}\{\mu\}^T[K_2]\{\mu\} + \{\hat{\eta}\}^T[K_3]\{\mu\}$$

$$-\{Q_4\}^T\{\hat{\eta}\} - \{Q_5\}^T\{\mu\}. \tag{11.21}$$

Since J is stationary, we require that

$$\frac{\partial J}{\partial \eta_j} = 0, \qquad j = 1, 2, \dots, E$$

and

$$\frac{\partial J}{\partial \mu_j} = 0, \qquad j = 1, 2, \dots, M$$

which leads to a set of linear algebraic equations

$$[K_1]\{\eta\} + [K_3]\{\mu\} = \{Q_4\}, \qquad [K_2]\{\mu\} + [K_3]^T\{\hat{\eta}\} = \{Q_5\}. \tag{11.22}$$

Elimination of $\{\mu\}$ gives a matrix equation for $\{\eta\}$

$$[K]\{\eta\} \equiv \big[[K_1] - [K_3][K_2]^{-1}[K_3]^T\big]\{\eta\} = \{Q_4\} + [K_3][K_2]^{-1}\{Q_5\} \tag{11.23}$$

The resultant matrix $[K]$ is symmetric; only half of its elements need to be stored in the computer.

By heuristic reasoning and numerical experiments in which comparisons with analytic solutions were made in several cases, the following empirical rules on the element size have been found:

1 The element size should be everywhere less than 10% of the incident wavelength.
2 At the neighborhood of sharp curvature, elements smaller than the radius of curvature should be used locally, otherwise local but not global errors can result.

The number of coefficients M needed in the superelement is easy to decide by trial and error; it usually increases for shorter waves.

To minimize the bandwidth of $[K]$, the numbering of nodes should be proceeded first along one ring, then along the next ring either toward or from C. When this is done, the semibandwidth is then roughly equal to the number of nodes on C and is independent of the number of series coefficients in the superelement. Because of the symmetry and bandedness of the matrix, the numerical solution of the algebraic equation is particularly efficient. Clearly, the smaller the circle C, the smaller the bandwidth, and the amount of computation is usually reduced even though more coefficients may be required in the superelement.

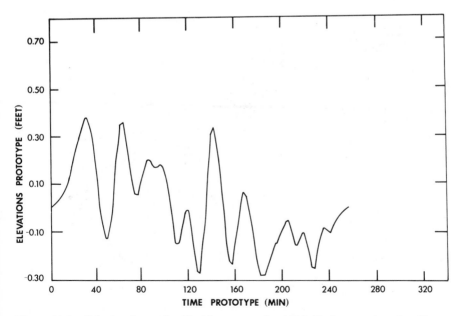

Figure 11.3 Calculated transient incident waves for 1964 Alaskan earthquake. (From Houston, 1978, *J. Phys. Ocean.* of American Meteorological Society.)

It should be remarked that the contour C need not be a circle; for an arbitrarily shaped contour numerical integration of the line integrals along C is needed and $[K_2]$ is full.

An interesting application of the hybrid-element method has been made by Houston (1978) on the response near the Hawaiian Islands to tsunamis caused by earthquakes near Chile in 1960 and Alaska in 1964. Based on a two-dimensional linearized long-wave equation including the earth's curvature, Houston used initial data estimated from the permanent displacement of the fault and calculated the waves up to a depth of 5000 m near the islands. The calculated wave history (shown in Fig. 11.3) was then Fourier analyzed into 18 harmonics with periods ranging from 14.5 min to 260 min. These harmonics were used as the incident waves. With the finite-element grid shown in Fig. 11.4 which has roughly 2500 nodes, the harmonic responses were calculated and superimposed in accordance with Eq. (2.56) to give the transient response

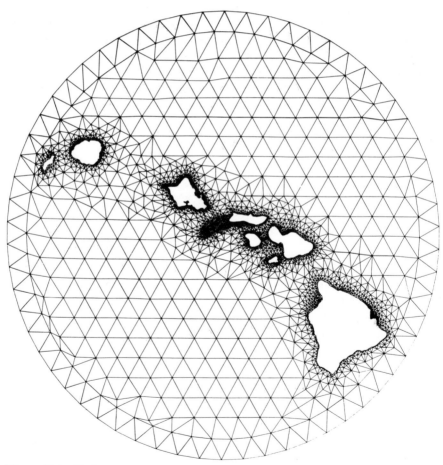

Figure 11.4 Typical finite-element grid surrounding Hawaiian islands. (From Houston, 1978, *J. Phys. Ocean.* of American Meterorological Society.)

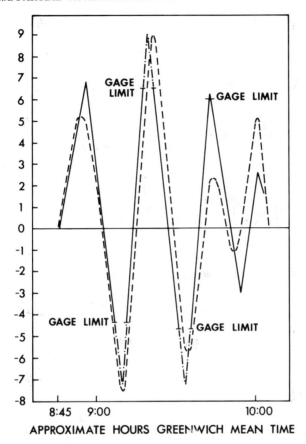

Figure 11.5 Calculated (dashed curve) versus measured (solid curve) tsunami records at Kahului, Maui for 1964 Alaskan earthquake. (From Houston, 1976, *J. Phys. Ocean.* of American Meteorological Society.)

at three wave stations: Kahului of Maui, Honolulu of Oahu, and Hilo of Hawaii. The agreement between theory and records is excellent. Figure 11.5 shows a sample comparison for the Alaskan earthquake with a 500-mile-long fault. The theoretical reason that the linear shallow-water equation is adequate for tsunami transocean propagation will be discussed in Chapter Eleven. This kind of calculation is useful for planning policies for tsunami flood insurance and can be applied to improve the speed and accuracy of tsunami warning systems.

APPENDIX 4.A: PARTIAL WAVE EXPANSION OF THE PLANE WAVE

Consider the product of the Taylor series expansions of $e^{zt/2}$ and $e^{-z/2t}$

$$e^{z(t-1/t)/2} = e^{zt/2}e^{-z/2t} = \left[\sum_{n=0}^{\infty}\frac{1}{n!}\left(\frac{zt}{2}\right)^n\right]\left[\sum_{m=0}^{\infty}\frac{1}{m!}\left(-\frac{z}{2t}\right)^m\right]. \quad \text{(A.1)}$$

Carrying out the multiplication and collecting the coefficients of the term t^n, we find that

$$e^{z(t-1/t)/2} = \sum_{n=-\infty}^{\infty} t^n \left[\frac{(z/2)^n}{n!} - \frac{(z/2)^{n+2}}{1!(n+1)!} + \frac{(z/2)^{n+4}}{2!(n+2)!} \right.$$

$$\left. + \cdots (-1)^r \frac{(z/2)^{n+2r}}{r!(n+r)!} \cdots \right].$$

Now the series in the brackets above is just the Bessel function of order n, $J_n(z)$; hence

$$e^{z(t-1/t)/2} = \sum_{n=-\infty}^{\infty} t^n J_n(z). \qquad (A.2)$$

If we substitute

$$t = ie^{i\theta}, \qquad \frac{1}{t} = -ie^{-i\theta}$$

into Eq. (A.2), then

$$e^{iz\cos\theta} = \sum_{-\infty}^{\infty} e^{in(\theta+\pi/2)} J_n(z). \qquad (A.3)$$

Now let us combine pairs of terms with equal but opposite n's and use the fact that

$$J_{-n}(z) = (-1)^n J_n(z).$$

Equation (A.3) can be written in an alternate form:

$$e^{iz\cos\theta} = \sum_{0}^{\infty} \varepsilon_n \cos n\theta (i)^n J_n(z) \qquad (A.4)$$

where the Jacobi symbols have been used.

For a plane wave, we let $z = kr$ in Eq. (A.4) and get

$$e^{ikx} = e^{ikr\cos\theta} = \sum_{0}^{\infty} \varepsilon_n \cos n\theta i^n J_n(kr). \qquad (A.5)$$

FIVE

Harbor Oscillations Excited by Incident Long Waves

5.1 INTRODUCTION

A harbor is a partially enclosed basin of water connected through one or more openings to the sea. Conventional harbors are built along a coast where a shielded area may be provided by natural indentations and/or by breakwaters protruding seaward from the coast, as sketched in Figs. 1.1a–1.1c. An artificial harbor can be far removed from the mainland; an example is the offshore harbor for the Atlantic Generating Stations, once planned by the New Jersey Public Service Gas and Electric Company, which would have enclosed two floating nuclear power plants by two giant breakwaters, Fig. 1.1d. There are also harbors on a small offshore island which may itself be near or far from the mainland, as in Fig. 1.1e.

Although a variety of external forcings can be responsible for significant oscillations within a harbor, the most studied forcing is caused by incident tsunamis which have typical periods from a few minutes to an hour and are originated from distant earthquakes. If the total duration of the tsunami is sufficiently long, oscillations excited in the harbor may persist for days, resulting in broken mooring lines, damaged fenders, hazards in berthing and loading or navigation through the entrance, and so on. Sometimes incoming ships have to wait outside the harbor until oscillations within subside, causing costly delays.

To understand roughly the physical mechanism of these oscillations, consider a harbor with the entrance in line with a long and straight coastline. Onshore waves are partly reflected and partly absorbed along the coast. A small portion is, however, diffracted through the entrance into the harbor and reflected repeatedly by the interior boundaries. Some of the reflected wave

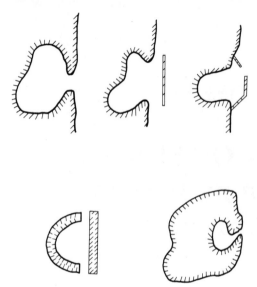

Figure 1.1   Variety of harbor configurations.

energy escapes the harbor and radiates to the ocean again, while some stays inside. If the wavetrain is of long duration, and the incident wave frequency is close to a standing wave frequency in the closed basin, resonance will occur in the basin so that a relatively weak incident wave can induce a large response in the harbor.

The peak amplitude at resonance can be limited by a number of mechanisms:

1   Radiation damping, associated with energy escaped seaward from the harbor entrance.
2   Frictional loss near the basin boundary and the harbor entrance.
3   Loss due to wave breaking on shallow beaches.
4   Finite-amplitude effects of energy transfer into higher harmonics.

Among these mechanisms, radiation damping is the most understood theoretically and was first treated in a pioneering paper by Miles and Munk (1961) for a rectangular harbor. Frictional losses occur along harbor boundaries and near breakwater tips at the entrance; these losses are harder to estimate and vary widely according to the property of the boundary. Reliable estimates require empirical information which is difficult to obtain by model tests because of the scale effects. Breaking is a phenomenon associated mostly with wind-generated waves on mild beaches and is not amenable to any theory at present. Fortunately, for very long tsunamis, breaking is usually unimportant.

In this chapter we shall ignore frictional and breaking losses and only include the effects of radiation damping. After the formulation, three elements of the harbor problem are discussed separately: standing waves in a basin, the concept of radiation damping, and diffraction through a gap. Next, for sinusoidal inputs and constant depth, we shall study the full problem coupling the ocean with harbors of various simple plan forms. Transients will be considered for a narrow bay. At the end of the chapter the general hybrid-element method of Section 4.11 will be modified for harbors of arbitrary depth and shape.

5.2 FORMULATION FOR HARBOR OSCILLATION PROBLEMS

For simplicity we make the following assumptions on the fluid motion: (i) inviscid fluid, (ii) irrotational flow, (iii) infinitesimal waves amplitude, (iv) very long wavelength compared to depth, and (v) lateral boundaries are perfectly reflective and vertical throughout the sea depth. The governing equations derived in Section 4.1 are applicable; for convenience they are recalled here. For transient motion the displacement satisfies the field equation

$$g\nabla \cdot (h\nabla\zeta) = \frac{\partial^2\zeta}{\partial t^2} \tag{2.1}$$

with the no-flux condition

$$h\frac{\partial\zeta}{\partial n} = 0 \tag{2.2}$$

on the lateral walls. For simple harmonic motion the spatial amplitude η of the free-surface displacement satisfies the field equation

$$\nabla \cdot (h\nabla\eta) + \frac{\omega^2}{g}\eta = 0 \tag{2.3}$$

subject to the no-flux condition

$$h\frac{\partial\eta}{\partial n} = 0 \tag{2.4}$$

on the lateral walls. For constant depth, Eq. (2.1) reduces to the classical wave equation, while Eq. (2.3) reduces to the Helmholtz equation

$$\nabla^2\eta + k^2\eta = 0 \tag{2.5}$$

where $\omega = (gh)^{-1/2}k$.

The radiation condition for sinusoidal motion can be stated explicitly if the topography far away from the harbor is simple. Consider a harbor on a coastline which is also perfectly reflective. Let Ω denote the region which includes the harbor and all the complex topography nearby, and let $\bar{\bar{\Omega}}$ be the remaining part of the ocean in which $h = \text{const}$ and the coastline $\bar{B}$ is straight (see Fig. 2.1). A plane incident wave may be expressed by

$$\eta^I = A \exp\left[ik(x \cos \theta_I + y \sin \theta_I)\right] \tag{2.6}$$

where A, k, and the direction θ_I are prescribed. The complete wave system in the ocean $\bar{\bar{\Omega}}$ may be split as

$$\eta = \eta^I + \eta^{I'} + \eta^S \tag{2.7}$$

where $\eta^{I'}$ denotes the reflected wave due to the straight coast without the local topography near the harbor, while η^S denotes the wave scattered by the local topography and radiated by the piston action at the entrance. Let the y axis coincide with the straight portion of the coast $\bar{B}$; the reflected wave is

$$\eta^{I'} = A \exp\left[ik(-x \cos \theta_I + y \sin \theta_I)\right] \tag{2.8}$$

so that on $\bar{B}$,

$$\frac{\partial}{\partial x}(\eta^I + \eta^{I'}) = 0. \tag{2.9}$$

Then the radiated/scattered wave cannot have any normal flux along the straight coast:

$$\frac{\partial}{\partial x}\eta^S = 0 \qquad \text{on } \bar{B}. \tag{2.10}$$

Furthermore, η^S must be outgoing at large distances,

$$(kr)^{1/2}\left(\frac{\partial}{\partial r} - ik\right)\eta^S \to 0, \qquad kr \to \infty. \tag{2.11}$$

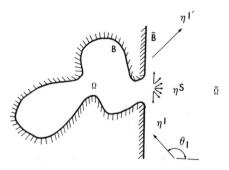

Figure 2.1 Definition sketch.

In the case of an offshore harbor many wavelengths away from a coast, one may simply omit the reflected wave $\eta^{I'}$ from Eq. (2.7). For other coastal configurations, or nonconstant depth in Ω, an explicit description of η^I and $\eta^{I'}$ may be a difficult task in itself.

When the depth is constant everywhere in Ω and in $\bar{\Omega}$, and all walls are vertical, the three-dimensional potential for arbitrary kh may be expressed as

$$\phi(x, y, z) = \frac{-ig\eta}{\omega} \frac{\cosh k(z + h)}{\cosh kh}.$$

Recall from Section 3.5 that η also satisfies the horizontal Helmholtz equation except that ω and k are related by $\omega^2 = gk \tanh kh$. Since the walls are vertical, the normal vector is in the horizontal plane, and the boundary condition is $\partial\eta/\partial n = 0$ on side walls. Thus, the boundary-value problems for long and short waves are formally the same. This mathematical analogy enables one to perform harbor experiments in deep water where nonlinear effects are easier to avoid.

5.3 NATURAL MODES IN A CLOSED BASIN OF SIMPLE FORM AND CONSTANT DEPTH

As a preliminary, it is useful to discuss the typical features of standing waves in a closed basin. For simplicity, we assume the depth to be constant. The boundary-value problem for η, which may now be taken as real, is defined by the homogeneous equations (2.5) and (2.4), and admits nontrivial solutions only when k equals certain eigenvalues. The corresponding ω's are called the *natural (or eigen) frequencies* and the corresponding η's the *natural (or eigen) modes*. Two elementary examples are discussed below.

5.3.1 A Rectangular Basin

Let the lateral boundaries of the basin be $x = 0, a$ and $y = 0, b$. The eigensolutions to Eq. (2.5) are found by separation of variables

$$\eta = A_{nm}\cos\frac{n\pi x}{a}\cos\frac{m\pi y}{b} \tag{3.1}$$

with $n, m = 0, 1, 2, 3, \ldots$. The corresponding eigenvalues are

$$k = k_{nm} = \left[\left(\frac{n\pi}{a}\right)^2 + \left(\frac{m\pi}{b}\right)^2\right]^{1/2}. \tag{3.2}$$

The natural periods are

$$T_{nm} = 2\pi/\omega_{nm} \tag{3.3}$$

where ω_{nm} are related to k_{nm} via the dispersion relation

$$\omega_{nm}^2 = ghk_{nm}^2. \tag{3.4}$$

If $a > b$, the lowest mode ($n = 1$, $m = 0$) has the lowest frequency and the longest period and is referred to as the *fundamental mode*. The corresponding motion is one dimensional.

If the ratio between the two sides is a rational number, that is, $a = pL$, $b = qL$ (p, q = integers),

$$k_{nm} = \left[\left(\frac{m}{p}\right)^2 + \left(\frac{n}{q}\right)^2\right]^{1/2} \frac{\pi}{L},$$

there are more than one set of (n, m) which correspond to the same eigen-frequency. This situation is called *degeneracy*.

Let us illustrate the spatial structure of the mode $(n, m) = (1, 1)$, that is,

$$\eta_{11} = A_{11}\cos\frac{\pi x}{a} \cos\frac{\pi y}{b}.$$

At the boundaries $x = 0, a$, and $y = 0, b$, the amplitude is maximum. On the other hand, the amplitude is zero along the *nodal lines* $x = a/2$ or $y = b/2$, which divide the basin into four rectangles. At a given instant two adjacent rectangles are opposite in phase. Thus, if two regions are above the mean water level, the other two are below and vice versa. In Fig. 3.1 the contour lines are shown.

For higher modes (n, m), the free surface is also similarly divided by n nodal lines along $x/a = \frac{1}{2}\pi, \frac{3}{2}\pi, \ldots, (n - \frac{1}{2})\pi$, as well as m nodal lines along $y/b = \frac{1}{2}\pi, \frac{3}{2}\pi, \ldots, (m - \frac{1}{2})\pi$.

5.3.2 A Circular Basin

Let the radius of the basin be a; polar coordinates (r, θ) may be chosen so that the origin is at the center. The governing Helmholtz equation may be written in the form of Eq. (9.10), Chapter Four. On the wall, $r = a$, the normal radial velocity component vanishes. Hence

$$\frac{\partial\eta}{\partial r} = 0. \tag{3.5}$$

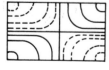

Figure 3.1 Free-surface contours of natural mode $\cos(\pi x/a)\cos(\pi y/b)$ in a rectangular basin.

A solution to the Helmholtz equation is, by separation of variables,

$$\eta = J_m(kr)(A_m\cos m\theta + B_m\sin m\theta) \tag{3.6}$$

where A_m and B_m are arbitrary constants. To satisfy the boundary condition we must have

$$J_m'(kr)\big|_{r=a} = J_m'(ka) = 0. \tag{3.7}$$

Now $J_m'(z)$ is an oscillatory function of z having an infinite number of zeroes. Denoting the nth zero of J_m' by j_{mn}': $J_m'(j_{mn}') = 0$, we have the eigenvalues,

$$k_{mn} = \frac{j_{mn}'}{a}, \qquad n = 1,2,3,\ldots, \quad m = 0,1,2,3,\ldots \tag{3.8}$$

The corresponding eigensolutions or natural modes are

$$\eta_{mn} = J_m(k_{mn}r)(A_{mn}\cos m\theta + B_{mn}\sin m\theta). \tag{3.9}$$

To illustrate the structure of a particular mode, we consider the free-surface variation for $\eta_{mn} = J_m(k_{mn}r)\cos m\theta$ with n, m fixed. Clearly, $\cos m\theta = 1$ when $m\theta = 0, 2\pi, 4\pi, \ldots, 2m\pi$, and -1 when $m\theta = \pi, 3\pi, 5\pi, \ldots, (2m-1)\pi$. Thus, $\theta = 0, \pi/m, 2\pi/m, 3\pi/m, \ldots$ are antinodal lines where the surface displacement is the greatest along a circle of given r. On the other hand, $\theta = \pi/2m, 3\pi/2m, 5\pi/2m, \ldots$ are the nodal lines where the displacement is zero. For a fixed θ, the curve $J_m(k_{mn}r)$ crosses the zero line exactly $n-1$ times within the range $r < a$ so that there are $n-1$ nodal rings; this fact is the consequence of the general Sturm's oscillation theorem in the theory of ordinary differential equations. The partition of the free surface into ups and downs is illustrated in Fig. 3.2. The values of these zeroes are available in Abramowitz and Stegun (1972) and are tabulated in Table 3.1. In ascending order, the indices (n, m) of the zeroes are $(0, 1)$, $(1, 1)$, $(2, 1)$, $(0, 2)$, (3.1), (4.1), $(1.2), \ldots$. For mass conservation, the mode $(0, 0)$ cannot exist in a completely closed basin.

Table 3.1 Values of j_{mn}' Such That $J_m'(j_{mn}') = 0$.

n \ m	0	1	2	3	4	5
1	0	1.84118	3.05424	4.20119	5.31755	6.41562
2	3.83171	5.33144	6.70713	8.01524	9.28240	10.51986
3	7.01559	8.53632	9.96947	11.34592	12.18190	13.98719
4	10.17346	11.70600	13.17037	14.58525	15.96411	17.31284

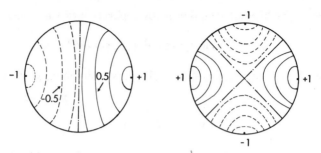

Figure 3.2 Contour lines of two natural modes in a circular basin: $J_1(k_{11}r)\cos\theta$ and $J_2(k_{21}r)\cos 2\theta$.

5.4   CONCEPT OF RADIATION DAMPING: A MODEL EXAMPLE

An important feature of wave diffraction in an infinite medium is that oscillations which originate from a finite region experience damping even when the medium is conservative. This damping is associated with the energy carried away to infinity by outgoing waves and is called *radiation damping*. To gain some idea about it let us study a very instructive model example of Carrier (1970) which has the physical features typical of a vibrating system coupled with propagating waves.

Consider a semi-infinite channel of uniform depth h and width b, Fig. 4.1. At $x = 0$ there rests a gate of mass M which is allowed to slide along the channel without friction. The gate is supported by a spring with elastic constant K. Assuming for simplicity that there is no leakage at $x = 0$, we wish to find the displacement $Xe^{-i\omega t}$ of the gate when there is an incident shallow-water wave of amplitude A and frequency ω from $x \sim +\infty$.

The water surface can be represented by

$$\zeta = \eta e^{-i\omega t} = A\left(e^{-ikx} + Re^{ikx}\right)e^{-i\omega t}$$

$$= A\left[e^{-ikx} + e^{ikx} + (R-1)e^{ikx}\right]e^{-i\omega t}. \qquad (4.1)$$

In the last brackets above the second term stands for the reflected wave when the gate is held stationary and the third term stands for the radiated wave due to the induced motion of the gate.

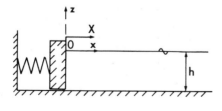

Figure 4.1 A spring-supported mass against water waves.

The equation of motion of the gate is

$$-M\omega^2 X = -KX - pbh \qquad (4.2)$$

where p is the hydrodynamic pressure per unit area at $x = 0$:

$$p = \rho g \eta = \rho g A (1 + R). \qquad (4.3)$$

Equations (4.2) and (4.3) can be combined to give

$$A(1 + R) = -\frac{K - M\omega^2}{\rho g b h} X. \qquad (4.4)$$

At $x = 0$ the fluid velocity $u(0) = (-ig/\omega)\eta_x(0)$ must equal the gate velocity $-i\omega X$; thus,

$$u(0) = -i\omega X = \frac{gkA}{\omega}(-1 + R). \qquad (4.5)$$

It is easy to deduce the solution from Eqs. (4.4) and (4.5)

$$\frac{X}{2A} = \frac{\rho g b h}{-K + M\omega^2 + i(\omega^2/k)\rho b h} = \frac{\rho g b h}{-K + M\omega^2 + i\omega(gh)^{1/2}(\rho b h)}. \qquad (4.6)$$

The radiated wave amplitude is

$$R - 1 = -2i\omega \left(\frac{h}{g}\right)^{1/2} \frac{X}{2A}.$$

Equation (4.6) may be compared with the usual mass–spring–dashpot system. Except for a constant of proportionality, the denominator in Eq. (4.6) may be called the *impedance*. The imaginary part (proportional to $\rho b h$) of the impedance plays the role of damping. To see this let us consider the system to be unforced. A nontrivial free vibration can still be described by Eq. (4.6) with $A = 0$ if we insist that the denominator vanishes, that is,

$$-K + M\omega^2 + i\omega(gh)^{1/2}(\rho b h) = 0 \qquad (4.7)$$

which is an eigenvalue condition with complex solutions for ω:

$$\omega = \pm \left\{ \omega_0^2 - \left[\frac{(gh)^{1/2}\rho b h}{2M}\right]^2 \right\}^{1/2} - \frac{i(gh)^{1/2}\rho b h}{2M}$$

where $\omega_0 = K/M$. Inserting either solution in the time factor $\exp(-i\omega t)$, we see that the oscillation decays exponentially at the rate proportional to

$$\frac{(gh)^{1/2}\rho bh}{2M}. \tag{4.8}$$

To trace the physical source of this damping, let us calculate the rate of work done by the radiated wave averaged over a cycle

$$\bar{E}_{rad} = \tfrac{1}{2}\operatorname{Re}[\,p_{rad}u^*\,]_{x=0}bh$$

$$= \tfrac{1}{2}bh\operatorname{Re}\rho gA(R-1)(-i\omega X)^*$$

$$= \tfrac{1}{2}\rho\frac{\omega^3}{k}\,|X|^2 = \tfrac{1}{2}\rho bh\big(\omega^2(gh)^{1/2}\big)\,|X|^2$$

after using Eq. (4.5) and $\omega = (gh)^{1/2}k$. This positive-definite quantity is clearly associated only with the damping term so that damping is due to the rate of work imparted by the radiated waves to the fluid. Therefore, we refer to the imaginary term in Eq. (4.6) as the *radiation damping*.

The response (4.6) may be written either as a function of ω:

$$\frac{X}{2A} = \left(\frac{\rho gbh}{M}\right)\left[\omega^2 - \frac{K}{M} + \frac{i\omega(gh)^{1/2}\rho bh}{M}\right]^{-1}, \tag{4.9}$$

or as a function of k:

$$\frac{X}{2A} = \left(-\frac{\rho b}{M}\right)\left(k^2 - \frac{K}{Mgh} + \frac{ik\rho bh}{M}\right)^{-1}. \tag{4.10}$$

In the complex k plane there are two poles located at

$$\pm\tilde{k} + i\hat{k} \tag{4.11}$$

with

$$\tilde{k} = k_0\left[1 - \left(\frac{\rho bh^2}{M}\right)^2\frac{Mg}{4Kh}\right]^{1/2}, \qquad k_0 \equiv \frac{\omega_0}{(gh)^{1/2}} \equiv \left(\frac{K}{M}\right)^{1/2}\frac{1}{(gh)^{1/2}}, \tag{4.12}$$

and

$$\hat{k} = -\frac{\rho bh}{2M} < 0. \tag{4.13}$$

Equation (4.10) then becomes

$$\frac{X}{2A} = \left(\frac{\rho b}{M}\right)(k - \tilde{k} - i\hat{k})^{-1}(k + \tilde{k} - i\hat{k})^{-1}. \qquad (4.14)$$

For small damping the two poles are only slightly below the real axis. In the physical problem, ω and k are both positive and real; the only pole of physical meaning is $\tilde{k} + i\hat{k}$. In its neighborhood, $|X|$ is large and Eq. (4.14) may be approximated by

$$\frac{X}{2A} \cong \left[\left(\frac{\rho b}{M}\right)\frac{1}{2\tilde{k}}\right](k - \tilde{k} - i\hat{k})^{-1}. \qquad (4.15)$$

The maximum of $|X/2A|^2$ is

$$\left|\frac{X}{2A}\right|^2_{max} = (\tilde{k}h)^{-2} \cong (k_0 h)^{-2}$$

which is attained near $k \cong \tilde{k}$. When $k - \tilde{k} = \pm\hat{k}$, the square response drops to one-half of the peak value, therefore $\hat{k}$ is a measure of the width of the response curve ($|X/2A|^2$ versus k). As in electric circuit theory, we may define the *quality factor* $\mathfrak{Q}$ by

$$\mathfrak{Q} = -\frac{\hat{k}}{\tilde{k}} \cong \frac{\rho b h}{2M}\left(\frac{M}{K}\right)^{1/2}(gh)^{1/2}. \qquad (4.16)$$

As the radiation damping $\hat{k}$ decreases, $\mathfrak{Q}$ decreases; the peak width of the response curve decreases, hence the shape of the curve sharpens. As is seen from Eq. (4.8), $\mathfrak{Q}\omega_0$ also corresponds to the rate of damping of free oscillations.

5.5   DIFFRACTION THROUGH A NARROW GAP

A harbor entrance is often just an opening along an otherwise long and thin breakwater. Transmission of waves through the opening is of obvious interest. For analytical simplicity, we assume that the breakwater is thin, vertical, and perfectly reflective, and the depth is constant so that the problem has an exact acoustical analogy.

Referring to Fig. 5.1, we consider normal incidence from $x > 0$. On the incidence side, $x > 0$, the total wave system consists of the incident wave, the reflected wave from a solid wall, and disturbances due to the fluid motion along the gap. On the transmission side, $x < 0$, there are only disturbances due to the motion along the gap. The gap acts as a piston in a baffle wall and radiates waves to infinity on both sides.

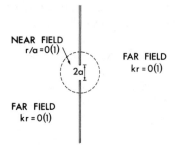

NEAR FIELD
$r/a = O(1)$

FAR FIELD
$kr = O(1)$

$2a$

FAR FIELD
$kr = O(1)$

Figure 5.1 A gap between two breakwaters.

While the boundary-value problem can be solved approximately for any gap width by the integral-equation method, we shall apply the method of *matched asymptotic expansions*, which is particularly convenient for a gap whose width is much less than the wavelength (Buchwald, 1971).

The intuitive ideas of the method have been explained in Section 4.2.2. Briefly, when various parts of the physical domain are governed by vastly different scales, we approximate the equations and the boundary conditions according to the local scales and seek solutions valid in these separate regions. The solution in one region usually does not satisfy the boundary conditions in the other, resulting in some indeterminancy. By the requirement that they match smoothly in some intermediate region, this determinancy is removed and the solution is obtained to a desired order.

Let us define the *far field* to be the region which is a few wavelengths away from the gap,

$$kr = O(1) \qquad \text{(the far field)}. \qquad (5.1)$$

Clearly, $1/k$ is the proper scale and all terms in the Helmholtz equation are equally important. At great distances from the gap, the radiated waves must satisfy the Helmholtz equation and the radiation condition. However, to the far-field observer, the gap is a very small region in the neighborhood of the origin. The radiated wave may be represented by superimposing solutions which are singular at the origin and cause no flux along the y axis:

$$\eta_{\pm}^{R} = \frac{\omega Q^{\pm}}{2g} H_0^{(1)}(kr) + \frac{\omega \mu^{\pm}}{2g} H_1^{(1)}(kr)\sin\theta + \cdots, \qquad x \gtrless 0. \qquad (5.2)$$

The total far-field solutions on both sides of the gap are

$$\eta_+ = 2A\cos kx + \eta_+^R, \quad x > 0, \qquad \text{and} \quad \eta_- = \eta_-^R, \quad x < 0. \qquad (5.3)$$

From the first term in the series of Eq. (5.2), the flux out of a semicircle of small radius around the origin is, for $x \lesssim 0$,

$$\text{flux} = \lim_{r \to 0} \pi r \left(\frac{-ig}{\omega} \frac{\omega Q^{\pm}}{2g} \right) \frac{\partial}{\partial r} H_0^{(1)}(kr) = Q^{\pm} .$$

Hence, the first term of Eq. (5.2) represents a source with a discharge rate of $Q^{\pm}$ into the half-plane $x \geq 0$. The subsequent terms represent doublet, quadrupole,..., and so on.

Near the junction, the length scale is the gap width; hence we may define a *near field* where

$$\frac{r}{a} = O(1). \tag{5.4}$$

In this region

$$\frac{k^2 \eta}{\nabla^2 \eta} = O(ka)^2$$

so that the flow is essentially governed by the Laplace equation

$$\nabla^2 \eta = 0 \tag{5.5}$$

with a relative error of order $O(ka)^2$. The no-flux condition must be satisfied on the solid walls. The radiation condition is no longer relevant and has to be discarded. Now Eq. (5.5) and the no-flux condition define a usual potential flow problem with time as a parameter only. Since η is harmonic, it may be taken as the real part of an analytic function W of the complex variable $z = x + jy$, that is,

$$\eta = \mathrm{Re}_j W(z) \tag{5.6}$$

where Re_j means the real part with respect to j, with i regarded as real. The solution of Eqs. (5.5) and (2.4) becomes the search for $W(z)$ that is analytic in the z plane with

$$\mathrm{Im}_j W(z) = \text{constant on the solid walls.} \tag{5.7}$$

For simple geometries, the solution is most effectively found by the technique of conformal mapping. In the present case, we use the Joukowski transformation in the airfoil theory

$$z = -\frac{ja}{2}\left(\tau + \frac{1}{\tau}\right) \tag{5.8}$$

to map the z plane outside the two breakwaters onto the upper half-plane of τ (see Fig. 5.2). In particular, the image of the solid wall ABD is the negative real τ axis and the image of $A'B'D'$ is the positive real τ axis. To satisfy the condition that $\mathrm{Im}_j W = 0$ on $A'B'D'$ and $\mathrm{Im}_j W = \text{const}$ on ABD, we adopt the solution

$$W(z) = C + M \ln \tau + C_1 \tau + C_2 \tau^2 + \cdots + C_{-1} \tau^{-1} + C_{-2} \tau^{-2} + \cdots \tag{5.9}$$

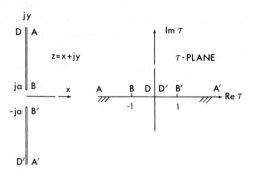

Figure 5.2 Mapping of the near field from the z plane (physical) to the upper half of the τ plane.

where the coefficients are real with respect to j but may be complex with respect to i. The coefficients $Q^\pm$ and $\mu^\pm$ in Eq. (5.2), as well as C, M, and C_1, C_{-1}, and so on will be found by matching the near and far fields.

Let us require that in an intermediate region, which appears to be close to the origin to the far-field observer ($kr \ll 1$), while very far away from the near-field observer ($r/a \gg 1$), the near- and far-field solutions should match smoothly. For $kr \ll 1$, the *inner expansion* of the total far field is

$$\eta_+ = 2A + \frac{i\omega Q^+}{g}\left(-\frac{i}{2} + \frac{1}{\pi}\ln\frac{\gamma kr}{2}\right) - \frac{\omega\mu^+}{2g}\frac{1}{r}\sin\theta + \cdots O(kr)^2\ln kr,$$

$$x > 0 \quad (5.10)$$

$$\eta_- = +\frac{i\omega Q^-}{g}\left(-\frac{i}{2} + \frac{1}{\pi}\ln\frac{\gamma kr}{2}\right) - \frac{\omega\mu^-}{2g}\frac{1}{r}\sin\theta + \cdots O(kr)^2\ln kr,$$

$$x < 0 \quad (5.11)$$

where $\ln\gamma = $ Euler's constant $= 0.5772157\ldots$. To approximate the near-field solution for $r/a \gg 1$, we must distinguish the two sides $x < 0$ and $x > 0$. On the side of $x > 0$, the region of $|z|/a \gg 1$ corresponds to $|\tau| \gg 1$ in the τ plane so that

$$\tau = \frac{2jz}{a}\left[1 + O\left(\frac{r}{a}\right)^{-2}\right] \quad (5.12)$$

from Eq. (5.8). If this is substituted into Eq. (5.6), the *outer expansion* of the near field η is obtained,

$$\eta \simeq \mathrm{Re}_j W \simeq \mathrm{Re}_j\left(C + M\ln\frac{2jz}{a}\right) + C_1\left(\frac{2jz}{a}\right) + \cdots + C_{-1}\left(\frac{a}{2jz}\right) + \cdots$$

$$= C + M\ln\frac{2r}{a} - C_1\left(\frac{2y}{a}\right) + \cdots - C_{-1}\left(\frac{a}{2r}\right)\sin\theta + \cdots \quad (5.13)$$

On the side of $x < 0$, the region of $|z|/a \gg 1$ corresponds to the origin in the τ plane. Hence from Eq. (5.8)

$$\tau = \frac{a}{2jz}\left[1 + O\left(\frac{z}{a}\right)^{-2}\right], \tag{5.14}$$

and the *outer expansion* of the near field is

$$\eta \simeq \mathrm{Re}_j W \simeq \mathrm{Re}_j\left\{C + M\ln\frac{a}{2jz} + C_1\left(\frac{a}{2jz}\right) + \cdots + C_{-1}\left(\frac{2jz}{a}\right) + \cdots\right\}$$

$$= C - M\ln\frac{2r}{a} - C_1\left(\frac{a}{2r}\right)\sin\theta + \cdots - C_{-1}\left(\frac{2y}{a}\right) + \cdots \tag{5.15}$$

We now equate Eqs. (5.10) and (5.13) to match η_+. From the coefficients of like terms several algebraic relations are found,

$$\text{(const):} \quad 2A + \frac{i\omega Q^+}{g}\left[-\frac{i}{2} + \frac{1}{\pi}\ln\frac{\gamma k}{2}\right] = C + M\ln\frac{2}{a} \tag{5.16a}$$

$$\text{(ln } r\text{):} \quad \frac{i\omega Q^+}{\pi g} = M \tag{5.16b}$$

$$(y)\text{:} \quad C_1 = 0 \tag{5.16c}$$

$$\left(\frac{1}{r}\sin\theta\right)\text{:} \quad C_{-1} = \frac{\omega\mu^+}{ga}. \tag{5.16d}$$

Matching η_- by equating Eqs. (5.11) and (5.15) similarly, we obtain

$$\text{(const):} \quad \frac{i\omega Q^-}{g}\left(-\frac{i}{2} + \frac{1}{\pi}\ln\frac{\gamma k}{2}\right) = C - M\ln\frac{2}{a} \tag{5.17a}$$

$$\text{(ln } r\text{):} \quad \frac{i\omega Q^-}{\pi g} = -M \tag{5.17b}$$

$$(y)\text{:} \quad C_{-1} = 0 \tag{5.17c}$$

$$\left(\frac{1}{r}\sin\theta\right)\text{:} \quad C_1 = \frac{\omega\mu^-}{ga}. \tag{5.17d}$$

Observe immediately that

$$C_1 = C_{-1} = 0, \tag{5.18a}$$

$$\mu^+ = \mu^- = 0. \tag{5.18b}$$

It can be shown that poles of higher order are likewise zero so that only the source is important at the leading order. Thus C_n, $n = \pm2, \pm3,\ldots$, are also zero and no nonzero powers of τ are needed in the inner solution to the present accuracy. These facts will be used in future analysis without further verification.

There are now only four unknowns remaining: $Q^\pm$, M, and C which can be solved to give

$$-\frac{i\omega Q^+}{g} = +\frac{i\omega Q^-}{g} = \frac{A}{-\frac{1}{2}i + (1/\pi)\ln(\gamma ka/4)}, \qquad (5.19a)$$

$$M = +\frac{i\omega Q^+}{\pi g} = -\frac{i\omega Q^-}{\pi g}, \qquad (5.19b)$$

$$C = A. \qquad (5.19c)$$

Combination of Eq. (15.19a) with Eq. (5.2) yields finally,

$$\eta_\pm^R \cong \frac{\pm i\frac{1}{2}AH_0^{(1)}(kr)}{-\frac{1}{2}i + (1/\pi)\ln(\gamma ka/4)}. \qquad (5.20)$$

Expanding the Hankel function for large kr, we have

$$\eta_\pm^R = \pm A\mathcal{C}\left(\frac{2}{\pi kr}\right)^{1/2} e^{ikr - i\pi/4} \qquad (5.21)$$

where

$$\mathcal{C} = \left(1 + \frac{2i}{\pi}\ln\frac{\gamma ka}{4}\right)^{-1}. \qquad (5.22)$$

The function $-\ln z$ approaches infinity very slowly as z diminishes. For example, $-\ln z = 2.0, 4.6, 6.9,\ldots$ for $z = 10^{-1}, 10^{-2}, 10^{-3}$, and so on. Thus, $-\ln(\gamma ka/4)$ is really not so large for practical ranges of ka, and $|\mathcal{C}|$ diminishes slowly as ka decreases, as shown below:

ka	1	0.1	0.01	0.001
$|\mathcal{C}|$	0.8890	0.4506	0.2786	0.1995

The persistence of transmission for very small ka is a typical result in potential theory and should be modified by real fluid effects of viscosity and separation at the tips.

To the present degree of approximation, the near field is dominated by a constant and a term proportional to $\ln r$. Physically, the former term represents a uniform rise and fall of the free surface and the latter indicates that the gap acts as a source to one side and a sink of equal magnitude to the other.

In closing, it should be mentioned that for arbitrary ka the diffraction problem can be solved by a number of approximate methods based on the method of integral equations. As ka increases the radiated wave amplitude $\mathcal{Q}$ has a more complicated dependence on θ. As will be shown later, significant resonance in a harbor occurs when the wavelength is at least comparable to the harbor dimensions, which usually far exceeds the width of the harbor entrance. Hence, we shall not discuss the gap problem further here.

5.6 SCATTERING BY A LONG AND NARROW CANAL OR A BAY

5.6.1 General Solution

Let us consider a narrow canal of width $2a$ open to the ocean. The geometry is depicted in Fig. 6.1. For long waves, $ka \ll 1$, the far field in the canal can only be one dimensional, this being a special case of Section 4.1.2. Therefore, the general far-field solution in the canal is

$$\eta_c = Be^{-ikx} + De^{ikx}, \qquad x < 0 \tag{6.1}$$

with the inner expansion

$$\eta_c = (B + D) + ik(-B + D)x + \cdots O(kx)^2 \qquad \text{for } |kx| \ll 1. \tag{6.2}$$

The far-field solution in the ocean is, as before,

$$\eta_0 \cong 2A \cos kx + \frac{\omega Q}{2g} H_0^{(1)}(kr) \tag{6.3}$$

with the inner expansion

$$\eta_0 = 2A + \frac{\omega Q}{2g}\left(1 + \frac{2i}{\pi} \ln \frac{\gamma kr}{2}\right) + O(kr), \qquad kr \ll 1. \tag{6.4}$$

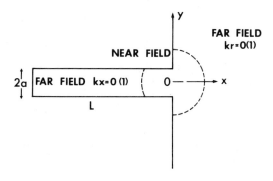

Figure 6.1 A narrow bay.

The near-field problem is that of a potential flow past a right-angled estuary; see Fig. 6.2. By the Schwarz–Christoffel transformation the physical region in the complex z plane ($z = x + jy$) can be mapped onto the upper half of the τ plane by

$$z = \frac{2a}{\pi}\left[-j(\tau^2 - 1)^{1/2} + \ln \frac{\tau}{(\tau^2 - 1)^{1/2} + j}\right] \tag{6.5}$$

(Kober, 1957, p. 155) with the images shown in Fig. 6.2. For single valuedness the square root $(\tau^2 - 1)^{1/2}$ is defined in the τ plane with a cut along the real axis $-1 \le \mathrm{Re}\,\tau \le 1$, and the branch is chosen so that $(\tau^2 - 1)^{1/2} \to \tau$ as $|\tau| \to \infty$. The logarithmic function $\ln \tau$ is defined with a cut along the positive real axis.

The near-field approximation has to be analytic in τ as before:

$$\eta = \mathrm{Re}_j W(\tau) = \mathrm{Re}_j(M \ln \tau + C) \tag{6.6}$$

with M and C real in j. Its outer expansion must be calculated by distinguishing the two sides $x \gtrless 0$. On the ocean side, $x > 0$, large $|z|/a$ corresponds to large $|\tau|$ (see Fig. 6.2). By expanding the right-hand side of Eq. (6.5) we have

$$z = \frac{2a}{\pi}\left[-j\tau + O\left(\frac{1}{\tau}\right)\right], \qquad -j\tau = \frac{\pi z}{2a}\left[1 + O\left(\frac{a}{z}\right)^2\right], \qquad x > 0. \tag{6.7}$$

Substituting Eq. (6.7) into Eq. (6.6), we obtain the outer expansion of the near field

$$\eta \cong \mathrm{Re}_j M \ln \frac{j\pi z}{2a} + C = M \ln \frac{\pi r}{2a} + C, \qquad x > 0. \tag{6.8}$$

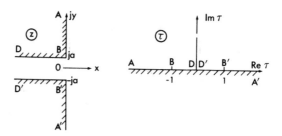

Figure 6.2   Mapping of the near field from the z plane to the upper half of the τ plane.

On the canal side, $x < 0$, large $|z/a|$ corresponds to small $|\tau|$. Since from Eq. (6.5)

$$\frac{\pi z}{2a} = 1 + \ln \tau - \ln 2j + O(\tau)^2$$

$$= \ln \frac{e\tau}{2j} + O(\tau)^2 \quad \text{or} \quad \tau \cong \frac{2j}{e} e^{\pi z/2a}, \qquad -\frac{x}{a} \gg 1,$$

we have

$$\ln \tau \cong \frac{\pi z}{2a} - \ln \frac{e}{2j}, \tag{6.9}$$

the error being exponentially small for $x/a \to -\infty$. The outer expansion of the near-field solution is, therefore,

$$\eta \cong M \frac{\pi x}{2a} - M \ln \frac{e}{2} + C, \qquad x < 0. \tag{6.10}$$

Matching of the inner and outer solutions on the canal side $x < 0$ gives

$$B + D = C - M \ln \frac{e}{2}, \tag{6.11a}$$

$$ik(-B + D) = \frac{\pi M}{2a}. \tag{6.11b}$$

Similarly, on the ocean side we obtain by matching Eqs. (6.4) and (6.8)

$$2A + \frac{\omega}{2g} Q \left(1 + \frac{2i}{\pi} \ln \frac{\gamma k}{2} \right) = C + M \ln \frac{\pi}{2a}, \tag{6.11c}$$

$$\frac{iQ\omega}{\pi g} = M. \tag{6.11d}$$

Thus far, there are four algebraic equations for five unknowns: B, D, C, M, and Q; A being prescribed. One more condition is needed which depends on the constraints at the far end of the canal. The following possibilities are of physical interest:

1 Wave scattering into an infinitely long channel with no reflection from the far end. Since only left-going waves are possible, $D = 0$, so that the channel solution is

$$\eta_c = B e^{-ikx}. \tag{6.12}$$

2   Wave incident from the far end of the channel and transmitted into the ocean. In this case D is given and $A = 0$.

3   Wave scattering into a long bay of length L where the far end $x = -L$ is highly reflective. Here we impose

$$\frac{\partial \eta_c}{\partial x} = 0, \qquad x = -L.$$

The appropriate outer solution is

$$\eta_c = E \cos k(x + L), \tag{6.13}$$

so that

$$B = \tfrac{1}{2} E e^{-ikL} \tag{6.14a}$$

$$D = \tfrac{1}{2} E e^{ikL}. \tag{6.14b}$$

The corresponding inner expansion is

$$\eta_c = E[\cos kL + (\sin kL)kx] + O(kx)^2. \tag{6.15}$$

The algebraic problems for the unknown coefficients can now be solved for each case. For instance, for the first problem (wave scattered into a long canal), we get

$$\frac{\omega Q}{2g} = \frac{2Aka}{[1 + ka + (2ika/\pi)\ln(2\gamma ka/\pi e)]}, \tag{6.16}$$

$$B = \frac{-2A}{[1 + ka + (2ika/\pi)\ln(2\gamma ka/\pi e)]}. \tag{6.17}$$

Again, C is concerned with the near field only and will not be recorded. Equation (6.16) gives the strength of the source radiating waves back to the infinite ocean and Eq. (6.17) gives the amplitude of the transmitted wave.

Exercise 6.1

Complete the solution for problem 2, wave propagation from the channel into the ocean, and discuss the result.

5.6.2   An Open Narrow Bay

The case of a narrow bay of finite length, that is, problem 3 above, illustrates many common features of harbor resonance, hence a detailed analysis is given here. A complete analysis was first given by Miles and Munk (1961) whose approximation at the entrance was slightly different from matched asymptotics (Ünlüata and Mei, 1973).

By combining Eqs. (6.14a) and (6.14b) with Eqs. (6.11a)–(6.11d), we obtain the bay response η_c and the discharge Q through the bay entrance:

$$\eta_c = \frac{2A \cos k(x + L)}{\cos kL + (2ka/\pi)\sin kL \ln(2\gamma ka/\pi e) - ika \sin kL} \qquad (6.18)$$

$$\frac{\omega Q}{2g} = \frac{-2Aika \sin kL}{\cos kL + (2ka/\pi)\sin kL \ln(2\gamma ka/\pi e) - ika \sin kL} \qquad (6.19)$$

where η_c refers to the far-field motion away from the entrance by a distance much greater than $2a$ but much less than the wavelength. Relative to the standing wave amplitude $2A$, an amplification factor $\mathcal{Q}$ may be defined,

$$\mathcal{Q} = \frac{1}{\cos kL + (2ka/\pi)\sin kL \ln(2\gamma ka/\pi e) - ika \sin kL} \qquad (6.20)$$

so that

$$\eta_c = 2A\mathcal{Q} \cos k(x + L). \qquad (6.21)$$

The plot of $|\mathcal{Q}|^2$ versus kL will be called the *response curve*, with ka being a parameter.

Since $ka \ll 1$, the response curve has a peak near the zeroes of $\cos kL$, that is,

$$\cos kL \cong 0, \qquad kL \cong k_n L = (n + \tfrac{1}{2})\pi, \qquad n = 0, 1, 2, \ldots$$

Because of the small terms of $O(ka)$ in Eq. (6.20), the resonant peaks are slightly shifted from these crude values. A better approximation is obtained by letting

$$k = k_n + \Delta$$

and expanding for small Δ:

$$\cos kL = -L\Delta \sin k_n L + O(\Delta^3), \qquad \sin kL = \sin k_n L + O(\Delta^2).$$

In the neighborhood of the nth resonant peak, the amplification factor is

$$\mathcal{Q} \cong \frac{1}{-\sin k_n L [L\Delta - (2k_n a/\pi)\ln(2\gamma k_n a/\pi e) + ik_n a]}$$

$$\cong \frac{1}{(-1)^{n+1}[(k - \tilde{k}_n)L + ik_n a]}, \qquad (6.22)$$

with

$$\tilde{k}_n \equiv k_n \left(1 + \frac{2}{\pi} \frac{a}{L} \ln \frac{2\gamma k_n a}{\pi e} \right) \tag{6.23}$$

which may be compared with Eq. (4.15) for the model example. Clearly, the peak is at $k = \tilde{k}_n$ and the shift of the peak is given by

$$\left(\tilde{k}_n - k_n \right) = \frac{2}{\pi} \frac{k_n a}{L} \ln \frac{2\gamma k_n a}{\pi e} < 0. \tag{6.24}$$

Around the peak, the square of the amplification factor is

$$|\mathcal{Q}|^2 = \frac{1}{|(\tilde{k}_n - \tilde{k}_n)L|^2 + (k_n a)^2}, \tag{6.25}$$

while the peak value is

$$|\mathcal{Q}|_{max} = \frac{1}{k_n a} = \frac{1}{\left(n + \frac{1}{2} \right) \pi a / L}. \tag{6.26}$$

Thus, the height of successive resonant peaks decreases with the mode number n.

For mode n, the plot of $\mathcal{Q}$ versus kL is approximately symmetrical about the peak. At the values

$$\left(k - \tilde{k}_n \right)L = \pm k_n a,$$

$|\mathcal{Q}|^2$ is reduced by half. Thus, $k_n a$ is a measure of both the peak height and the half-width of the resonance curve. The corresponding wave profile in the bay is roughly proportional to

$$\cos\left[\left(n + \frac{1}{2} \right) \pi \left(\frac{x}{L} + 1 \right) \right].$$

In particular, at the lowest mode $n = 0$, the bay length is about one-quarter of the wavelength so that the harbor entrance is very close to the first node. Compare two bays of equal length L but different width $2a$. The narrower harbor has the small shift $(k_n - \tilde{k}_n)L$ and the resonant peak is sharper and higher. In the limit of $ka \to 0$ the radiation damping diminishes to zero and the peak height becomes infinite. Since the width of the resonant peak in the response diagram also diminishes with a, the incident wave must be precisely tuned to the peak frequency in order to resonate the harbor. If the tuning is slightly off, the response is greatly reduced. The feature that the resonant response increases with narrowing entrance does not always agree with practical experience and is one aspect of the *harbor paradox* termed by Miles and Munk (1961). This paradox can be removed by considering friction at the harbor entrance and/or nonlinearity, both of which will be considered in later chapters.

From Eq. (6.19) the discharge per unit depth at the harbor entrance Q, which is essentially the amplitude of the radiated waves, is also maximum at a resonant peak. The maximum values of Q are obtained by letting the real part in the denominator vanish so that

$$\max |Q| = \frac{2Ag}{\omega_n}$$

where $\omega_n = k_n(gh)^{1/2}$. The resonant discharge is smaller for a higher mode.

Notice that at the nth peak the free surface has an apparent node at $x = \Delta L$ so that

$$\cos \tilde{k}_n(\Delta L + L) = 0,$$

or

$$(k_n + \Delta k_n)(\Delta L + L) = (n + \tfrac{1}{2})\pi,$$

or

$$\frac{\Delta L}{L} \cong -\frac{\Delta k_n}{k_n} = -\frac{2}{\pi}\frac{a}{L}\ln\frac{2\gamma k_n a}{\pi e} > 0,$$

which decreases with a/L and with n. Thus, the effective length of the bay is greater than the actual L. This increase in length can be thought of as the added inertia of the ocean water near the entrance.

The analytical result Eq. (6.20) is accurate as long as kb is small. For a special case treated by finite elements and shown later in Fig. 10.3, the present theory is quantitatively satisfactory only for the lowest (quarter wavelength) mode.

Exercise 6.2

Study the mutual influence of two straight and narrow canals of finite lengths opened perpendicularly to the same straight coast. Consider the angle of incidence to be arbitrary (Mei and Foda, 1979, where a mathematically similar problem of elastic SH waves incident on open cavities is treated).

Exercise 6.3

Study the oscillation in a semicircular canal of narrow width $2a$, with both ends open to the same straight coast. Consider the angle of incidence to be arbitrary (Mei and Foda, 1979).

Exercise 6.4   *A Modelling Effect in Harbor Resonance (Rogers and Mei, 1977)*

In harbor experiments the ocean is limited by the finite size of the test basin. Typically, the wavemaker is at a finite distance L' from the coast. Show by the method of images that the effect of the wavemaker at $x = L'$ can be approximately accounted for by letting the far-field solution in the ocean be

$$\eta = 2A \cos kx + \frac{\omega Q}{2g} \left\{ H_0^{(1)}(kr) + \sum_{n=1}^{\infty} \left[H_0^{(1)}(k \,|\, \mathbf{r} - 2nL'\mathbf{e}_x \,|) \right.\right.$$

$$\left.\left. + H_0^{(1)}(k \,|\, \mathbf{r} + 2nL'\mathbf{e}_x \,|) \right] \right\}$$

for $0 < x < L'$ and $kL' = m\pi$. Near the entrance for $kr \ll 1$ the correction due to finite L' is

$$e = \frac{\omega Q}{g} \sum_{n=1}^{\infty} H_0^{(1)}(2nkL').$$

For large kL', the Hankel functions may be replaced by their asymptotic approximations so that

$$e \cong \frac{\omega Q}{g} \left(\frac{2}{\pi} \right)^{1/2} e^{-i\pi/4} \sum_{n=1}^{\infty} \frac{e^{inZ}}{(nZ)^{1/2}}, \qquad Z = 2kL'.$$

For $Z = 2kL' \gg 1$ the series can be approximated by an integral in the following manner:

$$\sum_{n=1}^{\infty} f(n) = \sum_{n=1}^{\infty} f(n)\,\Delta n \qquad (\text{since } \Delta n = 1)$$

$$= \sum_{n=1}^{\infty} f(\sigma Z) Z \,\Delta\sigma \qquad \text{with } \frac{n}{Z} = \sigma$$

$$\cong \int_{1/Z}^{\infty} f(\sigma Z) Z \, d\sigma.$$

Express the integral as a Fresnel integral and show that $e \sim O(kL')^{-3/2}$, hence give your own criterion on how large a wave tank should be to model an infinite ocean.

5.7   A RECTANGULAR HARBOR WITH A NARROW ENTRANCE

In addition to the physical features deduced in Section 5.6, a harbor with comparable dimensions in both horizontal directions has a new mode of oscillation where the free surface within the harbor rises and falls in unison. This phenomenon is familiar in acoustics and may be demonstrated by a simple analysis. In reference to Fig. 7.1, consider a basin of surface area S open to the infinite ocean through a channel of length L and width a, where L is assumed to be sufficiently long so that the added hydrodynamic length is negligible by comparison. Let the free-surface amplitude at A be ζ and the velocity in the channel be U. Continuity requires that

$$\frac{\partial \zeta}{\partial t} S = - Uah.$$

When resonance occurs in the basin, ζ is small at the estuary B; the pressure gradient between A and B is approximately

$$-\frac{\Delta p}{\Delta x} = \frac{p_A - p_B}{L} = \frac{\rho g \zeta}{L}.$$

The momentum equation for the channel water is

$$\frac{\partial U}{\partial t} = \frac{g\zeta}{L}.$$

Combining the mass and momentum equations by eliminating U, we get

$$S\frac{\partial^2 \zeta}{\partial t^2} + gz\frac{a}{L}\zeta = 0,$$

which resembles a mass–spring system, and has a natural mode with the natural frequency

$$\omega = \left(\frac{gha}{SL}\right)^{1/2}.$$

The corresponding characteristic wavenumber is, in dimensionless form, $kS^{1/2} = (a/L)^{1/2}$ and is very small. This mode of oscillation is called the *Helmholtz mode* in acoustics and the *pumping mode* in harbor engineering literature. Clearly, the existence of the Helmholtz mode is associated with the finite harbor area. Since the narrow bay in Section 5.6 corresponds to an oscillation with a spring but without mass, there is no Helmholtz mode.

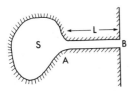

Figure 7.1

We now turn to the special example of a rectangular harbor for a detailed analysis. This example was first studied by Miles and Munk (1961) with revisions by Garrett (1970). Use of matched asymptotics was made by Ünlüata and Mei (1973). Let the sides of the harbor be B and L as shown in Fig. 7.2. The harbor entrance is a gap in an otherwise thin and straight breakwater aligned with the coast. The gap width is assumed to be small compared to the wavelength $ka \ll 1$.

Assuming for simplicity that the incidence is normal, the outer solution for the ocean is again given by Eq. (6.3):

$$\eta_0 = 2A \cos kx - \frac{i\omega}{g} Q_0 \left[\frac{i}{2} H_0^{(1)}(kr) \right], \qquad r^2 = x^2 + y^2 \qquad (7.1)$$

where the coordinate system has its origin at the center of the entrance. The inner expansion of η_0 is recorded here for convenience,

$$\eta_0 = 2A + i\frac{\omega}{g} Q_0 \left(\frac{-i}{2} + \frac{1}{\pi} \ln \frac{\gamma k}{2} \right) + i\frac{\omega}{g\pi} Q_0 \ln r + O(kr \ln kr). \qquad (7.2)$$

The inner solution near the entrance is the potential flow past a gap and has the following two-term outer expansion (cf. Eqs. (5.13) and (5.15)):

$$\eta_E = C \mp M \ln \frac{a}{2} + M \left(\begin{array}{l} +\ln r \\ -\ln r_1 \end{array} \right), \qquad \begin{array}{ll} x > 0 & (7.3a) \\ x < 0 \ (x_1 > 0). & (7.3b) \end{array}$$

To describe the interior of the harbor, it is convenient to use a different coordinate system (x_1, y_1) where the origin coincides with a corner of the basin so that

$$x = -x_1, \quad y = y_1' - y_1, \quad r_1^2 = x_1^2 + (y_1 - y_1')^2. \qquad (7.4)$$

The center of the harbor entrance is at $y_1 = y_1'$ (see Fig. 7.2).

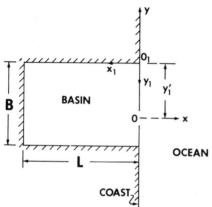

Figure 7.2 A rectangular harbor behind a straight coast.

5.7.1 Solution by Matched Asymptotic Expansions

Let it be understood that *in the harbor* the coordinate system (x_1, y_1) is used. However, for brevity, the subscripts $(\ )_1$ will be omitted.

To the leading order the outer solution for the harbor is the field due to a pulsating source of unknown strength Q_H at the point $x = 0$, $y = y'$. Let $G(x, y; y')$ be the solution corresponding to

$$\nabla^2 G + k^2 G = 0, \tag{7.5a}$$

$$\frac{\partial G}{\partial y} = 0, \qquad\qquad 0 < x < L, \quad y = 0, B, \tag{7.5b}$$

$$\frac{\partial G}{\partial x} = 0, \qquad\qquad x = L \tag{7.5c}$$

$$= \delta(y - y'), \qquad x = 0, \quad 0 < y < \beta. \tag{7.5d}$$

Since G represents the solution for a point source of unit discharge, it follows that

$$\eta_H = \frac{i\omega}{g} Q_H G(x, y; y') \tag{7.6}$$

is the desired outer solution in the harbor basin. The function of G is a kind of Green's function; its solution is derived in Appendix 5.A. We only give the following result:

$$G(x, y; y') = \sum_{n=0}^{\infty} X_n(x) Y_n(y) Y_n(y') \tag{7.7}$$

where

$$X_n(x) = \frac{\varepsilon_n \cos K_n(x - L)}{K_n B \sin K_n L}, \tag{7.8a}$$

$$Y_n(y) = \cos\left(\frac{n\pi y}{B}\right), \tag{7.8b}$$

$$K_n = \left[k^2 - \left(\frac{n\pi}{B}\right)^2\right]^{1/2}, \tag{7.8c}$$

and ε_n is again the Jacobi symbol. The inner expansion, however, needs a little work. The series for G converges slowly as it stands since

$$K_n \sim i\left(\frac{n\pi}{B}\right),$$

for n large and

$$\frac{\varepsilon_n \cos K_n(x - L)}{K_n B \sin K_n L} = \frac{2 \cosh[n\pi(x - L)/B]}{n\pi \sinh(n\pi L/B)} + O\left(\frac{1}{n^3}\right)$$

$$= -\frac{2}{n\pi} e^{-n\pi x/B} + O\left(\frac{1}{n^3}\right).$$

The nth term dies out only as fast as $1/n$. A usual trick to speed up the convergence is to try summing the series composed of the leading approximation of each term:

$$\tilde{G} = \sum_1^\infty -\frac{2}{n\pi} e^{-n\pi x/B} Y_n(y) Y_n(y'). \tag{7.9}$$

The remaining series

$$X_0 Y_0(y) Y_0(y') + \sum_{n=1}^\infty \left(X_n + \frac{2}{n\pi} e^{-n\pi x/B} \right) Y_n(y) Y_n(y') \tag{7.10}$$

then converges much faster (as $1/n^3$) (see Kantorovich and Krylov, 1964, p. 79ff for examples of this technique). The summation turns out to be possible and is detailed in Appendix 5.B; we only give the following result:

$$\tilde{G} = \frac{1}{2\pi} \ln |1 - e^{-Z_s}|^2 |1 - e^{-Z_s'}|^2, \tag{7.11}$$

where

$$Z_s = \frac{\pi}{B} [x + j(y - y')], \qquad Z_s' = \frac{\pi}{B} [x + j(y + y')]. \tag{7.12}$$

Note that Z_s is the normalized complex distance from the field point (x, y) to the source $(0, y')$, and Z_s' is the normalized complex distance from (x, y) to the mirror image of the source located at $(0, -y')$. Very close to the mouth, $r/B \ll 1$, we have $|Z_s| \ll 1$. Since

$$Z_s' = Z_s + \frac{2j\pi y'}{B},$$

$$1 - e^{-Z_s} = Z_s[1 + O(Z_s)],$$

$$1 - e^{-Z_s'} = (1 - e^{-2j\pi y'/B})[1 + O(Z_s)], \tag{7.13}$$

it follows that

$$|1 - e^{-Z_s}|^2 = \left(\frac{\pi r}{B}\right)^2\left[1 + O\left(\frac{r}{B}\right)\right],$$

$$|1 - e^{-Z_s}|^2 = 4\sin^2\frac{\pi y'}{B}\left[1 + O\left(\frac{r}{B}\right)\right]. \tag{7.14}$$

Substituting these formulas into Eq. (7.11), we get

$$\tilde{G} = \frac{1}{\pi}\ln\left(\frac{2\pi r}{B}\sin\frac{\pi y'}{B}\right) + O\left(\frac{r}{B}\right) \tag{7.15}$$

which is logarithmically singular as $r \to 0$. This is an expected result since $r = 0$ is the source point. From Eq. (7.15) the flux through an infinitesimal half-circle surrounding the source point on the side $x > 0$ is unity. Thus, $\tilde{G}$ represents the singular part of the Green function, and the residual series in Eq. (7.10) must be regular at the source point $r = 0$. The leading inner expansion of G is, therefore,

$$G(x, y; y') \cong \frac{1}{\pi}\ln\left(\frac{2\pi r}{B}\sin\frac{\pi y'}{B}\right) + F, \tag{7.16}$$

where F is the value of the residual series evaluated at the source point

$$F = \sum_{n=1}^{\infty}\left(\frac{\varepsilon_n\cos K_n L}{K_n B\sin K_n L} + \frac{2}{n\pi}\right)Y_n(y')Y_n(y') + \frac{\cos kL}{kB\sin kL}. \tag{7.17}$$

Finally, the inner expansion of the outer solution is

$$\eta_H \cong i\frac{\omega}{g}Q_H\left[\frac{1}{\pi}\ln\left(\frac{2\pi}{B}\sin\frac{\pi y'}{B}\right) + F\right] + i\frac{\omega}{g}\frac{Q_H}{\pi}\ln r. \tag{7.18}$$

Now we can perform the matching. On the ocean side the constant terms and the $\ln r$ terms in Eqs. (7.2) and (7.3a) must match separately; two equations are obtained

$$C - M\ln\frac{a}{2} = 2A + i\frac{\omega}{g}Q_0\left(\frac{-i}{2} + \frac{1}{\pi}\ln\frac{\gamma k}{2}\right), \tag{7.19}$$

and

$$M = i\frac{\omega}{g}\frac{Q_0}{\pi}. \tag{7.20}$$

Similarly, by matching Eqs. (7.3b) and (7.18) on the harbor side $x < 0$

$(x_1 > 0)$, we obtain

$$C + M \ln \frac{a}{2} = i \frac{\omega}{g} Q_H \left[\frac{1}{\pi} \ln \left(\frac{2\pi}{B} \sin \frac{\pi y'}{B} \right) + F \right], \tag{7.21}$$

$$- M = i \frac{\omega}{g} \frac{Q_H}{\pi}. \tag{7.22}$$

These four algebraic equations (7.19)–(7.22) can be easily solved for the unknowns C, Q_0, Q_H, and M. An immediate result is that $Q_0 = -Q_H$ which could have been anticipated on grounds of continuity. The most important result is

$$i \frac{\omega}{g} Q_0 = \frac{-i\omega Q_H}{g} = -2A \left[\frac{-i}{2} + F - I \right]^{-1}, \tag{7.23}$$

where

$$I = \frac{1}{\pi} \ln \left[\frac{4B}{(\pi \gamma k a^2 \sin(\pi y'/B))} \right]. \tag{7.24}$$

Finally, the far-field solution in the harbor is

$$\eta_H = \frac{-2A}{-i/2 + F - I} \sum_n X_n(x) Y_n(y) Y_n(y') \tag{7.25}$$

which can be used to calculate numerically the harbor response at almost all points except within a small region a wavelength from the entrance.

5.7.2 Resonant Spectrum and Response for Non-Helmholtz Modes

In order to enhance the physical understanding of the numerical results to be presented later, it is useful to examine the formulas (7.23) and (7.25) approximately.

When the incident wavenumber k is close to one of the natural modes of the closed basin, $k_{nm} = [(n\pi/B)^2 + (m\pi/L)^2]^{1/2}$, resonance should be expected. In the neighborhood of k_{nm} let

$$k = k_{nm} + \Delta, \tag{7.26}$$

and assume that

$$\frac{\Delta}{k_{nm}} \ll 1. \tag{7.27}$$

From Eq. (7.8c) we have

$$K_n L = L \left[(k_{nm} + \Delta)^2 - \left(\frac{n\pi}{B} \right)^2 \right]^{1/2}$$

$$\cong m\pi + \frac{k_{nm} L^2 \Delta}{m\pi}, \qquad n = 0, 1, 2, \ldots, \quad m = 1, 2, 3, \ldots \quad (7.28a)$$

or

$$\cong L(2k_{n0}\Delta)^{1/2} \qquad m = 0, \ n = 1, 2, 3, \ldots \quad (7.28b)$$

Note that $\sin K_n L$ is nearly zero for either Eq. (7.28a) or (7.28b). In the series for G or F the nth term is much greater than all the rest. The dominant term of the series F is

$$F \cong \frac{\varepsilon_n \cos K_n L}{K_n B \sin K_n L} \cos^2 \frac{n\pi y'}{B} \cong \frac{c}{\Delta}, \qquad \text{with } c = \frac{\varepsilon_n \varepsilon_m}{2k_{nm} BL} \cos^2 \frac{n\pi y'}{B}$$

$$(7.29)$$

when at least one of the indices n or m is not zero, whereas G is approximated by

$$X_n Y_n(y) Y_n(y') \cong \frac{c}{\Delta} \frac{\cos(m\pi/L)(x - L)}{\cos m\pi} Y_n(y) Y_n(y').$$

Equation (7.25) gives the harbor response

$$\eta_H \cong \frac{-2A}{-i/2 + c/\Delta - I(k_{nm})} \frac{c}{\Delta} \frac{\cos(m\pi/L)(x - L)}{\cos m\pi} Y_n(y) Y_n(y')$$

$$(7.30)$$

where

$$I(k_{nm}) \equiv [I(k)]_{k=k_{nm}} \qquad (7.31)$$

is logarithmically large for small $k_{nm}a$. It is important to point out that when $\Delta = 0$, that is, $k = k_{nm}$, the right-hand side of Eq. (7.30) approaches

$$-2A \frac{\cos(m\pi/L)(x - L)}{\cos m\pi} Y_n(y) Y_n(y')$$

which is not large. Thus, the resonant mode does not coincide with the natural mode of the closed basin. The amplification factor $\mathcal{R}$ for mode (n, m) can be defined by

$$\frac{\eta_H}{2A} = \mathcal{Q}\frac{\cos(m\pi/L)(x-L)}{\cos m\pi}Y_n(y)Y_n(y'), \tag{7.32a}$$

where

$$\mathcal{Q} = \frac{c/I}{\Delta - c/I + \frac{1}{2}i\,\Delta/I}. \tag{7.32b}[§]$$

Equation (7.32b) has the same form as Eq. (4.15) for the model example. Therefore, the term $i\Delta/2I$ is associated with radiation damping which depends on the frequency shift Δ. Consider the square of the modal amplification factor

$$|\mathcal{Q}|^2 = \frac{(c/I)^2}{(\Delta - c/I)^2 + \Delta^2/4I^2}. \tag{7.33}$$

The minimum of the denominator occurs at

$$\Delta_{\text{peak}} = \frac{c}{I}\left(1 + \frac{1}{4I^2}\right)^{-1} \cong \frac{c}{I}.$$

Hence, the resonant wavenumber is slightly greater than the natural value k_{nm},

$$\tilde{k}_{nm} \cong k_{nm} + \frac{c}{I}. \tag{7.34}$$

The correction c/I decreases as the width of the harbor mouth decreases. The peak value of $|\mathcal{Q}|^2$ is

$$|\mathcal{Q}|^2_{\text{max}} \cong 4I^2. \tag{7.35}$$

It is easy to check that when

$$\Delta \cong \frac{c}{I} \pm \frac{c}{2I^2},$$

the square response is reduced to half the peak value. Thus, $c/2I^2$ is essentially the half-width of the resonant peak in the plot of $|\mathcal{Q}|^2$ versus k. By the definition of I, a reduction of the entrance width results in an increase in I, hence an increase in $\mathcal{Q}_{\text{max}}$ and a decrease in the width of the resonant peak. Although narrowing the harbor entrance reduces the change of precise tuning for resonance, the peak response, if perfectly tuned, is heightened. This

[§]Note that $\mathcal{Q}$ has a simple pole in the complex k plane; the pole is located slightly below the real k axis.

behavior is related to the radiation damping which corresponds to the term $i\Delta/I$ in Eq. (7.32). Being proportional to $c/2I^2$ at resonance, the radiation damping diminishes sharply as a decreases. Thus, energy escapes the harbor with greater difficulty, and amplification is understandably more severe. This result is again likely at variance with the intuition of the designer who would normally narrow the entrance for better protection and is a feature of the *harbor paradox*. Note that the sharpening of a resonant peak is such that the area under the curve is roughly

$$| \mathcal{C} |^2_{\max} \frac{c}{2I^2} = 2c \tag{7.36}$$

which is independent of the entrance width and decreases with higher resonant mode (increasing n or m) (Garrett, 1970). If the incident waves are a stationary random process with the spectrum $S_I(k)$, it may be shown that the mean square response in the harbor is proportional to

$$\int_0^\infty S_I(k) \, | \mathcal{C} |^2 \, (k) \, dk$$

as in the case of simple oscillators. To this integral the resonant peak at k_{nm} makes a contribution approximately equal to the product of $S_I(k_{nm})$ and the area of the $| \mathcal{C} |^2$ curve under the peak. Equation (7.36) then implies that the mean square response is unchanged as the harbor entrance narrows. This result is another feature of the *harbor paradox*.

By combining Eq. (7.29) with Eq. (7.23) the discharge per unit depth near a resonant peak through the harbor entrance is found

$$\frac{i\omega Q_0}{g} = \left(\frac{-2A}{-i/2 + c/\Delta - I} \right). \tag{7.37}$$

Note that when the entrance width a vanishes, $k = k_{nm}$ and $Q_0 = 0$ so that pressure is transmitted across the entrance but mass is not. At resonance the discharge is simply $4A$; the corresponding averaged current velocity through the entrance U_E is

$$| U_E | = \left| \frac{Q_0}{2a} \right|_{\max} = \frac{2gA}{\omega_{nm} a} = \frac{gAT_{nm}}{\pi a} \tag{7.38}$$

where ω_{nm} is a resonant frequency $= (gh)^{1/2} k_{nm}$, and T_{nm} is the corresponding period. As a numerical example we take $a = 200$ m, $A = 0.2$ m, and $T_{nm} = 10$ min, then

$$| U_E | = 6 \text{ m/s}.$$

If the local breakwater thickness is estimated to be 10 m, the maximum local

Reynolds number is 6×10^7. As the flow is oscillatory, the instantaneous Reynolds number varies from 0 to $O(10^8)$. In reality there must be a significant energy loss due to vortex shedding and turbulence. Equation (7.38) also shows that $|U_E|$ is inversely proportional to the entrance width a, suggesting that for narrower entrance width real fluid effects should be more important. Dissipation at the entrance will be pursued in Chapter Six.

5.7.3 The Helmholtz Mode

The preceding analysis [in particular, Eq. (7.28)] does not apply when $n = m = 0$ so that $k_{00} = 0$. Since X_0 and Y_0 are constants, the free surface rises and falls in unison throughout the basin, and hence corresponds to the *Helmholtz* or the *pumping mode*. From Eq. (7.8c) we have

$$\Delta = k \quad \text{and} \quad K_n = k = \Delta \tag{7.39}$$

instead of Eq. (7.28). The leading terms in the series for F and G are associated with the term $n = 0$, and the harbor response is

$$\eta_H \cong \frac{-2A}{-i/2 + 1/k^2BL - I} \frac{1}{k^2BL}. \tag{7.40}$$

The resonant wavenumber $\tilde{k}_{00}$ is approximately given by the root of the transcendental equation

$$\frac{1}{k^2BL} - I = 0 \quad \text{or} \quad k^2BL = \frac{1}{I}, \qquad k \equiv \tilde{k}_{00}, \tag{7.41}$$

since I depends on K. As a decreases, I increases so that k decreases. The square amplification factor is

$$|\mathcal{Q}|^2 = \frac{\left(1/k^2BL\right)^2}{\frac{1}{4} + \left(1/k^2BL - I\right)^2}. \tag{7.42}$$

The peak value of $|\mathcal{Q}|^2$ is approximately $4I^2$ when Eq. (7.41) is satisfied. The half-width of the peak is obtained by letting

$$\frac{1}{k^2BL} - I = \frac{1}{4} \quad \text{or} \quad k - \tilde{k}_{00} = \pm\frac{1}{4}\frac{1}{\left(BLI^3\right)^{1/2}}. \tag{7.43}$$

As a decreases, $\mathcal{Q}^2_{\max}$ increases and the width of the peak decreases. However, the area under the peak of the curve of $|\mathcal{Q}|^2$ versus k is proportional to

$$\frac{1}{4}\frac{4I^2}{(BLI^3)^{1/2}} = \left(\frac{I}{BL}\right)^{1/2}, \tag{7.44}$$

which increases, although mildly, with decreasing a. The severity of the harbor paradox is worse for the Helmholtz mode than for non-Helmholtz modes, suggesting that friction loss is more important at the entrance, as will be discussed in detail in Chapter Six.

5.7.4 Numerical Results and Experiments

The response of a square harbor has been computed from Eq. (7.25) for a wide range of wavenumbers by Ünlüata and Mei (1973); these results are consistent with those of Miles (1971) who used a different approximate analysis. Two different entrance widths are studied; see Fig. 7.3. The effects of reducing a are clearly in agreement with the analysis in Sections 5.7.2 and 5.7.3.

Approximate analytical theories for a circular harbor have been worked out by Miles (1971) and Lee (1971). In addition, Lee performed experiments which were in excellent agreement with the linear theory. The only significant discrepancy appeared for the lowest resonant peaks, near which friction was likely important; see Fig. 7.4. It should be pointed out that Lee's experiments were made in very deep water, $kh \gg 1$, and comparison with the long-wave theory was based on the analogy for linear theories with constant depth and vertical side walls. However, in the laboratory, nonlinearity cannot be easily avoided in shallow water, and discrepancies between shallow-water experiments and the linearized long-wave theory can be expected to be large.

5.7.5 Effects of Finite Entry Channel

For a harbor with a single basin, Carrier, Shaw, and Miyata (1971) found that the finite length of the entry channel, or the finite thickness of the breakwater at the entrance, has qualitatively the same effect as a narrowed entrance. This conclusion can also be proven analytically by matched asymptotics. For a junction whose thickness $2d$ is of the same order of magnitude as the width $2a$, the near-field solution can be pursued by means of the Schwarz–Christoffel transformation. The results have been given by Davey (1944) in terms of elliptic integrals and have been used by Guiney, Noye, and Tuck (1972) for the transmission of deep-water waves through a narrow slit. Ünlüata and Mei (1973, in the context of harbor oscillations) and Tuck (1975, in a survey of wave transmission through small holes) have shown that all results obtained for a thin gap can be reinterpreted for a thick-walled gap if an effective width a_e is introduced to replace the actual width. The effective width a_e is given parametrically (through v) by the following relations:

$$\frac{a_e}{a} = 2pv^{1/2}, \qquad \frac{d}{a} = -\frac{p}{2}(K'v'^2 - 2K' - 2E') \tag{7.45}$$

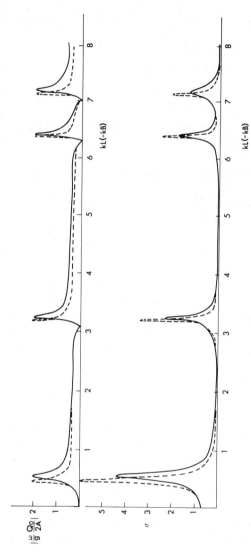

Figure 7.3 Root mean square response σ and normalized flux intensity $|(\omega/g)(Q_0/2A)|$ at the harbor mouth. $2a/B = 3 \times 10^{-2}$ (solid curve); $2a/B = 0.585 \times 10^{-2}$ (dashed curve). σ is defined in Eq. (9.5).

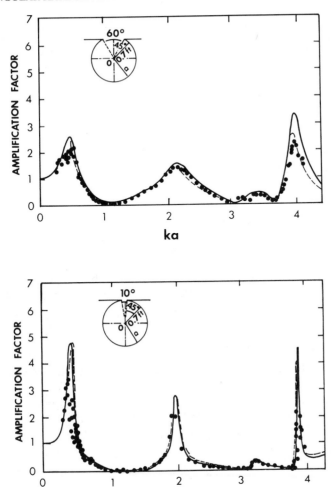

Figure 7.4   Response curve at $r = 0.7$ ft. $\theta = 45°$ of a circular harbor of radius $= 0.75$ ft. Solid and dashed curves are by two different theories; $\bigcirc$: experiment. (*a*) 60° opening; (*b*) 10° opening. (From Lee, 1971, *J. Fluid Mech*. Reproduced by permission of Cambridge University Press.)

where

$$p = \left(2E - v'^2 K\right)^{-1}$$

$$v' = \left(1 - v^2\right)^{1/2}$$

$$\left.\begin{array}{l} E = E(v) \\ K = K(v) \end{array}\right\} = \text{complete elliptic integrals of the} \left\{\begin{array}{l}\text{first}\\\text{second}\end{array}\right\} \text{kind}$$

$$E' \equiv E(v'), \qquad K' \equiv K(v').$$

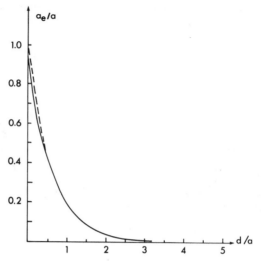

Figure 7.5 The ratio a_e/a of effective width to actual width of a junction as a function of the thickness-to-width ratio d/a. Equation (7.45): solid curve; Eq. (7.46): dashed curve. (From Mei and Ünlüata, 1978.)

We caution that the symbols E and K are conventional in the literature of elliptic integrals and do not have the same meaning as similar symbols used elsewhere in this section.

Details of the analysis leading to Eqs. (7.45) are rather complicated and can be found in Ünlüata and Mei (1973). We simply state that a_e/a decreases monotonically with d/a. A rather explicit approximation to Eqs. (7.45) valid for large d/a is

$$\frac{a_e}{a} \cong \frac{8}{\pi} \exp\left[-\left(\frac{\pi d}{2a} + 1\right)\right] \tag{7.46}$$

which is very accurate for $d/a > 0.5$. In fact, even for $d/a = 0$, Eq. (7.46) gives a fairly good result: $a_e/a = 0.937$, as shown in Fig. 7.5.

5.8 THE EFFECT OF PROTRUDING BREAKWATER

The coastline near the harbor entrance is often not a straight line, due either to the natural topography or to breakwaters protruding seaward. The latter is a common configuration for new and small harbors on shallow coasts. Physically, the protrusion alters the scattered waves when the harbor mouth is closed; therefore, the forcing agent at the mouth is different from that of a straight coastline. The radiation pattern is also altered, for now the harbor mouth is like a loudspeaker mounted on a protrusion from a baffle wall. The

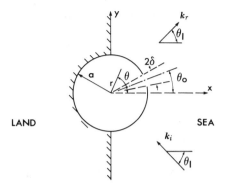

Figure 8.1 A circular harbor with protruding breakwater.

shape of the protrusion and the location of the harbor mouth become new factors which must be considered in design and/or operations.

To illustrate the effects of protrusion, we follow Mei and Petroni (1973) and consider a circular harbor with half of its area indented behind a straight coast. The breakwater is a semicircular arc with an opening centered at $\theta = \theta_0$. The harbor mouth subtends an angle 2δ. The geometry is shown in Fig. 8.1.

5.8.1 Representation of Solution

Consider the representation of the solution outside the harbor. We shall account for the straight coast first, then the semicircular breakwater, and finally the harbor mouth. Let the incidence angle be θ_I as shown in Fig. 8.1.

A perfectly reflective coast along the y axis introduces a reflected wave $\eta^{I'}$ in addition to the incoming wave η^I. In terms of partial waves, the sum is

$$\eta^I + \eta^{I'} = A \left[e^{ikr\cos(\theta + \theta_I)} + e^{ikr\cos(\theta - \theta_I)} \right]$$

$$= A \sum_m \varepsilon_m \left[(-i)^m \cos m(\theta + \theta_I) + (i)^m \cos m(\theta - \theta_I) \right] J_m(kr)$$

$$= 2A \sum_m \varepsilon_m \left(\cos \frac{m\pi}{2} \cos m\theta_I \cos m\theta + i \sin \frac{m\pi}{2} \sin m\theta_I \sin m\theta \right) J_m(kr),$$

$$m = 0, 1, 2, \ldots \quad (8.1)$$

It is easy to verify that there is no normal velocity on the y axis, that is,

$$\frac{1}{r} \frac{\partial}{\partial \theta} (\eta^I + \eta^{I'}) = 0, \qquad \theta = \pm \frac{\pi}{2}. \quad (8.2)$$

The presence of a solid circular cylinder centered at the origin creates scattered waves radiating outward to infinity. Thus, we must add to Eq. (8.1) terms which are proportional to $H_m(kr)(\genfrac{}{}{0pt}{}{\cos m\theta}{\sin m\theta})$, where H_m are the Hankel functions of the first kind. The coefficients must be chosen so that the sum of $\eta^I + \eta^{I'}$ and the scattered η^S satisfy

$$\frac{\partial \eta^0}{\partial r} = 0, \qquad r = a, \tag{8.3}$$

where

$$\eta^0 = \eta^I + \eta^{I'} + \eta^S \tag{8.4}$$

is the solution to the diffraction problem for a circular peninsula on a straight coast. The result is

$$\eta^0(r, \theta) = A \sum_m 2\varepsilon_m \left(\cos \frac{m\pi}{2} \cos m\theta_I \cos m\theta + i \sin \frac{m\pi}{2} \sin m\theta_I \sin m\theta \right)$$

$$\times \left[J_m(kr) - \frac{J_m'(ka)}{H_m'(ka)} H_m(kr) \right], \qquad r \geq a \tag{8.5}$$

in which $(\ )'$ denotes differentiation with respect to the argument. Note that the boundary condition on the coastline is still satisfied. The preceding solution can be thought of as two plane waves incident symmetrically from the opposite sides of the y axis and scattered by a circular cylinder in an open sea. The acoustic counterpart for one incident wave is well known (Morse and Feshbach, 1953, Vol. II, pp. 1387ff).

To complete the wave field outside the harbor, we must further correct for the piston action at the harbor entrance. The corresponding displacement can be formally given as

$$\eta^R = \sum_m \frac{H_m(kr)}{kaH_m'(ka)} (A_m \cos m\theta + B_m \sin m\theta), \qquad R \geq a \tag{8.6}$$

in which A_m and B_m are to be found.

In summary, the total field outside the harbor is

$$\eta_0 = \eta^0 + \eta^R, \qquad r \geq a, \qquad |\theta| < \tfrac{1}{2}\pi. \tag{8.7}$$

The displacement inside the harbor, η_H, must satisfy the Helmholtz equation so that the formal solution is

$$\eta_H(r, \theta) = \sum_m \frac{J_m(kr)}{kJ_m'(ka)} (C_m \cos m\theta + D_m \sin m\theta) \tag{8.8}$$

in which C_m and D_m are yet unknown.

5.8.2 Reduction to an Integral Equation

The unknown coefficients A_m, B_m, C_m, and D_m must be fixed so that the surface height (i.e., pressure) and the normal surface slope (i.e., normal velocity) are continuous at every point of the harbor entrance

$$\eta_0 = \eta_H, \tag{8.9}$$

$$\frac{\partial \eta_0}{\partial r} = \frac{\partial \eta_H}{\partial r}, \qquad r = a, \quad |\theta - \theta_0| \le \delta. \tag{8.10}$$

Let the surface slope in the radial direction at the entrance be formally $F(\theta)$

$$\frac{\partial \eta_0}{\partial r} = \frac{\partial \eta_H}{\partial r} = F(\theta) \tag{8.11}$$

which differs from zero only across the entrance $|\theta - \theta_0| \le \delta$. Applying Eq. (8.11) to Eq. (8.8) and using the theory of Fourier series, we can express the coefficients C_m and D_m as

$$C_m = \frac{\varepsilon_m}{2\pi} \int_M F(\theta') \cos m\theta' \, d\theta', \tag{8.12a}$$

$$D_m = \frac{\varepsilon_m}{2\pi} \int_M F(\theta') \sin m\theta' \, d\theta', \tag{8.12b}$$

in which M stands for harbor *mouth*.

Now apply Eq. (8.11) to Eq. (8.7). Some care is required for A_m and B_m as the conditions $\partial \eta_0 / \partial \theta = 0$ on $\theta = \pm \pi/2$, and $r > a$ must not be violated. The physical region outside the harbor is to the right of $x = 0$ (i.e., $|\theta| < \pi/2$). However, as far as the outside of the harbor is concerned, the problem is equivalent to an offshore circular harbor with two mouths symmetrically located with respect to the y axis, attacked by two symmetrically incident waves at the angles θ_I and $\pi - \theta_I$. The corresponding boundary condition is then

$$\frac{\partial \eta_0}{\partial r} = \frac{\partial \eta^R}{\partial r} = F(\theta), \qquad 0 \le \theta \le 2\pi, \quad r = a \tag{8.13a}$$

where

$$F(\theta) = F(\pi - \theta). \tag{8.13b}$$

With this condition the no-flux condition on the coast (y axis) is assured by symmetry.

Applying Eq. (8.13) to Eq. (8.6) for the complete range of $0 \leq \theta \leq 2\pi$, and using the symmetry property of $F(\theta)$, Eq. (8.13b), we obtain the Fourier coefficients

$$A_m = \frac{\varepsilon_m}{\pi} \cos^2 \frac{m\pi}{2} \int_M d\theta' F(\theta') \cos m\theta', \qquad (8.14a)$$

$$B_m = \frac{\varepsilon_m}{\pi} \sin^2 \frac{m\pi}{2} \int_M d\theta' F(\theta') \sin m\theta'. \qquad (8.14b)$$

Thus, $A_m = 0$ for $m =$ odd, and $B_m = 0$ for $m =$ even.

In terms of the surface slope, $F(\theta)$, we get from Eqs. (8.6) and (8.7),

$$\eta_0 = \eta^0 + a \sum_m \frac{\varepsilon_m}{\pi} \frac{H_m(kr)}{kaH'_m(ka)} \left[\cos^2 \frac{m\pi}{2} \cos m\theta \int_M du\, F(u) \cos mu \right.$$

$$\left. + \sin^2 \frac{m\pi}{2} \sin m\theta \int_M du\, F(u) \sin mu \right], \qquad r > a,$$

$$(8.15)$$

and from Eq. (8.8)

$$\eta_H = a \sum_m \frac{\varepsilon_m}{2\pi} \frac{J_m(kr)}{kaJ'_m(ka)} \left[\cos m\theta \int_M du\, F(u) \cos mu \right.$$

$$\left. + \sin m\theta \int_M du\, F(u) \sin mu \right], \qquad r < a. \quad (8.16)$$

Finally, condition (8.9) is invoked to match the surface displacement for all points at the harbor entrance $r = a$, $|\theta - \theta_0| \leq \delta$, leading to an integral equation for $F(\theta)$:

$$\int_M du\, F(u) K(\theta \,|\, u) = \frac{1}{a} \eta^0(a, \theta), \qquad |\theta - \theta_0| \leq \delta \qquad (8.17a)$$

in which the kernel K is

$$K(\theta \,|\, u) = K(u \,|\, \theta) = \sum_m \frac{\varepsilon_m}{2\pi} \frac{J_m(ka)}{kaJ'_m(ka)} (\cos m\theta \cos mu + \sin m\theta \sin mu)$$

$$- \sum_m \frac{\varepsilon_m}{\pi} \frac{H_m(ka)}{kaH'_m(ka)}$$

$$\times \left(\cos^2 \frac{m\pi}{2} \cos m\theta \cos mu + \sin^2 \frac{m\pi}{2} \sin m\theta \sin mu \right). \quad (8.17b)$$

Note the important property that the kernel is symmetric with respect to the interchange of u and θ.

The Wronskian identity (9.20), Chapter Four, can be used to rewrite the right-hand side of (8.17a)

$$\frac{1}{a}\eta^0(a, \theta) = \frac{A}{a}\sum_m \frac{2i}{\pi ka}\frac{2\varepsilon_m}{H'_m(ka)}$$

$$\times \left(\cos\frac{m\pi}{2}\cos m\theta_I\cos m\theta + i\sin\frac{m\pi}{2}\sin m\theta_I\sin m\theta \right),$$

$$|\theta - \theta_0| < \delta. \quad (8.18)$$

The crux of the problem is to solve $F(\theta)$ from Eq. (8.17) for every θ within $|\theta - \theta_0| \le \delta$. This can be done by a variety of numerical procedures, most of which lead to a set of finite algebraic equations. Alternatively, by recasting the integral equation as a variational principle, one may obtain a simple but optimal approximation whose numerical accuracy is best for narrow mouths, as has been demonstrated in the straight coast problem by Miles and Munk (1961). The latter approach is adopted below.

5.8.2 Approximate Solution by Variational Method

It can be shown that solving the integral equation (8.17) is equivalent to finding the extremum of the following functional (the proof is given in Appendix 5.C):

$$J[F(\theta)] = \frac{1}{2}\int_M\int F(\theta)K(\theta\,|\,u)F(u)\,d\theta\,du - \frac{1}{a}\int_M\eta^0(a, \theta)F(\theta)\,d\theta.$$

$$(8.19)$$

Although this variational principle can be used as the basis of the finite-element approximation, we take a less numerical approach and assume F to possess a certain form with a multiplicative parameter f_0, that is,

$$f(\theta) = f_0 f(\theta) \qquad (8.20)$$

with $f(\theta)$ prescribed; then

$$J = \frac{f_0^2}{2}\int_M\int f(\theta)K(\theta\,|\,u)f(u)\,d\theta\,du - \frac{f_0}{a}\int_M\eta^0(a, \theta)f(\theta)\,d\theta. \quad (8.21)$$

For J to be stationary, f_0 must be chosen such that $dJ/df_0 = 0$; thus

$$f_0 = \frac{(1/a)\int_M \eta^0(a,\theta)f(\theta)\,d\theta}{\int_M\int f(\theta)K(\theta\,|\,u)f(u)\,d\theta\,du}. \tag{8.22}$$

One reasonable choice for $f(\theta)$ is that

$$f = \frac{1}{\pi}\left[\delta^2 - (\theta - \theta_0)^2\right]^{-1/2} \tag{8.23}$$

which has the correct singularity at the tips and is particularly appropriate for a narrow mouth ($\ll$ wavelength). Intuitively, in the neighborhood of the narrow mouth, the term $k^2\eta$ can be neglected from the Helmholtz equation and Eq. (8.23) should be a good quasistatic approximation. For a wide mouth, Eq. (8.23) is not adequate and other methods must be used.

The identity

$$\frac{1}{\pi}\int_{\theta_0-\delta}^{\theta_0+\delta}\left[\delta^2 - (\theta - \theta_0)^2\right]^{-1/2}\begin{pmatrix}\cos m\theta \\ \sin m\theta\end{pmatrix} d\theta = \begin{pmatrix}\cos m\theta_0 \\ \sin m\theta_0\end{pmatrix}J_0(m\delta)$$

$$\tag{8.24}$$

changes the numerator of Eq. (8.22) to

$$\frac{A}{a}N \equiv \frac{1}{a}\int_M \eta^0(a,\theta)f(\theta)\,d\theta = \frac{A}{a}\sum_m \frac{2i}{\pi ka}\frac{2\varepsilon_m J_0(m\delta)}{H_m'(ka)}$$

$$\times\left(\cos\frac{m\pi}{2}\cos m\theta_I\cos m\theta_0 + i\sin\frac{m\pi}{2}\sin m\theta_I\sin m\theta_0\right). \tag{8.25}$$

Since η^0 is the wave pressure at the entrance in the absence of the opening, N is a weighted average of the forcing pressure. Note in particular the reciprocity between θ_0 and θ_I, that is, N is symmetrical with respect to the interchange of θ_0 and θ_I. Furthermore, the denominator of Eq. (8.22) is

$$D \equiv \int_M\int f(\theta)K(\theta\,|\,u)f(\theta)\,d\theta\,du = \sum_m \frac{\varepsilon_m}{2\pi}\frac{J_m(ka)J_0^2(m\delta)}{kaJ_m'(ka)}$$

$$- \sum_m \frac{\varepsilon_m}{2\pi}\frac{H_m(ka)J_0^2(m\delta)}{kaH_m'(ka)}(1 + \cos m\pi\cos 2m\theta_0). \tag{8.26}$$

The first series in Eq. (8.26) appears also in the case of a straight coast but the second series is now different. Substituting Eqs. (8.20) and (8.23) into Eq.

(8.16), and using Eqs. (8.22) and (8.24)–(8.26), we obtain the harbor response

$$\eta_H = A \frac{N}{D} \sum_m \frac{\varepsilon_m}{2\pi} \frac{J_0(m\delta)}{kaJ_m'(ka)} J_m(kr)\cos[m(\theta - \theta_0)]. \qquad (8.27)$$

The necessary numerical calculations merely involve summation of series.

Qualitative behavior of η_H near the natural modes can be examined analytically as seen in previous sections, and can be found in Mei and Petroni (1973).

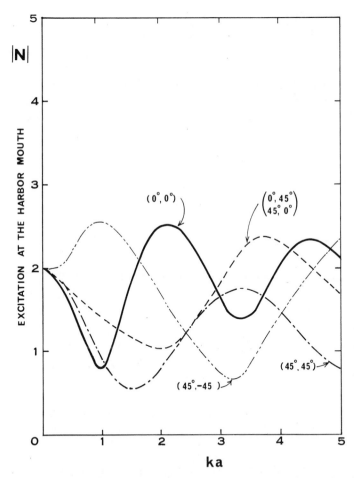

Figure 8.2 Excitation at the harbor mouth for an opening $2\delta = 10°$. Numbers in a parenthesis show (θ_0, θ_I) in degrees. (From Mei and Petroni, 1973, *J. Waterway, Port, Coastal and Ocean Div.* Reproduced by permission of American Society of Civil Engineers.)

5.8.4 Numerical Results

The excitation N at the mouth is plotted as a function of ka for several values of incidence angle θ_I and mouth position θ_0, in Fig. 8.2. In contrast with a straight coast, the most important new feature is the fluctuation with respect to ka, due to the complicated diffraction process. For a given ka and θ_I which characterize the incident wave, it is possible to orient the harbor mouth θ_0 so that the forcing is small.

Let us define

$$A_m = \frac{N}{D} \frac{\varepsilon_m}{2} \frac{J_0(m\delta)J_m(ka)}{kaJ'_m(ka)} \tag{8.28}$$

as the amplification factor for the modes with angular dependence $\cos m(\theta - \theta_0)$. Near a zero of $J'_m(ka)$, that is, $ka \cong j'_{ms}$, $s = 1, 2, \ldots, A_m$ is large and the standing wave mode (m, s) is resonated. From Section 5.3 the first few values of j'_{ms} are 0 (0, 1), 0.9184118 (1, 1), 3.05424 (2, 1), 3.83170 (0, 2), 4.20119 (3, 1), 5.33144 (1, 2)..., where the numbers in the parentheses refer to the index pair (m, s). Note that the mode $(0, 1)$ corresponds to the Helmholtz mode. Because

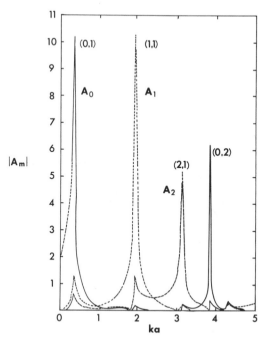

Figure 8.3 Amplification coefficient for mode m with $2\delta = 10°$, $\theta_0 = 0°$, $\theta_I = 0°$. (From Mei and Petroni, 1973, *J. Waterway, Port, Coastal and Ocean Div*. Reproduced by permission of American Society of Civil Engineers.)

of radiation damping, D is complex. The peak frequency is slightly shifted and A_m is finite at resonance. Figure 8.3 is a typical plot for $|A_m|$.

For any given set of ka, a, θ_0, and δ, the free-surface displacement at the point (r, θ) within the harbor can be calculated as soon as the amplification factors A_m are calculated. As a convenient measure of the overall response, the mean square elevation, averaged spatially over the entire area of the basin, is useful. From Eq. (8.27), the mean square

$$\overline{|\eta_H|^2} = \frac{1}{\pi a^2} \int_0^{2\pi} d\theta \int_0^a dr\, r\, |\eta_H|^2 = \frac{A^2}{\pi a^2 k^2} \sum_m \frac{|A_m|^2}{J_m^2(ka)} \frac{2\pi}{\varepsilon_m} \int_0^{ka} z J_m^2(z)\, dz$$

$$= \frac{4A^2}{(ka)^2} \sum_m \frac{1}{\varepsilon_m} \frac{|A_m|^2}{J_m^2(ka)} \sum_{n=0}^{m} (m + 2n + 1) J_{m+2n+1}^2(ka) \tag{8.29}$$

is obtained. The last integral is evaluated with the help of an identity in Abramowitz and Stegun (1972, p. 484).

For the following range of parameters: wavelength, $0 < ka < 5$; half-opening angle of harbor entrance, $\delta = 5°$; entrance position, $\theta_0 = 0°, 45°$; and the direction of incidence, $\theta_I = 0°, \pm 45°$, the root mean square of the harbor response $(|\eta_H|^2)^{1/2}$ is shown in Figs. 8.4a–8.4e.

Qualitatively, the response curves look the same in all cases with the peaks occurring at the expected places. However, the heights of the peaks differ quantitatively for different values of θ_0 and θ_I. As an example, consider the resonant mode (2, 1) near $ka = 3.1$. The peak heights decrease according to the

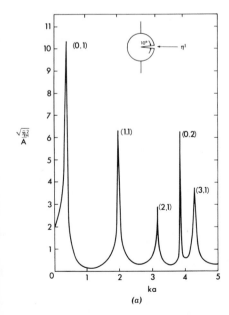

(a)

Figure 8.4 Mean harbor response for $2\delta = 10°$. (a) $\theta_0 = 0°$, $\theta_I = 0°$; (b) $\theta_0 = 0°$, $\theta_I = -45°$; (c) $\theta_0 = 45°$, $\theta_I = 0°$; (d) $\theta_0 = 45°$, $\theta_I = -45°$. (e) $\theta_0 = 45°$, $\theta_I = 45°$.

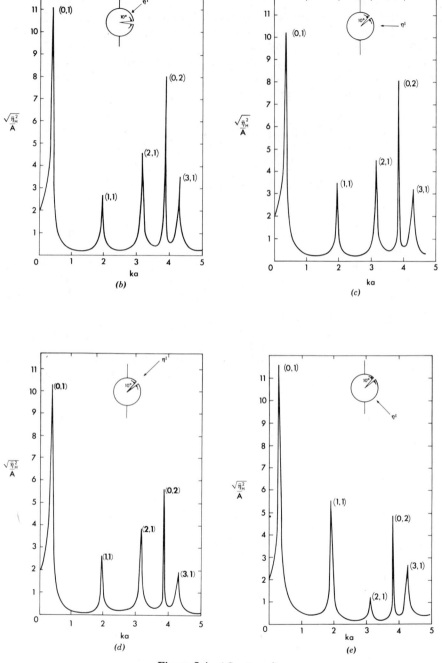

Figure 8.4 (*Continued*)

order ($\theta_0 = 45°$, $\theta_I = 0$), ($\theta_0 = 45°$, $\theta_I = 45°$), ($\theta_0 = 0$, $\theta_I = 0$), and ($\theta_0 = 45°$, $\theta_I = -45°$); the forcing N at the mouth is also in the same order as shown in Fig. 8.3. Thus, if the design wavelength is close to a particularly dangerous mode, proper siting of the entrance can reduce the input and the response.

Note that the Helmholtz mode is hardly affected by θ_0 and θ_I; this is consistent with Fig. 8.2 where $N \approx 2$ for small ka.

Comparing Figs. 8.4b (for $\theta_0 = 0$; $\theta_I = 45°$) and 8.4c (for $\theta_0 = 45°$, $\theta_I = 0°$), we see that the modes with $m = 0, 2$, namely, $(0, 1)$, $(0, 2)$, and $(2, 1)$, have the same peak height. This occurs because N is symmetric with respect to the interchange of θ_0 and θ_I [cf. Eq. (8.25)], while D is independent of θ_I[cf. Eq. (8.26)] and has the same value for the two sets of θ_0 and m.

Local responses may be calculated from Eq. (8.27) but are not pursued here.

5.9   A HARBOR WITH COUPLED BASINS

The total area of some harbors consists of two large basins connected by a narrow passage; the Long Beach Harbor in California is such an example. Several new features arise because of the added degree of freedom. For two equal circular basins with centers lying on a line normal to the coast, numerical and laboratory experiments were first conducted by Lee and Raichlen (1972) whose results showed a systematic doubling of resonant peaks in contrast to a harbor with only one basin. Further analytical studies were made by Mei and Ünlüata (1978) for narrow openings, using the method of matched asymptotics as in Section 5.7. Though straightforward, the analysis is necessarily lengthy and we only summarize here the approximate results for two equal rectangular basins. The configuration is shown in Fig. 9.1 where the widths of the harbor

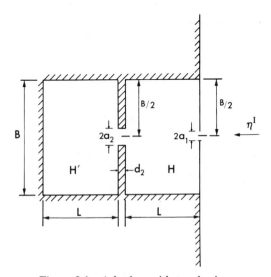

Figure 9.1   A harbor with two basins.

entrance and the inter-basin opening are denoted by $2a_1$ and $2a_2$, respectively.

Corresponding to the natural mode k_{nm} of a closed basin with n and m not both zero, there are two resonant peaks $\tilde{k}_{nm}^+$ and $\tilde{k}_{nm}^-$ with

$$\tilde{k}_{nm}^{\pm} \cong k_{nm} + c\left\{\frac{1}{I'} + \frac{1}{2I} \pm \left[\left(\frac{1}{I'}\right)^2 + \left(\frac{1}{2I}\right)^2\right]^{1/2}\right\}_{k=k_{nm}}, \qquad (9.1)$$

where

$$c = \frac{\varepsilon_n \varepsilon_m}{2k_{nm}BL}, \qquad (9.2)$$

and

$$I = \frac{1}{\pi} \ln \frac{4B}{\pi\gamma ka_1^2}, \qquad (9.3a)$$

$$I' = \frac{1}{\pi} \ln \frac{B^2}{\pi^2 a_2^2}. \qquad (9.3b)$$

The separation between the two peaks is

$$\tilde{k}_{nm}^+ - \tilde{k}_{nm}^- \cong 2c\left[\left(\frac{1}{I'}\right)^2 + \left(\frac{1}{2I}\right)^2\right]^{1/2}. \qquad (9.4)$$

When either a_1 or a_2 increases, I or I' decreases, hence the peaks move further apart. Corresponding to k_{00}, there are also two Helmholtz modes with similar dependence on the openings.

If we define σ_H^2 as the mean square amplification factor of the outer basin

$$\sigma_H^2 = \frac{1}{2BL} \int_0^L dx \int_0^B dy \frac{|\eta_H|^2}{4A^2}, \qquad (9.5)$$

and define $\sigma_H^{2'}$ similarly for the inner basin, it can be shown that

$$\frac{(\tilde{\sigma}_H^{2'})^+}{(\tilde{\sigma}_H^2)^+} \cong \frac{(\sigma_H^2)^-}{(\tilde{\sigma}_H^{2'})^-} \cong \left\{\frac{1}{\beta}\left[\left(\beta^2 + \frac{1}{4}\right)^{1/2} - \frac{1}{2}\right]\right\}_{k_{nm}}^2 \qquad (9.6)$$

where $\beta \equiv I/I'$ increases with decreasing harbor entrance ($2a_1$) or increasing inter-basin opening ($2a_2$). Since β lies between 0 and ∞, the following ordering is true

$$(\tilde{\sigma}_H^{2'})^- > (\tilde{\sigma}_H^2)^- > (\tilde{\sigma}_H^2)^+ > (\tilde{\sigma}_H^{2'})^+. \qquad (9.7)$$

For very large β, the ratios of mean squares in Eq. (9.6) tend to unity; the basin responses for both peaks are equalized. However, for small β these ratios become as small as 2β; the contrast of the basin responses then increases for both modes $\tilde{k}_{nm}^{\pm}$. In particular this implies that the inner basin becomes less protected for the lower mode $\tilde{k}_{nm}^{-}$. Clearly, this is a *paradox* associated with the coupling of basins in the context of the inviscid theory.

For the pair of Helmholtz modes corresponding to k_{00}, the ordering relation and the dependence on β is qualitatively the same. In addition, the two basins are out of phase for $\tilde{k}_{00}^{+}$ but in phase for $\tilde{k}_{00}^{-}$.

Numerical results based on the matched asymptotic theory are shown in Fig. 9.2 for two equal square basins. The width of the harbor entrance is fixed at $2a_1/B = 3 \times 10^{-2}$. In Fig. 9.2a we take $a_1 = a_2$ and the breakwaters have zero thickness. Within the computed range $0 < kB < 8$, the distinct natural modes of one basin H or H' are: $k_{01}B = \pi = 3.1415$, $k_{02}B = k_{20}B = 2\pi = 6.2833$, $k_{21}B = \sqrt{5}\,\pi = 7.0248$, which correspond to the second, third, and fourth pairs of peaks respectively, the first pair being the Helmholtz modes. Clearly, the ordering of the first three pairs of peaks obey Eq. (9.7). The

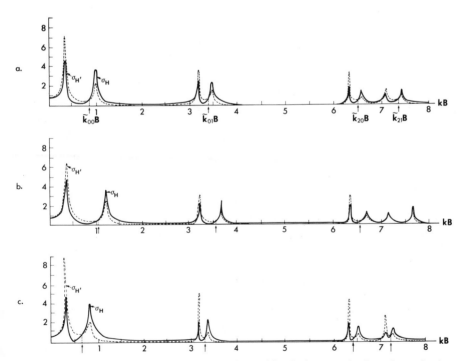

Figure 9.2 Root mean square responses of two identical square basins: Outer basin: solid curve; inner basin: dashed curve. $2a_1/B = 3 \times 10^{-2}$. (a) $a_1 = a_2$, $d_1 = d_2 = 0$; (b) $a_2 = 4a_1$, $d_1 = d_2 = 0$; (c) $a_1 = a_2$, $d_1 = 0$, $d_2 = 2a_1$. (From Mei and Ünlüata, 1976, *J. Eng. Math.*, Reproduced by permission of Sijthoff and Noordhoff.)

ordering of the last pair of peaks is only in partial agreement with Eq. (9.7) because the parameter $\tilde{k}a$ is no longer sufficiently small (> 0.107).

In Fig. 9.2b the inter-basin opening is changed to $a_2 = 4a_1$; the thickness of all breakwaters is still zero. Let us examine the lowest three pairs of peaks. In comparison with Fig. 9.2a the separation between a pair $\tilde{k}^{\pm}$ is indeed increased, and for the same mode the difference between the response of the basins is reduced, in accordance with Eq. (9.6). Note that for the fourth and highest pair of peaks the ordering rule Eq. (9.7) has deteriorated further, for now $\tilde{k}a_2 > 0.426$.

In Fig. 9.2c we keep $a_1 = a_2$ but increase the thickness of the breakwaters which divide the two basins, from zero to $d_2 = 2a_2$. In accordance with the results in Section 5.7.5 the effective width a_{2e} is reduced. In comparison with Fig. 9.2a the highest pair of peaks now obey the ordering rule since ($\tilde{k}a_{2e}$) is reduced to ~ 0.027. Moreover, the separation between pairs of peaks at $\tilde{k}^{\pm}$ decreases, while for the same mode ($\tilde{k}^{+}$ or $\tilde{k}^{-}$) the difference between the resonant responses of the basins is increased.

These features are consistent with the numerical and experimental findings of Lee and Raichlen (1972).

5.10 A NUMERICAL METHOD FOR HARBORS OF COMPLEX GEOMETRY

For harbors of constant depth but arbitrary plan form, numerical solutions have been obtained by Hwang and Tuck (1970) and Lee (1970) by the method of integral equations. Extension of integral equations for varying depth is rather complicated and expensive (Lautenbacher, 1970; Mattioli, 1978). The hybrid-element method (HEM) discussed in Section 4.11 can be modified to account for the coastline and other special features of harbors and is particularly well suited for variable harbor depth.

The coastline causes a reflected wave and modifies the scattered waves. Assume for simplicity that all topographical irregularities are within a contour C, and the coast is otherwise straight and coincides with the x axis, as shown in Fig. 10.1. In the superelement $\bar{\Omega}$ outside C, the depth is assumed to be a constant. Now the total wave in $\bar{\Omega}$ must consist of an incident wave, a reflected

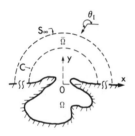

Figure 10.1 Harbor with a straight coast.

wave due to the straight coast, and a scattered wave. Thus,

$$\eta = \eta^I + \eta^S,$$

where

$$\eta^I = e^{ikr\cos(\theta - \theta_I)} + e^{ikr\cos(\theta + \theta_I)}$$

$$= 2 \sum_{n=0}^{\infty} \varepsilon_n(i)^n J_n(kr)\cos n\theta_I \cos n\theta, \tag{10.1}$$

and

$$\eta^S = \sum_{n=0}^{\infty} \alpha_n H_n(kr)\cos n\theta \quad \text{so that} \quad \frac{1}{r}\frac{\partial \eta}{\partial \theta} = 0, \qquad \theta = 0, \pi. \tag{10.2}$$

The same stationary functional (11.4), Chapter Four, still holds as long as one uses η^I and η^S as given here. The computational procedure remains the same.

Figure 10.2 shows the finite-element grid for a rectangular bay which was already studied by using matched asymptotics in Section 5.6.2. The response at

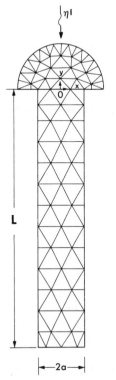

Figure 10.2 Network of finite elements for a rectangular bay. (From Mei and Chen, 1975, *Proc., Symp. on Modelling Techniques.* Reproduced by permission of American Society of Civil Engineers.)

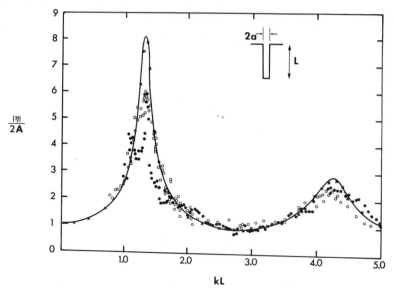

Figure 10.3 Amplification factor $|\eta|/2A$ at the inner end of an open rectangular bay, Solid curve: solution by integral equation (Lee, 1969); $\times\times\times$: solution by hybrid elements; $\bigcirc\bigcirc\bigcirc$: experiment by Lee (1969); $\bullet\bullet\bullet$: experiment by Ippen and Goda (1963). In both experiments, $2a = 2.38$ in., $L = 12.25$ in., and $h = 10.13$ in. (From Mei and Chen, 1975, *Proc., Symp. on Modelling Techniques.* Reproduced by permission of American Society of Civil Engineers.)

the inland end of the bay is shown in Fig. 10.3 for comparison with experiments performed in rather deep water. Discrepancy near the lowest peak suggests the importance of friction losses at the entrance.

If there is a thin breakwater, the velocity near the tip is very high so that local gradients are large. It is inefficient to increase the number of finite elements around the tip because ordinary interpolating functions cannot represent the singularity adequately. However, the hybrid element idea can again be applied by inserting a superelement Ω' centered at the tip (see Fig. 10.4). Within Ω', an analytical solution η' is used so that the singular behavior is accounted for exactly. The functional of Eq. (11.4), Chapter Four, must be modified by adding the following integral:

$$-\int_{C'} \left(\tfrac{1}{2}\eta' - \eta\right)\frac{\partial\eta'}{\partial n}\,ds, \tag{10.3}$$

where C' is the boundary of Ω', and **n** is the unit normal to C' pointing out of Ω'. Matching of η and $\partial\eta/\partial n$ is guaranteed as natural boundary conditions.

Figure 10.4 Neighborhood of the tip of a thin breakwater.

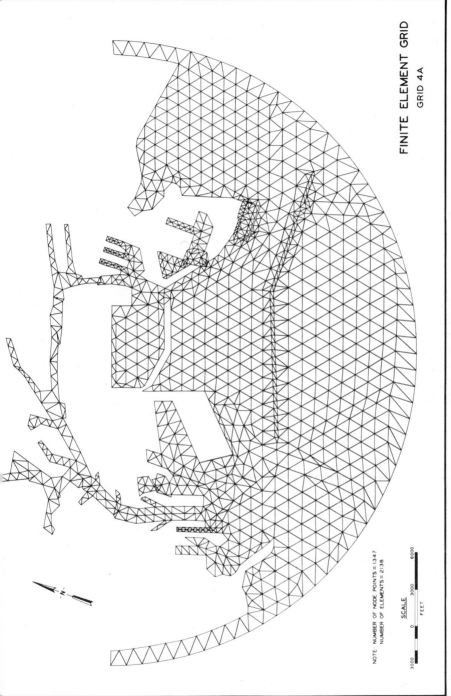

Figure 10.5 Finite-element grid for Long Beach Harbor. (From Houston, 1976, *Report, U.S. Army Waterways Experiment Station.*)

NOTE NUMBER OF NODE POINTS = 1347
NUMBER OF ELEMENTS = 2138

FINITE ELEMENT GRID

GRID 4A

For a thin breakwater the curve C' may be taken as a circle small enough so that the depth within is approximately constant. The proper form of η' is

$$\eta' = \sum_{n=0}^{\infty} \gamma_n J_{n/2}(kr')\cos\frac{n\theta'}{2}, \tag{10.4}$$

where r' and θ' are the local polar coordinates shown in Fig. 10.4. Equation (10.4) satisfies the Helmholtz equation and the no-flux conditions on the walls:

$$\frac{1}{r'}\frac{\partial\eta}{\partial\theta'} = 0 \quad \text{on} \quad \theta' = 0, 2\pi. \tag{10.5}$$

In addition, since

$$J_{1/2}(kr') = \left(\frac{2}{kr'}\right)^{1/2}\sin kr',$$

Eq. (10.4) has the right singular behavior. For other wedge-shaped corners with the wedge angle equaling a rational multiple of 2π, analytical representation similar to Eq. (10.4) can also be obtained (Chen and Mei, 1974).

Because of the versatility in treating arbitrary depth and boundary shape, the *hybrid element method* is a very powerful method for harbor studies. Indeed, its inception was in response to the proposed but now aborted offshore harbor project for the New Jersey Public Service Gas and Electric Company (Kehnemuyi and Nichols, 1973). Houston (1976) has made use of it in planning new piers and terminals in Long Beach Harbor; a sample finite-element grid is shown in Fig. 10.5.

5.11   HARBOR RESPONSE TO TRANSIENT INCIDENT WAVE

In Section 4.1.3 we have pointed out that the transient response in shallow water can be obtained by Fourier integration of the simple harmonic response. Let us now apply this procedure to the transient response in a harbor. Our analysis is adapted from Carrier (1970) who treated the elementary model of Section 5.4 in detail.

Let the transient incident wave be described by

$$\zeta^I(x, t) = \int_{-\infty}^{\infty} d\omega\, A_0(\omega)e^{-ikx-i\omega t} \tag{11.1}$$

where $A_0(-\omega) = A_0^*(\omega)$ for real ζ^I. The incident reflected wave system due to the straight coastline at $x = 0$ must be

$$\zeta^I + \zeta^{I'} = \int_{-\infty}^{\infty} d\omega\, 2A_0(\omega)(\cos kx)e^{-i\omega t}. \tag{11.2}$$

The harbor response may be written

$$\zeta_H = 2 \int_{-\infty}^{\infty} d\omega\, A_0(\omega) e^{-i\omega t} \eta_H(x, y, \omega) \tag{11.3}$$

where η_H is the frequency response to an incident wavetrain of unit amplitude. The path of integration should be slightly above the real axis so that $\zeta_H \to 0$ as $t \to -\infty$. Although all previous results were for positive and real ω and k, the results for negative and real ω can be inferred from the former by changing the sign of i, implying that

$$\eta_H(x, -\omega) = \eta_H^*(x, \omega). \tag{11.4}$$

Note first that to obtain the transient response by Fourier superposition it is, in principle, necessary to know $\eta_H(x, y, \omega)$ for the entire range of frequencies $-\infty < \omega < \infty$. This would require a numerical solution unrestricted for narrow entrance. Fortunately, tsunami inputs are usually of long periods and the harbor response is significant only in the lowest few modes; inaccuracy in the high-frequency range should not be essential. Therefore, it is reasonable to expect that the approximate long-wave theory developed in previous sections may be used without incurring gross error.

Consider a transient wave packet with a carrier frequency ω_0 and a slowly varying Gaussian envelope so that at $x = 0$

$$\zeta^I + \zeta^{I'} = 2Be^{-\Omega^2 t^2} \cos \omega_0 t, \qquad x = 0, \tag{11.5}$$

where

$$\frac{\omega_0}{\Omega} \gg 1. \tag{11.6}$$

The peak of the envelope strikes the coast at $t = 0$. The amplitude spectrum is easily found:

$$A_0(\omega) = \frac{B}{\pi^{1/2}} \frac{1}{2\Omega} \left\{ \exp\left[-\left(\frac{\omega - \omega_0}{2\Omega} \right)^2 \right] + \exp\left[-\left(\frac{\omega + \omega_0}{2\Omega} \right)^2 \right] \right\}.$$

Making use of Eq. (11.4), we can verify that

$$\zeta_H = \frac{B}{\pi^{1/2}} \frac{1}{\Omega} \mathrm{Re} \int_{-\infty}^{\infty} \left\{ \exp\left[-\left(\frac{\omega - \omega_0}{2\Omega} \right)^2 \right] \right\} \eta_H(x, \omega) e^{-i\omega t}\, d\omega,$$

or, equivalently,

$$\zeta_H = \frac{B}{\pi^{1/2}} \frac{1}{K} \mathrm{Re} \int_{-\infty}^{\infty} \left\{ \exp\left[-\left(\frac{k - k_0}{2K} \right)^2 \right] \right\} \eta_H(x, k) e^{-ik\tau}\, dk, \tag{11.7}$$

where

$$(k, k_0, K) = (\omega, \omega_0, \Omega)(gh)^{-1/2} \quad \text{and} \quad \tau = (gh)^{1/2}t. \qquad (11.8)$$

Equation (11.6) implies that

$$\frac{K}{k_0} \ll 1, \qquad (11.9)$$

so that the important part of the spectrum is narrow.

For simplicity only the rectangular bay in Section 5.5.2 is discussed here, but the analysis is much the same for other harbors where analytical solutions are available (see Carrier, 1970; Risser, 1976). Let us confine our attention to the point $x = -L$, that is, the landward end of the bay:

$$\zeta_H = \frac{B}{K\pi^{1/2}} \, \text{Re} \int_{-\infty}^{\infty} \mathcal{C}(k) \exp\left[-\frac{(k - k_0)^2}{4K^2} - ik\tau \right] dk, \qquad (11.10)$$

where

$$\mathcal{C} = \left[\cos kL + \frac{2ka}{\pi} \sin kL \ln \frac{2\gamma ka}{\pi e} - ika \sin kL \right]^{-1} \qquad (11.11)$$

is the amplification factor from Eq. (6.20). The preceding integral can be evaluated numerically. However, an analytical study for the present simple case here is both physically informative and useful in guiding numerical work. First of all, $\mathcal{C}(k)$ has a logarithmic branch point at $k = 0$. Since the path of integration is slightly above the real k axis, the branch cut must lie beneath the path and may be chosen as shown in Fig. 11.1. As a check recall that for negative and real ω one may replace i by $-i$ in Eq. (6.20) with the result

$$\mathcal{C}(\omega) = \left[\cos kL + \frac{2ka}{\pi} \sin kL \ln \frac{2\gamma ka}{\pi e} + ika \sin ka \right]^{-1},$$

$$k = \omega(gh)^{-1/2} < 0. \qquad (11.12)$$

Equation (11.12) can also be obtained from Eq. (6.20) by replacing k by $ke^{i\pi}$

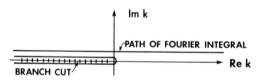

Figure 11.1   The complex plane for Fourier integral.

(not by $ke^{-i\pi}$). This particular choice of phase is consistent with the position of the branch cut.

Next, note that $\mathcal{C}(k)$ has poles in the lower half of the complex k plane at

$$\bar{k}_n = \pm \tilde{k}_n + i\hat{k}_n, \qquad \hat{k}_n < 0,$$

where $\tilde{k}_n$ corresponds to the nth resonant mode and $\hat{k}_n$ to the radiation damping rate.

The integral in Eq. (11.10) can be analyzed by the asymptotic *method of steepest descent*. Leaving the general exposition to many existing texts on applied mathematics (e.g., Carrier, Krook, and Pearson, 1966), we shall only explain this method for the problem at hand. Consider the phase function of the exponential in Eq. (11.10)

$$g(k) = -\frac{(k - k_0)^2}{4K^2} - ik\tau \qquad (11.13)$$

as an analytic function of k. Let $k = \alpha + i\beta$, then the contour lines of Re g and Im g are given by

$$\text{Re } g = \text{const:} \qquad (\alpha - k_0)^2 - (\beta + 2K^2\tau)^2 = \text{const} \qquad (11.14a)$$

$$\text{Im } g = \text{const:} \qquad (\alpha - k_0)(\beta + 2K^2\tau) = \text{const} \qquad (11.14b)$$

which are hyperbolas; contours of Re g are as shown in Fig. 11.2 for $\tau < 0$ and Fig. 11.3 for $\tau > 0$. The center of the hyperbolas is the point S where

$$k = k_0 - 2iK^2\tau = k_0 - 2iK\Omega t. \qquad (11.15)$$

Relative to the point S, the topography of Re g falls both to the east and west and rises both to the north and south; point S is therefore called the *saddle point*.[†] Because g is analytic, the contours of Im g are orthogonal to Re $g = $ const and are the paths along which Re g changes most rapidly; we call these the paths of the *steepest descent*. The strategy is to deform the original path to the steepest path so that the integrand is of significant magnitude only over a small stretch of the path.

Before the peak strikes the harbor mouth, $\tau < 0$, the saddle point is in the first quadrant. A closed rectangular contour is introduced in Fig. 11.2 with a horizontal path I_1 passing from the west valley over the saddle and down to the east valley. The integral along the two short vertical stretches at Re $k \to \pm \infty$ vanishes because of the exponential in the integrand. By Cauchy's theorem, $I_0 = I_1$. Now along path I_1, Im $g = 0$ and Re g drops off most rapidly on both sides of S; I_1 is a path of the *steepest descent*. Clearly, the most important

[†]In the most general situation a saddle point is defined by $dg/dz = 0$. Recall that an analytic function cannot have an extremum in the region of analyticity.

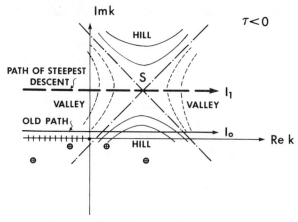

Figure 11.2　Topography near the saddle point for $\tau < 0$ (before the pulse strikes the coast). Vertical scale is exaggerated, $-2K^2\tau \ll k_0$, $K/k_0 \ll 1$.

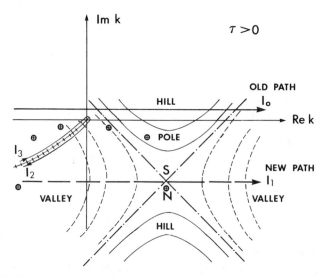

Figure 11.3　Topography near the saddle point for $\tau > 0$ (after the pulse strikes the coast). Branch cuts I_2 and I_3 follow the steepest path. N refers to a resonant pole.

contribution is from the neighborhood of S itself, where the rate of change of Re g is zero. Along I_1, let

$$\sigma = k - k_0 + 2iK^2\tau, \tag{11.16}$$

then

$$\zeta_H = \frac{B}{K\pi^{1/2}} \, \text{Re}\left\{ e^{-K^2\tau^2 - ik_0\tau} \int_{-\infty}^{\infty} d\sigma \exp\left(-\frac{\sigma^2}{4K^2}\right) \mathcal{Q}\Big|_{k=k_0-2iK^2\tau+\sigma} \right\}. \tag{11.17}$$

Because of the Gaussian factor $\exp(-K^2\tau^2)$ we need only confine our attention to $K\tau = O(1)$. Since $K/k_0 \ll 1$, the leading term in Eq. (11.17) is

$$\zeta_H \cong \frac{B}{K\pi^{1/2}} \operatorname{Re}\left\{ e^{-K^2\tau^2 - ik_0\tau}\mathcal{Q}(k_0)\int_{-\infty}^{\infty} d\sigma \exp\left(-\frac{\sigma^2}{4K^2}\right)\right\}$$

$$= 2B\operatorname{Re}\left\{\mathcal{Q}(k_0)e^{-\Omega^2 t^2}e^{-i\omega_0 t}\right\}, \qquad t,\tau < 0. \tag{11.18}$$

Thus, before the peak of the envelope strikes, the harbor responds passively with the same carrier frequency and similar envelope as the input.

Equation (11.18) is valid whenever the saddle point is at a distance greater than $O(K)$ from a pole of $\mathcal{Q}(k)$. Now all the poles of $\mathcal{Q}(k)$ correspond to resonances and have negative imaginary parts, $\hat{k}_n < 0$. If the lowest resonant mode has a $-\hat{k}_1 > O(K)$, that is, when the duration of the incident wave packet is much greater than the radiation damping time scale, Eq. (11.18) is valid up to $t = 0$ when the peak of the incident wave envelope strikes the coast. If the envelope is infinitely long, the peak of the envelope arrives at the coast only asymptotically in time. The input is effectively sinusoidal and the steady-state response is obtained by letting $\Omega \to 0$, while keeping t finite in Eq. (11.18); the result is, of course, already known.

After the peak of the incident packet strikes the coast, τ becomes positive and the saddle point S moves to the fourth quadrant. The most interesting case is when

$$k_0 = \tilde{k}_N, \tag{11.19}$$

that is, the incident wave has the same frequency as one of the lowest resonant modes N. For $K\tau \leq O(1)$ so that the wave packet is still appreciable, there is a time interval when S is just above the Nth pole, that is, $2K^2\tau < -\hat{k}_N$. In order to shift the integration path to a horizontal line passing through S, we must account for the contributions from the branch cut and from the poles lying above S and corresponding to modes with lower frequencies than the incident wave. Referring to Fig. 11.3, let us deform the branch cut so that it follows a path of the steepest descent. By Cauchy's theorem

$$\zeta_H = \zeta_H(I_1) + \zeta_H(I_2) + \zeta_H(I_3) - \frac{B}{K\pi^{1/2}}\operatorname{Re} 2\pi i \sum_n \text{residues}, \tag{11.20}$$

where $\zeta_H(I_\alpha)$ stands for Eq. (11.10) with I_α as the integration path. Along I_2 and I_3 which are the steepest paths, the neighborhood of $k = 0$ contributes the most to the integral. Because of the exponentially small factor $\exp(-k_0^2/4K^2)$, the corresponding integrals $\zeta_H(I_2)$ and $\zeta_H(I_3)$ are not important.

The residue for mode n can be obtained by first using Eq. (6.22) which is valid near the pole

$$\mathcal{Q} \cong (-1)^{n+1} \frac{1}{\left(k - \tilde{k}_n\right)L + ik_n a} \tag{11.21}$$

$$= \frac{1}{(-1)^{n+1}L} \frac{1}{k - \left(\tilde{k}_n + i\hat{k}_n\right)} \qquad \text{with } \hat{k}_n = -\frac{k_n a}{L} < 0, \tag{11.22}$$

where $\tilde{k}_n$ is given by Eq. (6.23). The residue is

$$\frac{B}{K\pi^{1/2}} \frac{-2\pi i}{(-)^{n+1}L} e^{-i\tilde{k}_n \tau} e^{\hat{k}_n \tau} \exp\left[-\frac{\left(\tilde{k}_n - k_0 + i\hat{k}_n\right)^2}{4K^2}\right], \tag{11.23}$$

which is also exponentially small for all $n \neq N$.

We now examine the term $\zeta_H(I_1)$. Extra care is needed since the saddle point S is close to the pole $\hat{k}_N + i\hat{k}_N$. Again using Eq. (11.22) we obtain from Eq. (11.10) that

$$\zeta_H(I_1) = \frac{B}{K\pi^{1/2}} \operatorname{Re} \int_{-\infty}^{\infty} \frac{\exp\left[-(k - k_0)^2/4K^2 - ik\tau\right]}{\left[k - \left(\hat{k}_N + i\hat{k}_N\right)\right](-)^{N+1}L} dk$$

which can also be written

$$\zeta_H(I_1) \cong \frac{B}{K\pi^{1/2}} \operatorname{Re} \frac{e^{-K^2\tau^2} e^{-ik_N\tau}}{(-)^{N+1}L} \int_{-\infty}^{\infty} \frac{e^{-\sigma^2/4K^2} d\sigma}{\sigma - i\left(2K^2\tau + \hat{k}_N\right)} \tag{11.24}$$

after the use of Eqs. (11.16) and (11.19). The integral above may be evaluated exactly (Carrier, 1970) as is detailed in Appendix 5.D. In particular, for $2K^2\tau + \hat{k}_N < 0$, the saddle point S is still above the pole,

$$\zeta_H(I_1) \cong \frac{B}{K\pi^{1/2}} \operatorname{Re} \frac{-i\pi e^{-i\tilde{k}_N\tau}}{(-)^{N+1}L} \left\{ e^{\hat{k}_N^2/4K^2} e^{\hat{k}_N\tau} \left[1 + \operatorname{erf}\left(\frac{2K^2\tau + \hat{k}_N}{2K}\right)\right]\right\}.$$

$$\tag{11.25}$$

When time has increased further so that $2K^2\tau + \hat{k}_N$ is positive, the integral along I_1 becomes

$$\zeta_H(I_1) = \frac{B}{K\pi^{1/2}} \operatorname{Re} \frac{i\pi e^{-i\tilde{k}_N\tau}}{(-)^{N+1}L} \left\{ e^{\hat{k}_N^2/4K^2} e^{\hat{k}_N\tau} \left[1 - \operatorname{erf}\left(\frac{2K^2\tau + \hat{k}_N}{2K}\right)\right]\right\}$$

$$\tag{11.26}$$

(see Appendix 5.D). We must now add the residue from pole N:

$$\mathrm{Re}\,\frac{-2iB\pi^{1/2}}{(-)^{N+1}KL}e^{i\tilde{k}_N\tau}e^{\hat{k}_N\tau}e^{\hat{k}_N^2/4K^2},\tag{11.27}$$

which follows from Eq. (11.23). The combined result is also given by Eq. (11.25). It is evident that the bay is excited at the frequency of mode N and attenuates according to the time scale of radiation damping $\propto|\hat{k}_N|^{-1}$. The maximum amplitude in the bay increases with increasing duration of the incident wave group (decreasing K). For a shorter wave group (larger K) the maximum not only is smaller but also occurs later.

The preceding analysis can be extended to harbor basins which are truly two dimensional. There the Helmholtz mode has the lowest resonant frequency and damping rate and may be excited by a simple pulse with nearly zero carrier frequency $\omega \approx 0$. Carrier and Shaw (1970) integrated numerically the harmonic response to get the transient response in a rectangular harbor with $B = 600$ ft, $L = 100$ ft, and $h = 21$ ft, which approximately represents Barber's Point Harbor in Oahu, Hawaii. For a very long pulse (pulse duration is 6.4 times the Helmholtz mode period), they found the response to be passive as shown in Fig. 11.4a. The reason is that the Fourier spectrum of a flat pulse is very sharp with $k_0 = 0$ and very small K; the residue from the Helmholtz pole is very small. However, for a very short pulse, the Helmholtz mode is excited, as shown in Fig. 11.4b. This excitation occurs because the Fourier spectrum of a short incident pulse is quite broad (K large) so that the terrain near the

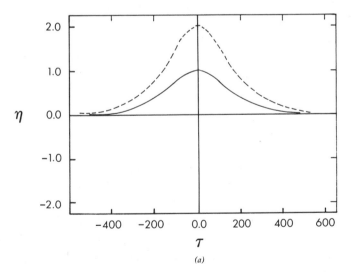

(a)

Figure 11.4 Transient responses to a single crested wave: (a) long pulse, (b) short pulse. Solid curve: incident wave; dashed curve: harbor reponse. (From Carrier and Shaw, 1970, *Tsunamis in the Pacific Ocean* edited by W. M. Adams. Reproduced by permission of University Press of Hawaii.)

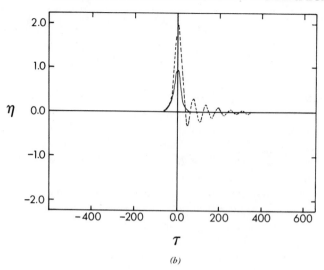

(b)

Figure 11.4   (*Continued*)

saddle is mild, and the residue from the Helmholtz pole is considerable. Finally, Fig. 11.5 shows the response of the Helmholtz mode excited by an incident wave packet. The growth to resonance for $t < 0$ and the reverberation and attenuation for $t > 0$ are qualitatively consistent with the analytical predictions of a narrow bay.

For two coupled square basins, analytical and numerical studies have been carried out by Risser (1976). In Fig. 11.6 a sample result by Risser is shown for the harbor system of Fig. 9.1a with $h = 20$ m and $B = L = 1000$ m. The carrier frequency of the wave packet is assumed to be $kB = 3.289$ which corresponds to the third lowest peak ($k_{01}^{-} B$ or $k_{10}^{-} B$). In physical terms the

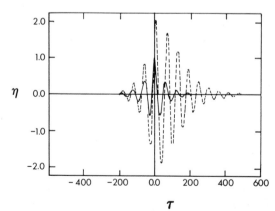

Figure 11.5   Transient response to a wave packet. (From Carrier and Shaw, 1970, *Tsunamis in the Pacific Ocean* edited by W. M. Adams. Reproduced by permission of University Press of Hawaii.)

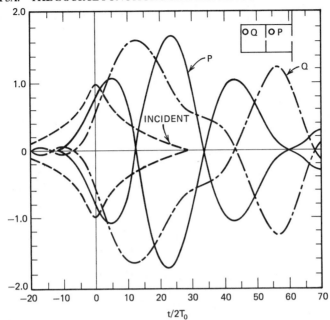

Figure 11.6 Sample transient responses in two identical square basins coupled in series. $2a_1 = 2a_2 = 3B \times 10^{-2}$, $k = 3.289 \times 10^{-3}$, $T_0 = 136$ s. Only the envelopes are shown (from Risser, 1976).

incident wave period is $T_0 = 136$ s. For a wave packet of duration $= 30T_0$, the envelopes of the responses at points P and Q as marked in the inset of Fig. 11.6 are shown. It is interesting that there is an oscillatory interchange of energy between the two basins and that the inner basin can have slightly higher response than the outer.

APPENDIX 5.A: THE SOURCE FUNCTION FOR A RECTANGULAR BASIN

In this appendix we deduce the source function $G(x, y; y')$ which satisfies the Helmholtz equation

$$(\nabla^2 + k^2)G = 0, \tag{A.1}$$

and the boundary conditions

$$\frac{\partial G}{\partial y} = 0 \qquad \text{on } y = 0, B, \quad 0 < x < L, \tag{A.2}$$

$$\frac{\partial G}{\partial x} = \begin{cases} 0, & x = L, \\ \delta(y - y'), & x = 0. \end{cases} \begin{matrix} \text{(A.3)} \\ \text{(A.4)} \end{matrix}$$

Thus, G represents a source on the wall at $x = 0$, $y = y'$.

We assume the solution to be in the form

$$G = \sum_{n=0}^{\infty} X_n(x)\cos \frac{n\pi y}{B}, \tag{A.5}$$

and note that it satisfies the zero-flux condition (A.2). Upon substituting Eq. (A.5) into Eq. (A.1), we have

$$\left(\frac{d^2}{dx^2} + K_n^2 \right) X_n = 0, \tag{A.6}$$

wherein

$$K_n^2 = k^2 - \left(\frac{n\pi}{B} \right)^2. \tag{A.7}$$

The solution of Eq. (A.6) satisfying Eq. (A.3) is simply

$$X_n = A_n \cos K_n(x - L). \tag{A.8}$$

To determine the unknown coefficients A_n in Eq. (A.8), we first substitute Eq. (A.8) into Eq. (A.5) and invoke the condition (A.4):

$$\sum_{n=0}^{\infty} A_n K_n \sin K_n L \cos \frac{n\pi y}{B} = \delta(y - y'). \tag{A.9}$$

Second, we multiply both sides of Eq. (A.9) by $\cos(n\pi y/B)$ and integrate with respect to y from 0 to B. By means of the orthogonality relation

$$\int_0^B \cos \frac{n\pi y}{B} \cos \frac{m\pi y}{B} dy = \frac{B\delta_{mn}}{\varepsilon_n}$$

where

$$\varepsilon_0 = 1, \qquad \varepsilon_n = 2, \qquad n = 1, 2, 3, \ldots, \tag{A.10}$$

the coefficients are found to be

$$A_n = \frac{\varepsilon_n}{BK_n \sin K_n L} \cos\left(\frac{n\pi y'}{B} \right). \tag{A.11}$$

Finally, the source function is given by

$$G(x, y; y') = \sum_{n=0}^{\infty} \frac{\varepsilon_n}{BK_n \sin K_n L} \cos K_n(x - L)\cos \frac{n\pi y}{B} \cos \frac{n\pi y'}{B}. \tag{A.12}$$

APPENDIX 5.B: SUMMATION OF THE G̃ SERIES

The series to be summed is

$$\tilde{G} = \sum_{n=1}^{\infty} \tilde{G}_n, \tag{B.1}$$

where

$$\tilde{G}_n = \tilde{X}_n(x) Y_n(y) Y_n(y'), \tag{B.2}$$

and

$$\tilde{X}_n = -\frac{2}{n\pi} e^{-n\pi x/B}, \tag{B.3}$$

$$Y_n = \cos \frac{n\pi y}{B}. \tag{B.4}$$

We let

$$\xi = \frac{\pi x}{B}, \qquad \eta = \frac{\pi y}{B}, \qquad \eta' = \frac{\pi y_0}{B}, \tag{B.5}$$

and note, by using trigonometric identities, that

$$Y_n(y) Y_n(y') = \cos n\eta \cos n\eta'$$
$$= \tfrac{1}{4}\{[e^{-jn(\eta-\eta')} + e^{-jn(\eta+\eta')}] + *\}, \tag{B.6}$$

where (*) denotes the complex conjugate of the preceding terms in curly parentheses. Upon substituting Eqs. (B.6) and (B.3) into Eq. (B.2), we have

$$\tilde{G}_n = -\frac{1}{2\pi}\left[\left(\frac{e^{-nZ_s}}{n} + \frac{e^{-nZ_s'}}{n}\right) + *\right], \tag{B.7}$$

where

$$Z_s = \xi + j(\eta - \eta'), \qquad Z_s' = \xi + j(\eta + \eta'). \tag{B.8}$$

The substitution of Eq. (B.7) into Eq. (B.1) yields four infinite series, each of which can be summed in closed form by the following formula (Collin, 1960, p. 579):

$$\sum_{n=1}^{\infty} \frac{e^{-ns}}{n} = -\ln(1 - e^{-s}), \tag{B.9}$$

yielding the result that

$$\tilde{G} = \sum_{n=1}^{\infty} \tilde{G}_n = \frac{1}{2\pi} \left\{ \left[\ln(1 - e^{-Z_s}) + \ln(1 - e^{-Z_s'}) \right] + * \right\}, \quad (B.10)$$

or

$$\tilde{G} = \frac{1}{2\pi} \ln \left\{ |1 - e^{-Z_s}|^2 \, |1 - e^{-Z_s'}|^2 \right\}. \quad (B.11)$$

APPENDIX 5.C: PROOF OF A VARIATIONAL PRINCIPLE

The functional J is defined as

$$J = \frac{1}{2} \int_M \int F(\theta) K(\theta \,|\, \theta') F(\theta') \, d\theta \, d\theta' - \frac{1}{a} \int_M F(\theta) \eta^0(\theta) \, d\theta \quad (C.1)$$

in which $F(\theta)$ is the solution to the integral equation

$$\int_M F(\theta') K(\theta \,|\, \theta') \, d\theta' = \frac{1}{a} \eta^0(\theta), \qquad \theta \text{ in } M \quad (C.2)$$

with a symmetric kernel

$$K(\theta \,|\, \theta') = K(\theta' \,|\, \theta). \quad (C.3)$$

Let us now prove that Eq. (C.2) is equivalent to the vanishing of the first variation of J, that is,

$$\delta J = 0. \quad (C.4)$$

Let an approximate solution be denoted by $\bar{F}$ which differs from the true solution F by δF, that is,

$$\bar{F} = F + \delta F \quad (C.5)$$

Thus, the approximate value of J is

$$\bar{J} = \frac{1}{2} \int_M \int (F + \delta F) K(F' + \delta F') \, d\theta \, d\theta' - \frac{1}{a} \int_M (F + \delta F) \eta^0 \, d\theta$$

$$= \frac{1}{2} \int_M \int [FF' + F'(\delta F) + F(\delta F')] K \, d\theta \, d\theta'$$

$$- \frac{1}{a} \int_M (F + \delta F) \eta^0 \, d\theta + O(\delta F)^2$$

$$= J + \frac{1}{2} \int_M \int (F' \delta F + F \delta F') K \, d\theta \, d\theta' - \frac{1}{a} \int_M \delta F \eta^0 \, d\theta + O(\delta F)^2 \quad (C.6)$$

in which the shorthand notation $F' = F(\theta')$ is used. The first variation of J is

$$\delta J = \bar{J} - J = \frac{1}{2} \int_M \int (F' \, \delta F + F \, \delta F') K \, d\theta \, d\theta' - \frac{1}{a} \int_M \delta F \, \eta^0 \, d\theta. \quad \text{(C.7)}$$

Due to the symmetry of K [cf. Eq. (C.3)],

$$\int_M \int F' \, \delta F K \, d\theta \, d\theta' = \int_M \int F \, \delta F' K \, d\theta \, d\theta', \quad \text{(C.8)}$$

it follows that

$$\delta J = \int_M d\theta \, F \int_M d\theta' \left[F(\theta') K(\theta \mid \theta') - \frac{1}{a} \eta^0(\theta') \right]. \quad \text{(C.9)}$$

Thus, the integral equation (C.2) implies Eq. (C.4). Conversely, if Eq. (C.4) is true for small but arbitrary δF, then Eq. (C.2) must be true.

APPENDIX 5.D: EVALUATION OF AN INTEGRAL

Carrier (1970) has given the following result:

$$J = \int_{-\infty}^{\infty} \frac{d\sigma \, e^{-\sigma^2/4K^2}}{\sigma + i\gamma} = -(\text{sgn } \gamma) i\pi e^{\gamma^2/4K^2} \left[1 - \text{erf}\left(\frac{|\gamma|}{2K} \right) \right], \quad \text{(D.1)}$$

which can be derived by contour integration (Risser, 1976). Consider first that $\gamma > 0$, and introduce $z = \sigma + i\gamma$. It follows that

$$J = \int_{-\infty+i\gamma}^{\infty+i\gamma} \frac{dz}{z} \exp\left[-\frac{(z^2 - 2i\gamma z - \gamma^2)}{4K^2} \right]. \quad \text{(D.2)}$$

The pole is now at the origin of the z plane while the path of integration is above the real z axis. By Cauchy's theorem, the integration path can be replaced by the real z axis indented above the origin.

Now break the integral into two parts: a principal-valued integral J_P

$$J_P = e^{\gamma^2/4K^2} \fint_{-\infty}^{\infty} \frac{dz}{z} e^{-(z^2 - 2i\gamma z)/4K^2}, \quad \text{(D.3)}$$

and an integral along the indentation:

$$J_\varepsilon = e^{\gamma^2/4K^2} \lim_{\varepsilon \to 0} \int_{\theta=\pi}^{\theta=0} \frac{d\varepsilon e^{i\theta}}{\varepsilon e^{i\theta}} = -i\pi e^{\gamma^2/4K^2}. \quad \text{(D.4)}$$

Denoting the principal-valued integral in (D.3) by F,

$$F = \oint_{-\infty}^{\infty} \frac{dz}{z} e^{-(z^2 - 2i\gamma z)/4K^2}, \tag{D.5}$$

we find

$$\frac{\partial F}{\partial \gamma} = \frac{i}{2K^2} \int_{-\infty}^{\infty} dz\, e^{-(z^2 - 2i\gamma z)/4K^2} = \frac{i}{2K^2} \int_{-\infty}^{\infty} dz\, e^{-(z - i\gamma)^2/4K^2} e^{-\gamma^2/4K^2}$$

$$= \frac{i}{2K^2} 2K\pi^{1/2} e^{-\gamma^2/4K^2} = \frac{i\pi^{1/2}}{K} e^{-\gamma^2/4K^2}. \tag{D.6}$$

When $\gamma = 0$, the integrand in Eq. (D.5) is odd in z; hence $F(\gamma = 0) = 0$. We may integrate Eq. (D.6) to get

$$F = \frac{i\pi^{1/2}}{K} \int_0^\gamma e^{-\gamma^2/4K^2}\, d\gamma = i\pi \frac{2}{\pi^{1/2}} \int_0^{\gamma/2K} e^{-\sigma^2}\, d\sigma = i\pi \operatorname{erf}\left(\frac{\gamma}{2K}\right). \tag{D.7}$$

Adding up J_p and J_ε, we get

$$J = J_p + J_\varepsilon = -i\pi e^{\gamma^2/4K^2}\left[1 - \operatorname{erf}\left(\frac{\gamma}{2K}\right)\right]. \tag{D.8}$$

If $\gamma < 0$, we write $\gamma = -|\gamma|$ which amounts to replacing i by $-i$ in the integrand of Eq. (D.1), hence the change of sign.

SIX

Effects of Head Loss at a Constriction on the Scattering of Long Waves: Hydraulic Theory

In the idealized theory of wave scattering in an inviscid fluid, it is usually assumed that the fluid flow is always tangential to the solid boundary of a wall or a structure. In reality, however, an adverse pressure gradient and viscosity can decelerate the fluid near a sharp convex corner, forcing the flow to separate and to form eddies of high vorticity and causing significant energy loss. This natural tendency is the basis of perforated panels on room walls to absorb sound energy. Jarlan (1965) introduced this idea to coastal engineering and patented the design of a caisson breakwater which has a perforated front wall but a solid back wall. Dissipation is enhanced by the water jets falling through the holes when the surface elevations on both sides differ. Breakwaters of similar types have been constructed at Baie Comeau Harbor and Chandler Harbor in Quebec, Canada, and at Roscoff Harbor in France (Richey and Sollitt, 1969). A more recent and dramatic example is the North Sea Ekofisk oil storage tank which is surrounded by a circular perforated breakwater with a diameter of approximately 92 m in water of 70 m (Gerwick and Hognestad, 1973). A breakwater consisting of a row of circular piles of 2 m diameter spaced with 0.5 cm gaps has been in use in Osaka Harbor, Japan (Hayashi, Kano and Shirai 1966).

In all these designs, flow separation due to sudden contraction and expansion is the primary physical feature. Now flow separation around a small cylinder is a related subject important in offshore structures and much experimental research has been done for isolated cylinders with smooth or sharp-edged boundaries (see, for example, Sarpkaya and Issacson, 1981). It is known from the experiments that the Strouhal number $U/\omega a$ (or equivalently the Keulegan–Carpenter number UT/a for the pioneering work of G. H. Keulegan and L. H. Carpenter, 1956, on oscillatory flows) is an important parameter, where U is the velocity amplitude and a the body dimension. According to Graham (1980), there appears to be at least two distinct regimes within the range of $2 < UT/a < 100$. For $UT/a > 20$ (the number depends on the cylinder cross-section), a limited wake containing a number of vortices extends downstream of the point of separation. With increasing UT/a, the wake lengthens and resembles more and more the steady stream situation of a Kármán vortex street. However, for $UT/a < 20$, vortices are shed from the separation points of the cylinder; each vortex is swept back by the reverse flow to the other side of the cylinder to pair up with a successive vortex of the opposite sign. This vortex pair is then convected away from the body at large angles ($\sim 45°$) to the incident flow. A friction loss formula which is quadratic in the local velocity [see Eq. (1.17) later] is satisfactory only for high values of UT/a. Unfortunately, a similar criterion is still lacking for gaps and holes. Since the Keulegan–Carpenter number can be rather large[†] for usual breakwater dimensions and wave periods, the quadratic loss formula provides at least a crude estimate until more experimental data become available. For one-dimensional scattering problems, such a semiempirical theory has been developed by Hayashi, Kano, and Shirai (1966), Terrett, Osorio, and Lean (1968), and others. Their reasoning has been further elaborated by Mei, Liu, and Ippen (1974) as described below.

6.1   ONE-DIMENSIONAL SCATTERING BY A SLOTTED OR PERFORATED BREAKWATER

6.1.1   The Field Equations

We limit ourselves to the study of small-amplitude waves in shallow water. Since the local velocity in the neighborhood of a sudden constriction can be large, let us include nonlinearity and begin with the equations of Airy, already given as Eqs. (5.11) and (5.12), Chapter Three:

$$\frac{\partial \zeta}{\partial t} + \nabla \cdot (\zeta + h)\mathbf{u} = 0, \tag{1.1}$$

$$\frac{\partial \mathbf{u}}{\partial t} + \mathbf{u} \cdot \nabla \mathbf{u} = -g\nabla\zeta. \tag{1.2}$$

[†]For tsunamis past a breakwater, we may take $U = 1$ m/s, $T = 3600$ s, and $a =$ breakwater thickness $= 10$ m, then $UT/a = 360$. For wind waves attacking a perforated breakwater we take $U = 3$ m/s for the velocity through the holes, $T = 10$ s, $a = 0.5$ m, then $UT/a = 60$.

Consider a thin barrier with vertical slots of width $2b$. The center-to-center spacing between any adjacent slots is $2a$. Periodicity allows us to consider a channel whose side walls coincide with the center lines of two adjacent vertical piles, as sketched in Fig. 1.1. The water depth h is assumed to be constant. The incoming wave is long, of low amplitude, and normally incident; hence it must satisfy

$$\frac{\partial \zeta}{\partial t} + h\frac{\partial u}{\partial x} = 0, \tag{1.3}$$

$$\frac{\partial u}{\partial t} + g\frac{\partial \zeta}{\partial x} = 0, \tag{1.4}$$

which are the limits of Eqs. (1.1) and (1.2) for $A/h \ll 1$. More explicitly, the incident wave may be written

$$\zeta^I = \tfrac{1}{2}A\big[e^{i(kx-\omega t)} + e^{-i(kx-\omega t)}\big], \tag{1.5}$$

$$u^I = \frac{gk}{\omega}\zeta^I. \tag{1.6}$$

It is assumed that the incident wave is not only long compared to the depth, but is even longer compared to the channel width, so that $ka \ll 1$.

Referring to Fig. 1.1, let the region at a distance $O(k^{-1})$ away from the constriction be called the *far field*. Because $ka \ll 1$, the flow is one dimensional and governed by Eqs. (1.3) and (1.4) on both sides of the barrier. Their solutions must be joined across the barrier by certain matching conditions which depend on the *near field*, defined as the neighborhood of $O(a)$ around the barrier.

6.1.2 The Matching Conditions and the Near Field

For sufficiently sharp constriction and moderate amplitude, flow is separated downstream of the barrier. A jet is formed which expands and collides with jets

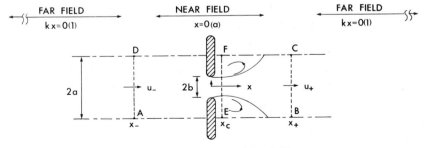

Figure 1.1 Near and far fields.

from adjacent slots to create two eddy zones. Further downstream the flow becomes nearly one dimensional again. We denote the outer limits of the near field by x_- and x_+ as shown in Fig. 1.1. The geometry is reversed with the reversal of flow direction.

In this complex near field the continuity equation (1.1) may be simplified. Even when the gaps or holes are absent, the wave amplitude can be $2A$ at most, hence ζ may be neglected relative to h. Furthermore, between x_- and x_+ the horizontal length scale is a, $\zeta = O(A)$, $t = O(\omega^{-1})$, and $u \gtrsim O((A/h)(gh)^{1/2})$; it follows that

$$\frac{\partial \zeta}{\partial t} \Big/ \nabla \cdot (h\mathbf{u}) \lesssim O(ka) \ll 1.$$

The continuity equation (1.1) becomes

$$\nabla \cdot \mathbf{u} \cong 0. \tag{1.7}$$

Outside the region of turbulent eddies the inviscid momentum equation applies. Since the velocity through the slots can be large, we retain the convective inertia term, that is, the entire Eq. (1.2) which is valid outside the eddy zone.

We now take the hydraulic approach by examining the differences of the cross-sectional averages between x_- and x_+. Upon integrating Eq. (1.7), it is obvious that

$$u_- S = u_c S_c = u_+ S, \tag{1.8}$$

where S is the gross area of the channel. S_c is the area at the *vena contracta* and is related to the net opening area S_0 by the empirical discharge coefficient c

$$S_c = cS_0; \tag{1.9}$$

u_c is the mean velocity at the *vena contracta*.

Outside the eddy zone, it is consistent with Eq. (1.2) to take $\mathbf{u}$ as irrotational so that $\mathbf{u} = \nabla\Phi$. A Bernoulli equation then holds:

$$\frac{\partial \Phi}{\partial t} + \frac{\mathbf{u}^2}{2} + g\zeta = \text{const.} \tag{1.10}$$

Applying Eq. (1.10) between x_- and x_c which is at the *vena contracta*, see Fig. (1.1), we get

$$\frac{\partial}{\partial t}(\Phi_c - \Phi_-) + \tfrac{1}{2}(u_c^2 - u_-^2) + g(\zeta_c - \zeta_-) = 0, \tag{1.11}$$

assuming that at both stations the transverse variation is negligible.

Downstream of the barrier we apply global momentum conservation to the control volume $EBCF$ in Fig. 1.1. In the wake of the solid barrier the mean fluid velocity is negligible and the free-surface height, hence the dynamic pressure, is essentially uniform in y, and is equal to that in the jet. Thus, the pressure force along EF is $\rho g \zeta_c S$. The balance of total momentum requires that

$$\rho g S(\zeta_c - \zeta_+) + \rho(u_c^2 S_c - u_+^2 S) = \rho \frac{\partial}{\partial t} \int_{x_c}^{x_+} S_J u \, dx, \qquad (1.12)$$

where S_J is the cross-section area of the jet. Subtracting Eq. (1.12) from Eq. (1.11) and invoking Eq. (1.8), we get

$$(\zeta_- - \zeta_+) = \frac{1}{2g} u_+^2 \left(\frac{S}{cS_0} - 1 \right)^2 + \frac{1}{g} \frac{\partial}{\partial t} \left[\int_{x_-}^{x_c} u \, dx + \int_{x_c}^{x_+} u \left(\frac{S_J}{S} \right) dx \right],$$

$$(1.13)$$

where the first integral above follows from the definition of Φ. If we introduce the loss coefficient f

$$f = \left(\frac{S}{cS_0} - 1 \right)^2, \qquad (1.14)$$

and the length L

$$L u_+ = \int_{x_-}^{x_c} u \, dx + \int_{x_c}^{x_+} u \left(\frac{S_J}{S} \right) dx, \qquad (1.15)$$

Eq. (1.13) may be rewritten

$$\zeta_- - \zeta_+ = \frac{f}{2g} u_+^2 + \frac{L}{g} \frac{\partial u_+}{\partial t}, \qquad u_+ > 0. \qquad (1.16)$$

If the argument is repeated for $u_+ < 0$, a negative sign appears in front of the first term on the right of Eq. (1.16). Accounting for both flow directions, we have

$$\zeta_- - \zeta_+ = \frac{f}{2g} u_+ |u_+| + \frac{L}{g} \frac{\partial u_+}{\partial t}. \qquad (1.17)$$

Once the coefficients f and L are determined empirically, Eqs. (1.8) and (1.17) provide the boundary conditions for the far-field solutions on two sides of the barrier. Since the far-field length scale is $O(k^{-1})$, these matching conditions can be approximately applied to the far-field solutions by letting $x \to \pm 0$.

If there is a solid wall at $x = l$, as in the case of a caisson with one wall perforated, the boundary condition $\partial \zeta_+ / \partial x = 0$ at $x = l$ must be added. If the region extends to $x \to \pm \infty$, then the scattered waves must be outgoing from $x = 0$ (radiation condition). In practice, the width of the caisson l is typically of order $O(10 \text{ m})$ and is shorter than a design wavelength. Use of Eqs. (1.9) and (1.17) for the perforated wall is therefore not entirely legitimate from a theoretical point of view.

As an aside, a formula may be deduced for the force acting on the solid barrier. Consider the fluid in $ABCD$ and outside the body and its wake. The net effect of pressure distribution on the upstream face of the body and in its wake is to produce a force $-F$ against the fluid in the control volume. Hence from momentum balance

$$S\left[\rho g(\zeta_- - \zeta_+) + \rho\left(u_-^2 - u_+^2\right)\right] - F = \rho \frac{\partial}{\partial t}\left[\int_{x_-}^{0} dx\, uS + \int_0^{x_+} dx\, uS_J\right],$$

it follows from Eqs. (1.8), (1.15), and (1.17) that

$$F = S\left[\rho g(\zeta_- - \zeta_+) - \rho L \frac{\partial u_+}{\partial t}\right]$$

$$= \tfrac{1}{2}\rho f S u_+ |u_+| . \tag{1.18}$$

6.1.3   The Coefficients f and L

It is well-known that in steady flows past sharp-edged orifices, the discharge coefficient c (thus f) depends primarily on the orifice geometry, if the Reynolds number is sufficiently high so that separation is a clear-cut feature. For a sharp-edged orifice, an empirical formula is

$$c = 0.6 + 0.4\left(\frac{S}{S_0}\right)^3 . \tag{1.19}$$

For thick or rounded edges, the discharge coefficient c is much closer to unity. According to this formula, c varies between 0.6 and 1. As the passing fluid slows down, the frequency of vortex shedding decreases. Thus, f and c should vary with the instantaneous velocity and acceleration, hence with Reynolds and Strouhal numbers. Since comprehensive experimental data on f and c are not available for this kind of oscillatory flows, the steady-state values are customarily used in the engineering literature. On the other hand, the length L is the most difficult to estimate. While Hayashi et al. (1966) neglected L altogether, Terrett et al. (1968) chose a constant value to fit the experiments. In the absence of separation, the boundary-value problem is linear; the corresponding length, denoted by L_0 herein, can be calculated by a two-dimensional theory. In fact, for long waves, L_0 is related to the transmission and reflection coefficients T and R as will now be shown. In the far field where the flow is one dimensional, the free-surface displacement is given by

$$\zeta_- = Ae^{i(kx-\omega t)} + ARe^{-i(kx-\omega t)}, \quad x < 0, \tag{1.20}$$

$$\zeta_+ = ATe^{i(kx-\omega t)}, \quad\quad\quad\quad x > 0. \tag{1.21}$$

Evaluating ζ_- at x_-, ζ_+ at x_+, and noting that $|kx_-|$, $kx_+ \ll 1$, we obtain

$$\zeta_- - \zeta_+ \simeq A(1 + R - T)e^{-i\omega t}. \tag{1.22}$$

At x_+ the velocity field on the right is

$$u_+ = \frac{gk}{\omega}\zeta_+ \simeq \frac{gk}{\omega}ATe^{-i\omega t}. \tag{1.23}$$

With Eq. (1.23), Eq. (1.22) may be written in the form

$$\zeta_- - \zeta_+ = \left[\frac{T - (1 + R)}{ikT}\right]\frac{1}{g}\frac{\partial u_+}{\partial t} = \frac{L_0}{g}\frac{\partial u_+}{\partial t}, \tag{1.24}$$

where L_0 is given by

$$L_0 = \left[\frac{T - (1 + R)}{ikT}\right]. \tag{1.25}$$

After multiplying by $\rho g S$, we can interpret Eq. (1.24) as Newton's second law for a mass of $\rho S L_0$ subject to the net force $\rho g S(\zeta_- - \zeta_+)$; the effect of the orifice is equivalent to adding a mass of $\rho S L_0$ at the section $x = 0$. The transmission and reflection coefficients must be found by a locally two-dimensional theory. Now long waves of small amplitude are exactly analogous to sound waves; analytical results known for several acoustic orifices (a slot in a rectangular duct, a circular hole in a circular pipe, and so on) may be applied here. As an example, for a thin two-dimensional slot, we have

$$\frac{L_0}{2a} \simeq \frac{1}{\pi}\ln\frac{1}{2}\left(\tan\frac{\pi b}{4a} + \cot\frac{\pi b}{4a}\right), \quad ka \ll 1 \tag{1.26}$$

(Morse and Ingard, 1968). Note that for long waves L_0 depends on k in the above approximation. The derivation of this formula is left as an exercise of matched asymptotics. In the limits of small and large gaps, we get

$$\frac{L_0}{2a} \simeq \frac{1}{\pi}\ln\frac{2}{\pi}\frac{a}{b} = \frac{1}{\pi}\ln\frac{2}{\pi}\frac{S}{S_0}, \quad \frac{b}{a} \ll 1,$$

$$\frac{L_0}{2a} \simeq \frac{\pi}{8}\left(\frac{a - b}{a}\right); \quad\quad \frac{a - b}{a} \ll 1. \tag{1.27}$$

Even for quite large S/S_0, $\ln(S/S_0)$ remains practically of $O(1)$; the ratio

$$\frac{L_0}{g\zeta}\frac{\partial u_+}{\partial t} = O(kL_0) \tag{1.28}$$

is very small for long waves. Thus, $\zeta_- \simeq \zeta_+$ and the constriction is ineffective. This result is related to the deduction in Section 5.5 that the transmission coefficient is nearly unity in the irrotational theory, unless S/S_0 is an enormous number.

With separation, L is obviously less tractable theoretically. Compared with the unseparated flow, separation reduces the curvature of the local streamlines around the gap. The local acceleration, which is responsible for the hydrodynamic reaction and hence the apparent mass, should be reduced also. In other words, we can expect the inviscid L_0 to be the upper bound of L. Now compare the inertia term to the friction loss term in Eq. (1.17)

$$\alpha = \frac{(L/g)(\partial u_+/\partial t)}{(f/2g)u_+|u_+|} = O\left(\frac{kL}{\frac{1}{2}fA/h}\right) = O\left(\frac{2kaL/2a}{\frac{1}{2}fA/h}\right).$$

Using L_0 in Eq. (1.26) for L, one sees that the aforementioned ratio is only important for relatively short waves. As the area ratio S/S_0 increases, $L/2a \sim \ln(S/S_0)$ and $f \sim (S/S_0)^2$, so that the preceding ratio diminishes rapidly as

$$\alpha \simeq \frac{(4/\pi)ka\ln(2/\pi)(S/S_0)}{f(A/h)(S/S_0)^2}. \tag{1.29}$$

Thus, for small gaps or holes, friction loss dominates for small but practical wave amplitudes. As a numerical example, assume that $A = 1$ m, $h = 10$ m, $T = 2\pi/\omega = 10$ s, $k = \omega/(gh)^{1/2} = 0.06/$m, and $c = 1$. For the perforated breakwater of the North Sea oil storage tank, $S/S_0 \simeq 2$ with hole diameter ≈ 1 m. We may take for estimates $a = 1$ m and $b = 0.5$ m as the equivalent channel and gap widths, respectively. For the Osaka pile breakwater, the values are close to $a = 1$ m, $b = 25$ mm, and $S/S_0 = 40$. The apparent orifice length is calculated according to Eq. (1.26) and the ratio α by Eq. (1.29) as given in Table 1.1. Note that even for wind waves, α is quite small and diminishes for larger amplitudes or longer waves, or both. Thus, in practical calculations there is an ample range of circumstances in which the inertia term can be ignored altogether.

Table 1.1   Ratio α According to Eq. (1.29).

a(m)	b(m)	$\dfrac{b}{a}$ or $\dfrac{S_0}{S}$	$\dfrac{L}{2a}$	f	$\dfrac{1}{2}\dfrac{fA}{h}$	α
1	0.5	0.5	0.22	1	00.5	0.1332
1	0.025	0.025	2.32	1.521	76	0.0018

Exercise 1.1

Use comformal mapping and matched asymptotics to verify Eq. (1.26).

6.1.4 Equivalent Linearization

The quadratic friction term in Eq. (1.17) makes the entire problem nonlinear, and the response to a simple harmonic input should contain many harmonics. If the response is dominated by the first harmonic as the input, as should be checked *a posteriori*, then the so-called *equivalent linearization* may be applied. Let the friction term be expressed in a linear form $c_e u$, that is,

$$\zeta_- - \zeta_+ = c_e u, \tag{1.30}$$

where c_e denotes the equivalent friction coefficient. We shall choose c_e in such a way that the mean square of the error

$$e = \frac{f}{2g} u|u| - c_e u \tag{1.31}$$

is minimum. Averaged over a period, the mean square is

$$\overline{e^2} = \overline{\left(\frac{f}{2g} u|u|\right)^2} - \frac{f}{g} c_e \overline{u^2|u|} + c_e^2 \overline{u^2}. \tag{1.32}$$

The minimum occurs when $\partial \overline{e^2}/\partial c_e = 0$, which determines the optimal c_e

$$c_e = \frac{f}{2g} \frac{\overline{u^2|u|}}{\overline{u^2}}. \tag{1.33}$$

The equivalent friction coefficient now depends on u, and is not known before the solution is completed. Alternatively, Eq. (1.33) can be obtained by requiring that the nonlinear friction force and the equivalent linear friction give the same energy loss per period. Approximating u by a simple harmonic, that is,

$$u \cong \tfrac{1}{2}\left(U_0 e^{-i\omega t} + U_0^* e^{-i\omega t}\right) = |U_0| \cos \omega(t+\tau) \tag{1.34}$$

where τ is the phase of U_0, that is, $U_0 = |U_0| e^{-i\omega \tau}$, we have

$$\overline{u^2} = \frac{\omega}{2\pi} \int_0^{2\pi/\omega} dt \, |U_0|^2 \cos^2 \omega(t+\tau) = |U_0|^2 \frac{1}{2\pi} \int_0^{2\pi} d\sigma \cos^2 \sigma,$$

$$= |U_0|^2 \pi$$

$$\overline{u^2|u|} = \frac{1}{2\pi} \int_0^{2\pi} d\sigma \cos^2 \sigma \, |\cos \sigma| \cdot |U_0|^3 = \tfrac{8}{3} |U_0|^3.$$

It follows that

$$c_e = \frac{f}{2g} \frac{8}{3\pi} |U_0|,$$ (1.35)

which depends on the amplitude of the motion.

6.1.5 Approximate and Exact Solutions

We first derive the approximate solution by using Eq. (1.30) instead of Eq. (1.17). The solution can be written in the following form:

$$\zeta_- = Ae^{-i\omega t}[e^{+ikx} + Re^{-ikx}],$$

$$\qquad\qquad\qquad\qquad\qquad\qquad x < 0 \qquad (1.36)$$

$$u_- = \frac{gk}{\omega} Ae^{-i\omega t}[e^{+ikx} - Re^{-ikx}],$$

$$\zeta_+ = ATe^{-i\omega t + ikx},$$

$$\qquad\qquad\qquad\qquad\qquad\qquad x > 0. \qquad (1.37)$$

$$u_+ = \frac{gk}{\omega} ATe^{-i\omega t + ikx},$$

Invoking continuity of velocity (1.8), we obtain

$$T = \frac{\omega}{gk} \frac{U_0}{A} = \frac{U_0}{(A/h)(gh)^{1/2}}$$ (1.38)

$$R = 1 - T = 1 - \frac{\omega}{gk} \frac{U_0}{A}.$$ (1.39)

Applying the head-loss condition (1.30), we have

$$A = \left(\frac{c_e}{2} + \frac{\omega}{gk}\right) U_0$$

which becomes

$$A = \frac{\omega}{gk} U_0 + \frac{2}{3\pi} \frac{f}{g} U_0 |U_0|$$ (1.40)

upon inserting Eq. (1.35). The phases of U_0 and A_0 are equal and may be taken to be zero, that is, $U_0 = |U_0|$. Equation (1.40) is a quadratic equation for U_0, which may be solved:

$$U_0 = \frac{A}{h}(gh)^{1/2}T = \frac{A}{h}(gh)^{1/2} \frac{(1 + 2\beta)^{1/2} - 1}{\beta}$$ (1.41)

where $\beta = (4/3\pi)(fA/h)$ (Hayashi et al., 1966).

For small-amplitude waves or a wide opening, β is much less than unity. Taylor expansion of the right-hand side of Eq. (1.41) gives

$$U_0 = \frac{A}{h}(gh)^{1/2}\left[1 - \tfrac{1}{2}\beta + O(\beta^2)\right].$$

Thus,

$$c_e \cong \frac{f}{2g}\frac{8}{3\pi}\left(\frac{A}{h}(gh)^{1/2}\right)\left(1 - \frac{1}{2}\frac{4}{3\pi}\frac{fA}{h}\right), \qquad (1.42)$$

$$T \cong (1 - \tfrac{1}{2}\beta), \qquad R \cong \tfrac{1}{2}\beta.$$

For small openings, $S_0/S \ll 1$, f is large according to Eq. (1.14); Eq. (1.41) may be approximated for large β by

$$U_0 = \frac{A}{h}(gh)^{1/2}\left(\frac{2}{\beta}\right)^{1/2} = \left(\frac{3\pi}{2f}gA\right)^{1/2} \cong \left(\frac{3\pi}{2}\right)^{1/2}\frac{cS_0}{S}(gA)^{1/2}. \quad (1.43)$$

The above limit may be deduced more directly as follows. By assuming that the reflection is nearly total, one has a standing wave with amplitude $2A$ on the side $x < 0$. The maximum difference of the free surfaces on both sides is $2A$ which gives the discharge velocity $(2gA)^{1/2}$ through a tiny hole, according to Torricelli's law. The discharge velocity defined by averaging over the gross area of S is then $U_0 = (2gA)^{1/2}S_0/S$, which corresponds to $c = 2/(3\pi)^{1/2} = 0.65$ in Eq. (1.43), the theoretical range of c being $0.6 < c < 1$. Finally, the corresponding transmission coefficient is

$$T \cong \left(\frac{2}{\beta}\right)^{1/2} = \left(\frac{3\pi}{2}\frac{h}{fA}\right)^{1/2} = \left(\frac{3\pi}{2}\frac{h}{A}\right)^{1/2}\frac{cS_0}{S}. \qquad (1.44)$$

The scattering coefficients T and R are plotted in Figs. 1.2a and 1.2b.

Özsoy (1977) compared the experiments of Hayashi et al. (1966) for a closely spaced pile breakwater with the theory of this section. With f given by Eq. (1.14), he found that the agreement in transmission and reflection coefficients was rather good (Figs. 1.2a and 1.2b). Özsoy also performed experiments for vertical slots in a thin barrier ($b/a = 0.052, 0.103, 0.162, 0.441$, and $d/2b \leq 0.133$ where $d =$ thickness, $2b =$ slot width, and $2a = 0.87$ m). The empirical coefficient f has a significant scatter for a fixed b/a (see Fig. 1.3), suggesting that other parameters such as the Strouhal number might be important. Further information of interest may be found in Özsoy.

The present problem with the nonlinear boundary condition (1.17) without $(L/g)(\partial u/\partial t)$ has been solved exactly by Mei, Liu, and Ippen (1974). We include it here to show that although odd higher harmonics exist, the fundamental harmonic dominates in practice and is quite accurately given by the method of equivalent linearization.

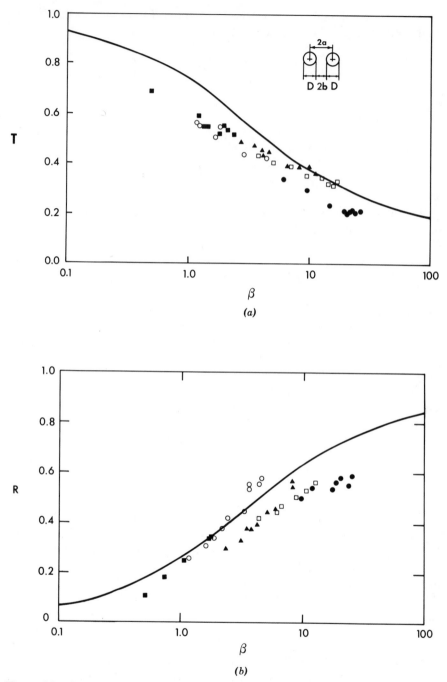

Figure 1.2 Comparison of theory (solid curve), Eqs. (1.14) & (1.41), with experiments by Hayashi et al. (1966). Experiments were performed for various b/a: (●: 0.055; □: 0.075; ▲: 0.091; ○: 0.141; ■: 0.182). (a) T; (b) R (from Özsoy, 1977).

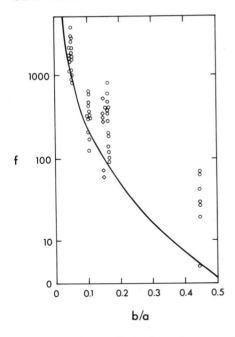

Figure 1.3 The friction coefficient as a function of b/a. (From Özsoy, 1977).

Due to the nonlinear boundary condition we propose the solution

$$\zeta_- = \zeta_+ + \frac{1}{2} \sum_\infty^\infty A_m e^{-im(kx+\omega t)}, \qquad (1.45)$$
$$x < 0,$$

$$u_- = u_I - \sum_{m=\infty}^\infty \frac{gk}{2\omega} A_m e^{-im(kx+\omega t)}, \qquad (1.46)$$

$$\zeta_+ = \frac{1}{2} \sum B_m e^{im(kx-\omega t)}, \qquad (1.47)$$
$$x > 0.$$

$$u_+ = \frac{gk}{2\omega} \sum B_m e^{im(kx-\omega t)}, \qquad (1.48)$$

By time averaging the governing conditions, it may be shown that $\bar{\zeta}$ is zero if $\bar{u}$ is assumed to be zero at one end. Thus, there is no zeroth harmonic in the series above. For every harmonic we require that

$$A_{-m} = A_m^*, \qquad B_{-m} = B_m^*, \qquad (1.49)$$

so that all physical quantities are real.

By matching the velocity according to Eq. (1.8), it follows that

$$\begin{aligned} B_m &= -A_m & \text{for } m \neq 1, \\ B_1 &= A_1 + 1 & \text{for } m = 1. \end{aligned} \qquad (1.50)$$

Since

$$\zeta_- = \frac{\omega}{gk}\left(u_I + \frac{gk}{2\omega}\sum A_m e^{-im\omega t}\right) = \frac{\omega}{gh}\left[2u_I - \left(u_I - \frac{gk}{2\omega}\sum A_m e^{-im\omega t}\right)\right]$$

$$= \frac{\omega}{gk}(2u_I - u_+),$$

at $x = 0 -$, and

$$\zeta_+ = \frac{\omega}{gk}u_+ \,,$$

at $x = 0 +$, it follows from Eq. (1.17) with $L = 0$ that

$$\frac{f}{2g}u_+ |u_+| + 2\frac{\omega}{gk}u_0 = 2\frac{\omega}{gk}u_I = 2A\cos\omega t.$$

Clearly, the signs of u_0 and u_I are the same, that is, u_0 and $u_I(0, t)$ are always in phase. The above equation then gives

$$\frac{f}{2g}|u_+|^2 + \frac{2\omega}{gk}|u_+| = 2A|\cos\omega t| \,. \tag{1.51}$$

In terms of the dimensionless variable W defined by

$$u_+(0, t) = A\frac{gk}{\omega}W = \frac{A}{h}(gh)^{1/2}W(t), \tag{1.52}$$

the solution to Eq. (1.51) is

$$|W| = \frac{(1 + 2\beta'|\cos\omega t|)^{1/2} - 1}{\beta'}, \tag{1.53}$$

where

$$\beta' = \frac{fA}{2h} = \frac{3\pi}{8}\beta. \tag{1.54}$$

When W is expressed as a Fourier series

$$W = \frac{1}{2}\sum T_m e^{-im\omega t}, \tag{1.55}$$

the Fourier coefficient must be

$$\frac{T_m}{2} = \frac{1}{2\pi}\int_0^{2\pi} d\tau\, e^{im\tau}\text{sgn}(\cos\tau)|W(\tau)| \,.$$

The results are

$$T_m = 0, \qquad m = 2, 4, 6, \ldots, \text{even}, \tag{1.56}$$

$$\tfrac{1}{2} T_m = \frac{2}{\pi} \left[\frac{(-1)^{(m+3)/2}}{\beta m} - \frac{M_m(\beta')}{\beta'} \right], \qquad m = 1, 3, 5, \ldots, \text{odd}, \tag{1.57}$$

where

$$M_m(\beta') = \int_0^{\pi/2} d\tau \cos m\tau (1 + 2\beta' \cos \tau)^{1/2}, \tag{1.58}$$

which may be expressed in elliptic integrals but is readily integrated numerically.

By combining Eqs. (1.46), (1.50), (1.52), and (1.55), we get

$$B_m = A T_m, \tag{1.59}$$

so that T_m is the transmission coefficient of the mth harmonic. The reflected coefficient for the mth harmonic is

$$R_1 = 1 - T_1, \qquad R_m = -T_m, \tag{1.60}$$

Table 1.2   Transmission Coefficients as a Function of
$\beta' = fA/2h, T_m$: mth **Harmonic**

β'	T_1		T_3
	Exact	Approximate	Exact
0.0	1	1	0
0.1	0.9601	0.9608	−0.0052
0.2	0.9271	0.9290	−0.0120
0.3	0.8978	0.8975	−0.0169
0.4	0.8719	0.8712	−0.0207
0.5	0.8486	0.8476	−0.0238
0.6	0.8276	0.8262	−0.0264
0.7	0.8084	0.8067	−0.0285
0.8	0.7907	0.7888	−0.0304
0.9	0.7744	0.7722	−0.0319
1.0	0.7593	0.7569	−0.0332
2.0	0.6498	0.6459	−0.0400
3.0	0.5813	0.5766	−0.0418
4.0	0.5326	0.5275	−0.0421
5.0	0.4954	0.4902	−0.0418

where $A_m = AR_m$. Table 1.2 shows the calculated first and third harmonics by the exact theory, and the first harmonic by the equivalent linearization for $1 < \beta < 5$. The smallness of the third harmonic and the efficiency of the approximate theory are evident.

Exercise 1.2

Consider a caisson breakwater which is composed of two parallel walls $x = 0$ and $x = l$. The wall $x = 0$ faces the incoming waves normally and is perforated with the area ratio S_0/S. Use the equivalent linear friction formula (1.30) to find the reflection coefficient. Discuss the effect of l.

6.2   EFFECT OF ENTRANCE LOSS ON HARBOR OSCILLATIONS

In Chapter Five where real fluid effects were ignored, the resonant response in a harbor was found to increase for decreasing entrance width. However, experiments by Lee (1971) confirmed this trend only for a relatively wide entrance and showed the reduction of the entrance width ultimately reduced the peak response. This discrepancy suggests the importance of friction loss at the harbor entrance. Indeed, Japanese engineers have successfully utilized friction to diminish tsunami effects in Ofunato Bay by narrowing the entrance with two transverse breakwaters. In a research conducted for the Ofunato project, Ito (1970) and Horikawa and Nishimura (1970) found experimentally that entrance friction practically eliminated the quarter-wave mode in the long bay. They also developed a theoretical model which incorporated the hydraulic loss formula (1.17) without the apparent inertia, that is,

$$\zeta^- - \zeta^+ = \frac{f}{2g} u |u| . \tag{2.1}$$

While more experimental information is needed for two-dimensional problems with a constriction, the simple formula (2.1) with an estimated constant f appears to yield reasonable global predictions. Based on the same assumptions, Ünlüata and Mei (1975) studied the problem analytically for the simple rectangular harbor with a centered entrance, and Miles and Lee (1975) for the Helmholtz mode of general harbor shape. The theory of Ünlüata and Mei, simplified by ignoring all higher harmonics, is presented below.

6.2.1   The Boundary-Value Problem

For analytical convenience we consider a rectangular harbor with a centered entrance as depicted in Fig. 7.2, Chapter Five.

In the ocean $x > 0$, we separate the radiated waves from the normally incident and the reflected waves

$$\eta_0 = 2A \cos kx + \eta^R . \tag{2.2}$$

The radiated wave η^R satisfies

$$\nabla^2\eta^R + k^2\eta^R = 0, \tag{2.3}$$

$$\frac{\partial\eta^R}{\partial x} = 0, \qquad\qquad |y| > a \tag{2.4}$$

$$\text{on } x = 0,$$

$$\frac{\partial\eta^R}{\partial x} = \frac{i\omega}{g}U(y), \qquad |y| < a \tag{2.5}$$

and must behave as outgoing waves at infinity. The velocity across the harbor entrance is denoted by $U(y) = |U(y)|e^{-i\omega\tau}$ where τ is the phase of U. In the harbor $x < 0$, the displacement amplitude η_H is governed by

$$\nabla^2\eta_H + k^2\eta_H = 0, \tag{2.6}$$

$$\frac{\partial\eta_H}{\partial x} = 0, \qquad\qquad x = -L, \quad |y| < B \tag{2.7}$$

$$\frac{\partial\eta_H}{\partial y} = 0, \qquad\qquad y = \pm\tfrac{1}{2}B, \quad -L < x < 0 \tag{2.8}$$

$$\frac{\partial\eta_H}{\partial x} = 0, \qquad\qquad a < |y| < \frac{B}{2}, \tag{2.9}$$

$$x = 0.$$

$$= \frac{i\omega}{g}U(y), \qquad |y| < a, \tag{2.10}$$

Together, Eqs. (2.9) and (2.10) imply the continuity of normal velocity across the entrance. In addition we have at $x = 0$, $|y| < a$,

$$\eta_H - \eta_0 = c_e U, \qquad c_e = \frac{8}{3\pi}\frac{f}{2g}|U|. \tag{2.11}$$

The solution can be formally written as a superposition of sources:

$$\eta_0 = \int_{-a}^{a} U(y')G_0(x, y|y')\,dy' + 2A\cos kx, \tag{2.12}$$

$$\eta_H = \int_{-a}^{a} U(y')G_H(x, y|y')\,dy', \tag{2.13}$$

where the source functions are

$$G_0 = -\frac{i\omega}{g}\frac{i}{2}H_0^{(1)}(kr),\tag{2.14}$$

$$G_H = -\frac{i\omega}{g}\left[\frac{\cos k(x+L)}{kh\sin kL} + 2\sum_{n=1}^{\infty}\frac{\cos K_n(x+L)}{K_n B\sin K_n L}\cos\frac{2n\pi y}{B}\cos\frac{2n\pi y'}{B}\right],$$

$$\tag{2.15}$$

with $K_n = [k^2 - (2n\pi/B)^2]^{1/2}$. Equations (2.14) and (2.15) are essentially the same as Eqs. (7.1) and (7.7), Chapter Five, except for a factor $-i\omega/g$. Substituting Eqs. (2.12) and (2.13) into Eq. (2.11), we get

$$\int_{-a}^{a} M(y\,|\,y')U(y)\,dy - 2A = c_e U(y) = \frac{f}{2g}\frac{8}{3\pi}|U|U = \frac{f}{2g}\frac{8}{3\pi}|U|^2 e^{-i\omega\tau}$$

$$\tag{2.16}$$

where

$$M(y\,|\,y') = G_H(0,\,y\,|\,y') - G_0(0,\,y\,|\,y').\tag{2.17}$$

Equation (2.16) is a nonlinear integral equation which can be solved numerically for $U(y)$. Because of the uncertainties in f, we shall be contented with a gross estimate by assuming that U is constant in y for $|y| < a$ and try to satisfy Eq. (2.16) only on the average, that is,

$$U\int\int_{-a}^{a} M(y\,|\,y')\,dy\,dy' - 4aA = a\frac{f}{g}\frac{8}{3\pi}|U|^2 e^{-i\omega\tau}.\tag{2.18}$$

For a small gap in a perfect fluid, $U(y)$ is well approximated by $(\text{const})(a^2 - y^2)^{-1/2}$. With flow separation, however, $U(y)$ should no longer be singular at the tips. For this reason and for mathematical simplicity, a uniform velocity distribution is adopted. Needless to say, this approximation cannot give the correct U in detail, and the inaccuracy is not easy to ascertain. Since it is already known that the harbor response for a narrow entrance is related to the total flux through the entrance, a gross global error is not likely to occur using this approximation. With

$$\frac{2a^2\omega}{g}D = -\int\int_{-a}^{a} M\,dy\,dy',\tag{2.19}$$

Eq. (2.18) may be rearranged as follows:

$$-\left(\frac{f}{\omega a}S\,|\,U|\right)^2 + \left(\frac{f}{\omega a}S\,|\,U|\right)D = \frac{2fA}{h}\frac{S}{(ka)^2}\exp(-i\omega\tau),\tag{2.20}$$

where use has been made of $\omega^2 = gk^2h$. Taking the square of the magnitude, we get

$$W^4 + 2(\operatorname{Re} D)W^3 + |D|^2W^2 - \left[\frac{4\beta}{(ka)^2}\right]^2 = 0, \qquad (2.21)$$

in which Re D is the real part of D,

$$W = \frac{fS}{\omega a}|U|, \quad \text{and} \quad \beta = \frac{fA}{2h}S = \frac{2fA}{3\pi h}. \qquad (2.22)$$

Equation (2.21) is a quartic equation for W, which can be solved numerically. Afterward the phase $\omega\tau$ follows from Eq. (2.20) and the solution for U is complete. Finally, the value of U is substituted into Eq. (2.13) to give the harbor response.

6.2.2 Local and Mean Square Response in the Harbor

From Eqs. (2.13) and (2.16), the response at a point (x, y) in the harbor is

$$\eta(x, y) = \frac{(A/a)\int_{-a}^{a} G_H(x, y|y')\,dy'}{(1/4a^2)\iint_{-a}^{a} M(y|y')\,dy\,dy' - (fS/2ga)|U|}. \qquad (2.23)$$

Since the integral

$$\frac{1}{2a}\int_{-a}^{a} G_H\,dy'$$

$$= -\frac{i\omega}{g}\left[\frac{\cos k(x+L)}{kb\sin kL} + 2\sum_{n=1}^{\infty}\frac{\cos K_n(x+L)}{K_n B\sin K_n L}\frac{\sin n\alpha}{n\alpha}\cos\frac{2n\pi y}{B}\right],$$

$$(2.24)$$

in which

$$\alpha \equiv \frac{2\pi a}{B}, \qquad (2.25)$$

is the response to an oscillating piston with uniform velocity $1/2a$ at the entrance (thus, unit total discharge per unit depth), the remaining factor

$$Q \equiv \frac{2A}{(1/4a^2)\iint_{-a}^{a} M\,dy\,dy' - fS|U|/2ga} \qquad (2.26)$$

in Eq. (2.23) represents the amplitude of the discharge across the entrance. We introduce the normalized mean square response as follows:

$$\sigma^2 = \frac{1}{2} \frac{1}{BL} \int_{-L}^{0} dx \int_{-B/2}^{B/2} dy \left| \frac{\eta_H}{2A} \right|^2$$

$$= \frac{1}{2} \frac{|Q/2A|^2}{BL} \int_{-L}^{0} dx \int_{-B/2}^{B/2} dy \left| \frac{1}{2a} \int_{-a}^{a} dy' G_H \right|^2. \tag{2.27}$$

After evaluating all the integrals, one obtains

$$\sigma^2 = \frac{1}{4} \left| \frac{Q}{2A} \frac{\omega}{g} \right|^2 F, \tag{2.28}$$

in which

$$F = \frac{1}{(kB \sin kL)^2} \left(1 + \frac{\sin 2kL}{2kL} \right)$$

$$+ 2 \sum_{n=1}^{\infty} \left(\frac{(\sin n\alpha)/n\alpha}{K_n B \sin K_n B} \right)^2 \left(1 + \frac{\sin K_n L}{2 K_n L} \right). \tag{2.29}$$

Various aspects of these general formulas are examined in the following subsections.

6.2.3 Approximations for Narrow Entrance

As may be anticipated intuitively, the effect of head loss is most important for narrow entrances and near resonant peaks. Hence further considerations will be restricted to $ka \ll 1$.

By a common technique of improving the convergence of a series, it can be shown that

$$\frac{1}{4a^2} \iint_{-a}^{a} M \, dy \, dy' \simeq -\frac{i\omega}{g} \left(-\frac{i}{2} + F - I \right) + O(k^2 a^2 \ln ka), \tag{2.30}$$

in which

$$F \equiv \frac{\cot kL}{kB} + 2 \sum_{n=1}^{\infty} \left(\frac{\cot K_n L}{K_n B} + \frac{1}{2n\pi} \right) \left(\frac{\sin n\alpha}{n\alpha} \right)^2, \tag{2.31}$$

and

$$I = -\left(\ln \frac{\pi k a^2 \gamma}{4B} + \ln 16 - 3 \right), \tag{2.32}$$

with $\ln \gamma = 0.5772157 =$ Euler's constant. Details of the derivation are given in Appendix 6.A. The approximate discharge per unit depth is

$$Q \simeq \frac{2A}{-(i\omega/g)\left[-\frac{1}{2}i(1 + W) + F - I\right]}, \qquad W = \frac{fS|U|}{a\omega}, \qquad (2.33)$$

which can be combined with Eq. (2.28) for the approximate σ and with Eq. (2.23) for the approximate η_H. The result should be compared with Eq. (7.23), Chapter Five.

The effect of head loss enters the theory explicitly only through the factor $W = fS|U|/\omega a$ in the entrance discharge Q [cf. Eq. (2.33)]. In the absence of friction ($f = 0$), the resonance features in the harbor have been studied in Chapter Five. In particular, the term $-\frac{1}{2}i$ in brackets in Eq. (2.33) (with $W = 0$) corresponds to radiation damping. Clearly, the term $(-\frac{1}{2}i)(1 + W)$ corresponds to the sum of radiation and friction damping at the entrance. For a narrow entrance and $f = W = 0$, radiation damping should be weak so that resonance occurs near the natural modes of the completely closed basin, that is,

$$k = k_{mn} = \left[\left(\frac{m\pi}{L}\right)^2 + \left(\frac{2n\pi}{B}\right)^2\right]^{1/2}, \qquad m = 0, 1, 2, 3, \ldots; n = 0, 1, 2, 3, \ldots$$
$$(2.34)$$

Now, when $W \neq 0$ but

$$W \leqq O(1), \qquad (2.35)$$

the friction loss, and thus the total damping, is also weak; the resonant peaks should still be near k_{mn} and the neighborhoods of the resonant peaks can be examined in the manner of Section 5.7. For brevity, only the peaks themselves, when they are well isolated, are studied in the next subsection.

6.2.4 Small Radiation and Friction Damping

Near the natural mode $k = k_{mn}$, the magnitude of Q is, from Eq. (2.33),

$$|Q| = \frac{2A(\omega/g)^{-1}}{\left[|\frac{1}{2}(1 + W)|^2 + (F - I)^2\right]^{1/2}}. \qquad (2.36)$$

Since F is large for $k \simeq k_{mn}$ and I is logarithmically large for small ka^2/B, the maxima of $|Q|$ occur approximately where

$$F - I = 0, \qquad (2.37)$$

provided that $W \leqq O(1)$. Let $(\tilde{\ })$ denote quantities evaluated at the resonant peaks. In particular, the real roots of Eq. (2.37) will be designated by $\tilde{k}_{mn}$. Since F and I are independent of f, the locations of the resonant peaks are not strongly affected by friction losses. The corresponding maximum discharge is

$$|\tilde{Q}| = \left(\frac{4A(\omega/g)^{1/2}}{1 + W} \right)_{\omega = \tilde{\omega}_{mn}}. \qquad (2.38)$$

Since

$$|\tilde{Q}|_{f=0} = \left(\frac{4A}{\omega/g} \right)_{\omega = \tilde{\omega}_{mn}} \qquad (2.39)$$

for $f = 0$, the reduction ratio for peak discharge at the entrance is

$$\frac{|\tilde{Q}|}{|\tilde{Q}|_{f=0}} = \frac{1}{(1 + W)_{\omega = \tilde{\omega}_{mn}}}. \qquad (2.40)$$

In view of Eq. (2.28), Eq. (2.40) is also the reduction ratio for the root mean square response at resonance since F and the resonant wavenumbers are approximately independent of f.

The value of W at resonance remains to be found. From Eqs. (2.30) and (2.19) it follows that for a narrow entrance and at resonance

$$\frac{2a^2\omega}{g} D = -\iint_{-a}^{a} M \, dy \, dy' \simeq \frac{2a^2\omega}{g}, \qquad \text{or} \qquad D \simeq 1. \qquad (2.41)$$

Hence,

$$\omega\tau = 0 \qquad (2.42)$$

from Eq. (2.20) and W can be solved from Eq. (2.21):

$$W = \frac{fS|U|}{\omega a} = \frac{1}{2}\left[-1 + \left(1 + \frac{16\beta}{(ka)^2} \right)^{1/2} \right]_{k=\tilde{k}_{mn}} \qquad (2.43)$$

in which $\beta = 2fA/3\pi h$ [cf. Eq. (2.22)]. Note that the original condition (2.35) implies that

$$\frac{16\beta}{(ka)^2} \leqq O(1). \qquad (2.44)$$

After Eq. (2.43) is substituted into Eq. (2.40), the reduction factor is found to be

$$\frac{\tilde{\sigma}}{\tilde{\sigma}_{f=0}} \cong \frac{\tilde{Q}}{\tilde{Q}_{f=0}} = \left(\frac{2}{1 + \left(1 + 16\beta / (ka)^2\right)^{1/2}} \right)_{k=\tilde{k}_{mn}}, \qquad (2.45)$$

where the value of $\tilde{k}_{mn}$ may be estimated by the natural wavenumbers of the closed basin when n and m are not both zero (non-Helmholtz mode). For the Helmholtz mode, $\tilde{k}_{00}$ may be estimated by the inviscid value.

It can be concluded from Eq. (2.45) that the reduction of resonant peaks by entrance loss is more pronounced for increasing $16\beta / (\tilde{k}a)^2$, that is, for (i) larger f, (ii) larger amplitude, (iii) longer waves or lower resonant modes, or (iv) narrower entrance. The wisdom in the Ofunato breakwater design is evident. Items (iii) and (iv) are also consistent with the experimental observations by Lee (1971) for a circular harbor.

With regard to the parameter $16\beta / (\tilde{k}a)^2$, it should be pointed out that the loss coefficient f may depend on the Strouhal and Reynolds numbers and on the geometry of the breakwater tips at the entrance. Ito (1970) suggests that the empirical value $f = 1.5$ gives reasonable results for the Ofunato tsunami breakwater. For reference, note that when $A = 0.5$ m, $h = 10$ m and f is taken to be 1, then $\beta = 10^{-2}$. Now take a square basin with $B = L$; the lowest few

Table 2.1 Reduction Factor According to Eq. (2.45).

m, n		(0, 0)	(1, 0)	(2, 0), (0, 1)	(1, 1)
$\tilde{k}B$		0.55	π	2π	$\sqrt{5}\,\pi$
$\tilde{k}a = \dfrac{2a}{B}\dfrac{\tilde{k}B}{2}$		0.825×10^{-2}	4.71×10^{-2}	9.42×10^{-2}	10.53×10^{-2}
	$\beta = 10^{-2}$	2350[†]	72.1[†]	18.0[†]	14.43[†]
	$\beta = 10^{-3}$	23.5[†]	0.721	0.18	1.44
$\dfrac{16\beta}{(\tilde{k}a)^2}$	$\beta = 10^{-4}$	23.5[†]	0.721	0.18	0.144
	$\beta = 10^{-5}$	2.35	0.0721	0.018	0.0144
	$\beta = 10^{-2}$	0.04[†]	0.209[†]	0.373[†]	0.406[†]
	$\beta = 10^{-3}$	0.122[†]	0.517	0.748	0.781
$\dfrac{\tilde{\sigma}}{\tilde{\sigma}_{f=0}}$	$\beta = 10^{-4}$	0.336[†]	0.865	0.959	0.966
	$\beta = 10^{-5}$	0.707	0.983	0.996	0.996

[a] From Mei, Liu, and Ippen (1974). J. *Waterway, Port, Coastal and Ocean Division*. Reproduced by permission of the American Society of Civil Engineers.

natural modes of the closed basin are:

$$k_{10}L = \pi, \qquad k_{01}L = k_{20}L = 2\pi, \qquad \text{and} \qquad k_{11}L = 5^{1/2}\pi.$$

For a narrow entrance with $2a/B = 3 \times 10^{-2}$, the reduction factor and the parameter $16\beta/(\tilde{k}a)^2$ are listed in Table 2.1 for a range of β no greater than 10^{-2}. The values marked with a † violate the assumption that $16\beta/(\tilde{k}a)^2 \leq O(1)$ and the calculated reduction factor is not reliable quantitatively; a different approximation is needed.

6.2.5 Large Friction Damping

Table 2.1 shows that the value of $16\beta/(\tilde{k}a)^2$ can be very large for the lowest resonant mode or for narrow entrance. From Eq. (2.21) the value of $W = fS|U|/\omega a$ is also large and can be approximated to the leading order by

$$W \simeq \frac{(4\beta)^{1/2}}{ka}, \tag{2.46}$$

or

$$|U| \simeq \left(\frac{2gA}{fS}\right)^{1/2}. \tag{2.47}$$

Note that the velocity U is proportional to $(2gA)^{1/2}$, as in the elementary Torricelli's law. The corresponding discharge per unit depth through the entrance is

$$|Q| \simeq 2a\left(\frac{2gA}{fS}\right)^{1/2}, \tag{2.48}$$

which diminishes with the entrance width $2a$. Combining this result with Eq. (2.28), it may be concluded that sufficiently large head loss removes the harbor paradox, that is, the response ultimately diminishes with the entrance width.

6.2.6 Numerical Results for General W

As shown in Table 2.1, the assumption of small friction damping $W \leq O(1)$ is not appropriate for the Helmholtz mode, and an accurate solution of W from the quartic equation (2.21) is necessary. Since the real part of D is proportional to the total damping, the coefficients of W^4, W^3, and W^2 in Eq. (2.21) are positive; one and only one real and positive solution can exist. After solving for W numerically the root mean square response for the fundamental harmonic can be calculated from Eq. (2.28) without the assumption of $ka \ll 1$. The normalized flux and the root mean square response are plotted in Fig. 2.1 for

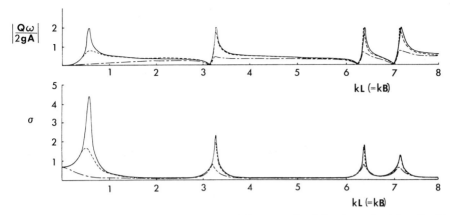

Figure 2.1 Root mean square response σ and the normalized flux intensity $|\,Q\omega/2gA\,|$ of the fundamental harmonic as function of kL ($= kB$). Normalized entrance width $2a/B = 3 \times 10^{-2}$. $\beta = 0$: solid curve; $\beta = 10^{-4}$: dashed curve, $\beta = 10^{-2}$: dash–dot curve. (From Ünlüata and Mei, 1975, *J. Waterway, Port, Coastal and Ocean Division*. Reproduced by permission of American Society of Civil Engineers.)

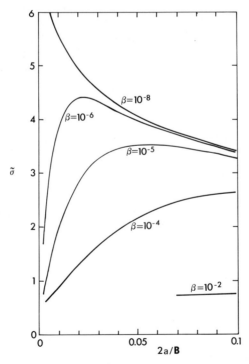

Figure 2.2 Mean resonant amplification $\tilde{\sigma}$ of the fundamental harmonic for the Helmholtz mode. (From Ünlüala and Mei, 1975, *J. Waterway, Port, Coastal and Ocean Division*. Reproduced by permission of American Society of Civil Engineers.)

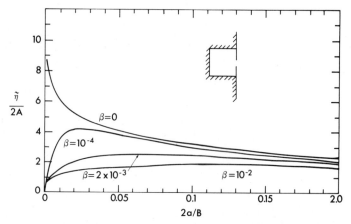

Figure 2.3 Resonant amplification $\tilde{\sigma}$ of the fundamental harmonic at the corner $x = -B, y = \frac{1}{2}B$ for the first mode $k_{01}B = \pi$ (or $\eta \propto \cos(2\pi y/B)$). (From Ünlüata and Mei, 1975 *J. Waterway, Port, Coastal and Ocean Division.* Reproduced by permission of American Society of Civil Engineers.)

the range $0 < kL < 8$ for $\beta = 10^{-2}$, 10^{-4}, and 0. For comparison, the inviscid theory which predicts the harbor paradox is also shown. In Figs. 2.2 and 2.3, the amplification ratios for the Helmholtz mode ($m = n = 0$) and for the mode $m = 1$, $n = 0$ are plotted as functions of the normalized entrance width $2a/B$. Again, the harbor response reduces with $2a/B$ for a fixed β.

APPENDIX 6.A: APPROXIMATIONS OF AN INTEGRAL FOR $ka \ll 1$

The integral in question is

$$\frac{1}{4a^2} \iint_{-a}^{a} dy\, dy'\, M(y\,|\,y') = \frac{1}{4a^2} \iint_{-a}^{a} [G_H(0, y, y') - G_0(0, y\,|\,y')]\, dy\, dy'$$

$$= \mathcal{I}_H - \mathcal{I}_0. \tag{A.1}$$

From Eq. (2.14) it follows that

$$\mathcal{I}_0 \equiv \frac{1}{4a^2} \iint_{-a}^{a} G_0(0, y\,|\,y')\, dy\, dy' = \frac{1}{4a^2}\frac{i\omega}{g}\left(-\frac{i}{2}\right)\int_{-a}^{a} H_0^{(1)}(k\,|\,y - y'\,|)\, dy\, dy',$$

$$= \frac{\omega}{2g} \iint_{-1/2}^{1/2} H_0^{(1)}[(2ka)\,|\,\xi - \xi'\,|]\, d\xi\, d\xi', \tag{A.2}$$

after the transformation $y = 2a\xi$, $y' = 2a\xi'$. Approximating $H_0^{(1)}$ for small ka, one has

$$\mathcal{I}_0 = \frac{\omega}{2g} \int_{-1/2}^{1/2} \left[1 + \frac{2i}{\pi} \ln \left(\gamma ka \, | \xi - \xi' | \right) \right] d\xi \, d\xi' + O(ka)^2 \ln(ka), \quad (A.3)$$

where $\ln \gamma = 0.577\,2157 = $ Euler's constant. Thus,

$$\mathcal{I}_0 = \frac{\omega}{2g} \left[\left(1 + \frac{2i}{\pi} \ln \gamma ka \right) + \frac{2i}{\pi} \int_{-1/2}^{1/2} \ln | \xi - \xi' | \, d\xi \, d\xi' \right] \quad (A.4)$$

$$= \frac{\omega}{2g} \left[1 + \frac{2i}{\pi} \left(\ln \gamma ka - \frac{3}{2} \right) \right] + O\left[(ka)^2 \ln ka \right], \quad (A.5)$$

where the following identity has been used:

$$\iint_{-1/2}^{1/2} \ln | \xi - \xi' | \, d\xi \, d\xi' = -\tfrac{3}{2}. \quad (A.6)$$

From Eq. (2.15) it follows that

$$\mathcal{I}_H = \frac{1}{4a^2} \iint_{-a}^{a} G_H(0, y | y') \, dy'$$

$$= -\frac{i\omega}{g} \frac{1}{4a^2} \iint_{-a}^{a} dy \, dy' \left(\frac{\cot kL}{kB} + 2 \sum_{n=1}^{\infty} \frac{\cot K_n L}{K_n B} \cos \frac{2n\pi y}{B} \cos \frac{2n\pi y'}{B} \right)$$

$$= -\frac{i\omega}{g} \left[\frac{\cot kL}{kB} + 2 \sum_{n=1}^{\infty} \frac{\cot K_n L}{K_n B} \left(\frac{\sin n\alpha}{n\alpha} \right)^2 \right], \quad (A.7)$$

where $\alpha \equiv 2\pi a/B$. Consider the series above. Since the harbor dimensions have been assumed to be comparable to the incident wavelength we must have

$$\alpha = \frac{2\pi a}{B} \ll 1.$$

Now for large n, the nth term in the last series of Eq. (A.7) approaches

$$\frac{\cot K_n L}{K_n B} \left(\frac{\sin n\alpha}{n\alpha} \right)^2 \rightarrow -\frac{1}{2n\pi} \left(\frac{\sin n\alpha}{n\alpha} \right)^2;$$

the series, to be denoted by Σ, may be written

$$\Sigma = -\frac{1}{\pi} \sum_{n=1}^{\infty} \frac{1}{n} \left(\frac{\sin n\alpha}{n\alpha} \right)^2 + 2 \sum_{n=1}^{\infty} \left(\frac{\cot K_n L}{K_n B} + \frac{1}{2n\pi} \right) \left(\frac{\sin n\alpha}{n\alpha} \right)^2.$$

$$(A.8)$$

The second series on the right-hand side above now converges rapidly; its sum
with the remaining term in Eq. (A.1) will be denoted by F:

$$F = \frac{\cot kL}{kB} + 2 \sum_{n=1}^{\infty} \left(\frac{\cot K_n L}{K_n B} + \frac{1}{2n\pi} \right) \left(\frac{\sin n\alpha}{n\alpha} \right)^2. \qquad (A.9)$$

The first series on the right-hand side of Eq. (A.8) may be approximately
summed in closed form. By the following rearrangement,

$$F' \equiv -\frac{1}{\pi} \sum_{1}^{\infty} \frac{1}{n} \left(\frac{\sin n\alpha}{n\alpha} \right)^2 = \frac{1}{2\pi\alpha^2} \sum_{1}^{\infty} \frac{1}{n^3} (\cos 2n\alpha - 1),$$

it can be found from Collin (1960, p. 579) that

$$\sum_{1}^{\infty} \frac{\cos nz}{n^3} = \frac{z^2}{2} \ln z - \frac{3}{4} z^2 + O(z^4) + \sum_{1}^{\infty} \frac{1}{n^3}.$$

Hence

$$F' = \frac{1}{2\pi\alpha^2} \left[\frac{(2\alpha)^2}{2} \ln 2\alpha - \frac{3}{4} (2\alpha)^2 + O(\alpha)^4 \right]$$

$$= \frac{1}{\pi} \left(\ln 2\alpha - \frac{3}{2} \right) \left[1 + O(\alpha^2) \right]$$

$$= \frac{1}{\pi} \left(\ln \frac{4\pi a}{B} - \frac{3}{2} \right) \left[1 + O\left(\frac{2\pi a}{B} \right)^2 \right].$$

Substituting into Eq. (A.8) and combining with Eqs. (A.7) and (A.9), one gets

$$\frac{1}{4a^2} \int_{-a}^{a} M \, dy \, dy'$$

$$= \mathcal{G}_H - \mathcal{G}_0 \simeq -\frac{i\omega}{g} \left\{ F + \frac{1}{\pi} \left(\ln \frac{4\pi a}{B} - \frac{3}{2} \right) \right.$$

$$+ O\left[(ka)^2 \ln ka \right] - \frac{i}{2} + \frac{1}{\pi} \left(\ln \gamma ka - \frac{3}{2} \right) \Big\}$$

$$= -\frac{i\omega}{g} \left\{ -\frac{i}{2} + F + \left[\ln \left(\frac{\pi ka^2 \gamma}{4B} \right) + \ln 16 - 3 \right] + O(ka)^2 \ln(ka) \right\}.$$

$$(A.10)$$

Since $\ln 16 - 3 = -0.2274$ and $ka^2/B \ll 1$, the square bracket is negative and will be denoted by

$$I = -\left[\ln\left(\frac{\pi ka^2 \gamma}{4B}\right) + \ln 16 - 3\right]. \qquad (A.11)$$

In summary, the integral in Eq. (A.1) is

$$\frac{1}{4a^2}\iint_{-a}^{a} M(y\,|\,y')\,dy\,dy' \simeq -\frac{i\omega}{g}\left(-\frac{i}{2} + F - I\right) + O(k^2 a^2 \ln ka).$$

$$(A.12)$$

SEVEN

Floating Body Dynamics: Diffraction and Radiation by Large Bodies

7.1 INTRODUCTION

Ships, buoys, barges, floating docks, breakwaters, submersibles supporting oil drilling rigs, and so on, are all structures whose safety and performance depend on their response to waves. In a calm sea, the body weight, the buoyancy force, and possible forces from external constraints such as tension legs, keep the body in static equilibrium. In waves, the presence of a sufficiently large body causes diffraction (scattering) of waves. The body must absorb some of the incident wave momentum and therefore must suffer a dynamic force. If the constraints, such as the mooring lines, are not sufficiently rigid, the body oscillates, hence further radiates waves, and experiences reacting forces from the surrounding fluid and from the constraint. Since the reacting forces depend on the motion of the body itself, the body, the constraint, and the surrounding water are dynamically coupled in the presence of incoming waves.

In modern offshore oil exploration, the gravity-type structure is frequently used in relatively shallow seas. This massive structure sits on the sea bottom and serves both as a storage tank and a support for the drilling deck above the sea level. The intensity of wave pressure has a direct effect on the stability of the seabed near the structure. Moreover, wave forces on the structure are transmitted to the seabed, causing stresses and deformation in both solids. Thus, the design of gravity structure involves the dynamic interaction of three media (water, structure, and seabed). A comprehensive discussion involving all three elements at once is too complex to be discussed here; we shall limit ourselves to the interaction of water with a rigid body.

First, a few remarks must be made on what is a *large* body. There are at least three relevant length scales in wave–body interaction: the characteristic body dimension a, the wavelength $2\pi/k$, and the wave amplitude A. Among these scales two ratios may be formed, for example, ka and A/a. If $ka \geq O(1)$, a body is regarded as large; its presence alters the pattern of wave propagation significantly and produces diffraction. Ships, submersibles, and underwater storage tanks fall into this category. For small bodies ($ka \ll 1$), such as the structural members of a drilling tower, diffraction is of minor importance. When A/a is sufficiently large, the local velocity gradient near the small body augments the effect of viscosity and induces flow separation and vortex shedding, leading to the so-called *form drag*. At present, the inviscid linearized diffraction theory has been fairly well developed for $A/a \ll 1$ and $ka = O(1)$ with considerable experimental confirmation. The case of $A/a \geq O(1)$ and $ka \ll 1$ has been the subject of intensive experimental studies (Sarpkaya and Issacson, 1981), but is not easily describable on purely theoretical ground. The intermediate case of $A/a \geq O(1)$ and $ka = O(1)$ involves both separation and nonlinear diffraction and is the most difficult and least explored area of all.

It is useful to recall from the classical viscous flow theory why A/a plays a role in flow separation. If a circular cylinder of radius a starts to accelerate from rest with a uniform acceleration b, the inviscid flow just outside the cylinder is potential and is given by

$$U(x, t) = t2b \sin \theta, \qquad (1.1)$$

where U is the tangential velocity along the cylinder and θ is the angle of a point along the cylinder measured from the forward stagnation point. From the solution of the viscous boundary layer along the cylinder surface, it is known that the boundary layer begins to separate at the rear stagnation point $\theta = \pi$ after the critical time t_c

$$t_c = \left(1.04 \frac{a}{b}\right)^{1/2} \qquad (1.2)$$

(Schlichting, 1968, p. 407). If the acceleration is continued, the rear region of separation expands toward the forward stagnation point; eddies are generated and shed downstream. The above information may be used to give an order-of-magnitude estimate for an oscillating cylinder in a calm fluid or an oscillating flow around a stationary cylinder. Let the amplitude of the oscillating velocity be U_0 so that the acceleration is $b = O(\omega U_0)$. The separation time is

$$t_c = O\left(\left(\frac{a}{\omega U_0}\right)^{1/2}\right). \qquad (1.3)$$

Over a period $2\pi/\omega$, there is no separation if

$$\frac{2\pi}{\omega} < t_c \quad \text{or} \quad \frac{2\pi}{\omega} < c\left(\frac{a}{\omega U_0}\right)^{1/2} \qquad \text{where } c = O(1). \qquad (1.4)$$

This criterion may be rewritten in terms of the Strouhal number, which is just $1/2\pi$ times the Keulegan–Carpenter number $U_0 T/a$:

$$\frac{U_0}{\omega a} < \left(\frac{c}{2\pi}\right)^2. \tag{1.5}$$

Since $U_0 = O(\omega A)$ in water waves where A is the wave amplitude, the criterion may be expressed

$$\frac{A}{a} < \left(\frac{c}{2\pi}\right)^2. \tag{1.6}$$

Thus, the parameter A/a governs the phenomenon of separation. Keulegan and Carpenter (1956) found experimentally that for a circular cylinder the value $A/a = 1$ is sufficiently large for flow separation and form drag is important. Since A is also the measure of the particle orbit in waves, a convenient rule of thumb is that no separation occurs if the orbital diameter is much less than the cylinder diameter.

Although large floating or submerged structures must often be designed to withstand storm waves with typical amplitudes as high as 15 m, little is known theoretically of the diffraction of large amplitude waves. Some of the difficulties of nonlinearity may be appreciated from the example given in Section 12.10. On the other hand, wave–body interaction at infinitesimal amplitudes is a richly developed subject, which has been serving the field of naval architecture with great success. Its applications in offshore technology have stimulated further advances in recent years. An acquaintance with the elements of linearized theory is now quite essential to the rational design of a costly new project. A comprehensive survey of the relevant topics may be found in the classic work of Wehausen and Laitone (1960), and in Wehausen (1971). Some of this material, and additional topics of more recent origin, are expounded here.

In Section 7.2 the governing equations for a partially constrained floating rigid body are derived for a body of arbitrary shape. The equations of static equilibrium and of small-amplitude motion are obtained as the zeroth- and the first-order approximations, respectively. The main body of this chapter is devoted to simple harmonic motions. Thanks to linearity, the fluid motion may be decoupled from that of the rigid body. A survey of the theoretical aspects of wave scattering and radiation is given in Sections 7.3–7.6 which include the derivations of the general reciprocity theorems for a single body. The principles of a versatile numerical method of hybrid elements which extends the method described in Chapter Four are then briefly described in Section 7.7, followed by some remarks on the alternative methods of integral equations in Section 7.8. Applications of the theoretical tools are demonstrated via the topic of wave-power absorption in Section 7.9. General formulas for the second-order drift forces are deduced in Section 7.10 by using the first-order results. Finally, we develop in Section 7.11 the general relationships between the simple harmonic response and the responses to transient waves.

7.2 LINEARIZED EQUATIONS OF MOTION FOR A CONSTRAINED FLOATING BODY

Details of the derivation presented below are due to John (1949) whose formal approach not only leads systematically to the complete first-order theory but also shows how higher-order extension may be made.

As in the case of the free surface, there are *kinematic* and *dynamic* conditions on the wetted body surface S_B, relating the motions of the body and of the surrounding water.

7.2.1 The Kinematic Condition

Let the instantaneous position of S_B be described by $z = f(x, y, t)$. Continuity of normal velocity requires that

$$\Phi_x f_x + \Phi_y f_y + f_t = \Phi_z, \qquad z = f(x, y, t). \tag{2.1}$$

For small-amplitude motions, we expand f in powers of the wave slope $\varepsilon = kA$ which is expected to characterize the body motion also,

$$z = f^{(0)}(x, y) + \varepsilon f^{(1)}(x, y, t) + \varepsilon^2 f^{(2)}(x, y, t) + \cdots \tag{2.2}$$

where $f^{(0)}(x, y)$ corresponds to the rest position of S_B, that is, $S_B^{(0)}$. Likewise the velocity potential may also be expanded:

$$\Phi = \varepsilon \Phi^{(1)} + \varepsilon^2 \Phi^{(2)} + \cdots . \tag{2.3}$$

Since the body motion is small, any function evaluated on S_B may be expanded about $S_B^{(0)}$, that is, $z = f^{(0)}(x, y)$. To the order $O(\varepsilon)$, Eq. (2.1) gives

$$\phi_x^{(1)} f_x^{(0)} + \phi_y^{(1)} f_y^{(0)} + f_t^{(1)} = \phi_z^{(1)} \qquad \text{on } z = f^{(0)}(x, y). \tag{2.4}$$

It is necessary to find $f^{(1)}$. Let the center of rotation of the rigid body be Q which has the following moving coordinate:

$$\mathbf{X}(t) = \mathbf{X}^{(0)} + \varepsilon \mathbf{X}^{(1)}(t) + \varepsilon^2 \mathbf{X}^{(2)}(t) + \cdots , \qquad \mathbf{X} = (X, Y, Z), \qquad \text{and so on,} \tag{2.5}$$

where $\mathbf{X}^{(0)}$ is the rest position of Q independent of t. For generality, Q need not coincide with the center of mass of the body. We introduce $\bar{\mathbf{x}}$ to be the coordinate system fixed with the body such that $\bar{\mathbf{x}} = \mathbf{x}$ when the body is at its rest position. If the angular displacement of the body is $\varepsilon\boldsymbol{\theta}^{(1)}(t)$ with the components $\varepsilon\alpha$, $\varepsilon\beta$, and $\varepsilon\gamma$ about axes parallel to x, y, and z, then the two coordinate systems $\mathbf{x}$ and $\bar{\mathbf{x}}$ are related, to the first order, by

$$\mathbf{x} = \bar{\mathbf{x}} + \varepsilon\big[\mathbf{X}^{(1)} + \boldsymbol{\theta}^{(1)} \times (\bar{\mathbf{x}} - \mathbf{X}^{(0)})\big] + O(\varepsilon^2). \tag{2.6}$$

To the same accuracy $\bar{\mathbf{x}}$ may be solved in terms of $\mathbf{x}$ to give

$$\bar{\mathbf{x}} = \mathbf{x} - \varepsilon\big[\mathbf{X}^{(1)} + \boldsymbol{\theta}^{(1)} \times (\mathbf{x} - \mathbf{X}^{(0)})\big] + O(\varepsilon^2), \tag{2.7a}$$

or, in component form,

$$\bar{x} = x - \varepsilon\big[X^{(1)} + \beta(z - Z^{(0)}) - \gamma(y - Y^{(0)})\big], \tag{2.7b}$$

$$\bar{y} = y - \varepsilon\big[Y^{(1)} + \gamma(x - X^{(0)}) - \alpha(z - Z^{(0)})\big], \tag{2.7c}$$

$$\bar{z} = z - \varepsilon\big[Z^{(1)} + \alpha(y - Y^{(0)}) - \beta(x - X^{(0)})\big]. \tag{2.7d}$$

Since $\mathbf{x} = \bar{\mathbf{x}}$ at rest by definition, the following is true:

$$\bar{z} = f^{(0)}(\bar{x}, \bar{y}). \tag{2.8}$$

Upon substituting Eq. (2.7) into Eq. (2.8), expanding about $S_B^{(0)}$, and then comparing the result with Eq. (2.2), we get

$$
\begin{aligned}
f^{(1)} = {}& Z^{(1)} + \alpha(y - Y^{(0)}) - \beta(x - X^{(0)}) \\
& - f_x^{(0)}\big[X^{(1)} + \beta(z - Z^{(0)}) - \gamma(y - Y^{(0)})\big] \\
& - f_y^{(0)}\big[Y^{(1)} + \gamma(x - X^{(0)}) - \alpha(z - Z^{(0)})\big].
\end{aligned} \tag{2.9}
$$

Finally, the combination of Eqs. (2.9) and (2.4) yields the first-order kinematic condition:

$$
\begin{aligned}
-\Phi_x^{(1)}f_x^{(0)} - \Phi_y^{(1)}f_y^{(0)} + \Phi_z^{(1)} = {}& -f_x^{(0)}\big[X_t^{(1)} + \beta_t(z - Z^{(0)}) - \gamma_t(y - Y^{(0)})\big] \\
& - f_y^{(0)}\big[Y_t^{(1)} + \gamma_t(x - X^{(0)}) - \alpha_t(z - Z^{(0)})\big] \\
& + Z_t^{(1)} + \alpha_t(y - Y^{(0)}) - \beta_t(x - X^{(0)}). \quad (2.10)
\end{aligned}
$$

There are more compact ways of expressing Eq. (2.10). Since the unit normal vector $\mathbf{n}$ pointing into the body is

$$\mathbf{n} = \left(-f_x^{(0)}, -f_y^{(0)}, 1\right)\left[1 + \left(f_x^{(0)}\right)^2 + \left(f_y^{(0)}\right)^2\right]^{-1/2}, \tag{2.11}$$

Eq. (2.10) may be written

$$\frac{\partial \Phi^{(1)}}{\partial n} = \mathbf{X}_t^{(1)} \cdot \mathbf{n} + \left[\boldsymbol{\theta}_t^{(1)} \times (\mathbf{x} - \mathbf{X}^{(0)}) \right] \cdot \mathbf{n}$$

$$= \mathbf{X}_t^{(1)} \cdot \mathbf{n} + \boldsymbol{\theta}_t^{(1)} \cdot \left[(\mathbf{x} - \mathbf{X}^{(0)}) \times \mathbf{n} \right]. \tag{2.12}$$

Alternatively, we may introduce the six-dimensional generalized displacement vector $\{X_\alpha\}$ and the generalized normal vector $\{n_\alpha\}$ defined by

$$\{X_\alpha\}^T = \{X^{(1)}, Y^{(1)}, Z^{(1)}, \alpha, \beta, \gamma\} = \{\mathbf{X}^{(1)}, \boldsymbol{\theta}^{(1)}\},$$

$$\{n_\alpha\}^T = \{n_1, n_2, n_3, -[n_2(z - Z^{(0)}) - n_3(y - Y^{(0)})],$$

$$-[n_3(x - X^{(0)}) - n_1(z - Z^{(0)})],$$

$$-[n_1(y - Y^{(0)}) - n_2(x - X^{(0)})]\}$$

$$= \{\mathbf{n}, (\mathbf{x} - \mathbf{X}^{(0)}) \times \mathbf{n}\}, \tag{2.13}$$

where $\{\ \}$ denotes a column vector and $\{\ \}^T$ its transpose, and the subscript α ranges over $1, 2, \ldots, 6$. Equation (2.12) becomes

$$\frac{\partial \Phi^{(1)}}{\partial n} = \sum_{\alpha=1}^{6} (X_\alpha)_t n_\alpha = \{X\}_t^T \{n\}. \tag{2.14}$$

Exercise 2.1

Show that the order $O(\varepsilon^2)$ correction to the kinematic boundary condition is

$$\phi_x^{(2)} f_x^{(0)} + \phi_y^{(2)} f_y^{(0)} + f_t^{(2)} = \phi_z^{(2)} - \left(\phi_x^{(1)} f_x^{(1)} + \phi_y^{(1)} f_y^{(1)} \right)$$

$$+ f^{(1)} \left(\phi_{zz}^{(1)} - \phi_{xz}^{(1)} f_x^{(0)} - \phi_{yz}^{(1)} f_y^{(0)} \right).$$

However, to get $f^{(2)}$ it is necessary to improve Eq. (2.6) which is good only for infinitesimal rotation. For finite rotation, the concept of Euler's angles is needed (Goldstein, 1950, Chapter 4).

We turn next to the dynamic conditions.

7.2.2   Conservation of Linear Momentum

Let M be the mass of the entire floating body, part of which may be above the free surface, and $\mathbf{x}^c(t)$ be the position of the center of mass. Conservation of linear momentum requires that

$$Mx_{tt}^c = \iint_{S_B} P\mathbf{n}\, dS - Mg\mathbf{e}_3 + \mathbf{F}, \tag{2.15}$$

where S_B is the instantaneous wetted body surface. $\mathbf{F}$ denotes the constraining force from external support, such as mooring lines, tension legs, and so on, and consists also of a static and a dynamic part:

$$\mathbf{F}(t) = \mathbf{F}^{(0)} + \varepsilon \mathbf{F}^{(1)}(t) + \cdots. \tag{2.16}$$

Up to the first order, Eq. (2.7) applies to $\mathbf{x}^c(t)$ so that

$$\mathbf{x}^c = \bar{\mathbf{x}}^c + \varepsilon\left[\mathbf{X}^{(1)} + \boldsymbol{\theta}^{(1)} \times (\bar{\mathbf{x}}^c - \mathbf{X}^{(0)})\right] + O(\varepsilon^2).$$

The left-hand side of Eq. (2.15) becomes

$$Mx_{tt}^c = \varepsilon M\left[\mathbf{X}_{tt}^{(1)} + \boldsymbol{\theta}_{tt}^{(1)} \times (\bar{\mathbf{x}}^c - \mathbf{X}^{(0)})\right] + O(\varepsilon^2). \tag{2.17}$$

On the right-hand side of Eq. (2.15), the linearized Bernoulli equation

$$P = -\rho g f - \varepsilon \rho \Phi_t^{(1)} + O(\varepsilon^2) \tag{2.18}$$

is used in the first integral. The second term above gives

$$-\varepsilon \iint_{S_B^{(0)}} \rho \Phi_t^{(1)} \mathbf{n}\, dS + O(\varepsilon^2), \tag{2.19}$$

where S_B has been approximated by $S_B^{(0)}$. To consider the buoyancy term $-\rho g f$ we assume for simplicity that the horizontal cross section of the body decreases in area with depth, although all the final results can be shown to remain valid for more complex geometries. Now on the instantaneous body surface S_B,

$$\mathbf{n}\, dS = (-f_x, -f_y, 1)\, dx\, dy,$$

and the domain of integration S_B may be replaced by the part of the water surface cut out by S_B, that is, S_A (see Fig. 2.1). The vertical component of buoyancy is

$$-\rho g \iint_{S_B} f n_3\, dS = -\rho g \iint_{S_A} (f^{(0)} + \varepsilon f^{(1)})\, dx\, dy.$$

Since S_A differs from its equilibrium counterpart $S_A^{(0)}$ by $O(\varepsilon)$, and since $f^{(0)} = 0$ in the equilibrium free surface, the surface S_A may be replaced by $S_A^{(0)}$

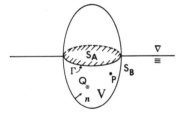

Figure 2.1 A floating body.

with an error of $O(\varepsilon^2)$. By partial integration, we have

$$\iint_{S_A^{(0)}} f_x^{(0)} \, dx \, dy = \oint_\Gamma [\, f^{(0)}]_{x_-}^{x_+} \, dy,$$

where Γ is the edge of $S_A^{(0)}$, that is, the water line, and x_+ and x_- refer to the points on Γ intersected by a line of constant y. Since $f^{(0)} = 0$ on Γ, the right-hand side integral in the equation above vanishes; it follows by using Eq. (2.9) that

$$-\rho g \iint_{S_B} fn_3 \, dS = -\rho g \iint_{S_A^{(0)}} f^{(0)} \, dx \, dy - \varepsilon \rho g \iint_{S_A^{(0)}} [\, Z^{(1)} + \alpha(\, y - Y^{(0)})$$

$$- \beta(x - X^{(0)})] \, dx \, dy + O(\varepsilon^2). \tag{2.20}$$

The first integral on the right is just the submerged volume $V^{(0)}$ (static displacement). When Eqs. (2.17), (2.19), and (2.20) are combined with Eq. (2.15), terms of different orders may be separated. At the zero-order, we have

$$Mg = \rho g V^{(0)} + \mathbb{F}_3^{(0)}, \tag{2.21}$$

which is just Archimedes' law.

Let $A^{(0)}$ be the area of $S_A^{(0)}$ and

$$I_1^A = \iint_{S_A^{(0)}} (x - X^{(0)}) \, dx \, dy, \qquad I_2^A = \iint_{S_A^{(0)}} (y - Y^{(0)}) \, dx \, dy \tag{2.22}$$

be the moments of inertia of the cut plane $S_A^{(0)}$, then the linearized z momentum equation at $O(\varepsilon)$ is

$$M[Z_{tt}^{(1)} + \alpha_{tt}(\bar{y}^c - Y^{(0)}) - \beta_{tt}(\bar{x}^c - X^{(0)})] = -\rho \iint_{S_B^{(0)}} \Phi_t^{(1)} n_3 \, dS + \mathbb{F}_3^{(1)}$$

$$- \rho g (I_2^A \alpha - I_1^A \beta + Z^{(1)} A^{(0)}). \tag{2.23a}$$

If the floating body is totally immersed, $S_A^{(0)}$ vanishes and buoyancy does not affect the dynamic equilibrium.

Consider next the x and y components. Note that

$$\rho g \iint_{S_B} fn_1 \, dS = -\rho g \iint_{S_B} ff_x \, dx \, dy = -\rho g \oint_\Gamma \left[\tfrac{1}{2} f^2\right]_{x_-}^{x_+} dy = O(\varepsilon^2),$$

and

$$\rho g \iint_{S_B} fn_2 \, dS = -\rho g \iint_{S_A} ff_y \, dx = -\rho g \oint_\Gamma \left[\tfrac{1}{2} f^2\right]_{y_-}^{y_+} dx = O(\varepsilon^2).$$

Thus, buoyancy has a negligible effect in the horizontal directions. It follows from Eq. (2.15) that at the first order

$$M\left[X_{tt}^{(1)} + \beta_{tt}(\bar{z}^c - Z^{(0)}) - \gamma_{tt}(\bar{y}^c - Y^{(0)}) \right] = -\rho \iint_{S_B^{(0)}} \Phi_t^{(1)} n_1 \, dS + \mathbb{F}_1^{(1)},$$

$$(2.23b)$$

$$M\left[Y_{tt}^{(1)} + \gamma_{tt}(\bar{x}^c - X^{(0)}) - \alpha_{tt}(\bar{z}^c - Z^{(0)}) \right] = -\rho \iint_{S_B^{(0)}} \Phi_t^{(1)} n_2 \, dS + \mathbb{F}_2^{(1)}.$$

$$(2.23c)$$

The zeroth order (static) balance of horizontal momentum is trivially satisfied.

7.2.3   Conservation of Angular Momentum

Let V_b be the volume of the whole body including the part above the free surface. The rate of change of angular momentum of the body about the axes passing through the fixed origin 0 is

$$\frac{d\mathbf{L}}{dt} = \frac{d}{dt} \iiint_{V_b} \mathbf{x} \times \mathbf{x}_t \, dm = \iiint_{V_b} \mathbf{x} \times \mathbf{x}_{tt} \, dm,$$

where dm is the body mass per unit volume and $\mathbf{x}(t)$ denotes the instantaneous position of a point in the rigid body. Conservation of angular momentum requires that

$$\iiint_{V_b} \mathbf{x} \times \mathbf{x}_{tt} \, dm = \iint_{S_B} \mathbf{x} \times P\mathbf{n} \, dS + \mathbf{x}^c \times (-Mg\mathbf{e}_3) + \mathbf{T}, \qquad (2.24)$$

where $\mathbf{T}$ is the constraining torque due to external influences. Note that by definition $\mathbf{x}^c$ is the center of mass of the body, that is,

$$\iiint_{V_b} \mathbf{x} \, dm = M\mathbf{x}^c. \tag{2.25}$$

The equation of linear momentum (2.15) may be rewritten

$$\iiint_{V_b} \mathbf{x}_{tt} \, dm = \iint_{S_B} P\mathbf{n} \, dS - Mg\mathbf{e}_3 + \mathbf{F}. \tag{2.26}$$

Taking the cross product of $\mathbf{X}$ with Eq. (2.26) and subtracting the result from Eq. (2.24), we get

$$\frac{d\mathbf{L}^Q}{dt} \equiv \iiint_{V_b} (\mathbf{x} - \mathbf{X}) \times \mathbf{x}_{tt} \, dm = \iint_{S_B} (\mathbf{x} - \mathbf{X}) \times P\mathbf{n} \, dS$$

$$+ (\mathbf{x}^c - \mathbf{X}) \times (-Mg\mathbf{e}_3) + (\mathbf{T} + \mathbf{F} \times \mathbf{X}) \tag{2.27}$$

which represents the conservation of angular momentum with respect to the center of rotation Q.

Consider the zero-order balance. The left-hand side of Eq. (2.27) is ineffective. On the right-hand side, the constraining torque is simple:

$$\mathbf{T} + \mathbf{F} \times \mathbf{X} = \left(\mathbf{T}^{(0)} + \mathbf{F}^{(0)} \times \mathbf{X}^{(0)}\right) + \varepsilon\left(\mathbf{T}^{(1)} + \mathbf{F}^{(1)} \times \mathbf{X}^{(0)}\right.$$

$$\left. + \mathbf{F}^{(0)} \times \mathbf{X}^{(1)}\right) + O(\varepsilon^2).$$

For convenience we denote

$$\mathfrak{T}^{(0)} = \mathbf{T}^{(0)} + \mathbf{F}^{(0)} \times \mathbf{X}^{(0)} \quad \text{and} \quad \mathfrak{T}^{(1)} = \mathbf{T}^{(1)} + \mathbf{F}^{(1)} \times \mathbf{X}^{(0)},$$

so that

$$\mathbf{T} + \mathbf{F} \times \mathbf{X} = \mathfrak{T}^{(0)} + \varepsilon\mathfrak{T}^{(1)} + \varepsilon\mathbf{F}^{(0)} \times \mathbf{X}^{(1)}. \tag{2.28}$$

When Eq. (2.28) is substituted into Eq. (2.27), it is easily seen that to the zeroth order

$$-\rho g \iint_{S_B^{(0)}} (\bar{\mathbf{x}} - \mathbf{X}^{(0)}) \times \mathbf{n} f^{(0)} \, dS + (\bar{\mathbf{x}}^c - \mathbf{X}^{(0)}) \times (-Mg\mathbf{e}_3) + \mathfrak{T}^{(0)} = 0.$$

$$\tag{2.29}$$

Being over the rest surface $S_B^{(0)}$, $\bar{\mathbf{x}}$ in the integral above may be replaced by $\mathbf{x}$. Using the argument leading to Eq. (2.20), we may write the x component of the same integral as

$$-\rho g\left[\iint_{S_A^{(0)}}(y - Y^{(0)})f^{(0)}\,dx\,dy + \iint_{S_A^{(0)}}f^{(0)}(f^{(0)} - Z^{(0)})f_y^{(0)}\,dx\,dy\right].$$

The second integral above vanishes after partial integration in y. On the other hand, since $f^{(0)} = -\int_{f^{(0)}}^0 dz$ the first integral may be rewritten

$$\rho g\iiint_{V^{(0)}}(y - Y^{(0)})\,dx\,dy\,dz \equiv \rho g I_2^V \equiv \rho g V^{(0)}(y^V - Y^{(0)}), \quad (2.30)$$

where $V^{(0)}$ is the static submerged volume, I_2^V its first moment about the plane $y = Y^{(0)}$, and y^V the y coordinate of the *center of buoyancy*. In terms of these symbols, we get from Eq. (2.29)

$$\rho g I_2^V - Mg(y^c - Y^{(0)}) + \mathcal{T}_1^{(0)} = 0. \quad (2.31)$$

Similarly, for the y component we may define I_1^V and x^V by

$$\rho g I_1^V \equiv \rho g V^{(0)}(x^V - X^{(0)}) \equiv \rho g\iiint_{V^{(0)}}(x - X^{(0)})\,dx\,dy\,dz, \quad (2.32)$$

and obtain

$$-\rho g I_1^V + Mg(\bar{x}^c - X^{(0)}) + \mathcal{T}_2^{(0)} = 0. \quad (2.33)$$

If there is no external contraint, $\mathcal{T}^{(0)} \equiv 0$; we must have $Mg = \rho g V^{(0)}$. It follows from Eqs. (2.31) and (2.33) that $x^c = x^V$ and $y^c = y^V$, that is, the centers of mass and of buoyancy must lie on the same vertical line.

We leave it for the reader to verify that the z component of the integral in Eq. (2.29) vanishes identically so that the corresponding static balance is trivial:

$$\mathcal{T}_3^{(0)} = 0. \quad (2.34)$$

Now consider the order $O(\varepsilon)$. On the left-hand side of Eq. (2.27) we may replace $\mathbf{x} - \mathbf{X}$ by $\bar{\mathbf{x}} - \mathbf{X}^{(0)}$ and invoke Eq. (2.6):

$$\mathbf{x}_{tt} = \varepsilon\left[\mathbf{X}_{tt}^{(1)} + \boldsymbol{\theta}_{tt}^{(1)} \times (\bar{\mathbf{x}} - \mathbf{X}^{(0)})\right].$$

Equation (2.27) becomes

$$\frac{d\mathbf{L}^{\varrho}}{dt} \equiv \varepsilon\left\{\left[\iiint_{V_b}(\bar{\mathbf{x}} - \mathbf{X}^{(0)})\,dm\right] \times \mathbf{X}_{tt}^{(1)} + \boldsymbol{\theta}_{tt}^{(1)}\iiint_{V_b}(\bar{\mathbf{x}} - \mathbf{X}^{(0)})^2\,dm\right.$$

$$\left. -\left[\boldsymbol{\theta}_{tt}^{(1)} \cdot \iiint_{V_b}(\bar{\mathbf{x}} - \mathbf{X}^{(0)})\right](\bar{\mathbf{x}} - \mathbf{X}^{(0)})\,dm\right\}. \quad (2.35)$$

Let us define the first and second moments of inertia as follows:

$$I_1^b = \iiint_{V_b} (\bar{x} - X^{(0)})\, dm \equiv M(\bar{x}^c - X^{(0)}),$$

$$I_{11}^b = \iiint_{V_b} (\bar{x} - X^{(0)})^2\, dm, \tag{2.36}$$

$$I_{12}^b = \iiint_{V_b} (\bar{x} - X^{(0)})(\bar{y} - Y^{(0)})\, dm.$$

Other moments $I_2^b, I_3^b, I_{22}^b, \ldots, I_{33}^b, \ldots$, and so on, are similarly defined. In terms of these moments, the left-hand side of Eq. (2.27) becomes, in component form,

$$\frac{dL_1^Q}{dt} = \varepsilon\{ I_2^b Z_{tt}^{(1)} - I_3^b Y_{tt}^{(1)} + (I_{22}^b + I_{33}^b)\alpha_{tt} - I_{21}^b \beta_{tt} - I_{31}^b \gamma_{tt}\},$$

$$\frac{dL_2^Q}{dt} = \varepsilon\{ I_3^b X_{tt}^{(1)} - I_1^b Z_{tt}^{(1)} + (I_{33}^b + I_{11}^b)\beta_{tt} - I_{32}^b \gamma_{tt} - I_{12}^b \alpha_{tt}\}, \tag{2.37}$$

$$\frac{dL_3^Q}{dt} = \varepsilon\{ I_1^b Y_{tt}^{(1)} - I_2^b X_{tt}^{(1)} + (I_{11}^b + I_{22}^b)\gamma_{tt} - I_{13}^b \alpha_{tt} - I_{23}^b \beta_{tt}\}.$$

There are no terms of zeroth order.

On the other hand, the right-hand side of Eq. (2.27) gives

$$-\varepsilon \iint_{S_B^{(0)}} \rho\Phi_t^{(1)}(\mathbf{x} - \mathbf{X}^{(0)}) \times \mathbf{n}\, dS$$

$$-\rho g\left\{ \iint_{S_B} f(\mathbf{x} - \mathbf{X}) \times \mathbf{n}\, dS - \iint_{S_B^{(0)}} f^{(0)}(\mathbf{x} - \mathbf{X}^{(0)}) \times \mathbf{n}\, dS \right\}$$

$$+\varepsilon\left[\boldsymbol{\theta}^{(1)} \times (\bar{\mathbf{x}}^c - \mathbf{X}^{(0)})\right] \times (-Mg\mathbf{e}_3) + \varepsilon(\mathfrak{J}^{(1)} + \mathbf{F}^{(0)} \times \mathbf{X}^{(1)}). \tag{2.38}$$

These terms represent torques of various physical origin: the first integral is due to hydrodynamics, the second from buoyancy, and the third from inertia, while the remaining terms are from the constraint. We now treat them separately.

Hydrodynamic Torque

The generalized normal (2.13b) enables us to express the components of the hydrodynamic torque

$$-\varepsilon \iint_{S_B^{(0)}} \rho\Phi_t^{(1)} n_\alpha\, dS, \qquad \alpha = 4, 5, 6. \tag{2.39}$$

Buoyancy Torque

Written in component form, the buoyancy torque is

$$
-\rho g \iint_{S_B} f(\mathbf{x} - \mathbf{X}) \times \mathbf{n}\, dS = -\rho g \iint_{S_B} f\{[(y - Y)n_3 - (z - Z)n_2]\mathbf{e}_1
$$

$$
+ [(z - Z)n_1 - (x - X)n_3]\mathbf{e}_2 + [(x - X)n_2 - (y - Y)n_1]\mathbf{e}_3\}\, dS
$$

$$
= -\rho g \iint_{S_A} f\{[(y - Y) + (f - Z)f_y]\mathbf{e}_1 + [-(f - Z)f_x - (x - X)]\mathbf{e}_2
$$

$$
+ [-(x - X)f_y + (y - Y)f_x]\mathbf{e}_3\}\, dx\, dy.
$$

After partial integration and noting $f = \varepsilon f^{(1)}$ on the edge of S_A, we find that the terms with f_x and f_y are of order $O(\varepsilon^2)$; hence

$$
-\rho g \iint_{S_B} f(\mathbf{x} - \mathbf{X}) \times \mathbf{n}\, dS = -\rho g \iint_{S_A} [f(y - Y)\mathbf{e}_1 - f(x - X)\mathbf{e}_2]\, dx\, dy.
$$

$$
(2.40)
$$

The second integral in $\{\ \}$ of Eq. (2.38) obviously takes a form similar to Eq. (2.40). Consider the x component of the same $\{\ \}$:

$$
\{\ \} = -\rho g \iint_{S_A} f(y - Y)\, dx\, dy - \rho g \iint_{S_A^{(0)}} f^{(0)}(y - Y^{(0)})\, dx\, dy
$$

$$
= -\varepsilon \rho g \iint_{S_A^{(0)}} [f^{(1)}(y - Y^{(0)}) - f^{(0)}Y^{(1)}]\, dx\, dy
$$

$$
= -\varepsilon \rho g \iint_{S_A^{(0)}} [Z^{(1)} + \alpha(y - Y^{(0)}) - \beta(x - X^{(0)})](y - Y^{(0)})\, dx\, dy
$$

$$
+ \varepsilon \rho g \iint_{S_A^{(0)}} f_x^{(0)}[X^{(1)} + \beta(z - Z^{(0)}) - \gamma(y - Y^{(0)})](y - Y^{(0)})\, dx\, dy
$$

$$
+ \varepsilon \rho g \iint_{S_A^{(0)}} f_y^{(0)}[Y^{(1)} + \gamma(x - X^{(0)}) - \alpha(z - Z^{(0)})](y - Y^{(0)})\, dx\, dy
$$

$$
+ \varepsilon \rho g Y^{(1)} \iint_{S_A^{(0)}} f^{(0)}\, dx\, dy.
$$

After partial integration, the second integral above vanishes; the third and fourth integrals combine to give

$$-\varepsilon\rho g \iint_{S_A^{(0)}} f^{(0)}\big[\gamma(x - X^{(0)}) - \alpha(z - Z^{(0)})\big]\, dx\, dy$$

$$= +\varepsilon\rho g \iiint_{V^{(0)}} \big[\gamma(x - X^{(0)}) - \alpha(z - Z^{(0)})\big]\, dx\, dy\, dz,$$

where use is made of $f^{(0)} = -\int_{f^{(0)}}^{0} dz$. Introducing the second moments

$$I_{22}^A = \iint_{S_A^{(0)}} (y - Y^{(0)})^2\, dx\, dy,$$

$$I_{12}^A = \iint_{S_A^{(0)}} (x - X^{(0)})(y - Y^{(0)})\, dx\, dy, \qquad \text{and so on,} \qquad (2.41)$$

we get the x component of the buoyancy torque:

$$-\varepsilon\rho g\big[Z^{(1)}I_2^A + \alpha I_{22}^A - \beta I_{12}^A - \gamma I_1^V + \alpha I_3^V\big], \qquad (2.42)$$

and, by similar arguments, the y component:

$$-\varepsilon\rho g\big[Z^{(1)}I_1^A + \alpha I_{21}^A - \beta I_{11}^A - \beta I_3^V + \gamma I_2^V\big]. \qquad (2.43)$$

There is no z component.

Inertia Torque

Expanding the triple vector product in Eq. (2.38), we get

$$\mathbf{e}_1(-Mg)\big[\gamma(\bar{x}^c - X^{(0)}) - \alpha(\bar{z}^c - Z^{(0)})\big]$$

$$+\mathbf{e}_2(Mg)\big[\beta(\bar{z}^c - Z^{(0)}) - \gamma(\bar{y}^c - Y^{(0)})\big]. \qquad (2.44)$$

Constraining Torque

$$\mathbf{e}_1\big(\mathfrak{T}_1^{(1)} + \mathbb{F}_2^{(0)}Z^{(1)} - \mathbb{F}_3^{(0)}Y^{(1)}\big)$$

$$+\mathbf{e}_2\big(\mathfrak{T}_2^{(1)} + \mathbb{F}_3^{(0)}X^{(1)} - \mathbb{F}_1^{(0)}Z^{(1)}\big) + \mathbf{e}_3\big(\mathfrak{T}_3^{(1)} + \mathbb{F}_1^{(0)}Y^{(1)} - \mathbb{F}_2^{(0)}X^{(1)}\big). \qquad (2.45)$$

Equations (2.39) and (2.42)–(2.45) may be combined with Eq. (2.38) and then with Eq. (2.37) to give the conservation equations of angular momentum, as follows: x component

$$I_2^b Z_{tt}^{(1)} - I_3^b Y_{tt}^{(1)} + \left(I_{22}^b + I_{33}^b\right)\alpha_{tt} - I_{21}^b \beta_{tt} - I_{13}^b \gamma_{tt}$$

$$= -\rho \iint_{S_B^{(0)}} \Phi_t^{(1)} n_4 \, dS - \rho g \left\{ Z^{(1)} I_2^A + \alpha\left(I_{22}^A + I_3^V\right) - \beta I_{12}^A - \gamma I_1^V \right\}$$

$$+ Mg \left[\alpha(\bar{z}^c - Z^{(0)}) - \gamma(\bar{x}^c - X^{(0)}) \right] + \mathcal{J}_1^{(1)} + \mathbb{F}_2^{(0)} Z^{(1)} - \mathbb{F}_3^{(0)} Y^{(1)},$$

$$(2.46a)$$

y component

$$I_3^b X_{tt}^{(1)} - I_1^b Z_{tt}^{(1)} + \left(I_{33}^b + I_{11}^b\right)\beta_{tt} - I_{32}^b \gamma_{tt} - I_{12}^b \alpha_{tt}$$

$$= -\rho \iint_{S_B^{(0)}} \Phi_t^{(1)} n_5 \, dS + \rho g \left\{ Z^{(1)} I_1^A + \alpha I_{21}^A + \beta\left(-I_{11}^A - I_3^V\right) + \gamma I_2^V \right\}$$

$$+ Mg \left[(\bar{z}^c - Z^{(0)})\beta - (\bar{y}^c - Y^{(0)})\gamma \right] + \mathcal{J}_2^{(1)} + \mathbb{F}_3^{(0)} X^{(1)} - \mathbb{F}_1^{(0)} Z^{(1)},$$

$$(2.46b)$$

z component

$$I_1^b Y_{tt}^{(1)} - I_2^b X_{tt}^{(1)} + \left(I_{11}^b + I_{22}^b\right)\gamma_{tt} - I_{13}^b \alpha_{tt} - I_{23}^b \beta_{tt}$$

$$= -\rho \iint_{S_B^{(0)}} \Phi_t^{(1)} n_6 \, dS + \mathcal{J}_3^{(1)} + \mathbb{F}_1^{(0)} Y^{(1)} - \mathbb{F}_2^{(0)} X^{(1)}. \qquad (2.46c)$$

Equations (2.23) and (2.46) can be regarded as the dynamic boundary conditions for the fluid problem. In the original derivation of John (1950), the reference point Q is assumed to coincide with the center of mass. The minor extension presented here is due to Serman (1978).

7.2.4 Summary of Dynamic Equations for a Floating Body in Matrix Form

The linear system of Eq. (2.22) and (2.46) can be summarized in matrix form in terms of the generalized coordinates (2.13a)

$$[M]\{\ddot{X}\} + [C]\{X\} = -\rho \iint_{S_B^{(0)}} dS \, \Phi_t^{(1)}\{n\} + \{\mathbb{F}\}, \qquad (2.47)$$

where an overhead dot denotes the time derivative and $\{\mathbb{F}\}$ is the generalized dynamic force of constraint:

$$\left\{ \mathbb{F}^{(1)}, \, \mathcal{J}^{(1)} + \mathbb{F}^{(0)} \times \mathbf{X}^{(1)} \right\}^T. \qquad (2.48)$$

The mass matrix $[M]$ and the buoyancy restoring force matrix $[C]$ are given respectively by Eqs. (2.49) and (2.50) shown on the next page.

$$[M] = \begin{bmatrix}
M & 0 & 0 & 0 & M(\bar{z}^c - Z^{(0)}) & -M(\bar{y}^c - Y^{(0)}) \\
0 & M & 0 & -M(\bar{z}^c - Z^{(0)}) & 0 & M(\bar{x}^c - X^{(0)}) \\
0 & 0 & M & M(\bar{y}^c - Y^{(0)}) & -M(\bar{x}^c - X^{(0)}) & 0 \\
0 & -M(\bar{z}^c - Z^{(0)}) & M(\bar{y}^c - Y^{(0)}) & (I_{22}^b + I_{33}^b) & -I_{21}^b & -I_{13}^b \\
M(\bar{z}^c - Z^{(0)}) & 0 & -M(\bar{x}^c - X^{(0)}) & -I_{12}^b & I_{33}^b + I_{11}^b & -I_{32}^b \\
-M(\bar{y}^c - Y^{(0)}) & M(\bar{x}^c - X^{(0)}) & 0 & -I_{13}^b & -I_{23}^b & I_{11}^b + I_{22}^b
\end{bmatrix} . \quad (2.49)$$

$$[C] = \begin{bmatrix}
0 & 0 & 0 & 0 & 0 & 0 \\
0 & 0 & 0 & 0 & 0 & 0 \\
0 & 0 & \rho g A & \rho g I_2^A & \rho g I_1^A & 0 \\
0 & \mathbb{F}_3^{(0)} & \begin{pmatrix} \rho g I_2^A - \mathbb{F}_2^{(0)} \end{pmatrix} & \begin{pmatrix} \rho g(I_{22}^A + I_3^V) \\ -Mg(\bar{z}^c - Z^{(0)}) \end{pmatrix} & \begin{pmatrix} \rho g(I_{11}^A + I_3^V) \\ -Mg(\bar{z}^c - Z^{(0)}) \end{pmatrix} & \begin{pmatrix} -\rho g I_1^V \\ +Mg(\bar{x}^c - X^{(0)}) \end{pmatrix} \\
-\mathbb{F}_3^{(0)} & 0 & \begin{pmatrix} -\rho g I_1^A + \mathbb{F}_1^{(0)} \end{pmatrix} & -\rho g I_{21}^A & -\rho g I_{12}^A & \begin{pmatrix} -\rho g I_2^V \\ +Mg(\bar{y}^c - Y^{(0)}) \end{pmatrix} \\
\mathbb{F}_2^{(0)} & -\mathbb{F}_1^{(0)} & 0 & 0 & 0 & 0
\end{bmatrix} . \quad (2.50)$$

It is obvious that the matrix $[M]$ is symmetric. If the center of mass coincides with the center of rotation, then $\bar{\mathbf{x}}^c - \mathbf{X}^{(0)} = 0$, and many terms in $[M]$ and $[C]$ vanish identically. Furthermore, when there is no constraint, $I_1^V = I_2^V = 0$ by Eqs. (2.31) and (2.33); $[C]$ is also symmetric.

In some situations the contraining force $\mathbb{F}^{(1)}$ and torque $\mathbf{T}^{(1)}$ may depend on the body displacement; equations governing the dynamics of the constraints must then be added.

The Two-Dimensional Limit

For a long horizontal cylinder with normally incident waves from one side, the motion can be described in the cross-sectional plane of x and z. It is only necessary to restrict our attention to unit length in the y direction.

For the rigid body there are only two translational modes x and z and one rotational mode β about the y axis. The relevant generalized coordinates are 1, 3, and 5. For a right-handed coordinate system, the negative y axis points out of the paper so that positive rotations are clockwise.

The linear momentum equations are reduced to

$$M\left[X_{tt}^{(1)} + \beta_{tt}(\bar{z}^c - Z^{(0)}) \right] = -\rho \int_{S_B^{(0)}} \Phi_t^{(1)} n_1 \, dS + \mathbb{F}_1^{(1)}, \tag{2.51a}$$

$$M\left[Z_{tt}^{(1)} - \beta_{tt}(\bar{x}^c - X^{(0)}) \right] = -\rho \int_{S_B^{(0)}} \Phi_t^{(1)} n_3 \, dS + \mathbb{F}_3^{(1)} + \rho g I_1^A \beta - \rho g Z^{(1)} A, \tag{2.51b}$$

and the angular momentum about the y axis is

$$I_3^b X_{tt}^{(1)} - I_1^b Z_{tt}^{(1)} + \left(I_{33}^b + I_{11}^b \right)\beta_{tt}$$

$$= -\rho \int_{S_B^{(0)}} \Phi_t^{(1)} n_5 \, dS + \rho g\left\{ Z^{(1)} I_1^A - \beta\left(I_{11}^A + I_3^V \right) \right\}$$

$$+ Mg(\bar{z}^c - Z^{(0)})\beta + \mathcal{T}_2^{(1)} + \mathbb{F}_3^{(0)} X^{(1)} - \mathbb{F}_1^{(0)} Z^{(1)}, \tag{2.52}$$

where

$$I_1^A = \int_{S_A^{(0)}} (x - X^{(0)}) \, dx, \qquad I_1^b = \iint_{V_b} (x - X^{(0)}) \, dm,$$

$$I_{11}^A = \int_{S_A^{(0)}} (x - X^{(0)})^2 \, dx, \qquad I_{11}^b = \iint_{V_b} (x - X^{(0)})^2 \, dm. \tag{2.53}$$

$$I_3^V = \iint_{V^{(0)}} (z - Z^{(0)}) \, dx \, dz.$$

In the expressions above, A is the length of the water line S_A, that is, the segment of the x axis which is displaced by the body, V is the submerged cross-section area, and V_b is the entire body cross section. The corresponding matrix equation is

$$[M]\frac{d^2}{dt^2}\begin{Bmatrix} X^{(1)} \\ Z^{(1)} \\ \beta \end{Bmatrix} + [C]\begin{Bmatrix} X^{(1)} \\ Z^{(1)} \\ \beta \end{Bmatrix} = -\rho\int_{S_B^{(0)}} dS\, \Phi_t^{(1)}\begin{Bmatrix} n_1 \\ n_3 \\ n_5 \end{Bmatrix}$$

$$+ \begin{Bmatrix} \mathbb{F}_1^{(1)} \\ \mathbb{F}_3^{(1)} \\ \mathcal{T}_2^{(1)} + \mathbb{F}_3^{(0)}X_1^{(1)} - \mathbb{F}_1^{(0)}X_3^{(1)} \end{Bmatrix},$$

$$(2.54)$$

where

$$[M] = \begin{bmatrix} M & 0 & M(\bar{z}^c - Z^{(0)}) \\ 0 & M & -M(\bar{x}^c - X^{(0)}) \\ M(\bar{z}^c - Z^{(0)}) & -M(\bar{x}^c - X^{(0)}) & I_{11}^b + I_{33}^b \end{bmatrix},$$

$$(2.55)$$

and

$$[C] = \begin{bmatrix} 0 & 0 & 0 \\ 0 & \rho g A & -\rho g I_1^A \\ -\mathbb{F}_3^{(0)} & -\rho g I_1^A + \mathbb{F}_1^{(0)} & \begin{pmatrix} \rho g(I_{11}^A + I_3^V) \\ -Mg(\bar{z}^c - Z^{(0)}) \end{pmatrix} \end{bmatrix}.$$

$$(2.56)$$

where the inertia terms refer to the unit length of the cylinder.

So far our derivation has been rather formal so that no first-order term could escape our attention. For the relatively simple case of two dimensions, it is instructive to reexamine some of the terms heuristically. Consider the torque caused by rotation β. Referring to Fig. 2.2 for the boundary element dS, a positive (clockwise) β will induce an added buoyancy force of the magnitude

$$-\rho g[\boldsymbol{\theta}^{(1)} \times (\mathbf{x} - \mathbf{X}^{(0)})] \cdot \mathbf{n}\, dS = -\beta[(z - Z^{(0)})n_1 - (x - X^{(0)})n_3]\rho g\, dS$$

$$(2.57)$$

in the vertical direction. The total restoring moment is

$$-\rho g\beta\int_{S_B^{(0)}}[(z - Z^{(0)})n_1 - (x - X^{(0)})n_3](x - X^{(0)})\, dS.$$

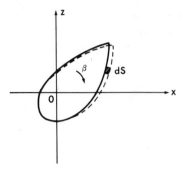

Figure 2.2

Since $n_1 \, dS = dz$ and $n_3 \, dS = -dx$, this moment can also be expressed

$$-\rho g \beta \left[\int_{S_B^{(0)}} (x - X^{(0)})(z - Z^{(0)}) \, dz + \int_{S_B^{(0)}} (x - X^{(0)})^2 \, dx \right]. \qquad (2.58)$$

Noting that $(x - X^{(0)}) = \int_{X^{(0)}}^{x} dx$, we may rewrite the first integral in Eq. (2.58) as

$$\iint_{V^{(0)}} (z - Z^{(0)}) \, dx \, dz = I_3^V,$$

and the second integral in Eq. (2.58) as

$$\int_{S_B^{(0)}} (x - X^{(0)})^2 \, dx = I_{11}^A.$$

Thus, Eq. (2.58) is in agreement with the β terms in Eq. (2.52c) and the appearance of I_3^V and I_{11}^A is verified by an elementary consideration.

In subsequent sections all superscripts (0) and (1) will be omitted for brevity.

7.3 SIMPLE HARMONIC MOTION

7.3.1 Decomposition into Diffraction and Radiation Problems

As in simpler vibrating systems governed by linear ordinary differential equations with constant coefficients, the most basic task is to study the frequency response to a simple harmonic excitation. Furthermore, it is convenient to decouple the hydrodynamics from the body dynamics by the following device (Haskind, 1944). Let us introduce the complex amplitude

$$\{ \Phi, \dot{X}_\alpha \}^T = \mathrm{Re} \{ \phi, V_\alpha \}^T e^{-i\omega t}, \qquad (3.1)$$

where V_α denotes the amplitude of the generalized body velocity, and the decomposition

$$\phi = \phi^D + \sum_\alpha V_\alpha \phi_\alpha, \qquad \alpha = 1, 2, \ldots, 6. \tag{3.2}$$

The potential ϕ^D is governed by the following conditions:

$$\nabla^2 \phi^D = 0 \qquad \text{in the fluid,} \tag{3.3}$$

$$\frac{\partial \phi^D}{\partial z} - \frac{\omega^2}{g} \phi^D = 0 \qquad \text{on } S_F \ (z = 0), \tag{3.4}$$

$$\frac{\partial \phi^D}{\partial z} = 0 \qquad \text{on } B_0 \ (z = -h, \text{ sea bottom}), \tag{3.5}$$

$$\frac{\partial \phi^D}{\partial n} = 0 \qquad \text{on } S_B \text{ (body surface),} \tag{3.6}$$

$$\phi^D - \phi^I \qquad \text{outgoing at infinity.} \tag{3.7}$$

ϕ^I represents the incident wave potential. Thus, ϕ^D represents the *diffraction* potential when the body is held stationary in incoming waves. Using the generalized normal introduced in Eq. (2.4), we define ϕ_α to satisfy Eqs. (3.3)–(3.5) and

$$\frac{\partial \phi_\alpha}{\partial n} = n_\alpha \qquad \text{on } S_B, \tag{3.8}$$

$$\phi_\alpha \qquad \text{outgoing at infinity.} \tag{3.9}$$

It is obvious that the total potential ϕ of Eq. (3.2) satisfies

$$\frac{\partial \phi}{\partial n} = \sum_\alpha V_\alpha n_\alpha \qquad \text{on } S_B, \tag{3.10}$$

hence Φ satisfies Eq. (2.14). Each ϕ_α corresponds to a generalized mode of forced motion with unit body velocity and is a *radiation* potential.

The advantage of the decomposition is that the component problems for ϕ^D and ϕ_α involve only hydrodynamics and can be solved separately first. The results are then used in Eq. (2.47) or (2.54) to determine the body motion. The solution of ϕ^D and ϕ_α is usually the most difficult part of the entire task.

For very large structures, elastic deformation due to wave forces may be appreciable. If the motion of the body surface is decomposed into real-valued normal modes, then decomposition can still be introduced with suitable reinterpretation of the generalized normal n_α.

We shall always assume that sufficiently far from the body or variable depth, the sea bottom is horizontal at the depth h. We further assume a plane incident wave propagating in the direction $\theta = \theta_I$; the corresponding potential is

$$\phi^I = \frac{-igA}{\omega} \frac{\cosh k(z + h)}{\cosh kh} e^{ikr\cos(\theta - \theta_I)}. \tag{3.11}$$

7.3.2 Exciting and Restoring Forces; Added Mass and Radiation Damping for a Body of Arbitrary Shape

We now introduce some general definitions applicable to a body of arbitrary shape.

The αth component of the generalized hydrodynamic force on the body is

$$\mathscr{F}_\alpha = \iint_{S_B} Pn_\alpha \, dS = \mathrm{Re}\big(F_\alpha e^{-i\omega t}\big), \tag{3.12}$$

where

$$F_\alpha = i\rho\omega \iint_{S_B} \phi n_\alpha \, dS. \tag{3.13}$$

Substituting Eq. (3.2) into Eq. (3.13), we get

$$F_\alpha = i\rho\omega \iint_{S_B} \phi^D n_\alpha \, dS + \sum_\beta i\rho\omega V_\beta \iint_{S_B} \phi_\beta n_\alpha \, dS. \tag{3.14}$$

We also denote

$$F_\alpha^D = i\rho\omega \iint_{S_B} \phi^D n_\alpha \, dS, \qquad f_{\beta\alpha} = i\rho\omega \iint_{S_B} \phi_\beta n_\alpha \, dS \tag{3.15}$$

so that

$$F_\alpha = F_\alpha^D + \sum_\beta V_\beta f_{\beta\alpha}. \tag{3.16}$$

The vector $\{F_\alpha^D\}$ is the *exciting force* on a stationary body due to diffraction, while the matrix $[f_{\beta\alpha}]$ is called the *restoring force* matrix. The component $f_{\beta\alpha}$ represents the hydrodynamic reaction in direction α due to the normal mode β. Consider the effect of the β mode only. The restoring force is, without summing over β,

$$\text{Re}\left(V_\beta f_{\beta\alpha} e^{-i\omega t}\right) = \text{Re}\left[(\text{Re } f_{\beta\alpha} + i \text{ Im } f_{\beta\alpha}) V_\beta e^{-i\omega t}\right]$$

$$= \text{Re}\left[\left(i\rho\omega \iint_{S_B} \text{Re } \phi_\beta n_\alpha \, dS - \rho\omega \iint_{S_B} \text{Im } \phi_\beta n_\alpha \, dS\right) \cdot V_\beta e^{-i\omega t}\right]$$

$$= -\left(\rho \iint_{S_B} \text{Re } \phi_\beta n_\alpha \, dS\right) \text{Re } \frac{d}{dt}\left(V_\beta e^{-i\omega t}\right) - \left(\rho\omega \int_{S_B} \text{Im } \phi_\beta n_\alpha \, dS\right) \text{Re}\left(V_\beta e^{-i\omega t}\right)$$

$$= -\left(\rho \iint_{S_B} \text{Re } \phi_\beta n_\alpha \, dS\right) \ddot{X}_\beta - \left(\rho\omega \iint_{S_B} \text{Im } \phi_\beta n_\alpha \, dS\right) \dot{X}_\beta. \tag{3.17}$$

The first integral above is proportional to the body acceleration and is the hydrodynamic inertia. We therefore call the matrix

$$[\mu]: \qquad \mu_{\beta\alpha} = \rho \iint_{S_B} \text{Re } \phi_\beta n_\alpha \, dS = \frac{1}{\omega} \text{ Im } f_{\beta\alpha} \tag{3.18}$$

the *added mass matrix*. The second integral in Eq. (3.17) is proportional to the body velocity and is expected to lead to damping, as will be shown. We call the matrix

$$[\lambda]: \qquad \lambda_{\beta\alpha} = \rho\omega \iint_{S_B} \text{Im } \phi_\beta n_\alpha \, dS = -\text{Re } f_{\beta\alpha} \tag{3.19}$$

the *radiation damping matrix*. In terms of these matrices the restoring force due to all six modes is expressed as

$$\mathscr{F}_\alpha^R = -\sum_\beta \mu_{\beta\alpha} \ddot{X}_\beta - \sum_\beta \lambda_{\beta\alpha} \dot{X}_\beta, \qquad \text{so that } \mathscr{F}_\alpha = \mathscr{F}_\alpha^D + \mathscr{F}_\alpha^R. \tag{3.20}$$

To justify the name "damping matrix" for $[\lambda]$, let us consider the average rate of work done *by* the body *to* the fluid over a period:

$$\bar{E} = -\sum_\alpha \overline{\mathscr{F}_\alpha^R \dot{X}_\alpha} = \sum_{\alpha,\beta} \mu_{\beta\alpha} \overline{\ddot{X}_\beta \dot{X}_\alpha} + \sum_{\alpha,\beta} \lambda_{\beta\alpha} \overline{\dot{X}_\beta \dot{X}_\alpha}. \tag{3.21}$$

Because $\mu_{\alpha\beta} = \mu_{\beta\alpha}$, which will be proven later, the first term may be written

$$\frac{1}{2} \sum_{\alpha,\beta} \mu_{\beta\alpha} \overline{\left(\ddot{X}_\beta \dot{X}_\alpha + \dot{X}_\alpha \ddot{X}_\beta\right)} = \frac{1}{2} \sum_{\alpha,\beta} \mu_{\beta\alpha} \overline{\frac{d}{dt} \dot{X}_\beta \dot{X}_\alpha} = 0, \tag{3.22}$$

which vanishes due to periodicity. Thus,

$$\bar{E} = \sum_{\alpha,\beta} \lambda_{\beta\alpha} \, \dot{\overline{X}}_\beta \dot{X}_\alpha \tag{3.23}$$

and $[\lambda]$ is associated with the energy given away by the oscillating body, hence the name *radiation damping*.

Finally, with these definitions, the matrix equation for the rigid body (2.47) may be written

$$\left[-\omega^2([M] + [\mu]) + [C] - i\omega[\lambda]\right]\{\xi\} = \{F^D\} + \{\mathbb{F}\}, \tag{3.24}$$

where $\{\xi\}$ is the amplitude of $\{X\}$:

$$\{X\} = \mathrm{Re}\{\xi\}e^{-i\omega t}. \tag{3.25}$$

From Eqs. (2.49) and (2.50), $[M]$ and $[C]$ are known from the equilibrium geometry of the body. Solutions of the hydrodynamic boundary-value problems then give $[\mu]$, $[\lambda]$, and $\{F^D\}$. Once the information for the constraint is known, Eq. (3.24) may be solved for $\{\xi\}$.

7.4 FORMAL REPRESENTATIONS OF VELOCITY POTENTIAL WHEN $h = $ CONSTANT

7.4.1 Away from the Body

Let all the geometrical departures (bodies, topography, etc.) from a sea of constant depth be confined in an imaginary vertical cylinder of finite size. In the exterior of this cylinder the general solution can be represented analytically in the form of an eigenfunction expansion, as shown below.

Two Dimensions

By separation of variables, $\phi = \psi(x)f(z)$, it is easy to show that $\psi(x) = e^{\pm ikx}$ and that

$$f'' - k^2 f = 0, \qquad -h < z < 0 \tag{4.1a}$$

$$f' - \sigma f = 0, \qquad z = 0, \tag{4.1b}$$

$$f' = 0, \qquad z = -h \tag{4.1c}$$

with $\sigma = \omega^2/g$. This is an eigenvalue problem of the Sturm–Liouville type. The solution is proportional to $\cosh k(z + h)$ and the eigenvalue condition for k is the usual dispersion relation

$$\sigma = k \tanh kh. \tag{4.2}$$

Let us examine Eq. (4.2) graphically. There is a pair of real roots which are familiar; see Fig. 4.1. Both of these roots $\pm k$ correspond to the same normalized eigenfunction; hence only the positive real root needs to be considered in the future. For later convenience, we introduce the normalized eigenfunction:

$$f_0(z) = \frac{\sqrt{2}\,\cosh k(z+h)}{(h + \sigma^{-1}\sinh^2 kh)^{1/2}} \qquad \text{so that } \int_{-h}^{0} f_0^2(z)\,dz = 1. \qquad (4.3)$$

In addition, there are also imaginary eigenvalues $k = i\kappa$ corresponding to the real solutions of

$$\sigma = -\kappa \tan \kappa h . \qquad (4.4)$$

Graphically, these roots are the intersections of $-\sigma h/\kappa h$ and $\tan \kappa h$, as shown in Fig. 4.1b. Since $\tan \kappa h$ has infinitely many branches, there is an infinite number of discrete roots $\kappa = \pm k_n$. Again it is only necessary to consider

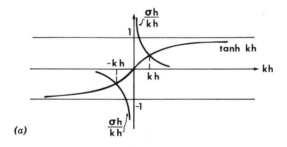

(a)

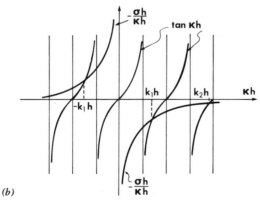

(b)

Figure 4.1 Graphical solution of dispersion relation (a) real solutions and (b) imaginary solutions.

positive k_n. From Fig. 4.1b, it is evident that

$$\frac{\pi}{2} < k_1 h < \pi, \quad \frac{3\pi}{2} < k_2 h < 2\pi, \quad (n - \tfrac{1}{2})\pi < k_n h < n\pi. \quad (4.5)$$

As n becomes large, $k_n h$ approaches $n\pi$. The corresponding normalized eigenfunctions are

$$f_n(z) = \frac{\sqrt{2}\,\cos k_n(z + h)}{\left(h - \sigma^{-1}\sin^2 k_n h\right)^{1/2}}, \qquad \int_{-h}^{0} f_n^2 \, dz = 1. \quad (4.6)$$

By straightforward integration and by using the eigenvalue condition (4.2) or (4.4), it can be shown that distinct eigenfunctions are orthogonal to one another, that is,

$$\int_{-h}^{0} f_m(z)f_n(z)\,dz = \delta_{mn}, \qquad m, n = 0, 1, 2, \ldots . \quad (4.7)$$

Kreisel (1949) has proven that the set $\{f_n\}$, $n = 0, 1, 2, \ldots$, is complete so that any function $G(z)$ in the interval $z \in [-h, 0]$ can be represented by a Fourier series based on $\{f_n\}$. Consequently, one can use the series

$$\phi(x, z) = a_0 f_0(z)e^{\pm ikx} + \sum_{n=1}^{\infty} b_n f_n(z)e^{\mp k_n x} \qquad \begin{cases} x_+ < x < \infty \\ x_- > x > -\infty \end{cases}$$

$$(4.8)$$

to represent the potential of a radiated or scattered wave. The first term corresponds to the propagating mode, while the series terms are only of local importance and are called the *evanescent* modes.

Three-Dimensional Eigensolutions in Cylindrical Polar Coordinates

The Laplace equation in polar coordinates is

$$\frac{1}{r}\frac{\partial}{\partial r}\left(r\frac{\partial \phi}{\partial r}\right) + \frac{1}{r^2}\frac{\partial^2 \phi}{\partial \theta^2} + \frac{\partial^2 \phi}{\partial z^2} = 0. \quad (4.9)$$

The eigensolution corresponding to the real and the imaginary eigenvalues are

$$\begin{Bmatrix} H_m^{(1)}(kr) \\ H_m^{(2)}(kr) \end{Bmatrix} \begin{Bmatrix} \cos m\theta \\ \sin m\theta \end{Bmatrix} f_0(z), \qquad m = 0, 1, 2, 3, \ldots$$

$$\begin{Bmatrix} I_m(k_n r) \\ K_m(k_n r) \end{Bmatrix} \begin{Bmatrix} \cos m\theta \\ \sin m\theta \end{Bmatrix} f_n(z), \qquad n = 1, 2, 3, \ldots \quad (4.10)$$

where $H_m^{(1)}$ and $H_m^{(2)}$ are the Hankel functions of the first and second kinds, and I_m and K_m are the modified Bessel functions of the first and second kinds, respectively. The most general representation which satisfies the radiation condition at infinity is

$$\phi = \sum_m H_m^{(1)}(kr)(\alpha_{0m}\cos m\theta + \beta_{0m}\sin m\theta)f_0(z)$$

$$+ \sum_m \sum_n K_m(k_n r)(\alpha_{nm}\cos m\theta + \beta_{nm}\sin m\theta)f_n(z). \qquad (4.11)$$

Because of the exponential attenuation with r, terms associated with $K_m(k_n r)$ are the evanescent modes.

From the general representations (4.8) and (4.11) we may express the radiation condition in more explicit mathematical forms as follows:

Two Dimensions: For $\phi = \phi^S \equiv \phi^D - \phi^I$, or ϕ_α

$$\phi \rightarrow -\frac{ig\mathcal{Q}_\pm}{\omega}\frac{\cosh k(z+h)}{\cosh kh}e^{\pm ikx} \qquad \text{as } kx \rightarrow \pm\infty, \qquad (4.12)$$

or equivalently

$$\frac{\partial\phi}{\partial x} \mp ik\phi \rightarrow 0 \qquad \text{as } kx \rightarrow \pm\infty. \qquad (4.13)$$

Three Dimensions: With the asymptotic formula for $H_m^{(1)}$, the wave potential can be written

$$\phi \sim -\frac{ig}{\omega}\sum_m H_m^{(1)}(kr)(\alpha'_{0m}\cos m\theta + \beta'_{0m}\sin m\theta)\frac{\cosh k(z+h)}{\cosh kh}$$

$$\sim \left\{\sum_m (\alpha'_{0m}\cos m\theta + \beta'_{0m}\sin m\theta)e^{-im\pi/2 - i\pi/4}\right\}$$

$$\cdot (-)\frac{ig}{\omega}\left(\frac{2}{\pi kr}\right)^{1/2}e^{ikr - i\pi/4}\frac{\cosh k(z+h)}{\cosh kh}. \qquad (4.14)$$

Let us denote the quantity inside { } above by $\mathcal{Q}(\theta)$; then

$$\phi \sim \frac{-ig\mathcal{Q}(\theta)}{\omega}\left(\frac{2}{\pi kr}\right)^{1/2}e^{ikr - i\pi/4}\frac{\cosh k(z+h)}{\cosh kh}, \qquad kr \rightarrow \infty.$$

$$(4.15)$$

$\mathcal{Q}(\theta)$ represents the angular variation of the radially spreading wave. By differentiation, the above statement can be expressed alternatively as

$$(kr)^{1/2}\left(\frac{\partial \phi}{\partial r} - ik\phi\right) \to 0, \qquad kr \to \infty \tag{4.16}$$

just as the case for shallow water waves.

Exercise 4.1

In a sea of constant depth, a row of equally spaced, identical bodies symmetrical about their axes parallel to x are fixed along the y axis. A train of plane waves is incident in the direction of positive x. Formulate the problem. Find the most general expressions for the velocity potential on the reflection and transmission sides. Give the explicit forms of the evanescent and the propagating modes. What is the effect of spacing on the number of propagating modes?

7.4.2   The Entire Fluid Domain

A formal solution for the whole fluid domain is possible by the use of Green's theorem and the so-called Green function.

For two twice-differentiable functions f and g, Green's theorem is

$$\iiint_{\Omega} (f\nabla^2 g - g\nabla^2 f)\, d\Omega = \iint_{\partial\Omega} \left(f\frac{\partial g}{\partial n} - g\frac{\partial f}{\partial n}\right) dS, \tag{4.17}$$

where Ω is a closed volume, $\partial\Omega$ its boundary, and $\mathbf{n}$ a unit normal to $\partial\Omega$ and outward from Ω.

The Green function $G(\mathbf{x}\,|\,\mathbf{x}_0)$ is defined to be the potential at any field point $\mathbf{x}$ due to an oscillating source of unit strength at $\mathbf{x}_0$. The governing conditions for G are

Two Dimensions: $\mathbf{x} = (x, z)$, $\mathbf{x}_0 = (x_0, z_0)$

$$\left(\frac{\partial^2}{\partial x^2} + \frac{\partial^2}{\partial z^2}\right) G = \delta(x - x_0)\delta(z - z_0) \qquad \text{in the fluid,} \tag{4.18}$$

$$\frac{\partial G}{\partial z} - \sigma G = 0, \qquad \sigma = \frac{\omega^2}{g}, \quad z = 0, \tag{4.19}$$

$$\frac{\partial G}{\partial z} = 0, \qquad z = -h, \tag{4.20}$$

$$\frac{\partial G}{\partial x} \mp ikG = 0, \qquad k\,|x - x_0| \sim \infty. \tag{4.21}$$

Three Dimensions: $\mathbf{x} = (x, y, z)$, $\mathbf{x}_0 = (x_0, y_0, z_0)$. Equations (4.19) and (4.20) are still valid but Eqs. (4.18) and (4.21) must be replaced by

$$\left(\frac{\partial^2}{\partial x^2} + \frac{\partial^2}{\partial y^2} + \frac{\partial^2}{\partial z^2} \right) G = \delta(x - x_0)\delta(y - y_0)\delta(z - z_0), \quad (4.22)$$

$$(kr)^{1/2}\left(\frac{\partial G}{\partial r} - ikG \right) \to 0, \quad kr \sim \infty. \quad (4.23)$$

The depth h is constant everywhere and the differentiations refer to $\mathbf{x}$ only.

Postponing the details of G, let us first apply Green's theorem to a radiation or a scattering potential and to the Green function, that is, $f = \phi$ where $\phi = \phi^R$ or ϕ^S and $g = G$. Consider only the three-dimensional case and let Ω be the control volume bounded by the free surface S_F, the body surface S_B, the sea bottom B_0, and a vertical circular cylinder S_∞ of great radius surrounding the body. For simplicity, the sea bottom for the radiation or the scattering problem is assumed to be horizontal everywhere, but it is not difficult to extend to the case where the bottom irregularity is confined in a finite region near the body. It follows from Eq. (4.17) that

$$\iiint_\Omega (\phi \nabla^2 G - G \nabla^2 \phi)\, d\Omega$$

$$= \left\{ \iint_{S_F} + \iint_{S_B} + \iint_{B_0} + \iint_{S_\infty} \right\} \left(\phi \frac{\partial G}{\partial n} - G \frac{\partial \phi}{\partial n} \right) dS.$$

The field equations for ϕ and G and the property of the δ function reduce the left-hand side to simply $\phi(\mathbf{x}_0)$. Because of the boundary conditions on ϕ and G, the surface integrals on S_F, B_0, and S_∞ vanish. Thus the preceding equation becomes

$$\phi(\mathbf{x}_0) = \iint_{S_B} \left(\phi \frac{\partial G}{\partial n} - G \frac{\partial \phi}{\partial n} \right) dS, \quad \mathbf{x}_0 \in \Omega, \notin S_B. \quad (4.24)$$

Now the Green function is symmetric with respect to the interchange of $\mathbf{x}$ and $\mathbf{x}_0$, that is,

$$G(\mathbf{x} \mid \mathbf{x}_0) = G(\mathbf{x}_0 \mid \mathbf{x}). \quad (4.25)$$

This fact may be proven straightforwardly by introducing another Green function with the source at $\mathbf{x}'$, $G(\mathbf{x} \mid \mathbf{x}')$, and applying Green's formula to $G(\mathbf{x} \mid \mathbf{x}_0)$ and $G(\mathbf{x} \mid \mathbf{x}')$. With Eq. (4.25) one can interchange $\mathbf{x}_0$ and $\mathbf{x}$ in Eq.

(4.24) to get

$$\phi(\mathbf{x}) = \iint_{S_B}\left[\phi(\mathbf{x}_0)\frac{\partial G}{\partial n_0} - G\frac{\partial \phi(\mathbf{x}_0)}{\partial n_0}\right]dS_0, \qquad \mathbf{x} \in \Omega, \notin S_B. \quad (4.26)$$

Thus, if ϕ and $\partial\phi/\partial n$ are known on the body surface, $\phi(\mathbf{x})$ is known everywhere. But in practice only the normal velocity $\partial\phi/\partial n$ is prescribed on the body, while $\phi(\mathbf{x}_0)$ is not known a priori. The representation (4.26) is therefore only formal.

While other applications of Eq. (4.26) will be discussed later, we shall derive here an asymptotic approximation of ϕ for $kr \gg 1$. For this purpose we need to obtain G explicitly. For three dimensions, one form of G is cited below, while further information may be found in Appendix 7.A.

$$G(\mathbf{x}\,|\,\mathbf{x}_0) = -\frac{i}{2}\frac{\sigma^2 - k^2}{h(k^2 - \sigma^2) + \sigma}\cosh k(z_0 + h)\cosh k(z + h)H_0^{(1)}(kR)$$

$$+ \frac{1}{\pi}\sum_{n=1}^{\infty}\frac{k_n^2 + \sigma^2}{h(k_n^2 + \sigma^2) - \sigma}\cos k_n(z_0 + h)\cos k_n(z + h)K_0(k_n R),$$

$$\tag{4.27}$$

where

$$R = \left[(x - x_0)^2 + (y - y_0)^2\right]^{1/2}.$$

For $kR \gg 1$ it is sufficient to keep the propagating mode only which may be written

$$G(\mathbf{x}\,|\,\mathbf{x}_0) \cong C_0\cosh k(z + h)\cosh k(z_0 + h)\left(\frac{2}{\pi kR}\right)^{1/2}e^{ikR - i\pi/4},$$

$$\tag{4.28}$$

where C_0 is a constant coefficient

$$C_0 = \frac{(-\tfrac{1}{2}i)(\sigma^2 - k^2)}{h(k^2 - \sigma^2) + \sigma}. \quad (4.29)$$

Using polar coordinates

$$(x, y) = r(\cos\theta, \sin\theta), \qquad (x_0, y_0) = r_0(\cos\theta_0, \sin\theta_0)$$

and approximating R for $r \gg r_0$

$$R = \left[r^2 + r_0^2 - 2rr_0\cos(\theta - \theta_0)\right]^{1/2} \cong r - r_0\cos(\theta - \theta_0),$$

we get

$$G(\mathbf{x}\,|\,\mathbf{x}_0) \cong C_0\cosh k(z + h)\cosh k(z_0 + h)\left(\frac{2}{\pi kr}\right)^{1/2} e^{ikr}e^{-ikr_0\cos(\theta - \theta_0)}e^{-i\pi/4}.$$

$$(4.30)$$

Substituting this into Eq. (4.26), we get

$$\phi(\mathbf{x}) \cong C_0\cosh^2 kh \frac{\cosh k(z + h)}{\cosh kh}\left(\frac{2}{\pi kr}\right)^{1/2} e^{ikr - i\pi/4}$$

$$\times \iint_{S_B} dS_0\left(\phi\frac{\partial}{\partial n_0} - \frac{\partial\phi}{\partial n_0}\right)\frac{\cosh k(z_0 + h)}{\cosh kh}e^{-ikr_0\cos(\theta - \theta_0)}, \quad (4.31)$$

which can also be written in the form of Eq. (4.15). This result gives a relation between the far-field amplitude $\mathcal{C}(\theta)$ and the values of ϕ and $\partial\phi/\partial n_0$ on S_B:

$$\mathcal{C}(\theta) = \frac{i\omega}{g} C_0\cosh^2 kh \iint_{S_B} dS_0\left(\phi\frac{\partial}{\partial n_0} - \frac{\partial\phi}{\partial n_0}\right)\frac{\cosh k(z_0 + h)}{\cosh kh}e^{-ikr_0\cos(\theta - \theta_0)}.$$

$$(4.32)$$

When ϕ and $\partial\phi/\partial n$ are known on the body, $\mathcal{C}(\theta)$ is obtainable by quadrature. In Russian literature the integral above is used to define the so-called Kochin's H function.

For two dimensions the Green function may be expressed as

$$G(\mathbf{x}\,|\,\mathbf{x}_0) = -\frac{i}{k}\left(h + \frac{1}{\sigma}\sinh^2 kh\right)^{-1} e^{ik|x - x_0|}\cosh k(z_0 + h)\cosh k(z + h)$$

$$+ \sum_{n=1}^{\infty}\frac{1}{k_n}\left(h - \frac{1}{\sigma}\sin^2 k_n h\right)^{-1} e^{-k_n|x - x_0|}\cos k_n(z_0 + h)\cos k_n(z + h).$$

$$(4.33)$$

The full derivation is given in Appendix 7.A. For $|kx| \gg 1$, only the propagating mode needs to be kept and Eq. (4.26) may be written

$$\phi(\mathbf{x}) \cong -\frac{i}{k}\frac{\cosh^2 khe^{\pm ikx}}{h + (1/\sigma)\sinh^2 kh}\frac{\cosh k(z+h)}{\cosh kh}$$

$$\times \int_{S_B} dS_0 \left(\phi\frac{\partial}{\partial n_0} - \frac{\partial\phi}{\partial n_0}\right)e^{\mp ikx_0}\frac{\cosh k(z_0+h)}{\cosh kh}. \qquad (4.34)$$

The wave amplitude at $x \sim \pm\infty$ is

$$\mathcal{Q}_{\pm} = \frac{\omega}{gk}\frac{\cosh^2 kh}{h + (1/\sigma)\sinh^2 kh}\int_{S_B} dS_0 \left(\phi\frac{\partial}{\partial n_0} - \frac{\partial\phi}{\partial n_0}\right)e^{\mp ikx_0}\frac{\cosh k(z_0+h)}{\cosh kh}.$$

$$(4.35)$$

In this form $\mathcal{Q}_{\pm}$ is essentially the two-dimensional Kochin H function.
We now leave generalities behind for a specific solution.

7.5 SCATTERING BY A VERTICAL CYLINDER WITH CIRCULAR CROSS SECTION

There are only a few geometries in the theory of water-wave diffraction where exact analytical solutions have been found. One of them is a vertical cylinder of circular cross section, extending from the sea bottom to the free surface. Without loss of generality, we take the incident wave to arrive from $x \sim -\infty$, as given by Eq. (3.8). The orthogonality property (4.7) can be used to show that only the propagating mode matters and the evanescent modes vanish identically.[†] Consequently, the total potential can be expressed

$$\phi = -\frac{ig}{\omega}\eta(x, y)\frac{\cosh k(z+h)}{\cosh kh}, \qquad (5.1)$$

where η is the free-surface displacement and satisfies the two-dimensional Helmholtz equation. The free surface of the incident wave is, in terms of partial waves,

$$\eta^I = Ae^{ikx} = Ae^{ikr\cos\theta}$$

$$= A\sum_{m=0}^{\infty}\varepsilon_m(i)^m J_m(kr)\cos m\theta. \qquad (5.2)$$

[†]This result holds for a vertical cylinder of arbitrary cross section.

The total free-surface displacement is given by

$$\eta = A \sum_{m=0}^{\infty} \varepsilon_m(i)^m \left\{ J_m(kr) - H_m(kr) \frac{J_m'(ka)}{H_m'(ka)} \right\} \cos m\theta, \qquad (5.3)$$

where $H_m \equiv H_m^{(1)}$ and $H_m'(s) \equiv dH_m/ds$ for the sake of brevity. The dynamic pressure may be calculated as

$$p(r, \theta, z) = i\omega\rho\phi = \rho g \eta \frac{\cosh k(z + h)}{\cosh kh}. \qquad (5.4)$$

In general, the pressure at any point on the cylinder can be calculated by summing up the infinite series

$$p(a, \theta, z) = \rho g A \frac{\cosh k(z + h)}{\cosh kh}$$

$$\times \sum_{m=0}^{\infty} \varepsilon_m(i)^m \left\{ J_m(ka) - H_m(ka) \frac{J_m'(ka)}{H_m'(ka)} \right\} \cos m\theta. \qquad (5.5)$$

In view of the Wronksian identity Eq. (9.20), Chapter Four, the pressure may be rewritten

$$p(a, \theta, z) = \rho g A \frac{\cosh k(z + h)}{\cosh kh} \sum_{m=0}^{\infty} \frac{2(i)^{(m+1)} \varepsilon_m \cos m\theta}{\pi ka H_m'(ka)}. \qquad (5.6)$$

The series also represents the normalized free-surface displacement around the cylinder. For small ka, the pressure p is rather uniform around the cylinder. As ka increases, the variation becomes more complex, as shown in Fig. 5.1.

Let us integrate the pressure on the cylinder. On a horizontal slice of a unit height, the force in the direction of wave propagation is

$$\frac{dF_x}{dz} = -a \int_0^{2\pi} p(a, \theta, z) \cos \theta \, d\theta.$$

By the orthogonality of cosines, only the term $m = 1$ in the series of Eq. (5.6) remains,

$$\frac{dF_x}{dz} = \frac{4A}{ka} \frac{\rho g a}{H_1'(ka)} \frac{\cosh k(z + h)}{\cosh kh}. \qquad (5.7)$$

The total horizontal force on the cylinder is then

$$F_x = \int_{-h}^0 \frac{dF_x}{dz} dz = \frac{4\rho g A a h}{ka H_1'(ka)} \frac{\tanh kh}{kh}. \qquad (5.8)$$

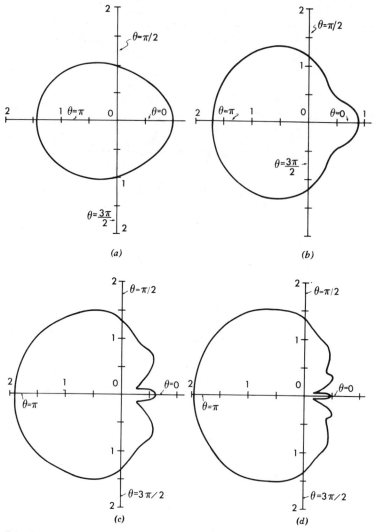

Figure 5.1 Polar distribution of run-up on a circular cylinder extending the entire sea depth. Incident wave is in the direction $\theta = 0$. (a) $ka = 0.5$, (b) $ka = 1.0$, (c) $ka = 3.0$, and (d) $ka = 5.0$.

The total moment about an axis parallel to y passing though the bottom of the cylinder is

$$M_y = -\int_{-h}^{0}(z+h)\frac{dF_x}{dz}dz = -\frac{4\rho gAah^2}{kaH_1'(ka)}\frac{kh\sinh kh - \cosh kh + 1}{(kh)^2\cosh kh},$$

$$(5.9)$$

which is positive counterclockwise. Formulas (5.8) and (5.9) were first given by McCamy and Fuchs (1954). To show the dependence of F_x and M_y on ka, it is only necessary to plot the magnitude and the phase of the factor $[kaH_1'(ka)]^{-1}$:

$$\left|\frac{1}{kaH_1'(ka)}\right| = (ka)^{-1}\left\{[J_1'(ka)]^2 + [Y_1'(ka)]^2\right\}^{-1/2}, \qquad (5.10)$$

and the phase $\tan \delta = Y_1'(ka)/J_1'(ka)$; see Fig. 5.2.

Let us introduce the inertia and drag coefficients per unit height, analogous respectively to the added mass and damping coefficients in the restoring forces on a body in forced radiation, as defined in more general terms in Section 7.3.2. For a unit horizontal slice of the cylinder we write

$$\text{Re}\left(\frac{dF_x}{dz}e^{-i\omega t}\right) = \rho\pi a^2\left(C_M\dot{U} + \omega C_D U\right), \qquad (5.11)$$

where U is the velocity of the incident wave at $x = 0$ in the absence of the cylinder,

$$U = \text{Re}\frac{\partial\phi^I}{\partial x}e^{-i\omega t} = \frac{gkA}{\omega}\frac{\cosh k(z+h)}{\cosh kh}\cos \omega t. \qquad (5.12)$$

From Eq. (5.7) it can be deduced that

$$\text{Re}\left(\frac{dF_x}{dz}e^{-i\omega t}\right) = \frac{4\rho gaA}{ka|H_1'(ka)|^2}\frac{\cosh k(z+h)}{\cosh kh}\left[J_1'(ka)\cos \omega t - Y_1'(ka)\sin \omega t\right].$$

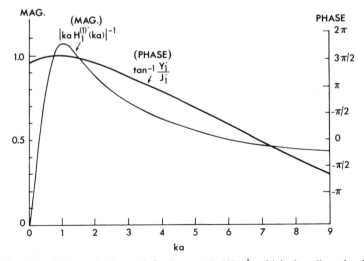

Figure 5.2 Magnitude and phase of the factor $(kaH_1')^{-1}$, which describes the dependence of force and moment on ka (cf. Eqs. (5.8) and (5.9)).

Upon comparing with Eq. (5.11), we obtain

$$C_M = \frac{4}{\pi} \frac{Y_1'(ka)}{(ka)^2 |H_1'(ka)|^2}, \tag{5.13a}$$

$$C_D = \frac{4}{\pi} \frac{J_1'(ka)}{(ka)^2 |H_1'(ka)|^2}. \tag{5.13b}$$

The inertia coefficient is plotted in Fig. 5.3. It is important that even in the absence of viscosity the cylinder experiences a drag (damping) force, which is due to the transport of energy toward infinity by scattered waves.

For a small cylinder we have approximately

$$C_M \cong 2, \tag{5.14}$$

and

$$C_D \cong \frac{\pi(ka)^2}{2}. \tag{5.15}$$

Thus, the drag coefficient due to waves is very small; in reality, viscous effects including vortex shedding are much more dominant. Note also that the apparent mass is twice that of the accelerating circular cylinder in an otherwise calm fluid without a free surface. To help understand this limit, we examine the neighborhood of the cylinder where $r = O(a)$. By keeping only the two

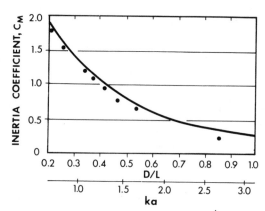

Figure 5.3 Inertia coefficient C_M by Eq. (5.13a) and by experiment. (From Charkrabarti and Tam, 1975, *J. Ship Res.* Reproduced by permission of the Society of Naval Architects and Marine Engineers.)

most important terms, $m = 0, 1$, in Eq. (5.3), we find

$$\phi \simeq -\frac{igA}{\omega}\frac{\cosh k(z+h)}{\cosh kh}\left\{1 + ik\left(r + \frac{a^2}{r}\right)\cos\theta\right\}$$

$$= -\frac{igA}{\omega}\frac{\cosh k(z+h)}{\cosh kh}\left\{1 + ik\left(x + \frac{a^2}{r}\cos\theta\right)\right\}. \qquad (5.16)$$

At a fixed z, this is just the potential for an oscillating flow past a stationary circular cylinder. The first term, ikx, corresponding to the incident wave in the absence of the cylinder, gives half of C_M, while the scattered waves, due to the presence of the cylinder, gives the other half.

Not too close to the small cylinder, the amplitude of the scattered wave is, from Eq. (5.3),

$$\frac{\eta^S}{A} = -H_0(kr)\frac{J'_0(ka)}{H'_0(ka)} - 2iH_1(kr)\frac{J'_1(ka)}{H'_1(ka)}\cos\theta + O(ka)^3$$

$$= \frac{\pi}{2}(ka)^2\left\{-\frac{i}{2}H_0(kr) - H_1(kr)\cos\theta\right\} + O(ka)^3. \qquad (5.17)$$

The scattered wave therefore consists of a radially symmetric wave and a dipole-like wave; their amplitudes are, however, very small $(O(ka)^2)$. Correspondingly, the damping force given by Eq. (5.15) is small.

The linearized theory here has been checked experimentally by Charkrabarti and Tam (1975), as shown in Fig. 5.3 for the coefficient C_M. Their measurements were taken for a cylinder of radius $a = 40.5$ in. in water depth of $h = 47.25$ in. The range of wave period was 1.0 to 3.5 s, while the incident wave amplitude was under 4.5 in.; the corresponding range of kA was $0.1 < kA < 0.38$, which included some steep waves. There was a good agreement between theory and experiment for $0.2 < ka < 0.65$. For still smaller radius or larger waves, viscous effect and flow separation became dominant which greatly affected both C_M and C_D.

Exercise 5.1

Solve the problem of a vertical circular cylinder forced to *sway* in the x direction, or to *roll* about the axis $x = 0$, $z = c$. The boundary condition on the cylinder is

$$\frac{\partial\phi}{\partial r} = U\cos\theta, \qquad \text{sway}$$

or

$$\frac{\partial\phi}{\partial r} = \Omega(c - z)\cos\theta, \qquad \text{roll}.$$

Find the added inertia and the damping force matrices by numerical summation of the series.

Exercise 5.2

Let a plane wavetrain be incident from $x \sim -\infty$ toward a vertical cylinder whose cross section is described by $r = a[1 + \varepsilon\xi(\theta)]$ with $\varepsilon \ll 1$. Show that, on a unit horizontal slice at height z,
  (a) the in-line wave force is

$$f_x \cong -a\int_0^{2\pi} d\theta\left\{\left[p^{(0)} + \varepsilon\left(\xi p^{(0)} + p^{(1)}\right)\right]\cos\theta + \varepsilon\xi' p^{(0)}\sin\theta\right\}_{r=a};$$

  (b) the transverse wave force is

$$f_y \cong -\varepsilon a\int_0^{2\pi} d\theta\left\{\left[\xi p^{(0)} + p^{(1)}\right]\sin\theta - \xi' p^{(0)}\cos\theta\right\}_{r=a};\ \text{and}$$

  (c) the moment about the z axis is

$$m_z \cong \varepsilon a^2\int_0^{2\pi} d\theta\,\xi' p^{(0)}(a, \theta),$$

where $p = p^{(0)} + \varepsilon p^{(1)} + \cdots$.
  For a slightly elliptical cylinder whose major axis is inclined at the angle α with the x axis, solve for $p^{(1)}$ explicitly and find f_x, f_y, and m_z.

7.6 GENERAL IDENTITIES FOR THE DIFFRACTION AND RADIATION OF SIMPLE HARMONIC WAVES

As seen in Section 4.7.4 and in many other wave problems, there are certain very general identities relating quantities of practical interest. These identities are useful for several reasons. First, they facilitate the theoretical understanding of the physical problem. Second, they provide necessary checks for an analytical or numerical theory. Third, they minimize the task of computing those quantities related by them. For constant h, many of these identities have been derived individually for a long time (Wehausen, 1971). Here we shall follow the systematic approach of Newman (1976) by selecting exhaustively various pairings of ϕ^D and ϕ_α for f and g in Green's formula (4.17).

  Let f and g be any two velocity potentials ϕ and ψ, and let the bounding surface $\partial\Omega$ be divided into the free surface S_F, the bottom B_0, the body S_B, and a vertical circular cylinder S_∞ with an arbitrarily large radius. The volume integral in Eq. (4.17) clearly vanishes. By invoking the appropriate boundary conditions, neither the sea bottom B_0 nor the free surface S_F gives any contribution to the surface integral; therefore,

$$\iint_{S_B+S_\infty} \left(\phi \frac{\partial \psi}{\partial n} - \psi \frac{\partial \phi}{\partial n} \right) dS = 0. \tag{6.1}$$

If we take $f = \phi$ but $g = \psi^*$ where $(\ )^*$ denotes the complex conjugate and use the fact that ψ^* satisfies the conjugate of the equations governing ψ, it follows that

$$\iint_{S_B+S_\infty} \left[\phi \frac{\partial \psi^*}{\partial n} - \psi^* \frac{\partial \phi}{\partial n} \right] dS = 0. \tag{6.2}$$

Further information may be obtained by specifying the boundary conditions on S_B and S_∞. A general result used repeatedly later is that the surface integral of S_∞ in Eq. (6.1) vanishes if ϕ and ψ are outgoing waves at infinity and satisfy Eq. (4.16), that is,

$$\iint_{S_\infty} \left(\phi \frac{\partial \psi}{\partial n} - \psi \frac{\partial \phi}{\partial n} \right) dS = 0. \tag{6.3}$$

Alternatively one may regard Eq. (6.3) as the weak radiation condition on ϕ if ψ satisfies the strong condition (4.16). As a consequence, Eq. (6.1) gives.

$$\iint_{S_B} \left(\phi \frac{\partial \psi}{\partial n} - \psi \frac{\partial \phi}{\partial n} \right) dS = 0. \tag{6.4}$$

We now specify ϕ and ψ even further.

7.6.1 Relations between Two Radiation Problems and Their Consequences

Let $\phi = \phi_\alpha$ and $\psi = \phi_\beta$ be the radiation potentials for two normal modes. Since they both satisfy Eq. (4.16), Eq. (6.4) applies. If use is made of the boundary condition on the body, we have

$$\iint_{S_B} \phi_\beta n_\alpha \, dS = \iint_{S_B} \phi_\alpha n_\beta \, dS. \tag{6.5}$$

In view of the definitions in Eqs. (3.15), (3.18), and (3.19), the restoring force, the added mass, and the damping matrices must be symmetric:

$$f_{\beta\alpha} = f_{\alpha\beta}, \qquad \mu_{\alpha\beta} = \mu_{\beta\alpha}, \qquad \lambda_{\alpha\beta} = \lambda_{\beta\alpha}. \tag{6.6}$$

It should be stressed that there is no restriction on the symmetry of the body.

Recall that the average rate of work done by the body to water as given by Eq. (3.23) was deduced after using the symmetry of $\lambda_{\alpha\beta}$. Now the working rate can be alternatively obtained in terms of the total radiation potential $\phi^R = \Sigma V_\alpha \phi_\alpha$ as follows:

$$\bar{E} = -\frac{\text{Re}}{2}\left[i\omega\rho \iint_{S_B} \phi^R \frac{\partial \phi^{R*}}{\partial n}\, dS\right] = \frac{\omega\rho}{2}\,\text{Im}\iint_{S_B} \phi^R \frac{\partial \phi^{R*}}{\partial n}\, dS$$

$$= \frac{\omega\rho}{2}\,\frac{1}{2i}\iint_{S_B}\left[\phi^R \frac{\partial \phi^{R*}}{\partial n} - \phi^{R*}\frac{\partial \phi^R}{\partial n}\right]\, dS.$$

From Eq. (6.2) the integral over S_B is the negative of a similar integral over S_∞; hence

$$\bar{E} = \frac{\rho\omega i}{4}\iint_{S_\infty}\left[\phi^R \frac{\partial \phi^{R*}}{\partial n} - \phi^{R*}\frac{\partial \phi^R}{\partial n}\right]\, dS. \qquad (6.7)$$

If the asymptotic expressions of ϕ^R and ϕ^{R*}, that is, Eq. (4.15) and its complex conjugate, are used in Eq. (6.7), we get

$$\bar{E} = \frac{i\rho\omega}{4}\int_0^{2\pi}\int_{-h}^0\left[-\frac{ig}{\omega}\left(\frac{2}{\pi kr}\right)^{1/2}\right]^2 2ik\,|\,\mathcal{Q}^R(\theta)\,|^2\frac{\cosh^2 k(z+h)}{\cosh^2 kh}\,r\,d\theta\,dz,$$

$$(6.8)$$

which is non-negative. Therefore, the radiation condition is consistent with the physical fact that energy can only flow from the rigid body to water when the body is forced externally to move.[†] Let us now return to Eq. (3.23). Since $\dot{X}_\alpha$ and $\dot{X}_\beta$ can be arbitrarily prescribed, the non-negativeness of $\bar{E}$ implies that the damping matrix is positive semidefinite. As a corollary, the diagonal terms of $[\lambda]$, λ_{11}, λ_{22}, and λ_{33}, are each individually positive semidefinite.

The radiation damping rate may also be related to the radiated amplitude in the far field. Putting $\phi = \phi_\alpha$ and $\psi^* = \phi_\beta^*$ in Eq. (6.2), we get

$$\iint_{S_B}\left(\phi_\alpha n_\beta - \phi_\beta^* n_\alpha\right)\, dS = -\iint_{S_\infty}\left(\phi_\alpha \frac{\partial \phi_\beta^*}{\partial n} - \phi_\beta^*\frac{\partial \phi_\alpha}{\partial n}\right)\, dS.$$

By virtue of the symmetry relation (6.5), the left side is seen to be

$$2i\,\text{Im}\iint_{S_B} \phi_\alpha n_\beta\, dS = \frac{2i\lambda_{\alpha\beta}}{\rho\omega}.$$

[†] There are special cases where an oscillating body does not radiate any energy for a particular frequency (e.g., Bessho, 1965; Frank, 1967; Kyozuka and Yoshida, 1981).

It follows after using the radiation condition that

$$\lambda_{\alpha\beta} = -\frac{\rho\omega}{2i} \iint_{S_\infty} \left(\phi_\alpha \frac{\partial \phi_\beta^*}{\partial n} - \phi_\beta^* \frac{2\phi_\alpha}{\partial n} \right) dS = \rho\omega k \iint_{S_\infty} \phi_\alpha \phi_\beta^* \, dS. \quad (6.9a)$$

In particular we have

$$\lambda_{\alpha\alpha} = -\rho\omega \, \text{Im} \iint_{S_\infty} \phi_\alpha \frac{\partial \phi_\alpha^*}{\partial n} \, dS = \rho\omega k \iint_{S_\infty} |\phi_\alpha|^2 \, dS > 0 \quad (6.9b)$$

for $\alpha = \beta$. The right-hand side of Eqs. (6.9a) and (6.9b) may be expressed in terms of far-field amplitudes, as shown below.

Two Dimensions

According to Eq. (4.12), the asymptotic behavior of ϕ_α is

$$\phi_\alpha \sim -\frac{ig\mathcal{Q}_\alpha^\pm}{\omega} e^{\pm ikx} \frac{\cosh k(z+h)}{\cosh kh}, \quad x \sim \pm\infty, \quad (6.10)$$

where $\mathcal{Q}_\alpha^\pm$ have the dimension of time. The right-hand side of Eq. (6.9a) consists of two integrals on $S_\infty^\pm$ at $x \sim \pm\infty$. By straightforward substitution, it can be shown that

$$\lambda_{\alpha\beta} = \rho g C_g \left(\mathcal{Q}_\alpha^- \mathcal{Q}_\beta^- {}^* + \mathcal{Q}_\alpha^+ \mathcal{Q}_\beta^+ {}^* \right). \quad (6.11)$$

Three Dimensions

The asymptotic behavior is, from Eq. (4.15),

$$\phi_\alpha = -\frac{ig\mathcal{Q}_\alpha(\theta)}{\omega} \frac{\cosh k(z+h)}{\cosh kh} \left(\frac{2}{\pi kr} \right)^{1/2} e^{ikr - i\pi/4}. \quad (6.12)$$

We leave it as an exercise to show from Eq. (6.9a) that

$$\lambda_{\alpha\beta} = \frac{2}{\pi k} \rho g C_g \int_0^{2\pi} \mathcal{Q}_\alpha(\theta) \mathcal{Q}_\beta^*(\theta) \, d\theta. \quad (6.13)$$

7.6.2 Relations between Two Diffraction Problems

Let $\phi = \phi^{(1)}$ and $\psi = \phi^{(2)}$ be two diffraction problems corresponding to different angles of incidence. As the normal velocity vanishes on the body, the surface integral on S_B in Eq. (6.1) vanishes. It follows that

$$\iint_{S_\infty} \left(\phi^{(1)} \frac{\partial \phi^{(2)}}{\partial n} - \phi^{(2)} \frac{\partial \phi^{(1)}}{\partial n} \right) dS = 0. \quad (6.14)$$

For the same reason, Eq. (6.2) gives

$$\iint_{S_\infty} \left(\phi^{(1)} \frac{\partial \phi^{(2)*}}{\partial n} - \phi^{(2)*} \frac{\partial \phi^{(1)}}{\partial n} \right) dS = 0. \tag{6.15}$$

A special case of Eq. (6.15) is obtained by letting $\phi^{(1)} = \phi^{(2)} = \phi$; then

$$\text{Im} \iint_{S_\infty} \phi \frac{\partial \phi^*}{\partial n} dS = 0. \tag{6.16}$$

The left side is proportional to the pressure work done on the surface S_∞. Hence Eq. (6.16) states the conservation of energy. Let us examine these results in terms of the far-field amplitudes.

Two Dimensions

Let $\phi^{(1)}$ ($\phi^{(2)}$) be the wave potential with an incident wave from left (right) to right (left). The asymptotic potentials are

$$\phi^{(1)} \sim \begin{Bmatrix} e^{ikx} + R_1 e^{-ikx} \\ \sim T_1 e^{ikx} \end{Bmatrix} \cdot \left(-\frac{igA}{\omega} \right) \frac{\cosh k\,(z+h\,)}{\cosh kh}, \qquad \begin{matrix} x \sim -\infty, \\ x \sim \infty, \end{matrix}$$

$$\tag{6.17}$$

and

$$\phi^{(2)} \sim \begin{Bmatrix} T_2 e^{-ikx} \\ \sim e^{-ikx} + R_2 e^{ikx} \end{Bmatrix} \cdot \left(-\frac{igA}{\omega} \right) \frac{\cosh k\,(z+h\,)}{\cosh kh}, \qquad \begin{matrix} x \sim -\infty, \\ x \sim +\infty \end{matrix}.$$

$$\tag{6.18}$$

Let $S_\infty^\pm$ be two vertical lines at $x \sim \pm\infty$. Then on $S_\infty^\pm$, $\partial/\partial n = \pm\partial/\partial x$. Upon substitution of Eqs. (6.17) and (6.18) into Eq. (6.14), we obtain, after a little algebra,

$$T_1 = T_2 \tag{6.19}$$

(Kreisel, 1949). The transmission coefficient is independent of the direction of the incident wave even though the body may not possess any symmetry at all! It should be emphasized that this result does not imply the equality of two problems and their solutions; only the far fields are involved here. Performing similar computations, we get from Eq. (6.15)

$$R_1 T_2^* + R_2^* T_1 = 0, \tag{6.20}$$

(Meyer, 1955; Newman, 1965), and from Eq. (6.16) for a single diffraction problem

$$|R^2| + |T^2| = 1. \tag{6.21}$$

Further inferences may be made. Using Eq. (6.19), Eq. (6.20) leads to

$$|R_1| = |R_2| \tag{6.22}$$

which was first obtained by Kreisel (1949) in a different way. Let δ_1^T and δ_2^R be the phase angles of the transmission and reflection coefficients as defined by

$$T_j = |T_j| e^{i\delta_j^T}, \qquad R_j = |R_j| e^{i\delta_j^R}, \qquad j = 1, 2. \tag{6.23}$$

It follows from Eqs. (6.19) and (6.20) that

$$\delta_1^T = \delta_2^T, \tag{6.24}$$

and

$$\delta_1^R + \delta_2^R = \delta_1^T + \delta_2^T \pm \pi \tag{6.25}$$

(Newman, 1965). In the special case of a symmetric body, $\delta_1^R = \delta_2^R$ and $\delta_1^T = \delta_2^T$; hence

$$\delta_1^R = \delta_1^T \pm \frac{\pi}{2}. \tag{6.26}$$

Equations (6.19)–(6.22) are extensions of similar results in Section 4.7.4 for long waves.

Three Dimensions

$$\phi^{(l)} \cong -\frac{igA}{\omega} \frac{\cosh k(z+h)}{\cosh kh} \left[e^{ikr\cos(\theta - \theta_l)} + \left(\frac{2}{\pi kr}\right)^{1/2} \mathcal{C}_l^S(\theta) e^{ikr - i\pi/4} \right],$$

$$l = 1, 2 \quad (6.27)$$

where $\mathcal{C}_l^S$ denotes the normalized scattered wave amplitude due to the incident wave in the direction θ_l. Substitution of Eq. (6.27) into Eqs. (6.14), (6.15), and (6.16) yields three relations governing $\mathcal{C}_l^S(\theta)$. Let us illustrate the analysis for the energy conservation theorem only. Due to Eq. (6.27), the left-hand side of

Eq. (6.16) reads

$$
2i \, \mathrm{Im} \int_0^{2\pi} d\theta \int_{-h}^0 dz \, \frac{\cosh^2 k(z+h)}{\cosh^2 kh} \left(\frac{gA}{\omega} \right)^2 r
$$

$$
\times \left[e^{ikr\cos(\theta-\theta_I)} + \left(\frac{2}{\pi kr} \right)^{1/2} \mathcal{Q}_I^S e^{ikr-i\pi/4} \right]
$$

$$
\times \left[-ik\cos(\theta-\theta_I)e^{-ikr\cos(\theta-\theta_I)} - ik\left(\frac{2}{\pi kr} \right)^{1/2} \mathcal{Q}_I^{S*} e^{-ikr+i\pi/4} \right].
$$

After multiplying out the integrand and omitting a constant factor, we get

$$
-\mathrm{Im} \, i \int_0^{2\pi} d\theta \left\{ \frac{2}{\pi} | \mathcal{Q}_I^S(\theta) |^2 + kr\cos(\theta-\theta_I) + \left(\frac{2}{\pi}kr \right)^{1/2} \mathcal{Q}_I^S \cos(\theta-\theta_I) \right.
$$

$$
\left. \times e^{ikr[1-\cos(\theta-\theta_I)]}e^{-i\pi/4} + \left(\frac{2}{\pi}kr \right)^{1/2} \mathcal{Q}_I^{S*} e^{-ikr[1-\cos(\theta-\theta_I)]}e^{i\pi/4} \right\} = 0.
$$

$$
(6.28)
$$

The term proportional to kr vanishes by periodicity. The fourth term may be combined with the third upon using the identity $\mathrm{Im} \, if = \mathrm{Im} \, if^*$; the combination gives

$$
e^{-i\pi/4} \left(\frac{2}{\pi}kr \right)^{1/2} \int_0^{2\pi} d\theta \, \mathcal{Q}_I^S e^{ikr[1-\cos(\theta-\theta_I)]} [1 + \cos(\theta-\theta_I)], \quad (6.29)
$$

which can be approximated for $kr \gg 1$ by the method of stationary phase. The stationary phase points occur at

$$
\frac{\partial}{\partial\theta} [1 - \cos(\theta-\theta_I)] = \sin(\theta-\theta_I) = 0 \quad \text{or} \quad \theta = \theta_I, \theta_I + \pi \quad (6.30)
$$

within the interval $[0, 2\pi]$. In the neighborhood of the first stationary phase point the integrand in Eq. (6.29) is roughly

$$
e^{ikr(\theta-\theta_I)^2/2} 2\mathcal{Q}_I^S(\theta_I). \quad (6.31)
$$

When the limits are approximated by $(-\infty, \infty)$, the integral in Eq. (6.29) can be evaluated:

$$
\mathcal{Q}_I^S(\theta_I) \int_{-\infty}^{\infty} e^{ikr\theta^2/2} \, d\theta = \left(\frac{2\pi}{kr} \right)^{1/2} e^{i\pi/4} \mathcal{Q}_I^S(\theta_I) \quad \text{as } kr \gg 1. \quad (6.32)
$$

Near the second stationary point the integrand of Eq. (6.29) vanishes. Finally, Eq. (6.28) becomes

$$\frac{1}{\pi} \int_0^{2\pi} |@_l^S(\theta)|^2 \, d\theta = -2 \operatorname{Re} @_l^S(\theta_l) \tag{6.33}$$

which was first derived by Maruo (1960). Similar relations are well known in quantum mechanics and other physical contexts as the *optical theorem*. Since $|@_l^S(\theta)|^2 \, d\theta$ is a measure of the scattered energy within the wedge $(\theta, \theta + d\theta)$, the integral on the left of Eq. (6.33) is a measure of the total scattered energy and is a feature of the body. The theorem implies that the total energy scattered can be alternatively obtained from the scattered wave amplitude in the forward direction alone. This result is important in experiments if the total scattered energy is of interest, as is the case in many branches of physics.

By a similar analysis for Eq. (6.14), and using the approximation

$$\int_0^{2\pi} @_1^S(\theta) e^{-i\pi/4} [1 - \cos(\theta - \theta_2)] e^{ikr[1+\cos(\theta-\theta_2)]} \, d\theta$$

$$= 2 \left(\frac{2\pi}{kr} \right)^{1/2} @_1^S(\theta_2 + \pi) \qquad \text{for } kr \gg 1,$$

we get

$$@_1^S(\theta_2 + \pi) = @_2^S(\theta_1 + \pi). \tag{6.34}$$

In particular, if $\theta_1 = 0$ and $\theta_2 = -\pi$, then $@_1^S(0) = @_2^S(\pi)$; this coincides with Eq. (6.19) which relates the transmission coefficients in two dimensions. In general, Eq. (6.35) states that the amplitude of the first scattered wave toward the second incident wave is equal to the amplitude of the second scattered wave toward the first incident wave.

We leave it for the reader to show from Eq. (6.15) that

$$-\pi \left[@_1^S(\theta_2) + @_2^{S*}(\theta_1) \right] = \int_0^{2\pi} d\theta \, @_1^S(\theta) @_2^{S*}(\theta) \tag{6.35}$$

of which Eq. (6.33) is just a special case.

Exercise 6.1

Show that for a two-dimensional rigid body induced to oscillate by an incident wave A from $x \sim -\infty$,

$$1 = \left| R + \frac{A_-}{A} \right|^2 + \left| T + \frac{A_+}{A} \right|^2 \tag{6.36}$$

in which R and T are the reflection and transmission coefficients when the body is held stationary, while A_+ and A_- are induced wave amplitudes toward $x \sim +\infty$ and $x \sim -\infty$, respectively.

Exercise 6.2

Consider a variable bottom which changes from one constant depth to another, that is, $h(x) \to h_+$ or h_- as $x \sim \infty$ or $-\infty$. A train of waves is incident obliquely from the left. What should the far-field expressions be for ϕ on both sides of the transition? Define the reflection (R) and transmission (T) coefficients, and derive an energy relation for R and T.

7.6.3   One Diffraction Problem and One Radiation Problem

Haskind–Hanaoka Theorem

There is a remarkable theorem due independently to Haskind (1957) and Hanaoka (1959), and popularized by Newman (1960), which relates the αth generalized component of the exciting force (due to diffraction) on a fixed body to the radiation potential of the αth normal mode of the same body; specifically,

$$F_\alpha^D = i\omega\rho \iint_{S_B} \left(\phi^I \frac{\partial \phi_\alpha}{\partial n} - \phi_\alpha \frac{\partial \phi^I}{\partial n} \right) dS = -i\omega\rho \iint_{S_\infty} \left[\phi^I \frac{\partial \phi_\alpha}{\partial n} - \phi_\alpha \frac{\partial \phi^I}{\partial n} \right] dS.$$

$$(6.37)$$

By definition,

$$F_\alpha^D = \iint_{S_B} pn_\alpha \, dS = i\omega\rho \iint_{S_B} (\phi^I + \phi^S)n_\alpha \, dS = i\omega\rho \iint_{S_B} (\phi^I + \phi^S) \frac{\partial \phi_\alpha}{\partial n} \, dS.$$

$$(6.38)$$

Because ϕ^S and ϕ_α are outgoing at infinity, Eq. (6.4) applies so that

$$F_\alpha^D = i\omega\rho \iint_{S_B} \left(\phi^I \frac{\partial \phi_\alpha}{\partial n} + \phi_\alpha \frac{\partial \phi^S}{\partial n} \right) dS. \qquad (6.39)$$

By the boundary condition $\partial \phi^S / \partial n = -\partial \phi^I / \partial n$, the first equality in Eq. (6.37) follows at once; the second equality then follows by letting $\phi = \phi^I$ and $\psi = \phi^S$ in Eq. (6.1).

We leave it for the reader to deduce from the proper asymptotic expressions the following explicit formulas:

Two Dimensions:

$$F_\alpha^D = -2\rho g A \mathcal{Q}_\alpha^- C_g. \qquad (6.40)$$

Three Dimensions:

$$F_\alpha^D = -\frac{4}{k}\rho g A \mathcal{C}_\alpha(\theta_I + \pi)C_g.$$ (6.41)

In both cases the exciting force is related to the wave amplitude in the direction opposite to the incident waves.

Exercise 6.3

Show from Eqs. (6.13) and (6.41) that in three dimensions the damping coefficients and the exciting force are related by

$$\lambda_{\alpha\alpha} = \frac{k/8\pi}{\rho g C_g |A|^2} \int_0^{2\pi} |F_\alpha^D(\theta)|^2 \, d\theta.$$ (6.42)

Bessho–Newman Relations

A still less obvious identity between radiation and scattering problems was discovered by Bessho (1967) for two dimensions, rediscovered and extended for three dimensions by Newman (1975, 1976).

Assuming that the normal velocity on S_B has the same phase everywhere, we may always redefine time so that

$$\frac{\partial \phi^R}{\partial n} = V_n = \text{real} \quad \text{or} \quad \frac{\partial}{\partial n}(\phi^R - \phi^{R*}) = 0 \quad \text{on} \quad S_B. \quad (6.43)$$

Because $\phi^I + \phi^S$ satisfies the same condition on S_B, it follows from Eq. (6.1) that

$$\iint_{S_\infty} \left[(\phi^R - \phi^{R*})\frac{\partial}{\partial n}(\phi^I + \phi^S) - (\phi^I + \phi^S)\frac{\partial}{\partial n}(\phi^R - \phi^{R*}) \right] dS = 0.$$

(6.44)

By applying Eq. (6.3) to ϕ^R and ϕ^S, we may rewrite Eq. (6.44)

$$\iint_{S_\infty} \left[(\phi^R - \phi^{R*})\frac{\partial \phi^I}{\partial n} - \phi^I \frac{\partial}{\partial n}(\phi^R - \phi^{R*}) \right] dS$$

$$= \iint_{S_\infty} \left(\phi^{R*}\frac{\partial \phi^S}{\partial n} - \phi^S \frac{\partial \phi^{R*}}{\partial n} \right) dS. \quad (6.45)$$

Equation (6.45) relates the far fields of ϕ^R and ϕ^S. More explicit implications follow.

Two Dimensions: By letting S_∞ be a large rectangular box and using Eq. (6.10) in Eq. (6.45), the two-dimensional Bessho–Newman relation can be obtained:

$$A_- - RA_-^* - TA_+^* = 0. \tag{6.46}$$

Consider now the special case of a body which is symmetrical about its vertical plane and is executing either a symmetric mode (heave) or an antisymmetric mode (roll or sway). For the symmetric mode, we have

$$A_+ = A_- = A_s = |A_s| e^{i\delta_s}, \tag{6.47}$$

and for the antisymmetric mode

$$A_+ = -A_- = |A_a| e^{i\delta_a}. \tag{6.48}$$

Substituting Eqs. (6.47) and (6.48) in turn into Eq. (6.46), we get

$$R + T = e^{2i\delta_s}, \qquad R - T = e^{2i\delta_s}. \tag{6.49}$$

Since both roll and sway are antisymmetric modes, we have the striking result that the phases of their radiated waves must satisfy

$$\delta_1 = \delta_5 \pm \pi. \tag{6.50}$$

Furthermore, for the same body and bottom geometry one may obtain the scattering coefficients R and T by solving two radiation problems (for two modes, say). With the help of the Haskind–Hanaoka relation for exciting forces, it is possible to obtain all the important global quantities in both radiation and scattering problems by solving the radiation problems alone!

Three Dimensions: Use of the asymptotic formulas for ϕ^S and ϕ^R transforms the right-hand side of Eq. (6.45) to:

$$C 2i \frac{2}{\pi} \int_0^{2\pi} d\theta \, @^S(\theta) @^{R*}(\theta), \tag{6.51}$$

where C is a constant multiplier. For the left-hand side of Eq. (6.45) we observe first that

$$\iint_{S_\infty} \left(\phi^R \frac{\partial \phi^I}{\partial n} - \phi^I \frac{\partial \phi^R}{\partial n} \right) dS = Ci(kr)^{1/2} \frac{2}{\pi} \int_0^{2\pi} d\theta \, @^R [1 - \cos(\theta - \theta_I)]$$

$$\times e^{ikr[\cos(\theta - \theta_I) + 1]}$$

$$= 4iC @^R(\theta_I + \pi) \tag{6.52}$$

after using Eq. (6.34); then by similar reasoning,

$$\iint_{S_\infty} \left(\phi^{R*} \frac{\partial \phi^I}{\partial n} - \phi^I \frac{\partial \phi^{R*}}{\partial n} \right) dS = 4iC\mathcal{Q}^{R*}(\theta_I). \tag{6.53}$$

Finally, from Eq. (6.45) we get the three-dimensional Bessho–Newman relations:

$$-\mathcal{Q}^{R*}(\theta_I) + \mathcal{Q}^R(\theta_I + \pi) = \frac{1}{\pi} \int_0^{2\pi} \mathcal{Q}^{R*}\mathcal{Q}^S \, d\theta. \tag{6.54}$$

Consider the special case of a body with rotational symmetry about the z axis. Without loss of generality we let $\theta_I = 0$. Now only three modes are distinct, that is, sway ($\alpha = 1$), heave ($\alpha = 3$), and roll about the y axis ($\alpha = 5$).

$$\mathcal{Q}_\alpha(\theta) = (\text{const})\cos\theta, \qquad \alpha = 1, 5,$$

$$\mathcal{Q}_3(\theta) = \text{const}. \tag{6.55}$$

For sway, we expect that $\mathcal{Q}^R = \mathcal{Q}_1(0)\cos\theta$. It follows from Eq. (6.54) that

$$-\mathcal{Q}_1^*(0) - \mathcal{Q}_1(0) = \frac{1}{\pi}\mathcal{Q}_1^*(0)\int_0^{2\pi} \mathcal{Q}^S \cos\theta \, d\theta, \tag{6.56}$$

or

$$-1 - \frac{\mathcal{Q}_1(0)}{\mathcal{Q}_1^*(0)} = \frac{1}{\pi} \int_0^{2\pi} \mathcal{Q}^S \cos\theta \, d\theta. \tag{6.57}$$

For roll, we also expect $\mathcal{Q}^R = \mathcal{Q}_5(0)\cos\theta$ although $\mathcal{Q}_1(0) \neq \mathcal{Q}_5(0)$. Equation (6.57) also applies if $\mathcal{Q}_1(0)$ is replaced by $\mathcal{Q}_5(0)$; therefore

$$\frac{\mathcal{Q}_1(0)}{\mathcal{Q}_1^*(0)} = \frac{\mathcal{Q}_5(0)}{\mathcal{Q}_5^*(0)}. \tag{6.58}$$

If we denote the phase angle of $\mathcal{Q}_\alpha(0)$ by δ_α, that is, $\mathcal{Q}_\alpha(0) = |\mathcal{Q}_\alpha(0)| e^{i\delta_\alpha}$, then the phase angles due to sway and roll are the same:

$$\delta_1 = \delta_5. \tag{6.59}$$

This result was first shown for a circular cylinder by Garrett (1970) and generalized in the present manner by Newman (1976). From Eq. (6.41) the phases of the sway exciting force and the roll exciting moment must also be the same. Similar results hold for a two-dimensional body symmetrical about its vertical axis.

For the heave mode, $\mathcal{Q}^R = \mathcal{Q}_3$, which is independent of θ. A similar application of Eq. (6.55) gives

$$e^{2i\delta_3} - 1 = \frac{i}{\pi} \int_0^{2\pi} \mathcal{Q}^S(\theta)\, d\theta. \tag{6.60}$$

All the general identities discussed in this section have been generalized for N bodies (Srokosz, 1980). These identities are useful in theoretical argument and can be used to check the correctness of calculations or to reduce the numerical work of calculating certain global quantities. Nevertheless, they do not change the basic need of an efficient technique for solving the typical hydrodynamic boundary-value problem; hence the next section.

7.7 NUMERICAL SOLUTION BY HYBRID ELEMENT METHOD

Two primary classes of numerical methods have been well developed for diffraction problems of this chapter. One class is based on finite elements, and the other on integral equations. In each class, there are variations in details, but the general spirits are exemplified by the two methods discussed briefly here and in the next section. Further information may be found in the survey by Mei (1978), Susbielles and Bratu (1981), Sarpkaya and Issacson (1981) and Yeung (1982). In particular, the second and third references contain the numerical and experimental results of a variety of geometries.

In this section we shall extend the hybrid element method of Section 4.11 for long waves in shallow water to water of arbitrary depth. As before, one of the main ideas is to employ the finite-element approximation near the body and analytical representation everywhere else. As the first step we must establish the variational principle which only involves integrals over a finite domain surrounding the body.

With reference to Fig. 7.1a let all bodies and bottom irregularities be localized within a vertical cylinder C of finite size. Beyond C the ocean depth is

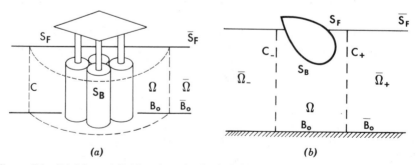

Figure 7.1 Division of fluid regions in the hybrid element method: (a) three dimensional; (b) two dimensional.

assumed to be constant everywhere. Overhead bars will be used to distinguish quantities outside C (i.e., in the superelement $\bar{\Omega}$) from those within C (i.e., in Ω). Let $\bar{\phi}$ satisfy *exactly* the Laplace equation, the boundary conditions on the free surface $\bar{S}_F$ and on the bottom $\bar{B}_0$, and the radiation condition at infinity. Since nothing is yet specified on C, $\bar{\phi}$ is so far unknown but may be formally represented in a variety of ways. One way is to use Green's function and express $\bar{\phi}$ as the superposition of sources on C with unknown source strength. Another way is to use eigenfunction expansions with unknown coefficients as in Eq. (4.8) for two dimensions and Eq. (4.11) for three dimensions. The latter course will be adopted here.

7.7.1 The Variational Formulation

Taking the radiation problem for demonstration, we shall show that the stationarity of the functional

$$J(\phi, \bar{\phi}) = \frac{1}{2} \iiint_\Omega (\nabla \phi)^2 \, d\Omega - \frac{\omega^2}{2g} \iint_{S_F} \phi^2 \, dS$$

$$- \iint_{S_B} V\phi \, dS + \iint_C \left(\tfrac{1}{2}\bar{\phi} - \phi \right) \frac{\partial \bar{\phi}}{\partial n} \, dS \tag{7.1}$$

implies, and is implied by, the original boundary-value problem. Equating the first variation to zero, we get,

$$\delta J = 0 = \iiint_\Omega \nabla \phi \cdot \nabla (\delta \phi) \, d\Omega - \frac{\omega^2}{g} \iint_{S_F} \phi \, \delta \phi \, dS - \iint_{S_B} V \, \delta \phi \, dS$$

$$+ \iint_C (\bar{\phi} - \phi) \frac{\partial \, \delta \bar{\phi}}{\partial n} \, dS + \tfrac{1}{2} \iint_C \left(\delta \bar{\phi} \frac{\partial \bar{\phi}}{\partial n} - \bar{\phi} \frac{\partial \delta \bar{\phi}}{\partial n} \right) \, dS - \iint_C \delta \phi \frac{\partial \bar{\phi}}{\partial n} \, dS \tag{7.2}$$

after some arrangement. Since $\delta \bar{\phi}$ satisfies the same conditions as $\bar{\phi}$, it may be shown by applying Green's formula to $\delta \bar{\phi}$ and $\bar{\phi}$ over $\bar{\Omega}$ that the last integral in Eq. (7.2) vanishes identically. By partial integration and Gauss' theorem, the first integral in Eq. (7.2) may be transformed to surface integrals, leading to the following:

$$\delta J = 0 = - \iiint_\Omega \delta \phi \nabla^2 \phi \, d\Omega + \iint_{S_F} \delta \phi \left(\frac{\partial \phi}{\partial z} - \frac{\omega^2}{g} \phi \right) \, dS + \iint_{B_0} \delta \phi \frac{\partial \phi}{\partial n} \, dS$$

$$+ \iint_{S_B} \delta \phi \left(\frac{\partial \phi}{\partial n} - V \right) \, dS + \iint_C (\bar{\phi} - \phi) \frac{\partial \delta \bar{\phi}}{\partial n} \, dS + \iint_C \delta \bar{\phi} \left(\frac{\partial \phi}{\partial n} - \frac{\partial \bar{\phi}}{\partial n} \right) \, dS. \tag{7.3}$$

In order for $\delta J = 0$ for arbitrary $\delta\phi$ and $\delta\bar\phi$, it is both necessary and sufficient that Laplace's equation be satisfied as the Euler–Lagrange equation, while all the other boundary conditions on S_F, S_B, and B_0, including the continuity of ϕ and $\partial\phi/\partial n$ across C, must be satisfied as *natural conditions*. Thus, the stationarity of J is equivalent to the boundary-value problem for the radiation potential.

Equation (7.2) without the last integral may be stated in another way. Since only first derivatives are involved, ϕ and $\delta\phi$ need only be piecewise linear in Ω. In mathematical language, ϕ and $\delta\phi$ are said to be in the Sobolev space $H^1(\Omega)$ defined by

$$f \in H^1(\Omega) \quad \text{if} \quad \iiint_\Omega \left[(\nabla f)^2 + f^2 \right] d\Omega < \infty.$$

Being analytic functions, $\bar\phi$ and $\delta\bar\phi$ are said to be in the space $C^\infty(\Omega)$. Thus, Eq. (7.2) may be stated as follows: Find $\phi \in H^1(\Omega)$ and $\bar\phi \in C^\infty(\Omega)$ such that for every $\psi \in H^1(\Omega)$:

$$-\iiint_\Omega \nabla\phi \cdot \nabla\psi \, d\Omega + \frac{\omega^2}{g} \iint_{S_F} \phi\psi \, dS + \iint_{S_B} V\psi \, dS + \iint_C \frac{\partial\bar\phi}{\partial n} \psi \, dS = 0,$$

(7.4)

and for every $\bar\psi \in C^\infty(\Omega)$:

$$\iint_C (\phi - \bar\phi) \frac{\partial\bar\psi}{\partial n} \, dS = 0, \qquad (7.5)$$

where we have written

$$\psi \equiv \delta\phi, \qquad \bar\psi = \delta\bar\phi. \qquad (7.6)$$

Equations (7.4) and (7.5) constitute the *weak formulation* of the problem (Aranha, Mei, and Yue, 1979).

Finally, the reader may verify that the proper functional for the diffraction problem is

$$J(\phi, \bar\phi) = \frac{1}{2} \iiint_\Omega (\nabla\phi)^2 - \frac{\omega^2}{2g} \iint_{S_F} \phi^2 \, dS + \iint_C \left[\left(\tfrac{1}{2}\bar\phi^S - \phi^S \right) \frac{\partial\bar\phi^S}{\partial n} - \bar\phi^S \frac{\partial\phi^I}{\partial n} \right] dS,$$

(7.7)

where $\phi^S = \phi - \phi^I$ and $\bar\phi^S = \bar\phi - \phi^I$.

Exercise 7.1

Use Eq. (7.7) to deduce the expansion coefficients in Eq. (5.3) for the circular cylinder, by making the entire fluid domain the superelement $\bar\Omega$.

7.7.2 The Approximate Solution

Equation (7.2) is now most convenient for setting up the discrete approximation. For simplicity, we shall demonstrate for the two-dimensional problem in
the vertical plane. As shown in Fig. 7.1b there are now two superelements, $\bar{\Omega}_-$
and $\bar{\Omega}_+$, to the left of C_- and to the right of C_+, respectively. In $\bar{\Omega}_-$ and $\bar{\Omega}_+$
we take,

$$\bar{\phi}^+ = A_0^\pm e^{\pm ikx} f_0^{(z)} + \sum_n A_n^\pm e^{\pm k_n x} f_n(z), \qquad x \in \bar{\Omega}_\pm . \tag{7.8}$$

The region Ω is divided into triangles to form a network with N nodes. Let us
introduce the pyramid function $F_i(x, z)$ which is linear in x and z in each
triangle, has the value 1 at the node (x_i, z_i), and vanishes at all other nodes,
that is,

$$F_i(x_j, z_j) = \delta_{ij}. \tag{7.9}$$

Clearly, F_i is in $H^1(\Omega)$. We now express

$$\phi = \sum_{j=1}^N \phi_j F_j(x, z), \tag{7.10a}$$

$$\psi = \sum_{i=1}^N \psi_i F_i(x, z), \tag{7.10b}$$

where ϕ_j is the nodal potential, ψ_i is the nodal weight, and

$$\bar{\psi}^\pm = B_0^\pm e^{\pm ikx} f_0(z) + \sum_{n=1}^\infty B_n^\pm e^{\mp k_n x} f_n(z). \tag{7.11}$$

Substituting into Eq. (7.4), we get from various integrals

$$\iint_\Omega \nabla\phi \cdot \nabla\psi \, d\Omega = \sum_i \sum_j \psi_i \left(\iint_\Omega \nabla F_i \cdot \nabla F_j \, d\Omega \right) \phi_j = \{\psi\}^T [K_1]\{\phi\},$$

$$\tag{7.12a}$$

$$-\frac{\omega^2}{g} \int_{S_F} \phi\psi \, dS = \sum_i \sum_j \psi_i \left(\int_{S_F} F_i F_j \, dS \right) \phi_j = \{\psi^F\}^T [K_2]\{\phi^F\}, \tag{7.12b}$$

$$-\int_{S_B} V\psi \, dS = -\sum_i \psi_i \left(\int_{S_B} V F_i \, dS \right) = -\{\psi^B\}^T \{V\}. \tag{7.12c}$$

After truncating the series (7.8) and (7.11) after M terms, we have

$$\bar{\phi}^{\pm} = \sum_{j=0}^{M-1} \bar{\phi}_j^{\pm} f_j(z), \tag{7.13a}$$

$$\bar{\psi}^{\pm} = \sum_{j=0}^{M-1} \bar{\psi}_j^{\pm} f_j(z), \tag{7.13b}$$

$$\frac{\partial \bar{\phi}^{\pm}}{\partial n} = \sum_{j=0}^{M-1} \sigma_j^{\pm} k_j \bar{\phi}_j^{\pm} f_j(z), \tag{7.13c}$$

$$\frac{\partial \bar{\psi}^{\pm}}{\partial n} = \sum_{j=0}^{M-1} \sigma_j^{\pm} k_j \bar{\psi}_j^{\pm} f_j(z), \tag{7.13d}$$

where $\sigma_0^{\pm} = \pm i$, $\sigma_j^{\pm} = \mp 1$, and $j = 1, 2, 3, \ldots$. If (ϕ_i^+, ψ_i^+) denote the N_+ nodal values on C_+, the remaining integrals on C_+ in Eq. (7.4) become

$$\int_{C_+} \frac{\partial \bar{\phi}}{\partial n} \psi \, dS = \sum_{i=1}^{N_+} \sum_{j=1}^{M-1} \psi_i^+ \left(\int_{C_+} \sigma_j^+ k_j f_j F_i \, dz \right) \bar{\phi}_j^+ = \{\psi^+\}^T [K_3^+]\{\bar{\phi}^+\},$$

$$\tag{7.14a}$$

$$\int_{C_+} \phi \frac{\partial \bar{\psi}}{\partial n} \, dS = \sum_{i=1}^{N_+} \sum_{j=1}^{M-1} \phi_i^+ \left(\int_{C_+} \sigma_j^+ k_j f_j F_i \, dz \right) \bar{\psi}_j^+ = \{\phi^+\}^T [K_3^+]\{\bar{\psi}^+\}$$

$$= \{\bar{\psi}^+\}^T [K_3^+]^T \{\phi^+\}, \tag{7.14b}$$

$$\int_{S_+} \bar{\phi} \frac{\partial \bar{\psi}}{\partial n} \, dS = \sum_{i=1}^{M-1} \sum_{j=1}^{M-1} \bar{\psi}_i^+ \left(\int_{S_+} \sigma_j^+ k_i f_i f_j \, dz \right) \bar{\phi}_j^+$$

$$= \sum_{i=1}^{M-1} \sum_{j=1}^{M-1} \bar{\psi}_i^+ (\sigma_j^+ k_i \delta_{ij}) \bar{\phi}_j^+ = \{\bar{\psi}^+\}^T [K_4^+]\{\bar{\phi}^+\}. \tag{7.14c}$$

Similar expressions may be obtained for integrals along C_-.

Using Eqs. (7.12) and (7.14), we assemble Eqs. (7.4) and (7.5) to give

$$\{\psi\}^T [K]\{\phi\} = \{\psi\}^T \{V\}, \tag{7.15}$$

where the transpose of the column vector $\{\phi\}$ is arranged as follows

$$\{\phi\}^T = \left\{ \{\bar{\phi}^-\}^T, \{\phi^-\}^T, \{\tilde{\phi}\}^T, \{\phi^+\}^T, \{\bar{\phi}^+\}^T \right\}$$

with $\{\tilde{\phi}\}$ corresponding to the nodes inside Ω and on S_B. The column vector

$\{\psi\}$ is arranged similarly. The column vector $\{V\}$ has nonzero entries only for the nodes on S_B. The global stiffness matrix $[K]$ is of the following form:

$$
\begin{bmatrix}
[K_4^-] & & & \\
[K_3^-] & & & \\
 & [K_1] + [K_2] & & \\
 & & [K_3^+] & \\
 & & [K_4^+] &
\end{bmatrix}
$$

which is symmetric. Since $\{\psi\}$ is arbitrary,

$$[K]\{\phi\} = \{V\} \tag{7.16}$$

which may be solved numerically.

To minimize the bandwidth of $[K]$, it is desirable to number the nodes downward along a vertical line, followed by the nodes on the next vertical line..., and to let $\{\bar{\phi}^-\}$ and $\{\bar{\phi}^+\}$ occupy the top and bottom positions, respectively, of the column vector $\{\phi\}$. In this arrangement, the semibandwidth is roughly equal to the number of nodes on a vertical line, that is, N_+ or N_-. Further details may be found in Bai and Yeung (1974) for two dimensions and Yue, Chen, and Mei (1976) for three dimensions.

When the sea depth becomes infinite, the eigenfunction expansion becomes inefficient and it is more advantageous to localize the finite element in all three dimensions and to employ Green's function for $\bar{\phi}$. These modifications have been accomplished by Seto and Yamamoto (1975) and Lenoir and Jami (1978).

7.7.3 A Theoretical Property of the Hybrid Element Method

In the weak formulation, let us choose among all admissible functions $\psi = \phi^*$. Equation (7.4) yields

$$
-\iiint_\Omega |\nabla\phi|^2 \, d\Omega + \frac{\omega^2}{g} \iint_{S_F} |\phi|^2 \, dS + \iint_{S_B} V\phi^* \, dS + \iint_C \frac{\partial\bar{\phi}}{\partial n}\phi^* \, dS = 0.
$$

$$\tag{7.17}$$

Since the first two integrals are purely real, we must have

$$\text{Im}\iint_{S_B} V\phi^* \, dS = -\text{Im}\iint_C \phi^* \frac{\partial\bar{\phi}}{\partial n} \, dS. \tag{7.18}$$

From $C^\infty(\bar{\Omega})$ let us choose $\bar{\psi} = \partial\bar{\phi}^*/\partial n$; Eq. (7.5) then implies

$$\iint_C \phi \frac{\partial\bar{\phi}^*}{\partial n} \, dS = \iint_C \bar{\phi} \frac{\partial\bar{\phi}^*}{\partial n} \, dS. \tag{7.19}$$

Using the complex conjugate of Eq. (7.19) in Eq. (7.18), we get

$$\text{Im} \iint_{S_B} V\phi^* \, dS = -\text{Im} \iint_C \bar{\phi}^* \frac{\partial \bar{\phi}}{\partial n} \, dS. \tag{7.20}$$

But this is formally the statement of energy conservation that the rate of work done by the body is equal to the rate of energy flux through the cylinder C. The remarkable point is that ϕ^* and $\bar{\phi}$ (and $\bar{\phi}^*$) correspond to the approximate solution whose accuracy depends on the discretization and the number of terms in the truncated series. Therefore, the present hybrid element method, which imposes continuity on C in a special way according to Eq. (7.5), preserves energy conservation. This property was observed in numerical experiments by Yue and Mei and proven by Aranha; it implies that the energy theorem *cannot* be used as a way to check the discretization error. Indeed, by proper selections of ϕ, ψ, and $\bar{\psi}$ one can further prove that *all* the global identities deduced in Section 7.6 are preserved in the weak formulation. Aranha, Mei, and Yue (1979) also gave numerical evidence of these properties by treating a given problem with two vastly different finite-element grids. They showed that while the two grids gave quite different answers at any point, the global identities were satisfied to an extremely high degree with a minute round-off error. Therefore, the relations in Section 7.6 are not *sufficient* to guarantee an accurate solution, although they are *necessary*, when the present hybrid element method is employed. Similar caution is warranted when other numerical methods are used.

7.7.4 A Numerical Example

As an application of the hybrid element method, we shall show some sample numerical results for a floating two-dimensional cylinder with the cross section of a tear drop. This cylinder was used by Salter as a device to extract energy from water waves, a topic which will be further discussed in Section 7.9. Here we shall only give the hydrodynamical quantities (Mynett, Serman, and Mei, 1979).

A sample finite-element grid for Salter's cam is shown in Fig. 7.2. Note that the lines C_+ and C_- are as close as one element away from the body. The results shown in Figs. 7.3a–7.3e correspond to a schematized Salter cam with a circular stern of radius a and a straight bow slanted at an angle $\theta = \pi/4$. The water depth is kept at a constant $h = 4a$.

For the diffraction problem the reflection and transmission coefficients are given in Fig. 7.3a for three depths of submergence, $s = 0$, $a/2$, and a. As $\omega(a/g)^{1/2}$ increases beyond 1, $|T| \to 1$ and $|R| \to 0$. Figures 7.3b–7.3d give the exciting forces F_1^D (horizontal), F_3^D (vertical), and F_5^D (moment about the y axis). Note that the largest forces occur around $\omega(a/g)^{1/2} = 1$.

For the radiation problem, the real quantities λ_{ij} and μ_{ij} are shown in Fig. 7.3e.

For three-dimensional examples and programming details, reference is made to Yue, Chen, and Mei (1976).

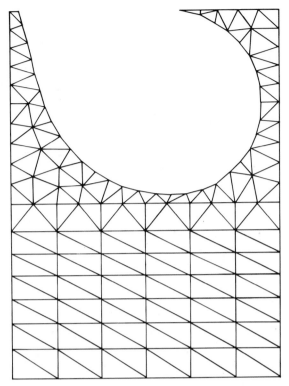

Figure 7.2 Sample finite-element grid for Salter's cam.

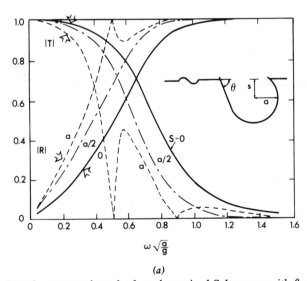

(a)

Figure 7.3 Sample computed results for schematized Salter cam with $\theta = \frac{1}{4}\pi$, $h = 4a$. *(a)* Transmission and reflection coefficients; *(b)* sway exciting force $\hat{F}_1^D = F_1^D / \rho g A a$; *(c)* heave exciting force $\hat{F}_3^D = F_3^D / \rho g A a$; *(d)* roll exciting moment $\hat{F}_5^D = F_5^D / \rho g A a^2$; *(e)* roll inertia and damping coefficients, $\hat{\mu}_{55} = \mu_{55} / \rho a^4$, $\hat{\lambda}_{55} = \lambda_{55} (\rho a^4 (g/a)^{1/2})^{-1}$. (From Mynett et al., 1979, *Appl. Ocean Res.* Reproduced by permission of CML Publications.)

337

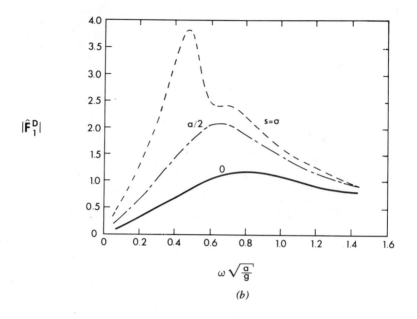

(b)

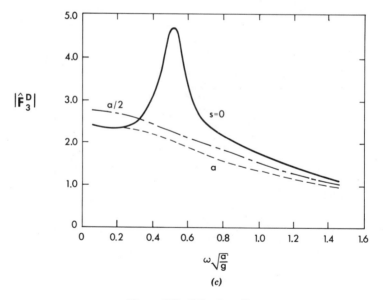

(c)

Figure 7.3 (*Continued*)

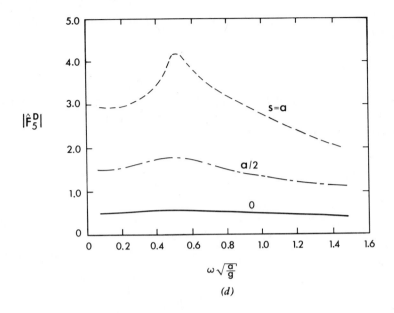

(d)

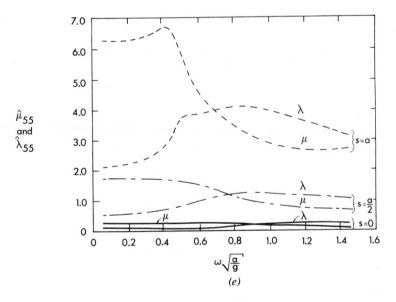

(e)

Figure 7.3 (*Continued*)

7.8   REMARKS ON THE NUMERICAL METHODS BY INTEGRAL EQUATIONS

7.8.1   The Integral Equations

Another powerful numerical technique for diffraction is the method of integral equations via Green's function, which was introduced in Section 7.4.2. Let us consider three dimensions and recall that

$$\phi(\mathbf{x}_0) = \iint_{S_B} \left(\phi \frac{\partial G}{\partial n} - G \frac{\partial \phi}{\partial n} \right) dS \qquad \text{if } \mathbf{x}_0 \in \Omega \text{ but not on } S_B. \quad (8.1)$$

If $\mathbf{x}_0$ is on S_B, we must proceed with caution and exclude from Ω a small hemisphere Ω_ε centered at $\mathbf{x}_0$ as shown in Fig. 8.1.

Let the surface of the ε-sphere be denoted by S_ε. Application of Green's theorem to ϕ and G over Ω (excluding Ω_ε) leads to

$$\iint_{S_B + S_\varepsilon} \left(\phi \frac{\partial G}{\partial n} - G \frac{\partial \phi}{\partial n} \right) dS = 0, \quad (8.2)$$

where the integral over S_B is to be interpreted as the principal value, a circular patch of radius ε being excluded. Over S_ε, ϕ is nearly constant and equal to $\phi(\mathbf{x}_0)$, and $\partial G/\partial n$ is much greater than G. Integrating the polar form of Eq. (4.22) for small r where $r = |\mathbf{x} - \mathbf{x}_0|$, we get

$$\frac{\partial G}{\partial r} \sim \frac{1}{4\pi r^2}, \qquad r \to 0. \quad (8.3)$$

Therefore, the integral over the hemisphere S_ε is

$$\lim_{\varepsilon \to 0} \iint_{S_\varepsilon} \left(\phi \frac{\partial G}{\partial n} - G \frac{\partial \phi}{\partial n} \right) dS = -\lim_{\varepsilon \to 0} \phi(\mathbf{x}_0) \iint_{S_\varepsilon} \frac{\partial G}{\partial r} dS = -\frac{1}{2} \phi(\mathbf{x}_0).$$

$$(8.4)$$

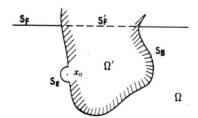

Figure 8.1

Putting Eq. (8.4) into Eq. (8.2), we get

$$\tfrac{1}{2}\phi(\mathbf{x}_0) = \iint_{S_B} \left(\phi \frac{\partial G}{\partial n} - G \frac{\partial \phi}{\partial n} \right) dS, \qquad \mathbf{x}_0 \text{ on } S_B. \tag{8.5}$$

Since $\partial \phi / \partial n$ is prescribed in a diffraction or radiation problem, Eq. (8.5) is a Fredholm integral equation of the second kind for $\phi(\mathbf{x}_0)$ ($\mathbf{x}_0 \in S_B$). By dividing S_B into discrete panels and approximating ϕ in each panel by a constant ϕ_i and then carrying out the integration, one obtains a system of algebraic equations for ϕ_i, $i = 1, 2, \ldots, N$, which can be solved for ϕ_i. Afterward, ϕ at any other point is given by Eq. (8.1) or (8.5).

In practice, higher-order interpolating functions may be used within each panel. More important, when $\mathbf{x}$ and $\mathbf{x}_0$ are close, the integration in Eq. (8.5) must be carried out to high accuracy; indeed, analytic integration is often necessary. While the advantage of this method is that the number of unknowns is usually small, being the ϕ_i on S_B, evaluation of the integrals in order to get the matrix coefficients is a laborious task both for the worker and for the computer (Jawson and Symm, 1977).

A closely related approach is the method of source distribution where one represents the potential everywhere in Ω by

$$\phi(\mathbf{x}_0) = \iint_{S_B} \gamma G \, dS_0 \tag{8.6}$$

with γ unknown. The boundary condition on the body requires that

$$\lim_{\mathbf{x} \to \mathbf{x}_0} \iint_{S_B} \gamma \frac{\partial G}{\partial n_0} dS_0 = U(\mathbf{x}_0) \tag{8.7}$$

as $\mathbf{x}$ approaches S_B from without. To carry out the limiting process we indent S_B inward by a hemispherical surface S_ε. Since,

$$\frac{\partial G}{\partial n_0} = -\frac{\partial G}{\partial r} \cong -\frac{1}{4\pi r^2} \tag{8.8}$$

on S_ε, we get the limit

$$-\tfrac{1}{2}\gamma(\mathbf{x}_0) + \iint_{S_B} \gamma \frac{\partial G}{\partial n} dS = U(\mathbf{x}_0). \tag{8.9}$$

This integral equation has the same kernel as Eq. (8.5).

7.8.2 Irregular Frequencies

The type of integral equations (8.5) or (8.9) is known to possess the so-called *irregular frequencies* and nontrivial eigensolutions if S_B intersects the free surface. The numerical consequence is that the approximate matrix equation becomes ill-conditioned. This drawback is strictly a feature of the kernel and does not imply that the original diffraction or radiation problem has eigensolutions, as is well known in acoustics (Lamb, 1932, Sec. 290). In the context of a floating body in water waves, John (1950) was the first to call attention to irregular frequencies, while Frank (1967) gave the first numerical evidence.

To see the origin of irregular frequencies we consider first a fictitious problem for the interior of S_B, that is, Ω' which intersects the free surface over the portion S_F'. Let the interior potential ϕ' satisfy

$$\nabla^2 \phi' = 0 \qquad \text{in } \Omega', \tag{8.10}$$

$$\frac{\partial \phi'}{\partial z} - \sigma \phi' = 0 \qquad z = 0 \text{ on } S_F'. \tag{8.11}$$

Using the same Green's function G as before which is defined everywhere in $\Omega + \Omega'$, and applying Green's theorem to ϕ' and G over Ω', we get

$$-\phi'(\mathbf{x}_0) = \iint_{S_B} \left(\phi' \frac{\partial G}{\partial n} - G \frac{\partial \phi'}{\partial n} \right) dS, \qquad \mathbf{x}_0 \in \Omega', \text{ not on } S_B. \tag{8.12}$$

The minus sign on the left is due to the fact that the unit normal $\mathbf{n}$ is from Ω to Ω'. It is also easy to deduce that

$$\begin{bmatrix} -\tfrac{1}{2}\phi'(\mathbf{x}_0) \\ 0 \end{bmatrix} = \iint_{S_B} \left(\phi' \frac{\partial G}{\partial n_0} - G \frac{\partial \phi'}{\partial n_0} \right) dS_0 \qquad \text{if } \mathbf{x}_0 \begin{bmatrix} \text{on } S_B \\ \in \Omega \end{bmatrix}. \qquad \begin{matrix} (8.13a) \\ (8.13b) \end{matrix}$$

Now subtracting Eq. (8.13b) from Eq. (8.1), we get

$$\phi(\mathbf{x}_0) = \iint_{S_B} \left[(\phi - \phi') \frac{\partial G}{\partial n} - \left(\frac{\partial \phi}{\partial n} - \frac{\partial \phi'}{\partial n} \right) G \right] dS \qquad \mathbf{x}_0 \text{ in } \Omega \text{ but not on } S_B. \tag{8.14}$$

If we impose that

$$\phi' = \phi \qquad \text{on } S_B \tag{8.15}$$

as the boundary condition for the interior problem, we recover the source representation (8.6) upon defining

$$\gamma = -\left(\frac{\partial \phi}{\partial n} - \frac{\partial \phi'}{\partial n} \right)_{S_B}. \tag{8.16}$$

Therefore, Eq. (8.6) and the integral equation (8.9) are related to the interior problem defined by Eqs. (8.10) and (8.11) and the Dirichlet condition (8.15). It may be shown that this interior problem has eigensolutions defined by the homogeneous problem

$$\nabla^2 \psi_m = 0 \qquad \text{in } \Omega', \tag{8.17}$$

$$\frac{\partial \psi_m}{\partial z} - \sigma_m \psi_m = 0 \qquad \text{on } S_F', \tag{8.18}$$

$$\psi_m = 0 \qquad \text{on } S_B, \tag{8.19}$$

where $\sigma_m = \omega_m^2/g$ and ω_m is the mth eigenfrequency. If $\omega = \omega_m$, the inhomogeneous problem for ϕ' does not have a unique solution. It follows from Eq. (8.16) that the source distribution is not unique; hence the integral equation method must fail. The irregular frequencies are precisely the eigenfrequencies of the fictitious interior problem with the Dirichlet condition on S_B.

A simple example serves to confirm the theoretical discussion above. Let a vertical circular cylinder in shallow water pulsate uniformly in all directions. The potential satisfying the Helmholtz equation, the radiation condition, and

$$\frac{\partial \phi}{\partial r} = U \qquad \text{on } r = a \tag{8.20}$$

is easily found to be

$$\phi = \frac{(U/k)H_0^{(1)}(kr)}{H_0^{(1)'}(ka)}. \tag{8.21}$$

On the other hand, the Green function is

$$G = \frac{i}{4} H_0^{(1)}(kR), \qquad R^2 = r^2 + r_0^2 - 2rr_0\cos(\theta - \theta_0) \tag{8.22}$$

which may be expanded to

$$G = \frac{i}{4} \sum_{n=0}^{\infty} \varepsilon_n J_n(kr) H_n^{(1)}(kr_0)\cos n(\theta - \theta_0) \tag{8.23}$$

by the addition theorem. The integral equation for the source strength is

$$-U = - \lim_{r_0 \to a} a \int_0^{2\pi} \left[\gamma \frac{\partial G}{\partial n_0} \right]_{r=a} d\theta$$

$$= -\frac{i}{4}\gamma ka \int_0^{2\pi} d\theta \sum_n \varepsilon_n J_n(ka) H_n^{(1)'}(ka)\cos n(\theta - \theta_0)$$

$$= -\frac{i}{4} 2\pi \gamma ka J_0(ka) H_0^{(1)'}(ka). \tag{8.24}$$

Hence

$$\gamma = -\frac{2iU}{\pi ka}\left[J_0(ka)H_0^{(1)\prime}(ka)\right]^{-1}. \tag{8.25}$$

Formally, we may substitute Eq. (8.25) into Eq. (8.6)

$$\phi(r,\theta) = a\int_0^{2\pi}\gamma G\,d\theta_0 \tag{8.26}$$

to recover Eq. (8.21). However, γ is clearly infinite at the zeroes of $J_0(ka)$, that is, at the eigenfrequencies of the interior Dirichlet problem.

There are ways to avoid irregular frequencies in solving these integral equations. Usually one introduces additional artificial unknowns and more conditions so as to improve the conditioning of the matrix equations. There are other integral equation methods which do not employ Green's function and hence do not share the same difficulties, for example, Yeung (1975). References may be made to Mei (1978) or Yeung (1982) and the literature cited therein.

In contrast, the hybrid element method of Section 7.7 does not seem to have irregular frequencies. Indeed, Aranha has shown theoretically that for long waves the method gives unique solutions for all ω (see Aranha et al., 1979). Yue (private communication) has experimented numerically for three-dimensional floating boxes of rectangular and circular shapes; no irregular frequencies have been encountered. By combining finite elements with singularity distributions, Lenoir and Jami (1978) have also devised a hybrid method which is free of irregular frequencies.

7.9 POWER ABSORPTION BY FLOATING BODIES

7.9.1 Introduction

To demonstrate the applications of the theory developed in this chapter, we shall discuss a novel subject of wave-power absorption by floating bodies.

The power flux within a unit crest length of a plane sine wavetrain in deep water is $\frac{1}{2}\rho g A^2 C_g = \rho g^2 A^2 T/8\pi$. By this simple formula the available power for 10-s-period swells is estimated below.

Amplitude A (m)	Wave Power (kW/m)
0.5	10
1	40
5	1000

If $A = 1$ m, the power along 25 km of the sea coast is 1000 MW and is comparable to the capacity of a typical conventional power plant. Of course, this estimate must be reduced by directivity of the wave spectrum and varies widely with season and locale. Although the economic feasibility and environmental consequences of wave power are still matters of contention, technical potentials and challenges have already spurred serious research in several countries (see Jansson, Lunde, and Rindby, 1979). Contemplated applications include small-scale power supplies for desolate islands, remote lighthouses, and desalinization of sea water (Pleass, 1978), and so on, or, more optimistically, integration with conventional power plants in a large grid system.

That power can be extracted from sea waves is readily seen by modifying the elementary example discussed in Section 5.4.

If a damper (i.e., a power takeoff device) is attached to the block shown in Fig. 4.1, Chapter Five, power is fed into the damper by the oscillating block. By proper adjustment of the spring and the damping rate, the net radiated waves to the right can be made to vanish so that all the incident wave power is removed by the damper. A detailed analysis for long waves can be worked out in the same manner as in Section 5.4. For better realism the spring may be replaced by a finite body of water filling the space between the block and the shore, if the distance between them is suitably chosen.

A large variety of designs have been proposed (see McCormick, 1981). In most of them the energy converter (turbine, generator) is on or directly connected to the structure. The designs may be classified into three types according to the gross geometry of the structure. With respect to plane incident waves, these are (a) beam–sea absorbers,[†] (b) omnidirectional absorbers, and (c) head–sea absorbers.[†] A beam–(head-) sea absorber is an elongated device with the longitudinal axis parallel (perpendicular) to the incident wave crests. A special case of the omnidirectional absorber is a *point absorber* which can be a buoy, or a resonant water column whose diameter is much smaller than the dominant wavelength.

Beam–Sea Absorbers

To save construction material, it is natural to use floating bodies whose draft is less than the water depth. Consider a cylinder which is symmetrical about its vertical axis and is allowed to have two modes of motion: heave and sway (or roll). Let R and T be the reflection and transmission coefficients, respectively, when the body is held fixed. The induced heave (sway) creates radiation potentials which are symmetric (antisymmetric) with respect to the vertical axis. The corresponding normalized radiated wave amplitudes at infinity are A_s (A_a) at $x \sim +\infty$ and A_s $(-A_a)$ at $x \sim -\infty$. If the induced motion is controlled (e.g., by dampers and springs) so that $R + A_s - A_a = 0$ and $T + A_s + A_a = 0$, then all the incident wave energy is absorbed.

[†](a) and (c) are called *terminators* and *attenuators*, respectively, by Professor M. J. French (1979) and in the United Kingdom.

By using a cam with a cross section in the shape of an inclined tear drop, Salter (1974) demonstrated that roll mode alone is sufficient to give high efficiency. This is partly due to the fact that for sufficiently short waves or large cylinder $T \approx 0$ and $|R| \approx 1$. Because of the inclined bow and the circular stern, the nodding cam radiates larger waves to the incidence than to the transmission side. By controlling the phase and magnitude of A_-, it is possible to make $R + A_-$ very small, leaving a large fraction of incident power for extraction. This device has been christened the *duck*, because of its pointed beak, round tail, and nodding motion.

Many other varieties of design are possible. For example, one may hinge one end of a floating pontoon on a sea wall and install a converter (e.g., a torsional damper) at the hinge. Proper control of the pontoon size and motion will enable complete removal of incident wave energy also. Instead of a fixed hinge, several pontoons may be hinged in a series and moored only at the bow; energy may then be extracted from the relative rotation of adjacent pontoons about the hinges. This is the two-dimensional version of the raft invented by G. Hagen (1975) and C. Cockerell (Wooley and Platts, 1975). Another device makes use of the fact that a submerged stationary circular cylinder does not reflect normally incident waves, while the transmitted waves only suffer a phase shift (Ursell, 1950). If the same cylinder is made to move along a circular orbit around its axis, then wave radiation will be in one direction only (Ogilvie, 1963). Because of these properties, a buoyant circular cylinder tied to a pair of taut moorings can be controlled to absorb all the incident wave energy (Evans et al., 1979).

There are also designs involving pneumatic mechanisms for power takeoff. For example, the oscillating water column of the National Engineering Laboratory, United Kingdom, has a line of caissons moored to the sea floor. The interior of the caisson is an air–water chamber open to the sea through a slot below the free surface on the incidence side. With proper chamber dimensions, the water level within may be tuned to resonance, thereby producing high-pressure air capable of driving, through a rectifier, a turbine in a duct.

Omnidirectional Absorbers

An extensively studied point absorber is a small vertical cylindrical buoy moored to the sea bottom (Budal and Falnes, 1975; Falnes and Budal, 1978). Work is done by the heaving buoy relative to the mooring rod or line. When properly tuned, energy can be absorbed from a sizable width of the wave crest much larger than the diameter of the buoy.

Another variety makes use of a resonant tube which can be either totally (Simon, 1981; see also Lighthill, 1979a for the beam–sea version) or partially submerged.

The so-called dam–atoll (Wirt and Higgins, 1979) uses a submerged circular island to focus the incoming wave rays to the top of the island, where special vanes guide the high water down to a vertical shaft which houses a turbine.

Head–Sea Absorbers

In contrast to the beam–sea absorbers, a slender body pointed normally to the wave crests suffers relatively small horizontal wave forces. The differences among various slender head–sea absorbers lie primarily in the details of power takeoff. In the floating raft system of G. Hagen and C. Cockerell each raft has several pontoons connected by hinges, and energy is derived from the relative rotation about the hinges. The Kaimei ship of M. Masuda (1979) has a series of vertical chambers open along the keel to the sea. In each chamber there is a turbine which is driven by the oscillatory water column. The system of M. J. French (1979) has many flexible air bags mounted on a slender floating frame. The varying pressure from the passing waves compresses the bags which then act as pumps to circulate air in a pipe connected to a turbine.

When the power converter is physically a part of the main structure placed in the sea, costs for energy storage and transmission to land are important concerns. To circumvent these costs, Mehlum and Stamnes (1979) introduced a submerged (beam–sea) lens along a line parallel to the incoming wave crests. Each section of the lens is designed to induce little reflection but to cause different phase lag in the transmitted wave. After passing the lens, the wave crests become concave and converge toward a focal point where a converter is installed. The lateral dimensions of the lens are comparable to other beam–sea devices.

As an introduction to an expanding literature, we now analyze a few simple devices.

7.9.2 A Two-Dimensional Beam–Sea Absorber—Salter's Cam (Duck)

Consider normal incidence and an infinitely long cam with a rigid shaft. The problem is two dimensional. Let $\bar{X}$ and $\bar{Z}$ denote the mean position of the center of the shaft (point Q), ξ_1 and ξ_3 the translations of the shaft, and ξ_5 the rotation about the shaft.

The equations of the cam are given by Eq. (3.24). There is now a reacting force due to the energy generator which transforms the mechanical energy of waves to another form. For generality, we assume that the generator exerts a force on the cam

$$\left(\omega^2\mu'_{\alpha\beta} - C'_{\alpha\beta} + i\omega\lambda'_{\alpha\beta}\right)\xi_\beta, \qquad \alpha, \beta = 1, 3, 5 \tag{9.1}$$

which is partly inertial ($\mu'_{\alpha\beta}$), partly elastic ($C'_{\alpha\beta}$), and partly damping ($\lambda'_{\alpha\beta}$). These primed matrices are the characteristics of the energy generator. If the force in the αth direction due to ξ_β is nonzero, off-diagonal terms of the matrices do not vanish, meaning that modes of the generator can be coupled. The equations of motion of the cam are, therefore,

$$\left[-\omega^2\left(M_{\alpha\beta} + \mu_{\alpha\beta} + \mu'_{\alpha\beta}\right) + \left(C_{\alpha\beta} + C'_{\alpha\beta}\right) - i\omega\left(\lambda_{\alpha\beta} + \lambda'_{\alpha\beta}\right)\right]\xi_\beta = F_\alpha^D + \mathbb{F}_\alpha,$$

$$\tag{9.2}$$

where F_α^D is the exciting force due to diffraction and $\mathbb{F}_\alpha$ the constraining force on the shaft, where the matrices $[M]$, $[C]$, $[\mu]$, and $[\lambda]$ have been defined in Sections 7.2 and 7.3.

When the shaft is fixed, $\xi_1 = \xi_3 = 0$, there is only one degree of freedom, that is, ξ_5 (roll). If we assume that the friction between the cam and the shaft is also included in $\lambda'_{\alpha\beta}$ of Eq. (9.1), then $\mathbb{F}_5 = 0$ and Eq. (9.2) may be simplified to

$$\left[-\omega^2 \mathfrak{M}_{55} + \bar{\mathcal{C}}_{55} - i\omega(\lambda_{55} + \lambda'_{55})\right]\xi_5 = F_5^D, \tag{9.3}$$

where

$$\mathfrak{M}_{55} = m_{55} + \mu_{55} + \mu'_{55}, \qquad \mathcal{C}_{55} = C_{55} + C'_{55}.$$

The solution is obvious. The known value of ξ_5 can be substituted into the remaining momentum equations:

$$\left[-\omega^2(m_{\alpha 5} + \mu_{\alpha 5} + \mu'_{\alpha 5}) + (C_{\alpha 5} + C'_{\alpha 5}) - i\omega(\lambda_{\alpha 5} + \lambda'_{\alpha 5})\right]\xi_5 = F_\alpha^D + \mathbb{F}_\alpha,$$

$$\alpha = 1, 3 \quad (9.4)$$

to determine the constraining force $\mathbb{F}_\alpha$, which is the negative of the force exerted by the cam on the shaft.

Equation (9.4) is formally identical to that describing a simple harmonic oscillator. Clearly, the power extracted is

$$\bar{E} = \tfrac{1}{2}\lambda'_{55}\omega^2 |\xi_5|^2 = \frac{1}{2} \frac{\lambda'_{55}\omega^2 |F_5^D|^2}{\left(\mathcal{C}_{55} - \mathfrak{M}_{55}\omega^2\right)^2 + (\lambda_{55} + \lambda'_{55})^2\omega^2}. \tag{9.5}$$

We emphasize that this total power removed from the waves includes both usable and wasted power due to friction and generator loss, and so on. If the incident wave frequency and the cam geometry are both fixed, then μ_{55}, λ_{55}, and F_5^D are fixed. Let us suppose that the cam inertia M_{55} and the damping rate λ'_{55} can be adjusted to maximize the power output. Therefore we require that

$$\frac{\partial P}{\partial \lambda'_{55}} = 0 \quad \text{and} \quad \frac{\partial P}{\partial\left(\mathcal{C}_{55} - \mathfrak{M}_{55}\omega^2\right)} = 0, \tag{9.6}$$

leading to the optimizing criteria

$$\mathcal{C}_{55} - \mathfrak{M}_{55}\omega^2 = 0, \tag{9.7}$$

and

$$\lambda'_{55} = \lambda_{55}. \tag{9.8}$$

According to the first criterion, the cam must be tuned to resonance, which is intuitively reasonable. The second criterion means that the effective extraction rate must equal the effective radiation damping rate, which is a result not easily anticipated. Substituting Eqs. (9.7) and (9.8) into Eq. (9.5), we get the optimum power

$$\bar{E}_{opt} = \frac{|F_5^D|^2}{8\lambda_{55}}. \tag{9.9}$$

By means of Eq. (6.11) and the Haskind–Hanaoka relation (6.40), the right-hand side of Eq. (9.9) may be represented in terms of the far-field amplitudes $\mathcal{Q}^{\pm}$ due to unit roll:

$$\bar{E}_{opt} = \frac{\frac{1}{2}\rho g A^2 C_g}{1 + |\mathcal{Q}^+|^2/|\mathcal{Q}^-|^2}. \tag{9.10}$$

Since the numerator is the power flux of the incident wave per unit crest length, the optimum efficiency is

$$E_{ff}^{opt} = \left(1 + \frac{|\mathcal{Q}^+|^2}{|\mathcal{Q}^-|^2}\right)^{-1}. \tag{9.11}$$

This formula (Evans, 1976; Mei, 1976; Newman, 1976) gives the maximum possible efficiency for energy removed from the waves. Through $\mathcal{Q}^{\pm}$, E_{ff}^{opt} is a function of frequency and the cam geometry. For a body symmetrical about its vertical axis, $|\mathcal{Q}^+| = |\mathcal{Q}^-|$; the maximum ideal efficiency is at most $\frac{1}{2}$. As $|\mathcal{Q}^+|/|\mathcal{Q}^-|$ decreases, this efficiency increases toward unity; this can be achieved by an asymmetrical cam, and Salter's cam with its circular stern and inclined bow certainly conforms with this principle.

Using the damping and added mass coefficients calculated by numerical solutions of the requisite scattering and radiation problems, Mynett, Serman, and Mei (1979) have obtained the following results for Salter's model cam profile shown in Fig. 7.2 which is slightly different from the profile used for Figs. 7.3a–7.3d. The depth of water is kept at $h = 4a$, where a is the radius of the cam stern.

Figure 9.1 shows the efficiency curves for a variety of extraction rates. Note first that if the cam is so small that for a given design frequency the resonance corresponds to $\hat{\omega} < 0.5$ where $\hat{\omega} = \omega(a/g)^{1/2}$, then not only is the peak efficiency low but the bandwidth of the efficiency curve is narrow. The sharpness is associated with the rapid decrease of radiation damping (see Fig. 7.3e for the qualitative trend).

For a sufficiently large cam the bandwidth of the efficiency curve easily spans over a range $k_0 < k < 2k_0$ which is the range in which most of the wind wave energy lies. Thus, Salter's cam with one degree of freedom is excellent in efficiency. Figure 9.2 gives the roll amplitude for various damping rates.

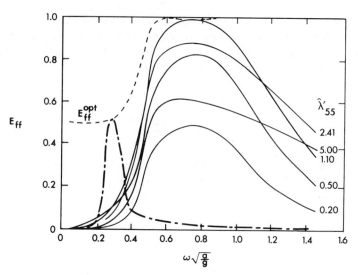

Figure 9.1 Efficiencies of Salter's cam on a fixed shaft. Optimum efficiency E_{ff}^{opt}. Solid curve: Inertia is optimized at $\omega(a/g)^{1/2} = 0.7$ ($\hat{I}_{55} = I_{55}/\rho a^4 = 1.9$) but extraction rates vary from below to above optimum ($\hat{\lambda}'_{55} = 1.10$) where $\hat{\lambda}'_{55} = \lambda_{55}/\rho a^4(g/a)^{1/2}$. Dash–dot curve: inertia and extraction rates are optimized at $\omega(a/g)^{1/2} = 0.25$ with $\hat{I}_{55} = 13$, and $\hat{\lambda}'_{55} = 0.4$ (From Mynett et al., 1979, *Appl. Ocean Res.* Reproduced by permission of CML Publications.)

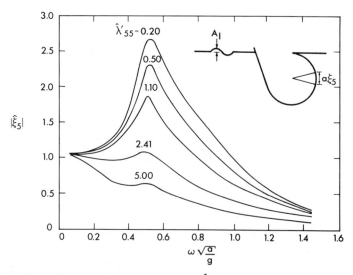

Figure 9.2 Normalized amplitude of roll angle $\hat{\xi}_5 \equiv (a\xi_5)/A$, corresponding to solid curves in Fig. 9.1. (From Mynett et al., 1979, *Appl. Ocean Res.* Reproduced by permission of CML Publications.)

The major drawback of a cam on a fixed axis is the large wave force which must be endured by the shaft (Figs. 9.3a and 9.3b). The maximum vertical and horizontal forces at resonance are of the order $(5\rho gaA)$ per unit cam length. Let $a = 10\ m$, $A = 1\ m$, and $\rho g = 10^4\ \text{kg/m}^3$. The peak force can be $f \sim 500$ tons/m. This large force is inevitable on any rigidly supported beam–sea absorber of similar draft.

Because a rigid support is costly, there are now attempts to allow for a flexible support with provisions to utilize each mode of motion for power production. For this purpose an intricate automatic control system is needed which is capable of varying the impedance of the power-takeoff mechanism in order to maximize the efficiency according to the changing sea spectrum. Indeed, Salter (1979) is now developing a system with a string of ducks which are slackly moored. Precessing gyros, controlled by microprocessors, are sealed within each duck. Energy is extracted from the precession relative to the moving duck. While automatic control may be a technologically innovative solution of some difficult problems unique to the sea, strong sentiments exist which prefer the use of simple devices at the expense of low efficiency. Which option is better only time will tell.

7.9.3 Optimum Efficiency of Three-Dimensional Absorbers

A complete study of three-dimensional absorbers regarding optimum and off-optimum performances, wave forces, and so on, requires considerable numerical computations for the scattering and radiation problems (see, e.g., Simon, 1981; Thomas, 1981). The detailed results vary with the design. We shall limit our discussion here to certain features regarding the optimum efficiency without specifying the power-takeoff system (Newman, 1976, 1979; Budal, 1977).

First, the power extracted by a body must equal the rate of work done by the dynamic pressure in water to the body, that is,

$$
\bar{E} = \iint_{S_B} \overline{p\frac{\partial \Phi}{\partial n}}\, dS = \frac{i\omega\rho}{4} \iint_{S_B} \left(\phi\frac{\partial \phi^*}{\partial n} - \phi^*\frac{\partial \phi}{\partial n} \right) dS
$$

$$
= -\frac{i\omega\rho}{4} \iint_{S_\infty} \left[\phi\frac{\partial \phi^*}{\partial n} - \phi^*\frac{\partial \phi}{\partial n} \right] dS. \tag{9.12}
$$

Let the total potential be decomposed to three parts: incident, scattered, and radiated:

$$
\phi = \phi^I + \phi^S + \phi^R. \tag{9.13}
$$

Let us define $\mathcal{Q}$ to be the amplitude factor of the total outgoing waves:

$$
\phi^S + \phi^R = -\frac{igA\mathcal{Q}}{\omega} \frac{\cosh k(z + h)}{\cosh kh} \left(\frac{2}{\pi kr} \right)^{1/2} e^{i(kr - \pi/4)}. \tag{9.14}
$$

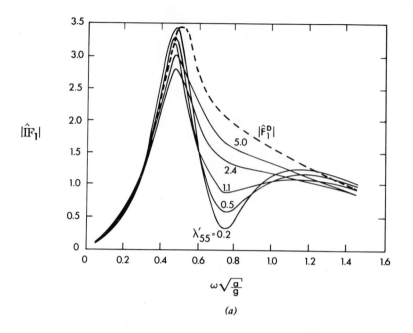

(a)

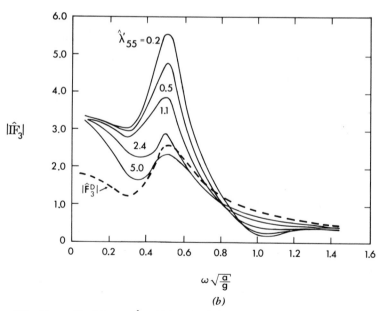

(b)

Figure 9.3 Normalized forces $\hat{F} = F/\rho g A a$. Solid curves correspond to solid curves in Fig. 9.1. (*a*) Horizontal exciting force $\hat{F}_1^D$ and horizontal constraining force $\hat{\mathbb{F}}_1$; (*b*) vertical exciting force $\hat{F}_3^D$ and vertical constraining force $\hat{\mathbb{F}}_3$ (From Mynett et al., 1979, *Appl. Ocean Res.* Reproduced by permission of CML Publications.)

It follows after familiar use of stationary phase that

$$\bar{E} = \frac{\rho g A^2}{k} C_g \left\{ -\frac{1}{\pi} \int_0^{2\pi} d\theta \, |\mathcal{Q}(\theta)|^2 - 2 \operatorname{Re} \mathcal{Q}(0) \right\} \qquad (9.15)$$

which is a generalized optical theorem.

We further define the *absorption* (or *capture*) width as the ratio of the absorbed power to the incident power per unit width of the crest, that is,

$$W = \bar{E} \left(\tfrac{1}{2} \rho g A^2 C_g \right)^{-1}, \qquad (9.16)$$

then the following dimensionless ratio

$$kW = -\frac{2}{\pi} \int_0^{2\pi} d\theta \, |\mathcal{Q}(\theta)|^2 - 4 \operatorname{Re} \mathcal{Q}(0) \qquad (9.17)$$

is clearly a measure of efficiency. For greatest efficiency, it is necessary to maximize the total scattered and radiated waves in the forward direction $\theta = 0$.

Newman has also deduced useful expressions for $\bar{E}$ and kW in terms of the radiation potentials only. To anticipate the behavior of articulated rafts or deformable bodies, he assumes that the normal velocity of the body surface can be expressed as the sum of real-valued normal modes $f_\alpha(x, y, z)$ on S_B, that is,

$$\frac{\partial \phi^R}{\partial n} = A \sum_\alpha V_\alpha f_\alpha(x, y, z) \qquad \text{on } S_B, \qquad (9.18)$$

with V_α being complex constants. The total potential can then be decomposed as

$$\Phi = A e^{-i\omega t} \left(\phi^D + \sum_\alpha V_\alpha \phi_\alpha \right), \qquad (9.19)$$

where

$$\frac{\partial \phi^D}{\partial n} = 0, \qquad \text{on } S_B, \qquad (9.20)$$

$$\frac{\partial \phi_\alpha}{\partial n} = f_\alpha, \qquad \text{on } S_B. \qquad (9.21)$$

The extracted energy is

$$\bar{E} = \iint_{S_B} \overline{p \frac{\partial \Phi}{\partial n}} \, dS$$

$$= \tfrac{1}{2}\rho\omega A^2 \mathrm{Re} \iint_{S_B} i\left(\phi^D + \sum_\alpha V_\alpha \phi_\alpha\right)\left(\sum_\beta V_\beta^* f_\beta\right) dS$$

$$= \tfrac{1}{2}\rho\omega A^2 \mathrm{Re}\left\{i\sum_\alpha V_\alpha^* \iint_{S_B} \phi^D f_\alpha \, dS + i\sum_\alpha \sum_\beta V_\alpha V_\beta^* \iint_{S_B} \phi_\alpha f_\beta \, dS\right\}. \quad (9.22)$$

For the first series above, the Haskind–Hanaoka theorem applies,

$$\iint_{S_B} \phi^D f_\alpha \, dS = \frac{4ig}{\omega k} C_g \mathcal{C}_\alpha(\pi), \quad (9.23)$$

which is simply Eq. (6.41) when n_α is replaced by f_α. Furthermore, the restoring force matrix is symmetric, that is,

$$\iint_{S_B} \phi_\alpha f_\beta \, dS = \iint_{S_B} \phi_\beta f_\alpha \, dS. \quad (9.24)$$

Now the double series in Eq. (9.22) may be manipulated by interchanging the indices to give

$$\mathrm{Re}\, i\sum_\alpha \sum_\beta V_\alpha V_\beta^* \iint_{S_B} \phi_\alpha f_\beta \, dS = -\mathrm{Re}\sum_\alpha \sum_\beta V_\alpha V_\beta^* \left(\mathrm{Im} \iint_{S_B} \phi_\alpha f_\beta \, dS\right).$$

Again by using the symmetry theorem (9.24), we obtain

$$\mathrm{Im} \iint_{S_B} \phi_\alpha f_\beta \, dS = \frac{1}{2i} \iint_{S_B} \left(\phi_\alpha \frac{\partial \phi_\beta^*}{\partial n} - \phi_\beta^* \frac{\partial \phi_\alpha}{\partial n}\right) dS$$

$$= \frac{1}{2i} \iint_{S_\infty} \left(\phi_\alpha \frac{\partial \phi_\beta^*}{\partial r} - \phi_\beta^* \frac{\partial \phi_\alpha}{\partial r}\right) dS$$

$$= \frac{2gC_g}{\pi\omega k} \int_0^{2\pi} d\theta \, \mathcal{C}_\alpha(\theta)\mathcal{C}_\beta^*(\theta)$$

which agrees with Eq. (6.13); therefore,

$$\mathrm{Re}\, i\sum_\alpha \sum_\beta V_\alpha V_\beta^* \iint_{S_B} \phi_\alpha f_\beta \, dS = -\frac{2gC_g}{\pi\omega k} \int_0^{2\pi} d\theta \left|\sum_\alpha V_\alpha \mathcal{C}_\alpha(\theta)\right|^2. \quad (9.25)$$

When Eqs. (8.24) and (9.25) are put in Eq. (9.22), the extracted power becomes

$$\bar{E} = -\frac{1}{2}\rho g A^2 \frac{C_g}{k} \left[\frac{2}{\pi} \int_0^{2\pi} d\theta \left| \sum_\alpha V_\alpha \mathcal{C}_\alpha \right|^2 - \mathrm{Re}\, 4 \sum_\alpha V_\alpha \mathcal{C}_\alpha^*(\pi) \right] \quad (9.26)$$

which is expressed only in terms of the radiation potentials. Accordingly, the efficiency is

$$kW = -\frac{2}{\pi} \int_0^{2\pi} d\theta \left| \sum_\alpha V_\alpha \mathcal{C}_\alpha(\theta) \right|^2 - 4\,\mathrm{Re}\, \sum_\alpha V_\alpha \mathcal{C}_\alpha^*(\pi). \quad (9.27)$$

If there is only one degree of freedom, the summation sign can be removed from Eq. (9.27). To maximize the efficiency the second term in Eq. (9.27) must be real and negative, that is,

$$V_\alpha \mathcal{C}_\alpha^*(\pi) = -|V_\alpha||\mathcal{C}_\alpha(\pi)|.$$

The maximum value is

$$(kW)_{\mathrm{opt}} = \frac{2\pi |\mathcal{C}_\alpha(\pi)|^2}{\int_0^{2\pi} |\mathcal{C}_\alpha(\theta)|^2\, d\theta}, \quad (9.28a)$$

which occurs when

$$|V_\alpha| = \frac{\pi |\mathcal{C}_\alpha(\pi)|^2}{\int_0^{2\pi} |\mathcal{C}_\alpha(\theta)|^2\, d\theta}. \quad (9.28b)$$

Equation (9.28) is the three-dimensional counterpart of Eq. (9.11) and implies that for one degree of freedom, focusing the radiated waves in the opposite sense of the incident wave can increase efficiency. We now seek more specific results.

Omnidirectional Absorbers

Consider an absorber which is an axially symmetric rigid body about an upright axis. There are only three modes $\alpha = 1$, 3, and 5. Substituting Eq. (6.55) into Eq. (9.26) and integrating, we get

$$\bar{E} = -\frac{1}{k}\rho g A^2 C_g \{ |V_1 \mathcal{C}_1(0) + V_5 \mathcal{C}_5(0)|^2 + 2|V_3 \mathcal{C}_3(0)|^2$$

$$+2\,\mathrm{Re}\left[V_1 \mathcal{C}_1(0) + V_5 \mathcal{C}_5(0) \right] - 2\,\mathrm{Re}\left[V_3 \mathcal{C}_3(0) \right]. \quad (9.29)$$

For maximum $\bar{E}$ we must orchestrate the motion so that $V_1 \mathcal{C}_1(0) + V_5 \mathcal{C}_5(0)$ is real and positive while $V_3 \mathcal{C}_3(0)$ is real and negative. The optimum criteria are

$$V_1 \mathcal{C}_1(0) + V_5 \mathcal{C}_5(0) = 1 \quad \text{and} \quad V_3 \mathcal{C}_3(0) = -\tfrac{1}{2}. \tag{9.30}$$

The corresponding optimum absorbed power and efficiency are

$$\bar{E}_{\text{opt}} = \frac{3}{2k} \rho g A^2 C_g, \tag{9.31a}$$

$$k W_{\text{opt}} = 3. \tag{9.31b}$$

When all three modes are optimized, the absorption width is about $\tfrac{1}{2}\lambda$, *independent* of the body size. If the axially symmetric buoy only heaves (such as a small buoy), Eq. (9.28) shows that $k W_{\text{opt}} = 1$ which is one-third of the possible maximum when all three are optimized. These results may be confirmed in another way. Recall that the incident wave may be expanded as

$$A e^{ikx} = A \sum_{n=0}^{\infty} \varepsilon_n (i)^n J_n(kr) \cos n\theta.$$

In the far field the nth partial wave mode may be approximated by

$$\cos n\theta \frac{A}{2} \left(\frac{2}{\pi k r} \right)^{1/2} \left\{ \exp\left[i\left(kr - \frac{\pi}{4} - \frac{n\pi}{2} \right) \right] + \exp\left[-i\left(kr - \frac{\pi}{4} - \frac{n\pi}{2} \right) \right] \right\}$$

which consists of both outgoing and incoming parts. Supposing that the body is only allowed to heave, its induced motion in the fluid can only be axially symmetric, that is, isotropic. With proper orchestration the radiated wave can be combined with the isotropic mode ($n = 0$) of the scattered wave to cancel the outgoing part of the isotropic mode in the incident wave. The converging part of the incident isotropic mode is available for absorption; the corresponding rate of energy influx through a circular cylindrical surface around the body is $\tfrac{1}{2}\rho g A^2 C_g/k$, which implies an absorption width of $W = 1/k$. Similarly, if the body can only sway or roll, the radiated wave is proportional to $\cos\theta$ only and can be used to cancel the outgoing part of the partial wave mode with $n = 1$. The energy influx from the converging part is then $\rho g A^2 C_g/k$, which implies $W = 2/k$.

It is interesting that the value of the optimum efficiency does not depend on the size of the buoy, suggesting that a small buoy can be as efficient as a larger one. However, a small buoy can only achieve the same optimum by oscillating at a very large amplitude, which may be demonstrated for the case of heave only $V_3 \neq 0$, $V_1 = V_5 = 0$. By the Haskind–Hanaoka theorem, Eq. (6.41), $\mathcal{C}_3(0)$ may be expressed in terms of the exciting force F_3^D

$$\mathcal{C}_3(0) = -\frac{k}{4} F_3^D \left(\rho g C_g A \right)^{-1}. \tag{9.32}$$

For a small buoy, F_3^D may be estimated by ignoring both diffraction and the spatial variation of pressure around the buoy, yielding

$$F_3^D \cong \rho g A \pi a^2, \qquad (9.33)$$

where a is the radius of the buoy at the water plane. Putting Eq. (9.33) into Eq. (9.32) and then into Eq. (9.30), we get

$$\mathcal{Q}_3 = -\frac{\pi k a^2}{4 C_g}, \qquad (9.34a)$$

$$A V_3 = \frac{2}{\pi} \frac{A}{a} \frac{C_g}{ka}. \qquad (9.34b)$$

Thus, for diminishing ka the heave velocity $A V_3$ must become very large for optimum absorption.

Furthermore, from Eq. (6.13) the radiation damping coefficient may be calculated:

$$\lambda_{33} \cong \frac{\rho g a^4 k}{4 C_g} \qquad (9.35)$$

which also diminishes with ka. By considerations similar to Section 7.9.2, the response curve of a small heaving buoy must be very sharply peaked. In order to achieve high efficiency for the broad frequency spectrum of sea waves, it is necessary to tune a small heaving buoy to optimum conditions for nearly every frequency in that spectrum. The following tuning method by phase control has been demonstrated in the laboratory by Falnes and Budal (1978) and Budal et al. (1979). A small buoy is used so that the resonant frequency is always far above the incident wave frequency. In order to maximize the power output, an electronic circuit is introduced to alternately lock or free the buoy so that its heave velocity is roughly in phase with the exciting force. Note that for a small buoy the exciting force is in phase with the incident wave surface, scattering being negligible. Wave height measured instantaneously in front of the buoy can be signaled back for optimum phase control.

In large scale applications, it is natural to envision an array of point absorbers. The total system is in general no longer omnidirectional and interference among neighboring buoys can either enhance or reduce the overall efficiency. Budal (1977), Falnes and Budal (1978), Budal et al. (1979), and Falnes (1980) have made a number of important theoretical and experimental contributions to this topic. In particular, Budal studied a single row of N small and identical buoys equally spaced at the distance d between centers, with the incident wave inclined at the angle α with respect to the line of centers. He assumed that all the buoys only heaved at equal amplitude, but different phase. When $kd = 0(1)$, he used an approximation for weak interaction and showed the total capture width to be

$$kW = Nq, \qquad (9.36a)$$

with q being the interaction factor

$$q = \left[1 + \frac{2}{N} \sum_{n=1}^{N} (N - n)\cos(nkd \sin \alpha) J_0(nkd) \right]^{-1}. \qquad (9.36b)$$

When $d \to \infty$, $q \to 1$; the total capture width is the sum of the individual capture width of N isolated bodies. Figure 9.4 shows the interference factor for normal incidence $\alpha = 0$ and for $N = 2$, 10, and ∞. Clearly, interference can be constructive so as to augment the capture width. In the limit of $N \to \infty$,

$$W = \frac{Nd \cos \alpha}{2} \qquad \text{for } kd < \frac{2\pi}{1 + \sin \alpha},$$

implying that the row of buoys absorbs half of the power incident on the projected length $Nd \cos \alpha$. Budal further studied two parallel rows and showed that all the incident power on the projected length could be absorbed.

Srokosz (1979) (see also Evans, 1979, and Falnes, 1980) relaxed the constraint of Budal and allowed the N point absorbers to have unequal motions. The optimum interaction factor q was found to be greater than that given by

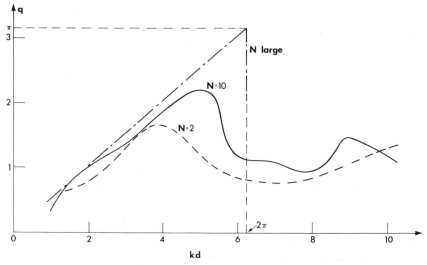

Figure 9.4 The interaction factor q [Eq. (9.36b)] for normal incidence and for two, ten, and an infinite number of bodies in a row. For small kd, Eq. (9.36b) is invalid. All bodies move with equal amplitudes. (From Budal, 1976, *J. Ship Res.* Reproduced by permission of The Society of Naval Architects and Marine Engineers.)

(9.36b) for low value of kd but the difference diminished for sufficiently large kd. For two absorbers the unconstrained result is

$$q = \frac{1 - J_0(kd)\cos(kd\sin\alpha)}{1 - J_0^2(kd)}. \tag{9.36c}$$

A sample comparison is shown in Fig. 9.5. Note that for long waves $kd \to 0$ the unconstrained theory gives different q for different incidences. These limits are of doubtful practical significance since the predicted amplitude of optimum motion would be infinite and inconsistent with the assumption of linearity.

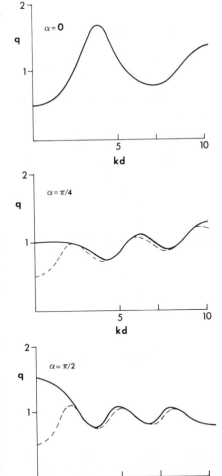

Figure 9.5 The interaction factor for and two bodies at various incidences: (a) $\alpha = 0$ normal incidence; (b) $\alpha = \frac{1}{4}\pi$ oblique incidence; (c) $\alpha = \frac{1}{2}\pi$ waves along the line of axis of bodies. Dashed curve: Budal's theory requiring two bodies moving at equal amplitude; solid curve: Srokoz's theory without Budal's constraint. (From Srokosz, 1979.)

A Head–Sea Absorber: Hagen–Cockerell Raft

Newman (1979) has made a penetrating study of the optimum efficiency of a slender Hagen–Cockerell raft which has fore-and-aft symmetry. Following him we let the longitudinal axis of the slender body be the x axis. In general, the vertical displacement Z of the raft may be decomposed into a number of modes $f_\alpha(x)$:

$$Z(x, t) = A \sum_\alpha Z_\alpha f_\alpha(x) e^{-i\omega t}, \tag{9.37}$$

where Z_α and f_α are dimensionless and f_α is real. Let us consider deep water only. $\mathcal{Q}_\alpha(\theta)$ may be related to the potential on the body by Eq. (4.32) which becomes, in the limit of $kh \to \infty$,

$$\mathcal{Q}_\alpha(\theta) = -\frac{\omega k}{2g} \iint_{S_B} dS \left(\phi_\alpha \frac{\partial}{\partial n} - \frac{\partial \phi_\alpha}{\partial n} \right) e^{kz} \exp(-ikx \cos\theta - iky \sin\theta).$$

$$\tag{9.38}$$

For a slender body with beam and draft small compared to the wavelength, ky, $kz \ll 1$ on S_B and the term $\partial \phi_\alpha / \partial n$ above dominates so that

$$\mathcal{Q}_\alpha(\theta) \cong -\frac{\omega k}{2g} \iint_{S_B} dS \frac{\partial \phi_\alpha}{\partial n} e^{-ikx \cos\theta}.$$

On the body

$$dS \frac{\partial \phi_\alpha}{\partial n} = -i\omega f_\alpha n_z \, dx \left[(dy)^2 + (dz)^2 \right]^{1/2} = -i\omega f_\alpha \, dx \, dy,$$

hence

$$\mathcal{Q}_\alpha(\theta) \cong -\frac{i}{2} k^2 \int_{-l/2}^{l/2} dx \, b(x) f_\alpha(x) e^{-ikx \cos\theta}, \tag{9.39}$$

where l is the body length. Assume for simplicity that $b = \text{const}$ and end effects are minor; Eq. (9.39) may be written

$$\mathcal{Q}_\alpha(\theta) = -\frac{i}{2} K^2 \left(\frac{2b}{l} \right) \int_{-1}^{1} dX \, f_\alpha(X) e^{-iKX \cos\theta} \tag{9.40}$$

after introducing $X = 2x/l$ and $K = \frac{1}{2}kl$. Clearly, $\mathcal{Q}_\alpha(\theta)$ is a function of $\cos\theta$. For a body with fore-and-aft symmetry, the motion can always be decomposed into modes which are either even or odd in X. Let f_α be even in X when α is an even integer, then

$$\mathcal{Q}_\alpha(\theta) = -iK^2\left(\frac{2b}{l}\right)\int_0^1 dX\, f_\alpha(X)\cos(KX\cos\theta) \qquad (9.41)$$

is even in $\cos\theta$ and imaginary. Similarly, let f_α be odd in X when α is an odd integer, then

$$\mathcal{Q}_\alpha(\theta) = K^2\left(\frac{2b}{l}\right)\int_0^1 dX\, f_\alpha(X)\sin(KX\cos\theta) \qquad (9.42)$$

is odd in $\cos\theta$ and real. For varying b, the product bf_α may be considered as the effective modal shape.

The even and odd symmetry of $\mathcal{Q}_\alpha$ implies that the absorption width may be decomposed into even and odd parts also:

$$kW = kW^e + kW^o, \qquad (9.43)$$

where

$$kW^e = -\frac{2}{\pi}\int_0^{2\pi} d\theta\left|\sum_\alpha Z_\alpha \mathcal{Q}_\alpha(\theta)\right|^2 - 4\,\mathrm{Re}\sum_\alpha\left[Z_\alpha \mathcal{Q}_\alpha^*(\pi)\right], \qquad (9.44)$$

with a similar formula for kW^o.

To extract energy from heave or pitch the raft must be dynamically coupled with a stationary structure which must endure a large wave force. For the sake of reducing structural costs, a slack mooring is likely preferable in design. One can then expect no absorption from these rigid body modes $\alpha = 0, 1$; therefore

$$kW_0 = -\frac{2}{\pi}\int_0^\infty d\theta\,|Z_0\mathcal{Q}_0(\theta)|^2 - 4\,\mathrm{Re}\left[Z_0\mathcal{Q}_0^*(\pi)\right] = 0, \qquad (9.45)$$

and a similar relation $kW_1 = 0$ holds. Subtracting Eq. (9.45) from kW^e, we get

$$kW^e = -\frac{2}{\pi}\int_0^{2\pi} d\theta\left|\sum_\alpha{}' Z_\alpha \mathcal{Q}_\alpha(\theta)\right|^2 - 4\,\mathrm{Re}\sum_\alpha{}'\left\{Z_\alpha \mathcal{Q}_\alpha^*(\pi)\right\}$$

$$-\frac{4}{\pi}\,\mathrm{Re}\int_0^{2\pi} d\theta\, Z_0^* \mathcal{Q}_0^*(\theta)\sum_\alpha{}' Z_\alpha \mathcal{Q}_\alpha(\theta), \qquad \alpha = 2, 4, 6, \ldots \quad (9.46)$$

where $\alpha = 0$ is excluded from the sum $\sum{}'$. Recall first that $\mathcal{Q}_\alpha$ is real for all even α. For optimum performance at higher modes $\alpha \geq 2$, the phases must be such that $Z_\alpha \mathcal{Q}_\alpha^*$ is real and negative. Now $\alpha = 0$ is a passive mode with $Z_0 = O(1)$ at most. Since $\mathcal{Q}_0 = O(2b/l)$ from Eq. (9.41), Eq. (9.45) implies that $\mathrm{Re}(Z_0\mathcal{Q}_0) = O(2b/l)^2$. Consequently, the last term in Eq. (9.46) may be neglected with a relative error of $O(2b/l)$. Thus,

$$kW^e \cong -\frac{2}{\pi}\int_0^{2\pi}d\theta\left|\sum_\alpha' Z_\alpha \mathcal{Q}_\alpha(\theta)\right|^2 - 4\,\mathrm{Re}\sum_\alpha'\left\{Z_\alpha \mathcal{Q}_\alpha(\pi)\right\}^*. \qquad (9.47)$$

A similar result may be obtained for kW^o.

If there is only one even or odd mode, the summation sign may be removed from Eq. (9.47). For optimum efficiency one must have

$$Z^e \mathcal{Q}^{e*}(\pi) = |Z^e|\,|\mathcal{Q}^e(\pi)|\,e^{i\pi}$$

so that

$$kW^e = -\frac{2}{\pi}|Z^e|^2\int_0^{2\pi}d\theta\,|\mathcal{Q}^e(\theta)|^2 + 4\,|Z^e|\,|\mathcal{Q}^e(\pi)|. \qquad (9.48)$$

A similar expression holds for the odd mode. Equations (9.28a) and (9.28b) hold for $(kW^e)_{\mathrm{opt}}$ and $(kW^o)_{\mathrm{opt}}$ when $\mathcal{Q}_\alpha$ is replaced by $\mathcal{Q}^e$ and $\mathcal{Q}^o$, respectively.

Now consider an articulated raft made up of three pontoons with two hinges located at $x = \pm\frac{1}{2}al$ (i.e., $X = \pm a$). There can only be four independent modes of transverse motion. Hence, aside from heave and pitch, there are only one even and one odd mode. The following modal shapes can be shown to be orthogonal to heave and pitch:

$$f^e = \begin{cases} 1, & |X|<a \\ 1 - \dfrac{2(|X|-a)}{(1-a)^2}, & a<|X|<1 \end{cases} \qquad (9.49a)$$

$$f^o = \frac{\pm a\left[2 + (a^2-3)\,|X|\right]}{(2+a\,)(1-a\,)^2}, \qquad \pm a \lessgtr X \lessgtr \pm 1. \qquad (9.49b)$$

The corresponding maximum displacements are

$$f_{\max}^e = f^e(1) = \frac{1+a}{1-a} \qquad (9.50a)$$

$$f_{\max}^o = \begin{cases} f^o(a) = a, & a < \sqrt{2} - 1 \\ f^o(1) = \dfrac{a(1+a)}{(2+a)(1-a)}, & a > \sqrt{2} - 1 \end{cases}. \qquad (9.50b)$$

A power-takeoff mechanism can, in principle, be designed for each mode. Now the amplitude functions may be evaluated explicitly:

$$\mathcal{Q}^e = \frac{2iK^2b}{l}\left\{\frac{1+a}{1-a}\frac{\sin(K\cos\theta)}{K\cos\theta} + \frac{2}{(1-a)^2(K\cos\theta)^2}\right.$$

$$\left. \times\left[\cos(K\cos\theta) - \cos(Ka\cos\theta)\right]\right\} \qquad (9.51a)$$

and

$$\mathcal{Q}^o = \frac{K^2b}{l}\left\{\frac{-2a}{(2+a)(1-a)^2K\cos\theta}\right.$$

$$\times\left[\frac{2\sin(Ka\cos\theta)}{Ka\cos\theta} + (a^2-3)\frac{\sin(K\cos\theta)}{K\cos\theta}\right.$$

$$\left.\left. +(1-a^2)\cos(K\cos\theta)\right]\right\}. \qquad (9.51b)$$

These functions may be substituted into Eq. (9.48) for the optimum absorption widths which are plotted as solid lines in Fig. 9.6. Although $W/l \uparrow \infty$ as $k \downarrow 0$, it may be shown that kW approaches a finite limit for long waves. Nevertheless Z^e and Z^o become unbounded, implying violent motion. A further condition that the vertical displacement must not exceed a certain limit,

$$\beta = \frac{|Z^{e,o}|b}{l} < 0.2 \quad \text{or} \quad 0.4, \qquad (9.52)$$

may be imposed on Eq. (9.48) for the maximum absorption width. The corresponding curves for $(W/l)_{\text{opt}}$ are shown in Fig. 9.6 by long dashes for $\beta < 0.4$ and short dashes for $\beta < 0.2$. These constraints can be enforced by electronically controlling the impedances. The total optimum power is the sum of the contributions from the two modes. Within the practical range of $0.2 < a < 0.6$ the optimum W/l is insensitive to a.

Further studies on the impedance of the power-takeoff mechanism at the hinges have been made by Haren and Mei (1980). They have shown that it is necessary to have negative springs and even negative damping at the forward hinge so that the induced raft motion can radiate large waves in the forward direction. This result is in accordance with Eq. (9.15) since the scattered wave is negligible. Heuristically, the radiated waves should be made to cancel the diverging part of as many partial wave modes in the incident wave as possible.

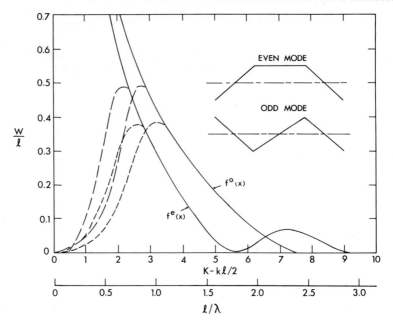

Figure 9.6 Absorption width ratios W/l for hinged raft ($a = 0.4$) with the modes f^e and f^o of optimum amplitude and phase. Broken curves are for the limited displacements defined by Eq. (9.52). (From Newman, 1979, *Appl. Ocean Res.* Reproduced by permission of CML Publications.)

For "infinite" absorption width the radiated waves must have the following far field:

$$\eta^R \cong - \left(\frac{2}{\pi kr} \right)^{1/2} e^{i(kr-\pi/4)} \frac{A}{2} \sum_{n=0} \varepsilon_n (i)^n e^{-in\pi/2} \cos n\theta$$

$$= - \left(\frac{2}{\pi kr} \right)^{1/2} e^{i(kr-\pi/4)} \frac{A}{2} \delta(\theta).$$

The δ-function dependence implies forward focusing. Of course, this limit cannot be achieved in practice for a raft of finite length.

Finally, Newman (1980, private communication) has also extended the slender body theory to the Kaimei ship with a discrete number of resonant chambers.

For a survey of many other theoretical aspects of wave-power absorption, reference should be made to Evans (1981).

7.10   DRIFT FORCES

Consider the time-averaged momentum flux of a plane progressive wave across a vertical plane

$$\overline{M} = \rho \, \overline{\int_{-h}^{\zeta} u^2 \, dz} \, .$$

If the linearized result is used in the integrand, $\overline{M}$ may be calculated to second-order accuracy in wave slope,

$$\overline{M} = \frac{EC_g}{C} + O(kA)^3, \tag{10.1}$$

where $E = \frac{1}{2}\rho g A^2$.

First, consider a two-dimensional obstacle in normally incident waves. If all the wave energy is absorbed by the body, as is the case of the wave-power devices discussed in Section 7.9, the mean momentum must also be totally absorbed by the body. Therefore, the body must experience a steady force EC_g/C. If the body reflects all the energy instead, then the reflected wave carries a momentum of EC_g/C in the opposite direction. The steady force on the body must be equal to the total rate of momentum change $2EC_g/C$, which is similar to a steady jet impinging normally on a wall. More generally, the body scatters waves by its presence and radiates waves by its motion. Let the amplitude of the left-going waves on the incidence side be $(R + \mathcal{C}_-)A$ and of the right-going waves on the transmission side be $(T + \mathcal{C}_+)A$. The net steady force on the body must be

$$\overline{F}_x = \frac{EC_g}{C}\left(1 + |R + \mathcal{C}_-|^2 - |T + \mathcal{C}_+|^2\right) \tag{10.2}$$

(Longuet-Higgins, 1977), and can be calculated as soon as R, T, $\mathcal{C}_-$, and $\mathcal{C}_+$ are known from the linearized theory. This steady force is important in designing moorings or dynamic positioning devices to prevent the body from drifting, hence it is called the *drift force*. Because of energy conservation

$$|R + \mathcal{C}_-|^2 + |T + \mathcal{C}_+|^2 + E_{ff} = 1, \tag{10.3}$$

where E_{ff} is the efficiency of energy absorption, the drift force may also be expressed as

$$\overline{F}_x = \frac{EC_g}{C}\left(E_{ff} + 2|R + \mathcal{C}_-|^2\right). \tag{10.4}$$

Thus, $\bar{F}_x$ is positive, that is, in the direction of the incident wave propagation, if $E_{ff} > 0$. However, if sufficient energy is supplied to the body so that

$$E_{ff} < -2 \, | \, R + \mathcal{Q}_- \, |^2, \tag{10.5}$$

then $\bar{F}_x$ becomes negative, that is, against the waves.

Figure 10.1 shows the horizontal drift force on a Salter duck under various conditions. Absorption of wave power is limited to the roll mode and is seen to reduce the drift force for a broad range of frequencies. In particular, at high ka, we expect $T \approx 0$ and the induced motion (hence $\mathcal{Q}_-$) to be small, so that with no absorption, $\bar{F}_x \cong 2EC_g/C$, as is confirmed by the dotted and dashed curves. When absorption is complete, $R + A_- \cong 0$ and $E_{ff} \approx 1$ so that $\bar{F}_x = EC_g/C$ which implies a 50% reduction; this result also agrees with the solid and the dash–dot curves in Fig. 10.1.

General formulas for drift forces on three-dimensional bodies were first derived by Maruo (1960) and extended to include the drift moment about the vertical axis by Newman (1967). The derivation is also based on momentum balance and is described below in detail in the manner of Newman.

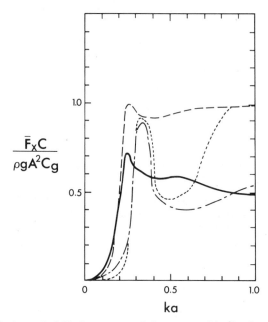

Figure 10.1 Horizontal drift force on a Salter cam with fixed axis: $a =$ radius of circular stern. No energy absorption: solid curve, roll only; dash–dot curve, roll and sway. With energy absorption: dashed curve, roll only; dotted curve, roll and away. (From Maeda, Tanaka, and Kinoshita, 1980, *Proc., 13th Symp. Naval Hydrody.* Reproduced by permission of U.S. Office of Naval Research.)

For any vector quantity per unit volume inside the moving volume V bounded by S, the following kinematic transport theorem (see, e.g., Batchelor, 1967, p. 135) is true:

$$\frac{d}{dt} \iiint_V \mathbf{G} \, dV = \iiint_V \frac{\partial \mathbf{G}}{\partial t} \, dV + \iint_S \mathbf{G} U_n \, dS, \tag{10.6}$$

where U_n denotes the normal velocity of S. Now let $\mathbf{G} = \rho \mathbf{u}$ be the linear momentum per unit volume and $\mathbf{M}$ be the total linear momentum in V. In index form we have

$$\frac{dM_i}{dt} = \rho \iiint_V \frac{\partial u_i}{\partial t} \, dV + \rho \iint_S u_i U_n \, dS. \tag{10.7}$$

If we use the Euler equation in the form

$$\frac{\partial u_i}{\partial t} = -\frac{\partial}{\partial x_i}\left(\frac{p}{\rho} + gz\right) - \frac{\partial}{\partial x_j}(u_i u_j),$$

the first integral on the right of Eq. (10.7) may be transformed to a surface integral via Gauss' theorem

$$\frac{dM_i}{dt} = -\iint_S (P + \rho g z \delta_{i3}) n_i \, dS - \iint_S \rho u_i (u_j n_j - U_n) \, dS. \tag{10.8}$$

In particular, the horizontal components are

$$\frac{d}{dt}\begin{pmatrix} M_x \\ M_y \end{pmatrix} = -\rho \iint_S \left[\frac{P}{\rho}\begin{pmatrix} n_x \\ n_y \end{pmatrix} + \begin{pmatrix} u \\ v \end{pmatrix}(\mathbf{u} \cdot \mathbf{n} - U_n)\right] dS. \tag{10.9}$$

Now let S be the union of the wetted body surface S_B, the free surface S_F, the horizontal sea bottom B_0, and the fixed vertical cylinder S_∞ at "infinity." On the material surfaces S_B, S_F, and B_0, $\mathbf{u} \cdot \mathbf{n} - U_n = 0$. On the free surface, $P = 0$, while on S_∞, $U_n = 0$. On the flat bottom B_0, $n_x = n_y = 0$. Therefore, the total force needed to hold the body is

$$\begin{pmatrix} F_x \\ F_y \end{pmatrix} \equiv \iint_{S_B} P \begin{pmatrix} n_x \\ n_y \end{pmatrix} dS = -\iint_{S_\infty} \left[P\begin{pmatrix} n_x \\ n_y \end{pmatrix} + \rho\begin{pmatrix} u \\ v \end{pmatrix}(\mathbf{u} \cdot \mathbf{n}) \right] dS - \frac{d}{dt}\begin{pmatrix} M_x \\ M_y \end{pmatrix}.$$

$$\tag{10.10}$$

When the time averages are taken and periodicity invoked, the last term on the right gives no contribution; the drift force components are

$$\left(\begin{matrix} \overline{F_x} \\ \overline{F_y} \end{matrix}\right) = -\overline{\iint_{S_\infty}\left[P\left(\begin{matrix} n_x \\ n_y \end{matrix}\right) + \rho\left(\begin{matrix} u \\ v \end{matrix}\right)(\mathbf{u}\cdot\mathbf{n})\right]dS}\,. \tag{10.11}$$

If S_∞ is further taken to be a circular cylinder of large radius R, then we may employ polar coordinates to write

$$\overline{F_x} = -\overline{\iint_{S_\infty}\left[P\cos\theta + \rho u_r(u_r\cos\theta - u_\theta\sin\theta)R\,d\theta\,dz\right]}\,,$$

$$\tag{10.12}$$

$$\overline{F_y} = -\overline{\iint_{S_\infty}\left[P\sin\theta + \rho u_r(u_r\sin\theta + u_\theta\cos\theta)R\,d\theta\,dz\right]}\,.$$

So far, the results above are exact; from here on only second order terms will be preserved. Using the Bernoulli equation, we get

$$-R\int_0^{2\pi}d\theta\left(\begin{matrix}\cos\theta \\ \sin\theta\end{matrix}\right)\overline{\int_{-h}^{\zeta}P\,dz} \cong R\int_0^{2\pi}d\theta\left(\begin{matrix}\cos\theta \\ \sin\theta\end{matrix}\right)\left\{\rho\int_{-h}^0 dz\overline{\left[\Phi_t + gz + \tfrac12(\nabla\Phi)^2\right]}\right.$$

$$\left.+\rho\overline{\int_0^{\zeta}dz\left[\Phi_t + gz + \tfrac12(\nabla\Phi)^2\right]}\right\}_{r=R}. \tag{10.13}$$

In the first integral on the right the time average can be performed directly on the integrand. The term $\overline{\Phi_t}$ is zero owing to periodicity, while the hydrostatic term ρgz does not contribute after integrating with respect to θ; the remaining term $\tfrac12(\nabla\Phi)^2$ can be calculated with an error of $O(kA)^3$ by using the linearized Φ. In the second integral, the contribution from the term $\tfrac12(\nabla\Phi)^2$ is $O(kA)^3$ and negligible; the remaining terms can be evaluated to $O(kA)^2$ as

$$\rho\overline{\int_0^{\zeta}(\Phi_t + gz)\,dz} = \rho\left(\overline{\Phi_t|_{z=0}\zeta + \tfrac12 g\zeta^2}\right) = -\frac{\rho}{2g}\,\overline{\Phi_t^2}\bigg|_{z=0}$$

$$= -\frac{\rho}{2g}\,\overline{\left(\mathrm{Re}\, i\omega\phi e^{-i\omega t}\right)^2}_{z=0} = -\frac{\rho\omega^2}{4g}\,|\phi|_{z=0}^2\,.$$

$$\tag{10.14}$$

It follows that

$$
\iint_{S_\infty} P\begin{pmatrix} \cos\theta \\ \sin\theta \end{pmatrix} R\, d\theta\, dz \cong \int_0^{2\pi} R\, d\theta \begin{pmatrix} \cos\theta \\ \sin\theta \end{pmatrix}
$$

$$
\times \left\{ \left[-\int_{-h}^0 dz \tfrac{1}{2}\, \overline{\rho(\nabla\Phi)^2} \right] + \frac{\rho\omega^2}{4g} \overline{|\phi|_{z=0}^2} \right\}.
$$

$$(10.15)$$

Substituting Eq. (10.15) into Eq. (10.12), we get

$$
\bar{F}_x = -\int_0^{2\pi} \rho R\, d\theta \left\{ \int_{-h}^0 dz \left\{ -\frac{1}{2}\left[\overline{\left(\frac{\partial\Phi}{\partial r}\right)^2} + \frac{1}{R^2}\overline{\left(\frac{\partial\Phi}{\partial\theta}\right)^2} + \overline{\left(\frac{\partial\Phi}{\partial z}\right)^2} \right]\cos\theta \right.\right.
$$

$$
\left.\left. + \overline{\left(\frac{\partial\Phi}{\partial r}\right)^2}\cos\theta - \frac{1}{R}\overline{\frac{\partial\Phi}{\partial r}\frac{\partial\Phi}{\partial\theta}}\sin\theta \right\} + \frac{\omega^2}{4g}\overline{|\phi|_{z=0}^2}\cos\theta \right\}_{r=R}, \quad (10.16)
$$

$$
\bar{F}_y = -\int_0^{2\pi} \rho R\, d\theta \left\{ \int_{-h}^0 dz \left\{ -\frac{1}{2}\left[\overline{\left(\frac{\partial\Phi}{\partial r}\right)^2} + \frac{1}{R^2}\overline{\left(\frac{\partial\Phi}{\partial\theta}\right)^2} + \overline{\left(\frac{\partial\Phi}{\partial z}\right)^2} \right]\sin\theta \right.\right.
$$

$$
\left.\left. + \overline{\left(\frac{\partial\Phi}{\partial r}\right)^2}\sin\theta + \frac{\cos\theta}{R}\overline{\left(\frac{\partial\Phi}{\partial r}\frac{\partial\Phi}{\partial\theta}\right)} \right\} + \frac{\omega^2}{4g}\overline{|\phi|_{z=0}^2}\sin\theta \right\}_{r=R}. \quad (10.17)
$$

We now make use of the asymptotic formula for large r,

$$
\Phi = \mathrm{Re}\left[(\phi^I + \tilde\phi)e^{-i\omega t} \right] \tag{10.18a}
$$

with

$$
\phi^I = -\frac{igA}{\omega}\frac{\cosh k(z+h)}{\cosh kh}e^{ikr\cos\theta}, \tag{10.18b}
$$

$$
\tilde\phi = -\frac{igA}{\omega}\frac{\cosh k(z+h)}{\cosh kh}\mathcal{C}(\theta)\left(\frac{2}{\pi kr}\right)^{1/2}e^{ikr-i\pi/4} \tag{10.18c}
$$

where $\tilde\phi$ represents the sum of the scattered and the radiated waves. It is useful to realize that in the quadratic terms in Eqs. (10.16) and (10.17), products involving only ϕ^I will not contribute, that is, there cannot be any drift force with incident waves alone. In the limit of $R \to \infty$, the terms multiplied by

$1/R^2$ vanish. Also, since only cross products of ϕ' and $\tilde{\phi}$ appear, the term

$$\frac{1}{R}\overline{\frac{\partial \Phi}{\partial r}\frac{\partial \Phi}{\partial \theta}} \sim R^{-3/2}$$

does not contribute to the integral. As a result, Eqs. (10.16) and (10.17) reduce to

$$\bar{F}_x = -\int_0^{2\pi} \rho R\, d\theta \cos\theta \left\{ \int_{-h}^0 dz\, \frac{1}{2}\left[\overline{\left(\frac{\partial \Phi}{\partial r}\right)^2} - \overline{\left(\frac{\partial \Phi}{\partial z}\right)^2} \right] + \frac{\omega^2}{4g}\overline{|\phi|^2}_{z=0} \right\}_{r=R},$$

$$(10.19)$$

$$\bar{F}_y = -\int_0^{2\pi} \rho R\, d\theta \sin\theta \left\{ \int_{-h}^0 dz\, \frac{1}{2}\left[\overline{\left(\frac{\partial \Phi}{\partial r}\right)^2} - \overline{\left(\frac{\partial \Phi}{\partial z}\right)^2} \right] + \frac{\omega^2}{4g}\overline{|\phi|^2}_{z=0} \right\}_{r=R}.$$

$$(10.20)$$

The remaining task involves the familiar use of the stationary phase method; we only record the final results here:

$$\bar{F}_x = -\frac{\rho g A^2}{k}\frac{C_g}{C}\left\{ \frac{1}{\pi}\int_0^{2\pi}\cos\theta\,|\mathcal{Q}(\theta)|^2\, d\theta + 2\,\mathrm{Re}\,\mathcal{Q}(0) \right\}, \quad (10.21)$$

$$\bar{F}_y = -\frac{\rho g A^2}{k}\frac{C_g}{C}\frac{1}{\pi}\int_0^{2\pi}\sin\theta\,|\mathcal{Q}(\theta)|^2\, d\theta. \quad (10.22)$$

It should be emphasized that $\mathcal{Q}$ combines scattering and radiation. Equation (10.21) can be rearranged by eliminating $2\,\mathrm{Re}\,\mathcal{Q}(0)$ with the help of the generalized optical theorem (9.15), yielding

$$\bar{F}_x = \frac{\rho g A^2 C_g}{kC}\left[\frac{\frac{1}{2}k\bar{\dot{E}}}{\frac{1}{2}\rho g A^2 C_g} + \frac{1}{\pi}\int_0^{2\pi}(1-\cos\theta)\,|\mathcal{Q}|^2\, d\theta \right] \quad (10.23)$$

where $\bar{\dot{E}}$ is the rate of power absorption by the body and the first term in the brackets is the efficiency of power absorption within a width λ/π of the wave front. The integral above is positive-definite, therefore, $\bar{F}_x$ is positive-definite if $\dot{E} \geq 0$.

The moment about the z axis on a stationary body in waves is defined as

$$T_z = \iint_{S_B} P(\mathbf{r}\times\mathbf{n})\cdot\mathbf{e}_z\, dS. \quad (10.24)$$

We leave it as an exercise to show that the change of angular momentum about

z is

$$- \iint_{S} \left[P(\mathbf{r} \times \mathbf{n}) \cdot \mathbf{e}_z + \rho(\mathbf{r} \times \mathbf{u}) \cdot \mathbf{e}_z(\mathbf{u} \cdot \mathbf{n} - U_n) \right] dS,$$

and that the mean drift moment is

$$\bar{T}_z = -\rho \iint_{S_\infty} \overline{\frac{\partial \Phi}{\partial r} \frac{\partial \Phi}{\partial \theta}} R \, d\theta \, dz$$

$$= \rho g A^2 \frac{1}{k^2} \frac{C_g}{C} \operatorname{Im}\left[\frac{d\mathcal{Q}^*}{d\theta} \Big|_{\theta=0} + \frac{1}{\pi} \int_0^{2\pi} \mathcal{Q}(\theta) \frac{d\mathcal{Q}^*(\theta)}{d\theta} d\theta \right]. \quad (10.25)$$

In summary, the drift forces can be inferred from the far field of the first-order theory.

The subject is being developed further for the *slowly varying* drift force in irregular seas. If the incident waves consist of two slightly different frequencies $\omega \pm \Delta\omega/2$, the second-order response will contain terms such as $e^{\pm 2i\omega t}$ and $e^{\pm i\Delta\omega t}$. The latter has a much longer time scale than $2\pi/\omega$. The corresponding long period forces can resonate the natural modes of the mooring lines attached to a floating body, thereby causing severe strain. For further information see Newman (1974) and Faltinsen and Løken (1979).

Exercise 10.1

For two dimensions, show from Eq. (10.11) that

$$\bar{F}_x = \left[\frac{\rho}{2} \int_{-h}^{0} (\bar{v}^2 - \bar{u}^2) \, dz - \frac{\rho\omega^2}{4g} |\phi|^2_{z=0} \right]_{x\sim-\infty}^{x\sim\infty}$$

and then rederive Eq. (10.2).

Exercise 10.2

For a Hagen–Cockerell wave absorber in head seas, the scattering amplitude $A^S(\theta)$ is negligible if the raft is very slender. Use the approximate estimates of the radiated waves A^e and A^o [cf. Eqs. (9.52a) and (9.52b)] to calculate the drift force in the direction of the incident waves.

7.11   PRINCIPLES OF CALCULATING THE TRANSIENT MOTION OF A FLOATING BODY

In this section we shall endeavor to show how the transient response of a floating body can, in principle, be calculated from the sinusoidal response for which general numerical methods are available. The material here[†] is based on the work of Cummins (1962), Ogilvie (1964), and Wehausen (1967).

7.11.1   Radiated Waves Caused by Impulsive Motion of a Floating Body

Let a generalized displacement in the α direction be given impulsively to the body at the instant $t = \tau$, so that the generalized body velocity is

$$\Delta V_\alpha = V_\alpha(\tau)\delta(t - \tau), \qquad V_\alpha = \dot{X}_\alpha. \tag{11.1}$$

Let Φ^Δ be the velocity potential of the radiated waves caused by the impulse. Φ^Δ must satisfy the Laplace equation in the fluid, the free-surface condition (2.11), Section 1.2, with $P_a = 0$, and $\partial\Phi^\Delta/\partial n = 0$ on the horizontal bottom ($z = -h$). The boundary condition on the body surface is

$$\frac{\partial\Phi^\Delta}{\partial n} = \Delta V_\alpha n_\alpha = V_\alpha n_\alpha \delta(t - \tau). \tag{11.2}$$

In addition we require boundedness at infinity

$$|\nabla\Phi^\Delta| \to 0 \qquad \text{as } r \to \infty, t < \infty, \tag{11.3}$$

and that there be no disturbance before $t = \tau$:

$$\Phi^\Delta = 0, \qquad t < \tau, \qquad \text{all } \mathbf{x}. \tag{11.4}$$

Cummins introduced the following decomposition:

$$\Phi^\Delta(\mathbf{x}, t) = V_\alpha(\tau)[\Omega_\alpha(\mathbf{x})\delta(t - \tau) + \Gamma_\alpha(\mathbf{x}, t - \tau)H(t - \tau)], \tag{11.5}$$

where H is the Heaviside step function. Summation over the repeated index α is implied. We now require Ω_α and Γ_α to satisfy the following conditions on the body:

$$\frac{\partial\Omega_\alpha}{\partial n} = n_\alpha \tag{11.6a}$$

$$\text{on } S_B.$$

$$\frac{\partial\Gamma_\alpha}{\partial n} = 0 \tag{11.6b}$$

Both Ω_α and Γ_α are further subjected to Laplace's equation and the bottom boundary condition. Care is needed in satisfying the remaining condition on the free surface. Substituting Eq. (11.5) into the free-surface condition (2.11),

---

[†]I have benefited from the survey by Serman (1978).

Section 1.2, and noting that $f(t)\delta(t) = f(0)\delta(t)$ for any smooth $f(t)$, we have

$$\frac{\partial^2 \Phi^\Delta}{\partial t^2} = \left\{ \ddot{\delta}\Omega_\alpha + \delta\Gamma_\alpha(\mathbf{x}, 0) + \delta\frac{\partial}{\partial t}\Gamma_\alpha(\mathbf{x}, t - \tau) + H\frac{\partial^2}{\partial t^2}\Gamma_\alpha(\mathbf{x}, t - \tau) \right\} V_\alpha,$$

(11.7a)

$$g\frac{\partial \Phi^\Delta}{\partial z} = g\left[\delta\frac{\partial\Omega_\alpha}{\partial z} + H\frac{\partial\Gamma_\alpha}{\partial z} \right] V_\alpha,$$

(11.7b)

where the arguments of δ, $\dot{\delta}$, $\ddot{\delta}$, and H are all $t - \tau$. After inserting Eqs. (11.7a) and (11.7b) into Eq. (2.11), Section 1.2, we use the fact that H, δ, $\dot{\delta}$, and $\ddot{\delta}$ are singularities of different orders, and equate to zero their coefficients individually. In this manner four conditions are obtained on the free surface. The initial-boundary-value problems for Ω_α and Γ_α may be summarized:

$$\nabla^2\Omega_\alpha = 0 \qquad \text{in fluid,} \tag{11.8a}$$

$$\Omega_\alpha(\mathbf{x}) = 0, \qquad z = 0, \tag{11.8b}$$

$$\frac{\partial\Omega_\alpha}{\partial z} = 0, \qquad z = -h, \tag{11.8c}$$

$$\frac{\partial\Omega_\alpha}{\partial n} = n_\alpha \qquad \text{on } S_B, \tag{11.8d}$$

$$\Omega_\alpha \to 0 \qquad |\mathbf{x}| \to \infty. \tag{11.8e}$$

The above boundary-value problem for Ω_α may be regarded as a radiation problem in the limit of infinite frequency. For Γ_α we have

$$\nabla^2\Gamma_\alpha = 0 \qquad \text{in the fluid,} \tag{11.9a}$$

$$\Gamma_\alpha = 0, \qquad z = 0, \quad t = t_0, \tag{11.9b}$$

$$\frac{\partial\Gamma_\alpha}{\partial t} + g\frac{\partial\Omega_\alpha}{\partial z} = 0, \qquad z = 0, \quad t = t_0, \tag{11.9c}$$

$$\frac{\partial^2\Gamma_\alpha}{\partial t^2} + g\frac{\partial\Gamma_\alpha}{\partial z} = 0, \qquad z = 0, \quad t_0 < t < \infty, \tag{11.9d}$$

$$\frac{\partial\Gamma_\alpha}{\partial z} = 0, \qquad z = -h, \tag{11.9e}$$

$$\frac{\partial\Gamma_\alpha}{\partial n} = 0, \qquad \text{on } S_B. \tag{11.9f}$$

Note that Γ_α is just a Cauchy–Poisson problem in the presence of a fixed body.

Let us try to obtain the response to a continuous $V(t)$ by integrating the effect of a succession of impulses over $-\infty < \tau < \infty$, that is, by integrating Eq. (11.5),

$$\Phi^R = \int_{-\infty}^{\infty} d\tau \, \Phi^\Delta = \Omega_\alpha(\mathbf{x}) V_\alpha(t) + \int_{-\infty}^{t} \Gamma_\alpha(\mathbf{x}, t - \tau) V_\alpha(\tau) \, d\tau \quad (11.10a)$$

$$= \Omega_\alpha(\mathbf{x}) V(t) + \int_{0}^{\infty} \Gamma_\alpha(\mathbf{x}, \tau) V_\alpha(t - \tau) \, d\tau. \quad (11.10b)$$

It is straightforward to verify that Φ^R satisfies the governing equation and all the boundary and initial conditions.

The hydrodynamic force on the body can be derived in terms of Ω_α and Γ_α by first taking the time derivatives of Φ^R

$$\frac{\partial \Phi^R}{\partial t} = \Omega_\alpha(\mathbf{x}) \dot{V}_\alpha(t) - \Gamma_\alpha(\mathbf{x}, 0) V_\alpha(t) + \Gamma_\alpha(\mathbf{x}, \infty) V_\alpha(-\infty)$$

$$+ \int_{-\infty}^{t} \Gamma_\alpha(\mathbf{x}, t - \tau) \dot{V}_\alpha(\tau) \, d\tau. \quad (11.11)$$

Because of Eq. (11.9b) and $V_\alpha(-\infty) = 0$, we simply have

$$\frac{\partial \Phi^R}{\partial t} = \Omega_\alpha(\mathbf{x}) \ddot{X}_\alpha(t) + \int_{-\infty}^{t} \Gamma_\alpha(\mathbf{x}, t - \tau) \ddot{X}_\alpha(\tau) \, d\tau, \quad (11.12)$$

where use is made of $V_\alpha = \dot{X}_\alpha$. The β component of the generalized restoring force reacting on the body is

$$\mathscr{F}_\beta^R(t) = -\left[\rho \iint_{S_B} \Omega_\alpha(\mathbf{x}) n_\beta \, dS \right] \ddot{X}_\alpha + \int_{-\infty}^{t} \iint_{S_B} \rho \Gamma_\alpha(\mathbf{x}, t - \tau) n_\beta \ddot{X}_\alpha(\tau) \, d\tau \, dS,$$

$$(11.13)$$

where n_β is the generalized normal. Following Wehausen, let us define

$$\mu_{\beta\alpha}(\infty) = \rho \iiint_{S_B} \Omega_\alpha(\mathbf{x}) n_\beta \, dS, \quad (11.14)$$

$$L_{\beta\alpha}(t) = \rho \iiint_{S_B} \Gamma_\alpha(\mathbf{x}, t) n_\beta \, dS, \quad (11.15)$$

which are both real, then

$$\mathscr{F}_\beta^R(t) = -\mu_{\beta\alpha}(\infty) \ddot{X}_\alpha(t) - \int_{-\infty}^{t} L_{\beta\alpha}(t - \tau) \ddot{X}_\alpha(\tau) \, d\tau. \quad (11.16)$$

7.11.2 Relation to the Frequency Response

Let

$$V_\alpha(t) = \operatorname{Re} \bar{V}_\alpha e^{-i\omega t}, \tag{11.17}$$

and suppose the motion to have begun from $t \sim -\infty$. Substituting Eq. (11.17) into Eq. (11.10b), we get

$$\Phi^R(\mathbf{x}, t) = \operatorname{Re}\left\{\left[\Omega_\alpha(\mathbf{x}) + \int_0^\infty \Gamma_\alpha(\mathbf{x}, \tau)e^{i\omega\tau}\,d\tau\right]\bar{V}_\alpha e^{-i\omega t}\right\}. \tag{11.18}$$

Now the quantity in [ ] must be the normalized simple harmonic response due to forced radiation with $\partial\phi_\alpha/\partial n = n_\alpha$ on the body; therefore,

$$\phi_\alpha(\mathbf{x}, \omega) = \operatorname{Re}\phi_\alpha + i\operatorname{Im}\phi_\alpha = \Omega_\alpha(\mathbf{x}) + \int_0^\infty \Gamma_\alpha(\mathbf{x}, \tau)e^{i\omega\tau}\,d\tau. \tag{11.19}$$

Since $\Gamma_\alpha = 0$ for $\tau \leq 0$, the integral above is just the Fourier transform of Γ_α. Once ϕ_α (and its special limit Ω_α) is solved for a broad range of frequencies by any of the modern numerical means, $\Gamma_\alpha(\mathbf{x}, t)$ can be obtained by inverse Fourier transform for which the numerical technique of fast Fourier transform is available. The β component of the restoring force on the body is, from Eq. (11.13),

$$\mathscr{F}_\beta^R = -\rho \iint_{S_B} \Phi_t^R n_\beta\,dS$$

$$= \operatorname{Re}\left\{\left[-\rho \iint_{S_B} \Omega_\alpha n_\beta\,dS - \rho \int_0^\infty d\tau\,e^{i\omega\tau}\iint_{S_B}\Gamma_\alpha n_\beta\,dS\right]\bar{V}_\alpha e^{-i\omega t}\right\}.$$

$$\tag{11.20}$$

If Eqs. (11.14) and (11.15) are substituted into Eq. (11.20) and the result is compared with the definitions of apparent mass and damping coefficients in Section 7.3.2, we conclude that

$$\mu_{\alpha\beta}(\omega) + \frac{i}{\omega}\lambda_{\alpha\beta}(\omega) = \mu_{\alpha\beta}(\infty) + \int_0^\infty L_{\alpha\beta}(t)e^{i\omega\tau}\,d\tau. \tag{11.21}$$

Separating the real and imaginary parts, we finally get

$$\mu_{\alpha\beta}(\omega) - \mu_{\alpha\beta}(\infty) = \int_0^\infty L_{\alpha\beta}(\tau)\cos\omega\tau\,d\tau, \tag{11.22}$$

$$\lambda_{\alpha\beta}(\omega) = \omega\int_0^\infty L_{\alpha\beta}(\tau)\sin\omega\tau\,d\tau \tag{11.23}$$

(Wehausen, 1967). Conversely, if the added mass or damping coefficients are known for all frequencies $0 \leq \omega < \infty$, $L_{\alpha\beta}(t)$ can be found by inverse cosine or sine transform. Also, since $L_{\alpha\beta}$ can be found from either Eq. (11.22) or (11.23), a relation should exist between the apparent mass and the damping coefficients. Such a relation is called the Kramers–Kronig relation (see Ogilvie, 1964) which also appears in other branches of physics.

7.11.3 Exciting Force Caused by Scattering of Transient Incident Waves

To complete the formulation of the dynamic problem of a body in transient incident waves it is necessary to know the exciting force on a stationary body caused by diffraction. Wehausen (1967) has shown that the exciting force can be obtained from the radiation problem through a generalized Haskind relation, as described below.

Let us define

$$B_\alpha(\mathbf{x}, t - \tau) = \Omega_\alpha(\mathbf{x}) + \int_\tau^t \Gamma_\alpha(\mathbf{x}, t') \, dt'. \tag{11.24}$$

From the governing conditions of Ω_α and Γ_α it can be shown that B_α satisfies the following conditions:

$$\nabla^2 B_\alpha = 0, \qquad \text{in the fluid,} \tag{11.25a}$$

$$\frac{\partial B_\alpha}{\partial n} = n_\alpha \qquad \text{on } S_B, \tag{11.25b}$$

$$\frac{\partial B_\alpha}{\partial z} = 0, \qquad z = -h, \tag{11.25c}$$

$$\frac{\partial^2 B_\alpha}{\partial t^2} + g \frac{\partial B_\alpha}{\partial z} = 0, \qquad z = 0, \text{ all } t > 0, \tag{11.25d}$$

$$\frac{\partial B_\alpha}{\partial t} = 0, \qquad z = 0, \quad t = 0, \tag{11.25e}$$

$$B_\alpha(\mathbf{x}, t) = 0, \qquad z = 0, \quad t = 0, \tag{11.25f}$$

$$B_\alpha(\mathbf{x}, t) = \Omega_\alpha(\mathbf{x}), \qquad t = 0, \text{ in the fluid,} \tag{11.25g}$$

$$B_\alpha = O(R^{-2}), \qquad R \to \infty, t < \infty. \tag{11.25h}$$

The last condition on the behavior of B_α at infinity requires mathematical arguments too lengthy to enter here (Stoker, 1957).

Next we let $\Phi^I(\mathbf{x}, t)$ denote the potential of the transient incident wave and $\Phi^S(\mathbf{x}, t)$ the scattered wave potential. Φ^I is originated in a finite region before striking the body. Φ^S satisfies Eqs. (11.25a)–(11.25h) provided that n_α is replaced by $-\partial\Phi^I/\partial n$ in Eq. (11.25b) and Ω_α is replaced by 0 in Eq. (11.25g).

Let us apply Green's theorem to $\partial\Phi^S(\mathbf{x}, \tau)/\partial\tau$ and $B_\alpha(\mathbf{x}, t - \tau)$ for the usual control volume bounded by S_F, S_B, and B_0 and a circular vertical cylinder S_∞ of great radius R. After the familiar use of the governing equations, we get

$$\iint_{S_B} \left(\frac{\partial\Phi^S}{\partial\tau} n_\alpha - \frac{\partial^2\Phi^S}{\partial\tau\,\partial n} B_\alpha \right) dS + \iint_{S_F} \left(\frac{\partial\Phi^S}{\partial\tau} \frac{\partial B_\alpha}{\partial z} - \frac{\partial^2\Phi^S}{\partial\tau\partial z} B_\alpha \right) dS = 0.$$

(11.26)

When the boundary conditions and $f_\tau(t - \tau) = -f_t(t - \tau)$ are used, the second integral may be expressed

$$-\frac{1}{g} \iint_{S_F} \left(\frac{\partial\Phi^S}{\partial\tau} \frac{\partial^2 B_\alpha}{\partial t^2} - \frac{\partial^3\Phi^S}{\partial\tau^3} B_\alpha \right) dS = \frac{1}{g} \iint_{S_F} \frac{\partial}{\partial\tau} \left(\frac{\partial\Phi^S}{\partial\tau} \frac{\partial B_\alpha}{\partial t} + \frac{\partial^2\Phi^S}{\partial\tau^2} B_\alpha \right) dS.$$

(11.27)

With this result Eq. (11.26) may be integrated with respect to τ from $-\infty$ to t, yielding

$$\int_{-\infty}^{t} dt \iint_{S_B} \left(\frac{\partial\Phi^S}{\partial\tau} n_\alpha - \frac{\partial^2\Phi^s}{\partial\tau\,\partial n} B_\alpha \right) dS = 0,$$

(11.28)

where the integrated terms vanish by virtue of Eqs. (11.25e) and (11.25f) for B_α and the fact that $\partial\Phi^S/\partial t$, $\partial^2\Phi^S/\partial t^2 \to 0$ as $t \to \infty$. Differentiating Eq. (11.28) with respect to t, and using

$$\frac{\partial\Phi^S}{\partial n} = -\frac{\partial\Phi^I}{\partial n} \qquad \text{on } S_B$$

(11.29)

along with Eqs. (11.24) and (11.25g), we find that

$$-\iint_{S_B} \frac{\partial\Phi^S}{\partial t} n_\alpha\, dS = \int_{-\infty}^{t} d\tau \iint_{S_B} \frac{\partial^2\Phi^I}{\partial\tau\,\partial n} \Gamma_\alpha(\mathbf{x}, t - \tau)\, dS + \iint_{S_B} \frac{\partial^2\Phi^I}{\partial t\,\partial n} \Omega_\alpha(\mathbf{x})\, dS.$$

(11.30)

Finally, the exciting (diffraction) force is

$$\mathcal{F}_\alpha^D(t) = -\rho \iint_{S_B} \left(\frac{\partial \Phi^I}{\partial t} + \frac{\partial \Phi^S}{\partial t} \right) n_\alpha \, dS$$

$$= -\rho \iint_{S_B} \left(\frac{\partial \Phi^I}{\partial t} n_\alpha - \frac{\partial^2 \Phi^I}{\partial t \, \partial n} \Omega_\alpha \right) dS$$

$$+ \rho \int_{-\infty}^{t} d\tau \iint_{S_B} \frac{\partial^2 \Phi^I}{\partial \tau \, \partial n}(\mathbf{x}, \tau) \Gamma_\alpha(\mathbf{x}, t - \tau) \, dS \qquad (11.31)$$

which is due to Wehausen (1967).

Since the right-hand side depends only on the incident wave and on Ω_α and Γ_α, which are defined for the radiation problem, Eq. (11.31) is an extended Haskind–Hanaoka relation and may be used to calculate the exciting force without solving the transient diffraction problem.

7.11.4 Linearized Equations of Transient Motion of a Floating Body

The governing equation (2.47) still holds, but the hydrodynamic restoring force is given by Eq. (11.16) and the exciting force by Eq. (11.31), with the result that

$$\left[M_{\alpha\beta} + \mu_{\alpha\beta}(\infty) \right] \ddot{X}_\beta + \int_{-\infty}^{t} L_{\alpha\beta}(t - \tau) \ddot{X}_\beta(\tau) \, d\tau + C_{\alpha\beta} X_\beta = \mathcal{F}_\alpha^D + \mathbb{F}_\alpha$$

$$(11.32)$$

which is a set of integro-differential equations to be solved for given initial position and velocity of the body.

From the above results it is evident that the transient response may be calculated, in principle, from sinusoidal responses numerically. On this basis an analytical theory has been worked out by Ursell (1964) and Maskell and Ursell (1970) for a two-dimensional circular cylinder freely floating on the free surface. At present all existing numerical methods are costly for high frequencies. However, this shortcoming is not fatal in practice since the most important responses usually occur in the range where the product of k and the body size is of order unity. Alternative numerical methods via transient Green's functions for initial-value problems are being developed for general bodies and may someday prove to be more effective. (See *Proceedings of the 3rd International Conference on Numerical Ship Hydrodynamics*, 1981, Paris).

APPENDIX 7.A: DERIVATION OF GREEN'S FUNCTION

For both two and three dimensions the Green functions have been deduced by John (1950) and by other authors. Only the details for two dimensions are explained here. Defining the Green function $G(x \mid x_0)$ by Eqs. (4.18)–(4.21), and taking the Fourier transform of Eqs. (4.18)–(4.20) with respect to x, we get

$$\frac{d^2\tilde{G}}{dz^2} - K^2\tilde{G} = \delta(z - z_0)e^{-iKx_0}, \qquad (A.1)$$

$$\frac{d\tilde{G}}{dz} - \sigma\tilde{G} = 0, \qquad z = 0, \qquad (A.2)$$

$$\frac{d\tilde{G}}{dz} = 0, \qquad z = -h, \qquad (A.3)$$

where K is the Fourier transform variable. An equivalent boundary-value problem is defined by

$$\frac{d^2\tilde{G}}{dz^2} - K^2\tilde{G} = 0, \qquad -h < z < z_0, \quad z_0 < z < 0, \qquad (A.4)$$

$$\tilde{G}\big|_{z=z_0^+} - \tilde{G}\big|_{z=z_0^-} = 0, \qquad (A.5a)$$

$$\frac{d\tilde{G}}{dz}\bigg|_{z=z_0^+} - \frac{d\tilde{G}}{dz}\bigg|_{z=z_0^-} = e^{-iKx_0}, \qquad (A.5b)$$

and by Eqs. (A.2) and (A.3). The second jump condition (A.5b) may be deduced by integrating Eq. (A.1) across z_0 and invoking Eq. (A.5a). In terms of

$$z_> = \max(z, z_0) \qquad \text{and} \qquad z_< = \min(z, z_0), \qquad (A.6)$$

the solution is easily expressed in a compact form

$$\tilde{G} = \frac{e^{-iKx_0}}{K} \frac{\sigma \sinh Kz_> + K\cosh Kz_>}{\sigma \cosh Kh - K\sinh Kh} \cosh K(z_< + h). \qquad (A.7)$$

In applying the Fourier inversion formula, an ambiguity arises along the real K axis where two poles exist at the real zeroes of

$$\sigma \cosh Kh - K\sinh Kh = 0 \quad \text{or} \quad K\tanh Kh = \sigma, \qquad (A.8)$$

that is, $K = \pm k$. Now the radiation condition must be invoked to define the improper integral. To be checked for correctness, we now choose the path Γ

which is indented above $-k$ and below $+k$ along two semicircles of small radii, as shown in Fig. A.1. Thus,

$$G(\mathbf{x}\,|\,\mathbf{x}_0) = \frac{1}{2\pi}\int_\Gamma dK\, e^{iK(x-x_0)}\left\{\frac{1}{K}\frac{\sigma\sinh Kz_> + K\cosh Kz_>}{\sigma\cosh Kh - K\sinh Kh}\cosh K(z_< + h)\right\}.$$

(A.9)

The above representation can be manipulated into several different forms. One of them is an eigenfunction expansion which is deduced by recognizing that the integral has imaginary poles at the imaginary zeroes of Eq. (A.8), that is, at $K = \pm ik_n$, $n = 1, 2, 3, \ldots$, where k_n is a positive real root of

$$k_n\tan k_n h = -\sigma.$$

(A.10)

For $x > x_0$ we introduce a closed semicircular contour in the upper half complex plane of K, as shown in Fig. A.1, so that the integrand diminishes as $\mathrm{Im}\,K \uparrow \infty$. By Jordan's lemma, the line integral along the semicircle vanishes in the limit of infinite radius. Therefore, the original integral is equal to the sum of residues from the poles at k and ik_n, $n = 1, 2, 3\ldots$. The residues may be calculated by repeated use of Eq. (A.8) and (A.10). Similarly for $x < x_0$, we introduce a closed contour in the lower half plane, capturing the residues at $-k$ and $-ik_n$. The combined result is

$$G = -\frac{i}{k}\left(h + \frac{1}{\sigma}\sinh^2 kh\right)^{-1}e^{ik|x-x_0|}\cosh k(z+h)\cosh k(z_0+h)$$

$$+ \sum_{n=1}^{\infty}\frac{1}{k_n}\left(h - \frac{1}{\sigma}\sin^2 k_n h\right)^{-1}e^{-k_n|x-x_0|}\cos k_n(z+h)\cos k_n(z_0+h).$$

(A.11)

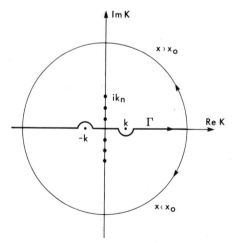

Figure A.1 Contours of integration.

Details are left as an exercise. Clearly, the first term represents the outgoing waves, while the series represents the evanescent modes. Thus, the radiation condition is satisfied. Indeed, any other indentation of the integral path would violate the radiation condition.

An alternate representation, which is important in the numerical method of integral equations, exhibits the singular part of the Green function explicitly. Near the source point (x_0, z_0) the Green function must be dominated by $(1/2\pi)\ln[(x - x_0)^2 + (z - z_0)^2]^{1/2}$. To extract this term we first use the symmetry of the bracket $\{ \}$ in Eq. (A.9) and of the contour in order to use only one-half of the path, that is,

$$G(\mathbf{x} \mid \mathbf{x}_0) = \frac{1}{\pi} \int_0^\infty \frac{dK}{K} \cos K(x - x_0)$$

$$\times \frac{\sigma \sinh Kz_> + K \cosh Kz_>}{\sigma \cosh Kh - K \sinh Kh} \cosh K(z_< + h), \qquad (A.12)$$

where the path of integration is the positive half of Γ in Fig. A.1. Let us define

$$r = \left[(x - x_0)^2 + (z - z_0)^2\right]^{1/2}, \qquad r' = \left[(x - x_0)^2 + (z + z_0 + 2h)^2\right]^{1/2},$$

$$(A.13)$$

where r' is the distance between the field point and the image of the source below the sea bottom. From the following identities of Laplace transform:

$$\int_0^\infty \frac{dK}{K} e^{-Kb}(1 - \cos Ka) = \ln\left[\frac{(a^2 + b^2)^{1/2}}{b}\right], \qquad (A.14)$$

$$\int_0^\infty \frac{dK}{K}(e^{-Kb} - e^{-Kh}) = \ln\left(\frac{h}{b}\right), \qquad (A.15)$$

it follows that

$$\ln\frac{r}{h} = \int_0^\infty \frac{dK}{K}(e^{-Kh} - e^{-Kb}\cos Ka), \qquad a = x - x_0, \quad b = z_> - z_<$$

$$(A.16)$$

with a similar formula for $\ln(r'/h)$. Substituting Eq. (A.16) into Eq. (A.12), we obtain

$$
2\pi G = \ln\frac{r}{h} + \ln\frac{r'}{h} + \int_0^\infty \frac{dK}{K}
$$

$$
\times\left\{\cos K(x - x_0)\left[2\frac{\sigma\sinh Kz_> + K\cosh Kz_>}{\sigma\cosh Kh - K\sinh Kh}\cosh K(z_< + h)\right.\right.
$$

$$
\left.\left. + e^{-K(z_< - z_>)} + e^{-K(z_> + z_< + 2h)}\right] - 2e^{-Kh}\right\}.
$$

After a little algebra and returning from $z_>$, $z_<$ to z and z_0, we get

$$
G = \frac{1}{2\pi}\ln\frac{r}{h} + \frac{1}{2\pi}\ln\frac{r'}{h} + 2\int_0^\infty \frac{dK}{K}
$$

$$
\times\left\{\frac{\cos K(x - x_0)}{\sigma\cosh Kh - K\sinh Kh}\right.
$$

$$
\left.\times\cosh K(z + h)\cosh K(z_0 + h) - e^{-Kh}\right\}, \qquad (A.17)
$$

which is the form deduced by John (1950).

The first two logarithmic terms above represent two sources (one in the fluid and one beneath the sea bottom); their sum satisfies the boundary condition at $z = -h$. The contour integral therefore represents the effect of the free surface. It may be easily checked that the contribution from the small semi-circle gives the outgoing wave, while the remaining principal-valued integral is of local importance only.

For infinite depth, the limit of Eq. (A.17) gives

$$
G(\mathbf{x}\,|\,\mathbf{x}_0) = \frac{1}{2\pi}\ln r + \int_0^\infty \frac{dK}{K}\left[\frac{\sigma + K}{\sigma - K}e^{K(z + z_0)}\cos K(x - x_0) - e^{-K}\right].
$$

$$
(A.18)
$$

For a three-dimensional ocean of constant depth, the Hankel transform may be used instead of the Fourier transform. A formula similar to Eq. (A.9) is obtained. One must then use some integral identities of Bessel functions to deduce alternate representations. The explanations for two dimensions should provide sufficient background for the reader to follow the details given in John (1950). Only certain key results are recorded here.

Corresponding to Eq. (A.12) the result can be expressed as

$$
G = -\frac{1}{2\pi}\int_0^\infty dK\, KJ_0(KR)\{\ \} \qquad (A.19)
$$

where the bracket { } is the same as that of Eq. (A.9). In series form G may be expressed as

$$G = -\frac{i}{2}\frac{k^2 - \sigma^2}{h(k^2 - \sigma^2) + \sigma}H_0^{(1)}(kR)\cosh k(z_0 + h)\cosh k(z + h)$$

$$+ \frac{1}{\pi}\sum_{n=1}^{\infty}\frac{k_n^2 + \sigma^2}{-h(k_n^2 + \sigma^2) + \sigma}K_0(k_n R)\cos k_n(z + h)\cos k_n(z_0 + h).$$

$$(A.20)$$

With the singular part exhibited, G may be written alternatively as

$$G = -\frac{1}{4\pi}\left(\frac{1}{r} + \frac{1}{r'}\right) + \frac{1}{4\pi}\int_0^{\infty}dK\, J_0(KR)\frac{2(\sigma + K)e^{-Kh}}{\sigma\cosh Kh - K\sinh Kh}$$

$$\times\cosh K(z + h)\cosh K(z_0 + h), \qquad\qquad (A.21)$$

where

$$r = \left[(x - x_0)^2 + (y - y_0)^2 + (z - z_0)^2\right]^{1/2},$$

$$r' = \left[(x - x_0)^2 + (y - y_0)^2 + (z + z_0 + 2h)^2\right]^{1/2}.$$

Finally, for infinite depth the three-dimensional Green's function is

$$G = -\frac{1}{4\pi r} + \frac{1}{4\pi}\int_0^{\infty}\frac{\sigma + K}{\sigma - K}e^{K(z+z_0)}J_0(KR)\,dK. \qquad (A.22)$$

EIGHT

Viscous Damping in Small-Amplitude Waves

8.1 INTRODUCTION

Except in Chapter Six, we have so far disregarded the effect of viscosity because it is usually so weak and confined in such thin layers as to exert very little influence on the wave motion over a few periods or wavelengths. However, over a long time compared to a characteristic wave period, or a long distance compared to a characteristic wavelength, the cumulative effect of viscosity on wave attenuation can be of first-order importance. In this chapter, theories are developed for the damping of infinitesimal waves by molecular viscosity. Within this framework the mechanism of energy transfer is examined in detail. A formal method of perturbation is also demonstrated for calculating viscous effects on both the amplitude and phase of the wave. The possible importance of air, dismissed in the past by most authors but pointed out recently by Dore (1978), is briefly discussed. Finally, a semiempirical theory of turbulent boundary layer near the sea bottom is described with a simple example of its applications.

8.2 LINEARIZED EQUATIONS OF VISCOUS FLOWS AND THE LAMINAR BOUNDARY LAYER

The full Navier–Stokes equations are given in Eqs. (1.1) and (1.2), Chapter One. For convenience we quote them here in index form. Let the rectangular coordinates be denoted by x_i, $i = 1, 2, 3$, with $x_1 = x$, $x_2 = y$, $x_3 = z$ and the corresponding velocity components be denoted by u_i. The equation of continuity is

$$\frac{\partial u_j}{\partial x_j} = 0, \tag{2.1}$$

where summation over the repeated indices is implied. The equation of momentum conservation is

$$\frac{\partial u_i}{\partial t} + u_j \frac{\partial u_i}{\partial x_j} = -g\delta_{i3} - \frac{1}{\rho}\frac{\partial p}{\partial x_i} + \frac{\partial \tau_{ij}}{\partial x_j}, \qquad i = 1, 2, 3, \qquad (2.2)$$

where τ_{ij} are the components of the viscous stress tensor

$$\tau_{ij} = 2\mu e_{ij}, \qquad e_{ij} = \frac{1}{2}\left(\frac{\partial u_i}{\partial x_j} + \frac{\partial u_j}{\partial x_i}\right). \qquad (2.3)$$

On a stationary solid the fluid velocity must vanish:

$$u_i = 0, \qquad i = 1, 2, 3. \qquad (2.4)$$

On a water surface $F = z - \zeta(x, y, t) = 0$ the kinematic condition still holds:

$$\frac{\partial \zeta}{\partial t} + u_1 \frac{\partial \zeta}{\partial x_1} + u_2 \frac{\partial \zeta}{\partial x_2} = u_3 \qquad z = \zeta. \qquad (2.5)$$

The dynamic boundary condition is the continuity of normal and tangential stresses:

$$\{-P\delta_{ij} + \tau_{ij}\}n_j = \{-P\delta_{ij} + \tau_{ij}\}_{\text{air}} n_j, \qquad z = \zeta. \qquad (2.6)$$

In this chapter the water surface will be assumed to be free of atmospheric pressure and stresses so that the right-hand side of Eq. (2.6) vanishes. In reality, surface contamination can complicate the above condition and affect wave damping considerably. It is also a common experience in the laboratory that fresh water gives a different damping rate from water that is visibly clean but has been in the tank for more than a day. This phenomenon is called *aging*. However, the physics of surface contamination and aging is not well understood and will not be pursued here.

To facilitate analysis, we assume infinitesimal amplitude and linearize Eq. (2.2),

$$\frac{\partial u_i}{\partial t} = -g\delta_{i3} - \frac{1}{\rho}\frac{\partial P}{\partial x_i} + \frac{\partial \tau_{ij}}{\partial x_j}. \qquad (2.7)$$

As is well known, any vector can be taken as the sum of an irrotational and a solenoidal vector (Morse and Feshbach, Vol. I, 1953, p. 53):

$$u_i = \frac{\partial \Phi}{\partial x_i} + U_i, \qquad (2.8)$$

with $\mathbf{U}$ being solenoidal, that is,

$$\frac{\partial U_i}{\partial x_i} = 0, \tag{2.9}$$

and $\nabla\Phi$ being irrotational. It follows from the continuity equation (2.1) that

$$\nabla^2\Phi = 0. \tag{2.10}$$

We now let

$$P = -\rho gz + p = -\rho gz - \rho\Phi_t, \tag{2.11}$$

where p is the dynamic pressure. Substituting Eq. (2.8) into Eq. (2.7) and using Eq. (2.11), we obtain

$$\frac{\partial U_i}{\partial t} = \nu\nabla^2 U_i. \tag{2.12}$$

The unknowns Φ and $\mathbf{U}$ are coupled by the boundary conditions, though not by the governing equations. In particular, there must be no slip on the solid wall S,

$$\frac{\partial\Phi}{\partial x_i} + U_i = 0 \qquad \text{on } S. \tag{2.13}$$

On the free surface, we apply the linearized boundary conditions on the mean free surface $z = 0$. The kinematic condition reads

$$\frac{\partial\zeta}{\partial t} = \Phi_z + W. \tag{2.14}$$

Dynamically, we also require the vanishing of the normal stress

$$\frac{\partial\Phi}{\partial t} + g\zeta + \frac{2\mu}{\rho}\frac{\partial w}{\partial z} = 0, \tag{2.15a}$$

and of the tangential stresses

$$\mu\left(\frac{\partial u}{\partial z} + \frac{\partial w}{\partial x}\right) = \mu\left(\frac{\partial v}{\partial z} + \frac{\partial w}{\partial y}\right) = 0. \tag{2.15b}$$

Strictly speaking, one may question the legitimacy of applying the stress-free conditions on $z = 0$ unless the wave amplitude is much smaller than the boundary layer thickness $\delta = O(\nu/\omega)^{1/2}$. However, since the stresses are related to the velocity gradient, viscosity has much less constraint on the free surface than near a solid wall where the velocity must vanish. In other words, the free-surface boundary layer is very weak. In fact, it will be shown shortly that the damping rate from the free-surface boundary layer is of the order $O(\mu^{3/2})$ as compared to $O(\mu^{1/2})$ from near the solid walls and $O(\mu)$ in the main body of the fluid. Hence, the error incurred from misplacing the free surface is only of the order $O(\mu^{3/2}kA)$ which is too small to be of concern here.

By taking the scalar product of the momentum equation (2.2) with $\mathbf{u}$ and integrating over *any* material volume whose boundary is $S_0(t)$, we obtain an equation stating the balance of mechanical energy inside the volume $V(t)$:

$$\iiint_V \frac{\partial}{\partial t} \frac{1}{2} \rho u_i u_i \, dV + \iint_{S_0} \frac{1}{2} \rho u_i u_i u_j n_j \, dS$$

$$= \iint_{S_0} \{-P\delta_{ij} + \tau_{ij}\} n_j u_i \, dS - \iiint_V \rho g \delta_{i3} u_i \, dV - \frac{1}{2\mu} \iiint_V \tau_{ij}\tau_{ij} \, dV.$$

$$(2.16)$$

The left-hand side is the total rate of energy change in V; the second term is the flux of energy across the boundary. On the right-hand side, the first integral represents the rate of working by surface stresses (pressure and viscous stress) acting on the boundary; the second integral is the rate of work done by body force throughout the volume, and the third integral is the rate of viscous dissipation (rate of working by viscous stresses on strain) throughout the volume.

The above energy equation is exact. Consistent with linearization, the second integral on the left may be neglected, being of the order $O(kA)$ higher than the rest. The mean position of S_0 may be used to replace the instantaneous bounding surface of V.

For later convenience we review the well-known theory of Stokes for an oscillatory boundary layer near a smooth wall. Let the inviscid velocity just outside the boundary layer near a flat wall be $U_I(t)$. Within the boundary layer the tangential velocity u is governed by

$$\frac{\partial u}{\partial t} = \frac{\partial U_I}{\partial t} + \nu \frac{\partial^2 u}{\partial z^2}, \tag{2.17}$$

where z is normal to the wall. On the wall ($z = 0$), $u = 0$. For large z, $u \sim U_I$. Assume U_I to be simple harmonic in time,

$$U_I = \mathrm{Re}\, U_0 e^{-i\omega t}, \tag{2.18}$$

then the tangential velocity in the boundary layer is given by

$$u = \mathrm{Re}\left(U_0 F_1(\xi) e^{-i\omega t}\right), \tag{2.19}$$

where

$$F_1(\xi) = 1 - e^{-(1-i)\xi},\tag{2.20}$$

with

$$\xi = \frac{z}{\delta}, \qquad \delta = \left(\frac{2\nu}{\omega}\right)^{1/2}.\tag{2.21}$$

The corresponding shear stress at the wall is

$$\tau_{xz}^B = \mu\frac{\partial u}{\partial z}\bigg|_0 = \mu\,\mathrm{Re}\left[\frac{U_0}{\delta}\sqrt{2}\,e^{-i(\omega t+\pi/4)}\right],\tag{2.22}$$

which is out of phase with the inviscid velocity U_I by $\frac{1}{4}\pi$.

8.3   DAMPING RATE AND THE PROCESS OF ENERGY TRANSFER

To understand the physical mechanism of wave damping, it is illuminating to examine the detailed process of energy transfer in the fluid. First, let us estimate the order of magnitude of the rate of dissipation in various parts of the fluid. We decompose the fluid velocity as in Eq. (2.8) into the rotational part $\mathbf{U}$ which depends directly on viscosity, and the irrotational part $\nabla\Phi$.

Outside of all boundary layers, it is reasonable to expect that $|\mathbf{U}|\ll\nabla\Phi$. There the rate of strain is dominated by the irrotational part whose velocity scale is ωA and length scale is either the basin dimension L or the wavelength $2\pi/k$, whichever is smaller. The energy dissipation rate is then

$$O\left(\frac{1}{\mu}\iiint_{R_I}\tau_{ij}^2\,dV\right) \sim \frac{\mu^2}{\mu}\frac{\omega^2 A^2 L^3}{L^2} \sim \mu\omega^2 A^2 L \propto \mu.\tag{3.1}$$

Inside a wall boundary layer R_ε (see Fig. 3.1), the tangential components of $\mathbf{U}$ and $\nabla\Phi$ are comparable, but the normal gradient of the tangential compo-

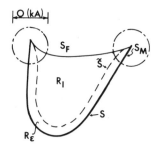

Figure 3.1 Division of fluid regions. The total fluid volume is defined by the instantaneous free surface S_F. Dashed lines designate the outer surface of the boundary layer.

nent of U is much greater and dominates the strain rate, so that

$$O\left(\frac{1}{\mu}\iiint_{R_\varepsilon}\tau_{ij}^2\,dV\right) \sim \frac{\mu^2}{\mu}\frac{\omega^2A^2}{\delta^2}L^2\delta \sim \mu\omega^2A^2L\cdot\frac{L}{\delta} \propto \mu^{1/2}. \qquad (3.2)$$

Consider the neighborhood of the free surface. In principle, a boundary layer R_F also exists there; its importance depends on the free-surface condition. Consider first a clean surface. Since U is negligible beneath the boundary layer, it cannot grow to an appreciable magnitude within a short distance of $O(\delta)$. Thus, the stress is mainly controlled by the potential velocity field which varies according to the length scale L or $2\pi/k$ and is of the same order as the stress in the main body of the fluid exterior to the boundary layer. Because of the small volume $O(\delta L^2)$ the rate of dissipation in the free-surface boundary layer is only

$$O\left(\frac{1}{\mu}\iiint_{R_F}\tau_{ij}^2\,dV\right) \sim \mu\left(\frac{\omega A}{L}\right)^2L^2\delta \sim \mu\omega^2A^2L\frac{\delta}{L} \propto \mu^{3/2}. \qquad (3.3)$$

At the other extreme, if the free surface is heavily contaminated, for example, by oil slicks, the fluid particles are immobilized so that the free surface may resemble an inextensible, though flexible, sheet. The stress in the free-surface boundary layer can then be as great as in the boundary layer near a solid wall. As stated in Section 8.2, surface contamination is assumed to be absent. Therefore, the largest energy dissipation takes place inside the wall boundary layers; only in very deep and unbounded water or for very short waves ($kh \gg 1$) is the dissipation in the main body of the fluid important.

With the energy flux term neglected on the left-hand side, Eq. (2.16) will now be averaged over a period T. We define the average as follows:

$$\bar{f}(t) = \frac{1}{T}\int_t^{t+T}f(t')\,dt' \qquad (3.4)$$

which has the important property that $\overline{\partial f/\partial t} = \partial\bar{f}/\partial t$. For generality, we take V to contain part of R_ε, R_F, and R_I, where R_I is the inviscid core.

From the second term on the right-hand side of Eq. (2.16), the work done by the body force is negligible everywhere except on the free surface and is given by

$$\overline{\iiint_V \rho g_i u_i\,dV} = -\rho g\overline{\iint_{S_F}\frac{\partial\zeta}{\partial t}\,dA\left[\int_0^\zeta dz\right]}$$

$$= \rho g\overline{\iint_{S_F}\frac{\partial\zeta}{\partial t}\zeta\,dS} = -\frac{\partial}{\partial t}\iint_{S_F}\frac{1}{2}\rho g\overline{\zeta^2}\,dS$$

which is just the negative of the rate of change of the potential energy. Combining the preceding term with the change of the kinetic energy in R, we have from Eq. (2.16)

$$\frac{\partial}{\partial t}\left[\iiint_V \frac{1}{2}\rho\,\overline{u_i u_i}\,dV + \frac{1}{2}\iint_{S_F}\rho g\,\overline{\zeta^2}\,ds\right]$$

$$= \iint_{S_0}\overline{(-p\delta_{ij} + \tau_{ij})n_j u_i}\,dS - \frac{1}{2\mu}\iiint_V \overline{\tau_{ij}^2}\,dV + O(A^3). \quad (3.5)$$

The left side represents the rate of change of the total kinetic and potential energy. On the right-hand side, the first integral gives the rates of working by the dynamic pressure p and by the viscous stresses, while the second integral gives the rate of dissipation.

From here on it is more convenient to employ dimensionless variables defined as follows:

$$(x', y', z') = k(x, y, z), \qquad t' = (gk)^{1/2}t$$

$$\mathbf{u}' = \frac{\mathbf{u}}{A(gk)^{1/2}}, \qquad p' = \frac{p}{\rho g A} \qquad\qquad (3.6)$$

$$\Phi' = \frac{\Phi k}{A(gk)^{1/2}}, \qquad \zeta' = \frac{\zeta}{A}$$

where all variables without primes are dimensional. A and k are the typical wave amplitude and wavenumber, respectively. After substitution, the primes are dropped for brevity. Let

$$\varepsilon = k\nu^{1/2}/(gk)^{1/4} \ll 1. \qquad\qquad (3.7)$$

Equation (3.5) becomes

$$\left\{\iiint_V \frac{\partial}{\partial t}\frac{\overline{u_i u_i}}{2}\,dV + \iint_{S_F}\frac{\partial}{\partial t}\overline{\zeta^2}\,dS\right\}$$
$$\underset{\text{(I)}}{}$$

$$= \iint_{S_0}\overline{-p u_i n_i}\,dS + 2\varepsilon^2\iint_{S_0}\overline{e_{ij}u_i n_j}\,dS - 2\varepsilon^2\iiint_V \overline{e_{ij}^2}\,dV. \qquad (3.8)$$
$$\underset{\text{(II)}}{}\qquad\qquad\underset{\text{(III)}}{}\qquad\qquad\underset{\text{(IV)}}{}$$

In this equation the free surface S_F may be excluded from the total S_0 which bounds V, due to the stress-free condition.

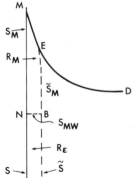

Figure 3.2 Enlarged view of the meniscus neighborhood. The line $NB(S_{MW})$ symbolizes the border between meniscus and wall boundary layers.

Referring to Figs. 3.1b and 3.2, we denote the meniscus boundary layer by R_M, whose dimensionless height is $O(kA)$. The solid boundaries of R_ε and R_M will be denoted by S and S_M, respectively, while the corresponding outer edges by $\tilde{S}$ and $\tilde{S}_M$, respectively. The precise physics in the neighborhood of the meniscus is a complex matter involving surface tension (Adams, 1941). In particular, the reversal of contact angles when the meniscus rises and falls is believed to contribute to damping by hysterisis (Miles, 1967), but the subject appears to be a poorly understood part of physical chemistry. We shall, therefore, not venture into hysterisis damping but postulate the following picture in the context of viscous fluid only. In the neighborhood of the meniscus the free surface attains its greatest height at the level M and the least height at the level N (Fig. 3.2). The wall below M is always wetted. The piece ME of the meniscus free surface changes from being nearly horizontal at the maximum rise to a thin film at the maximum fall. At any intermediate time during each period the meniscus consists of a thin viscous boundary layer $MNBE$ (of thickness $O(\varepsilon)$ and a height of $O(kA)$) near the wall, and a potential region whose boundary and volume change with time. The lower extreme of this boundary layer NB also forms the ceiling of the main wall boundary layer R_ε.

We now examine the leading terms in Eq. (3.8) for different control volumes (Mei and Liu, 1973).

8.3.1 The Entire Fluid

Let the volume V be the entire fluid volume R; hence the boundary surface consists of all wetted walls and the free surface:

$$V = R_I + R_\varepsilon + R_M, \qquad S_0 = S + S_M + S_F.$$

In Eq. (3.8), term (I) is well approximated by

$$\frac{1}{2} \frac{\partial}{\partial t} \left\{ \iiint_{R_I} \overline{u_i u_i}\, dV + \iint_{S_F} \overline{\zeta^2}\, dS \right\}$$

since $R_I = O(1) \gg R_\varepsilon + R_M = O(\varepsilon)$. The magnitude of $O(\partial/\partial t)$ is undetermined.

The sum of terms (II) and (III) vanishes on S_F because of the stress-free condition. Each term vanishes individually on S_W and S_M because $u_i = 0$.

Term (IV) $= O(\varepsilon)$ in the boundary layer since $e_{ij} = O(1/\varepsilon)$ and $R_\varepsilon + R_M = O(\varepsilon)$. The contribution from the essentially inviscid interior is $O(\varepsilon^2)$. Furthermore, the volume of the meniscus boundary layer is much smaller than that of the wall boundary layer ($R_M/R_\varepsilon = O(kA)$), while the straining rate is of the same order. Hence, we need only to account for the dissipation in the main wall boundary layer. In summary, we have, to $O(\varepsilon)$,

$$\frac{1}{2}\frac{\partial}{\partial t}\left\{ \iiint_{R_I} \overline{u_i u_i}\, dV + \iint_{S_F} \overline{\zeta^2}\, dS \right\} = 2\varepsilon^2 \iiint_{R_\varepsilon} \overline{e_{ij}^2}\, dV. \tag{3.9}$$

This formula is the basis of many existing damping theories and implies that both the time derivative and the dimensionless damping rate are $O(\varepsilon)$.

8.3.2 Meniscus Boundary Layer

Now consider the meniscus boundary layer with volume $V = R_M$ and bounding surface $S_M + \tilde{S}_M + S_{MW}$, where S_M is the side wall, $\tilde{S}_M$ is the outer edge of the boundary layer, and S_{MW} is the borderline with the main wall layer (see Fig. 3.2). Referring to Eq. (3.8), we find the following:

Term (I): Since $u_i = O(1)$, $R_M = O(\varepsilon kA)$ and $\partial(\ )/\partial t = O(\varepsilon)$, term I is $O(\varepsilon^2 kA)$.

Term (II): S_M gives no contribution since $u_i = 0$. On S_{MW}, $u_i n_i = w = O(1)$, $p = O(1)$, and $S_{MW} = O(\varepsilon)$; hence the integral over S_{MW} is of $O(\varepsilon)$. Use is made of the fact that only the hydrodynamic pressure p does work. On the outer edge $\tilde{S}_M$ of the meniscus boundary layer the tangential velocity is $w = O(1)$. Note that the tangential (vertical) length scale is $O(kA)$; hence by continuity the normal (horizontal) velocity is $u_i n_i = O(\varepsilon/kA)$. Since $p = -\Phi_t = O(1)$ and the area of $\tilde{S}_M$ is of $O(kA)$, we have

$$\iint_{\tilde{S}_M} \overline{pu_i n_i}\, dS = O(\varepsilon).$$

Term (III): On S_M, $u_i = 0$. On $\tilde{S}_M$, $e_{ij} = O(1)$. The integral on $S_M + \tilde{S}_M$ is $O(\varepsilon^2 kA)$. On S_{MW}, $e_{ij} = O(\varepsilon^{-1})$ and $u_i = O(1)$, and the area of $S_{MW} = O(\varepsilon)$; hence the integral is only of $O(\varepsilon^2)$.

Term (IV): As was estimated before, the dissipation rate in this volume is of $O(\varepsilon kA)$.

Thus, to $O(\varepsilon)$, Eq. (3.7) reduces to

$$\iint_{\tilde{S}_M} \overline{pu_i n_i}\, dS + \iint_{S_{MW}} \overline{pu_i n_i}\, dS = O(\varepsilon^2), \tag{3.10}$$

which means that, by pressure working on S_M, power is fed into the meniscus boundary layer from the inviscid core, and then transmitted essentially undiminished to the main wall boundary layer, also through pressure working on S_{MW}. Thus, the meniscus boundary layer serves as a channel of energy flow from waves to the main side-wall boundary layer!

8.3.3 Wall Boundary Layer

Take the wall layer bounded by the solid surface S, the outer edge of the boundary layer $\tilde{S}$, and the narrow strip bordering the free-surface meniscus boundary layer S_{MW}. Similar estimates give that term $(I) = O(\varepsilon^2)$, term $(II) = O(\varepsilon)$, term $(III) = O(\varepsilon^2)$, and term $(IV) = O(\varepsilon)$. In particular, the pressure working term has contributions from both $\tilde{S}$ and S_{MW}. The resulting energy budget is, to $O(\varepsilon)$, given by

$$- \iint_{\tilde{S}_{MW}} \overline{pw}\, dS - \iint_{S} \overline{pu_i}\, n_i\, dS - 2\varepsilon^2 \iiint_{R_\varepsilon} \overline{e_{ij}^2}\, dV = 0. \qquad (3.11)$$

Physically, pressure working on the ceiling and on the side balances the dissipation within.

If the water basin has sharp convex corners below the free surface, it is convenient to separate the wall layer into a side-wall layer R_W and a bottom layer R_B. Within the corner region whose volume is $O(\varepsilon^2)$ (see Fig. 3.3), the velocity field is essentially stagnant; all surface and volume integrals in Eq. (3.8) associated with the corner are at most of $O(\varepsilon^2)$ and can be ignored. Thus, the energy budget for the bottom boundary layer is

$$\iint_{\tilde{S}_B} \overline{pu_i}\, n_i\, dS - 2\varepsilon^2 \iiint_{R_B} \overline{e_{ij}^2}\, dV = O(\varepsilon^2), \qquad (3.12)$$

that is, dissipation within R_B is balanced by pressure working on the side $\tilde{S}_B$.

8.3.4 Interior Core

It can be shown that term $(I) = O(\varepsilon)$, term $(II) = O(\varepsilon)$, term $(III) = O(\varepsilon^2)$ since $e_{ij}, u_j = O(1)$, and term $(IV) = O(\varepsilon^2)$. Furthermore, the free-surface

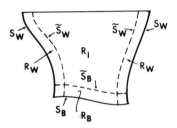

Figure 3.3 Basins with corners. The main boundary layer is separated into side-wall layer R_W and bottom layer R_B.

integral can be transformed into one for potential energy as before. Hence,

$$\frac{\partial}{\partial t}\left\{ \iiint_{R_I} \tfrac{1}{2}\overline{u_i u_i}\, dV + \iint_{S_F} \tfrac{1}{2}\overline{\zeta^2}\, dS \right\} = -\iint_{\tilde{S}+\tilde{S}_M} \overline{p u_i n_i}\, dS + O(\varepsilon^2). \quad (3.13)$$

It may be verified that Eqs. (3.9)–(3.12) are totally consistent with Eq. (3.13). Upon using Eq. (3.10), we can also write Eq. (3.13) as

$$\frac{\partial}{\partial t} \iiint_{R} \frac{1}{2}\overline{u_i u_i}\, dV + \iint_{S_F} \frac{1}{2}\overline{\zeta^2}\, dS = -\iint_{\tilde{S}} \overline{p u_i n_i}\, dS - \iint_{\tilde{S}_{MW}} \overline{pw}\, dS,$$

$$(3.14)$$

where R is the average volume of the entire fluid.

8.3.5 The Damping Rate

A general formula for the damping rate can now be easily derived. As a slight modification of Eq. (5.10), Chapter One, we can show that

$$a_l a_m = \tfrac{1}{2}\operatorname{Re}(A_l^* A_m)\big[\exp(2\omega^{(i)}t)\big](1 + O(\varepsilon))$$

$$= \tfrac{1}{2}\operatorname{Im}(i A_l^* A_m)\big[\exp(2\omega^{(i)}t)\big](1 + O(\varepsilon)), \qquad (3.15)$$

if

$$a_l(t) = \operatorname{Re} A_l \exp\big(-i(\omega^{(r)} + i\omega^{(i)})t\big), \qquad (3.16)$$

with $\omega^{(i)} = O(\varepsilon)\omega^{(r)}$. For the left-hand side of Eq. (3.14), the kinetic and potential energies may be calculated by the potential theory (Appendix 8.A) and are equal to each other. On the right-hand side the following formulas are true to the leading order:

$$[p]_{\tilde{S}} = -\left[\frac{\partial\phi}{\partial t}\right]_{\tilde{S}} = i\omega_0[\phi_0]_S e^{-i\omega t} + O(\varepsilon),$$

$$[p]_{\tilde{S}_{MW}} = i\omega_0[\phi_0]_{\tilde{S},\, z=0} e^{-i\omega t} + O(\varepsilon), \qquad (3.17)$$

$$[u_i n_i]_{\tilde{S}} = -[\mathbf{n}\cdot\mathbf{U}]_S e^{-i\omega t} + O(\varepsilon^2),$$

where $(\phi)_0$ refers to the inviscid leading order approximation, while $\{\mathbf{n}\cdot\mathbf{U}\}_S$ is the viscous correction of the normal velocity in the boundary layer, which will be explained more explicitly in Section 8.4. The term pw is of the order $O(\varepsilon)$ and can be obtained from continuity after the tangential velocity is found.

Upon substituting Eq. (3.17) into Eq. (3.14), we obtain

$$2\omega^{(i)} \iiint\limits_R |\nabla\phi_0|^2 \, dV = \text{Im}\left\{i\left[\iint\limits_S (i\omega_0\phi_0)^*(\mathbf{n}\cdot\mathbf{U})_S \, dS\right.\right.$$

$$\left.\left. -\frac{1}{\varepsilon}\iint\limits_{S_{MW}} (i\omega_0\phi_0)_S W \, dS\right]\right\},$$

which gives the dimensionless damping rate:

$$\text{Im}\,\frac{\omega^{(i)}}{\omega_0} = \frac{\text{Im}\left\{\iint\limits_S \phi_0^*[\mathbf{n}\cdot\mathbf{U}]_S \, dS + \iint\limits_{S_{MW}} \phi_0^* W \, dS\right\}}{2\iiint\limits_R |\nabla\phi_0|^2 \, dV}. \tag{3.18}$$

For a specific problem, the explicit value of the damping rate can be calculated from the potential solution and from the viscous correction in the boundary layers, both to the leading order.

Alternatively, one may start from Eq. (3.9) and obtain another formula for the damping rate. By the use of Eqs. (3.10), (3.11), and (3.12) the equivalence of the two formulas may be shown. It should be stressed that the meniscus term over S_{MW}, which is easily overlooked in intuitive reasoning, plays a crucial role.

8.4 DAMPING RATE BY A PERTURBATION ANALYSIS

In this section a more formal way of deducing Eq. (3.18) by a perturbation analysis is presented. The idea is to incorporate the slow rate of decay into a multiple-scale scheme and to make boundary-layer corrections iteratively (Johns, 1968; Dore, 1969; Greenspan, 1968; Mei and Liu, 1973). Let us first write the governing equations in Section 8.1 in terms of the dimensionless variables of Eq. (3.6):

$$\mathbf{u} = \nabla\Phi + \mathbf{U}, \tag{4.1}$$

$$\nabla^2\Phi = 0, \tag{4.2}$$

$$\nabla\cdot\mathbf{U} = 0, \tag{4.3}$$

$$\frac{\partial\mathbf{U}}{\partial t} = \varepsilon^2\nabla^2\mathbf{U}, \tag{4.4}$$

$$p = -\Phi_t. \tag{4.5}$$

The dimensionless boundary conditions are

$$\nabla \Phi + \mathbf{U} = 0, \qquad \text{on } S \equiv S_B + S_W \text{ (solid surface)}, \qquad (4.6)$$

$$\frac{\partial \zeta}{\partial t} = W + \frac{\partial \Phi}{\partial z}, \qquad (4.7)$$

$$\Phi_t + \zeta + 2\varepsilon \frac{\partial w}{\partial z} = 0, \qquad (4.8)$$

$$\varepsilon^2 \left(\frac{\partial u}{\partial z} + \frac{\partial w}{\partial x} \right) = \varepsilon^2 \left(\frac{\partial v}{\partial z} + \frac{\partial w}{\partial y} \right) = 0, \qquad \left. \begin{array}{c} \\ \end{array} \right\} \text{ on } z = 0, \qquad (4.9)$$

where the small parameter ε is defined by Eq. (3.7). Combining Eqs. (4.7)–(4.9), we have

$$\frac{\partial^2 \Phi}{\partial t^2} + \frac{\partial \Phi}{\partial z} + W + 2\varepsilon^2 \frac{\partial^2 w}{\partial t \, \partial z} = 0, \qquad z = 0. \qquad (4.10)$$

The tangential stress conditions are

$$\varepsilon^2 \left(2 \frac{\partial^2 \Phi}{\partial x \, \partial z} + \frac{\partial W}{\partial x} \right) + \varepsilon^2 \frac{\partial U}{\partial z} = 0, \qquad \varepsilon^2 \left(2 \frac{\partial^2 \Phi}{\partial y \, \partial z} + \frac{\partial \overline{W}}{\partial y} \right) + \varepsilon^2 \frac{\partial V}{\partial z} = 0.$$

$$(4.11)$$

Equation (4.11) suggests that $\partial U / \partial z$, $\partial V / \partial z$ can be of the order $O(1)$. However, U and V vanish outside the boundary layer and they can grow at most to be $O(\varepsilon)$ inside. From continuity, we must have $\partial W / \partial z = O(\varepsilon)$. Thus, the last term in Eq. (4.10) is of the order $O(\varepsilon^3)$ and can be neglected. Consequently, we have

$$\frac{\partial^2 \Phi}{\partial t^2} + \frac{\partial \Phi}{\partial z} + W = O(\varepsilon^3) \qquad z = 0. \qquad (4.12)$$

Since W is of the same order as the other terms only in a thin strip of the meniscus boundary layer and much less elsewhere, its global effect can only be felt at the order $O(\varepsilon)$.

Expecting that the rotational part $\mathbf{U}$ varies rapidly within the dimensionless distance $O(\varepsilon)$, we introduce a boundary-layer coordinate

$$\zeta = \frac{x_N}{\varepsilon} \qquad \text{so that} \quad \mathbf{U} = \mathbf{U}(\mathbf{x}_T, \zeta), \qquad (4.13)$$

where $\mathbf{x}_T$, x_N form a locally rectangular coordinate system with x_N being in the normal direction pointing *into* the fluid from the wall, hence, opposite to $\mathbf{n}$, as shown in Fig. 4.1.

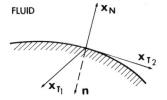

FLUID

Figure 4.1 Local coordinate system.

If the dimensional radius of curvature of the wall is comparable to $1/k$, then $\partial/\partial x_T = O(\varepsilon\,\partial/\partial x_N)$. The continuity equation in terms of the local coordinate is

$$-\frac{\partial \mathbf{n}\cdot \mathbf{U}}{\partial x_N} + \frac{\partial U_{T_1}}{\partial x_{T_1}} + \frac{\partial U_{T_2}}{\partial x_{T_2}} = 0, \qquad (4.14)$$

so,

$$-\frac{\partial \mathbf{n}\cdot \mathbf{U}}{\partial \xi} + \varepsilon\left(\frac{\partial U_{T_1}}{\partial x_{T_1}} + \frac{\partial U_{T_2}}{\partial x_{T_2}}\right) = 0. \qquad (4.15)$$

We now introduce perturbation expansions. Since there is a damping time scale of the order $O(T/\varepsilon)$ in addition to the wave period T, it is natural to employ the multiple-scale expansions with t and εt as independent variables. Equivalently, we may expand the dimensionless ω in powers of ε. Thus,

$$\Phi = \left[\phi_0(\mathbf{x}) + \varepsilon\phi_1(\mathbf{x}) + O(\varepsilon^2)\right]e^{-i\omega t}, \qquad (4.16a)$$

$$\mathbf{U} = \left[\mathbf{q}_0(\mathbf{x}_T, \xi) + \varepsilon\mathbf{q}_1(\mathbf{x}_T, \xi) + O(\varepsilon^2)\right]e^{-i\omega t}, \qquad (4.16b)$$

$$\omega = \omega_0 + \varepsilon\omega_1 + O(\varepsilon^2). \qquad (4.16c)$$

When Eq. (4.16) is substituted into Eqs. (4.2)–(4.10) and the orders are separated, a sequence of problems is obtained.
  (i) Inviscid solution at $O(\varepsilon^0)$:

$$\nabla^2\phi_0 = 0, \qquad (4.17a)$$

$$\mathbf{n}\cdot\nabla\phi_0 = 0 \qquad \text{on } S, \qquad (4.17b)$$

$$\frac{\partial\phi_0}{\partial z} - \omega_0^2\phi_0 = 0 \qquad \text{on } S_F, \quad z = 0. \qquad (4.17c)$$

Note that W is not present at this order because of its small area of effectiveness. This solution gives rise to a nonzero tangential velocity on S.

(ii) Boundary-layer correction of $O(\varepsilon^0)$:

$$\frac{\partial^2 \mathbf{q}_0}{\partial \xi^2} = -i\omega_0 \mathbf{q}_0, \tag{4.18a}$$

$$\mathbf{q}_0 = -\nabla \phi_0 \qquad \text{on } S, \tag{4.18b}$$

$$\mathbf{q}_0 \to 0 \qquad \xi \to \infty. \tag{4.18c}$$

Note that $\mathbf{q}_0$ is tangential to the wall S, that is, $\mathbf{n} \cdot \mathbf{q}_0 = 0$. The continuity equation gives

$$-\frac{\partial}{\partial \xi}(\mathbf{n} \cdot \mathbf{q}_1) + \left(\frac{\partial U_{T_1}}{\partial x_{T_1}} + \frac{\partial U_{T_2}}{\partial x_{T_2}}\right)_0 = 0 \tag{4.19}$$

from which $\mathbf{n} \cdot \mathbf{q}_1$ may be integrated from $\xi \sim \infty$ (outside the boundary layer) inward to ξ, subject to the condition that $\mathbf{n} \cdot \mathbf{q}_1 = 0$ at $\xi \sim \infty$. In general, $(\mathbf{n} \cdot \mathbf{q}_1)$ is nonzero at the wall S ($\xi = 0$) and must be cancelled by $\partial \phi_1 / \partial x_N$ at the order $O(\varepsilon)$.

(iii) Inviscid correction of $O(\varepsilon)$:

$$\nabla^2 \phi_1 = 0, \tag{4.20a}$$

$$\mathbf{n} \cdot \nabla \phi_1 = -[\mathbf{n} \cdot \mathbf{q}_1]_S \qquad \text{on } S, \tag{4.20b}$$

$$\frac{\partial \phi_1}{\partial z} - \omega_0^2 \phi_1 = 2\omega_0 \omega_1 \phi_0 - \frac{W_0}{\varepsilon} \qquad \text{on } S_F. \tag{4.20c}$$

Note that W_0/ε is effective only over an area of $O(\varepsilon)$.

Problem (i) is homogeneous and is an eigenvalue problem. The eigenfrequency ω_0 and the eigenfunction $\phi_0(x)$ can be found in principle.

Now Problem (ii) is the classical Stokes' problem of a plate oscillating in its own plane; the solution is similar to Eqs. (2.19)–(2.21). In the present dimensionless variables, the solution is

$$\mathbf{q}_0 = -[\nabla \phi_0]_S \Gamma(\xi), \qquad \Gamma(\xi) = \exp\left[-(1-i)(\omega_0/2)^{1/2} \xi\right] \tag{4.21}$$

which can be used to obtain $\mathbf{n} \cdot \mathbf{q}_1$ by integrating Eq. (4.19). Now Problem (iii) for ϕ_1 contains the unknown ω_1 which we are seeking. The homogeneous boundary-value problem is identical to that of ϕ_0 which has nontrivial solutions. By invoking the Fredholm alternative, that is, applying Green's formula to ϕ_0^* and ϕ_1

$$\iiint_R \left(\phi_0^* \nabla^2 \phi_1 - \phi_1 \nabla^2 \phi_0^*\right) dV = \iint_{S+S_F} \left(\phi_0^* \frac{\partial \phi_1}{\partial n} - \phi_1 \frac{\partial \phi_0^*}{\partial n}\right) dS,$$

and by using all boundary conditions on ϕ_0^* and ϕ_1, we obtain ω_1 immediately:

$$\omega_1 = \frac{\iint\limits_{S} \phi_0^* [\mathbf{n} \cdot \mathbf{q}_1]_S \, dS + \iint\limits_{S_F} \phi_0^* W_0 \, dS/\varepsilon}{2\omega_0 \iint\limits_{S_F} |\phi_0|^2 \, dS}. \tag{4.22}$$

An alternative form of Eq. (4.22) may be derived by noting that

$$\nabla \cdot (\phi_0^* \nabla \phi_0) = |\nabla \phi|^2 + \phi_0^* \nabla^2 \phi_0 = |\nabla \phi_0|^2.$$

Integrating the preceding equation over the entire volume R and using Gauss' theorem, we obtain

$$\iiint\limits_{R} |\nabla \phi_0|^2 \, dV = \iint\limits_{S+S_F} \phi_0^* \frac{\partial \phi_0}{\partial n} \, dS = \iint\limits_{S_F} \phi_0^* \frac{\partial \phi_0}{\partial z} \, dS + \iint\limits_{S} \phi_0^* \frac{\partial \phi_0}{\partial n} \, dS.$$

From the boundary conditions on S_F and S, it follows that

$$\iiint\limits_{R} |\nabla \phi_0|^2 \, dV = \omega_0^2 \iint\limits_{S_F} |\phi_0|^2 \, dS.$$

Since $W_0 = 0$ except in S_{MW}, Eq. (4.22) becomes

$$\frac{\omega_1}{\omega_0} = \frac{1}{2} \frac{\iint\limits_{S} \phi_0^* [\mathbf{n} \cdot \mathbf{q}_1]_S \, dS + \iint\limits_{S_{MW}} (\phi_0^* W_0 \, dS/\varepsilon)}{\iiint\limits_{R} |\nabla \phi_0|^2 \, dV}. \tag{4.23}$$

Finally we have, in *physical variables*, $\omega = \omega_0 + \omega_1$ and

$$\frac{\omega_1}{\omega_0} = \frac{\iint\limits_{S} \phi_0^* [\mathbf{n} \cdot \mathbf{q}_1]_S \, dS + \iint\limits_{S_{MW}} \phi_0^* W_0 \, dS}{2 \iiint\limits_{R} |\nabla \phi_0|^2 \, dV}. \tag{4.24}$$

In particular, the imaginary part of Eq. (4.23) is precisely Eq. (3.18).

The perturbation analysis gives as a bonus the real part of ω_1 which represents a shift of the eigenfrequency due to viscosity.

The case of progressive waves in an infinitely long channel of uniform cross section (see Fig. 4.2) can be worked out in a manner similar to that for standing waves. If the progressive waves are strictly sinusoidal in time as in the case of a laboratory wave flume, we should expect them to attenuate in the

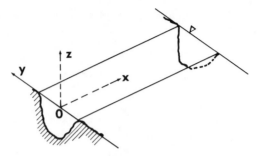

Figure 4.2

direction of propagation. The spatial rate of attenuation can be derived for arbitrary cross section by following the steps suggested in the next exercise.

Exercise 4.1   Boundary-Layer Effect on Progressive Waves in a Uniform Channel

Rewrite the linearized governing equations in terms of the following dimensionless variables

$$(x, y, z) = \left(\frac{\omega^2}{g}\right)(x', y', z'), \qquad t = \omega t',$$

$$u = \frac{u'}{\omega A}, \qquad \phi = \phi'\left(\frac{\omega}{gA}\right), \qquad \zeta = \frac{\zeta'}{A}. \tag{4.25}$$

Define the small parameter

$$\varepsilon = \frac{\omega^2}{g}\left(\frac{\nu}{\omega}\right)^{1/2}, \tag{4.26}$$

and assume the expansions

$$\Phi(x, y, z, t) = \left[\phi_0(y, z) + \varepsilon\phi_1(y, z) + \cdots\right]e^{i(kx-t)}, \tag{4.27a}$$

$$\mathbf{U} = \left[\mathbf{q}_0(y, z) + \varepsilon\mathbf{q}_1(y, z) + \cdots\right]e^{i(kx-t)}, \tag{4.27b}$$

$$k = k_0 + \varepsilon k_1 + \cdots, \tag{4.27c}$$

then show that

$$k_1 = \frac{-\displaystyle\iint_S \phi_0^*[\mathbf{n} \cdot \mathbf{q}_1]_S \, dS - \iint_{S_{MW}} \phi_0^* W_0 \, dS/\varepsilon}{2k\displaystyle\iiint_R |\phi_0|^2 \, dS}, \tag{4.28}$$

where the numerator involves line integrals and the denominator is an area integral. Here S is the wetted wall contour, S_F is the contour of the free surface, and R is the total water area of a cross section of the channel.

8.5 DETAILS FOR STANDING WAVES IN A CIRCULAR BASIN

Consider a circular basin of radius a' and depth h', and choose a polar-coordinate system with the origin at the center of the free surface. The first-order potential solution for a typical mode (m, n) with $m, n = 1, 2, 3, \ldots$, is, in physical variables,

$$\phi'(r', \theta, z') = \frac{-iA\omega_0'}{k'} \frac{\cosh k'(z' + h')}{\sinh k'h'} J_n(k'r')\sin n\theta, \tag{5.1}$$

where $k' = k'_{mn}$ is the mth root, $m = 1, 2, 3, \ldots$, of

$$J_n'(k'a') \equiv \frac{d}{dk'a'} J_n(k'a') = 0, \tag{5.2}$$

and

$$\omega_0'^2 = gk'\tanh k'h'. \tag{5.3}$$

For nondimensionalization it is convenient to use $k' = k'_{mn}$ as the scaling wavenumber so that in dimensionless variables Eqs. (5.1)–(5.3) become

$$\phi_0 = -i\omega_0 \frac{\cosh(z + h)}{\sinh h} J_n(r)\sin n\theta, \tag{5.4}$$

$$J_n'(a) = 0, \tag{5.5}$$

$$\omega_0 = \tanh h, \tag{5.6}$$

where

$$a = k'_{mn}a', \quad h = k'_{mn}h', \quad \text{and} \quad \omega_0 = \frac{\omega_0'}{(gk'_{mn})^{1/2}}. \tag{5.7}$$

Let U, V, and W denote the dimensionless components of the rotational velocity in the r, θ, and z directions, respectively. Divide the boundary layer into two parts, that is, the side wall and the bottom layer, designated by the subscripts W and B, and let the boundary-layer coordinates be

$$\xi_W = \frac{a - r}{\varepsilon} \quad \text{and} \quad \xi_B = \frac{z + h}{\varepsilon}. \tag{5.8}$$

The first-order boundary layer solutions are

$$U_{0W} = 0, \tag{5.9a}$$

$$V_{0W} = -\left(\frac{1}{r}\frac{\partial \phi_0}{\partial \theta}\right)_{r=a} \Gamma(\xi_W) \tag{5.9b}$$

$$= i\omega_0 \frac{n}{a} J_n(a)\frac{\cosh(z+h)}{\sinh h}\cosh n\theta \Gamma(\xi_W),$$

$$W_{0W} = -\left(\frac{\partial \phi_0}{\partial z}\right)_{r=a} \Gamma(\xi_W)$$

$$= i\omega_0 J_n(a)\frac{\sinh(z+h)}{\sinh h}\sin n\theta \Gamma(\xi_W) \tag{5.9c}$$

near the side wall, and

$$U_{0B} = -\left(\frac{\partial \phi_0}{\partial r}\right)_{z=-h} \Gamma(\xi_B)$$

$$= \frac{i\omega_0}{\sinh h} J_n'(r)\sin n\theta \Gamma(\xi_B), \tag{5.10a}$$

$$V_{0B} = -\left(\frac{1}{r}\frac{\partial \phi_0}{\partial \theta}\right)_{z=-h} \Gamma(\xi_B)$$

$$= \frac{i\omega_0}{\sinh h}\frac{n}{r} J_n(r)\cos n\theta \Gamma(\xi_B), \tag{5.10b}$$

$$W_{0B} = 0 \tag{5.10c}$$

near the bottom, with $\Gamma(\xi)$ being Stokes' solution as defined in Eq. (4.21). The induced velocity normal to the side-wall boundary layer is obtained by integrating the continuity equation (in polar coordinates)

$$[\mathbf{n}\cdot\mathbf{q}_1]_{S_W} = U_{1W} = \int_\infty^{\xi_W}\left(\frac{1}{a}\frac{\partial V_{0W}}{\partial \theta} + \frac{\partial W_{0W}}{\partial z}\right)d\xi_W$$

$$= -(1-i)\left(\frac{\omega_0}{2}\right)^{1/2}\left(\frac{n^2}{a^2} - 1\right)\frac{\cosh(z+h)}{\sinh h}\sin n\theta J_n(a)\Gamma(\xi_W),$$

$$\tag{5.11}$$

where use is made of the fact that

$$-i\omega_0 \int_\infty^\xi \Gamma(\xi)\,d\zeta = -(1-i)\left(\frac{\omega_0}{2}\right)^{1/2}\Gamma(\xi).$$

Similarly, the normal velocity in the bottom boundary layer is

$$[\mathbf{n} \cdot \mathbf{q}_1]_{S_B} = W_{1B} = \int_{\infty}^{\xi_B} \left(\frac{\partial U_{0B}}{\partial r} + \frac{U_{0B}}{r} + \frac{1}{r}\frac{\partial V_{0B}}{\partial \theta} \right) d\xi_B$$

$$= \frac{i\omega_0}{\sinh h} \sin n\theta \left(J_n'' + \frac{J_n'}{r} - \frac{n^2}{r^2} \right) \int_{\infty}^{\xi_B} \Gamma(\xi_B)\, d\xi_B$$

$$= -(1-i)\left(\frac{\omega_0}{2}\right)^{1/2} \frac{\sin n\theta}{\sinh h} J_n(r)\Gamma(\xi_B). \tag{5.12}$$

Now the S integral in Eq. (4.22) or (4.23) is composed of two parts

$$\iint_S \phi_0^*[\mathbf{n} \cdot \mathbf{q}_1]\, dS = \iint_{S_W} \phi_0^*[\mathbf{n} \cdot \mathbf{q}_1]\, dS + \iint_{S_B} \phi_0^*[\mathbf{n} \cdot \mathbf{q}_1]\, dS$$

$$= \int_{-h}^{0} dz \int_{0}^{2\pi} \left[i\omega_0 \frac{\cosh(z+h)}{\sin h} J_n(a)\sin n\theta \right]$$

$$\times \left[-(1-i)\left(\frac{\omega_0}{2}\right)^{1/2}\left(\frac{n^2}{a^2} - 1\right) \right.$$

$$\times \left. \frac{\cosh(z+h)}{\sinh h} J_n(a)\sin n\theta \right] a\, d\theta$$

$$+ \int_{0}^{2\pi} d\theta \int_{0}^{a} \left[i\omega_0 J_n(r)\frac{\sin n\theta}{\sinh h} \right]$$

$$\times \left[-(1-i)\left(\frac{\omega_0}{2}\right)^{1/2} J_n(r)\frac{\sin n\theta}{\sinh h} \right] r\, d\theta. \tag{5.13}$$

If the orthogonality property

$$\int_{0}^{a} r J_n^2(r)\, dr = \frac{1}{2}(a^2 - n^2)J_n^2(a) \qquad \text{if } J_n'(a) = 0, \tag{5.14}$$

is used, Eq. (5.13) becomes

$$\iint_S \phi_0^*[\mathbf{n} \cdot \mathbf{q}]\, dS = -(1+i)\omega_0\left(\frac{\omega_0}{2}\right)^{1/2}\left(\frac{n^2}{a^2} - 1\right)\frac{aJ_n^2(a)}{\sinh^2 h}\frac{\pi}{4}(\sinh 2h + 2h)$$

$$- (1+i)\omega_0\left(\frac{\omega_0}{2}\right)^{1/2}\frac{a^2-n^2}{2}\frac{\pi J_n^2(a)}{\sinh^2 h}$$

$$= -(1+i)\omega_0\left(\frac{\omega_0}{2}\right)^{1/2}\frac{\pi}{4}\frac{J_n^2(a)}{\sinh^2 h}(a^2 - n^2)$$

$$\times \left[2 - \frac{1}{a}(\sinh 2h + 2h) \right]. \tag{5.15}$$

From Eqs. (5.4) and (5.9c) the second integral in Eq. (4.22) along the free surface is

$$\iint_{\tilde{S}_{MW}} \phi_0^* W_0 \frac{dS}{\varepsilon} = -\int_0^{2\pi} a\, d\theta \int_0^\infty d\xi_W \left[i\omega_0 \frac{\cosh h}{\sinh h} J_n(a)\sin n\theta \right]$$

$$\cdot \left[-i\omega_0 J_n(a)\sin n\theta \right] \Gamma(\xi_W) = \pi a J_n^2(a) \frac{1+i}{2(\frac{1}{2}\omega_0)^{1/2}}, \quad (5.16)$$

where the dispersion relation has been applied.

Finally, the denominator of Eq. (4.22) is

$$\iint_{\tilde{S}_F} |\phi_0|^2\, dS = \int_0^{2\pi} d\theta \int_0^a \omega_0^2 \frac{\cosh^2 h}{\sinh^2 h} \sin^2 n\theta\, r J_n^2(r)\, dr$$

$$= \frac{\pi}{\omega_0^2} \frac{1}{2}(a^2 - n^2) J_n^2(a). \quad (5.17)$$

When Eqs. (5.15)–(5.17) are substituted into Eq. (4.22), the dimensionless frequency correction ω_1 is obtained:

$$\omega_1 = -(1+i)\left(\frac{\omega_0}{2}\right)^{1/2} \left[\frac{a^2+n^2}{2a(a^2-n^2)} + \left(1-\frac{h}{a}\right)\frac{1}{\sinh 2h} \right] \quad (5.18)$$

where the dispersion relation is again applied.

Returning to dimensional variables, we now have

$$\omega = \omega_0 \left\{ 1 - k'\delta\frac{1+i}{2} \left[\frac{(k'a')^2+n^2}{2k'a'((k'a')^2-n^2)} + \left(1-\frac{h'}{a'}\right)\frac{1}{\sinh 2kh'} \right] \right\}$$

$$(5.19)$$

where $\delta = (2\nu/\omega_0)^{1/2}$ is the boundary-layer thickness and ω_0 and k' are the eigenfrequency and the eigen wavenumber, respectively, of a specific mode (m, n). The imaginary part which gives the damping rate was first derived by Case and Parkinson (1957) by using Eq. (3.9). If $k'h'$ is large, the second term in Eq. (5.19) becomes insignificant. On the other hand, for a shallow basin where $k'h'$ and h'/a' are small, the bottom boundary layer becomes dominant.

It is interesting to examine the energy details for a side-wall boundary layer, using the explicit solution to check Eq. (3.11).

The average work done by the pressure on the strip of surface S_{MW} is, omitting the factor $\exp[-2\omega^{(i)}t]$,

$$\text{Re} \int_0^{2\pi} a\, d\theta \int_0^\infty [p_0^*]_{z=0,\, r=a}[w]_{z=0}\, d\zeta_w$$

$$= \text{Re}\, a \int_0^{2\pi} d\theta [-i\omega_0\phi_0^*]_{z=0,\, r=a} \int_0^\infty [W_{0w}]_{z=0}\, d\zeta_w$$

$$= -\pi a \left(\frac{\omega_0}{2}\right)^{1/2} J_n^2(a). \tag{5.20}$$

The dynamic pressure and the potential part of the velocity are out of phase to the present order of approximation. The work done by the pressure on the interface between the inviscid interior and the side-wall layer is

$$\text{Re} \int_h^0 dz \int_0^{2\pi} a \left[p_0^*\left(-\frac{\partial \phi_1}{\partial r}\right)\right]_{r=a} d\theta = \text{Re} \int_{-h}^0 dz \int_0^{2\pi} a[-i\omega_0\phi_0^* U_{1W}]_{r=a}\, d\theta$$

$$= -\frac{\pi a}{2}\left(\frac{\omega_0}{2}\right)^{1/2}\left(1 + \frac{2h}{\sinh 2h}\right)\left(\frac{n^2}{a^2} - 1\right) J_n^2(a).$$

$$\tag{5.21}$$

Note that the outward normal points toward the z axis. Finally, the average rate of viscous dissipation in the side-wall layer is

$$\text{Re} \int_{-h}^0 dz \int_0^{2\pi} a\, d\theta \int_0^\infty \left(\left|\frac{\partial V_{0W}}{\partial \zeta_W}\right|^2 + \left|\frac{\partial W_{0W}}{\partial \zeta_W}\right|^2 \right) d\zeta_W$$

$$= \pi a \left(\frac{\omega_0}{2}\right)^{1/2} J_n^2(a)\left[\frac{1}{2}\left(\frac{n^2}{a^2} - 1\right)\left(1 + \frac{2h}{\sinh 2h}\right) + 1\right]. \tag{5.22}$$

The three energy terms in Eq. (3.11) add up precisely to zero, as estimated by Eq. (3.10).

Note from Eq. (5.14) that $n^2 < a^2 = (k'_{mn}a')^2$ for all modes. We conclude from Eq. (5.21) that the side-wall layer *receives* power from waves through the meniscus boundary layer above, spends only a part of it on internal dissipation, and gives up the rest to the inviscid interior!

A similar calculation for the bottom layer confirms Eq. (3.12) with no surprises.

Other interesting examples are left as exercises.

Exercise 5.1

Deduce for a progressive wave advancing down a uniform rectangular channel, that

$$k_1 = \frac{1+i}{2^{1/2}} \frac{k_0}{b} \left(\frac{2k_0 b + \sinh 2k_0 h}{2k_0 h + \sinh 2k_0 h} \right) \tag{5.23}$$

(Hunt, 1952) where $2b$ is the dimensionless width, h is the dimensionless depth, and $k_0 \tanh k_0 h = 1$. Discuss separately the energy balance in the boundary layers near the side wall and near the bottom.

Exercise 5.2

Get the frequency correction due to viscosity for the standing wave in a rectangular basin

$$\phi_0 = \frac{-i\omega_0}{\sinh h} \cosh(z+h) \cos \frac{n\pi x}{a} \cos \frac{m\pi y}{b}. \tag{5.24}$$

The physical wavenumber used for normalization is

$$k_{mn}^2 = \left(\frac{n\pi}{a'} \right)^2 + \left(\frac{m\pi}{b'} \right)^2. \tag{5.25}$$

Show that

$$\omega_1 = -(1+i)\left(\frac{\omega_0}{2} \right)^{1/2} \left\{ \frac{1}{a} \left[2 - \left(\frac{n\pi}{a} \right)^2 \right] \right.$$
$$\left. + \frac{1}{b} \left[2 - \left(\frac{m\pi}{b} \right)^2 \right] + \frac{1}{\sinh 2h} \left[1 - 2\frac{h}{a}\left(\frac{n\pi}{a} \right)^2 - 2\frac{h}{b}\left(\frac{m\pi}{b} \right)^2 \right] \right\}$$
$$\tag{5.26}$$

(Keulegan, 1959). Discuss separately the energy balance in the boundary layers near the side wall and near the bottom.

8.6 THE EFFECT OF AIR ON THE DAMPING OF DEEP WATER WAVES

An additional factor, which has not received much attention until recently (Dore, 1978), is the presence of air above the water surface. Despite its relatively small density and viscosity, air, which must be moved by the waves, can contribute as much dissipation as water if the water depth is great and the side-wall effects are negligible.

The physical argument, due to Dore, is as follows. Let the physical properties of air be distinguished by primes. In deep water waves the rate of dissipation in water is of the order $O(\mu U^2/\lambda)$, where U is the typical orbital velocity in the waves. On the free surface, the tangential motion of the water particles induces a Stokes boundary layer in air. Because of the much smaller density of air, the motion of water particles near the free surface is hardly affected and the free-surface boundary layer in water remains ineffective. The rate of dissipation in the air boundary layer, whose thickness is $\delta' = (2\nu'/\omega)^{1/2}$, is of the order $O(\mu'U'^2/\delta')$. The ratio of the two dissipation rates is

$$O\left(\frac{\mu'\lambda}{\mu\delta'}\right). \tag{6.1a}$$

Although $\mu'/\mu \ll 1$, λ/δ' is much greater than unity for sufficiently long waves so that the ratio (6.1a) can be of order unity. Using definitions and the dispersion relation $\omega^2 = gk$, we may rewrite ratio (6.1a)

$$\frac{\mu'}{\mu}\left(\frac{\rho'}{\mu'}\right)^{1/2}\left(\frac{g}{\lambda}\right)^4\lambda. \tag{6.1b}$$

Since the following values are representative

<table>
<tr><td></td><td>water</td><td>air</td></tr>
<tr><td></td><td>$\mu = 1.3 \times 10^{-2}$ g/cm $-$ s</td><td>$\mu' = 1.76 \times 10^{-4}$ g/cm $-$ s</td></tr>
<tr><td></td><td>$\rho = 1$ g/cm^3</td><td>$\rho' = 1.25 \times 10^{-3}$ g/cm^3</td></tr>
</table>

the ratio (6.1b) is small only if λ is much less than 10 cm. Thus, in a natural environment where the range of interest is $\lambda > 10$ cm, air cannot be ignored.

If one only wishes to find the damping rate, the simplest procedure is to start from the potential-theory result for water and calculate the boundary layer in the air to ensure that there is no slip between air and water. The rate of dissipation within this boundary layer may be added to that within the main body of water to give the total damping rate. This approach was used by Dore. However, we sketch below a perturbation analysis with a view to obtaining the damping rate as well as the real frequency shift. With two fluids the algebra is lengthy, but straightforward.

Defining $\varepsilon = k(\nu/\omega)^{1/2}$, $\varepsilon' = k(\nu'/\omega)^{1/2}$, we first note that $\mu'/\mu = 10^{-2} \ll 1$ and $\rho'/\rho = 10^{-3} \ll 1$, but that

$$\frac{\varepsilon'}{\varepsilon} = \left(\frac{\nu'}{\nu}\right)^{1/2} = \left(\frac{\mu'\rho}{\mu\rho'}\right)^{1/2} = O(1).$$

Let us assume for the convenience of ordering that

$$O(\varepsilon) = O(\varepsilon') = O(\bar{\varepsilon}),$$

$$\alpha = \frac{\mu'}{\mu} = O(\bar{\varepsilon}) \quad \text{and} \quad \beta = \frac{\rho'}{\rho} = O(\bar{\varepsilon}). \tag{6.2}$$

In terms of the normalized variables defined in Eq. (3.6), the linearized dimensionless equations are

$$\mathbf{u} = \nabla\Phi + \mathbf{U}, \qquad \mathbf{u}' = \nabla\Phi' + \mathbf{U}', \tag{6.3}$$

with

$$\nabla^2\Phi = 0, \qquad \nabla^2\Phi' = 0, \tag{6.4}$$

$$\frac{\partial\mathbf{U}}{\partial t} = \varepsilon\nabla^2\mathbf{U}, \qquad \frac{\partial\mathbf{U}'}{\partial t} = \varepsilon'^2\nabla^2\mathbf{U}, \tag{6.5}$$

$$\nabla\cdot\mathbf{U} = 0, \qquad \nabla\cdot\mathbf{U}' = 0. \tag{6.6}$$

The kinematic boundary conditions on the mean free surface $z = 0$ are

$$\frac{\partial\zeta}{\partial t} = W + \frac{\partial\Phi}{\partial z}, \tag{6.7a}$$

$$\frac{\partial\zeta}{\partial t} = W' + \frac{\partial\Phi'}{\partial z}, \tag{6.7b}$$

$$U = U'. \tag{6.8}$$

The dynamic stress conditions on the free surface are as follows:
normal:

$$\frac{\partial\Phi}{\partial t} + \zeta + 2\varepsilon^2\left(\frac{\partial W}{\partial z} + \frac{\partial^2\Phi}{\partial z^2}\right) = \beta\left[\frac{\partial\Phi'}{\partial t} + \zeta + 2\varepsilon'^2\left(\frac{\partial W'}{\partial z} + \frac{\partial^2\Phi'}{\partial z^2}\right)\right], \tag{6.9}$$

tangential:

$$2\frac{\partial^2\Phi}{\partial x\,\partial z} + \frac{\partial U}{\partial z} + \frac{\partial W}{\partial x} = \alpha\left(2\frac{\partial^2\Phi'}{\partial x\,\partial z} + \frac{\partial U'}{\partial z} + \frac{\partial W'}{\partial x}\right). \tag{6.10}$$

In addition, $\Phi, \Phi' \to 0$ and $\mathbf{U}, \mathbf{U}' \to 0$, when $z \to -\infty, \infty$, respectively.

We now assume the following expansions:

$$\Phi = \left(\phi_0 + \bar{\varepsilon}\phi_1 + \bar{\varepsilon}^2\phi_2 + \cdots\right)e^{i\theta}, \tag{6.11}$$

$$U = \left(U_0 + \bar{\varepsilon}U_1 + \bar{\varepsilon}^2U_2 + \cdots\right)e^{i\theta}, \tag{6.12}$$

$$W = \left(\bar{\varepsilon}W_1 + \bar{\varepsilon}^2W_2 + \cdots\right)e^{i\theta}, \tag{6.13}$$

$$\zeta = \left(\zeta_0 + \bar{\varepsilon}\zeta_1 + \bar{\varepsilon}^2\zeta_2 + \cdots\right)e^{i\theta}, \tag{6.14}$$

and similar expansions for Φ', U', and W', with

$$\theta = x - \omega t, \tag{6.15}$$

$$\omega = \omega_0 + \bar{\varepsilon}\omega_1 + \bar{\varepsilon}^2\omega_2 + \cdots, \tag{6.16}$$

and

$$\phi_n = \phi_n(x, z)\phi'_n = \phi'_n(x, z) \quad \text{and} \quad U_n = U_n\left(\frac{x, z}{\varepsilon}\right), \quad U'_n = U'_n\left(x, \frac{z}{\varepsilon'}\right). \tag{6.17}$$

The details of the perturbation analysis are very similar to those of Section 8.4. In the following chart we only indicate the procedure of obtaining ω_2:

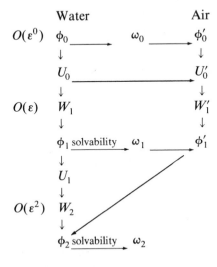

At each order, the potential parts ϕ_n and ϕ'_n are subject to the normal components of the kinematic and dynamic conditions, whereas the tangential components of the rotational velocity are subject to the tangential components of the same conditions. The vertical components W_n and W_n' follow from continuity. We leave it for the reader to verify that

$$\omega_0 = 1, \qquad \bar{\varepsilon}\omega_1 = -\beta,$$

and

$$
\bar{\varepsilon}^2\omega_2 = \frac{1}{2}\left[\beta^2 - \frac{1}{\sqrt{2}}\left(\alpha\frac{\varepsilon^2}{\varepsilon'} + \beta\varepsilon'\right)\right] - i\left[2\varepsilon^2 + \frac{1}{\sqrt{2}}\left(\alpha\frac{\varepsilon^2}{\varepsilon'} + \beta\varepsilon'\right)\right]
$$

$$
= \frac{1}{2}\left[\left(\frac{\rho'}{\rho}\right)^2 - \sqrt{2}\,\varepsilon\left(\frac{\mu'\rho'}{\mu\rho}\right)^{1/2}\right] - i\left[2\varepsilon^2 + \sqrt{2}\,\varepsilon\left(\frac{\mu'\rho'}{\mu\rho}\right)^{1/2}\right]. \qquad (6.18)
$$

The imaginary part of $\bar{\varepsilon}^2\omega_2$ gives the damping rate, as found by Dore. To have some numerical ideas, Dore has computed the time required for the wave amplitude to decrease by a factor of e^{-1}, for two examples:

Wavelength	Air–Water	Vacuum—Water
1 m	75.9 min	161.8 min
100 m	30.5 days	3.1 years

The importance of air is evident. A more complete picture is seen in Fig. 6.1. The real part of the frequency change which shows the effect on the phase is

$$
\mathrm{Re}\!\left(\bar{\varepsilon}\omega_1 + \bar{\varepsilon}^2\omega_2\right) = \frac{\rho'}{\rho} + \frac{1}{2}\left[\left(\frac{\rho'}{\rho}\right)^2 - \sqrt{2}\,\varepsilon\left(\frac{\mu'\rho'}{\mu\rho}\right)^{1/2}\right]. \qquad (6.19)
$$

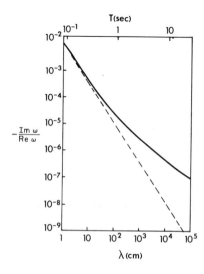

Figure 6.1 Viscous attenuation rates of surface waves in deep water with a clean surface. Solid curve: air–water interface; Dashed curve: vacuum–water interface. (From Dore, 1978, *Geophys. Astrophys. Fluid Dyn.* Reproduced by permission of Gordon and Breach Scientific Publishers.)

In open seas, the coupling of air and water may involve a turbulent transfer of energy and momentum near the interface; this is an important part of the subject of wave generation by wind (Phillips, 1977). The intermittent breaking of large crests also contributes to the damping of random sea waves. However, these aspects are too complicated to be entered here.

8.7 THE TURBULENT BOUNDARY LAYER NEAR A ROUGH BOTTOM

8.7.1 The Boundary-Layer Structure

When the wave amplitude becomes large, the laminar Stokes layer near a smooth wall can be unstable to infinitesimal disturbance and transition to turbulence is possible. According to Li (1954) who has conducted experiments with an oscillatory plate, the critical condition is $A_0/\delta = 280$, where A_0 is the orbital amplitude just outside the boundary layer. In nature the roughness at the sea bottom enhances turbulence. Jonsson (1966) has given the following empirical criteria for turbulence:

$$\mathsf{R}_E = \frac{\omega A_0^2}{\nu} > 1.26 \times 10^4 \quad \text{and} \quad \frac{A_0}{k_N} \geq \frac{4}{\pi}(2\,\mathsf{R}_E)^{1/2}$$

where R_E is the Reynolds number and k_N is the typical roughness height.[†] Assuming fully developed turbulence, Kajiura (1964) has proposed a semi–theoretical model. Specifically, an eddy viscosity ν_e is assumed to relate the stress to the velocity gradient within the bottom boundary layer, that is,

$$\frac{\tau_{xz}}{\rho} = \frac{\tau}{\rho} = \nu_e \frac{\partial u}{\partial z}, \tag{7.1}$$

where

$$\nu_e = \kappa |u_*| z. \tag{7.2}$$

The coefficient κ is the Kármán constant, which is approximately equal to 0.4 for steady boundary layers in clean water, while u_* is the friction velocity, which is formally related to the bottom stress τ^B by

$$\tau^B = \rho |u_*| u_*. \tag{7.3}$$

By definition, u_* depends on time also. When Eq. (7.1) is combined with the

[†]The subscript N is for Nikuradse whose experiments on roughness in steady turbulent flows are well known.

momentum equation

$$\frac{\partial u}{\partial t} = \frac{\partial U_I}{\partial t} + \frac{\partial}{\partial z}\left(\kappa\,|\,u_*\,|\,z\frac{\partial u}{\partial z}\right), \tag{7.4}$$

the mathematics of solving u becomes nonlinear. One simplification is to replace $|\,u_*\,|$ above by a representative constant. For example, Kajiura invokes the idea of equivalent linearization and introduces the constant $\tilde{u}_*$ such that

$$\tau^B = \rho\tilde{u}_* u_* \tag{7.5}$$

gives the same rate of dissipation as Eq. (7.3). Assuming u_* to be sinusoidal with amplitude $\hat{u}_*$, we can easily verify that

$$\tilde{u}_* = \frac{8}{3\pi}\hat{u}_*. \tag{7.6}$$

Alternatively, Grant (1977), and Grant and Madsen (1979a) assume

$$\tau^B = \rho\bar{u}_* u_*, \tag{7.7}$$

where $\bar{u}_*$ is defined by the average bottom stress $\bar{\tau}^B$ according to

$$\bar{u}_* = \left(\frac{\bar{\tau}^B}{\rho}\right)^{1/2}. \tag{7.8}$$

From Eq. (7.6) and the average of Eq. (7.3),

$$\bar{\tau}^B = \rho\,\overline{|\,u_*\,|\,u_*} = \tfrac{1}{2}\rho\hat{u}_*^2 = \tfrac{1}{2}\hat{\tau}^B, \tag{7.9}$$

where $\hat{\tau}^B$ denotes the amplitude of τ^B. It follows that

$$\sqrt{2}\,\bar{u}_* = \hat{u}_* = \frac{3\pi}{8}\tilde{u}_*. \tag{7.10}$$

We may now solve Eq. (7.4) for u subject to the boundary conditions

$$u \to U_I, \qquad z \sim \infty, \tag{7.11}$$

$$u = 0, \qquad z = z_0, \tag{7.12}$$

where z_0 denotes the effective position of the bottom and is an empirical constant which depends on the roughness. The solution for a simple harmonic wave is

$$u = \mathrm{Re}\left[U_0\bar{e}^{i\omega t}\left(1 - \frac{\ker 2\zeta^{1/2} - i\,\ker 2\zeta^{1/2}}{\ker 2\zeta_0^{1/2} - i\,\ker 2\zeta_0^{1/2}}\right)\right], \tag{7.13}$$

where

$$\zeta = \frac{z}{l}, \qquad \zeta_0 = \frac{z_0}{l}, \qquad l = \frac{\kappa \bar{u}_*}{\omega}, \qquad (7.14)$$

and $\mathrm{ker}(x)$ and $\mathrm{kei}(x)$ are the real and imaginary parts of $K_0(xe^{-i\pi/4})$, respectively. The above solution is essentially the same as Kajiura (1964) if $\bar{u}_*$ is replaced by $(3\pi/8(2)^{1/2})\bar{u}_*$ (Grant, 1977; Grant and Madsen, 1979a).

The friction velocity, so far unknown, may be found by combining Eqs. (7.1) and (7.2) and the original definition (7.3); the result is

$$u_* = \kappa z \left. \frac{\partial u}{\partial z} \right|_{z=0}. \qquad (7.15)$$

Using the approximation

$$\mathrm{ker}\, 2\zeta^{1/2} - i\,\mathrm{kei}\, 2\zeta^{1/2} = -\frac{1}{2}\ln\zeta - 0.5772 + \frac{i}{4} + O(\zeta\ln\zeta) \qquad (7.16)$$

for small ζ, we obtain

$$u_* = \mathrm{Re}\left[\frac{\kappa}{2} U_0 e^{-i\omega t}\left(\mathrm{ker}\, 2\zeta_0^{1/2} - i\,\mathrm{ker}\, 2\zeta_0^{1/2}\right)^{-1}\right], \qquad (7.17)$$

which defines both the magnitude and the phase of u_*. In particular, the magnitude is

$$\hat{u}_* = \sqrt{2}\,\bar{u}_* = \frac{\kappa U_0}{\left[\left(\mathrm{ker}\, 2\zeta_0^{1/2}\right)^2 + \left(\mathrm{kei}\, 2\zeta_0^{1/2}\right)^2\right]^{1/2}}. \qquad (7.18)$$

As soon as z_0 is prescribed, Eq. (7.18) gives $\bar{u}_*$ implicitly, after which u is completely known. For a natural sea bottom, however, it is not an easy matter to estimate z_0 because the bottom roughness depends not only on the sand grains but also on the ripples.

Extensive experiments on artificially roughened bottoms have been performed by Jonsson (1966) (see also Jonsson and Carlsen, 1976) and Horikawa and Watanabe (1968). These authors measured u as a function of x and t and calculated ν_e by

$$\nu_e = -\left[\int_\infty^z dz\, \frac{\partial}{\partial t}(u - U_I)\bigg/\frac{\partial u}{\partial z}\right]. \qquad (7.19)$$

At any fixed instant of the wave period the eddy viscosity was found to increase from zero to a maximum value and then to decrease again with height. At some instants, however, negative eddy viscosity was found over part of the height. The magnitude of ν_e could be as much as 100 cm^2/s, and 20–40 cm^2/s was the typical range. At a fixed z, ν_e oscillated in time at twice the fundamental frequency. Horikawa and Watanabe also found that the measured velocity typically contained higher harmonics. The eddy viscosity, calculated from Eq. (7.19) by using the first harmonic only, varied widely over a wave period, the typical range being -15 cm²/s to $+15$ cm²/s. The temporal dependence of ν_e is not yet well understood and does not strictly conform with the theory given here.

Grant (1977) and Grant and Madsen (1979a) have made a careful comparison of the theoretical (hence only first harmonic) velocity profile with the available experiments of Jonsson. With z_0 properly selected, they found Eq. (7.13) to give a fairly good prediction for the magnitude but not the phase of u. Furthermore, they have extended this type of theory to include currents (Grant and Madsen, 1979a, b).

8.7.2 The Friction Coefficient

A practical motivation for studying bottom stresses is to find the friction coefficient f_w defined by

$$\tau^B = \tfrac{1}{2} f_w \rho \, | \, U_I | \, U_I, \tag{7.20}$$

which may be used to find the effect on the wave field outside the boundary layer.

Taking the maximum of both sides of Eq. (7.20) and using Eq. (7.10), we get

$$\bar{u}_* = \tfrac{1}{2} f_w^{1/2} U_0 \tag{7.21}$$

so that Eq. (7.20) is also an implicit equation for f_w in terms of U_0. Grant (1977) lets

$$U_0 = A_0 \omega \quad \text{and} \quad z_0 = \frac{k_N}{30}$$

for fully rough turbulent flows, where A_0 is the inviscid orbital amplitude just above the boundary layer and k_N is the characteristic bed roughness height.[†] Upon further using the approximation (7.16) for small ζ_0, he has shown that

$$\frac{1}{4.06 f_w^{1/2}} + \log_{10} \frac{1}{4 f_w^{1/2}} = \log_{10} \frac{A_0}{k_N} - 0.325, \tag{7.22}$$

which is of the form given by Jonsson (1966) except for the last constant being -0.08, and implied by Kajiura (1964). Figure 7.1 shows the variations of f_w versus A_0/k_N according to Eq. (7.22) [Jonsson (1966) and Kajiura (1968) who

---

[†] It must be pointed out that in natural surroundings, k_N cannot be defined unequivocally.

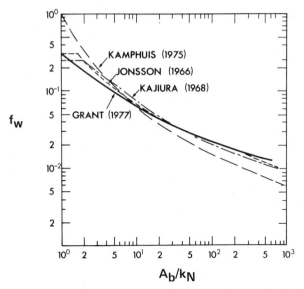

Figure 7.1 Comparison of wave friction factor formulas of Kamphuis (1975), Jonsson (1966), Kajiura (1967), and Grant (1977) (from Grant, 1977).

used a more complex mode of ν_e]. The empirical curve of Kamphuis (1975) is also shown.

The length scale l is clearly a characteristic length of the boundary layer. Grant and Madsen found that at the height $2l$ the velocity u reached 90% of the inviscid value U_l. Thus, they defined the turbulent boundary-layer thickness δ_t to be $2l$ which implied from Eqs. (7.14) and (7.21) that

$$\frac{\delta_t}{A_0} = 0.4 f_w^{1/2}. \tag{7.23}$$

This ratio varied from $0.04 \sim 0.4$ for $A_0/k_N = 10^3 \sim 1$.

8.7.3 Bottom Friction on the Damping of Standing Shallow-Water Waves in a Basin

As an application of the foregoing results, we consider the effect of friction on standing waves in shallow water.

Let us recall the continuity equation

$$\frac{\partial \zeta}{\partial t} + \nabla \cdot (\mathbf{u} h) = 0, \tag{7.24}$$

and add the bottom stress to the momentum equation

$$\frac{\partial \mathbf{u}}{\partial t} = -g\nabla\zeta - \frac{\tau^B}{\rho h}. \tag{7.25}$$

Multiplying Eq. (7.25) by $\mathbf{u}h$, we get

$$\frac{\partial}{\partial t}\left(\tfrac{1}{2}\mathbf{u}^2 h\right) = -g\nabla\cdot(\mathbf{u}h\zeta) + g\zeta\nabla\cdot(\mathbf{u}h) - \frac{\tau^B\cdot\mathbf{u}}{\rho}$$

$$= -g\nabla\cdot(\mathbf{u}h\zeta) - \frac{\partial}{\partial t}\left(\frac{g\zeta^2}{2}\right) - \frac{\tau^B\cdot\mathbf{u}}{\rho}$$

after using Eq. (7.24). Integrating over the entire basin area S and invoking the no-flux boundary condition, we get

$$\frac{\partial}{\partial t}\iint_S \left(\frac{\rho}{2}\mathbf{u}^2 h + \frac{\rho}{2}g\zeta^2\right)dx\,dy = -\iint_S \tau^B\cdot\mathbf{u}\,dx\,dy. \tag{7.26}$$

Since the integral on the left is the total energy E, Eq. (7.26) simply states that the energy decay rate is equal to the rate of work done at the bottom.

Using Eq. (7.20), invoking equivalent linearization and neglecting possible phase difference between τ^B and $\mathbf{u}$, we get

$$\tau^B = \frac{4\rho}{3\pi}\frac{f_w U_0}{h_0}\mathbf{u}h, \tag{7.27}$$

where $U_0 h/h_0$ is the amplitude of $\mathbf{u}$ which, in general, depends on $\mathbf{x}$. For estimation we shall simply redefine U_0 as a constant typical of $\mathbf{u}$. Substituting Eq. (7.27) into Eq. (7.26) and taking the time average over a wave period, we find

$$\frac{\partial \bar{E}}{\partial t} = \frac{4\rho}{3\pi}\frac{f_w U_0}{h_0}\iint_S \overline{\mathbf{u}^2}h\,dx\,dy = \frac{4}{3\pi}\frac{f_w U_0}{h_0}\bar{E}, \tag{7.28}$$

where we have assumed that the friction is small enough, so that the kinetic and potential energies are equal as in the inviscid limit (equipartition theorem, Appendix 8.A). Equation (7.28) implies that energy damps out according to

$$\bar{E}(t) = \bar{E}(0)e^{-\alpha t} \quad \text{with } \alpha = \frac{4}{3\pi}\frac{f_w U_0}{h_0}. \tag{7.29}$$

To have a quantitative idea of the damping rate we take $h = 10$ m, $U_0 = (A/h)(gh)^{1/2} \cong 0.1$ m/s and $f = 0.025$. It follows that $\alpha \sim 10^{-4}$ s^{-1}.

It is now opportune to assess the possible importance of bottom friction in harbor resonance. For the open bay of Section 5.6, the radiation decay rate can be estimated from Eq. (6.22), Section 5.6, to be $\omega a/L$ where a is the half-width and L the length of the bay. If we take $a = 100$ m, $L = 5000$ m, and $h = 10$ m, the lowest resonant wavenumber is $k_0 = \pi/2L$ so that $\omega = (gh)^{1/2}k_0 = (\pi/2L)(gh)^{1/2} = \pi \times 10^{-3}$ s^{-1}. The corresponding radiation damping rate is about $\pi/5 \times 10^{-4}$ s^{-1} which is comparable to the bottom friction. Thus, in practical computations of harbor oscillations, bottom friction may be of importance.

In principle, U_0 represents the local velocity amplitude and is unknown *a priori*. Therefore, to include turbulent bottom friction in a general wave problem such as harbor oscillations would involve nonlinear mathematics and massive computations. In practice, one does not have sufficient information of the bottom roughness to warrant such an effort; the coefficient $f_w U_0/h_0$ in Eq. (7.27) is often taken to be a constant instead.

Over a natural sea bottom there are other physical factors which may contribute to the damping of waves. An important contribution to the bottom roughness is the sand ripples formed naturally by waves. According to an experimental study by Vitale (1979), ripples may amount to 25–34 percent of the total friction on a natural seabed. A theoretical model which accounts for energy losses from vortex shedding at the sharp fixed ripple crests has been advanced by Longuet-Higgins (1981). Smith and McLean (1977) and Grant and Madsen (1982) have developed semiempirical theories which include both the form drag and the movement of the sand ripples. Another factor is the scattering by topographical irregularities with dimensions much larger than sand grains or ripples but much less than a wavelength. Theoretical estimates based on a random process model have been given by Long (1973). In addition, water percolates into the sea bottom and loses energy due to friction within the pores. This kind of loss is, however, difficult to estimate; various hypotheses regarding the solid skeleton (rigid, elastic, viscoelastic, porous or nonporous, etc.) have been proposed. Since the pertinence of the hypotheses must vary widely from one locale to another, we do not pursue it here.

APPENDIX 8.A: AN EQUIPARTITION THEOREM

For simple harmonic infinitesimal waves

$$\mathbf{u} = \nabla\phi e^{-i\omega t}, \tag{A.1}$$

the time-averaged kinetic energy is

$$\frac{1}{2}\iiint_V \rho\,\overline{u_i u_i}\,dV = \frac{1}{4}\iiint_V \rho(\nabla\phi\cdot\nabla\phi^*)\,dV. \tag{A.2}$$

The potential energy is

$$\frac{1}{2} \iint\limits_{S_F} \rho g \overline{\zeta^2} \, dS = \frac{\rho g}{4} \iint\limits_{S_F} |\eta|^2 \, dS = \frac{\rho}{4} \frac{\omega^2}{g} \iint\limits_{S_F} |\phi|^2 \, dS \qquad (A.3)$$

after the kinematic boundary condition is invoked. By virtue of an identity derived on p. 399 (see the equation preceding Eq. (4.23)), the right-hand-sides of Eqs. (A.2) and (A.3) are equal, implying that

$$\frac{1}{2} \rho \iiint\limits_{V} \overline{u_i u_i} \, dV = \frac{1}{2} \iint\limits_{S_F} \rho g \overline{\zeta^2} \, dS = \frac{1}{2} \overline{E}, \qquad (A.4)$$

Thus, the total energy is equally divided between kinetic and potential energies. This is an example of the equipartition theorem which is valid in many other conservative physical systems.

The reader should try to check the equipartition theorem in progressive waves by using the explicit formulas for u, v, and ζ.

NINE

Mass Transport Due to Viscosity

9.1 INTRODUCTION

In addition to pure fluctuations, waves can induce currents which do not change their directions during a time long in comparison with a wave period. Although the current velocity is often weak, its persistence can result in the transport of bottom sediments. In order to understand the changes of the sea bottom, it is necessary to acquire a good understanding of currents generated by waves.

Two different types of currents which owe their existence to wave fluctuations are of interest to us. In this chapter we examine the *mass transport* within the boundary layer near the sea bottom when the wave field above is essentially inviscid and irrotational. In the next chapter we shall examine the longshore current and its variations in and near the surf (breaking) zone of a gently sloping shore. In both cases the driving mechanism is the steady momentum flux due to the convective inertia of the waves, and dissipation plays a central role in maintaining the steady drift.

It was first discovered in acoustics that steady currents could be induced near a solid wall adjacent to an oscillating fluid. This phenomenon of *acoustic streaming* makes standing waves in tubes visible by the accumulation of dust particles at the nodes. Rayleigh (1883) first analyzed the phenomenon theoretically and determined Eulerian streaming in the boundary layer. Extension to water waves propagating in one direction was made by Longuet-Higgins (1953) who pointed out that Stokes drift had to be added to Eulerian streaming velocity in order to find the particle (Lagrangian) drift which is now known as the *mass transport velocity*. Hunt and Johns (1963) deduced formulas for the induced Eulerian streaming in two-dimensional waves and for Lagrangian drift at the outer edge of the boundary layer. Carter, Liu, and Mei (1973) gave Lagrangian drift throughout the boundary layer. While all of the theories above have been based on the assumption of constant viscosity, Longuet-Higgins (1958) and Johns (1970) have proposed theories of vertically varying eddy viscosity. Because of our incomplete knowledge of the turbulent boundary layer and because the variable eddy viscosity theory has not produced qualitatively different results, we shall only discuss the model of constant viscosity in this chapter.

9.2   MASS TRANSPORT NEAR THE SEA BOTTOM—GENERAL THEORY

It is convenient to adopt a local coordinate system (x, y, z) with x, y in the plane of the boundary layer and z pointing normally into the inviscid region as shown in Fig. 2.1. For small bottom slopes, z is nearly in the vertical direction. By continuity, the normal velocity component w is small so that w/u and $w/v = O(k\delta)$, where k is the typical wavenumber and δ is the boundary-layer thickness. Let A be the typical orbital amplitude near the bottom, then

$$\frac{\nu(\partial^2/\partial x^2 + \partial^2/\partial y^2)\begin{bmatrix} u \\ v \end{bmatrix}}{\dfrac{\partial}{\partial t}\begin{bmatrix} u \\ v \end{bmatrix}} \sim O(k\delta)^2, \qquad \frac{\left(u\dfrac{\partial}{\partial x} + v\dfrac{\partial}{\partial y}\right)\begin{bmatrix} u \\ v \end{bmatrix}}{\dfrac{\partial}{\partial t}\begin{bmatrix} u \\ v \end{bmatrix}} \sim O(kA).$$

As long as $1 \gg kA \gg (k\delta)^2$, Navier—Stokes equations can be approximated by

$$\frac{\partial u}{\partial x} + \frac{\partial v}{\partial y} + \frac{\partial w}{\partial z} = 0, \tag{2.1a}$$

$$\frac{\partial u}{\partial t} + u\frac{\partial u}{\partial x} + v\frac{\partial u}{\partial y} + w\frac{\partial u}{\partial z} = -\frac{1}{\rho}\frac{\partial P}{\partial x} + \nu\frac{\partial^2 u}{\partial z^2}, \tag{2.1b}$$

$$\frac{\partial v}{\partial t} + u\frac{\partial v}{\partial x} + v\frac{\partial v}{\partial y} + w\frac{\partial v}{\partial z} = -\frac{1}{\rho}\frac{\partial P}{\partial y} + \nu\frac{\partial^2 v}{\partial z^2}, \tag{2.1c}$$

$$0 = -\frac{1}{\rho}\frac{\partial P}{\partial z} - g. \tag{2.1d}$$

In the last equation of vertical momentum, we have ignored terms of the order

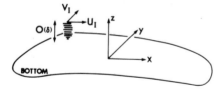

Figure 2.1   Local coordinates near the bottom.

$O(k\delta)$. Thus, $p = P - \rho gz$ does not vary in the direction normal to the boundary layer and must be the same as its value in the inviscid flow field just outside. Consequently,

$$-\frac{1}{\rho}\frac{\partial p}{\partial x} = \frac{\partial U_I}{\partial t} + U_I\frac{\partial U_I}{\partial x} + V_I\frac{\partial U_I}{\partial y}, \tag{2.2a}$$

$$-\frac{1}{\rho}\frac{\partial p}{\partial y} = \frac{\partial V_I}{\partial t} + U_I\frac{\partial V_I}{\partial x} + V_I\frac{\partial V_I}{\partial y}, \tag{2.2b}$$

where U_I and V_I are the tangential components of the inviscid velocity field at the wall, with W_I being zero. In this section U_I and V_I are assumed to be prescribed. Let the total velocity components in the boundary layer u and v be expanded as a perturbation series with kA as the implied small parameter:

$$u = u_1 + u_2 + \cdots \tag{2.3a}$$

$$v = v_1 + v_2 + \cdots \tag{2.3b}$$

where $u_1 = O(\omega A)$ and $u_2 = O(\omega kA^2)$, and so on. At the first order, Eqs. (2.1b) and (2.1c) give

$$\frac{\partial u_1}{\partial t} = \frac{\partial U_I}{\partial t} + \nu\frac{\partial^2 u_1}{\partial z^2}, \tag{2.4a}$$

$$\frac{\partial v_1}{\partial t} = \frac{\partial V_I}{\partial t} + \nu\frac{\partial^2 v_1}{\partial z^2}. \tag{2.4b}$$

The boundary conditions are

$$(u_1, v_1) = 0 \qquad \text{on } z = 0, \tag{2.5}$$

$$(u_1, v_1) \rightarrow U_I, V_I, \qquad \frac{z}{\delta} \gg 1. \tag{2.6}$$

For simple harmonic motion, we write

$$U_I(x, y, t) = \text{Re}\left[U_0(x, y)e^{-i\omega t}\right] \tag{2.7a}$$

$$V_I(x, y, t) = \text{Re}\left[V_0(x, y)e^{-i\omega t}\right]. \tag{2.7b}$$

Since x and y are just parameters in the boundary-value problem, u_1 and v_1 are again given by the Stokes solution:

$$\begin{pmatrix} u_1 \\ v_1 \end{pmatrix} = \text{Re}\left[\begin{pmatrix} U_0 \\ V_0 \end{pmatrix} F_1(\xi)e^{-i\omega t}\right], \tag{2.8}$$

where F_1 is given by $F_1 = 1 - e^{-(1-i)\xi}$, $\xi = z/\delta$. From the continuity equation, the normal component is found to be

$$w_1 = -\delta \int_0^\xi d\xi \left(\frac{\partial u_1}{\partial x} + \frac{\partial v_1}{\partial y} \right) = -\delta \operatorname{Re}\left\{ \left(\frac{\partial U_0}{\partial x} + \frac{\partial V_0}{\partial y} \right) e^{-i\omega t} \int_0^\xi F_1(\xi)\, d\xi \right\}$$

$$= \operatorname{Re}\left\{ \delta \left(\frac{\partial U_0}{\partial x} + \frac{\partial V_0}{\partial y} \right) e^{-i\omega t} \left[\frac{1+i}{2} [1 - e^{-(1-i)\xi}] - \xi \right] \right\}. \tag{2.9}$$

Although w_1 is of the order $O(k\delta)$ times u_1 and v_1, which is insignificant by itself, its effects through the terms $w_1 \partial u_1/\partial z$ and $w_1 \partial v_1/\partial z$ are as important as the other convection terms and cannot be overlooked. Now at the next order, the equations are

$$\frac{\partial u_2}{\partial t} - \nu \frac{\partial^2 u_2}{\partial z^2} = U_I \frac{\partial U_I}{\partial x} + V_I \frac{\partial U_I}{\partial y} - \left(u_1 \frac{\partial u_1}{\partial x} + v_1 \frac{\partial u_1}{\partial y} + w_1 \frac{\partial u_1}{\partial z} \right), \tag{2.10a}$$

$$\frac{\partial v_2}{\partial t} - \nu \frac{\partial^2 v_2}{\partial \dot{z}^2} = U_I \frac{\partial V_I}{\partial x} + V_I \frac{\partial V_I}{\partial y} - \left(u_1 \frac{\partial v_1}{\partial x} + v_1 \frac{\partial v_1}{\partial y} + w_1 \frac{\partial v_1}{\partial z} \right). \tag{2.10b}$$

Equations (2.10a) and (2.10b) can be used to obtain the complete second-order velocity field. Because the forcing terms on the right of Eqs. (2.10a) and (2.10b) contain zeroth and second harmonics, we must expect that u_2 and v_2 contain the same harmonics in response, that is,

$$u_2 = \overline{u_2} + \operatorname{Re} \tilde{u}_2 e^{-2i\omega t}, \qquad v_2 = \overline{v_2} + \operatorname{Re} \tilde{v}_2 e^{-2i\omega t}, \tag{2.11}$$

where $\overline{u}_2$ and $\overline{v}_2$ do not vary with time and are referred to as the *induced-streaming* velocities. In the terminology of electrical circuits, $\overline{u}_2$ and $\overline{v}_2$ are the DC (direct current) as opposed to the time-varying AC (alternating current) components. While the AC parts $\tilde{u}_2$ and $\tilde{v}_2$ modify the oscillatory velocity field, their influence is overshadowed by u_1 and v_1. The components $\overline{u}_2$ and $\overline{v}_2$ are new, however, and are the leading contribution to the steady motion. Although small, they are responsible for the steady drifting of fluid particles after many periods. Let us concentrate on these mean values by taking the averages of (2.10a) and (2.10b):

$$-\nu \frac{\partial^2 \overline{u}_2}{\partial z^2} = \overline{U_I \frac{\partial U_I}{\partial x}} + \overline{V_I \frac{\partial U_I}{\partial y}} \ominus \left(\overline{u_1 \frac{\partial u_1}{\partial x}} + \overline{v_1 \frac{\partial u_1}{\partial y}} + \overline{w_1 \frac{\partial u_1}{\partial z}} \right), \tag{2.11a}$$

$$-\nu \frac{\partial^2 \overline{v}_2}{\partial z^2} = \overline{U_I \frac{\partial V_I}{\partial x}} + \overline{V_I \frac{\partial V_I}{\partial y}} - \left(\overline{u_1 \frac{\partial v_1}{\partial x}} + \overline{v_1 \frac{\partial v_1}{\partial y}} + \overline{w_1 \frac{\partial v_1}{\partial z}} \right). \tag{2.11b}$$

Before proceeding further let us use the continuity equation for $\mathbf{u}_1$ to rewrite

$$u_{1i}\frac{\partial}{\partial x_i}u_{1j} = \frac{\partial}{\partial x_i}(u_{1i}u_{1j}), \qquad i = 1,2,3,$$

thus

$$-\nu\frac{\partial^2 \bar{u}_2}{\partial z^2} = \overline{U_I\frac{\partial U_I}{\partial x}} + \overline{V_I\frac{\partial U_I}{\partial y}} - \left(\frac{\partial}{\partial x}\overline{u_1u_1} + \frac{\partial}{\partial y}\overline{u_1v_1} + \frac{\partial}{\partial z}\overline{u_1w_1}\right), \quad (2.12a)$$

$$-\nu\frac{\partial^2 \bar{v}_2}{\partial z^2} = \overline{U_I\frac{\partial V_I}{\partial x}} + \overline{V_I\frac{\partial V_I}{\partial y}} - \left(\frac{\partial}{\partial x}\overline{u_1v_1} + \frac{\partial}{\partial y}\overline{v_1v_1} + \frac{\partial}{\partial z}\overline{v_1w_1}\right). \quad (2.12b)$$

The terms $\overline{u_1u_1}$, $\overline{u_1v_1}$, and so forth, are just the components of the Reynolds stress tensor representing momentum fluxes due to wave fluctuations. The physical picture is now clear, that is, the mean streaming current $(\bar{u}_2, \bar{v}_2)$ arises because a mean shear stress field must be present in order to balance the mean dynamic pressure field and the Reynolds stress field. Equations (2.12a) and (2.12b) may be integrated straightforwardly with the boundary conditions that there be no velocity at the wall and no stress at the outer edge of the boundary layer,

$$(\bar{u}_2, \bar{v}_2) = 0, \qquad z = 0; \qquad (2.13a)$$

$$\frac{\partial}{\partial z}(\bar{u}_2, \bar{v}_2) \to 0, \qquad \frac{z}{\delta} \to \infty. \qquad (2.13b)$$

After some algebra, the following results are obtained:

$$\bar{u} = \bar{u}_2 = -\frac{1}{\omega}\text{Re}\left[F_2 U_0\frac{\partial U_0^*}{\partial x} + F_3 V_0\frac{\partial U_0^*}{\partial y} + F_4 U_0\frac{\partial V_0^*}{\partial y}\right], \qquad (2.14a)$$

$$\bar{v} = \bar{v}_2 = -\frac{1}{\omega}\text{Re}\left[F_2 V_0\frac{\partial V_0^*}{\partial y} + F_3 U_0\frac{\partial V_0^*}{\partial x} + F_4 V_0\frac{\partial U_0^*}{\partial x}\right], \qquad (2.14b)$$

where U_0 and V_0 are the amplitudes of U_I and V_I as defined in Eq. (2.7), and

$$F_2 = -\frac{1}{2}(1 - 3i)e^{(-1+i)\xi} - \frac{i}{2}e^{-(1+i)\xi} - \frac{1+i}{4}e^{-2\xi}$$
$$+ \frac{1}{2}(1 + i)\xi e^{(-1+i)\xi} + \frac{3}{4}(1 - i), \qquad (2.15a)$$

$$F_3 = \frac{1}{2}ie^{(-1+i)\xi} - \frac{i}{2}e^{-(1+i)\xi} - \frac{1}{4}e^{-2\xi} + \frac{1}{4}, \qquad (2.15b)$$

$$F_4 = -\frac{1}{2}(1 - 2i)e^{(-1+i)\xi} + \frac{1+i}{2}\xi e^{-(1-i)\xi}$$
$$- \frac{i}{4}e^{-2\xi} + \frac{1}{4}(2 - 3i). \qquad (2.15c)$$

The present two-dimensional formulas were first derived by Hunt and Johns (1963).

The results just given are Eulerian streaming velocities. To infer the motion of a marked fluid particle, it is necessary to calculate Lagrangian velocity. Let $x(x_0, t)$ be the position of the particle which was at point x_0 when $t = t_0$, namely,

$$x(x_0, t_0) = x_0.$$

At any subsequent time t, the position and velocity of the particle are related by

$$\frac{dx}{dt} \equiv u_L(x_0, t) \tag{2.16}$$

where u_L denotes Lagrangian velocity of the particle. When a particle arrives at x at the time t, its Lagrangian velocity must be the same as its Eulerian velocity $u(x, t)$, that is,

$$u_L(x_0, t) = u(x, t).$$

From Eq. (2.16), the particle position at time t is

$$x = x_0 + \int_{t_0}^{t} dt' u_L(x_0, t'),$$

so that

$$u_L(x_0, t) = u\left[x_0 + \int_{t_0}^{t} dt' u_L(x_0, t'), t\right].$$

For sufficiently small $t - t_0$ (say within a few periods), the integral, being the particle displacement, is of the order $O(A)$ and is comparable to the orbital size. Hence, we may expand u as a Taylor series:

$$u_L(x_0, t) \simeq u(x_0, t) + (x - x_0) \cdot \nabla_0 u(x_0, t) + \cdots$$

$$\simeq u(x_0, t) + \left[\int_{t_0}^{t} dt' u_L(x_0, t')\right] \cdot \nabla_0 u(x_0, t) + \cdots,$$

where $\nabla_0 = (\partial/\partial x_0, \partial/\partial y_0, \partial/\partial z_0)$. Clearly, to the first order u_L and u are the same so that we may replace u_L by u in the integral with second-order accuracy. The streaming velocity is

$$\bar{u}_L(x_0, t) = \bar{u}(x_0, t) + \overline{\left[\int_{t_0}^{t} dt' u(x_0, t')\right] \cdot \nabla_0 u(x_0, t)} + \cdots \tag{2.17}$$

Thus, for sufficiently small time, the particle velocity is equal to its initial velocity plus a correction due to the fact that the particle moves in an environment where the velocity field varies. The correction is proportional to the distance traveled and to the spatial rate of change of the local velocity field. Now $\bar{\mathbf{u}} = \bar{\mathbf{u}}_2$ and the quadratic terms are both of second order; hence, in the second term $\mathbf{u}$ may be approximated by $\mathbf{u}_1$. Finally, since all spatial variables are $\mathbf{x}_0$, the subscripts $(\;)_0$ may be dropped. The components of $\bar{\mathbf{u}}_L$ are then

$$\bar{u}_L = \bar{u}_2 + \overline{\left(\int^t u_1\, dt'\right)\frac{\partial u_1}{\partial x}} + \overline{\left(\int^t v_1\, dt'\right)\frac{\partial u_1}{\partial y}} + \overline{\left(\int^t w_1\, dt'\right)\frac{\partial u_1}{\partial z}},$$

(2.18a)

$$\bar{v}_L = \bar{v}_2 + \overline{\left(\int^t u_1\, dt'\right)\frac{\partial v_1}{\partial x}} + \overline{\left(\int^t v_1\, dt'\right)\frac{\partial v_1}{\partial y}} + \overline{\left(\int^t w_1\, dt'\right)\frac{\partial v_1}{\partial z}};$$

(2.18b)

the lower limits of integration are immaterial. Equations (2.18a) and (2.18b) give the components of the *mass transport velocity* of a marked fluid particle which is located at (x, y, z) at time t.

The difference between Lagrangian and Eulerian drifts is represented by the integral terms above, or in vector form, by

$$\overline{\left[\int^t \mathbf{u}_1(\mathbf{x}, t')\, dt'\right] \cdot \nabla \mathbf{u}_1(\mathbf{x}, t)},$$ ' (2.19)

which is known as *Stokes' drift*.

With the known expressions of u_1, v_1, and w_1 in Eqs. (2.8) and (2.9), various terms in Stokes' drift can be evaluated

$$\overline{\left(\int^t u_1\, dt'\right)\frac{\partial u_1}{\partial x}} = \mathrm{Re}\,\frac{i}{2\omega}\left\{U_0\frac{\partial U_0^*}{\partial x}[1 - e^{-(1-i)\xi} - e^{-(1+i)\xi} + e^{-2\xi}]\right\},$$

$$\overline{\left(\int^t v_1\, dt'\right)\frac{\partial u_1}{\partial y}} = \mathrm{Re}\,\frac{i}{2\omega}\left\{V_0\frac{\partial U_0^*}{\partial y}[1 - e^{-(1-i)\xi} - e^{-(1+i)\xi} + e^{-2\xi}]\right\},$$

$$\overline{\left(\int^t w_1\, dt'\right)\frac{\partial u_1}{\partial z}} = \mathrm{Re}\,\frac{1}{2\omega}\left\{\left[U_0\frac{\partial U_0^*}{\partial x} + U_0\frac{\partial V_0^*}{\partial y}\right]\right.$$

$$\left.\times\left[e^{-2\xi} - e^{-(1-i)\xi} + (1+i)\xi e^{-(1-i)\xi}\right]\right\}.$$

The other terms may be obtained by interchanging u_1 and v_1. Combining

Stokes' with Eulerian mean drift, we finally obtain Lagrangian mass transport velocities:

$$\bar{u}_L = \frac{1}{4\omega} \, \mathrm{Re}\left\{ F_5 U_0 \frac{\partial U_0^*}{\partial x} + F_6 V_0 \frac{\partial U_0^*}{\partial y} + F_7 U_0 \frac{\partial V_0^*}{\partial y} \right\}, \qquad (2.20a)$$

$$\bar{v}_L = \frac{1}{4\omega} \, \mathrm{Re}\left\{ F_5 V_0 \frac{\partial V_0^*}{\partial y} + F_6 U_0 \frac{\partial V_0^*}{\partial x} + F_7 V_0 \frac{\partial U_0^*}{\partial x} \right\}, \qquad (2.20b)$$

where

$$F_5 = -8ie^{-(1-i)\xi} + 3(1+i)e^{-2\xi} - 3 + 5i, \qquad (2.21a)$$

$$F_6 = -4ie^{-(1-i)\xi} + (1+2i)e^{-2\xi} - 1 + 2i, \qquad (2.21b)$$

$$F_7 = -4ie^{-(1-i)\xi} + (2+i)e^{-2\xi} - 2 + 3i. \qquad (2.21c)$$

These formulas were derived by Carter, Liu, and Mei (1973) from Hunt and Johns (1963). In particular, at the outer edge of the boundary layer

$$F_5 \to -3 + 5i, \qquad F_6 \to -1 + 2i, \qquad F_7 \to -2 + 3i,$$

so that the mass transport velocity at $z \gg \delta$ is

$$\bar{u}_L = -\frac{1}{4\omega} \, \mathrm{Re}\left[3U_0 \frac{\partial U_0^*}{\partial x} + V_0 \frac{\partial V_0^*}{\partial y} + 2U_0 \frac{\partial V_0^*}{\partial y} \right.$$
$$\left. -i\left(5U_0 \frac{\partial U^*}{\partial x} + 3U_0 \frac{\partial V^*}{\partial y} + 2V_0 \frac{\partial U_0^*}{\partial y} \right) \right], \qquad (2.22a)$$

$$\bar{v}_L = -\frac{1}{4\omega} \, \mathrm{Re}\left[3V_0 \frac{\partial V_0^*}{\partial y} + U_0 \frac{\partial V_0^*}{\partial y} + 2V_0 \frac{\partial U_0^*}{\partial x} \right.$$
$$\left. -i\left(5V_0 \frac{\partial U_0^*}{\partial y} + 3V_0 \frac{\partial U_0^*}{\partial x} + 2U_0 \frac{\partial V_0^*}{\partial x} \right) \right] \qquad (2.22b)$$

(Hunt and Johns, 1963). The one-dimensional version for $u \neq 0$ and $v = 0$ was first given by Longuet-Higgins (1953). Thus far, Eqs. (2.20) and (2.21) are applicable to any oscillatory flows, water waves being a special case.

Although the mass transport velocities owe their existence to viscosity, their limits at the outer edge of the boundary layer do not depend on ν. Also, as soon as the inviscid wave field is known, it is, in principle, a straightforward matter of numerical differentiation to obtain the mass transport in the boundary layer. We now study two special cases and their possible qualitative implications on sediment transport.

9.3 BOTTOM MASS TRANSPORT UNDER A LONG CRESTED WAVE

Consider a wave system described by

$$\zeta = \operatorname{Re} A[e^{-i(kx+\omega t)} + Re^{-i(kx-\omega t)}], \tag{3.1}$$

which represents the sum of an incident wave and a reflected wave, R being the reflection coefficient. Without loss of generality we may take A and R to be real since the phase of a complex R may be eliminated by redefining x. The corresponding velocity potential is

$$\Phi = \operatorname{Re}\left[-\frac{igA}{\omega}\frac{\cosh k(z+h)}{\cosh kh}(e^{-ikx} + Re^{ikx})e^{-i\omega t}\right]. \tag{3.2}$$

This potential includes the standing waves as a special case ($R = 1$) and may also be applied to shoaling waves on a mildly sloping beach if R is taken to be zero and kx is replaced by $\int k\,dx$. At the sea bottom $z = -h$, the inviscid tangential velocity is

$$U_I = \operatorname{Re} U_0 e^{-i\omega t} = \operatorname{Re}[\Phi_x(x, -h, t)],$$

$$= \operatorname{Re}\left[\frac{-\omega A}{\sinh kh}(e^{-ikx} - Re^{ikx})e^{-i\omega t}\right], \qquad V_I = 0 \tag{3.3}$$

hence,

$$U_0 = \frac{-\omega A}{\sinh kh}(e^{-ikx} - Re^{ikx}), \qquad V_0 = 0. \tag{3.4}$$

Lagrangian mass transport in the boundary layer follows from Eqs. (2.20a) and (2.21a)

$$\bar{u}_L = \operatorname{Re}\left(\frac{1}{4\omega}F_5 U_0 \frac{\partial U_0^*}{\partial x}\right)$$

$$= \frac{k\omega A^2}{4\sinh^2 kh}\big[(1 - R^2)(8e^{-\xi}\cos\xi - 3e^{-\xi} - 5)$$

$$+ 2R\sin 2kx(8e^{-\xi}\sin\xi + 3e^{-2\xi} - 3)\big], \tag{3.5}$$

which is due to Longuet-Higgins (1953). For several values of R the profiles of $\bar{u}_L$ are shown in Figs. 3.1a–3.1c.

The special case of a purely progressive wave corresponds to $R = 0$:

$$\bar{u}_L = \frac{k\omega A^2}{4\sinh^2 kh}(8e^{-\xi}\cos\xi - 3e^{-\xi} - 5), \tag{3.6}$$

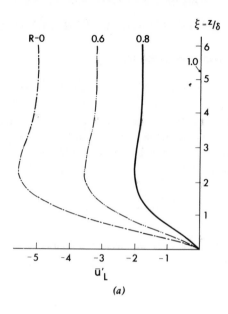

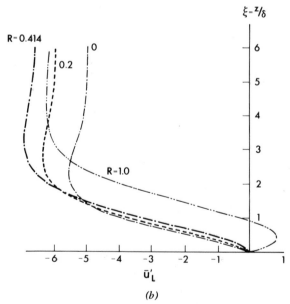

Figure 3.1 Mass transport profile within Stokes boundary layer near the bottom for various instants and reflection coefficients. Envelope maxima are at $2kx = 0$ and 2π and envelope minima at $2kx = \pi$. Dimensionless $\bar{u}'_L = \bar{u}_L(4\sinh^2 kh/\omega kA^2)$. (a) $2kx = 0$; (b) $2kx = \pi$; (c) $2kx = \frac{3}{2}\pi$. (From Carter et al., 1973, *J. Waterway, Port, Coastal and Ocean Div.* Reproduced by permission of American Society of Civil Engineers.)

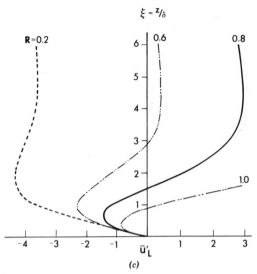

$\xi = z/\delta$

(c)

Figure 3.1 (*Continued*).

which varies monotonically from zero at $\xi = 0$ to

$$\bar{u}_L(\infty) = -\frac{5k\omega A^2}{4 \sinh^2 kh} \qquad \text{as } \xi \to \infty. \tag{3.7}$$

On the other hand, in front of a long and straight sea wall the reflection is nearly complete so that $R = 1$. In this case

$$\bar{u}_L = \frac{k\omega A^2}{4 \sinh^2 kh} 2 \sin 2kx \left(8e^{-\xi}\sin \xi + 3e^{-2\xi} - 3\right), \tag{3.8}$$

where the vertical variation described by the parenthesis changes sign at $\xi \approx 1$, see Fig. 3.1*b*. The sign of $\bar{u}_L$ now depends on kx and ξ. In Fig. 3.2 we sketch the free surface of a standing wave with nodes at $kx = \frac{1}{2}\pi, \frac{3}{2}\pi, \ldots$ and antinodes at $kx = 0, \pi, 2\pi, \ldots$ Above $\xi \approx 1$, that is, in the top part of the boundary layer, the mass transport always converges toward the antinodes and diverges away from the nodes, while below $\xi \approx 1$, that is, in the bottom part of the boundary layer, the reverse is true. From the curves for $R = 1$ in Fig. 3.1*b* the net mass flux above the level $\xi = 1$ certainly exceeds that below, implying that the streak lines must be as sketched in Fig. 3.2*d*. Near the antinodes some fluid sinks below $\xi \approx 1$ to form closed circulating cells within the boundary layer, while other fluid escapes upward. A more complete picture involves the introduction of a thicker boundary layer which is caused by the convective inertia of the streaming velocity but will not be discussed here (Mei, Liu, and Carter, 1972; Dore, 1976).

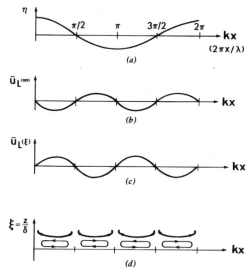

Figure 3.2 Schematic variation of mass transport velocity beneath a standing wave.

The mass transport velocity near the bottom ($\xi \approx 0$) is likely to be influential for the motion of heavy sediments which roll on the bottom. Near the bottom, $\bar{u}_L$ is in the direction of the incident wave for a purely progressive wave ($R = 0$) and is alternately along and opposite to the incident wave for a completely reflected wave ($R = 1$). It is therefore interesting to examine the effect of intermediate R. Reversal of sign in $\bar{u}_L(x, \xi)$ near the bottom requires that the profile of $\bar{u}_L$ has a slope greater than zero, that is,

$$\frac{\partial \bar{u}_L}{\partial \xi}(x, 0) > 0, \qquad \xi = 0. \tag{3.9}$$

Because of Eq. (3.5), Eq. (3.9) implies

$$R^2 + 2R \sin kx - 1 > 0.$$

For a fixed x the critical value R_c is

$$R_c = -\sin 2kx + (1 + \sin^2 2kx)^{1/2}. \tag{3.10}$$

If $R > R_c$, reversal of direction occurs. Since $0 < R < 1$, an acceptable solution of R_c can occur only when $\sin 2kx > 0$ or $0 < 2kx < \pi$. Accordingly, the variation of R_c versus $2kx$ is shown in Fig. 3.3. The lowest R_c is $2^{1/2} - 1 = 0.414$ which occurs at $2kx = \frac{1}{2}\pi$ or $x/\lambda = \frac{1}{4}$. For any $0.414 < R < 1$ reversal of $\bar{u}_L$ occurs in the range of $x_1 < x < x_2$ where $R = R_c$ at $x = x_1$ and x_2. The corresponding vertical variations of $\bar{u}_L(x, \xi)$ can be seen in Figs. 3.1a–3.1c.

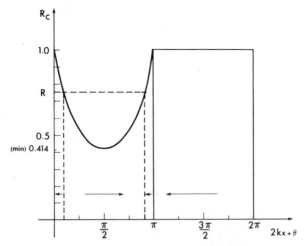

Figure 3.3 Critical reflection coefficient for reversal of bottom mass transport velocity. (From Carter, Liu, and Mei, 1973, *J. Waterway, Port, Coastal and Ocean Div.* Reproduced by permission of American Society of Civil Engineers.)

In the case of partial reflection $0 < R < 1$, the total wave can be viewed as a progressive wave with a modulated envelope

$$\zeta = \mathrm{Re}\{[A(1 + Re^{-2ikx})]e^{-i(kx+\omega t)}\}$$

$$= \mathrm{Re}\, \mathcal{Q}(x)e^{-i(kx+\omega t - \theta_{\mathcal{Q}}(x))} \qquad (3.11)$$

where $\theta_{\mathcal{Q}}$ is the phase of the square bracket above and

$$\mathcal{Q} = A[(1 + R^2) + 2R\cos 2kx]^{1/2}. \qquad (3.12)$$

The amplitude $\mathcal{Q}$ of the envelope is the greatest $[= A(1 + R)]$ at $2kx = 0, \pm 2\pi, \pm 4\pi, \ldots$ and the smallest $[= A(1 - R)]$ at $2kx = \pm\pi, \pm 3\pi, \pm 5\pi, \ldots$ Thus, if $R > 0.414$, reversal of $\bar{u}_L$ occurs within the quarter-wavelength in front of every peak $\mathcal{Q}_{\max}$ of the envelope.

The mass transport near the bottom of the boundary layer should be particularly relevant to heavy particles which roll on the bottom as *bed load*. On many beaches of the world the sand size distribution is commonly in the range from 0.05 mm to 10 mm (pebbles). As pointed out in Section 8.7 the eddy viscosity is typically 10–100 cm^2/s for a hydrodynamically rough bottom. The scale of the boundary-layer thickness $\delta = (2\nu/\omega)^{1/2}$ can be estimated to be roughly 6–18 cm for a wave of period 10s and is larger than most sand grains. Once the sand grains are mobilized by the first-order oscillatory velocity of the nearby fluid, the heavy ones will likely move, on the average, in the same general direction as the mass transport velocity near the bottom; the magnitude of the velocity of the sand particles themselves is so far too difficult to predict. For beaches of mild slope and for normally incident waves of short periods, reflection is small and the direction of net sand movement should be onshore. If reflection is large as in the vicinity of a sea wall, a cliff, near a beach of steep slope, or in the presence of a large offshore sand bar, $\bar{u}_L$ can be opposite in direction to the incident waves and can cause deposition on the sea bottom along parallel lines spaced at half-wavelengths apart. Dunes parallel to the coasts can, therefore, be formed at the same spacings. Such periodic dunes can be found on the coast of Florida (see Plate 1, from Lau and Travis, 1973).

In the case of large reflection ($R > 0.414$) the sign change of $\bar{u}_L$ should also have consequences on *sediment sorting*. Because fine particles are more likely to

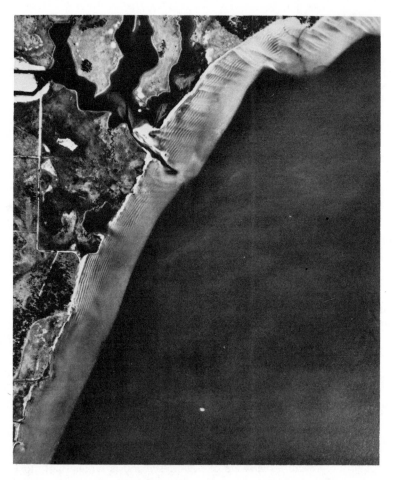

Plate 1   Aerial view of submarine longshore bars Escambia Bay, Florida (From Lau and Travis, 1973, *J. Geophys. Res.* Reproduced by permission of American Geophysical Union.)

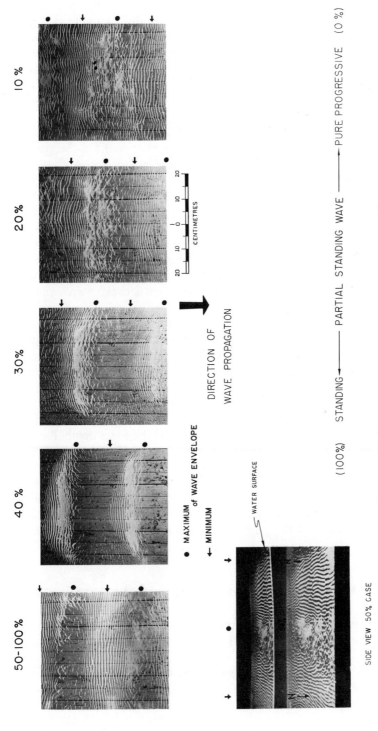

Plate 2 Sediment morphology for variable beach reflectivity. (From Carter, Liu, and Mei, 1973, *J. Waterway, Port, Coastal and Ocean Div.* Reproduced by permission of American Society of Civil Engineers.)

be transported in suspension, they should be influenced by the value of $\bar{u}_L$ near the top of the boundary layer. Therefore, they may drift in opposite directions to the heavier particles rolling beneath. In a standing wave the light particles should be deposited around the antinodes and the heavy particles around the nodes no matter what the initial distribution of sediment size is. This *sorting* has been observed in the laboratory by Noda (1968) De Best and Bijker (1971), and Carter, Liu, and Mei (1973). In nature, finer sand is found near dune troughs and coarser sand near dune crests (Inman, 1957).

Bagnold (1946) and Herbich, Murphy, and Van Weele (1965) experimented with sand beds under a standing wave. Dunes were found to be present at precisely the intervals of half of the length of the surface waves. Carter, et al. (1973) performed experiments for partially standing waves over a solid bottom sprinkled sparcely with sand; they found the same initial tendency of scour and deposition on a bottom covered with a thick layer of sand. For $R > 0.50$, the tendency toward large-scale depositions became visible near the peaks of the wave envelope (Plate 2), thereby supporting the relevance of the present theory to the initial development of a flat sandy bottom. Once grown to a sufficient size, the dunes will alter the wave field; the conclusions of this section are no longer appropriate. The interaction between waves and dunes has not received much study.

As seen in Plate 2, ripples with wavelengths much less than the surface wavelength are a prominant feature of a sandy bottom. The mechanics of their creation is not well understood except that they are most likely the results of an instability mechanism caused by the first-order oscillatory fluid velocity near the bottom. Observations by Inman (1957), Zenkovich (1967), and others show that the ripple dimensions, amplitude, and wavelength depend strongly on the fluid orbital velocity (hence, the depth of water) and the grain size. Generally, the coarser the grain and/or the greater the water depth, the steeper the ripple profiles. In shallow water the ripples are flat-troughed as a series of solitary waves. Furthermore, the coarser sand is more abundant near the ripple crests while the finer sand is found near the trough. While the formation of sand ripples in waves awaits further research, steady streaming in an oscillating flow over a fixed rippled wall has been studied both theoretically and experimentally by Kaneko and Honji (1979) who found two circulation cells stacking above a ripple wave.

9.4   BOTTOM MASS TRANSPORT NEAR A SMALL STRUCTURE

Since the simple wave system in Section 9.3 already reveals many interesting features in the mass transport and its possible implications on the sand-bed configuration, it is now tempting to work out some cases which correspond to more complex wave systems. In principle, this is a straightforward but largely numerical problem; once the inviscid wave field (U_0, V_0) is known the mass transport is obtained from Eqs. (2.20a) and (2.20b) by differentiation. For illustration we shall study in this section only a three-dimensional problem where analytical information is possible (Lamoure and Mei, 1977).

There are many coastal structures, man-made or natural, whose governing horizontal dimension a, say, is much less than the prevailing wavelength $(\sim 1/k)$. The head of a breakwater and the piles of an oil drilling rig are such instances. Near these bodies where the local curvature is large, eddy formation or flow separation can exist when the wave amplitudes are sufficiently large. The corresponding flow field is impossible to deduce purely theoretically. However, for low-amplitude swells which are present most of the time, separation can be insignificant near the sea bottom and a potential theory may be applied to the waves. The assumption of small structure $ka \ll 1$ then simplifies the calculation of waves near the body, as demonstrated in Section 7.4 for a vertical pile of circular cross section. In addition, we shall assume for simplicity that all the solid lateral boundaries are vertical and the sea bottom is horizontal. The inviscid potential may be related to the free-surface displacement η by

$$\phi = -\frac{ig\eta(x, y)}{\omega} \frac{\cosh k(z + h)}{\cosh kh} e^{-i\omega t}, \qquad (4.1)$$

where η is governed by the Helmholtz equation. In the near field of the structure, that is, in the neighborhood within the radius of $O(a)$, the proper horizontal length scale of motion is a, so that the horizontal Laplacian in the Helmholtz equation is much more important

$$\frac{k^2\eta}{\nabla^2 n} \sim O(ka)^2 \ll 1 \qquad \text{where } \nabla = \left(\frac{\partial}{\partial x}, \frac{\partial}{\partial y}\right).$$

It follows that

$$\nabla^2\eta \simeq 0. \qquad (4.2)$$

Thus, the inviscid flow in the near field is harmonic in the horizontal plane, and many solutions in classical hydrodynamics which are solved by methods of conformal mapping can be employed here for analytical purposes.

The solution in the near field is indeterminate by a complex constant factor, which must be found by matching with the solution in the *far field* about a wavelength away from the structure. The details of the matching process are unimportant for present purposes. Since the complex constant factor is common to both U_0 and V_0, the phases of U_0 and V_0 cancel out in Eqs. (2.20a) and (2.20b).[†] Hence we can discuss the variation of mass transport by taking U_0

[†] These formulas do not hold within the distance of $O(\delta)$ from the vertical wall.

and V_0 to be real without any loss of generality, that is,

$$\bar{u}_L = \frac{1}{4\omega} \operatorname{Re}\left[U_0 \frac{\partial U_0}{\partial x} (F_5 - F_7) + F_6 V_0 \frac{\partial U_0}{\partial y} \right],$$

$$\bar{v}_L = \frac{1}{4\omega} \operatorname{Re}\left[F_6 U_0 \frac{\partial V_0}{\partial x} + (F_5 - F_7) V_0 \frac{\partial V_0}{\partial y} \right],$$

where

$$(U_0, V_0) = -\frac{ig\nabla\eta}{\omega} \operatorname{sech} kh . \tag{4.3}$$

Since $F_5 - F_7 = F_6$ from Eq. (2.21), we have

$$\bar{u}_L = \frac{1}{4\omega} \operatorname{Re}\left[F_6 \left(U_0 \frac{\partial U_0}{\partial x} + V_0 \frac{\partial V_0}{\partial y} \right) \right], \tag{4.4a}$$

$$\bar{v}_L = \frac{1}{4\omega} \operatorname{Re}\left[F_6 \left(U_0 \frac{\partial V_0}{\partial x} + V_0 \frac{\partial V_0}{\partial y} \right) \right], \tag{4.4b}$$

or, in vector form,

$$\bar{\mathbf{u}}_L = \frac{1}{4\omega} \operatorname{Re} F_6 \mathbf{U}_0 \cdot \nabla \mathbf{U}_0$$

$$= \frac{1}{4\omega} \nabla \frac{|\mathbf{U}_0|^2}{2} (4e^{-\xi}\sin\xi + e^{-2\xi} - 1). \tag{4.5}$$

Note that the strength of the mass transport velocity is proportional to the horizontal gradient of the inviscid velocity. Furthermore, the factor for vertical variation inside the boundary layer is

$$\operatorname{Re} F_6 = 4e^{-\xi}\sin\xi + e^{-2\xi} - 1 \simeq 2\xi \qquad \text{for } \xi \ll 1,$$
$$\simeq -1 \qquad\qquad\qquad \text{for } \xi \gg 1.$$

Thus, there is a sign reversal within the boundary layer. Very close to the bottom

$$\bar{\mathbf{u}}_L \simeq \frac{\xi}{4\omega} \nabla |\mathbf{U}_0|^2, \qquad \xi \ll 1, \tag{4.6}$$

so that the mass transport is in the direction of increasing $|\mathbf{U}_0|^2$. In particular, fluid must drift toward the convex corner of a vertical structure. On the other

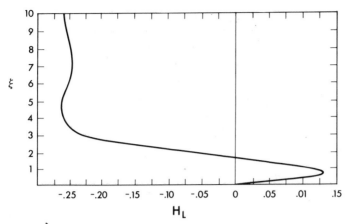

Figure 4.1 Vertical variation of Lagrangian mass transport in gravity waves near a small body; $\xi = z/\delta$. (From Lamoure and Mei, 1977, *J. Fluid Mechs.* Reproduced by permission of Cambridge University Press.)

hand, near the top of the boundary layer

$$\bar{\mathbf{u}}_L \simeq \frac{-1}{8\omega}\, \nabla\,|\,\mathbf{U}_0\,|^2, \tag{4.7}$$

so that the mass transport points away from the convex corner. The variation of Re F_6 for all ξ is plotted in Fig. 4.1.

The above simple conclusion suggests that heavy sediments, once mobilized to roll on the bottom, may be attracted toward the neighborhood of high velocity. More specifically, for a pile in the presence of waves from $x \sim -\infty$ ($\theta = \pi$) deposition is the largest near the two extremities on the two sides of the x axis, as shown in Fig. 4.2. By similar reasoning, heavy sediments would tend to deposit near the tip of a breakwater. Deposition would also be pronounced at the neck of a narrow harbor entrance, tending to close the harbor.

Why is $\bar{\mathbf{u}}_L$ parallel to $\nabla^2\mathbf{U}_0^2$? Consider a vertical column of fluid of unit square cross section with its base somewhere inside the boundary layer and its top at the outer edge. The mean shear stress acting at the base is $\mu\partial\bar{u}/\partial z$. The net normal stresses acting on a unit area of the vertical sides are the mean pressure gradient $\frac{1}{2}\rho\nabla\overline{U_I^2}$ and the gradient of the normal Reynolds stress $-(1/\rho)\rho\nabla\overline{\mathbf{u}^2}$. Because $w \cong 0$, the shear components of the Reynolds stress are negligible. Equating the forces, we have

$$\mu\frac{\partial\bar{u}}{\partial z} = \frac{\rho}{2}\int_0^\infty \overline{\nabla(U_I^2 - \mathbf{u}^2)}\,dz.$$

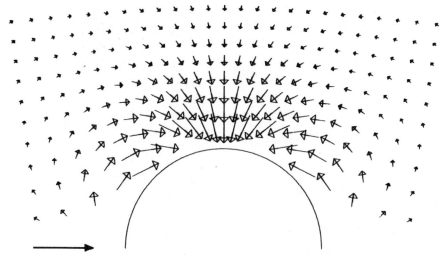

Figure 4.2 Mass transport velocity vectors in gravity waves near the solid bottom in the presence of a circular cylinder. (From Lamoure and Mei, 1977, *J. Fluid Mechs.* Reproduced by permission of Cambridge University Press.)

From the boundary-layer solution, the integral on the right is coplanar with and proportional to ∇U_0^2; the same must be true for $\bar{u}$ and $\bar{u}_L$.

The qualitative prediction of Eq. (4.7) has been verified experimentally by Lamoure and Mei (1977); see Plate 3. We caution that, in nature, the presence of solid particles in fluid not only produces turbulence but affects the nature of turbulence by their own dynamics. Hence the mechanics of littoral drift by waves is far more complex than the mass transport of pure fluid.

Other beach topographies have been attributed to the mass transport under complex wave systems by Holman and Bowen (1982).

Exercise 4.1

A plane incident wave of amplitude A propagates in the positive x direction in a sea of constant depth. A small circular pile is located at the origin. Use the fact from Section 7.3 that

$$\zeta \simeq \operatorname{Re} A e^{-i\omega t}\left[1 + ik\cos\theta\left(r + \frac{a^2}{r}\right)\right], \qquad \text{for } r = O(a)$$

to deduce the following formulas for the mass transport velocity in the bottom boundary layer near the pile:

$$\bar{u}_L = \frac{4}{a}\left(\frac{gkA}{\omega\cosh kh}\right)^2\left(\frac{a}{r}\right)^3\left\{ \begin{array}{c} \left(-\dfrac{a^2}{r^2} + \cos 2\theta\right)\mathbf{e}_r \\[2mm] \sin 2\theta\,\mathbf{e}_\theta \end{array} \right\}\frac{1}{2\omega}\operatorname{Re} F_6. \quad (4.8)$$

Make some computations to verify Fig. 4.2.

Plate 3   Sediment accumulation (see dark areas) near two sides of a circular pile. Wave crests are parallel to the ripple marks. Prior to wave motion, there were no sand particles in an annular area next to the cylinder. (From Lamoure and Mei, 1977, *J. Fluid Mechs*. Reproduced by permission of Cambridge University Press.)

9.5   REMARKS ON INDUCED STREAMING OUTSIDE THE STOKES BOUNDARY LAYER

One of the main features of the preceding deductions is that Eulerian streaming velocity approaches a finite value at the outer edge of the Stokes boundary layer. It is interesting to ask what the effect of this current is on the essentially inviscid core. Does this current induce further current throughout the entire water depth by viscous diffusion?

In studying the boundary layers near oscillating bodies, Stuart (1966) suggested that if the amplitude of motion was sufficiently large, the convective inertia of the induced current became important in the average momentum balance and had to be counteracted by a new shear stress in a relatively thick boundary layer. Riley (1967) and Wang (1968) developed this subject further. In particular, Wang analyzed systematically the variety of possibilities outside the Stokes layer. The following is a synopsis of Wang's reasoning.

In terms of physical variables, denoted with primes, the equations of motion are

$$\nabla' \cdot \mathbf{u}' = 0, \tag{5.1}$$

$$\frac{\partial \mathbf{\Omega}'}{\partial t'} - \nabla' \times (\mathbf{u}' \times \mathbf{\Omega}') = -\nu \nabla' \times (\nabla \times \mathbf{\Omega}'), \tag{5.2}$$

where $\mathbf{\Omega}' = \nabla' \times \mathbf{u}'$ is the vorticity vector. Note that the Navier–Stokes equation has been replaced by the vorticity equation (5.2).

Let the unknowns be split into a mean (DC) and a fluctuating (AC) part as follows:

$$\mathbf{u}' = \bar{\mathbf{u}}' + \tilde{\mathbf{u}}', \tag{5.3}$$

$$\mathbf{\Omega}' = \bar{\mathbf{\Omega}}' + \tilde{\mathbf{\Omega}}', \tag{5.4}$$

with $(\bar{\ })$ representing the time mean over a period and $(\tilde{\ })$ representing the fluctuations. Substituting into Eqs. (5.1) and (5.2) and taking time averages, we obtain equations governing the mean motion:

$$\nabla' \cdot \bar{\mathbf{u}}' = 0, \tag{5.5}$$

$$\nabla' \times \overline{(\tilde{\mathbf{u}}' \times \tilde{\mathbf{\Omega}}')} + \nabla' \times (\bar{\mathbf{u}}' \times \bar{\mathbf{\Omega}}') = \nu \nabla' \times (\nabla \times \bar{\mathbf{\Omega}}'). \tag{5.6}$$

In Eq. (5.6) the first term corresponds to the Reynolds stress contribution, which is the *steady part* of self-interaction of two oscillating components. This term is nonzero if $\tilde{\mathbf{u}}'$ and $\tilde{\mathbf{\Omega}}'$ are not always out of phase by $\frac{1}{2}\pi$.

Subtracting Eqs. (5.5) and (5.6) from Eqs. (5.1) and (5.2), we obtain the equations governing the fluctuations:

$$\nabla' \cdot \tilde{\mathbf{u}}' = 0, \tag{5.7}$$

$$\frac{\partial \tilde{\mathbf{\Omega}}'}{\partial t'} - \nabla'(\bar{\mathbf{u}}' \times \tilde{\mathbf{\Omega}}') - \nabla' \times (\tilde{\mathbf{u}}' \times \bar{\mathbf{\Omega}}') - \nabla' \times \overbrace{(\tilde{\mathbf{u}}' \times \tilde{\mathbf{\Omega}}')}$$

$$= -\nu \nabla' \times (\nabla \times \tilde{\mathbf{\Omega}}'), \quad (5.8)$$

where the last term on the left of Eq. (5.8) is the *unsteady part* of self-interaction. We now normalize the variables as follows:

$$\begin{aligned}
x &= kx', & t &= \omega t' \\
\tilde{\mathbf{u}} &= \frac{\tilde{\mathbf{u}}'}{\omega A}, & \bar{\mathbf{u}} &= \frac{\bar{\mathbf{u}}'}{\omega k A^2}, \\
\tilde{\mathbf{\Omega}} &= \frac{\tilde{\mathbf{\Omega}}'}{\omega k A}, & \bar{\mathbf{\Omega}} &= \frac{\bar{\mathbf{\Omega}}'}{\omega k^2 A^2},
\end{aligned} \tag{5.9}$$

where ω is the wave frequency, k is the wavenumber, and A is the wave amplitude. The scale of the fluctuation velocity is the fluid velocity in waves and the scale of the mean drift is that of the mass transport velocity (cf. Eq. (3.5)). In dimensionless form Eqs. (5.5)–(5.8) become

$$\nabla \cdot \bar{\mathbf{u}} = 0, \tag{5.10}$$

$$\underset{\text{(Reynolds stress)}}{(kA)^{-2} \nabla \times \overline{(\tilde{\mathbf{u}} \times \tilde{\mathbf{\Omega}})}} + \underset{\left(\substack{\text{steady} \\ \text{convection}}\right)}{\nabla \times (\bar{\mathbf{u}} \times \bar{\mathbf{\Omega}})} = \underset{\text{(diffusion)}}{\frac{\delta^2}{a^2} \nabla \times (\nabla \times \bar{\mathbf{\Omega}})}, \tag{5.11}$$

for the steady streaming, where

$$\delta = \left(\frac{2\nu}{\omega} \right)^{1/2} \tag{5.12}$$

and

$$\nabla \cdot \tilde{\mathbf{u}} = 0, \tag{5.13}$$

$$\frac{\partial \tilde{\mathbf{\Omega}}}{\partial t} - (kA)^2 \left[\nabla \times (\bar{\mathbf{u}} \times \tilde{\mathbf{\Omega}}) + \nabla \times (\tilde{\mathbf{u}} \times \bar{\mathbf{\Omega}}) \right] - (kA) \left[\nabla \times (\widetilde{\tilde{\mathbf{u}} \times \tilde{\mathbf{\Omega}}}) \right]$$

$$= - (k\delta)^2 \nabla \times \nabla \times \tilde{\mathbf{\Omega}}, \tag{5.14}$$

for the fluctuations.

It is now easy to delineate various regions and their proper approximations. In most laboratory setups, the wave slope and the viscosity are small,

$$kA \ll 1, \qquad k\delta \ll 1. \tag{5.15}$$

Therefore, for the fluctuations the convective term may be ignored from Eq. (5.14). Viscous diffusion can be ignored except near a solid wall within a layer of thickness $O(\delta)$. Within this layer the normal derivative dominates and Eq. (5.14) is approximately

$$\frac{\partial \tilde{\mathbf{\Omega}}}{\partial t} = (k\delta)^2 \frac{\partial^2 \tilde{\mathbf{\Omega}}}{\partial n^2}, \tag{5.16}$$

which governs the structure of the Stokes layer $O(\delta)$.

Now examine Eq. (5.11) for the steady streaming inside the Stokes layer. The Reynolds stress due to the mean motion is unimportant, and $\partial/\partial n \sim 1/k\delta$. Hence the Reynolds stress due to fluctuations must be balanced by a steady viscous stress:

$$\left(\frac{1}{kA} \right)^2 \cdot \nabla \times \overline{(\tilde{\mathbf{u}} \times \tilde{\mathbf{\Omega}})} = - \left(\frac{\delta}{A} \right)^2 \frac{\partial^2 \overline{\mathbf{\Omega}}}{\partial n^2}. \tag{5.17}$$

Outside the Stokes layer, since the cross product of the fluctuating parts $\tilde{\mathbf{u}}$, $\tilde{\mathbf{\Omega}}$ is known to be exponentially small, we have from Eq. (5.11)

$$\nabla \times (\bar{\mathbf{u}} \times \overline{\mathbf{\Omega}}) = \left(\frac{\delta}{A} \right)^2 \nabla \times (\nabla \times \overline{\mathbf{\Omega}}) \tag{5.18}$$

which leads to several possibilities:

(i) $\delta/A \gg 1$: The amplitude is very much smaller than the Stokes-layer thickness. The approximation

$$\nabla \times (\nabla \times \overline{\mathbf{\Omega}}) \simeq 0 \tag{5.19}$$

can be made which corresponds to small-Reynolds-number (creeping) flows. While unrealistic in practical situations, this approximation is the easiest to analyze.

(ii) $\delta/A \ll 1$: The amplitude is much greater than the Stokes-layer thickness, which corresponds to high-Reynolds-number flows. Viscosity and inertia are both important within a layer of thickness

$$O(\delta_1) = O\left(\frac{\delta}{kA}\right). \tag{5.20}$$

Within this layer the normal derivative again dominates so that

$$\nabla \times (\bar{\mathbf{u}} \times \bar{\mathbf{\Omega}}) \cong -(k\delta_1)^2 \frac{\partial^2 \bar{\mathbf{\Omega}}}{\partial n^2}, \tag{5.21}$$

which corresponds to the Blasius equation in the classical laminar boundary-layer theory. Vorticity is expected to be confined within this second boundary layer and diminishes to zero outside it (Stuart, 1966).

(iii) $\delta/A = O(1)$: Equation (5.18) must be solved exactly.

The case of standing waves has been analyzed by Mei, Liu, and Carter (1972) and Dore (1976) for the double structure of boundary layers, corresponding to Case (ii) above.

One aspect of two-dimensional progressive waves in a channel of constant depth deserves special remark. In earlier theoretical papers (e.g., Longuet-Higgins, 1953), the amplitude attentuation along the channel (x direction) was totally ignored, and the induced streaming was strictly horizontal and independent of x,

$$\bar{u} = \bar{u}(z) \propto A^2 f(z), \qquad \bar{v} = 0.$$

Under these assumptions the induced streaming is governed by Eq. (5.19) even when $\delta/A \ll 1$ and there is no double-layer structure. However, the attenuation rate in the longitudinal direction is not exactly zero and is known to be

$$\frac{1}{kA} \frac{dA}{dx'} = O(k\delta). \tag{5.22}$$

If the vertical length scale is taken to be $1/k$ for the time being, the order of magnitude of the terms in Eq. (5.18) is as follows:

$$\nabla \times (\bar{\mathbf{u}}' \times \bar{\mathbf{\Omega}}') = \nu \nabla \times (\nabla \times \bar{\mathbf{U}}'),$$

$$(k^2\delta)(\omega kA^2)(\omega k^2 A^2) \sim (\nu)(k^2)(\omega k^2 A^2) \tag{5.23}$$

or

$$O\left(\frac{\text{viscous diffusion}}{\text{inertia}}\right) = \frac{k\delta}{(kA)^2}. \tag{5.24}$$

For inertia terms to be negligible it is necessary that $k\delta \gg (kA)^2$ and not that $\delta/A \gg 1$. In many laboratory tanks with smooth bottoms, the ratio (5.24) is usually less than $O(1)$; for example, take $\nu = 10^{-2} \text{ cm}^2/\text{s}$, $k = 2\pi/50$ (wavelength $= 50$ cm), $\omega = 2\pi/T = 4\pi$ ($T = \frac{1}{2}$ s), and $A = 1 \sim 5$ cm, then $\delta = (10^{-2}/4\pi)^{1/2} = 0.03$ cm and

$$\frac{k\delta}{(kA)^2} = 0.05 \sim 0.25.$$

For very small $k\delta/(kA)^2$, there can be secondary boundary layers near the free surface and the solid bottom, of the thickness

$$\delta_2 = O\left(\frac{1}{k} \frac{(k\delta)^{1/2}}{kA}\right),$$

which is rather thick and may eventually grow to be comparable to the water depth. Furthermore, over a rough or sediment-laden bottom the eddy viscosity which is many times larger than the laminar viscosity can increase the boundary-layer thickness δ_2 even further. Convective inertia of the induced streaming can then become as important as lateral diffusion throughout the entire depth.

The details of secondary boundary layers in water waves are complicated (Dore, 1977). In the following section we only investigate the classic case of creeping flow theory due to Longuet-Higgins (1953).

9.6 CREEPING FLOW THEORY OF MASS TRANSPORT IN A CHANNEL OF FINITE DEPTH

In his original analysis, Longuet-Higgins (1953) employed a curvilinear coordinate system embedding the free surface so that the boundary layer could be measured from the instantaneous free surface. An alternative is to employ the Lagrangian coordinate system in terms of which the free-surface condition can be conveniently stated (Ünlüata and Mei, 1970). Let a and c be the initial horizontal and vertical coordinates of a fluid particle. In the a-c plane, the position of the free surface is simply $c = 0$ at all times and is no longer unknown. We refer the readers to Monin and Yaglom (1971) for a derivation of the Navier–Stokes equations in Lagrangian form and only quote the results for two dimensions.
Continuity:

$$\frac{\partial(x, z)}{\partial(a, c)} \equiv \begin{vmatrix} \dfrac{\partial x}{\partial a} & \dfrac{\partial x}{\partial c} \\ \dfrac{\partial z}{\partial a} & \dfrac{\partial z}{\partial c} \end{vmatrix} = 1, \tag{6.1}$$

x momentum:

$$x_{tt} = -\frac{1}{\rho}\frac{\partial(P, z)}{\partial(a, c)} + \nu\nabla^2 x_t, \tag{6.2}$$

z momentum:

$$z_{tt} + g = -\frac{1}{\rho}\frac{\partial(P, x)}{\partial(a, c)} + \nu\nabla^2 z_t, \tag{6.3}$$

where the Laplacian ∇^2, which is an Eulerian operator, can be transformed to

$$\nabla^2 x_t = \frac{\partial(\partial(x_t, z)/\partial(a, c), z)}{\partial(a, c)} + \frac{\partial(x, \partial(x, x_t)/\partial(a, c))}{\partial(a, c)}. \tag{6.4}$$

Vanishing of the normal and tangential stresses on the free surface can be stated exactly as

$$(-P + \tau_{zz})x_a + \tau_{xz}z_a = 0 \tag{6.5a}$$

$$\text{on } c = 0,$$

$$\tau_{zx}x_a - (-P + \tau_{xx})z_a = 0 \tag{6.5b}$$

where

$$\tau_{zz} = 2\mu[x_a z_{tc} - z_{ta}x_c], \tag{6.5c}$$

$$\tau_{xx} = 2\mu[x_{ta}z_c - x_{tc}z_a], \tag{6.5d}$$

$$\tau_{xz} = \mu[x_a x_{tc} + z_{ta}z_c - z_{tc}z_a - x_c x_{ta}]. \tag{6.5e}$$

The usual kinematic condition is accounted for by the very fact that $c = 0$ always refers to the free surface.

Pierson (1962) first applied Eqs. (6.1)–(6.3) to small-amplitude waves by introducing the perturbation series

$$x = a + x_1 + x_2 + \cdots, \tag{6.6a}$$

$$z = c + z_1 + z_2 + \cdots, \tag{6.6b}$$

$$P = (p_0 - \rho gc) + p_1 + p_2 + \cdots, \tag{6.6c}$$

where

$$(x_1, z_1, p_1) \sim O(kA), \qquad (x_2, z_2, p_2) \sim O(kA)^2.$$

Thus, these series are valid for a small time range in which the net displacement is small. The following equations are obtained at the order $O(kA)$:

$$x_{1ta} + z_{1tc} = 0, \tag{6.7}$$

$$x_{1tt} + \frac{p_{1a}}{\rho} = \nu\nabla_L^2 x_{1t}, \tag{6.8a}$$

$$z_{1tt} + gz_{1c} + \frac{p_{1c}}{\rho} = \nu\nabla_L^2 z_{1t}, \tag{6.8b}$$

where ∇_L^2 denotes the Laplacian for Lagrangian coordinates

$$\nabla_L^2 = \frac{\partial^2}{\partial a^2} + \frac{\partial^2}{\partial c^2}.$$

On the free surface, the normal and tangential stresses vanish:

$$p_1 = 2\rho\nu z_{1tc}, \tag{6.9a}$$
$$c = 0.$$
$$x_{1tc} + z_{1ta} = 0, \tag{6.9b}$$

On the bottom the no-slip condition is

$$x_{1t} = z_{1t} = 0, \qquad c = -h. \tag{6.10}$$

Formally, the first-order problem is identical to the linearized Navier–Stokes equations in Eulerian form if we identify

$$x_{1t} = u_1, \qquad z_{1t} = v_1. \tag{6.11}$$

Let

$$x_1 = x_1^p + x_1^r, \qquad z_1 = z_1^p + z_1^r, \tag{6.12}$$

where the potential part is governed by

$$\left(x_{1t}^p, z_{1t}^p\right) = \nabla_L\Phi, \tag{6.13}$$

$$\Phi_t = -\frac{p_1}{\rho} + gz_1, \tag{6.14}$$

$$\nabla_L^2\Phi = 0, \tag{6.15}$$

and the rotational part is governed by

$$x_{1tt}^r = \nu\nabla_L^2 x_t^r, \tag{6.16a}$$

$$z_{1tt}^r = \nu\nabla_L^2 z_t^r, \tag{6.16b}$$

$$x_{1ta}^r + z_{1tc}^r = 0. \tag{6.17}$$

The results in Section 9.3 may now be used. In the core outside the Stokes layers the solution is

$$\begin{pmatrix} x_1^p \\ z_1^p \end{pmatrix} = \frac{Ae^{i\theta}}{\sinh kh} \begin{pmatrix} i\cosh k(c+h) \\ \sinh k(c+h) \end{pmatrix},$$

(6.18)

with

$$\theta = ka - \omega t.$$

(6.19)

In the bottom boundary layer we have

$$x_1 = \frac{iA}{\sinh kh}\left\{ 1 - \exp\left[-(1-i)\frac{c+h}{\delta} \right] \right\} e^{i\theta},$$

(6.20)

$$z_1 = \frac{k\,\delta A}{(1-i)\sinh kh}\left\{ \exp\left[-(1-i)\frac{c+h}{\delta} \right] - 1 \right\} e^{i\theta}.$$

(6.21)

In the free-surface boundary layer, which was ignored previously, the potential part does not change significantly. Equation (6.16a) is approximately

$$\frac{\partial}{\partial t} x_{1t}^r = \nu \frac{\partial^2}{\partial c^2} x_{1t}^r.$$

(6.22)

The shear stress condition (6.9) on $c = 0$ now implies

$$\frac{\partial x_{1t}^r}{\partial c} = -\left(\frac{\partial}{\partial c} x_{1t}^p + \frac{\partial}{\partial a} z_{1t}^p \right) = -2\omega kAe^{i\theta}.$$

(6.23)

With the further requirement that

$$x_{1t}^r \to 0, \qquad \frac{c}{\delta} \to -\infty,$$

(6.24)

x_1^r is easily found to be

$$x_1^r = -\frac{2iAk\delta}{1-i} e^{i\theta} \exp\left[(1-i)\frac{c}{\delta} \right],$$

(6.25)

hence by continuity

$$z_1^r = O(k\delta \cdot x_1^r) = O(k\delta)^2.$$

(6.26)

The total solution in the free-surface layer is:

$$x_1 = iA \left\{ \coth kh - \frac{2k\delta}{(1-i)} \exp\left[(1-i)\frac{c}{\delta}\right] \right\} e^{i\theta}, \qquad (6.27)$$

$$\frac{\partial x_1}{\partial c} = ikA \left\{ 1 - 2\exp\left[(1-i)\frac{c}{\delta}\right] \right\} e^{i\theta}, \qquad (6.28)$$

$$z_1 = A \left\{ 1 + O(k\delta)^2 \right\} e^{i\theta}. \qquad (6.29)$$

Note that x_1' itself is insignificant but gives rise to $O(1)$ vorticity $(\partial x_1 / \partial c)$.

The second-order equations have been worked out by Pierson; only the horizontal component is needed here:

$$x_{2tt} + gz_{2a} + \frac{p_{2a}}{\rho} - \nu\nabla_L^2 x_{2t} = G, \qquad (6.30a)$$

where

$$G = -x_{1tt}x_{1a} - z_{1tt}z_{1a} + \nu\{(x_{1taa} + 3x_{1tcc})x_{1a} + 2x_{1taa}z_{1c}$$

$$+ x_{1ta}(z_{1ac} - x_{1cc}) - 2x_{1tac}(z_{1a} + x_{1c})$$

$$- x_{1tc}(z_{1aa} - x_{1ac}) + z_{1a}(z_{1taa} + z_{1tcc})\}. \qquad (6.30b)$$

We now take the time average of Eq. (6.30) and ignore the spatial attentuation in the direction of the waves, that is, $\bar{z}_{2a} = 0$. Since $\bar{x}_{2tt} = 0$, we have

$$-\nu\frac{d^2\bar{x}_{2t}}{dc^2} = -\frac{1}{\rho}\frac{\partial\bar{p}_2}{\partial a} + \bar{G} \qquad (6.31)$$

which is valid for all c. Denoting

$$x_1 = AX(c)e^{i\theta}, \qquad z_1 = AZ(c)e^{i\theta} \qquad (6.32)$$

where X and Z can be inferred for different regions, we obtain from Eq. (6.30b), after some algebra, that

$$\bar{G} = \frac{A^2}{2}\nu\omega k \, \mathrm{Re}\left\{ k^2\left(|X|^2 + |Z|^2\right) - 4X^*\frac{d^2X}{dc^2} - ik\frac{d}{dc}(X^*Z) \right.$$

$$\left. - 3\left|\frac{dX}{dc}\right|^2 - Z^*\frac{d^2Z}{dc^2} \right\}. \qquad (6.33)$$

Inside the free-surface boundary layer only the terms

$$4X^* \frac{d^2X}{dc^2} \quad \text{and} \quad 3\left|\frac{dX}{dc}\right|^2$$

dominate; Eq. (6.31) can be substantially simplified. On the free surface, the condition (6.5b) on the tangential stress gives

$$x_{2tc} + z_{2ta} = 3x_{1ta}z_{1a} - 2x_{1tc}x_{1a} + x_{1c}x_{1ta}, \qquad c = 0. \qquad (6.34)$$

After Eqs. (6.27) and (6.29) are used, the time average of Eq. (6.34) yields, happily,

$$\frac{d\bar{x}_{2t}}{dc} = O(k^2\delta^2\omega k^2 A^2). \qquad (6.35)$$

With this boundary condition, Eq. (6.31) can be integrated. In particular, the result leads to

$$\frac{d\bar{x}_{2t}}{dc} \to 4\omega k^2 A^2 \coth kh, \qquad \frac{c}{\delta} \to -\infty \qquad (6.36)$$

at the outer edge of the free-surface layer.

Either by a similar analysis, or by invoking the known result of Eq. (3.7), we have

$$\bar{x}_{2t} \to \frac{5\omega kA^2}{4} \sinh^2 kh, \qquad \frac{c+h}{\delta} \to \infty \qquad (6.37)$$

at the outer edge of the bottom layer.

Now in the core region between the two boundary layers, Eq. (6.31) takes the simple form

$$\nu \frac{d^2\bar{x}_{2t}}{dc^2} = \frac{\bar{p}_{2a}}{\rho} + 2\nu\omega k^3 A \frac{\cosh 2(c+h)}{\sinh^2 kh}, \qquad (6.38)$$

since $\bar{G}$ can be calculated from the first-order potential solution. Equation (6.38) is readily solved, subject to Eqs. (6.35) and (6.36) at $c = 0$ and $c = -h$, respectively; the result is

$$\bar{x}_{2t} = \frac{1}{\mu}\frac{\partial p_2}{\partial a}(c^2 - h^2) + \frac{\omega kA^2}{\sinh^2 kh}\left[\frac{3}{4} + k(c+h)\sinh 2kh\right.$$

$$\left. + \frac{1}{2}\cosh 2k(c+h)\right] \qquad (6.39)$$

which is Lagrangian mass transport outside the two boundary layers. The mean horizontal pressure gradient $\partial \bar{p}_2/\partial a$ can be determined by imposing one more condition. For example, Longuet-Higgins requires that the net mass flux throughout the depth be zero, a situation realizable in a tank of finite length; $\partial \bar{p}_2/\partial a$ is then readily found. The corresponding mass transport velocity is

$$\bar{x}_{2t} = \frac{A^2 \omega k}{4 \sinh^2 kh}\left[3 + 2\cosh 2k(c+h) + 3\left(\frac{\sinh 2kh}{2kh} + \frac{3}{2}\right)\left(\frac{c^2}{h^2} - 1\right)\right.$$

$$\left. + kh \sinh 2kh\left(1 + 4\frac{c}{h} + 3\frac{c^2}{h^2}\right)\right]. \qquad (6.40)$$

A typical plot is shown in Fig. 6.1.

In a laboratory tank of length L, the state of zero net flow is reached when a return flow is established throughout the length of the tank; the time required is $T_1 \sim O(L/\omega^2 kA^2)$. Now the time required for viscous diffusion to reach the entire depth is $T_2 = O(h^2/\nu)$. The ratio

$$\frac{T_2}{T_1} = \frac{h^2 \omega k A^2}{L\nu} = \frac{(kh)^2}{2kL}\left(\frac{A}{\delta}\right)^2$$

is of the order $O(1)$ in normal laboratory tanks. Thus, when vorticity is diffused throughout the depth, the return mass transport is just about established in most of the length. However, Eq. (6.40) has not been convincingly verified in experiments, possibly due to the effect of secondary boundary layers. In an ocean it is more appropriate to assume $\partial \bar{p}_2/\partial a = 0$; Eq. (6.39) without $\partial \bar{p}_2/\partial a$ then applies, if a suitable eddy viscosity is estimated for ν.

Note that the results for infinite depth cannot be obtained by taking the limit of Eq. (6.37) or (6.38). This is not surprising, since the theory is constructed on the basis that the attenuating length scale in the direction of wave propagation is infinite (i.e., $\gg h$). One cannot take the limit of $h \to \infty$ without first accounting for the effects of attenuation and Stuart's double

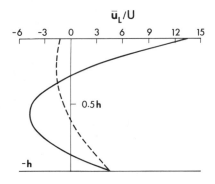

Figure 6.1 Comparison of mass transport velocity profiles for $kh = 0.5$, $T = 10$ s. Solid curve: Dore (1978, air–water); dashed curve: Longuet-Higgins (1953, vacuum–water). $U = A\omega^2 k(\sinh kh)^{-1}$. (From Dore, 1978, *Geophys. Astrophys. Fluid Dyn.* Reproduced by permission of Gordon and Breach Scientific Publishers.)

boundary layers. These remedies have been studied by Dore (1977), and Liu and Davis (1977).

Finally, the presence of air, which was shown in Section 8.6 to have a pronounced effect on wave damping, can also be important in the mass transport. By recalculating the boundary layers near the free surface, Dore (1978) found that Eq. (6.35) must be modified. As a result Eq. (6.40) had to be corrected by adding the following expression:

$$\frac{1}{4h}\left(\frac{\omega}{2}\right)^{3/2}\frac{1}{k}(1 + \coth kh)^2\left(\frac{\rho'u'}{\rho\mu}\right)^{1/2}\frac{1}{\nu^{1/2}}(c + h)(3c + h) \quad (6.41)$$

where ρ' and μ' refer to air. For a typical frequency and depth the resulting velocity profile is compared to Eq. (6.38), in Fig. 6.1.

9.7   FURTHER REFERENCES

As pointed out earlier, Longuet-Higgins' theory for the mass transport in a sea of finite depth leads to an unbounded limit for infinite depth. As an alternative to Stuart's second boundary layer, it turns out that earth rotation can give a finite limit. While irrelevant in the laboratory, this new physical factor cannot be ruled out in an ocean where distances of hundreds of kilometers are involved. Madsen (1978) and Huang (1978) showed independently that the wind-induced current in the Ekman layer near the sea surface could interact and modify the wave-induced mass transport in a nontrivial way. In particular, the mass transport was found to be confined within the Ekman layer. Madsen assumed that the effect of wind (hence, of air) was only to provide a steady driving shear stress to maintain the wave motion. Possible direct effect of air on wave-induced Reynolds' stresses in water was, however, not taken into account. In view of Dore's conclusion cited in Section 9.6, further study, which may yield results useful in oil spills, is desirable.

The effect of earth rotation is of first-order influence to tidal waves whose frequencies are comparable to the angular velocity of the earth. To the coastal engineer, the tide-induced mass transport may be of interest in affecting the silting and scour in an estuary, near a barrier island, and so on. Pertinent theoretical studies have been made by Hunt and Johns (1963), Longuet-Higgins (1970c), Moore (1970), and Huthnance (1981). Lamoure and Mei (1977) examined the bottom mass transport near small bodies and found for the northern hemisphere that the mass transport near the sea bottom had a counterclockwise tilt and varied in a spiraling manner in the Ekman layer.

TEN

Currents Induced
by Breaking Waves

10.1 INTRODUCTION

One of the most important reasons for studying nearshore currents is to help understand the interaction between the shoreline and the fluid motion in the sea. In the past, geographers and geomorphologists accumulated a good deal of descriptive knowledge on the formation and evolution of coastlines (Johnson, 1919; King, 1959; Zenkovich, 1967; Shepard and Wanless, 1973). Since the loss of sand and the recession of beaches can be undesirable to parks, residential areas, and coastal highways, and the accumulation of silt near a harbor entrance or a river inlet can be hazardous to navigation or flood control, it is the task of engineers to grapple with the mysteries of beach processes in order to plan proper defenses.

Most of the guidelines for coastal protection practices have been based on observational studies both in the field and in the laboratory (see, for example, *Shore Protection Manual*, U.S. Army Coastal Engineering Research Center, 1977, Vols. I–III). These studies have been and will undoubtedly remain indispensible in increasing our knowledge of shore processes. In recent years, much progress has also been made in the development of semiempirical theories on both the causative currents and the resultant sediment motion. The whole problem of shoreline dynamics is extremely complex, due in part to the turbulent interaction between sediments and fluid flow and in part to the large variety of currents that can be present (wind waves, tides, storm surges, river currents, and large-scale ocean circulation caused by wind, thermal or salinity gradients, etc.). Among these currents a prominent role is played by the breaking-wave-induced *longshore current* and its relatives, to which our attention is directed in this chapter.

With reference to Fig. 1.1, let us first give a bird's-eye description of the physical processes on a straight beach. A train of periodic waves with frequency

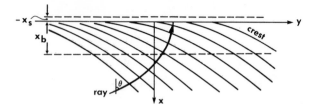

Figure 1.1   Waves incident on an infinitely long plane beach.

ω and amplitude A_∞ is incident from deep water $x \sim +\infty$ at the incidence angle θ_∞. When waves first feel the bottom, refraction causes the x component of the wavelength to decrease so that the wave rays tend to strike the shore normally; the wave amplitude varies such that energy is roughly conserved between two adjacent rays. In sufficiently shallow water the wave profile steepens and eventually breaks at a certain line $x = x_b$, where the depth is $h_b = h(x_b)$. This line of first breaking is called the *breaker line*, which divides the *shoaling* zone seaward of x_b and the *breaker zone* or the *surf zone* shoreward of x_b. Kinematically, the wave rays proceed in roughly the same direction as if there were no breaking. Once inside the surf zone, turbulence prevails so that the breaking wave is dissipated and diminishes in height. When the shore is reached, the wave crest finally exhausts itself in the form of a thin sheet rushing up the beach. After the highest climb, the sheet of fluid retreats seaward by gravity, encounters, and is swallowed by the next onrushing wave crest. Aside from this to-and-fro motion, however, there is also a less conspicuous drift along the shore in the direction of the longshore component of the incidence waves. This steady drift, called the *longshore current*, can be significant in magnitude, say $O(1\ \mathrm{m/s})$. Since the turbulent fluctuations in the surf zone dislodge many sand particles from their rest position and mix them with the fluid, the steady longshore drift becomes a powerful mechanism for sand transport, and, in turn, for the evolution of the beach.

In the earliest analytical theories on longshore currents, focus was centered on the total longshore flux in the entire surf zone. Putnam, Munk, and Traylor (1949) assumed that the crests of a periodic wavetrain were so well separated that they were practically independent of each other. The momentum flux of a solitary crest incident obliquely was averaged over the actual period to give the mean momentum flux entering the surf zone. After breaking, the crest dissipated into a much slower current along the shore and carried away certain momentum flux. By equating the longshore component of the net momentum flux into a control volume to the bottom friction experienced by this volume, they obtained the total longshore discharge. Eagleson (1965) discarded the assumption of solitary crests and modified the above argument by taking time averages of a periodic wavetrain described by the small-amplitude theory. He also considered the growth of the longshore current from a headland or an infinitely long surf zone barrier where the longshore current was zero. In both cases, only the mean across the surf zone was derived. In arriving at explicit results, empirical assumptions on breaking waves and bottom friction were introduced.

A significant step forward toward the variation of longshore current across the surf zone was made independently by Bowen (1969), Longuet-Higgins (1970a, b), and Thornton (1970). An important common feature of their theories is the use of the *radiation stresses* which represent the time averages of the local horizontal momentum fluxes. In regions where no breaking occurs (e.g., away from the surf zone) these stresses can be obtained theoretically as a second-order quantity from the first-order small-amplitude wave theory. In the surf zone where breaking and turbulence prevail, no theory exists for describing either the oscillatory waves or the radiation stresses. However, by hypothesizing some closure relations between the radiation stresses and mean-flow quantities, a semiempirical framework is obtained for calculating the mean flow in the surf zone.

The goal of this chapter is to introduce these recent advances. We first deduce the averaged equations of motion and then the radiation stresses. After a discussion of the empirical knowledge of breaking waves, semiempirical theories are presented for the uniform longshore current on a straight beach. Two examples of more complex wave fields are also discussed. Finally, some relevant references are given.

10.2 DEPTH AND TIME-AVERAGED EQUATIONS FOR THE MEAN MOTION

The main objective of this section is to obtain conservation laws of mass and horizontal momentum for the mean current field by arguments similar to those leading to the Reynolds equations for the mean turbulent flow (Longuet-Higgins and Stewart, 1962; Phillips, 1977).

For the convenience of vertical integration, the vertical and horizontal directions are distinguished. Specifically, we denote the vertical velocity by w and the vertical coordinate by z, the horizontal velocity components by u_i ($i = 1, 2$; $u_1 = u$, $u_2 = v$), and the horizontal coordinates by x_i ($i = 1, 2$; $x_1 = x$, $x_2 = y$). We define the mean velocity U_i ($i = 1, 2$) by integrating u_i over the instantaneous water depth and then over the time period T, that is,

$$U_i(x, y, t) = \frac{1}{\bar{\zeta} + h} \overline{\int_{-h}^{\zeta} u_i \, dz}, \qquad i = 1, 2, \tag{2.1}$$

where $\zeta(x, y, t)$ is the free-surface displacement, $\bar{\zeta}$ is its time mean, and $h(x, y)$ is the still water depth. Physically, $\rho U_i(\bar{\zeta} + h)$ is the mean rate of mass flux across a vertical plane of unit width along $x_i = $ const. The vector (U_1, U_2) may, therefore, be called the *mass flux velocity*, which depends only on the horizontal coordinates and the long time scale. Denoting the deviation from the mean by $\tilde{u}_i$, we have

$$u_i = U_i + \tilde{u}_i(x, y, z, t). \tag{2.2}^\dagger$$

It follows from the definition that

$$\overline{\int_{-h}^{\zeta} \tilde{u}_i \, dz} = 0. \tag{2.3}$$

10.2.1　Averaged Equation of Mass Conversion

Integrating the local continuity equation over the depth, we get

$$\int_{-h}^{\zeta} \frac{\partial u_i}{\partial x_i} dz + [w]_\zeta - [w]_{-h} = 0, \tag{2.4}$$

where $[w]_\zeta$ denotes the value of w at $z = \zeta$ and $[w]_{-h}$ denotes the value of w on the bottom. With Leibniz's rule, Eq. (2.4) can be written

$$\frac{\partial}{\partial x_i} \int_{-h}^{\zeta} u_i \, dz + \left[-u_i \frac{\partial \zeta}{\partial x_i} + w \right]_\zeta - \left[u_i \frac{\partial h}{\partial x_i} + w \right]_{-h} = 0. \tag{2.5}$$

On the free surface, the kinematic boundary condition

$$\frac{\partial \zeta}{\partial t} + u_i \frac{\partial \zeta}{\partial x_i} = w, \qquad z = \zeta, \tag{2.6}$$

may be recalled. For simplicity, we assume a rigid sea bottom at $z = -h$, then

$$u_i = w = 0 \quad \cdot \quad \text{for real fluid}, \tag{2.7}$$

or

$$\left[u_i \frac{\partial h}{\partial x_i} + w \right]_{-h} = 0 \qquad \text{for inviscid fluid}. \tag{2.8}$$

Because of Eqs. (2.6) and (2.7) or Eq. (2.8), Eq. (2.5) becomes

$$\frac{\partial \zeta}{\partial t} + \frac{\partial}{\partial x_i} \int_{-h}^{\zeta} u_i \, dz = 0. \tag{2.9}$$

†This definition differs slightly from that of Phillips (1977, p. 45).

Upon taking the time average, we get

$$\frac{\partial \bar{\zeta}}{\partial t} + \frac{\partial}{\partial x_i}\left[U_i(\bar{\zeta} + h)\right] = 0 \tag{2.10}$$

exactly; the same result was deduced before [see Eq. (5.12), Section 2.5] as an approximation.

10.2.2 Averaged Equations of Momentum Conservation

In order to include the laboratory case of smooth and rigid bottoms, we allow both viscous and turbulent stresses to be present. For a rough and rigid bottom, the viscous stresses are unimportant almost everywhere and may be discarded from the following discussion.

Let us first rearrange the Navier–Stokes equations of motion as follows: horizontal momentum ($j = 1, 2$):

$$\frac{\partial u_j}{\partial t} + \frac{\partial u_i u_j}{\partial x_i} + \frac{\partial u_j w}{\partial z} = \frac{1}{\rho}\frac{\partial}{\partial x_i}\left(-P\delta_{ij} + \tau_{ij}\right) + \frac{1}{\rho}\frac{\partial \tau_{j3}}{\partial z}, \tag{2.11}$$

vertical momentum:

$$\frac{\partial w}{\partial t} + \frac{\partial u_i w}{\partial x_i} + \frac{\partial w^2}{\partial z} = -\frac{1}{\rho}\frac{\partial}{\partial z}(P + \rho g z) + \frac{1}{\rho}\frac{\partial \tau_{i3}}{\partial x_i} + \frac{1}{\rho}\frac{\partial \tau_{33}}{\partial z}, \tag{2.12}$$

where τ_{ij}, τ_{j3}, and τ_{33} are the components of the viscous stress tensor. When the horizontal momentum equation (2.11) is integrated vertically, terms on the left-hand side give

$$\int_{-h}^{\zeta} dz\,\frac{\partial \rho u_j}{\partial t} = \frac{\partial}{\partial t}\int_{-h}^{\zeta} \rho u_j\,dz - \rho[u_j]_{\zeta}\frac{\partial \zeta}{\partial t},$$

$$\int_{-h}^{\zeta} dz\,\frac{\partial \rho u_i u_j}{\partial x_i} = \frac{\partial}{\partial x_i}\int_{-h}^{\zeta} \rho u_i u_j\,dz - \rho[u_i u_j]_{\zeta}\frac{\partial \zeta}{\partial x_i} - \rho[u_i u_j]_{-h}\frac{\partial h}{\partial x_i},$$

$$\int_{-h}^{\zeta} dz\,\frac{\partial \rho u_i w}{\partial z} = \rho[u_i w]_{\zeta} - \rho[u_i w]_{-h}.$$

Summing up these terms and invoking the kinematic boundary conditions on the free surface and on the bottom, that is, Eqs. (2.6) and (2.7), we obtain for the left-hand side of Eq. (2.11)

$$\text{L.H.S.} = \frac{\partial}{\partial t}\int_{-h}^{\zeta} \rho u_i\,dz + \frac{\partial}{\partial x_i}\int_{-h}^{\zeta} (\rho u_i u_j)\,dz. \tag{2.13}$$

Similarly, the right-hand side of Eq. (2.11) gives

$$\text{R.H.S.} = \int_{-h}^{\zeta} \frac{\partial}{\partial x_i} (-P\delta_{ij} + \tau_{ij}) \, dz + \int_{-h}^{\zeta} \frac{\partial \tau_{j3}}{\partial z} \, dz$$

$$= \frac{\partial}{\partial x_i} \int_{-h}^{\zeta} (-P\delta_{ij} + \tau_{ij}) \, dz - \left[-P\delta_{ij} + \tau_{ij} \right]_{\zeta} \frac{\partial \zeta}{\partial x_i}$$

$$- \left[-P\delta_{ij} + \tau_{ij} \right]_{-h} \frac{\partial h}{\partial x_i} + \left[\tau_{j3} \right]_{\zeta} - \left[\tau_{j3} \right]_{-h}. \qquad (2.14)$$

On the free surface, the atmospheric force per unit area must balance the fluid stresses. This balance implies that in the jth direction

$$\left[-P\delta_{ij} + \tau_{ij} \right] n_i + \tau_{j3} n_3 = \tau_j^F, \qquad z = \zeta, \quad j = 1, 2, \qquad (2.15)$$

where τ_j^F is the externally applied horizontal stress component which is defined to include both atmospheric pressure and shear stress given from meteorological data and $\mathbf{n} = (n_1, n_2, n_3)$ is the unit outward normal vector. On the free surface described by $F(x, y, z, t) = z - \zeta(x, y, t) = 0$, the unit outward normal is $\mathbf{n} = \nabla F / |\nabla F|$ where $\nabla F = (-\partial \zeta/\partial x, -\partial \zeta/\partial y, 1)$. Hence, Eq. (2.15) can be written

$$- \left[-P\delta_{ij} + \tau_{ij} \right]_{\zeta} \frac{\partial \zeta}{\partial x_i} + \left[\tau_{j3} \right]_{\zeta} = \tau_j^F |\nabla F|, \qquad z = \zeta. \qquad (2.16)$$

On the bottom $B(x, y, z) = z + h(x, y) = 0$ we denote the total horizontal shear stress by τ_j^B, that is,

$$\tau_{ij} n_i + \tau_{j3} n_3 = \tau_j^B, \qquad z = -h(x, y), \quad j = 1, 2. \qquad (2.17)$$

Since the unit outward normal to the bottom is $\mathbf{n} = -\nabla B / |\nabla B|$ where $\nabla B = (\partial h/\partial x, \partial h/\partial y, 1)$, Eq. (2.17) may be expressed as

$$\tau_{ij} \frac{\partial h}{\partial x_i} + \tau_{j3} = -\tau_j^B |\nabla B|, \qquad z = -h. \qquad (2.18)$$

With Eqs. (2.16) and (2.18), Eq. (2.14) becomes

$$\text{R.H.S.} = \frac{\partial}{\partial x_i} \int_{-h}^{\zeta} \left[-P\delta_{ij} + \tau_{ij} \right] dz + [P]_{-h} \frac{\partial h}{\partial x_i} + \left[\tau_j^F |\nabla F| - \tau_j^B |\nabla B| \right].$$

$$(2.19)$$

Equating Eqs. (2.13) and (2.19), we have

$$\frac{\partial}{\partial t}\int_{-h}^{\zeta}\rho u_j\,dz + \frac{\partial}{\partial x_i}\int_{-h}^{\zeta}\rho u_i u_j\,dz = [P]_{-h}\frac{\partial h}{\partial x_j} + \frac{\partial}{\partial x_i}\int_{-h}^{\zeta}[-P\delta_{ij}+\tau_{ij}]\,dz$$

$$+\left[\tau_j^F\,|\,\nabla F| - \tau_j^B\,|\,\nabla B|\right]. \qquad (2.20)$$

Physically, Eq. (2.20) represents the momentum balance in a vertical column of fluid of height $\zeta + h$ and unit cross section ($dx = 1$, $dy = 1$). On the left-hand side the terms are successively the acceleration and the net momentum flux through the vertical sides of the column. On the right, the terms represent, in the sequence shown, the pressure by the bottom to the fluid, the net surface stresses on the vertical sides, and the surface stress at the free surface and the sea bottom.

The time average of the left-hand side of Eq. (2.20) may be written

$$\rho(\bar\zeta + h)\left(\frac{\partial U_j}{\partial t} + U_i\frac{\partial U_j}{\partial x_i}\right) + \rho\frac{\partial}{\partial x_i}\overline{\left(\int_{-h}^{\zeta}\tilde u_i\tilde u_j\,dz\right)} \qquad (2.21)$$

after using Eq. (2.10). On the right-hand side of Eq. (2.20) we define $\bar p$ as the mean dynamic pressure at the bottom, that is,

$$\bar p = [\bar P]_{-h} - \rho g(\bar\zeta + h), \qquad (2.22)$$

so that

$$[\bar P]_{-h}\frac{\partial h}{\partial x_j} = \bar p\frac{\partial h}{\partial x_j} + \frac{\partial}{\partial x_j}\left[\frac{1}{2}\rho g(\bar\zeta + h)^2\right] - \rho g(\bar\zeta + h)\frac{\partial\bar\zeta}{\partial x_j}. \qquad (2.23)$$

Substituting Eqs. (2.21) and (2.23) into the time average of Eq. (2.20), we obtain finally for $j = 1, 2$

$$\rho(\bar\zeta + h)\left[\frac{\partial U_j}{\partial t} + U_i\frac{\partial U_j}{\partial x_i}\right] = \bar p\frac{\partial h}{\partial x_j} - \rho g(\bar\zeta + h)\frac{\partial\bar\zeta}{\partial x_j}$$

$$+ \frac{\partial}{\partial x_i}\left\{-S_{ij} + \overline{\int_{-h}^{\zeta}\tau_{ij}\,dz}\right\} + \overline{\tau_j^F\,|\,\nabla F|} - \overline{\tau_j^B\,|\,\nabla B|}, \qquad (2.24)$$

where the following definition has been introduced

$$S_{ij} = \overline{\int_{-h}^{\zeta}[P\delta_{ij} + \rho\tilde u_i\tilde u_j]\,dz} - \frac{\rho g}{2}(\bar\zeta + h)^2\delta_{ij}. \qquad (2.25)$$

Physically, S_{ij} is the (i, j) component of the stress tensor representing the *excess momentum fluxes.*[†] Since

$$\frac{\rho g}{2}(\bar{\zeta} + h)^2 = \int_{-h}^{\bar{\zeta}} \rho g(\bar{\zeta} - z)\, dz$$

is just the total mean hydrostatic pressure over the mean depth, S_{ij} may be written

$$S_{ij} = \left[\overline{\int_{-h}^{\zeta} P\, dz} - \int_{-h}^{\bar{\zeta}} \rho g(\bar{\zeta} - z)\, dz \right] \delta_{ij} + \overline{\int_{-h}^{\zeta} \rho \tilde{u}_i \tilde{u}_j\, dz}\, . \qquad (2.26)$$

Thus, S_{ij} represents the sum of the ith component of the excess hydrodynamic pressure on, and the net momentum flux across, a surface normal to the jth direction. Also on the right-hand side of Eq. (2.24), the quantity $\int_{-h}^{\zeta} \tau_{ij}\, dz$ corresponds to the mean momentum flux tensor caused by fluctuations of molecular scale. $\bar{p}\, \partial h / \partial x_j$ is the mean hydrodynamic reaction at the sea bottom and $\rho g\, \partial \bar{\zeta} / \partial x$ is the hydrostatic pressure gradient due to the mean sea level.

Thus far, Eqs. (2.10) and (2.24) are *exact*. In particular, the orders of magnitude of the mean current and the waves are yet arbitrary. Practical applications of these equations sometimes permit (or even demand) certain simplifications as will be shown later.

10.2.3   Some Preliminary Simplifications

Let us introduce certain simplifying assumptions before proceeding further.

First, let all mean quantities be steady in time,

$$\frac{\partial \bar{\zeta}}{\partial t} = \frac{\partial U_i}{\partial t} = 0, \qquad (2.27)$$

and let there be no atmospheric disturbance on the free surface,

$$\bar{\tau}_j^F = 0. \qquad (2.28)$$

These two assumptions preclude the study of the transient effects of wind which may have very direct influence on the short-term evolution of beaches. Little work has been done in this regard.

Let us now discuss further approximations for Eq. (2.26) when the following quantities are small: (i) viscosity, (ii) bottom slope, $|\nabla h| \leqq O(kA)$, and (iii) wave slope kA.

[†] The definition of S_{ij} is slightly different from Phillips [1977, Eq. (3.6.12)]. The difference is of the fourth order for infinitesimal waves.

Viscous Stress Terms

The integrated viscous stress terms, being the product of molecular viscosity and the horizontal gradients of the horizontal velocity, are of the order

$$\overline{\int_{-h}^{\zeta} \tau_{ij} \, dz} \sim O(\mu k \omega A h). \tag{2.29}$$

Since S_{ij} is of the order $O(\rho(\omega A)^2 h)$, we get

$$\frac{\overline{\int_{-h}^{\zeta} \tau_{ij} \, dz}}{S_{ij}} = O\left(\frac{k\nu}{\omega A}\right) = O(\mathsf{R}_\mathsf{E}^{-1}) \tag{2.30}$$

where

$$\mathsf{R}_\mathsf{E} \equiv \frac{\omega A}{k\nu} = \frac{A}{k\delta^2}$$

is the Reynolds number based on the wave orbital velocity and the wavelength. Under practical circumstances the above ratio is very small and the integral of viscous stress τ_{ij} is negligible.

The bottom stress $\overline{\tau}_j^B$ is, however, not necessarily small. In the present chapter the bottom slope of the beach will always be regarded as small,

$$|\nabla B| = 1 + O(\nabla h)^2,$$

and we may approximate

$$\overline{\tau}_j^B |\nabla B| = \overline{\tau}_j^B [1 + O(\nabla h)^2] \tag{2.31}$$

in Eq. (2.24). Recall from Section 8.9 that the $\overline{\tau}_j^B$ can be related to the local orbital velocity if there is a turbulent boundary layer.

Dynamic Pressure $\overline{p} \, \partial h / \partial x_j$

To estimate the mean dynamic pressure, we integrate the vertical momentum equation (2.12) and employ Leibniz's rule as before,

$$[P]_z = \rho g (\zeta - z) + \rho \left[\frac{\partial}{\partial t} \int_z^{\zeta} w \, dz + \frac{\partial}{\partial x_i} \int_z^{\zeta} u_i w \, dz \right]$$

$$- \rho \left[w \left(\frac{\partial \zeta}{\partial t} + u_i \frac{\partial \zeta}{\partial x_i} - w \right) \right]_{\zeta} - \rho [w^2]_z - \frac{\partial}{\partial x_i} \int_z^{\zeta} \tau_{i3} \, dz$$

$$- \left[-P + \tau_{33} - \tau_{i3} \frac{\partial \zeta}{\partial x_i} \right]_{\zeta} + [\tau_{33}]_z, \tag{2.32}$$

where $[P]_z$ means P evaluated at an arbitrary height z. The free-surface boundary terms vanish by Eqs. (2.6) and (2.16) if the atmospheric pressure is zero. The time average of the mean water pressure at any z is, therefore,

$$[\bar{P}]_z = \rho g(\bar{\zeta} - z) + \frac{\partial}{\partial x_i} \overline{\int_z^\zeta \rho u_i w \, dz} - \rho[\overline{w^2}]_z - \frac{\partial}{\partial x_i} \overline{\int_z^\zeta \tau_{i3} \, dz} + [\bar{\tau}_{33}]_z.$$

$$(2.33)$$

The integrated molecular viscous term is of the order $O(\mathsf{R}_\mathsf{E}^{-1})$ relative to the others. From continuity, the term $[\bar{\tau}_{33}]_z$ may be estimated by

$$[\bar{\tau}_{33}]_z = \mu \frac{\partial \bar{w}}{\partial z} \sim O\left(\mu \frac{\partial \bar{u}_i}{\partial x_i}\right)$$

and is also negligible. Thus, to a good approximation, we have

$$[\bar{P}]_z \simeq \left\{ \rho g(\bar{\zeta} - z) + \frac{\partial}{\partial x_i} \overline{\int_z^\zeta \rho u_i w \, dz} - \rho[\overline{w^2}]_z \right\} \{1 + O(\mathsf{R}_\mathsf{E}^{-1})\},$$

$$(2.34)$$

implying that viscosity does not have any direct influence on $[\bar{P}]_z$. The same is true for $[\bar{P}]_{-h}$

$$[\bar{P}]_{-h} \simeq \rho g(\bar{\zeta} + h) + \frac{\partial}{\partial x_i} \overline{\int_{-h}^\zeta \rho u_i w \, dz}. \qquad (2.35)$$

Combining Eqs. (2.35) and (2.22), we have simply

$$\bar{p} \frac{\partial h}{\partial x_j} = \left\{ \frac{\partial h}{\partial x_j} \frac{\partial}{\partial x_i} \overline{\int_{-h}^\zeta \rho u_i w \, dz} \right\} [1 + O(\mathsf{R}_\mathsf{E}^{-1}, |\nabla h|)]. \qquad (2.36)$$

Finally, let us compare the preceding term with other terms in Eq. (2.24). For short-crested waves we should expect the length scale of variations of U_i, $\bar{\zeta}$, S_{ij}, and so on, to be k^{-1} so that the largest terms among

$$\rho(\bar{\zeta} + h)U_i \frac{\partial U_j}{\partial x_i}, \qquad \rho g(\bar{\zeta} + h) \frac{\partial \bar{\zeta}}{\partial x_j}, \qquad \frac{\partial S_{ij}}{\partial x_i}$$

are of the order $O(kh\rho \omega^2 A^2)$. Now because of the factor ∇h,

$$\bar{p} \frac{\partial h}{\partial x_i} = O\left(|\nabla h| \rho (\omega A)^2 kh\right) \qquad (2.37)$$

is unimportant and can be ignored.

For long-crested waves refracted by the sea bottom, all mean quantities are expected to vary slowly in horizontal directions as the depth h itself; hence $\bar{p}\,\partial h/\partial x_i$ is again $O(|\nabla h|)$ times smaller than the largest remaining terms and may be ignored.

The Excess Momentum Flux Tensor—The Radiation Stresses for Small Fluctuations

In the exact definition for S_{ij}, Eq. (2.25), let $\tilde{u}_i$ and $\tilde{\zeta} \equiv \zeta - \bar{\zeta}$ be further decomposed into wave and turbulent fluctuations, which will be distinguished by single and double primes, respectively; thus

$$\tilde{u}_i = u_i' + u_i'', \qquad \tilde{\zeta} = \zeta - \bar{\zeta} = \zeta' + \zeta''. \tag{2.38}$$

Clearly, $\tilde{\zeta} = O(A)$. If we assume that the characteristic time scales of the two fluctuations are vastly different, then u_i' and u_j'' are uncorrelated

$$\overline{u_i' u_j''} = 0, \qquad \overline{\zeta' \zeta''} = 0. \tag{2.39}$$

The second term in Eq. (2.25) may now be approximated as

$$\rho \,\overline{\int_{-h}^{\zeta} \tilde{u}_i \tilde{u}_j \, dz} = \rho \int_{-h}^{\bar{\zeta}} \overline{\tilde{u}_i \tilde{u}_j} + O(\rho h \omega^2 A^3)$$

$$\cong \rho \int_{-h}^{\bar{\zeta}} \overline{u_i' u_j'} \, dz + \rho \int_{-h}^{\bar{\zeta}} \overline{u_i'' u_j''} \, dz. \tag{2.40}$$

To obtain a more explicit formula for the remaining terms in Eq. (2.25), we integrate Eq. (2.32) vertically and then average the result with respect to time:

$$\overline{\int_{-h}^{\zeta} P \, dz} = \rho g \,\overline{\frac{(\zeta + h)^2}{2}} + \rho \,\overline{\int_{-h}^{\zeta} dz \,\frac{\partial}{\partial t} \int_{z}^{\zeta} w \, dz}$$

$$+ \rho \,\overline{\int_{-h}^{\zeta} dz \,\frac{\partial}{\partial x_i} \int_{z}^{\zeta} u_i w \, dz} - \rho \,\overline{\int_{-h}^{\zeta} w^2 \, dz}. \tag{2.41}$$

Again, the integrated viscous stresses have been ignored. The first term on the right of Eq. (2.41) may be rewritten

$$\frac{\rho g}{2} \,\overline{(\zeta + h)^2} = \frac{\rho g}{2} \left[(\bar{\zeta} + h)^2 + \overline{\zeta'^2} + \overline{\zeta''^2} \right]. \tag{2.42}$$

The second term in Eq. (2.41) vanishes upon partial integration,

$$\overline{\int_{-h}^{\zeta} dz \,\frac{\partial}{\partial t} \int_{z}^{\zeta} w \, dz'} = \frac{\partial}{\partial t} \,\overline{\int_{-h}^{\zeta} dz \int_{z}^{\zeta} w \, dz'} = 0. \tag{2.43}$$

The averaged vertical velocity W is, by definition,

$$W = \frac{1}{\bar{\zeta}+h} \overline{\int_{-h}^{\bar{\zeta}} w \, dz} = \frac{1}{\bar{\zeta}+h} \left(\int_{-h}^{\bar{\zeta}} \overline{w} \, dz + \overline{\int_{\bar{\zeta}}^{\zeta} w \, dz} \right).$$

Assume that the horizontal scales of depth and U_i are comparable. Then either for a strong current $[O(U_i) = (gh)^{1/2}]$ with $\nabla h/kh < O(kA)$ or for a weak current $[O(U_i) = \omega A]$ with $\nabla h/kh = O(kA)$, the mean vertical current W is much smaller than w' or w'' by a factor $O(\nabla h)$ or $O(kA)$. It follows that

$$\overline{\int_{-h}^{\zeta} dz \frac{\partial}{\partial x_i} \int_z^{\zeta} u_i w \, dz} \cong \int_{-h}^{\bar{\zeta}} dz \frac{\partial}{\partial x_i} \int_z^{\bar{\zeta}} \left[\overline{u_i' w'} + \overline{u_i'' w''} \right] dz, \qquad (2.44)$$

and

$$-\rho \overline{\int_{-h}^{\zeta} w^2 \, dz} \cong -\rho \int_{-h}^{\bar{\zeta}} \left(\overline{w'^2} + \overline{w''^2} \right) dz, \qquad (2.45)$$

with a relative error of $O(|\nabla h|, kA)$. When Eqs. (2.42)–(2.45) are substituted into Eq. (2.41) and then into Eq. (2.25), the result is

$$S_{ij} = S_{ij}' + S_{ij}'', \qquad (2.46)$$

where

$$S_{ij}' = \rho \left\{ \frac{g \overline{\zeta'^2}}{2} + \int_{-h}^{\bar{\zeta}} dz \frac{\partial}{\partial x_i} \int_z^{\bar{\zeta}} \overline{u_i' w'} \, dz' - \int_{-h}^{\bar{\zeta}} \overline{w'^2} \, dz \right\} \delta_{ij} + \rho \int_{-h}^{\bar{\zeta}} \overline{u_i' u_j'} \, dz.$$

$$(2.47)$$

Similarly, S_{ij}'' is defined with all ( )' being replaced by ( )''.

$$S_{ij}'' = \rho \left\{ \frac{g \overline{\zeta''^2}}{2} + \int_{-h}^{\bar{\zeta}} dz \frac{\partial}{\partial x_i} \int_z^{\bar{\zeta}} \overline{u_i'' w''} \, dz' - \int_{-h}^{\bar{\zeta}} \overline{w''^2} \, dz \right\} \delta_{ij} + \rho \int_{-h}^{\bar{\zeta}} \overline{u_i'' u_j''} \, dz.$$

$$(2.48)$$

S_{ij}' and S_{ij}'' are, respectively, the excess momentum flux tensors due to wave and turbulent fluctuations. S_{ij}' is called the *radiation stress* tensor by Longuet-Higgins and Stewart (1962, 1964). Since the last term in S_{ij}''

$$\rho \int_{-h}^{\bar{\zeta}} \overline{u_i'' u_j''} \, dz$$

is the negative of the integrated Reynolds stress due to horizontal turbulent fluctuations, S''_{ij} is an extension of the Reynolds stress, with additional contributions from vertical fluctuations.

Away from the surf zone, turbulence is usually unimportant so that $S''_{ij} \ll S'_{ij}$; the wave radiation stress S'_{ij} can be evaluated by using the first-order wave theory, as will be shown shortly. Inside the surf zone, waves and turbulence can be equally strong; nonlinearity becomes important and u'_i may even be comparable to the phase velocity. While Eq. (2.25) still holds, definitions (2.47) and (2.48), which are approximate for small kA, should be modified. More important, neither stress can be deduced theoretically, and empirical closure conditions are needed.

In contrast to Eq. (2.48), Phillips (1977) did not include the diagonal terms in the curly brackets. This formal difference is, however, only of academic significance in view of the uncertainties in the closure hypothesis to be introduced for S''_{ij}.

10.2.4   Summary of Approximate Averaged Equations

The approximate equations for the mean motion may now be collected below:

$$\frac{\partial}{\partial x_i}\left[U_i(\bar{\zeta} + h)\right] = 0, \qquad (2.50)$$

and

$$U_i\frac{\partial U_j}{\partial x_i} = -g\frac{\partial \bar{\zeta}}{\partial x_j} - \frac{1}{\rho(\bar{\zeta} + h)}\frac{\partial}{\partial x_i}\left(S'_{ij} + S''_{ij}\right) - \frac{\bar{\tau}_j^B}{\rho(\bar{\zeta} + h)}, \qquad (2.51)$$

where S'_{ij} and S''_{ij} are given by Eqs. (2.47) and (2.48).

In the shoaling zone and away from lateral boundaries, such as breakwaters and jetties, turbulence is negligible so that S''_{ij} may be further neglected. If the mean current is of the second order in wave slope, then the inertia term in Eq. (2.51) is of the fourth order and may also be ignored. If there is no mean current, the mean bottom friction vanishes. Equation (2.51) may then be simplified to a statement of static balance,

$$0 \cong -g\frac{\partial \bar{\zeta}}{\partial x_j} - \frac{1}{\rho h}\frac{\partial S'_{ij}}{\partial x_i}. \qquad (2.52)$$

Thus, for any pure wave field, for example, the scattered field near a solid body, the variation of mean sea level is in static equilibrium with the divergence of the radiation stress field. It also follows from Eqs. (2.51) and (2.52) that the mean sea level $\bar{\zeta}$ is a second-order quantity $O(kA^2)$ for weak current and small-amplitude waves.

10.3 RADIATION STRESSES IN THE SHOALING ZONE—SMALL-AMPLITUDE WAVES ON CONSTANT OR NEARLY CONSTANT DEPTH

In this section we derive some explicit formulas for the wave-induced radiation stresses S'_{ij} in the shoaling zone. Consider first the more practical case of weak current where U_i is comparable to the wave velocity field (u'_i, w'). Due to the absence of breaking, turbulence must be rather insignificant except near the bottom. Consequently, the wave and the current fields should not affect each other at the first order in wave slope, and the linearized theory for waves is still applicable to the leading order. In particular, in a region of nearly constant depth and several wavelengths away from local scatterers, the wave potential can be expressed by Eq. (4.1), Chapter Nine, while wave-induced free-surface displacement ζ' is

$$\zeta'(x, y, t) = \operatorname{Re} \eta(x, y)e^{-i\omega t}$$

with η governed by the Helmholtz equation. Keeping only terms of the second order in wave slope, the upper limits of integration in Eq. (2.47) may be replaced by 0. The various terms involved are evaluated individually as follows:

$$\mathrm{I} = \rho \int_{-h}^{0} \overline{u'_i u'_j} \, dz = \frac{\rho g}{4} \operatorname{Re}\left(\frac{\partial \eta}{\partial x_i} \frac{\partial \eta^*}{\partial x_j}\right) \frac{1}{k^2}\left[1 + \frac{2kh}{\sinh 2kh}\right], \qquad (3.1)$$

$$\mathrm{II} = \tfrac{1}{2}\rho g\, \overline{\zeta'^2} = \tfrac{1}{4}\rho g\,|\eta|^2, \qquad (3.2)$$

$$\mathrm{III} = \rho \int_{-h}^{0} dz\, \frac{\partial}{\partial x_i} \int_{z}^{0} \overline{u'_i w'} \, dz$$

$$= \frac{\rho g}{4} \operatorname{Re} \frac{\partial}{\partial x_i}\left(\eta^* \frac{\partial \eta}{\partial x_i}\right)\frac{1}{2k^2}[2kh\cosh 2kh - 1], \qquad i = 1,2; \quad (3.3)$$

$$\mathrm{IV} = -\rho \int_{-h}^{0} \overline{w'^2} \, dz = \frac{\rho g}{4}|\eta|^2\left(\frac{2kh}{\sinh 2kh} - 1\right). \qquad (3.4)$$

Summing up the four integrals and using the Helmholtz equation, we obtain

$$S'_{ij} = \mathrm{I} + \mathrm{II} + \mathrm{III} + \mathrm{IV}$$

$$= \frac{\rho g}{4}\left\{\operatorname{Re}\left[\frac{\partial \eta}{\partial x_i} \frac{\partial \eta^*}{\partial x_j}\right]\frac{1}{k^2}\left(1 + \frac{2kh}{\sinh 2kh}\right)\right.$$

$$\left. + \delta_{ij}\left[|\eta|^2\frac{2kh}{\sinh 2kh} + \frac{2kh\cosh 2kh - 1}{2k^2}\left(\left|\frac{\partial \eta}{\partial x}\right|^2 + \left|\frac{\partial \eta}{\partial y}\right|^2 - k^2|\eta|^2\right)\right]\right\}$$

$$\tag{3.5}$$

(Mei, 1973).

The following two limits are easily found

1 Deep water $kh \gg 1$, $\omega^2 = gk$.

$$\frac{S_{xx}}{h} = \frac{S_{yy}}{h} = \frac{\rho g}{4} \frac{1}{k^2} (|\nabla \eta|^2 - k^2 |\eta|^2), \tag{3.6a}$$

$$\frac{S_{xy}}{h} = \frac{S_{yx}}{h} = 0. \tag{3.6b}$$

2 Shallow water $kh \ll 1$, $\omega = (gh)^{1/2} k$.

$$S_{xx} = \frac{\rho g}{2} \left(\frac{1}{k^2} \left| \frac{\partial \eta}{\partial x} \right|^2 + |\eta|^2 \right), \tag{3.7a}$$

$$S_{yy} = \frac{\rho g}{2} \left(\frac{1}{k^2} \left| \frac{\partial \eta}{\partial y} \right|^2 + |\eta|^2 \right), \tag{3.7b}$$

$$S_{xy} = S_{yx} = \frac{\rho g}{2} \frac{1}{k^2} \operatorname{Re} \frac{\partial \eta}{\partial x} \frac{\partial \eta^*}{\partial y}. \tag{3.7c}$$

For the special example of a progressive wave in the direction θ with respect to the x axis,

$$\eta = A e^{i\psi}, \qquad A \text{ real}, \tag{3.8}$$

where

$$\psi = k_1 x_1 + k_2 x_2, \tag{3.9}$$

$$k_1 = k \cos \theta, \qquad k_2 = k \sin \theta, \qquad x_1 = x, \qquad x_2 = y.$$

It is straightforward to show that the radiation stresses are

$$S'_{ij} = \frac{\rho g A^2}{4} \left\{ \frac{k_i k_j}{k^2} \left(1 + \frac{2kh}{\sinh 2kh} \right) + \delta_{ij} \frac{2kh}{\sinh 2kh} \right\}$$

$$= \frac{E}{2} \left\{ \frac{k_i k_j}{k^2} \frac{2C_g}{C} + \delta_{ij} \left(\frac{2C_g}{C} - 1 \right) \right\} \tag{3.10}$$

(Longuet-Higgins and Stewart, 1962, 1964). More explicitly, the components of the radiation stresses may be written

$$S'_{xx} = S'_{11} = \frac{E}{2}\left[\frac{2C_g}{C}\cos^2\theta + \left(\frac{2C_g}{C} - 1\right)\right],\qquad (3.11a)$$

$$S'_{yy} = S'_{22} = \frac{E}{2}\left[\frac{2C_g}{C}\sin^2\theta + \left(\frac{2C_g}{C} - 1\right)\right],\qquad (3.11b)$$

$$S'_{xy} = S'_{yx} = S'_{12} = S'_{21} = E\frac{C_g}{C}\sin\theta\cos\theta.\qquad (3.11c)$$

In very deep water $C_g/C \to \frac{1}{2}$ so that

$$S'_{xx} \cong \frac{E}{2}\cos^2\theta,\qquad (3.12a)$$

$$S'_{yy} \simeq \frac{E}{2}\sin^2\theta,\qquad (3.12b)$$

$$S'_{xy} = S'_{yx} \cong \frac{E}{2}\sin\theta\cos\theta,\qquad (3.12c)$$

while in very shallow water, $C_g/C \to 1$, so that

$$S'_{xx} \cong \frac{E}{2}[2\cos^2\theta + 1],\qquad (3.13a)$$

$$S'_{yy} \cong \frac{E}{2}(2\sin^2\theta + 1),\qquad (3.13b)$$

$$S'_{xy} = S'_{yx} \cong E\sin\theta\cos\theta.\qquad (3.13c)$$

Applications of Eqs. (3.10)–(3.13) will be found in the next section.

Although Eqs. (3.11)–(3.13) are derived for strictly constant depth, they are a good first approximation for slowly varying depth if A and k are interpreted as the local values corresponding to the local h.

For a strong current on a nearly constant depth, the linearized equations governing the leading order wave field can be inferred from Eqs. (6.17), (6.19), (6.20), and (6.23), Chapter Three, by ignoring nonlinearity and spatial variations of U_i and h. It can be shown that formulas (3.5) and (3.10) still hold if h is replaced by $h + \bar{\zeta}$, where $\bar{\zeta}$ is the mean sea level induced by the strong current according to Eq. (6.13), and ω and k must satisfy Eqs. (6.29) and (6.33), all of Chapter Three.

10.4   EMPIRICAL KNOWLEDGE OF BREAKING WAVES

At present, theoretical information on breaking waves on a sloping beach is still inadequate, especially with regard to breaking-induced turbulence. A succinct summary of the empirical knowledge of sinusoidal waves normally incident on a plane beach is given in Battjes (1974 a, b), from which much of the present section is extracted. The film entitled *Breaking Waves* made by Kjeldsen and Olsen (1971) is also very informative.

10.4.1   Breaking of Standing Waves on a Slope

On a smooth plane beach, the parameters which govern wave breaking are the wave slope and the beach slope $s \tan \alpha$. For sufficiently large s or sufficiently low amplitudes, an incident wave does not break and is completely reflected. When s decreases and/or kA increases, a threshold is reached where breaking begins. Irribarren and Nogales (1949) and others have found empirically that the single dimensionless parameter

$$\xi = \frac{s}{(H/\lambda_\infty)^{1/2}} = s\left(\frac{\pi}{k_\infty A}\right)^{1/2} \tag{4.1}$$

plays an important role, where H is the total height of the breaking wave and λ_∞ is the wavelength in deep water. The critical value is roughly

$$\xi_c \simeq \frac{4}{\pi^{1/2}} \simeq 2.3 \quad \text{or} \quad k_\infty A = \left(\frac{\pi}{4}\right)^2 s^2. \tag{4.2}$$

If $\xi < 2.3$, waves break and the reflection coefficient reduces to below unity. A heuristic explanation for the parameter ξ has been proposed by Munk and Wimbush (1969). Their reasoning is that the fluid acceleration downward along the slope cannot exceed the acceleration of a free-falling particle, that is, $g \sin \alpha$, without causing breaking. For a standing wave of local amplitude A, the maximum vertical fluid acceleration may be estimated as $\omega^2 A$; the maximum acceleration along the slope is roughly $\omega^2 A/\sin \alpha$. Hence, the critical value for breaking is

$$\frac{\omega^2 A}{\sin \alpha} = g \sin \alpha \quad \text{or} \quad k_\infty A = \sin^2 \alpha.$$

For small slope $s = \tan \alpha \simeq \sin \alpha$,

$$\xi_c \simeq \frac{s}{(H/\lambda_\infty)^{1/2}} = \frac{s\pi^{1/2}}{(k_\infty A)^{1/2}} = \pi^{1/2} = 1.772 \tag{4.3}$$

which is essentially the same as Eq. (4.2) for small slope except for the factor

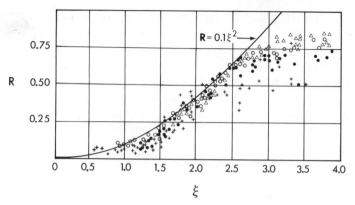

Figure 4.1 Reflection coefficient of a beach as a function of surf parameter $\xi = s(k_\infty A/\pi)^{-1/2}$ where A is the local amplitude. $+$: $s = 0.1$; $\bullet$: $s = 0.15$; $\bigcirc$: $s = 0.20$; and $\triangle$: $s = 0.30$ (from Battjes, 1974a, based on Moraes' data).

$(4/\pi)^2 = 1.62$. As will be shown in Chapter Eleven, Eq. (4.3) is the same as the theoretical deduction of Carrier and Greenspan (1957) based on Airy's nonlinear approximation and the criterion that the free surface is vertical at the breaking point. While the quantitative agreement may be accidental, it is reassuring that the role of ξ as the governing parameter has theoretical support.

As ξ decreases beyond the critical value, reflection from the slope decreases. Moraes (1970) has performed extensive experiments for wave reflection for various incoming waves and beach slopes. From these data Battjes (1974a) found that the reflection coefficient could be expressed as the function of ξ alone as shown in Fig. 4.1. This, of course, gives further significance to the quantity ξ which has been called the *surf parameter* by Battjes.

10.4.2 Types of Breakers on Mild Beaches

Three types of breakers may be crudely distinguished when ξ is below ξ_c (Galvin, 1968).[†] The *collapsing breaker* is associated with a large beach slope and flat incident waves. A small amount of breaking occurs only at the instantaneous shoreline. Dissipation of wave energy is quite small and reflection is nearly complete. When the beach slope decreases and/or the incident wave slope increases, a crest becomes forward-leaning as it advances toward the shore; its amplitude grows so that the profile is quite asymmetric. The crest ultimately curls forward and forms a jet plunging into the trough ahead; this type of breaker is called a *plunging breaker*, or a *plunger*. Because of the air tunnel formed by the splashing crest, the front of the breaking wave is accompanied by much noise and turbulence. Shortly after the collapse of the tunnel, a traveling bore is formed which marches shoreward with continued

---

[†]Galvin further calls an essentially nonbreaking wave on a beach a *surging breaker*.

dissipation. If the beach slope is further decreased and/or the incident wave slope increased, the onset of breaking occurs at a greater distance offshore when the wave crest is still symmetric. Breaking is signalled by the presence of foam draping the forward side of the crest; the trough in front is not visibly disturbed. This is called a *spilling breaker* in which dissipation takes place with a less violent appearance. Galvin classifies the breaker type on a quantitative scale:

$$\xi_b > 0.4 \qquad \text{spilling}$$

$$0.4 < \xi_b < 2.4 \qquad \text{plunging}$$

$$\xi_b > 2.0 \qquad \text{collapsing}$$

where ξ_b is the value of ξ at the breaker line. It must be emphasized that transition from one type to another is always gradual so that the numerical values marking the border lines cannot be precise. Theoretically, all three types may be regarded as plunging with the collapsing breakers having the narrowest surf zone and the spilling breakers having a negligible overhanging crest.

10.4.3 Maximum Wave Height

The maximum wave height corresponds to the limiting amplitude of a crest before breaking. There is as yet no simple theory for predicting the maximum wave height on a sloping beach. For a strictly horizontal bottom, there are perturbation theories of the Stokes type for periodic progressive waves (see Chapter Twelve). By taking the Stokes wave solution and assuming that at the threshold of breaking the fluid velocity at the crest equals the phase speed, Miche (1951) calculated the maximum wave height

$$\left(\frac{H}{\lambda}\right)_{\max} = \left(\frac{kH}{2\pi}\right)_{\max} = 0.14\tanh kh. \tag{4.4}$$

For very shallow water $kh \ll 1$, Eq. (4.4) gives

$$\frac{H_{\max}}{h} = 0.88, \tag{4.5}$$

which turns out to be in good agreement with experiments. Because the Stokes theory implies a symmetrical wave profile, the above result is more relevant to spilling breakers. Now breaking is a manifestation of extreme nonlinearity; a perturbation theory of a few orders is hardly satisfactory. An exact numerical theory of Schwartz (1974) for maximum wave height indicates, nevertheless, that Miche's estimate is quite good.

Experimental results for maximum wave height on a sloping bottom obtained by Iversen, Goda, Bowen et al., and Battjes have been collected by

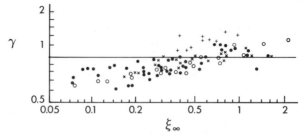

Figure 4.2 Breaker height to depth ratio as a function of $\xi_\infty = s(k_\infty A_\infty/\pi)^{-1/2}$ (from Battjes (1974a)). Sources of data are: ●: Iversen; ○: Goda; +: Bowen et al.; and ×: Battjes (1974a).

Battjes (1974a) as shown in Fig. 4.2. The ratio of height to depth at breaking is denoted by γ. The empirical range

$$\gamma = \frac{H_b}{h_b} = 0.7\text{--}1.2 \tag{4.6}$$

is roughly comparable to Eq. (4.5) and shows a weak dependence on the parameter ξ. In the absence of a satisfactory theory for wave breaking, these empirical results have been used for predicting the breaker line on a gently sloping beach in conjunction with a convenient theory for shoaling waves. From the theoretical relation between H and h before breaking, Eq. (4.6) may be invoked to calculate the breaking depth h_b, which, in turn, gives the position of the breaker line. As a crude estimate we take the linearized shoaling theory, Eq. (3.9), Chapter Three, to relate the wave amplitude and the depth for the entire shoaling zone. Near the breaker line, $kh \ll 1$; Eq. (3.10), Chapter Three, applies, namely,

$$\frac{A}{A_\infty} = (2|\cos\theta_\infty|)^{-1/2}\left(\frac{\omega^2 h}{g}\right)^{-1/4} \tag{4.7}$$

where A_0 and θ_0 have been replaced by A_∞ and θ_∞, respectively, and $k_\infty h_\infty \gg 1$ is assumed. Invoking $(A/h)_b = \frac{1}{2}(H/h)_b = \frac{1}{2}\gamma$, we get

$$\frac{\omega^2 h_b}{g} = \left(\frac{1}{\gamma}\frac{\omega^2 A_\infty}{g}\right)^{4/5}(2|\cos\alpha_\infty|)^{-1/2}. \tag{4.8}$$

From the observed data of H, Komar and Gaughan (1972) found that the best value for γ was 1.42 which was larger than that of Eq. (4.6). This discrepancy is likely due to the inadequacy of the linearized theory at the breaker line.

The ratio of breaker height to depth at the breaking line $\gamma = 0.7\text{--}1.2$ has been used also to approximate the ratio of height to depth of a breaking wave as it travels across the surf zone (Munk, 1949a). This simple statement describes empirically the diminishing of wave amplitude and has been adopted as a basic assumption in all longshore current theories.

10.5   THE STRUCTURE OF A UNIFORM LONGSHORE CURRENT ON A PLANE BEACH

Assume for simplicity that the beach has a plane bottom, that is, $h = sx$. Uniformity in the longshore direction of y implies that $\partial/\partial y = 0$. Hence the continuity equation is

$$\frac{\partial}{\partial x}\left[U(\bar{\zeta} + h)\right] = 0,$$

which leads to

$$U \equiv 0 \qquad \text{for all } x. \tag{5.1}$$

The mean velocity can only be alongshore, that is, $V = V(x)$.

In the offshore–onshore (x) direction, the bottom friction can be omitted from the momentum equation since $U = 0$; thus,

$$0 = -g\frac{\partial \bar{\zeta}}{\partial x} - \frac{1}{\rho(\bar{\zeta} + h)} \frac{\partial}{\partial x}(S'_{xx} + S''_{xx}). \tag{5.2}$$

In the alongshore (y) direction, the momentum equation reads

$$0 = -\frac{1}{\rho(\bar{\zeta} + h)}\left[\frac{\partial}{\partial x}(S'_{xy} + S''_{xy}) + \bar{\tau}_y^B\right]. \tag{5.3}$$

These formal equations are valid for both shoaling and surf zones. As in uniform pipe flows in ordinary fluid mechanics, the convective inertia terms are absent identically.

10.5.1   Shoaling Zone: $x > x_b$

In most of the shoaling zone the turbulence intensity is weak so that $S''_{xx} \ll S'_{xx}$. Furthermore, $\bar{\zeta}$ is of the second order in wave slope $O(kA^2)$ and is negligible compared to the still water depth. Hence, Eq. (5.2) can be written

$$0 \simeq -g\frac{\partial \bar{\zeta}}{\partial x} - \frac{1}{\rho h}\frac{\partial S'_{xx}}{\partial x}, \tag{5.4}$$

where S'_{xx} can be given approximately by Eq. (3.11a):

$$S'_{xx} = \frac{\rho g A^2}{4}\left[\frac{2C_g}{C}\cos^2\theta + \frac{2C_g}{C} - 1\right], \tag{5.5}$$

while C_g, C, θ, and A refer to the refracting waves implied by Eqs. (4.7) and (4.8). Qualitatively, S'_{xx} decreases as h decreases, that is, $\partial S'_{xx}/\partial x < 0$. It follows from Eq. (5.4) that $\partial \bar{\zeta}/\partial x > 0$, and $\bar{\zeta}$ steadily increases with x. Since it vanishes in very deep water, $\bar{\zeta}$ must be below the still water level; this variation is called the *set-down*. After some algebra, Eq. (5.4) may be integrated with respect to x to give $\bar{\zeta}$ (see Longuet-Higgins and Stewart, 1962, for normal incidence). The same result can also be derived simply by using the dynamic condition on the free surface. Approximating to the second order only, we have

$$g\zeta + \frac{\partial \Phi}{\partial t} + \zeta \frac{\partial^2 \Phi}{\partial t\, \partial z} + \frac{1}{2}\left(u'^2 + v'^2 + w'^2\right) \simeq 0, \qquad z = 0,$$

whose time average is

$$g\bar{\zeta} - \zeta \overline{\frac{\partial w'}{\partial t}} + \frac{1}{2}\left(\overline{u'^2} + \overline{v'^2} + \overline{w'^2}\right) = 0.$$

Using the kinematic condition on the free surface $\partial \zeta/\partial t = w'$, we get

$$g\bar{\zeta} = -\tfrac{1}{2}\left(\overline{u'^2} + \overline{v'^2} - \overline{w'^2}\right).$$

The linearized wave field may be recalled as

$$\Phi = \mathrm{Re}\, \frac{-igA}{\omega} \frac{\cosh k(z+h)}{\cosh kh} \exp\left[i\left(\int \alpha\, dx + \beta y - \omega t\right)\right].$$

Calculating the mean squares of the velocity components from this potential, we obtain

$$\bar{\zeta} = -\frac{kA^2}{2\sinh 2kh} \tag{5.6}$$

where derivatives of h have been ignored. Equation (5.6) is formally the same as if $h = \mathrm{const}$. Very near the breaker line $kh \ll 1$, Eq. (5.6) may be approximated by

$$\bar{\zeta} \simeq -\frac{A^2}{4h}. \tag{5.7}$$

For normal incidence, experiments by Bowen, Inman, and Simmons (1968) have confirmed Eq. (5.6) very well for nearly all points in the shoaling zone except very near the breaker line (see Fig. 5.1) where discrepancies are likely caused by nonlinearity and turbulence through S''_{xx}.

Now let us consider the alongshore (y) momentum. In the shoaling zone and far away from the breaker line, turbulence is negligible and there is no

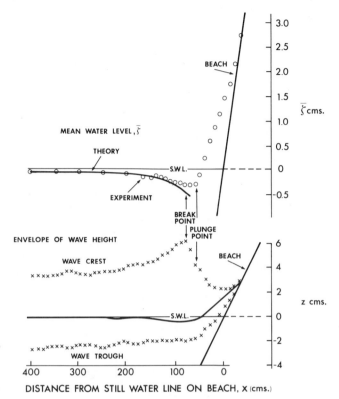

Figure 5.1 Comparison of experiments with theory for set-down and set-up on a plane beach. Data: wave period $= 1.14$ s; deep water wave height $H_\infty = 6.45$ cm; breaker height $H_b = 8.55$ m; beach slope $= 0.082$ (from Bowen et al., 1968, *J. Geophys. Res.*).

current; therefore,

$$\frac{\partial S'_{xy}}{\partial x} = 0 \quad \text{or} \quad S'_{xy} = \text{const.} \tag{5.8}$$

This result is a special case of the more general conclusion (2.52). As a check note that $EC_g \cos \theta$ is the constant rate of energy flux into a unit distance of y, and $(\sin \theta)/C$ is constant by Snell's law; therefore S'_{xy} is constant by Eq. (3.11c).

Near the breaker line there is a lateral transfer of turbulent momentum through the integrated Reynolds stress:

$$S''_{xy} = \rho \int_{-h}^{\bar{\zeta}} \overline{u''v''} \, dz. \tag{5.9}$$

If the modification of waves by turbulence diffused out of the surf zone is ignored, it follows from Eq. (5.3) that

$$0 = -\frac{\partial S''_{xy}}{\partial x} - \bar{\tau}_y^B.$$ (5.10)

This equation must be supplemented with further hypotheses on S''_{xy} and $\bar{\tau}_y^B$ as will be discussed later.

10.5.2 Surf Zone: $x < x_b$

Although formal equations are available (see Eqs. (2.47) and (2.48)), the stresses S'_{ij} and S''_{ij} are both unknown in the surf zone due to breaking-induced turbulence. This is a familiar situation in usual turbulent flows and closure hypotheses are needed in order to render the problem determinate. For the radiation stresses S'_{xx} and S'_{xy}, Bowen (1969), Longuet-Higgins (1970a, b), and Thornton (1970) introduced the important hypothesis that the relations (3.11)–(3.13) between S'_{ij} and $E = \frac{1}{2}\rho g A^2$ be formally valid even though they are derived for small-amplitude nonbreaking waves. In addition, the breaking-wave amplitude is assumed to be related to the local depth $(\bar{\zeta} + h)$ by

$$A = \frac{\gamma}{2}(\bar{\zeta} + h)$$ (5.11)

where γ is the same empirical constant given in Eq. (4.6).

We now apply this new hypothesis to the x-momentum equation in the surf zone. Using the shallow water approximation, we get

$$S'_{xx} = \tfrac{3}{16}\gamma^2\rho g(\bar{\zeta} + h)^2.$$ (5.12)

The cited authors did not include the turbulent stress S''_{xx} defined by Eq. (2.48); this omission is not easy to assess since S'_{xx} and S''_{xx} have not been measured separately.[†] Combining Eq. (5.12) with Eq. (5.4), we get the following differential equation:

$$0 = -\frac{\partial \bar{\zeta}}{\partial x} - \frac{3}{8}\gamma^2\frac{\partial}{\partial x}(\bar{\zeta} + h),$$

which immediately leads to

$$\frac{\partial \bar{\zeta}}{\partial x} = \frac{-\tfrac{3}{8}\gamma^2\,\partial h/\partial x}{1 + \tfrac{3}{8}\gamma^2}.$$ (5.13)

Integrating with respect to x and matching $\bar{\zeta}$ with the mean sea level $\bar{\zeta}_b$ at the

---

[†]Recall that in a turbulent pipe flow, terms analogous to S''_{xx} are present which alter the mean pressure corresponding to $\bar{\zeta}$ here.

breaker line, we get

$$\bar{\zeta} - \bar{\zeta}_b = \frac{\frac{3}{8}\gamma^2(h_b - h)}{1 + \frac{3}{8}\gamma^2}. \tag{5.14}$$

A crude estimate of $\bar{\zeta}_b$ can be obtained by using Eq. (4.8) and $(A/h)_b = \frac{1}{2}\gamma$ in Eq. (5.7). With this result the total mean depth may be written

$$\bar{\zeta} + h = \frac{h - h_s}{1 + \frac{3}{8}\gamma^2}, \tag{5.15}$$

where

$$h_s = -\left[\left(1 + \frac{3}{8}\gamma^2\right)\bar{\zeta}_b + \frac{3}{8}\gamma^2 h_b\right] \tag{5.16}$$

is the still water depth of the mean shoreline at which $(\bar{\zeta} + h)_s = 0$.

As the incident wave amplitude A_∞ increases, h_b increases. While the change in $\bar{\zeta}_b$ is small, the right-hand side of Eq. (5.14) increases. Hence, the set-up $\bar{\zeta}$ increases. Equation (5.13) has been verified experimentally for a plane beach by Bowen, Inman, and Simmons (1968); see Fig. 5.1. It may now be concluded that the momentum balance in the x direction results in a static set-down in the shoaling zone and a set-up in the surf zone.

With the mean sea-level known, the radiation shear stress in the shallow surf zone is

$$S'_{xy} \simeq \frac{1}{16}\rho g\gamma^2(\bar{\zeta} + h)^2\sin 2\theta \tag{5.17}$$

when Eq. (5.11) is combined with Eq. (3.13c). From Eq. (2.51), S'_{xy} now acts as the forcing term in the longshore direction which must be balanced by bottom friction and the turbulent Reynolds stress. At this stage empirical relations must be added to express $\bar{\tau}_y^B$ and S''_{xy} in terms of the mean current $V(x)$. We follow Longuet-Higgins (1970a) in this regard.

In open-channel hydraulics *bottom friction* is usually related to the total mean velocity excluding u''_i:

$$\bar{\tau}^B = \frac{f}{2}\rho\overline{|\mathbf{U} + \mathbf{u}'|(\mathbf{U} + \mathbf{u}')}, \qquad f = \text{friction coefficient}, \tag{5.18}$$

where $\mathbf{u}'$ is the wave velocity field evaluated at $z = -h$. Thus, in the y direction the mean bottom friction is

$$\bar{\tau}_y^B = \frac{f}{2}\rho\overline{|\mathbf{U} + \mathbf{u}'|(V + v')}. \tag{5.19}$$

For reasonably small incidence angle the wave vector in shallow water is essentially in the x direction, namely, $v' \ll u'$, while the shear current is in the y direction so that the total velocity is approximately

$$\mathbf{U} + \mathbf{u}' \cong u'\mathbf{e}_x + V\mathbf{e}_y,$$

and

$$\overline{\tau_y^B} \cong \frac{f}{2}\rho \overline{(u'^2 + V^2)^{1/2}} V.$$

Now $u' = U_0\cos \omega(t + t_0)$ where U_0 is the amplitude of the orbital (oscillatory) velocity at the bottom. In the surf zone the magnitudes of V and U_0 are usually comparable and the average

$$\overline{(U_0^2\cos^2\omega t + V^2)^{1/2}} = \frac{1}{T}\int_0^T (V^2 + U_0^2\cos^2\omega t)^{1/2}\, dt$$

$$= \frac{1}{T}\int_0^T ((V^2 + U_0^2) - U_0^2\sin^2\omega t)^{1/2}\, dt \quad (5.20)$$

may be evaluated in terms of the complete elliptic integral of the second kind (Jonsson, Skovgaard, and Jacobsen, 1974). Longuet-Higgins assumes, however, that $V \ll U_0$ so that

$$\overline{(u^2 + V^2)^{1/2}} \simeq \overline{|u'|} = U_0 \overline{|\cos \omega t|} = \frac{2}{\pi}U_0.$$

Consequently, the bottom friction becomes

$$\overline{\tau_y^B} \simeq -\frac{1}{\pi}f\rho U_0 V. \qquad (5.21)$$

The orbital velocity amplitude in the shoaling zone near the breaker line may be estimated according to the linear shallow water theory,

$$U_0 = \frac{A}{h}(gh)^{1/2}.$$

In the surf zone, no wave theory is available, but an order-of-magnitude estimate can be made by replacing h with $\bar{\zeta} + h$, and by using Eq. (5.11):

$$U_0 = \frac{A}{\bar{\zeta} + h}(g(\bar{\zeta} + h))^{1/2} \simeq \frac{1}{2}\gamma[g(\bar{\zeta} + h)]^{1/2}. \qquad (5.22)$$

The mean depth for a plane beach may be inferred from Eq. (5.15):

$$\bar{\zeta} + h = s'x', \qquad (5.23)$$

where

$$s' = \frac{s}{1 + \frac{3}{8}\gamma^2}, \tag{5.24}$$

and

$$x' = x - x_s = x - \frac{h_s}{s} \tag{5.25}$$

is the offshore distance measured from the mean shoreline. Thus, from Eq. (5.21)

$$\bar{\tau}_y^B = \frac{1}{\pi} \frac{f}{2} \rho\gamma (gs'x')^{1/2} V. \tag{5.26}$$

Moreover, Longuet-Higgins assumes the same formula for the shoaling zone.

The integrated Reynolds stress or the turbulent radiation stress S''_{xy} is assumed to be of the following form:

$$-S''_{xy} = (\bar{\zeta} + h)\rho\nu_e \frac{\partial V}{\partial x}, \tag{5.27}$$

where ν_e is the eddy viscosity which has the dimension

$$[\nu_e] = [U][L].$$

Longuet-Higgins takes the velocity scale $[U]$ to be proportional to $(g(\bar{\zeta} + h))^{1/2}$ which is of the same order as the orbital velocity U_0. Analogous to the mixing length of a turbulent flow near a wall, $[L]$ may be taken to be proportional to the distance x' from the shore. Thus,

$$-S''_{xy} = N\rho g^{1/2} s'^{3/2} x'^{5/2} \frac{dV}{dx'} \tag{5.28}$$

after the use of Eq. (5.23); N is an empirical coefficient. Again the same expression is assumed for the shoaling zone, which is likely an overestimate because of the expected reduction of turbulence.

In summary, the momentum equation is

$$N\rho g^{1/2} s'^{3/2} \frac{d}{dx'}\left(x'^{5/2}\frac{dV}{dx'}\right) - \frac{1}{\pi}\frac{f}{2}\rho\gamma g^{1/2} S'^{1/2} x'^{1/2} V$$

$$= -\frac{5}{16} g^{3/2} s'^{5/2} \gamma^2 \left(\frac{\sin\theta}{C}\right)_\infty (x')^{3/2}, \quad 0 < x' < x'_b,$$

$$= 0, \qquad\qquad\qquad\qquad x' > x'_b, \tag{5.29}$$

which may be solved by requiring that V be bounded for all $0 < x' < \infty$.

The solution for the longshore current profile $V(x')$ can be simply obtained. The analytical details are straightforward and we shall only present the numerical results in Fig. 5.2. The results depend on the parameter

$$P = 2\pi \frac{s'N}{\gamma f} \tag{5.30}$$

which signifies the relative importance of lateral turbulent diffusion to bottom friction. Qualitatively, the longshore velocity decays outside the surf zone. The experimental results of Galvin and Eagleson (1965) scatter within the range $0.05 < P < 1.0$ in Fig. 5.2.

The profile for zero turbulent diffusion gives a useful guide for the order of magnitude of the longshore current, although the discontinuity at the breaker line is physically unrealistic. Taking $N = 0$, we have

$$
\begin{aligned}
V &= 0, & x > x_b, \\
V &= \frac{5\pi}{8} \frac{\gamma s'}{\rho f} g s' x' \left(\frac{\sin\theta}{C} \right)_b, & x < x_b,
\end{aligned}
\tag{5.31}
$$

which is a discontinuous distribution with the following maximum at the breaker line:

$$V_m = \frac{5\pi}{8} \frac{\gamma s'}{\rho f} \left(g(\bar{\zeta} + h)_b \right)^{1/2} \sin\theta_b. \tag{5.32}$$

Thus, V_m increases with increasing incidence angle, bottom slope, or incident wave amplitude through $(\bar{\zeta} + h)_b$. By using empirical data in open-channel

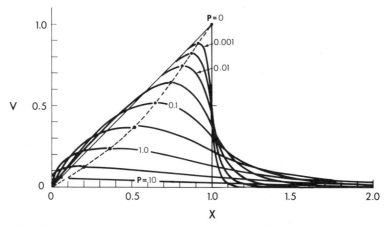

Figure 5.2 Theoretical form of the longshore current V/V_b as a function of $X = x/x_b$ and the lateral mixing parameter $P = 2\pi(s'N/\gamma f)$. (From Longuet-Higgins, 1970b, *J. Geophys. Res.*)

flows and assuming a sand grain diameter of 1 mm for bed roughness, Longuet-Higgins (1970a) estimated $f = 0.02$. Equation (5.32) then crudely agrees with both field and laboratory data if the breaker line depth h_b is reasonably estimated.

10.6 OTHER EMPIRICAL HYPOTHESES OR IMPROVEMENTS

From the simple example of a longshore current it is evident that one must resort to a number of empirical approximations which are based on limited observations. For example, that S'_{ij} has the same expression in both surf and shoaling zones can only be regarded as a physically plausible assumption. The hypothesis of Eq. (5.11) has some experimental support only for normally incident waves on a plane beach, but little is known for more complicated waves. Unfortunately, there appears to be no better replacements yet for these major assumptions. A few other questions, and attempted answers, are discussed below in order to reflect recent ideas on this subject.

10.6.1 Bottom Friction

The Friction Factor in the Joint Presence of Waves and Currents

Abundant experimental information exists for the friction coefficient f_c when the flow is unidirectional and steady. On the other hand, there are data for pure oscillatory flows without breaking waves, and semiempirical formulas for f_w are available [see Eq. (7.22), Chapter Eight]. These data are only appropriate for the shoaling zone. Jonsson, Skovgaard, and Jacobsen (1974) have proposed an interpolation formula for the combined case where both waves and current are present. They define the instantaneous f by

$$\tau^B = \tfrac{1}{2}\rho f\,|\mathfrak{U}|\,\mathfrak{U}, \qquad (6.1)$$

where $\mathfrak{U}$ is the vector sum of the current and the inviscid orbital velocity at the sea bottom. The instantaneous f is assumed to be

$$f = f_w + (f_c - f_w)\sin \mu, \qquad (6.2)$$

$\mu(t)$ being the angle between $\mathfrak{U}$ and the instantaneous wave orbital velocity.

In the special case where the current and the wave orbits are orthogonal, we have

$$\tan \mu = \frac{V}{U_0 \cos \omega t}. \qquad (6.3)$$

The time average of Eq. (6.1) leads to the integral in Eq. (5.20) which, in turn, gives

$$\bar{\tau}_y^B = \tfrac{1}{2} f_e \rho V^2,   \tag{6.4}$$

where the effective f_e is

$$f_e = f_c + \left\{ \frac{2}{\pi} \left[1 + \left(\frac{U_0}{V} \right)^2 \right]^{1/2} E(m) - 1 \right\} f_w,   \tag{6.5}$$

with $E(m)$ being the complete elliptic integral of the first kind and m its modulus.

$$E(m) = \int_0^{\pi/2} (1 - m \sin^2 \tau)^{1/2} \, d\tau   \qquad m = U_0^2 (V^2 + U_0^2)^{-1},   \tag{6.6}$$

For weak current $V/U_0 \ll 1$, it may be shown again that

$$\bar{\tau}_y^B = \frac{1}{\pi} f_w \rho U_0 V.   \tag{6.7}$$

No direct experimental verification of the general formula (6.2) is yet available. Moreover, adequate information on f_w is still missing for the surf zone. The situation is much better in the shoaling zone where a more rational theory has been proposed by Grant and Madsen (1979b) which extends the turbulent boundary-layer picture to waves with currents for any angle between them.

Large Angle of Incidence and Strong Current

Most existing laboratory experiments are performed with a fairly large incidence angle at the breaker line (10°–50°), and the measured longshore currents are often comparable to the local orbital velocity. Both these facts are not in accord with the assumptions made in Section 10.5. More specifically, if we use Eq. (5.31) for the longshore current velocity V and $\tfrac{1}{2}\gamma(g(\bar{\xi} + h))^{1/2}$ for the local orbital velocity U, then the assumption $V/u \ll 1$ implies a severe limitation on the incidence angle throughout the surf zone:

$$\sin \theta \ll \frac{4f(1 + 3\gamma^2/8)}{5\pi s} \cong \frac{0.315 f}{s}$$

(Liu and Dalrymple, 1977). If we take $f = 0.01$, $s = 0.1$ for laboratory experiments and $s = 0.01$ for natural beaches, then the theory (5.31) is limited to $\theta \ll 1.8°$ in the laboratory and $\theta \ll 18°$ in the field. Liu and Dalrymple (1978) reconsidered the problem by allowing large incidence angle and relatively strong longshore current, but ignoring lateral turbulence for simplicity. We

shall first discuss their theory of weak current but large angle of incidence.
   Let the total velocity be

$$\mathscr{U} = (\overline{U}\cos\alpha + u'\cos\theta)\mathbf{e}_x + (\overline{U}\sin\alpha + u'\sin\theta)\mathbf{e}_y, \tag{6.8}$$

where $\overline{U}$ and u' are the mean current and the oscillatory velocity, respectively, and α and θ are their inclinations with respect to the x axis. Let U_0 denote the magnitude of u', that is, $u' = U_0\cos\omega t$, and assume the current to be weak, that is, $\overline{U}/U_0 \ll 1$, then

$$|\mathscr{U}| \cong |u'| + \overline{U}\frac{|u'|}{u'}\cos(\theta - \alpha).$$

Putting this result into Eq. (6.1), taking averages, and using the fact that

$$\overline{|u'|} = \frac{2U_0}{\pi}, \qquad \overline{|u'|u'} = 0,$$

we obtain

$$\overline{\tau}^B = \rho f\left(\frac{U_0}{\pi}\right)\left\{[U(1 + \cos^2\theta) + V\sin\theta\cos\theta]\mathbf{e}_x\right.$$

$$\left. + [V(1 + \sin^2\theta) + U\sin\theta\cos\theta]\mathbf{e}_y\right\}, \tag{6.9}$$

where

$$U = \overline{U}\cos\alpha, \qquad V = \overline{U}\sin\alpha.$$

When the mean current and the orbital velocity are colinear, $\theta = \alpha = 0$, $V = 0$, and $\overline{U} = U$ so that

$$\overline{\tau}_B = 2f\left(\frac{U_0}{\pi}\right)U\mathbf{e}_x. \tag{6.10a}$$

On the other hand, if the two velocities are perpendicular, $\theta = 0$, $\alpha = \frac{1}{2}\pi$, then $U = 0$ and $\overline{u} = V$, so that

$$\overline{\tau}^B = \rho f\left(\frac{U_0}{\pi}\right)V\mathbf{e}_y \tag{6.10b}$$

which is the same as Eq. (6.7). The preceding formulas differ by a factor of 2, as first pointed out by Jonsson (1966).
   For a longshore current, $U = 0$ and $\alpha = \frac{1}{2}\pi$. Let θ be arbitrary so that

$$\overline{\tau}^B = \rho\frac{f}{2}\left(\frac{U_0}{\pi}\right)\left\{(V\sin 2\theta)\mathbf{e}_x + 2V(1 + \sin^2\theta)\mathbf{e}_y\right\}. \tag{6.11}$$

Using this and proceeding as in Section 10.5, one obtains

$$0 = -g\frac{d\bar{\zeta}}{dx} - \frac{g\gamma^2}{8(\bar{\zeta}+h)}\frac{d}{dx}\left[(\bar{\zeta}+h)^2(\tfrac{3}{2} - \sin^2\theta)\right]$$

$$- \frac{f}{2}\frac{\gamma}{\pi}g^{1/2}(\bar{\zeta}+h)^{-1/2}V\sin\theta\cos\theta \qquad (6.12)$$

in the x direction (offshore), and

$$0 = -\frac{g\gamma^2}{8(\bar{\zeta}+h)}\frac{d}{dx}\left[(\bar{\zeta}+h)^2\sin\theta\cos\theta\right]$$

$$- \frac{f}{2}\frac{\gamma}{\pi}g^{1/2}(\bar{\zeta}+h)^{-1/2}V(1+\sin^2\theta) \qquad (6.13)$$

in the y direction (alongshore). Note that the mean sea level $\bar{\zeta}$ is no longer independent of the longshore current. If Snell's law, $(\sin\theta)/C = ((\sin\theta)/C)_b$, and $C = [g(\bar{\zeta}+h)]^{1/2}$ are assumed, Eqs. (6.12) and (6.13) may be combined to give a single ordinary differential equation for $(\bar{\zeta}+h)$. The boundary condition is:

$$(\bar{\zeta}+h) = (\bar{\zeta}+h)_b. \qquad (6.14)$$

From the solution the mean shoreline position x_s can be immediately inferred, and the longshore current distribution V can be computed. The resulting formulas are omitted and only the current profiles are shown in Fig. 6.1. For a larger angle of incidence, the current velocity falls below the prediction [Eq. (5.32)] by Longuet-Higgins.

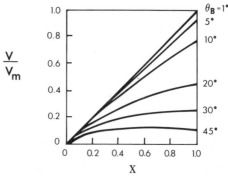

Figure 6.1 Nondimensional longshore current profiles by assuming weak current in bottom friction. Lateral diffusion is ignored. V_m is given by Eq. (5.32) and $X = (x - x_s)/(x_b - x_s)$. (From Liu and Dalrymple, 1978, *J. Marine Res.*)

Liu and Dalrymple further considered the effect of a strong current $|u'/U|$ ≥ 1. In general, the total velocity magnitude is

$$|\mathfrak{U}|=\left[(u')^2 + \overline{U}^2 + 2u'\overline{U}\cos(\theta - \alpha)\right]^{1/2}.$$

It turns out that the following three-term expansion for small $u'/\overline{U}$

$$|\mathfrak{U}|\cong \overline{U}\left\{1 + \left(\frac{u'}{U}\right)\cos(\theta - \alpha) + \frac{1}{2}\left(\frac{u'}{U}\right)^2 \sin^2(\theta - \alpha)\right\}^{1/2}$$

gives a fairly good numerical approximation for $|u'/U|< 1$ as long as $\theta - \alpha > 45°$, which is normally the case. The error term is $\frac{1}{2}(u'/U)^3\cos(\theta - \alpha)$ which further diminishes as $|\theta - \alpha|\to \frac{1}{2}\pi$. With this approximation the averaged bottom stress becomes, in general,

$$\overline{\tau}^B = \frac{\rho f}{4}\left\{\left[\frac{U_0^2}{2}\left(\frac{1}{2} \sin^2(\theta - \alpha)\cos \alpha + \cos(\theta - \alpha)\cos \theta\right) + \overline{U}\,U\right]\mathbf{e}_x\right.$$

$$\left. + \frac{U_0^2}{2}\left(\frac{1}{2} \sin^2(\theta - \alpha)\sin \alpha + \cos(\theta - \alpha)\sin \theta\right) + \overline{U}V\right]\mathbf{e}_y\right\}. \quad (6.15)$$

For a longshore current $\alpha = \frac{1}{2}\pi$, we have

$$\overline{\tau}^B = \frac{\rho f}{2}\left\{\left[\frac{U_0^2}{2} \sin 2\theta\right]\mathbf{e}_x + \left[V^2 + \frac{U_0^2}{4}(1 + \sin^2\theta)\right]\mathbf{e}_y\right\} \quad (6.16)$$

which may be inserted in the mean momentum equations for numerical solution. Sample results in Fig. 6.2 show a considerable reduction from the weak-current theory of Longuet-Higgins [see Eq. (5.32)].

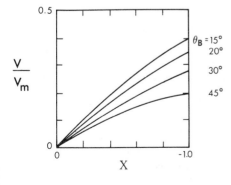

$\frac{V}{V_m}$

$\theta_B = 15°$
$20°$
$30°$
$45°$

X

Figure 6.2 Nondimensional longshore current profiles by assuming strong current in bottom friction. Lateral diffusion is ignored; bottom slope $s = 0.05$; $f = 0.025$. (From Liu and Dalrymple, 1978, *J. Marine Res.*)

10.6.2 Lateral Turbulent Diffusion S''_{xy}

In the theory of Bowen (1969), the simplest assumption is made that the eddy viscosity is constant in and outside the surf zone.

In the mixing length argument, Longuet-Higgins (1970b) estimates the velocity scale to be less than $0.1U_0 = 0.1\{\frac{1}{2}[g(\zeta + h)]^{1/2}\}$, and the mixing length scale to be $\kappa x'$ where $\kappa = 0.40$ is the Kármán constant; thus

$$N \leqq (0.1)\left(\frac{\gamma}{2}\right)\kappa \leqq 0.016. \tag{6.17}$$

Thornton (1970) and Jonsson, Skovgaard, and Jacobsen (1974) suggest that

$$-S''_{xy} = \rho\left(\frac{\gamma^2}{2\pi}s'\right)g^{1/2}s'^{3/2}x'^{5/2}\frac{dV}{dx}, \tag{6.18}$$

which implies an equivalent N of

$$N = \frac{\gamma^2}{2\pi}s' \simeq 0.1s'. \tag{6.19}$$

For $s' \sim 0.1$, Eq. (6.19) is not much different from Eq. (5.28) except that the former depends on the beach slope.

Battjes (1975) argues that turbulence in the surf zone is derived primarily from wave breaking and not from the horizontal shear in the mean longshore current. The rate of dissipation (into heat) of turbulent energy should be equal to the rate of loss of wave energy by breaking. The latter is approximately

$$\frac{d}{dx}(EC_g) = \frac{d}{dx}\left(\frac{1}{2}\rho g A^2 C_g\right) \simeq \frac{\gamma^2}{8}\rho g\frac{d}{dx}\left[h^2(gh)^{1/2}\right]$$

$$= \frac{5\gamma^2}{16}\rho g^{3/2}h^{3/2}\frac{dh}{dx}.$$

Now the rate of turbulent energy dissipation ε can be estimated by the fluctuating velocity q'' and the eddy size which must be limited by the depth h. By dimensional reasoning, we have $\varepsilon = q''^3/h$ per unit mass. Equating ε and dEC_g/dx in order of magnitude, we get an estimate of the turbulent velocity:

$$q'' \sim \left(\frac{5\gamma^2}{16}\frac{dh}{dx}\right)^{1/3}(gh)^{1/2}. \tag{6.20}$$

Battjes further assumes that the integrated Reynolds stress is related to the mean shear by

$$-S''_{xy} = -\rho\int_{-h}^{\bar{\zeta}}\overline{u''v''}\,dz \sim \rho h(q''h)\frac{dV}{dx}.$$

Substituting Eq. (6.20) for q'', he finds an expression for the eddy viscosity,

$$\nu_e \sim q''h = K\left(\frac{5}{16}\rho g^{3/2}h^{3/2}\frac{dh}{dx}\right)^{1/3}h \qquad (6.21)$$

where K is a constant of order unity. In order to compare Eq. (6.21) with Longuet–Higgins' assumption we express ν_e as

$$\nu_e = Nx(gh)^{1/2}.$$

The corresponding N for a plane beach with slope s is

$$N = \left(\tfrac{5}{16}\gamma^2\right)^{1/3}s^{4/3}K. \qquad (6.22)$$

This result differs slightly from that of Thornton and Jonsson et al. in the dependence of N on beach slope s. Available experiments are not yet sufficiently extensive to enable a choice among these alternative models.

Finally, observations by dye injection have shown that turbulence outside the surf zone is weak (Inman, Tait, and Nordstrom, 1971); therefore they do not support the assumption made in most theories that the eddy viscosities in and outside the surf zone are the same. Although this weakness is widely recognized, no satisfactory substitute is yet known.

10.7 CURRENTS BEHIND AN OFFSHORE BREAKWATER

The uniform longshore current along a straight beach is one dimensional with variations in the offshore–onshore direction only; the associated mathematics is especially simple since the convective acceleration of the mean current vanishes identically. We shall now discuss briefly two examples where the wave system, hence also the current, are two dimensional. Certain approximations will be made for the sake of mathematical expedience. Specifically, we shall omit convective inertia and turbulent diffusion. Further, we shall use very simple models for the bottom friction by ignoring the variation of the angle between the local mean current and the wave velocities. Also omitted is the possible modification of the waves by the mean current. Intuitively, waves are modified the most where there is a strong rip current (a narrow current in the offshore direction), and our omission is likely to give quantitatively poor results. Owing to the lack of definitive empirical information, an elaborate model will not be attempted here so that the results are of qualitative and order-of-magnitude significance only. The purpose of our examples in this section and the next is to illustrate how much a simple model can tell us about the current patterns, the difference in forcing mechanisms, and the possible inferences on littoral transport, in two interesting situations.

For the prevention of a given stretch of beach from excessive erosion, an offshore breakwater is sometimes built at a certain distance parallel to the shore to reduce the influence of the incident waves, as shown in Fig. 7.1. A well-known consequence is that, after a long time, the sandy beach may develop a cusp at the center of the wave shadow. If the breakwater is close enough to the shore the cusp head may eventually reach the back of the breakwater. This process is known in coastal geography to be partly responsible for the linking of an offshore island with the mainland and is called the *tombolo* effect. More on the offshore island will be discussed later. For the offshore breakwater, laboratory studies on a plane beach have been conducted by Sauvage de Saint Marc and Vincent (1955). On the bottom of a rigid beach, two opposing currents were observed to flow toward the center line of the shadow, and then to combine into a rip current toward the breakwater. When the bottom of the beach was covered with sand, the sand particles drifted towards and were deposited near the stagnation point along the initial shore. The deposition eventually grew to be a cusp. It is, therefore, of engineering and scientific interest to predict the current pattern behind the breakwater.

10.7.1 The Wave Field

Assume the bottom contours of the beach to be straight and parallel so that $h = h(x)$. Let the breakwater be of length B and be located outside the breaker line at $x = L$ from the still water shoreline ($x = 0$), where $L > x_b$. A normally incident wavetrain arrives from $x \sim \infty$. The waves in the shoaling zone must be first calculated by accounting for both diffraction and refraction; generally this is a numerical task. For this special case we follow the analytical approach of Liu and Mei (1974, 1976a) who extended the *parabolic approximation* to the case of varying depth.

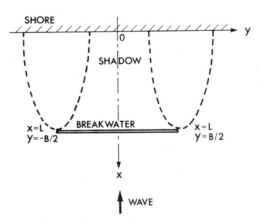

Figure 7.1 An offshore breakwater.

According to the geometrical optics approximation, the lines $y = \pm B/2$ and the breakwater divide the whole region into three zones. In the illuminated zone $|y| > B/2$, a plane progressive wave is dominant. In the reflection zone $|y| < B/2$, $x > L$, there is a standing wave due to the superposition of the incident and the reflected waves. All these waves are slowly refracted by the changing depth. In the shadow zone $|y| < B/2$, $0 < x < L$, no waves are present. This approximation is, of course, discontinuous near the boundaries of these zones. Boundary layers are needed to describe the transition smoothly by accounting for diffraction.

The exact linearized formulation of the diffraction problem calls for the solution of the boundary-value problem for the velocity potential $\Phi(x, y, z) = \phi(x, y, z)e^{-i\omega t}$, where ϕ satisfies Eqs. (5.4a)–(5.4b), Section 3.5, with $h = h(x)$. We assume that the depth varies slowly within a wavelength and define a small parameter μ such that

$$\mu = O\left(\frac{1}{kh}\frac{dh}{dx}\right) \ll 1. \tag{7.1}$$

Intuitively, the slow depth variation produces slow variations in the amplitude and phase of the propagating wave. Let us concentrate on the vicinity of the shadow boundary along $y - B/2$, $0 < x < L$ and introduce $\tilde{y} = y - B/2$. Far outside the shadow, the waves simply experience shoaling effects; the potential should approach

$$\phi \sim \psi(\mu x, z)\exp\left[-i\int k(\mu x)dx\right] \tag{7.2}$$

where $k(x)$ is the local wavenumber. If the breakwater is long enough so that the transition regions of the two shadow boundaries do not intersect, we expect that

$$\phi \sim 0 \tag{7.3}$$

far inside the shadow. Now we assume

$$\phi = \psi(\mu x, z)D\exp\left(-i\int k\,dx\right) \tag{7.4}$$

where the factor D accounts for diffraction and takes the following limits:

$$D \to 1 \qquad \text{as } \tilde{y} \to \infty \tag{7.5a}$$

and

$$D \to 0 \qquad \text{as } \tilde{y} \to -\infty. \tag{7.5b}$$

In the classical problem of tip diffraction in a uniform medium (corresponding to $h = $ const), it is known that the neighborhood of the shadow boundary is like a boundary layer. In particular, the variations in the transverse $(\tilde{y})$ direction are much more rapid than those in the longitudinal (x) direction, and that variations in both directions are slow sufficiently far away from the tip. Accepting this for the time being, we now let the diffraction factor D depend on x, $\tilde{y}$ slowly so that

$$D = D(\alpha x, \beta \tilde{y}) \qquad \text{with } \alpha \ll \beta \ll 1.$$

The precise nature of α and β will be determined later. Substituting Eq. (7.4) into Laplace's equation, we obtain

$$
\left\{ \frac{\partial^2 \psi}{\partial z^2} - k^2 \psi + \mu \left[2ik \frac{\partial \psi}{\partial(\mu x)} + i \frac{\partial k}{\partial(\mu x)} \psi \right] \right\} D
$$

$$
+ \left[\beta^2 \frac{\partial^2 D}{\partial(\beta \tilde{y})^2} - \alpha ik \frac{\partial D}{\partial(\alpha x)} \right] \psi = O(\alpha \mu, \alpha^2, \mu^2), \qquad (7.6)
$$

where the exponential factor has been omitted. The boundary conditions are

$$
\frac{\partial \psi}{\partial z} - \frac{\omega^2}{g} \psi = 0, \qquad z = 0, \qquad (7.7)
$$

and

$$
\frac{\partial \psi}{\partial z} = \mu ik \frac{dh}{d(\mu x)} \psi + O(\mu^2), \qquad z = -h(x). \qquad (7.8)
$$

Note that the boundary conditions do not depend on $\tilde{y}$. For $\tilde{y} \to \infty$, $D \to 1$, Eq. (7.6) reduces to

$$
\frac{\partial^2 \psi}{\partial z^2} - k^2 \psi = -\mu \left[2ik \frac{\partial \psi}{\partial(\mu x)} + i \frac{\partial k}{\partial(\mu x)} \psi \right] + O(\mu^2) \qquad (7.9)
$$

subject to the boundary conditions (7.7) and (7.8). This problem is a special case of the refraction theory, Chapter Three; the leading-order solution is

$$
\psi = -\frac{igA}{\omega} \frac{\cosh k(z + h)}{\cosh kh} [1 + O(\mu)] \qquad (7.10)
$$

where ω, k and A must satisfy Eqs. (1.15) and (3.9), Chapter Three, with

$\theta = \theta_0 = 0$. Because of Eq. (7.9), it follows from Eq. (7.6) that

$$\beta^2 \frac{\partial^2 D}{\partial(\beta\tilde{y})^2} - 2\alpha ik \frac{\partial D}{\partial(\alpha x)} \simeq 0. \tag{7.11}$$

Since the task of deducing the approximate equation is at last achieved, the small parameters μ, α, and β may now be dropped and the original variables x and y restored. Thus, Eq. (7.11) is simply

$$\frac{\partial^2 D}{\partial\tilde{y}^2} - ik\frac{\partial D}{\partial x} \simeq 0, \tag{7.12}$$

subject to the boundary conditions (7.5a) and (7.5b). The variable coefficient in Eq. (7.12) may be eliminated by introducing

$$\xi = k_0 \int_x^L \frac{dx}{k}, \tag{7.13}$$

where k_0 is the wavenumber at the breakwater $x = L$. Then,

$$\frac{\partial^2 D}{\partial\tilde{y}^2} + 2ik_0\frac{\partial D}{\partial\xi} = 0, \tag{7.14}$$

which is just the Schrödinger equation. Indeed the present boundary-value problem has been solved in Section 2.4.2 in a totally different context. For convenience the salient results are quoted here. Let

$$\sigma = \frac{k_0\tilde{y}}{(\pi k_0\xi)^{1/2}}, \tag{7.15}$$

then the diffraction factor D may be expressed

$$D = \frac{1}{2^{1/2}}e^{-i\pi/4}\{[\tfrac{1}{2} + C(\sigma)] + i[\tfrac{1}{2} + S(\sigma)]\}. \tag{7.16}$$

The magnitude of D is

$$|D| = \frac{1}{2^{1/2}}\{[\tfrac{1}{2} + C(\sigma)]^2 + [\tfrac{1}{2} + S(\sigma)]^2\}^{1/2}. \tag{7.17}$$

The diffraction factor is constant for $\sigma = $ const. Therefore, one may define the boundary of the transition zone by

$$(k_0\tilde{y})^2 = (\text{const})\pi k_0\xi = (\text{const})\pi k_0\int_x^L \frac{dx}{k} \qquad (7.18)$$

which is a sort of parabola.

From the solution for D, Eq. (7.18), it is clear that

$$\frac{1}{k_0 D}\frac{\partial D}{\partial x} = O\left(\frac{\partial \sigma}{\partial x}\right) = O(k_0\xi)^{-1},$$

$$\frac{1}{k_0 D}\frac{\partial D}{\partial \tilde{y}} = O\left(\frac{\partial \sigma}{\partial \tilde{y}}\right) = O(k_0\xi)^{-1/2}. \qquad (7.19)$$

Thus, variations in x and y directions are small only when $k_0\xi$ is large. In view of Eq. (7.13) the distance from the tip must be large. Because of these restrictions, the present approximate theory may be applied to the offshore breakwater of finite length only if $k_0 L$ and $k_0 B$ are both large so that the two shadow boundary layers do not overlap.

In summary, the potential in the shadow boundaries near $y = \pm\frac{1}{2}B, x < L$ is given approximately by

$$\Phi = \text{Re}\left(-\frac{ig\tilde{A}}{\omega}\frac{\cosh k(z+h)}{\cosh kh}\exp\left[-i\left(\int k\,dx + \omega t + \Theta\right)\right]\right), \qquad (7.20a)$$

where

$$\tilde{A} = A\left\{\frac{1}{2}\left[\frac{1}{2} + C(\sigma)\right]^2 + \frac{1}{2}\left[\frac{1}{2} + S(\sigma)\right]^2\right\}^{1/2}, \qquad (7.20b)$$

$$\Theta = \frac{\pi}{4} - \tan^{-1}\frac{\frac{1}{2} + S(\sigma)}{\frac{1}{2} + C(\sigma)}, \qquad (7.20c)$$

and

$$\sigma = \frac{k_0\left(y \pm \frac{1}{2}B\right)}{(\pi k_0\xi)^{1/2}}. \qquad (7.20d)$$

The diffraction factor $|D|$ is shown in Fig. 7.2 for the side $y > 0$.

Again we use the linearized theory to estimate the position of the breaker line $x = x_b(y)$. By invoking the hypothesis

$$\tilde{A}(x_b; y) = \frac{\gamma}{2}h(x_b),$$

which is an extrapolation of the empirical result established for strictly plane

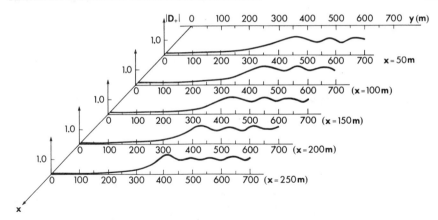

Figure 7.2 Variation of diffraction factor $|D_+|$ on the side of $x > 0$ for normal incidence on an offshore breakwater. The breakwater is at $x = 350$ m and has a length of 700 m. (From Liu and Mei, 1976a, *J. Geophys. Res.* Reproduced by permission of American Geophysical Union.)

waves, and by using Eq. (7.20b) for $\tilde{A}$, we can determine a crude breaker line. A further simplification may be made by taking $x = 0$ in the expression for ξ, that is,

$$\xi \to \xi_0 = k_0 \int_0^L \frac{dx}{k} .$$

The estimated breaker line is shown by dashes in Fig. 7.3, the important feature being the increase of the surf zone width from zero in the shadow to the finite limit far outside.

We emphasize that the quantitative prediction of wave transformation in the shoaling zone can certainly be improved. Indeed, a uniformly valid theory is now available (Liu, Lozano, and Pantazarus, 1979) which supports the present theory quite favorably.

10.7.2 The Mean Motion

With the convective inertia and turbulent diffusion S''_{ij} ignored for mathematical simplicity, the equations of motion are

$$\frac{\partial}{\partial x_i}[U_i(\bar{\xi} + h)] = 0, \tag{7.21}$$

$$0 = \rho g(\bar{\xi} + h)\frac{\partial \bar{\xi}}{\partial x_j} - \frac{\partial S'_{ij}}{\partial x_i} - \bar{\tau}_j^B. \tag{7.22}$$

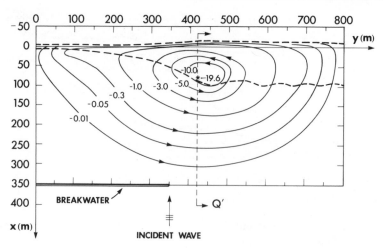

Figure 7.3 Streamline of mean current behind an offshore breakwater of normal incidence (breaker line is shown in dashes). (From Liu and Mei, 1976b, *J. Geophys. Res.* Reproduced by permission of American Geophysical Union.)

In the shoaling zone we approximate $\bar{\zeta} + h$ by h so that

$$\frac{\partial}{\partial x_i}(U_i h) = 0, \tag{7.23}$$

$$0 = -\rho g h \frac{\partial \bar{\zeta}}{\partial x_j} - \frac{\partial S'_{ij}}{\partial x_i} - \bar{\tau}_j^B. \tag{7.24}$$

As shown in Section 10.2.4 the current is expected to be significant in the surf zone but not in most of the shoaling zone. Thus, the mean sea level and the radiation stress should be in static equilibrium. Let us denote the corresponding mean sea level by $\bar{\zeta}_w$ so that

$$0 = -\rho g h \frac{\partial \bar{\zeta}_w}{\partial x_i} - \frac{\partial S'_{ij}}{\partial x_i}, \qquad x > x_b.$$

More specifically, since the wave in the shoaling zone is approximately a slowly modulated progressive wave normally incident toward the beach, the part $\bar{\zeta}_w$ is

$$\bar{\zeta}_w = -\frac{k\tilde{A}^2}{2\sinh 2kh}. \tag{7.25}$$

Very near the breaker line there must be some current in the shoaling zone due to continuity. As a consequence, a new adjustment of mean sea level $\bar{\zeta}_c$ is induced. Let the total mean sea level be expressed by

$$\bar{\zeta} = \bar{\zeta}_c + \bar{\zeta}_w,$$ (7.26)

where $\bar{\zeta}_c$ is related to the current through the bottom stress by

$$0 = -\rho g h \frac{\partial \bar{\zeta}_c}{\partial x_i} - \bar{\tau}_j^B.$$ (7.27)

In the surf zone it is no longer reasonable to ignore $\bar{\zeta}$ with respect to h; hence the problem is still nonlinear despite the omission of convective inertia. Now the radiation stresses are again assumed to be of the same form as in the shoaling zone. Since the waves are slowly modulated and normally incident progressive waves in shallow water, these stresses are given by

$$S'_{xx} \simeq \tfrac{3}{4}\rho g \tilde{A}^2,$$ (7.28a)

$$S'_{yy} = \tfrac{1}{4}\rho g \tilde{A}^2,$$ (7.28b)

$$S'_{xy} = S'_{yx} \simeq 0,$$ (7.28c)

[see Eq. (3.13)]. Note that while S'_{xy} is the only driving force in the longshore current, it disappears in normally incident waves. Again the assumption $\tilde{A} = \tfrac{1}{2}\gamma(\bar{\zeta} + h)$ is made so that the radiation stresses become

$$S'_{xx} = \tfrac{3}{16}\rho g \gamma^2 (\bar{\zeta} + h)^2,$$

$$S'_{yy} = \tfrac{1}{16}\rho g \gamma^2 (\bar{\zeta} + h)^2.$$ (7.29)

To complete the formulation, the following simple model for the bottom friction is introduced:

$$\bar{\tau}_i^B = \frac{f}{2}\rho U_0 U_i,$$ (7.30)

where U_0 is taken to be a constant typical of the orbital velocity in the shoaling zone

$$U_0 = \left(\frac{gkA}{\omega \cosh kh}\right)_{x_b}, \qquad x > x_b,$$ (7.31)

and the following orbital velocity in the surf zone

$$U_0 = \frac{\tilde{A}}{\bar{\zeta} + h}\left(g(\bar{\zeta} + h)\right)^{1/2} = \frac{\gamma}{2}\left(g(\bar{\zeta} + h)\right)^{1/2}, \qquad x < x_b.$$ (7.32)

With these assumptions the equations are closed. We add the boundary conditions that the mean flow is tangential to the mean shoreline ($\bar{\zeta} + h = 0$), to the line of symmetry ($y = 0$), and to the breakwater and vanishes at infinity. The mathematical problem is now determinate.

The above nonlinear boundary-value problem can only be solved numerically, as reported in Liu and Mei (1976b, see (1974) for further details). Only the calculated results are described here. Figure 7.3 presents the stream function ψ defined by

$$U(\bar{\zeta} + h) = -\frac{\partial \psi}{\partial y}, \qquad V(\bar{\zeta} + h) = \frac{\partial \psi}{\partial x}. \qquad (7.33)$$

The geometrical dimensions chosen for this sample case are $B = 700$ m, $L = 350$ m, and $dh/dx = 1/50$. The incident wave is assumed to be a swell of period $T = 10$ s and amplitude $A_\infty = 0.5$ m. For reference, the incident wavelength is 79 m and the amplitude is 0.52 m at the depth of the breakwater ($h = 7$ m). The empirical coefficients used are $f = 0.02$ and $\gamma = 0.8$.

The streamlines for the side $y > 0$ show a counterclockwise circulation. On the side $y < 0$ there is a symmetric cell in the clockwise direction. This result is certainly in agreement with known observations of currents and is consistent with the cusp or tombolo formation.

Figure 7.4 shows the corresponding mean sea level on the side of $y > 0$. Deep inside the shadow the mean sea level is zero; far outside the shadow the mean sea level has a set-up in the surf zone and a set-down in the shoaling zone, as expected.

For a better understanding of the physical mechanism of the circulation, we substitute assumptions (7.29), (7.30), and (7.32) into the momentum equation

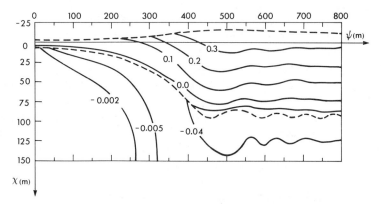

Figure 7.4 Contours of mean sea level displacement $\bar{\zeta}$ behind an offshore breakwater at normal incidence. Mean shore line and mean breaker line are shown in dashes. (From Liu and Mei, 1976b, *J. Geophys. Res.* Reproduced by permission of American Geophysical Union.)

(7.22), and then take the curl, yielding,

$$\frac{\partial}{\partial x}\left(\frac{V}{\bar{\zeta}+h}\right) - \frac{\partial}{\partial y}\left(\frac{U}{\bar{\zeta}+h}\right)$$
$$= \frac{2g}{f}\left\{\frac{h\bar{\zeta}_y - \frac{1}{4}\gamma^2(\bar{\zeta}+h)_x\bar{\zeta}_y}{\gamma g^{1/2}(\bar{\zeta}+h)^{3/2}} + \frac{\gamma\bar{\zeta}_{xy}}{2[g(\bar{\zeta}+h)]^{1/2}}\right\}. \quad (7.34)$$

The left-hand side of Eq. (7.34) is the potential vorticity in the mean flow; the right-hand side may be viewed as the forcing vorticity. Now the behavior of $\bar{\zeta}$ is easy to speculate, although it is a part of the solution. Because of the small numerical value of the factor $\frac{1}{4}\gamma^2 \simeq 0.16$ which is associated with S'_{xx} and S'_{yy}, the most important term is $h\bar{\zeta}_y$ in the first fraction on the right of Eq. (7.34). For $y > 0$, $\bar{\zeta}_y$ is certainly positive; the driving vorticity is positive and the responding current must be counterclockwise. Therefore, the gradient of mean sea level is the driving force; the radiation stresses act in an indirect way to produce the mean sea level.

If convective inertia were included, the streamlines would be more crowded around the x axis, resulting in a stronger rip current. Inclusion of the turbulent term S''_{xy} would push the center of the cell seaward where the speed is maximum. These effects have not yet been investigated quantitatively.

Although only the results for an offshore breakwater with normally incident waves are presented here, the following variations have also been studied by Liu and Mei (1976b): (i) oblique incidence: the added feature is that a uniform longshore current must be superposed, and (ii) oblique incidence on a long breakwater intersecting the shore normally. A cell is found around the shadow boundary, and flow is directed toward the breakwater near the shore and seaward along the shadow face of the breakwater. Laboratory evidence relevant to the second example has been reported by Shimano, Hom-ma, and Horikawa (1958) who performed experiments in a wave tank with a jetty intersecting the beach at an angle different from 90°. When the incident wave crests were parallel to the beach, flow was indeed found to be outward along the shadow side of the jetty. This tendency is qualitatively consistent with the results of the present section.

The formation of tombolos behind an offshore island is a common geographical phenomenon as discussed by Zenkovich (1967) whose sketches are reproduced here as Figs. 7.5a and 7.5b. For an island long in the direction parallel to the mainland coast, it is further possible to have two spits emanating from the mainland behind the island. The spits either form a lagoon first and then a single tombolo to link with the island (Fig. 7.5a) or extend to the island directly as twin tombolos (Fig. 7.5b). These varieties may be explained by the fact that breaking-induced currents are sufficiently strong to move heavy sediments only slightly outside, hence accumulation should occur slightly inside of the shadow boundaries. For a long island the shadows do not

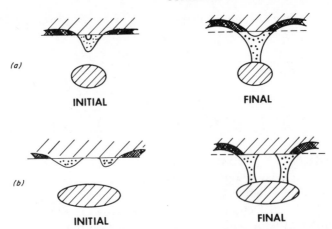

Figure 7.5 Formation of tombolos from the mainland toward the island when only the mainland beach is erodible (after Zenkovich, 1960, *Processes of Coastal Development*, Oliver and Boyd Publishers). (*a*) Small island and single tombolo; (*b*) long island and twin tombolos. Dots: area of deposition; double hatches: area of erosion; single hatches: undisturbed land.

overlap, so that there must be two initial spits which form new natural barriers for further deposition. If the initial spits are close, a lagoon forms first before reaching the island. For a still longer island twin tombolos result. So far we have tacitly assumed that only the shore of the mainland is erodible; the possible impact on an erodible island shore is the motivation of the next section.

10.8 CURRENTS AROUND A CONICAL ISLAND

Having discussed some effects of an offshore island on the mainland beach, we now turn to the effects of the beach on the island itself (Mei and Angelides, 1976). For simplicity in wave computation we select a conical island whose depth contours are concentric circles. From the shore ($r = b$) to the toe ($r = a$), the depth is a function of r only. Beyond the toe the sea depth is assumed to be constant (see Fig. 8.1). A plane wave is incident from the left $x \sim -\infty$ where $\theta = -\pi$. As in the last section, we first calculate the waves in the shoaling zone, then the breaker line, by invoking an empirical rule. Radiation stresses, calculated for the shoaling zone, are applied to the surf zone with the assumed closure relation between the breaking wave amplitude and the local mean depth. By neglecting convection and lateral turbulent diffusion, the mean circulation is obtained numerically from the average conservation equations of mass and momentum.

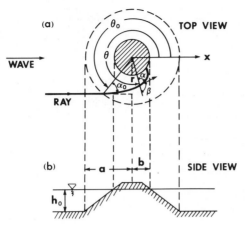

Figure 8.1 Definitions for a conical island.

10.8.1 The Wave Field

The island slope dh/dr is assumed to be so gentle that the ray (geometrical-optics) theory of refraction applies outside the surf zone. Because of symmetry, only the half to the right of the island, $\pi < \theta < 2\pi$, will be discussed. With reference to Fig. 8.1 the equation of the ray $r = r(\theta)$ is given by Eq. (4.4), Chapter Three. To illustrate the effect of the island size, we shall fix the wave frequency, the beach length $a - b$, the deep water depth h_0, and the slope dh/dr. Only the island radius b is left as a free parameter. Specifically, we take $T = 2\pi/\omega = 10$ s, $h_0 = 100$ ft, $dh/dr = 1/20$, $a - b = 2000$ ft. and take $b = 10,000$, 7370, and 5831 ft. for Cases I, II, and III, respectively.

After calculating the ray paths by numerical integration, the refracted wave amplitude A may be computed by invoking energy conservation in a ray tube. Assuming the linear refraction theory to be valid up to the breaker line, we introduce the familiar hypothesis that $(A/h)_b = \frac{1}{2}\gamma$, which determines the breaker line $r = r_b(\theta)$.

Figures 8.2a, b, and c show the geometry of the rays. Upon comparing these curves, it is evident that as the island size increases, the extent of the lee shore where no rays can enter increases. Case III is the threshold where the lee shore is just zero. In reality modifications due to diffraction must be present so that the wave intensity does not drop to zero without some further penetration. The existence of the lee shore for a large island is physically obvious, and suggests immediately that for a very large island there is no current caused by breaking in a portion of the lee.

10.8.2 The Mean Motion

The approximate mean equations of motion can be conveniently expressed in polar coordinates:

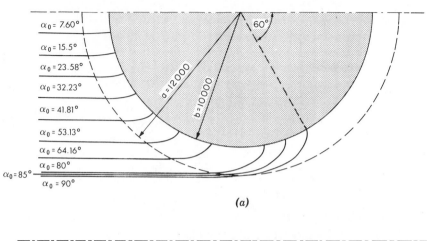

(a)

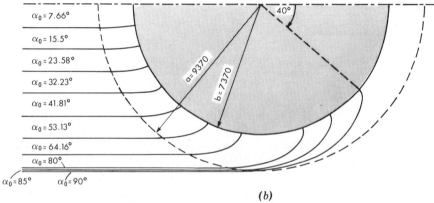

(b)

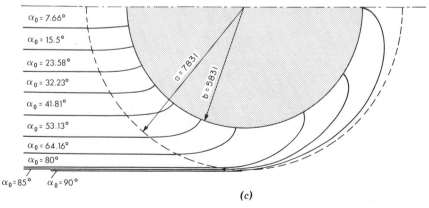

(c)

Figure 8.2 Geometry of rays near a conical island. (From Mei and Angelides, 1976, *Coastal Eng.* Reproduced by permission of Elsevier Scientific Publishing Co.) (*a*) Large island with a long lee shore; (*b*) intermediate island with a reduced lee shore; (*c*) small island with no lee shore.

498

Continuity:

$$\frac{\partial}{\partial r}\left[(\bar{\zeta}+h)U_r\right] + \frac{1}{r}\frac{\partial}{\partial \theta}\left[(\bar{\zeta}+h)U_\theta\right] + \frac{1}{r}U_r(\bar{\zeta}+h) = 0, \qquad (8.1)$$

r Momentum:

$$0 \simeq -\rho g(\bar{\zeta}+h)\frac{\partial \bar{\zeta}}{\partial r} - \frac{\partial S_{rr}}{\partial r} - \frac{f}{2}\rho U_0 U_r, \qquad (8.2)$$

θ Momentum:

$$0 \simeq -\rho g(\bar{\zeta}+h)\frac{1}{r}\frac{\partial \bar{\zeta}}{\partial \theta} - \frac{\partial S_{r\theta}}{\partial r} - \frac{f}{2}\rho U_0 U_\theta, \qquad (8.3)$$

where U_r and U_θ are the components of the mean current in the r and θ directions, respectively. Formal derivation of these equations is left as an exercise. Suffice it to say that the continuity equation is exact, and the momentum equations are approximately valid when

$$\frac{\partial}{\partial r} \gg \frac{1}{r} \quad \text{or} \quad \frac{1}{r}\frac{\partial}{\partial \theta}$$

which applies in the present example since the island is large and the radial range of the current is anticipated to be in and near the narrow surf zone. It is further reasonable to approximate the refracted waves locally by a plane progressive wave with incidence angle β. In shallow water, the important components of the radiation stress are

$$S'_{rr} \simeq \tfrac{1}{4}\rho g A^2(3 - 2\sin^2\beta), \qquad (8.4a)$$

$$S'_{r\theta} = S'_{\theta r} = \tfrac{1}{4}\rho g A^2 \sin^2\beta, \qquad (8.4b)$$

where β is the local incidence angle which differs for different rays and is a function of r and θ.

The closure condition $A = \gamma(\bar{\zeta}+h)/2$ and assumptions of U_0 similar to that of Section 10.7 are again introduced. The current velocities are replaced by the stream function ψ defined by

$$U_r = \frac{-1}{\bar{\zeta}+h}\frac{1}{r}\frac{\partial \psi}{\partial \theta}, \qquad (8.5a)$$

$$U_\theta = \frac{1}{\bar{\zeta}+h}\frac{\partial \psi}{\partial r}. \qquad (8.5b)$$

ψ must vanish on $\theta = \pi, 2\pi$, along the mean shoreline, and at $r \to \infty$. The mean motion is solved numerically as before. The streamlines calculated for an incident wave amplitude of $A_0 = 3$ ft are shown for three islands, Case I, II, and III in Fig. 8.3a, 8.3b, and 8.3c, respectively.

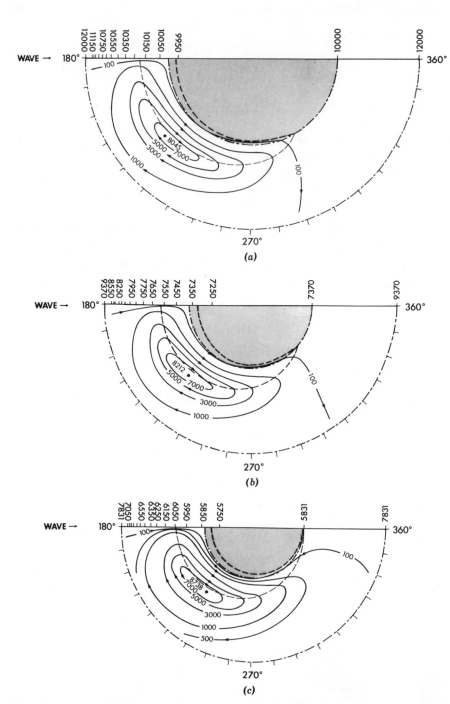

Figure 8.3 Streamlines of breaking-induced currents. (From Mei and Angelides, 1976, *Coastal Eng.* Reproduced by permission of Elsevier Scientific Publishing Co.) (*a*) A large island; (*b*) an intermediate island; (*c*) a small island.

The general picture is that there are two symmetric cells hugging the shores on each side of the island axis. Near the shoreline the flow is directed from the upwave side to the lee. Without significantly penetrating the lee shore the current returns at low velocities via the shoaling zone. The primary function of S'_{rr} is to establish the set-up in the surf zone, while the shear component $S'_{r\theta}$ provides the forcing of the longshore current. Thus, the driving mechanism here is roughly analogous to that of the uniform longshore current. The main difference is that the driving force varies in the longshore direction. One reason for such a difference is that $\partial S'_{r\theta}/\partial r$ depends on the local incidence angle β which increases with θ. The dominant term here is

$$\frac{g\gamma^2}{8}(\bar{\zeta}+h)\frac{\partial(\bar{\zeta}+h)}{\partial r}\sin 2\beta.$$

For a ray with a larger α_0, the value of β at the breaker line is also larger, that is, the incidence is more oblique. Numerically, it is found that $\partial S'_{r\theta}/\partial r$ always increases with θ from $\theta = \pi$ to the edge of the lee shore. Another reason is that the breaker zone width decreases from $\theta = \pi$ to the edge of the lee shore. Thus, the total driving force across the surf zone, which can be measured by the product of $\partial S'_{r\theta}/\partial r$ and the surf zone width, is zero at $\theta = \pi$, maximum at some intermediate θ, and zero again at the edge of the lee shore. Consequently, the center of the circulating cell is somewhere between the upwave point and the edge of the lee shore, as shown in Figs. 8.3a—8.3c.

As the island radius decreases from Cases I to III, ray convergence is more pronounced so that the current velocity is increased. The extent of the lee shore is also reduced, until in Case III the two longshore currents converge at the center of the lee and form a rip current leaving the island.

From these current patterns, one may venture to explain certain observations near an offshore island. On the upwave side, erosion prevails; sand is transported toward the island lee by longshore currents. Because of the large weight, sediments are deposited somewhere before reaching the edges of the lee shore. Therefore, if the island is large enough to have a finite lee shore, two sand spits may form, which may converge to form a lagoon if the mainland is very far, or may reach the mainland directly to form twin tombolos. If the island is so small that there is no lee shore, as in Case III, deposition occurs at

INITIAL   FINAL

Figure 8.4 Tombolo from a small island toward the mainland when only the island beach is erodible. (From Zenkovich, 1960, *Processes of Coastal Development*, Oliver and Boyd.)

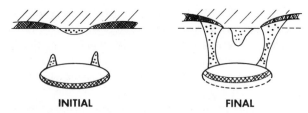

INITIAL   **FINAL**

Figure 8.5 Polygenetic formation of tombolos. (From Zenkovich, 1960, *Processes of Coastal Development*, Oliver and Boyd.)

the center of the lee and a single spit is formed. These features are known to happen when the mainland has a rocky coast (see Fig. 8.4) as was discussed by Zenkovich (1967).

In general the shores of both the island and the mainland are sandy. Spits can emanate from both of them; the tendency is shown in Fig. 8.5 and is called *polygenetic* by Zenkovich.

10.9 RELATED WORKS ON NEARSHORE CURRENTS

Since the introduction of radiation stresses, the literature on nearshore currents due to breaking waves has grown considerably. For uniform longshore currents, James (1974a, b) applied the theory of cnoidal waves and its approximation, the so-called *hyperbolic waves*, to allow for large amplitudes. Battjes (1972) developed a stochastic theory for longshore currents due to random waves; lateral turbulence was ignored.

On two-dimensional rip currents which are relevant for the formation of beach cusps, Arthur (1962) demonstrated the effect of nonlinear convective inertia of the current on the intensification of rips. Bowen (1969) and Bowen and Inman (1969) suggested that the presence of long period edge waves might be a cause of longshore variation of the mean sea-level, which, in turn, would induce periodic circulation cells. Sonu (1972) gave field evidence that rip currents could also be generated as a consequence of the periodic variation of beach contours. Noda (1974) developed a numerical model accounting for refraction and wave–current interaction for Sonu's problem. In the presence of periodic shoreline variation (shoals and embayments), rip currents can emanate from the shoals or from the embayments (Komar, 1971). Mei and Liu (1977) found that the ratio of the surf zone width to the longshore wavelength of the beach topography influenced the direction of the circulation cells. In nearly all of these theories, lateral turbulence was ignored. Extending Arthur's work, Tam (1973) used a boundary-layer approximation and treated the rip as a turbulent jet within which convective inertia was strong. The presence of the *rip head*, that is, the sudden expansion and dissipation into a big stagnant patch, was attributed to the sudden change of bottom slope. LeBlond and Tang (1974) studied the wave–current interaction for rips. Dalrymple (1975) gave experimental evidence of rip currents near a straight beach when two wavetrains were incident symmetrically from opposite sides of the shore-normal. Liu and Lennon (1978) applied the finite-element method for numerical modeling of nearshore currents.

Finally, a survey of the current state of sediment transport by the longshore current has been made by Ostendorf and Madsen (1979).

From the discussions in this chapter, it is evident that the idea of radiation stress has enabled us to study increasingly complex nearshore current problems with at least qualitative reliability. Formidable obstacles remain for a scientifically satisfactory understanding of the breaking waves, for which theory and observations are urgently needed.

ELEVEN

Nonlinear Long Waves in Shallow Water

11.1 DERIVATION AND CLASSIFICATION OF APPROXIMATE EQUATIONS

As pointed out in Chapter One, the linearized shallow-water approximation is useful only if the following two length ratios are small:

$$\mu \equiv kh \ll 1 \quad \text{and} \quad \varepsilon \equiv \frac{A}{h} \ll 1. \tag{1.1}$$

The second restriction is a severe one for many coastal problems; therefore, a nonlinear theory of shallow water waves is needed. The presence of two small parameters (three length scales) introduces new subtleties in the procedure of approximation since the magnitude of one ratio relative to the other is now important. Historically, two different theories, one by Airy and the other by Boussinesq (1877), and Korteweg and de Vries (1895), were separately developed which led to opposite conclusions regarding wave breaking on constant depth. The confusion was resolved in a fundamental paper by Ursell (1953) and further clarified by Lin and Clark (1959). In particular, Ursell has shown that the ratio

$$U_r = \frac{A}{h} \frac{1}{(kh)^2} = \frac{kA}{(kh)^3} = \frac{A\lambda^2}{h^3(2\pi)^2} \tag{1.2}$$

plays a central role in deciding the choice of approximations which correspond to very different physics. This ratio has since been widely referred to as the Ursell parameter and will be denoted by U_r in this book, although it also appeared in the earlier theory of Stokes.

For simplicity we shall illustrate the approximation procedure for constant depth, using the formalism of Benney (1966) and Peregrine (1967). The extension to variable depth is left as an exercise.

Since there are now two small parameters, it is advantageous to use dimensionless variables for the sake of clarity. The scales of the variables are suggested by the linearized theory:

$$(x', y') = k(x, y), \qquad z' = \frac{z}{h}, \qquad t' = k(gh)^{1/2}t$$

$$\zeta' = \frac{\zeta}{A}, \qquad \Phi' = \Phi\left[\frac{A}{kh}(gh)^{1/2}\right]^{-1}. \tag{1.3}$$

Note that the implied normalizations on velocity components are

$$(u, v) = \left(\frac{\partial}{\partial x}, \frac{\partial}{\partial y}\right)\Phi = \frac{A}{h}(gh)^{1/2}\left(\frac{\partial}{\partial x'}, \frac{\partial}{\partial y'}\right)\Phi' = \frac{A}{h}(gh)^{1/2}(u', v')$$

$$w = \frac{\partial\Phi}{\partial z} = \frac{1}{kh}\frac{A}{h}(gh)^{1/2}\frac{\partial\Phi'}{\partial z'} = \frac{1}{kh}\frac{A}{h}(gh)^{1/2}(w'). \tag{1.4}$$

The difference in scaling for horizontal and vertical components is required by continuity. The normalized equations are

$$\mu^2(\Phi'_{x'x'} + \Phi'_{y'y'}) + \Phi'_{z'z'} = 0, \qquad\qquad -1 < z' < \varepsilon\zeta' \tag{1.5}$$

$$\mu^2[\zeta'_{t'} + \varepsilon\Phi'_{x'}\zeta'_{x'} + \varepsilon\Phi'_{y'}\zeta'_{y'}] = \Phi'_{z'}, \qquad\qquad z' = \varepsilon\zeta' \tag{1.6}$$

$$\mu^2[\Phi'_{t'} + \zeta'] + \tfrac{1}{2}\varepsilon[\mu^2(\Phi'^2_{x'} + \Phi'^2_{y'}) + \Phi'^2_{z'}] = 0, \qquad z' = \varepsilon\zeta' \tag{1.7}$$

$$\Phi'_{z'} = 0, \qquad\qquad z = -1. \tag{1.8}$$

For convenience the primes will be dropped from here on. First, we only assume $\mu = kh$ to be small, leaving ε to be arbitrary for the time being. Since Φ is analytic, we may expand it as a power series in the vertical coordinate,

$$\Phi(x, y, z, t) = \sum_{n=0}^{\infty} (z + 1)^n\Phi_n \tag{1.9}$$

where $\Phi_n = \Phi_n(x, y, t)$, $n = 0, 1, 2, 3, \ldots$, whose orders of magnitude are yet unknown. Using ∇ to denote the horizontal gradient $(\partial/\partial x, \partial/\partial y)$, we first evaluate the derivatives

$$\nabla \Phi = \sum_0^\infty (z + 1)^n \nabla \Phi_n, \tag{1.10a}$$

$$\nabla^2 \Phi = \sum_0^\infty (z + 1)^n \nabla^2 \Phi_n, \tag{1.10b}$$

$$\frac{\partial \Phi}{\partial z} = \sum_0^\infty n(z + 1)^{n-1} \Phi_n = \sum_0^\infty (z + 1)^n (n + 1) \Phi_{n+1}, \tag{1.10c}$$

$$\frac{\partial^2 \Phi}{\partial z^2} = \sum_0^\infty (z + 1)^{n-1} (n + 1) n \Phi_{n+1} = \sum_0^\infty (z + 1)^n (n + 2)(n + 1) \Phi_{n+1}. \tag{1.10d}$$

Substituting Eqs. (1.10b and 1.10d) into the Laplace equation, we obtain

$$\mu^2 \nabla^2 \Phi + \frac{\partial^2}{\partial z^2} \Phi = \sum_{n=0}^\infty (z + 1)^n \left[\mu^2 \nabla^2 \Phi_n + (n + 2)(n + 1) \Phi_{n+2} \right] = 0. \tag{1.11}$$

Since z is arbitrary within $(-1, \varepsilon \zeta)$, the coefficient of each power of $(z + 1)$ must vanish, which yields a recursive relation

$$\Phi_{n+2} = \frac{-\mu^2 \nabla^2 \Phi_n}{(n + 2)(n + 1)}, \quad n = 0, 1, 2, \dots . \tag{1.12}$$

On the horizontal bottom Eq. (1.8) leads to

$$\Phi_1 \equiv 0$$

which implies from Eq. (1.12) that all Φ_n's with odd n vanish

$$\Phi_1 = \Phi_3 = \Phi_5 = \cdots 0. \tag{1.13}$$

For the even Φ_n's we have, specifically,

$$\Phi_2 = \frac{-\mu^2}{2 \cdot 1} \nabla^2 \Phi_0 = \frac{-\mu^2}{2!} \nabla^2 \Phi_0,$$

$$\Phi_4 = \frac{-\mu^2}{4 \cdot 3} \nabla^2 \Phi_2 = \frac{\mu^4}{4!} \nabla^2 \nabla^2 \Phi_0, \tag{1.14}$$

$$\Phi_6 = \frac{-\mu^2}{6 \cdot 5} \nabla^2 \Phi_4 = \frac{-\mu^6}{6!} \nabla^2 \nabla^2 \nabla^2 \Phi_0.$$

Since $\Phi_0 = O(\phi) = O(1)$, it may be concluded that $\Phi_2 = O(\mu^2)$, $\Phi_4 = O(\mu^4)$, and so on. Thus, with an error of $O(\mu^6)$ the potential is

$$\Phi = \Phi_0 - \frac{\mu^2}{2}(z + 1)^2 \nabla^2 \Phi_0 + \frac{\mu^4}{24}(z + 1)^4 \nabla^2 \nabla^2 \Phi_0 + O(\mu^6). \quad (1.15)$$

We must now make use of the boundary conditions on the free surface. Because of the multiplier μ^2 it is sufficient to keep $O(\mu^4)$ terms in Φ_t, Φ_x, and Φ_y.

$$\mu^2 \left[\frac{H_t}{\varepsilon} + \nabla H \cdot \left(\nabla \Phi_0 - \frac{\mu^2}{2} H^2 \nabla^2 \nabla \Phi_0 \right) \right]$$

$$= -\mu^2 H \nabla^2 \Phi_0 + \frac{\mu^4}{6} H^3 \nabla^2 \nabla^2 \Phi_0 + O(\mu^6), \quad (1.16)$$

$$\mu^2 \left[\Phi_{0_t} - \frac{\mu^2}{2} H^2 \nabla^2 \Phi_{0_t} + \zeta \right] + \tfrac{1}{2}\varepsilon\mu^2 \left[(\nabla \Phi_0)^2 - \mu^2 H^2 \nabla \Phi_0 \cdot \nabla^2 (\nabla \Phi_0) \right]$$

$$+ \tfrac{1}{2}\varepsilon\mu^4 H^2 (\nabla^2 \Phi_0)^2 = O(\mu^6), \quad (1.17)$$

where

$$H = 1 + \varepsilon\zeta \quad (1.18)$$

denotes the total depth. By defining

$$\mathbf{u}_0 = \nabla \Phi_0 \quad (1.19)$$

as the horizontal velocity at the bottom, Eq. (1.16) may be written

$$\frac{1}{\varepsilon} H_t + \nabla H \cdot \left(\mathbf{u}_0 - \frac{\mu^2}{2} H^2 \nabla^2 \mathbf{u}_0 \right) + H \nabla \cdot \mathbf{u}_0$$

$$- \frac{\mu^2}{6} H^3 \nabla^2 (\nabla \cdot \mathbf{u}_0) = O(\mu^4). \quad (1.20)$$

Taking the gradient of Eq. (1.17), we have

$$\mathbf{u}_{0_t} + \varepsilon \mathbf{u}_0 \cdot \nabla \mathbf{u}_0 + \frac{\nabla H}{\varepsilon}$$

$$+ \mu^2 \nabla \left[-\frac{\varepsilon}{2} H^2 \mathbf{u}_0 \cdot \nabla^2 \mathbf{u}_0 + \frac{\varepsilon}{2} H^2 (\nabla \cdot \mathbf{u}_0)^2 - \frac{1}{2} H^2 \nabla \cdot \mathbf{u}_{0_t} \right] = O(\mu^4).$$

$$(1.21)$$

Once ζ and $\mathbf{u}_0$ are solved, the actual velocity components are given by

$$(u, v) = \nabla \Phi = \mathbf{u}_0 - \frac{\mu^2}{2}(z + 1)^2 \nabla \nabla \cdot \mathbf{u}_0 + O(\mu^4), \qquad (1.22)$$

$$w = \frac{\partial \Phi}{\partial z} = -\mu^2(z + 1)\nabla^2 \Phi_0 = \mu^2(z + 1)\nabla \cdot \mathbf{u}_0 + O(\mu^4). \quad (1.23)$$

The pressure field may be obtained from Bernoulli's equation whose dimensionless form is, exactly,

$$-P = z + \varepsilon \left\{ \Phi_t + \frac{\varepsilon}{2}\left[(\nabla \Phi)^2 + \frac{1}{\mu^2}\Phi_z^2 \right] \right\}, \qquad (1.24)$$

where P has been normalized by $\rho g h$. The approximate pressure field follows by substituting Eqs. (1.21) and (1.22) into Eq. (1.24):

$$-P = z + \varepsilon \left\{ \left[\Phi_{0_t} - \frac{\mu^2}{2}(z + 1)^2 \nabla \cdot \mathbf{u}_{0_t} \right] \right.$$

$$\left. + \frac{\varepsilon}{2}\left[\mathbf{u}_0^2 - \mu^2(z + 1)^2 \mathbf{u}_0 \cdot \nabla^2 \mathbf{u}_0 + \mu^2(z + 1)^2(\nabla \cdot \mathbf{u}_0)^2 \right] \right\} + O(\mu^4).$$

Equation (1.17) can be used to eliminate Φ_{0_t}, so that

$$P = (\varepsilon \zeta - z) - \frac{\mu^2}{2}\left[H^2 - (z + 1)^2 \right]\left\{ \nabla \cdot \mathbf{u}_{0_t} + \varepsilon\left[\mathbf{u}_0 \cdot \nabla^2 \mathbf{u}_0 - (\nabla \cdot \mathbf{u}_0)^2 \right] \right\}$$

$$+ O(\mu^4). \qquad (1.25)$$

Instead of $\mathbf{u}_0$ we may introduce the depth-averaged horizontal velocity $\bar{\mathbf{u}}$ defined by

$$\bar{\mathbf{u}} = \frac{1}{H}\int_{-1}^{\varepsilon\zeta} dz\, \nabla\Phi = \frac{1}{H}\int_{-1}^{\varepsilon\zeta} dz\left(\mathbf{u}_0 - \frac{\mu^2}{2}(z + 1)^2 \nabla \nabla \cdot \mathbf{u}_0 + \cdots \right)$$

$$= \mathbf{u}_0 - \frac{\mu^2}{6}H^2\nabla^2\mathbf{u}_0 + O(\mu^4), \qquad (1.26)$$

which can be inverted to give

$$\mathbf{u}_0 = \bar{\mathbf{u}} + \frac{\mu^2}{6}H^2\nabla^2\bar{\mathbf{u}} + O(\mu^4). \qquad (1.27)$$

After Eq. (1.27) is substituted into Eq. (1.20), it follows at once that

$$H_t + \varepsilon \nabla \cdot (H\bar{\mathbf{u}}) = 0 \tag{1.28}$$

which is just the depth-averaged law of continuity and is, in fact, exact to all orders of (μ^2) [cf. Eq. (2.11), Chapter Ten]. Expressing $\mathbf{u}_0$ in terms of $\bar{\mathbf{u}}$ to the stated accuracy, we have

$$\bar{\mathbf{u}}_t + \varepsilon \bar{\mathbf{u}} \cdot \nabla \bar{\mathbf{u}} + \frac{\nabla H}{\varepsilon} + \frac{\mu^2}{6}(H^2 \nabla^2 \bar{\mathbf{u}})_t$$

$$+ \mu^2 \nabla \left\{ -\frac{\varepsilon}{3} H^2 \bar{\mathbf{u}} \cdot \nabla^2 \bar{\mathbf{u}} + \frac{\varepsilon}{2} H^2 (\nabla \cdot \bar{\mathbf{u}})^2 - \frac{H^2}{2} \nabla \cdot \bar{\mathbf{u}}_t \right\} = O(\mu^4). \tag{1.29}$$

It deserves emphasis that all the equations thus far are valid for arbitrary ε. Extension to higher orders in μ^2 is straightforward, although tedious.

Let us express Eqs. (1.28), (1.29), and (1.25) in physical variables

$$H_t + \nabla \cdot (H\bar{\mathbf{u}}) = 0, \tag{1.30}$$

$$\bar{\mathbf{u}}_t + \bar{\mathbf{u}} \cdot \nabla \bar{\mathbf{u}} + g\nabla H + \tfrac{1}{6}(H\nabla \nabla \cdot \bar{\mathbf{u}})_t$$

$$+ \nabla \left\{ -\frac{2}{3} H^2 \bar{\mathbf{u}} \cdot \nabla^2 \bar{\mathbf{u}} + \frac{H^2}{2}(\nabla \cdot \bar{\mathbf{u}})^2 - \frac{H^2}{2} \nabla \cdot \bar{\mathbf{u}}_t \right\} = 0, \tag{1.31}$$

and

$$P = \rho g(\zeta - z) - \tfrac{1}{2}\left[H^2 - (z + h)^2 \right]\left\{ \nabla \cdot \bar{\mathbf{u}}_t + \left[\bar{\mathbf{u}} \cdot \nabla^2 \bar{\mathbf{u}} - (\nabla \cdot \bar{\mathbf{u}})^2 \right]\right\}. \tag{1.32}$$

We now turn to limiting cases.

Airy's Theory for Very Long Waves: $\mu \to 0$, $\varepsilon = O(1)$

Airy's theory is the leading order approximation for very long waves of finite amplitude. By omitting terms proportional to μ^2 from Eqs. (1.25) and (1.29), we get, in physical variables,

$$\zeta_t + \nabla \cdot [(\zeta + h)\bar{\mathbf{u}}] = 0, \tag{1.33}$$

$$\bar{\mathbf{u}}_t + \bar{\mathbf{u}} \cdot \nabla \bar{\mathbf{u}} + g\nabla \zeta = 0, \tag{1.34}$$

$$P = \rho g(\zeta - z). \tag{1.35}$$

The preceding equations are actually valid for variable $h(x, y)$. A distinguishing feature of Airy's approximation is that the pressure is hydrostatic.

Boussinesq Theory: $O(\varepsilon) = O(\mu^2) < 1$

For weakly nonlinear and moderately long waves in shallow water, we approximate Eqs. (1.28), (1.29), and (1.25) to include terms of order $O(\varepsilon)$ and $O(\mu^2)$ only, obtaining

$$\zeta_t + \nabla \cdot [(\varepsilon\zeta + 1)\bar{u}] = 0, \tag{1.36}$$

$$\bar{u}_t + \varepsilon\bar{u} \cdot \nabla\bar{u} + \nabla\zeta - \frac{\mu^2}{3}\nabla\nabla \cdot \bar{u}_t = 0, \tag{1.37}$$

$$P = \varepsilon\zeta - z - \frac{\mu^2}{2}(-z^2 + 2z)\nabla \cdot \bar{u}_t. \tag{1.38}$$

In physical variables, they are

$$\zeta_t + \nabla \cdot [(\zeta + h)\bar{u}] = 0, \tag{1.39}$$

$$\bar{u}_t + \bar{u} \cdot \nabla\bar{u} + g\nabla\zeta - \frac{h^2}{3}\nabla\nabla \cdot \bar{u}_t = 0, \tag{1.40}$$

$$P = \rho g(\zeta - z) + \frac{\rho}{2}(2zh + z^2)\nabla \cdot \bar{u}_t. \tag{1.41}$$

Equations (1.36) and (1.37), or, equivalently, Eqs. (1.39) and (1.40), are called the Boussinesq equations. Note that the pressure field is no longer hydrostatic. Formally, Airy's and Boussinesq's theories differ by the linear term multiplied by μ^2 in Eq. (1.37). To see the physical significance of this term, let us examine the linearized versions of Eqs. (1.39) and (1.40) for one-dimensional infinitesimal waves

$$\zeta = Ae^{i(kx - \omega t)}, \qquad \bar{u} = Ue^{i(kx - \omega t)}.$$

When the exponential factors are cancelled, we get

$$-i\omega A + ikhU = 0,$$

$$-i\omega U + ikgA - \frac{h^2}{3}(ik)^2(-i\omega)U = 0,$$

which is a homogeneous set of equations for A and U. For a nontrivial solution the discriminant must vanish, that is,

$$\begin{vmatrix} -i\omega & ikh \\ igk & -i\omega\left(1 + \dfrac{k^2h^2}{3}\right) \end{vmatrix} = 0,$$

or

$$\omega^2 = \frac{ghk^2}{1 + \frac{1}{3}k^2h^2} = ghk^2\left(1 - \frac{k^2h^2}{3} + \cdots\right), \tag{1.42}$$

or

$$\text{phase velocity } C \cong (gh)^{1/2}\left(1 + \frac{k^2h^2}{3}\right)^{-1/2}$$

$$\cong (gh)^{1/2}\left(1 - \frac{k^2h^2}{3}\right)^{1/2}. \tag{1.43}$$

The term $\frac{1}{3}(kh)^2 = \frac{1}{3}\mu^2$ above arises from the term $(\frac{1}{3}h^2)u_{xxt}$ and represents frequency dispersion. Indeed Eq. (1.42) is precisely the two-term expansion of the familiar dispersion relation for arbitrary depth, $\omega^2 = gk \tanh kh$.

In summary, Boussinesq equations account for the effects of nonlinearity ε and dispersion μ^2 to the leading order. When $\varepsilon \gg \mu^2$, they reduce to Airy equations which are valid for all ε; when $\varepsilon \ll \mu^2$, they reduce to the linearized approximation with weak dispersion. When $\varepsilon \to 0$ and $\mu^2 \to 0$, the classical linearized wave equation is obtained.

Variable Depth

If the horizontal scale of depth variation is no greater than the typical wavelength, the procedure beginning from Eq. (1.9) may be extended. Now the bottom boundary condition is, in physical variables,

$$\Phi_z = -h_x\Phi_x - h_y\Phi_y, \qquad z = -h(x, y), \tag{1.44}$$

or in dimensionless variables

$$\Phi_z = -\mu^2\left(h_x\Phi_x + h_y\Phi_y\right), \qquad z = -h(x, y), \tag{1.45}$$

where the variable depth is normalized by the typical depth h_0 which is also used in defining the dimensionless variables of Eqs. (1.3) and (1.4). Instead of Eq. (1.9), one may assume

$$\Phi = \sum_{n=0}^{\infty} [z + h(x, y)]^n \Phi_n(x, y). \tag{1.46}$$

By substituting Eq. (1.46) into Laplace's equation and applying Eq. (1.45), we obtain a set of recursive relations among Φ_n. In particular, Φ_n's for odd n no longer vanish. Following arguments similar to those given here, Mei and LeMéhauté (1966, with corrections by Madsen (see Madsen and Mei, 1969)) obtained the Boussinesq equations for one-dimensional waves. For two dimensions the following equations were deduced by Peregrine (1967) in terms of $\bar{u}$ and ζ:

$$\zeta_t + \nabla \cdot [(h + \zeta)\bar{u}] = 0, \tag{1.47}$$

$$\frac{\partial \bar{u}}{\partial t} + \bar{u} \cdot \nabla \bar{u} + g\nabla\zeta = \frac{h}{2}\nabla[\nabla \cdot (h\bar{u}_t)] - \frac{h^2}{6}\nabla[\nabla \cdot \bar{u}_t]. \tag{1.48}$$

The algebra involved is straightforward but lengthy and may be left to the reader.

Exercise 1.1

Derive from Eqs. (1.5)–(1.7) and (1.45) the following alternative approximate equation for varying depth by keeping terms up to $O(\varepsilon^2)$ and $O(\mu^2)$ for $\varepsilon = O(\mu)$:

$$g\nabla \cdot (h\nabla\phi_0) - \phi_{0_{tt}} = \tfrac{1}{2}(\nabla\phi_0)_t^2 + \nabla \cdot \left\{\nabla\phi_{0_t} + \tfrac{1}{2}(\nabla\phi_0)^2\right\}$$

$$+ g\nabla \cdot \left(\frac{h^3}{6}\nabla\nabla^2\phi_0 - \frac{h^2}{2g}\nabla\phi_{0_{tt}}\right), \tag{1.49}$$

where $\phi_0(x, y, t)$ is the velocity potential at $z = 0$. In dimensionless form, the error is of the order $O(\varepsilon\mu^2, \mu^4)$ relative to the leading terms kept in Eq. (1.40). Begin by letting

$$\Phi = \sum_{n=0}^{\infty} \frac{z^n}{n!}\phi_n(x, y, t).$$

11.2 NONDISPERSIVE WAVES IN WATER OF CONSTANT DEPTH

11.2.1 Analogy to Gas Dynamics

When $(A/h)(kh)^{-2} \gg 1$, the approximate equations given by Eqs. (1.33) and (1.34) are analogous to those governing two-dimensional gas dynamics in an isentropic flow (Riabouchinsky, 1932). Let

$$\hat{\rho} = (\zeta + h) \tag{2.1}$$

denote the density of a fictitious compressible fluid. The continuity equation becomes

$$\frac{\partial\hat{\rho}}{\partial t} + \nabla \cdot (\hat{\rho}\bar{u}) = 0. \tag{2.2}$$

Defining a fictitious pressure by

$$\hat{p} = \frac{g}{2}(\zeta + h)^2 = \frac{g}{2}\hat{\rho}^2, \tag{2.3}$$

and using Eq. (2.1), we may rewrite the momentum equation in the form

$$\hat{\rho}\left[\frac{\partial \bar{u}}{\partial t} + \bar{u} \cdot \nabla \bar{u}\right] = -\nabla \hat{p}. \tag{2.4}$$

Equations (2.2) and (2.4) are just the continuity and momentum equations for a compressible fluid, while Eq. (2.3) is the equation of state for an isentropic flow of a perfect gas. In general, the equation of state for a calorically perfect gas is

$$\frac{p}{p_0} = \left(\frac{\rho}{\rho_0}\right)^{C_p/C_v} \exp\left[\frac{C_p(s - s_0)}{C_v(C_p - C_v)}\right], \tag{2.5}$$

where s is the entropy and p_0, ρ_0, and s_0 are the reference values of p, ρ, and s, respectively. C_p and C_v are specific heats under constant pressure and constant volume, respectively. Thus, Eq. (2.3) corresponds to a gas with $C_p/C_v = 2$, which does not exist in reality.

Because of this analogy, knowledge and methods established in gas dynamics can be transferred directly to long water waves. This aspect is most thoroughly exploited in Stoker (1957) on which much of the present section is based. It must be emphasized that while Eqs. (2.2) and (2.4) are exact in gas dynamics, they are only approximate in water waves when the horizontal scale of motion is much larger than the vertical scale. In particular, while the gas-dynamic theory succeeds in predicting shock formation, Airy's theory is basically unsuitable for predicting bore formation, a problem of enormous importance for which the full solution is not yet forthcoming. Therefore, care must be exercised in interpreting the results of Airy's theory.

11.2.2 Method of Characteristics for One-Dimensional Problems

As in one-dimensional gas dynamics, the *method of characteristics* is a very useful tool in Airy's shallow-water wave theory. The salient features of this method will be outlined only to the extent needed. A full account of the general method is available in Courant and Friedrichs (1949), Courant and Hilbert (1962), and Stoker (1957), among others.

With the overbars omitted for brevity, Eqs. (1.33) and (1.34) become

$$\frac{\partial \zeta}{\partial t} + \frac{\partial}{\partial x}[u(h + \zeta)] = 0, \tag{2.6}$$

$$\frac{\partial u}{\partial t} + u\frac{\partial u}{\partial x} + g\frac{\partial \zeta}{\partial x} = 0. \tag{2.7}$$

Let us rearrange the governing equations by introducing

$$C^2 = g(\zeta + h).$$ (2.8)

The continuity and momentum equations may be written

$$2C_t + 2uC_x + Cu_x = 0,$$ (2.9)

$$u_t + 2CC_x + uu_x = 0.$$ (2.10)

By adding and subtracting the two equations above, we get

$$\left[\frac{\partial}{\partial t} + (u + C)\frac{\partial}{\partial x} \right][u + 2C] = 0,$$ (2.11)

$$\left[\frac{\partial}{\partial t} + (u - C)\frac{\partial}{\partial x} \right][u - 2C] = 0.$$ (2.12)

Equations (2.11) and (2.12) describe the total rate of change of $u \pm 2C$ along the curves which are governed by

$$\frac{dx}{dt} = u \pm C.$$ (2.13)

These curves are called the *characteristics*. The solutions to $dx/dt = u + C$ are a one-parameter family of curves which we shall call C_+ characteristics, the parameter being the initial position of a curve at a given time t_0. Similarly, the solutions to $dx/dt = u - C$ are also a one-parameter family of curves, to be called C_- characteristics. A set of partial differential equations possessing real characteristics are said to be *hyperbolic*. Because the right-hand sides are zero, Eqs. (2.11) and (2.12) may be integrated to give

$$u + 2C = \text{const} \qquad \text{along } C_+ ,$$

$$u - 2C = \text{const} \qquad \text{along } C_- .$$ (2.14)

The quantities $u \pm 2C$ are called the *Riemann invariants*.

Let us explain how a typical initial-value problem can be solved in principle. Consider the example where $u(x, 0) = F(x)$ and $C(x, 0) = G(x)$ are given for $t = 0$, all x. The Riemann invariants are

$$\text{along } C_+ : \quad \frac{dx}{dt} = u + C, \quad u + 2C = F(x) + 2G(x) = R_+(x);$$

$$\text{along } C_- : \quad \frac{dx}{dt} = u - C, \quad u - 2C = F(x) - 2G(x) = R_-(x).$$

(2.15)

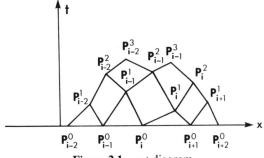

Figure 2.1 *x-t* diagram.

Consider the *x-t* diagram (Fig. 2.1). Divide the *x* axis into small intervals by grid points $x = x_1^0, \ldots, x_i^0, \ldots$; from the initial values $u(x_i^0, 0)$ and $C(x_i^0, 0)$ we calculate the initial stretch of the characteristics

$$C_+ : \quad x = x_i^0 + \left[u(x_i^0, 0) + C(x_i^0, 0) \right] t,$$

$$C_- : \quad x = x_i^0 + \left[u(x_i^0, 0) - C(x_i^0, 0) \right] t. \tag{2.16}$$

Consider a C_+ curve starting from x_i^0 and a C_- curve starting from a neighboring grid point x_{i+1}^0. Their intersecting point x_i^1, t_i^1 is found by solving the simultaneous equations

$$x_i^1 = x_i^0 + \left[u(x_i^0, 0) + C(x_i^0, 0) \right] t_i^1, \qquad\qquad C_+$$

$$x_i^1 = x_{i+1}^0 + \left[u(x_{i+1}^0, 0) - C(x_{i+1}^0, 0) \right] t_i^1, \qquad C_- . \tag{2.17}$$

Let us denote the intersecting point by $P_i^1 = (x_i^1, t_i^1)$. From all such neighboring points a row of new points $P_i^1(x_i^1, t_i^1)$, $i = 1, 2, \ldots$, is obtained. Now because P_i^1 is on the C_+ characteristic which is originated at $P_i^0(x_i^0, 0)$, we have

$$u(P_i^1) + 2C(P_i^1) = u(P_i^0) + 2C(P_i^0). \tag{2.18}$$

Because P_i^1 is also on the C_- characteristic originated at $P_{i+1}^0(x_{i+1}^0, 0)$, we have

$$u(P_i^1) - 2C(P_i^1) = u(P_{i+1}^0) - 2C(P_{i+1}^0). \tag{2.19}$$

Equations (2.18) and (2.19) can now be solved for $u(P_i^1)$ and $C(P_i^1)$. After obtaining the solution for all $i = 1, 2, 3 \ldots$ we have marched forward in time, although t_i^1 are in general not equal for all i. The points P_i^1 then form a new initial line from which the characteristics C_+ and C_- can be extended, and the solution procedure for u and C repeated, and so on. In this way a network of characteristic curves and the solution at each grid point are found.

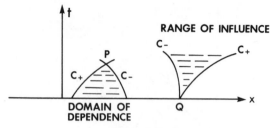

Figure 2.2

Through each point P passes a pair of characteristics C_+ and C_-. From the procedure just described it is clear that the solution at P depends only on the initial data between these two characteristics. As shown in Fig. 2.2, the triangular region, bounded by the initial curve and the two passing characteristics $C_\pm$ is called the *domain of dependence*. On the other hand, the initial data at point P_i^0 are felt by every later point in the sector bounded by the two characteristics $C_\pm$ initiated from P_i^0. This sector is called the *range of influence*.

Instead of an initial-value problem, consider an initial-boundary-value problem for the quadrant $x > 0$, $t > 0$; Fig. 2.3. Let the two characteristics $C_\pm$ emanating from the t axis point to the opposite sides of the t axis. Then only one boundary condition is needed on the t axis because the two unknowns u and C at any interior point P are uniquely determined by the two equations (Riemann invariants) along the two intersecting characteristics. For example, one may prescribe $u(0, t)$, while $C(0, t)$ must be compatible to the value determinable from the Riemann invariant along C_- going back to the x axis. In general, the initial curve can be anything in the x-t plane. If a section of the initial curve, say AO, is such that at every point both C_+ and C_- point into the region of interest as t increases, the curve AO is called *space-like*, and two initial conditions are needed. If a section OB is such that at every point only one (C_+ or C_-) points into the region of interest, then this section is called *time-like* and one boundary condition is required.

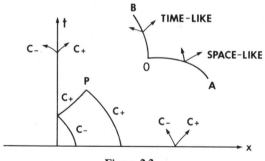

Figure 2.3

In principle, integration of the characteristics equations may be carried out graphically or numerically; the unknowns as well as the characteristics are found in the course of solution. In more general hyperbolic problems the characteristics equations corresponding to Eqs. (2.11) and (2.12) are inhomogeneous with the right-hand side depending on x, t, u, and C. There are no Riemann invariants and integration must be carried out numerically. Nevertheless, the partial differential equations are reduced to ordinary differential equations by means of characteristics.

In practice, the method of finite differences is a more convenient alternative. The basic aspects are amply described in Stoker (1957), while further extensions and improvements are a well-developed part of computational hydraulics. Interested readers should consult Abbott (1979) and current literature in hydraulics for further details. In the following subsections only analytic examples will be treated to explore the physical implications of nonlinearity.

11.2.3 Simple Waves and Constant States

As in gas dynamics, some useful concepts may be built on the basis of the following important theorem (Courant and Friedrichs, 1949):

If one of the C_+ (or C_-) characteristics is straight, then all other C_+ (or C_-) characteristics in the neighborhood must also be straight.

Consider two points A and B along the known straight characteristic C_+. From the Riemann invariant along C_+ it follows that

$$u(A) + 2C(A) = u(B) + 2C(B). \tag{2.20}$$

Since C_+ is straight, the following is true:

$$u(A) + C(A) = u(B) + C(B). \tag{2.21}$$

After subtracting Eq. (2.21) from Eq. (2.20), we see that

$$C(A) = C(B), \tag{2.22}$$

which implies

$$u(A) = u(B). \tag{2.23}$$

Thus, along a straight characteristic C_+, u and C are constants. Conversely, if u and C are constants along C_+, then C_+ is straight. Now let C'_+ be a neighboring characteristic of the C_+ family and C_- and C'_- be two characteristics of the C_- family. The intersecting points A, B, C, and D are shown in Fig. 2.4. Using the Riemann invariants along C'_+, C_-, and C'_-, respectively,

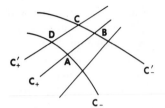

Figure 2.4

we have

$$u(C) + 2C(C) = u(D) + 2C(D), \tag{2.24}$$

$$u(D) - 2C(D) = u(A) - 2C(A), \tag{2.25}$$

$$u(C) - 2C(C) = u(B) - 2C(B). \tag{2.26}$$

Because of Eqs. (2.22) and (2.23) the last two equations give

$$u(C) - 2C(C) = u(D) - 2C(D). \tag{2.27}$$

It is evident from Eqs. (2.24) and (2.27) that

$$u(C) = u(D), \quad C(C) = C(D). \tag{2.28}$$

Thus, C'_+ is also straight, and the theorem is proved.

A region R in the x, t diagram is called a *simple wave* zone if all characteristics of one family are straight. Note that the straight characteristics are not necessarily parallel; the constant values of u and C are different along different characteristics. A special case of the simple wave is called the *constant state* where u and C are the same constants throughout. Both families of C_+ and C_- are then straight and parallel. It is clear that adjacent to a constant state there must be a simple wave.

Let us use these notions to study an initial-boundary-value problem.

11.2.4 Expansion and Compression Waves—Tendency of Breaking

In a linearized theory, the disturbances due to an initial depression on the free surface and to an initial elevation of the same shape are essentially the same except for a change of sign. The situation is quite different in a nonlinear problem. As an illustration, consider $\zeta(0, t)$ and hence $C(0, t)$ to be prescribed at the end $(x = 0)$ of a long channel $(x > 0)$. Figures 2.5a and 2.6a show $C(0, t)$ as a decreasing and increasing function of t, respectively. As shown in Figs. 2.5b and 2.6b, there is always an undisturbed zone DOB where

$$u = 0, \qquad C = C_0 = (gh)^{1/2} \tag{2.29}$$

which is a constant state.

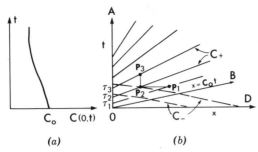

Figure 2.5 Expansion waves.

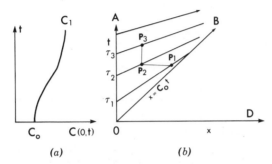

Figure 2.6 Compression waves.

Being next to a constant state, the sector BOA must be a simple wave where C_+ characteristics are straight. By using the Riemann invariant along a C_- which crosses the t axis at a given instant t, we get

$$u(0, t) - 2C(0, t) = 0 - 2C_0. \tag{2.30}$$

Therefore, the velocity at the channel estuary is not arbitrary but is given by

$$u(0, t) = 2C(0, t) - 2C_0. \tag{2.31}$$

Alternatively, one may prescribe $u(0, t)$; $C(0, t)$ is then implied by Eq. (2.31). Moreover, any C_+ intersecting the t axis at $t = \tau$ has the slope

$$\frac{dx}{dt} = u(0, \tau) + C(0, \tau) = 3C(0, \tau) - 2C_0. \tag{2.32}$$

From the Riemann invariant along a C_+ characteristic,

$$u + 2C = u(0, \tau) + 2C(0, \tau) = 4C(0, \tau) - 2C_0,$$

whereas from the Riemann invariant along a C_- characteristic,

$$u - 2C = -2C_0.$$

At the intersection point $P(x, t)$ of these two curves, we must have

$$C(x, t) = C(0, \tau) \tag{2.33a}$$

and

$$u(x, t) = 2[C(0, \tau) - C_0]. \tag{2.33b}$$

If $C(0, \tau)$ is a decreasing function of time, then the C_+ characteristics in zone AOB spread out as a fan. Referring to Fig. 2.5b, we compare $P_2(x, t_2)$ with two other points $P_1(x_1, t_2)$ and $P_3(x, t_3)$ with $x_1 > x$ and $t_3 > t_2$. It follows from Eq. (2.33a) that

$$C(P_3) - C(P_2) = C(0, \tau_3) - C(0, \tau_2) < 0,$$

$$C(P_2) - C(P_1) = C(0, \tau_2) - C(0, \tau_1) < 0.$$

Hence, the free-surface height increases with x for a fixed time. This physical situation is called an *expansion wave*.

Next, if $C(0, t)$ is an increasing function of time, then the neighboring C_+ characteristics in zone AOB tend to intersect. Referring to Fig. 2.6b, we compare $P_2(x, t_2)$ with two other points $P_1(x_1, t_2)$ and $P_3(x, t_3)$, with $x_1 > x$ and $t_3 > t_2$ before the intersection occurs. From Eq. (2.33a) we have

$$C(P_3) > C(P_2) > C(P_1)$$

instead. Thus, the free-surface height increases with time at a fixed x and decreases with x at a given t. As the two neighboring C_+ curves approach the intersection, the free-surface height becomes steeper. At the first moment of such an intersection, the free surface becomes locally vertical, and the wave profile *breaks*.

The preceding deduction on breaking requires that the free-surface slope be so large as to be incompatible with the basic assumption of Airy's theory, that is, $kA \sim \partial A/\partial x \ll 1$. Therefore, the implications near and after initial breaking cannot be reliable quantitatively. There are also cases where even the qualitative predictions are of doubtful validity. Consider, for example, the disturbance generated in a long wave tank by first pushing the wavemaker for a finite distance and then stopping it. The velocity $u(0, t)$ must increase from zero to a maximum and decrease to zero after a finite duration. From Eq. (2.31), $C(0, t)$ first increases from and then decreases to C_0. The C_+ characteristics emanating from the t axis must first lean toward and then away from their neighbors. After the wavemaker stops, a hump of finite length propagates on the free surface toward $x \sim \infty$. The front of the hump is a compression wave tending to steepen, while the lee is an expansion wave tending to flatten. According to Airy's approximation, an initial hump of finite amplitude would always break at the front. By a similar reasoning, waves generated by an oscillating flap started from rest would break near the front of every wave crest. These conclusions are, however, not always observed in experiments. Indeed, there is a considerable range of ε and μ where an initial hump or periodically driven waves does not break at all. Theoretically, the increase of free-surface slope implies that higher horizontal gradients, corresponding to frequency dispersion which is neglected in Airy's theory, become more and more important. Because of the large gradient near a breaking crest, a valid theory must be able to account for derivatives of all orders and full nonlinearity, which implies that the full equation must be solved exactly, at least locally. More research is needed in this difficult area.

While inadequate for the prediction of initial breaking, Airy's theory has relevance, however, to the events after breaking. Specifically, the propagation of a bore can be treated as a discontinuous (shock) solution within the framework of Airy's theory. A good example is the classical theory of hydraulic jumps in open channels. For a horizontal bottom, theories on bore propagation are thoroughly discussed in Stoker (1957).

11.3 NONBREAKING WAVES ON A SLOPE

For tsunamis with very long periods, the wave slope is so small that wave breaking is not important, that is, the shore is highly reflective. Of engineering interest is the quantity called *run-up*, that is, the maximum landward excursion of the shoreline. Since the water depth at the shoreline is zero, a nonlinear theory is needed. Airy's approximation, being valid for arbitrary A/h, is clearly the proper one to use.

Let us begin with Eqs. (1.47) and (1.48) without the dispersion terms and the overhead symbol (¯):

$$\frac{\partial \zeta}{\partial t} + \frac{\partial}{\partial x}\left[u(h(x) + \zeta)\right] = 0, \tag{3.1}$$

$$\frac{\partial u}{\partial t} + u\frac{\partial u}{\partial x} + g\frac{\partial \zeta}{\partial x} = 0, \tag{3.2}$$

which may be rewritten

$$(\zeta + h)_t + [u(\zeta + h)]_x = 0, \tag{3.3}$$

$$u_t + uu_x + g(\zeta + h)_x = gh_x. \tag{3.4}$$

With the definition

$$C^2 = g(\zeta + h), \tag{3.5}$$

Eqs. (3.3) and (3.4) may be further written

$$2C_t + 2uC_x + Cu_x = 0, \tag{3.6}$$

$$u_t + uu_x + 2CC_x = gh_x. \tag{3.7}$$

By adding and subtracting Eqs. (3.6) and (3.7), we arrive at a pair of characteristic equations:

$$\left[\frac{\partial}{\partial t} + (u \pm C)\frac{\partial}{\partial x}\right][u \pm 2C] = gh_x. \tag{3.8}$$

In general, Eq. (3.8) cannot be integrated to give Riemann invariants unless h_x is constant. For simplicity it is assumed from here on that

$$h = sx, \tag{3.9}$$

and that the sea is to the right of the slope. The Riemann invariants are

$$u + 2C - mt = \alpha \qquad \text{along } C_+ : \qquad \frac{dx}{dt} = u + C, \tag{3.10}$$

and

$$u - 2C - mt = \beta \qquad \text{along } C_- : \qquad \frac{dx}{dt} = u - C, \tag{3.11}$$

where $m = sg$. Since the values of α or β are different along different C_+ or C_- characteristics, the x, t plane may be covered with a new set of coordinate curves $\alpha = \text{const}$ and $\beta = \text{const}$. This suggests transforming the independent coordinates from x, t to α, β. Along a particular C_+ characteristic, only β is varying; hence the differential equation for C_+ may be written

$$x_\beta - (u + C)t_\beta = 0. \tag{3.12}$$

Likewise, the differential equation for C_- is

$$x_\alpha - (u - C)t_\alpha = 0. \tag{3.13}$$

Now Eqs. (3.12) and (3.13) are a set of partial differential equations for the dependent variables x and t as functions α and β. Since u and C are linearly related to the new independent variables α and β, these new equations are linear. Let us further introduce (λ, σ) as independent variables:

$$\frac{\lambda}{2} = \frac{\alpha + \beta}{2} = u - mt, \tag{3.14}$$

$$\frac{\sigma}{4} = \frac{\alpha - \beta}{4} = C, \tag{3.15}$$

so that

$$\frac{\partial}{\partial \alpha} = \frac{\partial \lambda}{\partial \alpha} \frac{\partial}{\partial \lambda} + \frac{\partial \sigma}{\partial \alpha} \frac{\partial}{\partial \sigma} = \frac{\partial}{\partial \lambda} + \frac{\partial}{\partial \sigma},$$

$$\frac{\partial}{\partial \beta} = \frac{\partial \lambda}{\partial \beta} \frac{\partial}{\partial \lambda} + \frac{\partial \sigma}{\partial \beta} \frac{\partial}{\partial \sigma} = \frac{\partial}{\partial \lambda} - \frac{\partial}{\partial \sigma}.$$

The equations for the characteristic curves become

$$-x_\lambda + x_\sigma - u(-t_\lambda + t_\sigma) - C(-t_\lambda + t_\sigma) = 0,$$

$$x_\lambda + x_\sigma - u(t_\lambda + t_\sigma) + C(t_\lambda + t_\sigma) = 0,$$

which may be added and subtracted to give

$$x_\sigma - ut_\sigma + Ct_\lambda = 0, \tag{3.16}$$

$$x_\lambda - ut_\lambda + Ct_\sigma = 0. \tag{3.17}$$

By cross-differentiation a single linear partial differential equation is obtained,

$$\sigma(t_{\lambda\lambda} - t_{\sigma\sigma}) - 3t_\sigma = 0 \tag{3.18}$$

(Stoker, 1948).

Carrier and Greenspan (1957) chose to use an auxiliary dependent variable ψ, as follows: With the help of Eqs. (3.14) and (3.15), t and C may be eliminated from Eq. (3.16) to yield

$$x_\sigma + \frac{u}{m}\left(\frac{\lambda}{2} - u\right)_\sigma - \frac{\sigma}{4m}\left(\frac{\lambda}{2} - u\right)_\lambda = 0,$$

which may be written

$$\left(x - \frac{u^2}{2m} - \frac{\sigma^2}{16m}\right)_\sigma + \frac{\sigma}{4m}u_\lambda = 0. \tag{3.19}$$

Similarly, Eq. (3.17) may be written

$$\left(x - \frac{u^2}{2m}\right)_\lambda + \frac{u}{2m} + \frac{\sigma u_\sigma}{4m} = 0. \tag{3.20}$$

Equation (3.19) suggests the introduction of a "stream function" ψ such that

$$\frac{-u}{m} = \frac{\psi_\sigma}{\sigma}, \tag{3.21}$$

$$x - \frac{u^2}{2m} - \frac{\sigma^2}{16m} = \frac{\psi_\lambda}{4}. \tag{3.22}$$

When the above substitutions are made into Eq. (3.20), a single equation for ψ is obtained

$$(\sigma\psi_\sigma)_\sigma - \sigma\psi_{\lambda\lambda} = 0, \tag{3.23}$$

which is equivalent to Eq. (3.18). Once ψ is found, u, C, x, and t are found, respectively, from Eqs. (3.21), (3.15), (3.22), and (3.14) in terms of the parameters λ and σ, which may be eliminated to give $u(x, t)$ and $C(x, t)$. The free-surface displacement is found from Eqs. (3.5), (3.19), and (3.15) to be:

$$\zeta = -s\left(x - \frac{\sigma^2}{16m}\right) = -s\left(\frac{u^2}{2m} + \frac{\psi_\lambda}{4}\right) \tag{3.24}$$

after Eq. (3.22) is taken into account.

It is useful to attach some physical significance to the new variables σ and λ. By definition, $\frac{1}{16}\sigma^2 = C^2 = g(\zeta + h)$ is proportional to the total depth, the instantaneous shoreline being at $\sigma = 0$. In the limit of deep water u becomes much less than C; it follows from Eq. (3.14) that $t \sim -\lambda/2m$ so that λ is closely related to time.

Because of the simplicity of Eq. (3.23), analytical solutions may be sought for some convenient initial and/or boundary conditions on ψ. The corresponding physical problem and solution are then found by inverse transformation. A simple example is discussed below.

11.3.1   Standing Waves of Finite Amplitude

Let us study a simple solution of the type

$$\psi = f(\sigma)\cos\left(\frac{\omega\lambda}{2m}\right). \tag{3.25}$$

The coefficient has been chosen so that ω represents frequency. Substituting Eq. (3.25) into Eq. (3.23), we get

$$\sigma^2 f'' + \sigma f' + \left(\frac{\omega}{2m}\right)^2 \sigma^2 f = 0. \tag{3.26}$$

A solution which is finite at $\sigma = 0$ is $J_0((\omega/2m)\sigma)$; thus,

$$\psi = \frac{8g}{\omega} A J_0\left(\frac{\omega}{2m}\sigma\right)\cos\left(\frac{\omega}{2m}\lambda\right). \tag{3.27}$$

At large distances

$$\sigma = 4C \cong 4(gh)^{1/2} = 4(gsx)^{1/2}, \qquad x > 0,$$

$$\lambda \sim -2mt,$$

so that

$$\psi \simeq \frac{8g}{\omega} A J_0 \left(2\omega \left(\frac{x}{sg} \right)^{1/2} \right) \cos \omega t. \tag{3.28}$$

It is easily shown that Eq. (3.28) is just the standing wave solution to the linearized equations

$$\frac{\partial \zeta}{\partial t} + \frac{\partial}{\partial x}(sxu) = 0,$$

$$\frac{\partial u}{\partial t} + g\frac{\partial \zeta}{\partial x} = 0. \tag{3.29}$$

The corresponding free surface is

$$\zeta_{\text{linear}} = -A J_0 \left(2\omega \left(\frac{x}{sg} \right)^{1/2} \right) \sin \omega t. \tag{3.30}$$

In Eq. (3.27), ψ is clearly a single-valued function of λ and σ. Is it also a single-valued function of x and t? In the theory of implicit functions it is well known that the mapping between (x, t) and (λ, σ) is not one-to-one only when the Jacobian

$$J = \frac{\partial(x, t)}{\partial(\sigma, \lambda)} = \begin{vmatrix} x_\sigma & t_\sigma \\ x_\lambda & t_\lambda \end{vmatrix} \tag{3.31}$$

vanishes at some λ and σ. Since $\partial C/\partial x$, which is proportional to the surface slope, is also given by

$$\frac{\partial C}{\partial x} = \frac{\begin{vmatrix} C_\sigma & t_\sigma \\ C_\lambda & t_\lambda \end{vmatrix}}{J},$$

the vanishing of J implies a vertical free surface, hence breaking. To gain some qualitative insight on the inception of breaking, we calculate J from Eq. (3.17),

$$J = (ut_\sigma - Ct_\lambda)t_\lambda - (ut_\lambda - Ct_\sigma)t_\sigma$$

$$= C(t_\sigma^2 - t_\lambda^2).$$

Since

$$t_\sigma = \frac{1}{m} u_\sigma = -\left(\frac{\psi_\sigma}{\sigma} \right)_\sigma,$$

$$t_\lambda = \frac{u_\lambda}{m} - \frac{1}{2m} = -\frac{\psi_{\sigma\lambda}}{\sigma} - \frac{1}{2m},$$

from Eq. (3.14), the vanishing of J implies

$$\pm \left(\frac{\psi_\sigma}{\sigma}\right)_\sigma + \left(\frac{\psi_{\sigma\lambda}}{\sigma} + \frac{1}{2m}\right) = 0. \tag{3.32}$$

The derivatives involved above can be found from Eq. (3.27). With the abbreviations $(\omega/2s)\sigma = \sigma'$ and $\omega\lambda/2s = \lambda'$, condition (3.32) becomes, after some algebra,

$$-\frac{8gA}{\omega}\left(\frac{\omega}{2m}\right)^3 m\left\{\pm\left[-\left(\sigma'J_0(\sigma') - 2J_1(\sigma')\right)\cos\lambda'\right] + \sigma'J_1(\sigma')\sin\lambda'\right\} = \tfrac{1}{2}\sigma'^2.$$

$$\tag{3.33}$$

For a fixed λ, the curly bracket on the left first increases with σ', then oscillates with diminishing amplitude about a mean which grows as $(\sigma')^{1/2}$, while the right-hand side increases as σ'^2. If the left-hand side is less than the right for small σ', then it remains so for all σ'. Since for small σ'

$$\left[-\sigma'J_0(\sigma') - 2J_1(\sigma')\right] \cong \frac{\sigma'^3}{8} + O(\sigma'^5),$$

$$\sigma'J_1(\sigma') = \frac{\sigma'^2}{2} + O(\sigma'^4),$$

the left-hand side of Eq. (3.33) is approximately

$$4gA\left(\frac{\omega}{2m}\right)^2\left[\pm\frac{\sigma'^3}{8}\cos\lambda' + \frac{\sigma'^2}{2}\sin\lambda'\right] + \cdots$$

which is always less than the right-hand side if

$$4gA\left(\frac{\omega}{2m}\right)^2 < 1. \tag{3.34a}$$

Thus, if

$$A < \frac{gs^2}{\omega^2} = \frac{1}{4\pi^2}gT^2s^2, \tag{3.34b}$$

no breaking will occur. Note that the limiting amplitudes increase with the squares of the period and of the bottom slope. This breaking criterion is meaningful only as a qualitative rather than quantitative guide. In fact, A should not be too close to this limiting value for the solution to be valid. It is nevertheless interesting that Eq. (3.34) is of the same form as the empirical law Eq. (4.2), Chapter Ten, in terms of the surf parameter.

Let us assume that Eq. (3.34) is satisfied and study the run-up of the shoreline. By taking $\sigma = 0$ in Eq. (3.22), we locate the instantaneous shoreline

$$x(0, \lambda) = \frac{u^2(0, \lambda)}{2m} + \frac{\psi_\lambda(0, \lambda)}{4}. \tag{3.35}$$

From Eqs. (3.21) and (3.27), we get

$$u(0, \lambda) = \frac{gA\omega}{m} \cos \frac{\omega\lambda}{2m}, \tag{3.36a}$$

$$\psi_\lambda(0, \lambda) = \frac{-4gA}{m} \sin \frac{\omega\lambda}{2m}, \tag{3.36b}$$

$$x(0, \lambda) = \frac{A}{s} \left[-\sin \frac{\omega\lambda}{2m} + \frac{gA}{2} \left(\frac{\omega}{2m} \right)^2 \cos^2 \frac{\omega\lambda}{2m} \right]. \tag{3.36c}$$

At the shoreline, t and λ are related by

$$t = -\frac{1}{m} \left(\frac{\lambda}{2} - u(0, \lambda) \right) = \frac{-1}{2m} \left[\lambda - \frac{2gA\omega}{m} \cos \frac{\omega\lambda}{2m} \right]. \tag{3.37}$$

Equations (3.36c) and (3.37) give the shoreline in terms of the parameter λ. Using condition (3.34a) in Eqs. (3.36), we find the run-up for waves just beginning to break at the shore,

$$-x \leq \frac{A}{s} \left(\sin \frac{\omega\lambda}{2m} - \frac{1}{8} \cos^2 \frac{\omega\lambda}{2m} \right). \tag{3.38}$$

The maximum run-up occurs when $\omega\lambda/2m = \frac{1}{2}\pi$, so that $x = -A/s$. This result can also be obtained from the linearized theory which is presumably invalid at the shore (Keller, 1961). From Eq. (3.30), the linear theory gives A as the maximum wave height at $x = 0$. Thus, by assuming the free surface to be horizontal near the shore, the maximum run-up is again $-A/s$. This coincidence gives us some hope that the linearized approximation may give a reasonable estimate for the run-up of nonbreaking waves on beaches of more general profile, for which a nonlinear analysis is too difficult.

The instantaneous free-surface profile and the flow field predicted by the present nonlinear theory are quite different from the linear theory; some computed results by Carrier and Greenspan (1957) are shown in Fig. 3.1.

11.3.2 Matching with Deep Water

In the sense of matched asymptotics, the results of Section 11.3.1 may be further used as the inner solution to be matched with a linear standing wave solution in the deep sea. This is possible because the outer limit of the

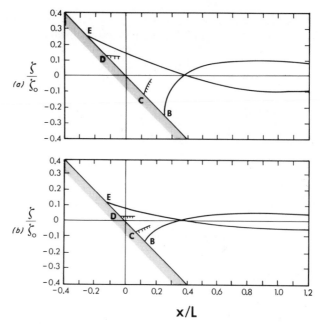

Figure 3.1 Standing wave profiles on a beach according to Eq. (3.27) at E, point of maximum run-up $\omega\lambda/2m = (\frac{1}{2})\pi$; B, $\omega\lambda/2m = \frac{3}{2}\pi$; C, D, intermediate times. (a) $A/\zeta_0 = 1$; (b) $A/\zeta_0 = \frac{1}{2}$ where $L = 4gs/\omega^2$, $\zeta_0 = 4gs^2/\omega^2$. (From Carrier and Greenspan, 1957, *J. Fluid Mech*. Reproduced by permission of Cambridge University Press.)

nonlinear Airy equations is linear and is the same as the inner limit ($kh \to 0$) of the linear WKB theory in gradually varying depth (Carrier, 1966; and Mei, 1966b).

Far from the beach the linearized theory should suffice. By adding two progressive waves propagating in opposite directions on a slowly varying bottom (Chapter Three), a standing wave is obtained,

$$\Phi = \frac{2gA^0}{\omega} \frac{\cosh k(z+h)}{\cosh kh} \cos \omega t \cos\left(\int^x k \, dx\right), \qquad (3.39)$$

$$\zeta = -2A^0 \sin \omega t \cos\left(\int^x k \, dx\right), \qquad (3.40)$$

where the lower limit of integration is, so far, arbitrary. The amplitude A^0 is given by A of Eq. (3.9), Chapter Three, with $\alpha = \alpha_0 = 0$. Let the reference point be far offshore ($x \sim \infty$) where the depth is great. Then in very shallow water

$$A^0 = A_\infty(k_\infty h)^{-1/4}2^{-1/2}, \qquad (3.41)$$

and

$$\int^x k\, dx = \frac{k_\infty}{s} \int^x \frac{dh}{h^{1/2}} = \frac{2\omega x^{1/2}}{(gs)^{1/2}} + \delta, \tag{3.42}$$

where $(\ )_\infty$ denotes quantities far offshore and δ is an integration constant. Hence, the inner approximation of Eq. (3.40) is

$$\zeta \cong -A_\infty 2^{1/2} \left(\frac{g}{\omega^2}\right)^{1/4} s^{-1/4} x^{-1/4} \sin \omega t \cos\left(\frac{2\omega x^{1/2}}{(gs)^{1/2}} + \delta\right). \tag{3.43}$$

On the other hand, the nonlinear solution valid near the shore becomes linear far offshore. Specifically, when $2\omega(x/sg)^{1/2} \gg 1$, the outer approximation is, from Eq. (3.30),

$$\zeta \cong -A J_0\left(2\omega\left(\frac{x}{sh}\right)^{1/2}\right) \sin \omega t$$

$$= -A\pi^{-1/2}\left(\frac{g}{\omega^2}\right)^{1/4} s^{1/4} x^{-1/4} \cos\left(2\omega\left(\frac{x}{sg}\right)^{1/2} - \frac{\pi}{4}\right) \sin \omega t. \tag{3.44}$$

Matching Eq. (3.43) with Eq. (3.44), we find

$$\delta = -\frac{\pi}{4}, \tag{3.45a}$$

and

$$A = A_\infty \left(\frac{2\pi}{s}\right)^{1/2}. \tag{3.45b}$$

Together Eqs. (3.27) and (3.40) give a uniformly valid theory for all depths.

Equations (3.45b) and (3.34b) can be combined to give the breaking criterion

$$k_\infty A = s^2 \quad \text{or} \quad k_\infty A_\infty = \frac{1}{(2\pi)^{1/2}} s^{5/2}. \tag{3.46}$$

With Eq. (3.45b) the empirical criterion (4.2), Chapter Ten, may be written

$$k_\infty A = \left(\frac{\pi}{4}\right)^2 s^2 \quad \text{or} \quad k_\infty A_\infty = \frac{1}{(2\pi)^{1/2}}\left(\frac{\pi}{4}\right)^2 s^{5/2}. \tag{3.47}$$

Based on a linearized theory of standing waves on a slope, Miche (1944) assumed breaking to begin when the free surface at the water edge was tangential to the sloping bottom. His criterion for the critical moment was derived in terms of $k_\infty A_\infty$ and may be expressed as

$$k_\infty A = 2s^2, \qquad k_\infty A_\infty = \left(\frac{2}{\pi}\right)^{1/2} s^{5/2} \qquad (3.48)$$

because of Eq. (3.45b). Recall from Eq. (4.1), Section 10.4, that $\xi = s(\pi/k_\infty A)^{1/2}$ is the surf parameter. Thus, the critical value of ξ is the largest in the experiments, intermediate by the nonlinear theory of Airy, and the smallest by the linearized theory of Miche. We hasten to emphasize that neither theory should be expected to be quantitatively correct, but both confirm the basic significance of the surf parameter. To bring the theory closer to experiments, it is likely that the boundary conditions must be exactly met near the breaking crest.

Despite the difficulties in predicting the incipient breaking, Airy's equations provide a useful framework for bore propagation. Assuming that the bore is already present, one may treat the motion away from the bore by Airy's equations and connect both sides of the bore by propoer shock conditions. Early studies of this kind may be found in Stoker (1957) for horizontal bottom and in Keller, Levine, and Whitham (1960) for a beach. Further analytical studies of a single bore and its run-up on a beach have been given by Ho and Meyer (1962), Shen and Meyer (1963a, b), and summarized by Meyer and Taylor (1972). Using finite differences, Hibberd and Peregrine (1979) have investigated a bore incident from a horizontal bottom toward a sloping beach. While these situations may be simulated in the laboratory, periodicity and breaking inception are two other aspects which are ever-present in nature and should be included in a more complete picture of bores climbing a beach.

11.3.3   Transient Responses to Initial Inputs

Another advantage of the linear equations (3.23) is that several interesting initial-value problems can be treated conveniently in an inverse way. A typical example is to let $\psi(\sigma, 0) = 0$ and $\psi_\lambda(\sigma, 0)$ be prescribed for $\lambda = 0$, $\sigma > 0$. The initial-value problem with finite ψ at $\sigma = 0$ may be obtained straightforwardly by Laplace transform. In view of Eqs. (3.14), (3.21) and (3.22), the initial conditions $\psi(\sigma, 0) = 0$ and $\psi_\lambda(\sigma, 0) \neq 0$ imply the physical conditions $u(x, 0) = 0$ and $\zeta(x, 0)$ at $t = 0$. Carrier and Greenspan studied a number of initial humps in this way.

Another interesting transformation which renders the nonlinear equations linear has been found by Tuck and Hwang (1972). Let

$$x^* = x + \frac{\zeta}{s}, \qquad t^* = t - \frac{u}{gs},$$

$$u^*(x^*, t^*) = u(x, t), \qquad \zeta^*(x^*, t^*) = \zeta + \frac{u^2}{2g}. \qquad (3.49)$$

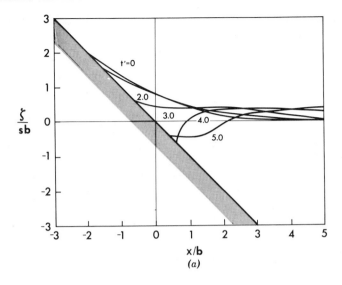

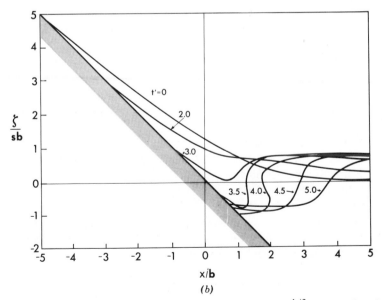

Figure 3.2 Transient waves on a beach where $t' = t(gs/b)^{1/2}$. (a) $a/sb = 2.0$; (b) $a/sb = 5.0$. (From Tuck and Hwang, 1972, *J. Fluid Mech*. Reproduced by permission of Cambridge University Press.)

Then it can be shown straightforwardly that $u(x^*, t^*)$ and $\zeta(x^*, t^*)$ satisfy the linearized equations

$$\frac{\partial \zeta^*}{\partial t^*} + s \frac{\partial}{\partial x^*} (x^* u^*) = 0, \tag{3.50}$$

$$\frac{\partial u^*}{\partial t^*} + g \frac{\partial \zeta^*}{\partial x^*} = 0. \tag{3.51}$$

Note that $sx^* = sx + \zeta$. Thus, one can also seek solutions in terms of the transformed variables, then find the physical problem later. A sample problem is to prescribe

$$\zeta^*(x^*, 0) = a e^{-x^*/b}, \tag{3.52}$$

and

$$u^*(x^*, 0) = 0. \tag{3.53}$$

The initial free surface is then given by

$$x = x^* - \frac{a}{s} e^{-x^*/b}, \tag{3.54}$$

or

$$x = -b \ln \frac{\zeta}{a} - \frac{\zeta}{s}. \tag{3.55}$$

The history of free-surface motion is shown in Figs. 3.2a and 3.2b.

11.4   SUBHARMONIC RESONANCE OF EDGE WAVES

In Section 4.8 linearized edge waves are found to be free modes trapped on a sloping beach. It was mentioned there that progressive edge waves may be generated directly by a storm traveling along the coast at a speed close to the phase velocity of an edge wave mode. The typical period of this kind of edge wave is related to the spatial extent of the storm area and is of the order of several hours. On the other hand, shorter edge waves of periods less than a few minutes are also of practical interest because they can induce either harbor seiches or rip currents which are responsible for *beach cusps*.

Can edge waves be excited by the ever-present incident swell?

If the coastline has an indentation, a linear resonance mechanism similar to harbor resonance is possible. Imagine a rectangular cove of length L with width B on an otherwise straight coast. Let the bottom of the cove be a beach of constant slope s. If the incident waves have the frequency ω which nearly satisfies the eigenvalue condition

$$\omega^2 = gks(2n + 1) \quad \text{with } k = \frac{2m\pi}{B}, \quad n, m = \text{integers},$$

and if the cove length L is $O(2\pi/k)$, then a trapped edge wave can certainly be excited in the cove. The finiteness of L makes it possible to have some radiation damping and enables the incident wave to supply energy to the edge wave.

In their studies of nearshore circulation, Bowen and Inman (1969) found field evidences of standing edge waves of periods comparable in order of magnitude to the period of the incoming swell. The amplified edge waves cause longshore modulation of the incident swell which may be sufficiently short to break near the shore. The periodic modulation affects the breaking wave height and the radiation stresses, leading to periodic cells of currents, which, in turn, lead to beach cusps. Motivated by these interests, Guza and Davis (1974) and Guza and Bowen (1976) made a systematic examination of the nonlinear mechanism of *subharmonic resonance* in which a standing edge wave of frequency ω was resonated by a normally incoming and reflected wave of frequency 2ω. Guza and Bowen employed Airy's shallow-water approximation as the basis of their theory. In addition to the initial instability of edge waves, the incident and reflected waves were found to leak energy by radiation due to quadratic nonlinearity. Consideration of cubic nonlinearity and of radiation damping enabled them to predict both the initial resonant growth and the final equilibrium amplitude. Their own experiments also strongly supported these findings.

Minzoni and Whitham (1977) further extended the theory of Guza et al. In particular, they began with the more complete but complicated theory of Friedrichs (1949) for normally incident and reflected waves on a beach of finite slope. Since most beach slopes are small, we shall follow Guza and Bowen in using the shallow-water equations

$$\zeta_t + \left[(sx + \zeta)\Phi_x\right]_x + \left[(sx + \zeta)\Phi_y\right]_y = 0, \tag{4.1}$$

$$\Phi_t + \tfrac{1}{2}\left(\Phi_x^2 + \Phi_y^2\right) + g\zeta = 0. \tag{4.2}$$

However, the more systematic perturbation arguments of Minzoni and Whitham will be applied, as in Rockliff (1978).

Eliminating ζ from Eqs. (4.1) and (4.2), we acquire a single nonlinear equation for Φ:

$$\mathcal{L}\Phi \equiv -\Phi_{tt} + sg\left[(x\Phi_x)_x + x\Phi_{yy}\right] = 2\left(\Phi_x\Phi_{xt} + \Phi_y\Phi_{yt}\right) + \Phi_t\left(\Phi_{xx} + \Phi_{yy}\right)$$

$$+ \tfrac{1}{2}\left(\Phi_x^2 + \Phi_y^2\right)\left(\Phi_{xx} + \Phi_{yy}\right) + \Phi_x^2\Phi_{xx} + \Phi_y^2\Phi_{yy} + 2\Phi_x\Phi_y\Phi_{xy}. \tag{4.3}$$

The linearized approximation of Eq. (4.3) admits two solutions. The first solution is the normally incident and reflected waves with frequency 2ω

$$\Phi_0 = -\frac{igA}{2\omega} J_0\left(4\omega\left(\frac{x}{gs}\right)^{1/2}\right)e^{-2i\omega t} + * \qquad (4.4)^\dagger$$

which is finite at the shore $x = 0$; ω can be arbitrary. The second solution is an edge wave mode of frequency ω (see Section 4.8). For simplicity, we shall only take the lowest mode (mode 0) which is described by

$$\Phi_e = -\frac{igB}{\omega} e^{-kx}\cos ky\, e^{-i\omega t} + *, \qquad (4.5)$$

subject to the dispersion relation (eigenvalue condition)

$$\omega = (gks)^{1/2}. \qquad (4.6)$$

If both Φ_0 and Φ_e are initially present, the quadratic term involving the pair (Φ_0, Φ_e) will, at the second order, give rise to a simple harmonic forcing term proportional to $G(x)\cos ky\, e^{-i\omega t}$ where $G(x)$ is a certain function of x. Unless $G(x)$ is in some sense orthogonal to the homogeneous solution e^{-kx}, the edge wave will be further resonated to great amplitude.

Let the linear terms on the left of Eq. (4.3) be denoted by $\mathcal{L}\Phi$ and the quadratic and cubic terms on the right be denoted symbolically by their possible combinations

$$\mathcal{L}\Phi_e + \mathcal{L}\Phi_0 = (\Phi_e, \Phi_e) + (\Phi_0, \Phi_0) + (\Phi_e, \Phi_0) + (\Phi_e, \Phi_e, \Phi_e)$$

$$+ (\Phi_0, \Phi_0, \Phi_0) + (\Phi_e, \Phi_0, \Phi_0) + (\Phi_0, \Phi_e, \Phi_e). \qquad (4.7)$$

The harmonics of the terms of the right are, respectively,

$$(0, \pm2) \quad (0, \pm4) \quad (\pm1, \pm3) \quad (\pm1, \pm3)$$

$$(\pm2, \pm6) \quad (\pm1, \pm3, \pm5) \quad (0, \pm2, \pm4). \qquad (4.8)$$

Let a and b denote the order of magnitude of A and B, respectively; the order of the terms on the right of Eq. (4.7) are, respectively,

$$(b^2), \quad (a^2), \quad (ab), \quad (b^3), \quad (a^3), \quad (a^2b), \quad (ab^2). \qquad (4.9)$$

Terms containing first harmonics are potentially resonance-forcing. Since $b \ll a$ at the initial stage of resonant growth, the third term on the right of Eq.

$\dagger(\ ) + *$ is the abbreviation for $(\ ) +$ the complex conjugate of $(\ )$.

(4.7) is of the order $O(ab)$ and is the most important term for resonating Φ_e. In response to this forcing, the amplitude B of the edge wave is no longer constant in time. From $-\partial^2\Phi_e/\partial t^2$, a term proportional to $\partial B/\partial t$, which represents the initial growth of B, is expected. Since $\partial B/\partial t$ must be comparable to the leading force term of $O(ab)$, the time scale of resonant growth must be $O(1/a)$. Consequently, two time scales are inherent in this problem: ω^{-1} and $(\omega ka)^{-1}$.

As resonance develops further, the edge wave is no longer small in comparison with the standing wave. Let us postulate the existence of a final equilibrium state where some cubic terms become as important as the quadratic terms. From Eqs. (4.8) and (4.9), we see that (Φ_e, Φ_0) can only be balanced by (Φ_e, Φ_e, Φ_e) which implies that $O(ab) = O(b^3)$ or $b = O(a^{1/2})$ at equilibrium. To account for this eventuality we allow Φ_e to be $O(a^{1/2})$, introduce the slow time $T = at$, and assume a multiple scale expansion as follows:

$$\Phi = \left[a^{1/2}\psi_1(x, y, T) + a^{3/2}\psi_2(x, y, T) + \cdots\right]e^{-i\omega t} + *$$

$$+ \left[a\phi_1(x, T) + a^2\phi_2(x, T) + \cdots\right]e^{-2i\omega t} + *. \tag{4.10}$$

The left-hand side of Eq. (4.3) becomes

$$a^{1/2}\left\{\omega^2\psi_1 + gs\left[(x\psi_{1x})_x + x\psi_{1yy}\right]\right\}e^{-i\omega t} + *$$
$$+ a\left\{4\omega^2\phi_1 + gs(x\phi_{1x})_x\right\}e^{-2i\omega t} + * \tag{4.11}$$
$$+ a^{3/2}\left\{\omega^2\psi_2 + gs\left[(x\psi_{2x})_x + x\psi_{2yy}\right] + 2i\omega\frac{\partial\psi_1}{\partial T}\right\}e^{-i\omega t} + *.$$

The right-hand side of Eq. (4.3) may be symbolically rewritten

$$a(\psi_1, \psi_1)e^{-2i\omega t} + * + a^{3/2}(\psi_1^*, \phi_1)e^{-i\omega t} + *$$

$$+ a^{3/2}(\psi_1^*, \psi_1, \psi_1)e^{-i\omega t} + *. \tag{4.12}$$

We now examine the perturbation equations sequentially and explicitly.

At $O(a^{1/2})$, the governing equation is

$$\omega^2\psi_1 + gs\left[(x\psi_{1x})_x + x\psi_{1yy}\right] = 0, \tag{4.13}$$

which has a homogeneous solution

$$\psi_1 = -\frac{igB}{\omega}e^{-kx}\cos ky \tag{4.14a}$$

with

$$\omega^2 = gks, \tag{4.14b}$$

that is, the lowest edge wave mode. For later purposes note that the factor $F = e^{-kx}$ which describes the x dependence satisfies

$$\omega^2 F + gs[(xF_x)_x - xk^2 F] = 0;$$

$$xF' = 0 \quad \text{at } x = 0; \qquad F \to 0, \quad x \sim \infty.$$

(4.15)

At $O(a)$, the governing equation is

$$4\omega^2 \phi_1 + gs(x\phi_{1x})_x = (\psi_1, \psi_1),$$

(4.16a)

where the quadratic forcing term is, after some algebra,

$$(\psi_1, \psi_1) = 2i\left(\frac{gk}{\omega}\right)^2 B^2 e^{-2kx}.$$

(4.16b)

The solution to Eq. (4.16) has both homogeneous and particular parts. The homogeneous solution is just the standing wave

$$\phi_1^h = -\frac{igA}{2\omega} J_0\left(4(kx)^{1/2}\right),$$

(4.17)

where A is the amplitude at the shore. The particular solution should be finite at $x = 0$ and outgoing at infinity. We leave it as an exercise to show that

$$\phi_1^p = i\frac{2\pi gk}{\omega s} B^2 G(\xi),$$

(4.18a)

where $\xi = kx$ and

$$G(\xi) = -E_2(\xi)J_0\left(4\xi^{1/2}\right) + E_1(\xi)Y_0\left(4\xi^{1/2}\right)$$

$$+ \left(E_2(\infty) - iE_1(\infty)\right)J_0\left(4\xi^{1/2}\right),$$

(4.18b)

with

$$E_1(\xi) = \int_0^\xi e^{-2\xi} J_0\left(4\xi^{1/2}\right) d\xi, \qquad E_2(\xi) = \int_0^\xi e^{-2\xi} Y_0\left(4\xi^{1/2}\right) d\xi. \quad (4.18c)$$

It is easy to check that ϕ_1^p satisfies the radiation condition.

Now we examine the $O(a^{3/2})$ problem. It is straightforward to calculate the cubic terms

$$(\psi_1^*, \psi_1, \psi_1) = -3i\left(\frac{g}{\omega}\right)^3 k^4 B^2 B^* e^{-3kx} \cos ky.$$

(4.19)

The quadratic terms are lengthier and can be written in the following form:

$$(\psi_1^*, \phi_1) = e^{-kx}\cos ky\left\{iAB^* \frac{g^2}{2\omega}\left(2k\frac{d}{dx}J_0 + \frac{d^2}{dx}J_0\right)\right.$$

$$\left. - i\frac{2\pi g^2 k}{\omega s}B^*B^2\left(2k\frac{dG}{dx} + \frac{d^2}{dx^2}G\right)\right\} \qquad (4.20)$$

without the factor $\cos ky$, the $O(a^{3/2})$ equation may be summarized as follows:

$$\omega^2 H + gs\left[(xH_x)_x - k^2 xH\right]$$

$$= \left[-2i\omega\frac{\partial B}{\partial T} + iAB^*\frac{g^2}{2\omega}\left(2k\frac{dJ_0}{dx} + \frac{d^2J_0}{dx^2}\right)\right.$$

$$\left. - i\frac{2\pi g^2 k}{\omega s}B^*B^2\left(2k\frac{dG}{dx} + \frac{d^2G}{dx^2}\right)\right]e^{-kx} - 3i\left(\frac{g}{\omega}\right)^2 k^4 B^2 B^* e^{-3kx},$$

$$(4.21)$$

where H is defined by

$$\psi_2 \equiv H(x)\cos ky. \qquad (4.22)$$

Since the homogeneous problem (4.15) has the nontrivial solution $F = e^{-kx}$, there must be a solvability condition for H (Fredholm alternative). This condition is derived as usual by multiplying Eq. (4.21) by F and Eq. (4.15) by H, and integrating the difference from 0 to ∞. After partial integration we obtain

$$\int_0^\infty dx\, e^{-kx}\{\text{RHS of Eq. (4.21)}\} = 0. \qquad (4.23)$$

This orthogonality condition gives a differential equation for B. With the change of variable $\xi = kx$, Eq. (4.23) becomes

$$\frac{\partial B}{\partial T} = \frac{iAB^*gk^2}{2\omega}\alpha - i\frac{2\pi gk^3}{\omega s}B^*B^2\beta - i\frac{3}{4}\frac{g^2}{\omega^3}k^4 B^* B^2, \qquad (4.24)$$

where α and β are a pair of definite integrals

$$\alpha = \int_0^\infty d\xi\, e^{-2\xi}\left(2\frac{dJ_0}{d\xi} + \frac{d^2J_0}{d\xi^2}\right) = 8E_1(\infty) = (2e^2)^{-1} = 0.06767,$$

$$(4.25a)$$

and

$$\beta = \int_0^\infty d\xi \, e^{-2\xi}\left(2\frac{dG}{d\xi} + \frac{d^2G}{d\xi^2}\right) = -\frac{1}{\pi} + E_G, \qquad (4.25b)$$

where

$$E_G = \int_0^\infty d\xi \, e^{-2\xi}G(\xi)$$

$$= -i(E_1(\infty))^2 + 2\int_0^\infty d\xi \, e^{-2\xi}Y_0(4\xi^{1/2})\int_0^\infty d\tau \, e^{-2\tau}J_0(4\tau^{1/2})$$

$$= 0.02862 - 0.0045782i.$$

The evaluation of the preceding integrals is detailed in Appendix 11.A.

Equation (4.24) describes the nonlinear evolution of the edge wave amplitude B and may be integrated numerically for a given A and an initial value $B(0)$. Its physical significance can be better seen by multiplying with B^* and adding the result to its complex conjugate, yielding

$$\frac{\partial|B|^2}{\partial T} = \frac{-gk^2\alpha A}{\omega}\,\mathrm{Im}\,B^{*2} + \frac{4\pi gk^3}{\omega s}|B|^4\mathrm{Im}\,\beta. \qquad (4.26)$$

This is an energy equation stating that the change of the edge wave is due to its interaction with the standing wave (the first term on the right) and the radiation of the second harmonic (the second term). Indeed, $\mathrm{Im}\,\beta$ is negative so that radiation causes damping, as expected.

During the initial state when B is still much less than A, the last two terms (proportional to B^*B^2) may be omitted from Eq. (4.24), leaving

$$\frac{\partial B}{\partial T} = \frac{4igk^2}{\omega}E_1(\infty)AB^*. \qquad (4.27)$$

Differentiating Eq. (4.27), we get

$$\frac{\partial^2 B}{\partial T^2} = |A|^2\left[\frac{4gk^2}{\omega}E_1(\infty)\right]^2 B, \qquad (4.28)$$

which can be solved to give

$$B(T) = B(0)\exp\left(\pm\frac{4gk^2}{\omega}|A|\,E_1(\infty)T\right). \qquad (4.29)$$

Hence the growth rate is proportional to $|A|$. Since $gk^3/\omega = \omega^5/g^2s^2$, the growth rate increases with higher frequency and smaller beach slope. This result agrees with the limit of the full theory at $s \to 0$ by Minzoni and Whitham.[†]

Consider now the nonlinear equilibrium state, when $\partial B / \partial T = 0$. From Eqs. (4.24) and (4.25) the equilibrium amplitude of the edge wave is found to be

$$|B| = \left[\frac{4E_1(\infty)s|A|}{k|16\pi E_G - \frac{5}{4}|} \right]^{1/2} . \tag{4.30}$$

The corresponding run-up, defined as the maximum excursion of the shore line, is

$$X_R = \frac{4|B|}{s} = 8\left(\frac{E_1(\infty)}{|16\pi E_G - \frac{5}{4}|} \right)^{1/2} \left(\frac{|A|}{ks} \right)^{1/2}$$

$$= 3.815\left(\frac{|A|}{ks} \right)^{1/2} = 3.815\left(\frac{|A|g}{\omega^2} \right)^{1/2} , \tag{4.31a}$$

which is equivalent to the limiting result ($s \to 0$) of Minzoni and Whitham (1977, Eq. (76)):

$$X_R = 5.4\left(\frac{ga_0}{\sigma^2} \right)^{1/2} \tag{4.31b}$$

where $a_0 = 2A$ and $\omega = 2\sigma$. This result also agrees with the theory of Guza and Bowen (1976). Model experiments were performed by the latter authors for a beach angle of $s = 6°$ and an incident wave period of $2\pi/2\omega = 2.7a$. For $2A$ ranging from 2 to 4 cm, the measured coefficient in Eq. (4.31b) was on the average 4.5 instead of 5.4, but in the experiments the standing waves ϕ_1 were very steep. In addition, the radiated second harmonic was also confirmed experimentally.

In view of Eq. (3.47), Section 11.3.2, the edge wave run-up can be expressed in terms of the deep-water amplitude A_∞ instead of A.

Guza and Bowen, and Minzoni and Whitham also discussed imperfect tuning and showed that a higher equilibrium amplitude could be achieved when phase matching was imperfect. For other issues their papers and Rockliff (1978) should be consulted.

[†]It may be remarked that while they give the same growth criterion for the leading order edge wave as the small-slope limit of the full theory, Airy's equations for a plane beach can lead to difficulties at the third order. Despite the solvability condition, Whitham (1976) found for a progressive edge wave of permanent form that the third-order solution became logarithmically unbounded with y as compared to the linear (first-order) solution. This nonuniformity has to be avoided either by using the full theory of arbitrary slope or by letting the sea depth approach a constant far offshore (Minzoni, 1976).

Exercise 4.1

Study the effect of imperfect resonance (detuning) when the forcing frequency is $\omega = (gks)^{1/2} + \kappa A$, where $\kappa = O(1)$. Show that when $\kappa \neq 0$, the initial growth rate is reduced but the final equilibrium amplitude is increased.

11.5   DISPERSIVE LONG WAVES OF PERMANENT FORM AND THE KORTEWEG–DE VRIES (KdV) EQUATION

In linear wave theories, solutions of the type $e^{ik(x-Ct)}$ are the most elementary; it is natural to inquire whether in nonlinear theories solutions which depend on x, t in the combination $(x - Ct)$ exist. Since these solutions represent waves propagating at constant speed without change of form, they are called the *permanent waves*. From earlier discussions, nonlinearity is known to steepen a crest, while dispersion tends to counteract this trend by "dispersing" into waves of different lengths. Permanent waves, if any, must therefore correspond to a dynamical equilibrium in which the two effects are in perfect balance.

The normalized equations of Boussinesq, Eqs. (1.36) and (1.37), can be written for one-dimensional waves as follows:

$$\zeta_t + \varepsilon(\zeta\Phi_x)_x + \Phi_{xx} = 0, \tag{5.1}$$

$$\Phi_t + \zeta + \frac{\varepsilon}{2}\Phi_x^2 = \frac{\mu^2}{3}\Phi_{xxt}, \tag{5.2}$$

upon substituting Φ_x for u. For brevity the symbols $(\ )'$ and $(\bar{\ })$ have been omitted. The spatial and time scales for normalization are k^{-1} and $(gh)^{1/2}k^{-1}$, respectively. Eliminating ζ from Eqs. (5.1) and (5.2), we obtain a single equation for Φ

$$\Phi_{tt} - \Phi_{xx} = \frac{\mu^2}{3}\Phi_{xxtt} - \varepsilon\left(\Phi_x^2 + \tfrac{1}{2}\Phi_t^2\right)_t, \tag{5.3}$$

where terms smaller than $O(\varepsilon, \mu^2)$ have been ignored. Equation (5.3) may also be called the Boussinesq equation. We now seek a solution of the following form:

$$\Phi = \Phi(\xi) \qquad \text{with } \xi = x - Ct. \tag{5.4}$$

Since

$$\frac{\partial}{\partial x} = \frac{d}{d\xi} \equiv (\ )' \quad \text{and} \quad \frac{\partial}{\partial t} = -C\frac{d}{d\xi} \equiv -C(\ )',$$

Eq. (5.3) becomes

$$(C^2 - 1)\Phi'' = \frac{\mu^2}{3}C^2\Phi'''' + \varepsilon C\left(1 + \frac{C^2}{2}\right)(\Phi'^2)'.$$

Now the preceding equation implies that $C = 1 + O(\varepsilon, \mu^2)$; hence, in all terms on the right-hand side C may be approximated by unity without affecting the accuracy. Integrating once with respect to ξ, we get

$$(C^2 - 1)\Phi' + A_1 = \frac{\mu^2}{3}\Phi''' + \frac{3\varepsilon}{2}(\Phi')^2.$$

To the leading order, $\zeta = -\Phi_t = \Phi'$; thus,

$$(C^2 - 1)\zeta + A_1 = \frac{\mu^2}{3}\zeta'' + \frac{3\varepsilon}{2}\zeta^2.$$

Finally, we multiply the above equation by ζ' and integrate once more to get

$$-\frac{\varepsilon}{2}\zeta^3 + (C^2 - 1)\frac{\zeta^2}{2} + A_1\zeta + A_2 = \frac{\mu^2}{6}\zeta'^2 \tag{5.5}$$

where the integration constants A_1 and A_2 are both of the order $O(\varepsilon, \mu^2)$.
Two cases will now be discussed.

11.5.1 Solitary Waves

A solitary wave[†] has a single crest whose amplitude diminishes to zero as $|\xi| \to \infty$. Since ζ, ζ', and ζ'' vanish at infinity, so should the constants A_1 and A_2. Equation (5.5) becomes simply

$$(\zeta')^2 = 3\zeta^2\left(\frac{C^2 - 1}{\varepsilon} - \zeta\right)\left(\frac{\varepsilon}{\mu^2}\right). \tag{5.6}$$

For the right-hand side to be positive we must have

$$C > 1 \quad \text{or} \quad C > (gh)^{1/2}$$

in physical variables; this wave speed is called *supercritical*. Furthermore, we must insist that $\zeta \le (C^2 - 1)/\varepsilon$. Hence $(C^2 - 1)/\varepsilon$ is just the maximum amplitude of the crest which is unity because of the normalization, that is,

$$C^2 = 1 + \varepsilon. \tag{5.7}$$

[†]Discovered by John Scott Russell in 1834.

In dimensional form Eq. (5.7) reads

$$C = (gh)^{1/2} \left(1 + \frac{A}{h}\right)^{1/2} = [g(h + A)]^{1/2} \tag{5.8}$$

which was first found by Rayleigh. Thus, the wave speed increases with amplitude.

With Eq. (5.7), Eq. (5.6) can be written

$$\frac{d\zeta}{d\xi} = \frac{(3\varepsilon)^{1/2}}{\mu} \zeta(1 - \zeta)^{1/2}$$

which can be integrated to give

$$\frac{\varepsilon^{1/2}}{\mu} 3^{1/2}(\xi - \xi_0) = -2 \tanh^{-1}(1 - \zeta)^{1/2},$$

or

$$\zeta = \operatorname{sech}^2 \frac{(3\varepsilon)^{1/2}}{2\mu}(\xi - \xi_0). \tag{5.9}$$

The corresponding profile is a solitary hill with the crest at $\xi = \xi_0$. The integration constant ξ_0 may be taken to be zero. In dimensional form the surface profile is

$$\zeta = A \operatorname{sech}^2 \frac{3^{1/2}}{2} \left(\frac{A}{h^3}\right)^{1/2} (x - Ct). \tag{5.10}$$

Thus, the higher the peak, the narrower the profile.

Solitary waves can be easily generated in a long tank by almost any kind of impulse. For experimental proof of this theory see Dailey and Stephans (1952a, b).

Major advances in numerical theory have been made in the past decade so that it is now possible to compute strongly nonlinear solitary waves. From the exact boundary conditions on the free surface, Byatt-Smith (1970) obtained an integral–differential equation which was then solved numerically. Fenton (1972) extended an expansion procedure due to Benjamin and Lighthill (1954) to the ninth order and made calculations comparable in accuracy to Byatt-Smith's as long as the amplitude was less than $\frac{3}{4}$ of the still-water depth. In a series of remarkable papers, Longuet-Higgins and associates made a most thorough investigation. By first deducing a number of exact relations on mass, momentum, energy, and circulation (Longuet-Higgins, 1974b), then carrying out numerical computations with the aid of either Padé approximants (Longuet-Higgins and Fenton, 1974) or an integral–differential equation

(Byatt-Smith and Longuet-Higgins, 1975), they uncovered a great deal of accurate information. Perhaps the most striking discovery is that the highest wave is not the most energetic. For fascinating details of both the approaches and the results, the reader is referred to these papers.

11.5.2 Cnoidal Waves

Besides the solitary wave just discussed, periodic permanent waves are possible. Let us first rewrite Eq. (5.5) as follows:

$$\frac{1}{3}\frac{\mu^2}{3}(\zeta')^2 = -\zeta^3 + \frac{C^2 - 1}{\varepsilon}\zeta^2 + B_1\zeta + B_2$$

$$= (\zeta_3 - \zeta)(\zeta - \zeta_2)(\zeta - \zeta_1) = P_3, \qquad (5.11)$$

where $\zeta_1 < \zeta_2 < \zeta_3$ are the three zeroes of the third-order polynomial P_3. Since the left-hand side of Eq. (5.11) is positive, the right-hand side must be positive also for a real solution. As shown in Fig. 5.1, ζ must lie between the two zeroes ζ_3 and ζ_2 which correspond to the heights of the crest and the trough, respectively. Their difference is the total dimensionless wave height

$$(\zeta_3 - \zeta_2) = H. \qquad (5.12)$$

Equation (5.11) can be integrated in terms of an elliptic integral by introducing

$$\zeta = \zeta_3\cos^2\phi + \zeta_2\sin^2\phi \qquad \text{with } \phi = \phi(\xi), \qquad (5.13)$$

and differentiating,

$$\zeta' = (\zeta_3 - \zeta_2)\phi'(-2\sin\phi\cos\phi). \qquad (5.14)$$

Equations (5.13) and (5.14) may be inserted into both sides of Eq. (5.11),

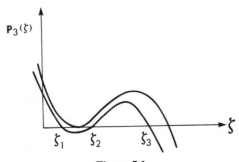

Figure 5.1

yielding

$$4\phi'^2(\zeta_3 - \zeta_2)^2\sin^2\phi\cos^2\phi = \frac{3\varepsilon}{\mu^2}\left[\zeta_3(1 - \cos^2\phi) - \zeta_2\sin^2\phi\right]$$

$$\cdot\left[-\zeta_3\cos^2\phi + \zeta_2(1 - \sin^2\phi)\right]$$

$$\cdot\left[\zeta_1 - \zeta_3(1 - \sin^2\phi) - \zeta_2\sin^2\phi\right]$$

$$= \frac{3\varepsilon}{\mu^2}(\zeta_3 - \zeta_2)^2\sin^2\phi\cos^2\phi\left[(\zeta_3 - \zeta_1) - (\zeta_3 - \zeta_2)\sin^2\phi\right],$$

$$(5.15)$$

or, after cancellation of common factors,

$$(\phi')^2 = \frac{3}{4}\frac{\varepsilon}{\mu^2}(\zeta_3 - \zeta_1)\left[1 - m\sin^2\phi\right], \qquad (5.16)$$

with

$$m = \frac{\zeta_3 - \zeta_2}{\zeta_3 - \zeta_1}, \qquad (5.17)$$

where $1 > m > 0$. Finally, we get

$$\int_0^\phi \frac{d\phi}{(1 - m\sin^2\phi)^{1/2}} \equiv F(\phi, m) = \pm\frac{(3\varepsilon)^{1/2}}{2\mu}(\zeta_3 - \zeta_1)^{1/2}(\xi - \xi_0).$$

The integral is the standard form for the incomplete elliptic integral of the first kind with modulus m. The above relation can be regarded as an implicit equation for ϕ as a function of ξ. More explicitly, we denote

$$\cos\phi = \text{Cn}\left[\frac{(3\varepsilon)^{1/2}}{2\mu}(\zeta_3 - \zeta_1)^{1/2}(\xi - \xi_0)\right],$$

$$\sin\phi = \text{Sn}\left[\frac{(3\varepsilon)^{1/2}}{2\mu}(\zeta_3 - \zeta_1)^{1/2}(\xi - \xi_0)\right],$$

where Cn and Sn are called the cosine-elliptic and the sine-elliptic functions. From Eq. (5.13) the surface height is then

$$\zeta = \zeta_2 + (\zeta_3 - \zeta_2)\text{Cn}^2\left[\frac{(3\varepsilon)^{1/2}}{2\mu}(\zeta_3 - \zeta_1)^{1/2}(\xi - \xi_0)\right], \qquad (5.18)$$

where ζ_2 is the level of the trough measured from the mean and is negative. In physical terms the surface height is

$$\zeta = \zeta_2 + (\zeta_3 - \zeta_2)\mathrm{Cn}^2\left[\frac{3^{1/2}}{2}\frac{(\zeta_3 - \zeta_2)^{1/2}}{h^{3/2}}(x - Ct - x_0)\right]. \quad (5.19)$$

Korteweg and de Vries (1895) coined the word cnoidal for the function Cn; thus Eq. (5.18) or (5.19) is now called the *cnoidal* wave.

Since $\cos\phi$ is periodic with period 2π, $\mathrm{Cn}(z)$ is, by definition, periodic with z with the period $4K$, where

$$K = F\left(\frac{\pi}{2}, m\right) = \int_0^{\pi/2}\frac{d\phi}{(1 - m\sin^2\phi)^{1/2}} \quad (5.20)$$

is the standard symbol for the complete elliptic integral of the first kind. Since Cn^2 must have the period $2K(m)$, the dimensionless wavelength λ of a cnoidal wave is given by

$$\frac{(3\varepsilon)^{1/2}}{2\mu}(\zeta_3 - \zeta_1)^{1/2}\lambda = 2K(m),$$

or

$$\lambda = \frac{4K(m)\mu}{[3(\zeta_3 - \zeta_1)\varepsilon]^{1/2}}. \quad (5.21)$$

The wavelength depends on the amplitude through m. Equation (5.18) may be rewritten

$$\zeta = \zeta_2 + (\zeta_3 - \zeta_2)\mathrm{Cn}^2\left[\frac{2K}{\lambda}(x - Ct)\right]. \quad (5.22)$$

The wave speed is found from Eq. (5.11) in terms of ζ_1, ζ_2, and ζ_3,

$$\frac{C^2 - 1}{\varepsilon} = \zeta_1 + \zeta_2 + \zeta_3,$$

or

$$C^2 = 1 + \varepsilon(\zeta_1 + \zeta_2 + \zeta_3). \quad (5.23)$$

In principle, the cnoidal wave is specified by three parameters ζ_1, ζ_2, and ζ_3, or equivalently by ζ_2, ζ_3, and the wavelength λ. For engineering usage, it is more convenient to replace ζ_1, ζ_2, and ζ_3 by the wavelength λ, the mean depth and the wave height H measured vertically from trough to crest. Let us define the mean depth so that the net area occupied by fluid within a wavelength is zero:

$$\int_0^\lambda \zeta \, d\xi = 0,$$

which implies

$$\int_0^\pi \left(\zeta_3 \cos^2\phi + \zeta_2 \sin^2\phi \right) \frac{d\xi}{d\phi} \, d\phi = 0$$

because of Eq. (5.13). From Eq. (5.17) and the square root of Eq. (5.16), the left-hand side of the preceding integral may be rewritten

$$\int_0^{\pi/2} d\phi \, \frac{\zeta_1 + (\zeta_3 - \zeta_1)(1 - m\sin^2\phi)}{(1 - m\sin^2\phi)^{1/2}} = 0,$$

where a constant multiplier has been dropped and the symmetry of $\sin^2\phi$ about $\phi = \frac{1}{2}\pi$ is used. Using Eq. (5.20) and the definition

$$E(m) = \int_0^{\pi/2} (1 - m\sin^2\phi)^{1/2} \, d\phi \qquad (5.24)$$

for the elliptic integral of the second kind, we have

$$\zeta_1 K(m) + (\zeta_3 - \zeta_1) E(m) = 0, \qquad (5.25)$$

or

$$\zeta_1 = -\frac{E}{K}(\zeta_3 - \zeta_1) = -\frac{\zeta_3 - \zeta_2}{m}\frac{E}{K} = -\frac{\cdot H}{m}\frac{E}{K}. \qquad (5.26)$$

It follows from Eq. (5.25) that

$$\zeta_3 = -\zeta_1 \left(\frac{K}{E} - 1 \right) = \frac{H}{m}\left(1 - \frac{E}{K} \right), \qquad (5.27)$$

and, finally,

$$\zeta_2 = \zeta_3 - H = \frac{H}{m}\left(1 - m - \frac{E}{K} \right). \qquad (5.28)$$

Thus, all three parameters ζ_1, ζ_2, and ζ_3 are expressed in terms of H and m, the mean depth of h being fixed. These expressions may be inserted into Eq. (5.23) for the dimensionless wave speed

$$C^2 = 1 + \varepsilon \frac{H}{m}\left[-m + 2 - \frac{3E}{K} \right], \qquad (5.29)$$

and into Eq. (5.21) for the dimensionless wavelength

$$\lambda = \frac{4K\mu}{(3\varepsilon)^{1/2}} \left(\frac{m}{H}\right)^{1/2}. \tag{5.30}$$

Now the dimensionless wave period is

$$T = \frac{\lambda}{C} = \frac{\left(4\mu/(3\varepsilon)^{1/2}\right)(m/H)^{1/2}K}{\left[1 + \varepsilon(H/m)(-m + 2 - 3E/K)\right]^{1/2}}. \tag{5.31}$$

To return to physical variables the following transformation is necessary:

$$x \to kx, \qquad t \to k(gh)^{1/2}t, \qquad C \to (gh)^{1/2}C,$$

$$\lambda \to k\lambda, \qquad H \to \frac{H}{A}.$$

By using the definitions $\mu = kh$ and $\varepsilon = A/h$, it is easy to obtain that, in physical variables,

$$C^2 = gh\left[1 + \frac{H}{h}\frac{1}{m}\left(-m + 2 - 3\frac{E}{K}\right)\right], \tag{5.32}$$

$$\lambda = 4Kh\left(\frac{m}{3H/h}\right)^{1/2}, \tag{5.33}$$

$$T \cong \left(\frac{h}{g}\right)^{1/2}\frac{4K\left(m/3H/h^{-1}\right)^{1/2}}{\left[1 + (H/h)(1/m)(-m + 2 - 3E/K)\right]^{1/2}}, \tag{5.34}$$

$$\zeta = \zeta_2 + H\operatorname{Cn}^2\left[\frac{2K}{\lambda}(x - Ct)\right]. \tag{5.35}$$

The parameter m can, in principle, be eliminated from any pair among Eqs. (5.32)–(5.34) so as to obtain relations of the type $C = C(T, H)$, $\lambda = \lambda(T, H)$, and so on. However, it is simpler to leave m as a parameter. Wiegel (1960) has plotted the wave profile ζ for various values of m ranging from 0 to almost 1; sample results are reproduced in Figs. 5.2 and 5.3.

For a better understanding of these curves it is helpful to check the following two limiting cases:

(i) $m \to 1$. In this limit $\zeta_2 \to \zeta_1$, $E(1) = 1$, and $K(1) \to \infty$; consequently, $\lambda \to \infty$ and $\operatorname{Cn}^2 u \to \operatorname{sech}^2 u$. However, from Eq. (5.30) the ratio K/λ approaches a finite limit so that

$$\zeta = H\operatorname{sech}^2\frac{3^{1/2}}{2}\left(\frac{H}{h^3}\right)^{1/2}(x - Ct)$$

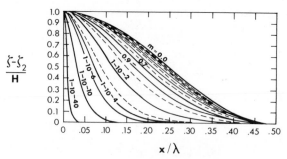

$$\frac{\zeta - \zeta_2}{H}$$

Figure 5.2 Surface profiles of cnoidal waves. (From Wiegel, 1960, *J. Fluid Mech.* Reproduced by permission of Cambridge University Press.)

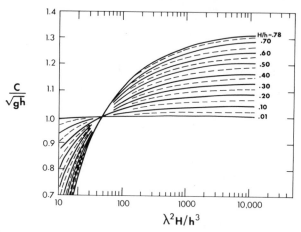

$$\frac{C}{\sqrt{gh}}$$

Figure 5.3 Dispersion relation of cnoidal waves. (From Wiegel, 1960, *J. Fluid Mech.* Reproduced by permission of Cambridge University Press.)

which is just Eq. (5.10). Thus, the solitary wave is the limit of the cnoidal wave with infinite wavelength. If the wavelength is kept fixed, the wave profile becomes isolated peaks, as is represented by the curve with $m \approx 1$ in Fig. 5.2. The wave speed becomes

$$C^2 = gh\left(1 + \frac{H}{h}\right).$$

This equation corresponds to the horizontal asymptotes in Fig. 5.3.

(ii) $m \to 0$. In this limit $\zeta_3 - \zeta_2 = H \to 0$, that is, the waves are infinitesimal. It can be easily shown that

$$C^2 \to gh, \qquad Cn(u \mid m) \to Cn(u \mid 0) \to \cos u, \qquad K \to \frac{\pi}{2},$$

and

$$\zeta = \zeta_2 + H \cos^2 \frac{\pi}{\lambda} (x - Ct).$$

Now $\zeta_2 = -a = -\frac{1}{2}H$ so that

$$\zeta = \frac{H}{2} \cos \frac{2\pi}{\lambda} (x - Ct)$$

which is the linearized sinusoidal wave, as is evident in Fig. 5.2. In Fig. 5.3 the wave speed is represented by the horizontal line $C/(gh)^{1/2} = 1$.

A fifth-order theory has also been worked out by Fenton (1979).

11.5.3   The Korteweg–de Vries (KdV) Equation

The permanent waves studied so far depend on the following variable:

$$\xi = x - Ct = x - t + O(\varepsilon)t$$

because of Eq. (5.7). Thus, an observer traveling at the linearized long-wave speed, which is unity in dimensionless variables, witnesses a slow variation in time. The dimensionless scale of the slow time variation is $1/\varepsilon$, which suggests the following variables for the more general transient evolution of nonlinear and dispersive long waves propagating in positive x direction:

$$\sigma = x - t, \qquad \tau = \varepsilon t. \tag{5.36}$$

In terms of these variables, the derivatives become

$$\frac{\partial}{\partial x} \rightarrow \frac{\partial}{\partial \sigma}, \qquad \frac{\partial}{\partial t} \rightarrow -\frac{\partial}{\partial \sigma} + \varepsilon \frac{\partial}{\partial \tau}.$$

By substituting these into the alternate Boussinesq equation (5.3), we get immediately

$$(\phi_\sigma)_\tau + \frac{3}{4} (\sigma_\sigma^2)_\sigma + \frac{\mu^2}{6\varepsilon} \phi_{\sigma\sigma\sigma\sigma} = O(\varepsilon, \mu^2).$$

To the leading order ϕ_σ may be replaced by ζ or u, that is,

$$\zeta_\tau + \frac{3}{2} \zeta \zeta_\sigma + \frac{\mu^2}{6\varepsilon} \zeta_{\sigma\sigma\sigma} \cong 0. \tag{5.37}$$

This is commonly called the Korteweg–de Vries (or KdV) equation. In physical variables and stationary coordinates, Eq. (5.37) takes the form

$$\frac{\partial \zeta}{\partial t} + (gh)^{1/2} \left(1 + \frac{3}{2} \frac{\zeta}{h}\right) \frac{\partial \zeta}{\partial x} + \frac{h^2}{6} (gh)^{1/2} \frac{\partial^3 \zeta}{\partial x^3} = 0. \tag{5.38}$$

In the stationary frame of reference the nonlinear and the dispersion terms are small quantities of order ε and μ^2, respectively, hence one may replace $(gh)^{1/2}\zeta_x$ by $-\zeta_t$ without affecting the accuracy. Possible alternative forms for Eq. (5.38) are:

$$\zeta_t + (gh)^{1/2}\zeta_x + \frac{3}{2}(gh)^{1/2}\frac{\zeta}{h}\left(\begin{array}{c} \zeta_x \\ -\dfrac{1}{(gh)^{1/2}}\zeta_t \end{array} \right)$$

$$+\frac{h^2}{6}(gh)^{1/2}\left[\begin{array}{c} \zeta_{xxx} \\ -\dfrac{1}{(gh)^{1/2}}\zeta_{xxt} \\ \dfrac{1}{gh}\zeta_{xtt} \\ -\dfrac{1}{(gh)^{3/2}}\zeta_{ttt} \end{array} \right] = 0.$$

$$\text{(5.39a)}$$
$$\text{(5.39b)}$$
$$\text{(5.39c)}$$
$$\text{(5.39d)}$$

It is easy to check that none of these forms will alter the linearized dispersion relation to the accuracy $O(\varepsilon, \mu^2)$. The permanent solitary and cnoidal waves can also be deduced from Eq. (5.37) by assuming $\zeta = \zeta(\sigma - C\tau)$; the details are left as an exercise.

11.6 NONLINEAR DISPERSIVE STANDING WAVES ON A HORIZONTAL BOTTOM

In studying the interaction between two trains of permanent waves propagating in two different directions, Benney and Luke (1964) found that the interaction effect remained $O(\varepsilon)$ times smaller than the primary waves, as long as $\tau = \varepsilon t = O(1)$ and the angle of intersection was greater than $O(\mu)$. This conclusion implies that two permanent nonlinear waves which do not travel in the same direction may be superimposed to give another leading-order solution to the Boussinesq equations. In particular, standing waves can be constructed by adding two opposite-going cnoidal waves. To show this result, it is convenient and sufficient to begin with Eq. (5.3).

We introduce the multiple scale expansion

$$\Phi = \phi^{(0)}(x, t; \tau) + \varepsilon\phi^{(1)}(x, t; \tau) + \cdots \tag{6.1}$$

into Eq. (5.3) and get

$$\phi_{tt}^{(0)} - \phi_{xx}^{(0)} = 0, \tag{6.2}$$

$$\phi_{tt}^{(1)} - \phi_{xx}^{(1)} = \frac{1}{3}\frac{\mu^2}{\varepsilon}\phi_{xxtt}^{(0)} - \left[\left(\phi_x^{(0)}\right)^2 + \tfrac{1}{2}\left(\phi_t^{(0)}\right)^2 \right]_t - 2\phi_{t\tau}^{(0)}. \tag{6.3}$$

The general solution to Eq. (6.2) is

$$\phi^{(0)} = \phi^+ (\sigma_+ ; \tau) + \phi^- (\sigma_- ; \tau), \tag{6.4}$$

where

$$\sigma_{\pm} = x \mp t. \tag{6.5}$$

Thus, ϕ^+ and ϕ^- travel to the right and left, respectively. Equation (6.3) may be written

$$-4 \frac{\partial^2 \phi^{(1)}}{\partial \sigma_+ \partial \sigma_-} = 2 \frac{\partial}{\partial \tau} \frac{\partial \phi^+}{\partial \sigma_+} + 3 \frac{\partial \phi^+}{\partial \sigma_+} \frac{\partial^2 \phi^+}{\partial \sigma_+^2} + \frac{1}{3} \frac{\mu^2}{\varepsilon} \frac{\partial^4 \phi^+}{\partial \sigma_+^4}$$

$$-2 \frac{\partial}{\partial \tau} \frac{\partial \phi^-}{\partial \sigma_-} - 3 \frac{\partial \phi^-}{\partial \sigma_-} \frac{\partial^2 \phi^-}{\partial \sigma_-^2} + \frac{1}{3} \frac{\mu^2}{\varepsilon} \frac{\partial^4 \phi^-}{\partial \sigma_-^4}$$

$$+ \frac{\partial^2 \phi^+}{\partial \sigma_+^2} \frac{\partial \phi^-}{\partial \sigma_-} - \frac{\partial \phi^+}{\partial \sigma_+} \frac{\partial^2 \phi^-}{\partial \sigma_-^2}. \tag{6.6}$$

For $\phi^{(1)}$ not to grow linearly with σ_+ or σ_-, the first and second lines on the right-hand side of Eq. (6.6) must vanish separately. Writing

$$\zeta^+ = -\phi_t^+ = \phi_{\sigma_+}^+ \quad \text{and} \quad \zeta^- = -\phi_t^- = -\phi_{\sigma_-}^-, \tag{6.7}$$

we obtain

$$\frac{\partial \zeta^+}{\partial \tau} + \frac{3}{2} \zeta^+ \frac{\partial \zeta^+}{\partial \sigma_+} + \frac{\mu^2}{6\varepsilon} \frac{\partial^3 \zeta^+}{\partial \sigma_+^3} = 0, \tag{6.8}$$

and

$$- \frac{\partial \zeta^-}{\partial \tau} + \frac{3}{2} \zeta^- \frac{\partial \zeta^-}{\partial \sigma_-} + \frac{\mu^2}{6\varepsilon} \frac{\partial^3 \zeta^-}{\partial \sigma_-^3} = 0. \tag{6.9}$$

Thus, the right- and left-going waves are decoupled and satisfy their own Korteweg–de Vries (KdV) equations separately. To the leading order, $\zeta^{(0)}$, which corresponds to $\phi^{(0)}$, may be constructed by superposition:

$$\zeta^{(0)} = \zeta^+ (\sigma_+ ; \tau) + \zeta^- (\sigma_- ; \tau). \tag{6.10}$$

Consider a one-dimensional tank in the spatial domain $0 \leq x \leq \pi$. If

$$\zeta^+ = F(-\sigma_+ ; \tau) \quad \text{and} \quad \zeta^- = F(\sigma_- ; \tau), \tag{6.11}$$

then the corresponding velocity field

$$u^{(0)} = -F(-\sigma_+ ; \tau) + F(\sigma_- ; \tau) \tag{6.12}$$

vanishes at the left wall $x = 0$ for all t. In order for $u^{(0)}$ to vanish also at $x = \pi$ we require

$$-F(t - \pi; \tau) + F(t + \pi; \tau) = 0,$$

which implies that $F(\sigma; \tau)$ is periodic in σ with the period 2π. A cnoidal wave of unit amplitude meets all these conditions:

$$F(\sigma; \tau) = f_2 + (f_3 - f_2)\mathrm{Cn}^2\left[\left(\frac{3\varepsilon}{2\mu^2}(f_3 - f_1)\right)^{1/2}(\sigma + \gamma\tau); m\right], \tag{6.13}$$

$$\frac{\varepsilon}{\mu^2} = \frac{4mK^2}{3\pi^2}, \tag{6.14a}$$

$$\gamma = \frac{1}{m}\left(1 - \frac{3}{2}\frac{E}{K}\right) - \frac{1}{2}, \tag{6.14b}$$

$$f_1 = -\frac{E}{mK}, \tag{6.14c}$$

$$f_3 = f_2 + 1 = \frac{1}{m}\left(1 - \frac{E}{K}\right). \tag{6.14d}$$

The standing wave is given by

$$\zeta^{(0)} = F((1 + \gamma\varepsilon)t - x) + F((1 + \gamma\varepsilon)t + x) \tag{6.15}$$

with the dimensionless period of

$$T = 2\pi(1 + \gamma\varepsilon)^{-1}. \tag{6.16}$$

The dimensionless crest-to-trough amplitude of the standing wave is 2. Given the following physical data: amplitude A_0, tank depth h, tank length L, and modal index n, the Ursell parameter

$$U_r = \frac{\varepsilon}{\mu^2} = \frac{A_0}{h}\left(\frac{L}{h}\right)^2\left(\frac{1}{n\pi}\right)^2$$

is known, and m and γ can be calculated from Eqs. (6.14a) and (6.14b), respectively. The remaining parameters f_1, f_2, and f_3 follow from Eqs. (6.14c) and (6.14d).

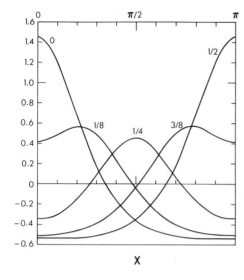

Figure 6.1 Instantaneous free surfaces of nonlinear standing waves within one-half of wavelength. Numbers near each profile indicate t/T for half a period (S. R. Rogers and C. C. Mei, 1975, in an unpublished manuscript).

Figure 6.1 shows the sample history of a standing wave within half of the tank for $n = 1$ and $U_r = 1.825$. Unlike the sloshing mode of a linear standing wave, the free surface is never horizontal and fixed nodes do not exist. Furthermore, a crest bounces back and forth between two end walls. As the Ursell number becomes very large, the bouncing crest approaches a solitary wave except very near the ends. These features are also present in the theory of nonlinear standing waves when kh is not small (Wehausen and Laitone, 1960, p. 665).

Lepelletier (1980) performed experiments by oscillating a shallow-water tank sinusoidally and horizontally at very low amplitudes, until steady resonance was reached. In the case where $kh = 0.314$, $A_0/h = 0.055 - 0.2$ (i.e., $U_r = 0.5-2$) the measurements agreed with the present theory very well. For two other cases with larger U_r (case (i) $kh = 0.157$, $A_0/h = 0.043-0.23$, $U_r = 1.72-9.32$; case (ii) $kh = 0.107$, $A_0/h = 0.04-0.075$, $U_r = 3.5-7.0$), appreciable quantitative discrepancies existed which were attributed to the strictly sinusoidal excitation.

Returning to Eq. (6.6), the next order correction is

$$\varepsilon \phi^{(1)} = \varepsilon \left[\phi_+^{(1)} + \phi_-^{(1)} - \frac{1}{4} \left(\phi^- \frac{\partial \phi^+}{\partial \sigma_+} - \phi^+ \frac{\partial \phi^-}{\partial \sigma_-} \right) \right] \tag{6.17}$$

where $\phi_+^{(1)}(\sigma_+)$ and $\phi_-^{(1)}(\sigma_-)$ are the homogeneous solutions to the KdV equations (6.8) and (6.9) and are indeterminate until cubic terms of the order $O(\varepsilon^2, \varepsilon\mu^2, \mu^4)$ are examined. The remaining terms in Eq. (6.17) represent the weak effect of interaction and can be straightforwardly calculated (Benney and Luke, 1964).

11.7 EVOLUTION OF AN INITIAL PULSE

In recent years, the KdV equation has been the object of intensive research, as it describes a wide range of nonlinear dispersive waves in different branches of physics. Extensive numerical studies of the KdV equation by Zabusky and Kruskal (1965) for periodic boundary conditions have shown that an initial hump goes through recurring stages of disintegration into, and recombination of, pulses, each of which has the properties of a solitary wave. These solitary waves travel at different speeds and may pass through one another. After the encounter, the solitary waves leave each other with the original features preserved; the only effect of the interaction is a change of phase. Because these features are common in the physics of particles such as electrons, protons, and so on, Zabusky and Kruskal coined the word *soliton* for a solitary wave.

In a landmark paper, Gardner, Greene, Kruskal, and Miura (1967) reported an analytical solution to the initial-value problem due to a disturbance of finite length in an infinite domain, $|x| < \infty$. Subsequent development stimulated by this work has had a profound impact on the study of nonlinear waves not only in fluids but also in numerous branches of physics and engineering. This section is intended as an introduction to this expanding literature. More extensive expositions may be found in Ablowitz and Segur (1981), Karpman (1975), Gardner et al. (1974), Lamb (1980), Scott et al. (1973), Miura (1974, 1976), and Whitham (1974).

The procedure of Gardner et al. may be regarded as an extension to the classical method of Fourier transform for linear partial differential equations (Ablowitz et al., 1974). Let $u(x, t)$ be governed by $u_t = \mathcal{L}u$ where $\mathcal{L}$ is a linear differential operator in x with constant coefficients, and let the initial data $u(x, 0)$ be prescribed for $|x| < \infty$. To find $u(x, t)$ for $t > 0$ we may take the exponential Fourier transform with respect to x and solve the initial-value problem for $\hat{u}(k, t)$ where k is the transform variable. Afterward the inverse Fourier transform is performed for $u(x, t)$. In the nonlinear problem here, the Fourier transform is replaced by the mapping into a linear eigenvalue problem (or direct scattering problem, step I). We then look for some information regarding the transient evolution of the eigenvalue problem from the given initial data (step II). Finally, an inverse mapping is performed to determine the transient solution (step III). The inverse mapping involves the so-called inverse scattering theory which is a linear problem. Therefore, by this indirect procedure the nonlinear problem is replaced by a series of linear problems.

In the following discussions, the original arguments of Gardner et al., are followed.

First we switch to their convention and write the KdV equation in the form

$$\eta_t - 6\eta\eta_x + \eta_{xxx} = 0 \tag{7.1}$$

which can be achieved by making the transformation

$$\zeta \to -4\eta, \qquad \sigma \to \frac{\mu}{(6\varepsilon)^{1/2}}x, \qquad \tau \to \frac{\mu}{(6\varepsilon)^{1/2}}t$$

to Eq. (5.37). The initial data $\eta(x, 0)$ is prescribed for $|x| < \infty$. For later use we express the solitary wave of amplitude $\frac{1}{2}f^2$ by

$$\eta = -\frac{f^2}{2} \operatorname{sech}^2 \frac{f}{2}(x - f^2 t).$$ (7.2)

For step I, the transformed problem turns out to center around the Schrödinger equation

$$\psi_{xx} + (\lambda - \eta)\psi = 0$$ (7.3)

where $\eta = \eta(x, t)$ with t being a parameter. In a typical quantum-mechanical problem, the *potential* η is prescribed, and the task is to find the eigenvalues λ and the eigenfunctions. Consider the case where $\eta < 0$ in a finite region, but vanishes sufficiently fast for $|x| \to \infty$. It is known that there are then two kinds of eigenvalue problems. On one hand the negative eigenvalues form a discrete spectrum, $\lambda = \lambda_n = -k_n^2$, $n = 1, 2, 3, \ldots$; the associated eigenfunctions ψ_n vanish at infinity, hence are called *bound states*. Their asymptotic behavior at infinity is

$$\begin{aligned}
\psi_n &\sim c_n e^{-k_n x}, & x &\sim \infty, \\
&\sim d_n e^{k_n x}, & x &\sim -\infty,
\end{aligned}$$ (7.4)

where c_n and d_n are the normalization constants chosen so that

$$\int_{-\infty}^{\infty} \psi_n^2 \, dx = 1 \qquad \text{all } n.$$ (7.5)

In general, k_n, c_n, and d_n depend on the parameter t. On the other hand, the positive eigenvalues $\lambda = k^2 > 0$ form a continuous spectrum and the eigenfunction for each k has the following asymptotic behavior:

$$\psi \sim a(k, t)e^{-ikx}, \qquad x \sim -\infty,$$ (7.6a)

$$\psi \sim e^{-ikx} + b(k, t)e^{ikx}, \qquad x \sim \infty.$$ (7.6b)

Clearly, for each of these k's, ψ describes a direct scattering problem where a is the transmission coefficient and b is the reflection coefficient; both coefficients depend on t.

For step II, a surprising discovery of Gardner et al. is that the eigenvalues of Eq. (7.3) do not vary with time if $\eta(x, t)$ in Eq. (7.3) satisfies the KdV equation (7.2) and vanishes sufficiently fast as $|x| \to \infty$. This result can be verified by solving η from Eq. (7.3) and substituting into Eq. (7.1) to get

$$\lambda_t \psi^2 + (\psi R_x - \psi_x R)_x = 0,$$ (7.7a)

where

$$R = \psi_t + \psi_{xxx} - 3(\eta + \lambda)\psi_x.$$ (7.7b)

Taking $\lambda = \lambda_n$ to be the nth discrete eigenvalue and integrating Eq. (7.7a) from $x \sim -\infty$ to $x \sim +\infty$, we obtain $\lambda_n = $ const by using Eq. (7.5). Consequently, λ_n may be determined from the initial data of η, that is, $\eta(x, 0)$. Afterward, the bound-state eigenfunctions $\psi_n(x, 0)$ and the normalization constants $c_n(0)$ can be calculated for $t = 0$. The continuous eigenvalue $\lambda = k^2$ ranges from 0 to ∞; this fact does not change with time, so λ may be taken as constant. The corresponding scattering problem for $\eta(x, 0)$ may be solved for the initial values $a(k, 0)$ and $b(k, 0)$.

We shall now show that $c_n(t)$ and $b(k, t)$ may be determined from their initial values $c_n(0)$ and $b(k, 0)$ without having to know $\psi(x, t)$ for all x. Equation (7.7a), which is valid for both discrete and continuous spectra, may be integrated twice to give

$$\psi_t + \psi_{xxx} - 3(\eta + \lambda)\psi_x = C\psi + D\psi \int^x \frac{dx}{\psi^2}. \tag{7.8}$$

If ψ is the eigenfunction of the discrete spectrum, ψ^{-2} is exponentially large and we must insist that $D = 0$. Multiplying Eq. (7.8) by ψ and integrating from $x \sim -\infty$ to $x \sim +\infty$, we get

$$\frac{\partial}{\partial t} \int_{-\infty}^{\infty} \frac{1}{2}\psi^2 \, dx + \int_{-\infty}^{\infty} (\psi\psi_{xx} - 2\psi_x^2 - 3\lambda\psi^2)_x \, dx = C \int_{-\infty}^{\infty} \psi^2 \, dx.$$

Clearly, the second integral vanishes for bound states and the first integral vanishes because of Eq. (7.5). Hence $C = 0$ and

$$\psi_t + \psi_{xxx} - 3(\eta + \lambda)\psi_x = 0. \tag{7.9}$$

For $x \sim +\infty$, $\eta \sim 0$ and ψ is given asymptotically by Eq. (7.4a); the above equation now implies that

$$\dot{c}_n - 4k_n^3 c_n = 0,$$

hence

$$c_n(t) = c_n(0)e^{4k_n^3 t}. \tag{7.10}$$

For the continuous spectrum we consider $x \sim -\infty$ and substitute Eq. (7.6a) (the transmission side) into Eq. (7.8) to get

$$a_t + 4ik^3 a = Ca + \frac{D}{a} \int^x e^{2ikx} \, dx. \tag{7.11}$$

Since a is independent of x, D must vanish so that

$$a_t + (4ik^3 - C)a = 0. \tag{7.12}$$

Consider next $x \sim +\infty$. By substituting Eq. (7.6b) (the incidence side) into Eq. (7.8) and separating the coefficients of $e^{\pm ikx}$, we find

$$C = 4ik^3 \tag{7.13a}$$

and

$$\dot{b} - 8ik^3 b = 0. \tag{7.13b}$$

From Eqs. (7.12), (7.13a) and (7.13b), the evolution of a and b are found:

$$a(k, t) = a(k, 0) \quad \text{and} \quad b(k, t) = b(k, 0)e^{8ik^3 t}. \tag{7.14}$$

Step III involves the inverse scattering problem for Eq. (7.3). According to Gelfand and Levitan (1955), $\eta(x, t)$ can be uniquely determined if the *scattering data* $c_n(t)$ and $b(k, t)$ are known. Specifically, we must first construct the kernel

$$B(x, t) = \frac{1}{2\pi} \int_{-\infty}^{\infty} b(k, t)e^{ikx}\, dk + \sum_{n=1}^{N} c_n^2(t)e^{-k_n x}, \tag{7.15}$$

and solve the Gelfand–Levitan integral equation[†] for K,

$$K(x, y, t) + B(x + y, t) + \int_{x}^{\infty} B(y + z, t)K(x, z, t)\, dz = 0. \tag{7.16}$$

Afterward $\eta(x, t)$ is given by

$$\eta(x, t) = -2\frac{d}{dx}K(x, x, t). \tag{7.17}$$

The basis of Eqs. (7.15)–(7.17) is a well-developed subject in mathematical physics but is too lengthy to be discussed here; interested readers may consult Lamb (1980) or Ablowitz and Segur (1981) for a thorough discussion.

If the initial potential leads to no reflection ($b(k, t) = 0$), so that only the discrete eigenvalues enter the kernel B, the linear Gelfand–Levitan integral equation can be solved explicitly (Gardner et al. 1967; Segur, 1973). Otherwise much information can still be deduced analytically.

Let us examine the asymptotic behavior for large x and t. Since our interest is only in $K(x, x, t)$ we need only consider

$$B(x + z, t) = \frac{1}{2\pi} \int_{-\infty}^{\infty} b(k, 0)e^{ik(x+z)-8ik^3 t}\, dk$$

$$+ \sum_{n=1}^{N} c_n^2(0)e^{[8k_n^3 t - k_n(x+z)]}. \tag{7.18}$$

[†]Also called the Marchenko equation (see Lamb, 1980).

For large t and $x \sim 4k_N^2 t$, where k_N is the largest of the discrete eigenvalues $(k_N > k_{N-1} > \cdots > k_2 > k_1)$, the Nth term dominates among the series terms and the major contribution to the integral of Eq. (7.16) comes from the neighborhood of the lower limit, that is, $z = x \sim 4k_N^2 t$. Furthermore, the integral in Eq. (7.18) may be shown by the method of stationary phase to diminish with t as $t^{-1/3}$ (Segur, 1973). Let B be approximated by the dominant term

$$B(x + z, t) = c_N^2(0)e^{-k_N(x+z)+8k_N^3 t}. \tag{7.19}$$

The Gelfand–Levitan equation becomes

$$K(x, x, t) + c_N^2(0)e^{8k_N^3 t - 2k_N x} + c_N^2(0)e^{8k_N^3 t - k_N x}\int_x^\infty e^{-k_N z}K(x, z, t)\, dz \cong 0. \tag{7.20}$$

Only the second variable in K pertains to the integral equation which can be solved by assuming

$$K(x, z, t) = L(x, t)e^{-k_N z}$$

so that

$$L + c_N^2(0)e^{8k_N^3 t}e^{-k_N x} + c_N^2(0)e^{8k_N^3 t}L\int_x^\infty e^{-2k_N z}\, dz = 0.$$

It follows that

$$K(x, x, t) = -\frac{c_N^2(0)e^{8k_N^3 t - 2k_N x}}{1 + \left(c_N^2(0)/2k_N\right)e^{8k_N^3 t - 2k_N x}}. \tag{7.21}$$

Finally, Eq. (7.17) gives

$$\eta = -2k_N^2 \operatorname{sech}^2\left(k_N(x - x_0) - 4k_N^3 t\right), \tag{7.22}$$

with

$$\frac{2k_N}{c_N^2(0)} \equiv e^{-2k_N x_0}.$$

In view of Eq. (7.2), Eq. (7.22) is a soliton of amplitude $2k_N^2$ and speed $4k_N^2$!

If we focus instead on large t but $x \sim 4k_n^2 t$ for any other n, the dominant term in B is the nth term in the series of Eq. (7.18). Evidently, a soliton of amplitude $2k_n^2$ should then travel at the speed $4k_n^2$. In summary, the initial hump eventually disintegrates into N solitons, each of which corresponds to a discrete eigenvalue of the initial "potential well." By a more elaborate analysis of the integral in Eq. (7.18), an oscillatory tail can be shown to follow the train of solitons. However, the lag increases with time so that the solitons are eventually alone at the front. This disintegration of an initial pulse to a train of solitons is also called *fission*.

The following results which follow the known properties of the Schrödinger equation are quoted below without proof:

1 If $\int \eta(x, 0)\, dx < 0$, which corresponds to a net hump, there is at least one bound state, hence one soliton.

2 Let η_0 and l be the characteristic height and length of the initial hump, then the number of bound states N is proportional to $\eta_0^{1/2} l$.

Rather than deducing these results in general terms, let us cite an exactly solvable example which is well known in quantum mechanics (Landau and Lifshitz, 1958, p. 70) and was first used in the present context by Zabusky (1968). If the initial profile can be expressed by

$$\eta(x, 0) = -p(p + 1)\operatorname{sech}^2 x, \qquad p > 0, \qquad (7.23)$$

the Schrödinger equation can be reduced to a hypergeometric equation. The bound-state eigenvalues are:

$$-\lambda_{N-1-n} = k_{N-1-n}^2 = (p - n)^2 = (c + N - 1 - n)^2 \qquad (7.24)$$

for $n = 0, 1, 2, \ldots, N - 1$ where c is the noninteger difference between $p + 1$ and N. The number N of bound states is the largest integer satisfying

$$N \leq p + 1. \qquad (7.25)$$

The success of Gardner et al. hinges, in part, on finding the proper linear eigenvalue problem [cf. (7.3)] which leads to constant eigenvalues. Their ingenuity has inspired further generalizations by Lax (1968), Zakharov and Shabat (1972), Ablowitz et al. (1974), and others for other equations governing nonlinear dispersive waves, all for pure initial-value problems on an infinite line. Apparently, similar treatments have not been successful for many other types of initial-boundary-value problems, for which numerical methods are now available for quantitative information (see, e.g., Zabusky (1968) and Fornberg and Whitham (1978)).

11.8 FISSION OF SOLITONS BY DECREASING DEPTH

In general, the effects of varying depth on long waves of finite amplitudes may be studied by solving numerically the partial differential equations (1.47) and (1.48) with appropriate initial and boundary conditions. For normal incidence on one-dimensional topography, Madsen and Mei (1969) and Madsen, Mei, and Savage (1970) expressed an equivalent set of equations in quasi-hyperbolic form and solved them by the method of characteristics. It was found numerically and confirmed experimentally that a soliton traveling from one constant depth to another constant but smaller depth, disintegrates into several solitons of varying sizes, trailed by an oscillatory tail. This *fission* is clearly related to the result of the last section. Analytical confirmation and extension were provided later by Tappert and Zabusky (1971), Johnson (1973) and Ono (1972).

The analytical theory is based on a KdV equation with variable coefficients under the assumption that the scale of depth variation $\mathcal{L}$ is $L\varepsilon^{-1}$, that is, much longer than wavelength L inherent in Eqs. (1.47) and (1.48). Let us, therefore, approximate Boussinesq equations further. After normalization and noting $h_x = O(\varepsilon)$, $h_{xx} = O(\varepsilon^2)$, and so on, we get

$$\zeta_t + [(\varepsilon\zeta + h)u]_x = 0, \tag{8.1}$$

$$u_t + \varepsilon u u_x + \zeta_x - \frac{\mu^2}{3}h^2 u_{xxt} = 0. \tag{8.2}$$

Let the following new variables be introduced:

$$X = \varepsilon x, \qquad \xi = \frac{1}{\varepsilon}\int^X h^{-1/2}\,dX - t \tag{8.3}$$

so that $h = h(X)$ with $h = 1$, $X \le X_0$, and ξ is the coordinate moving at the local linear wave speed. With the following changes

$$\frac{\partial}{\partial t} \to -\frac{\partial}{\partial\xi}, \qquad \frac{\partial}{\partial x} \to \varepsilon\frac{\partial}{\partial X} + h^{-1/2}\frac{\partial}{\partial\xi}, \tag{8.4}$$

Equations (8.1) and (8.2) become

$$-\zeta_\xi + \varepsilon h_X u + \varepsilon h u_X + \varepsilon\zeta_\xi h^{-1/2}u + \varepsilon\zeta h^{-1/2}u_\xi + h^{1/2}u_\xi = 0,$$

$$-h^{1/2}u_\xi + \varepsilon u u_\xi + \varepsilon h^{1/2}\zeta_X + \zeta_\xi + \frac{\mu^2}{3}h^{3/2}u_{\xi\xi\xi} = 0.$$

Adding the two equations above and using the leading order approximation

$u \cong h^{-1/2}\zeta$, we get, to the leading order,

$$2h^{1/2}\zeta_X + \frac{1}{2}\frac{h_X}{h^{1/2}}\zeta + \frac{3}{h}\zeta\zeta_\xi + \frac{\mu^2}{3\varepsilon}h\zeta_{\xi\xi\xi} = 0. \tag{8.5}$$

This extended KdV equation was deduced by Kakutani (1971) and Johnson (1973) and may be expressed in several forms. For example, let us apply the following transformation (Ono, 1972):

$$\zeta = -4h^2\psi, \qquad \tau = \frac{\mu^2}{6\varepsilon}\int_{X_0}^X h^{1/2}\,dX, \qquad \xi = \frac{\mu^2}{6\varepsilon}\sigma, \tag{8.6}$$

where the exponents of h are chosen so as to remove most of the variable coefficients. Equation (8.5) reduces to

$$\psi_\tau - 6\psi\psi_\sigma + \psi_{\sigma\sigma\sigma} + \nu(\tau)\psi = 0, \tag{8.7}$$

where

$$\nu = \frac{9}{4}\frac{h_X}{h^{3/2}}\left(\frac{6\varepsilon}{\mu^2}\right) \tag{8.8}$$

represents the effect of variable depth. Any finite-difference scheme devised for the KdV equation works for Eq. (8.7) as well (see Johnson, 1972).

To apply Eq. (8.7) to soliton fission by a shelf, Johnson and Ono proceeded along the following intuitive line inherent in the WKB approximation. A pair of integral laws are first derived which relate the initial and the end states of the transition. With the assumption that the soliton preserves its qualitative shape while climbing the transition, the two integral laws give the size and width of the pulse at the end of the transition, thereby specifying the initial data for the shallower shelf. The solution of the previous section can then be employed to find the eventual number of solitons.

Integrating Eq. (8.7) with respect to σ from $-\infty$ to ∞, we get

$$\frac{\partial}{\partial\tau}\int_{-\infty}^\infty \psi\,d\sigma + \left(-3\psi^2 + \psi_{\sigma\sigma}\right)_{-\infty}^\infty + \nu(\tau)\int_{-\infty}^\infty \psi\,d\sigma = 0.$$

If ψ and its derivatives are assumed to vanish at infinities, then

$$\left(\exp\int_0^\tau \nu\,d\tau\right)\int_{-\infty}^\infty \psi\,d\sigma = \text{const} \equiv \mathcal{I}. \tag{8.9}$$

Multiplying Eq. (8.7) by ψ and integrating with respect to σ, we get

$$\frac{\partial}{\partial\tau}\int_{-\infty}^\infty \frac{\psi^2}{2}\,d\sigma + \left[-2\psi^3 + \psi\psi_{\sigma\sigma} - \frac{1}{2}\psi_\sigma^2\right]_{-\infty}^\infty + 2\nu\int_{-\infty}^\infty \frac{\psi^2}{2}\,d\sigma = 0,$$

which may be integrated with respect to τ to give

$$\left(\exp 2 \int_0^\tau v\, d\tau\right) \int_{-\infty}^\infty \psi^2\, d\sigma = \text{const} = \mathscr{E}. \tag{8.10}$$

Equations (8.9) and (8.10) are two invariants of the approximate equation (8.7). From Eq. (8.8), it may be shown that

$$\exp \int_0^\tau v\, d\tau = h^{9/4},$$

hence $\mathscr{I}$ and $\mathscr{E}$ may be written

$$\mathscr{I} = h^{9/4} \int_{-\infty}^\infty \psi\, d\sigma \tag{8.11a}$$

and

$$\mathscr{E} = h^{9/2} \int_{-\infty}^\infty \psi^2\, d\sigma. \tag{8.11b}$$

Let us apply these invariants to a soliton climbing from a sea of unit depth on the left, over a smooth transition, to a horizontal shelf of smaller depth $h < 1$. We choose two stations X_0 and X_1 such that X_0 is at least half a wavelength before the transition, and X_1 is equally far after the transition; the corresponding τ values are $\tau = 0$ and $\tau = \tau_1$. For $L/\mathscr{L} = O(\varepsilon)$, it is reasonable to expect that the incident soliton retains roughly its pulse shape; hence

$$\psi = -\frac{\alpha^2}{2} \operatorname{sech}^2 \beta(\sigma - C\tau) \tag{8.12}$$

gives a good description. The parameters α and β are local height and length measures, being equal to α_0 and β_0 at $\tau = 0$, and to α_1 and β_1 at $\tau = \tau_1$. With these values, two algebraic equations are obtained from Eqs. (8.11a, b) for α_1 and β_1 which can be solved in terms of α_0 and β_0:

$$\alpha_1^2 = \alpha_0^2 h^{-9/4}, \qquad \beta_1 = \beta_0.$$

Now assume that the initial pulse is a soliton so that $\alpha_0 = 2$ and $\beta_0 = 1$. At the beginning of the shelf the pulse parameters must be $\alpha_1^2 = 4h^{-9/4}$ and $\beta_1 = 1$, or,

$$\psi = -2h^{-9/4} \operatorname{sech}^2(\sigma - \text{phase}) \tag{8.13}$$

which is no longer a soliton. Note from Eq. (8.6) that the peak amplitude changes according to

$$\frac{\zeta_1}{\zeta_0} \sim h^{-1/4} \tag{8.14}$$

(Ostrovsky and Pelinovskiy, 1970) which is the same as Green's law governing linear sinusoidal waves.

The known shape of the pulse at the beginning of the shelf, that is, Eq. (8.13), can be used as the initial data of the initial-value problem of Section 11.7. Using the results (7.23)–(7.25), we may conclude that N solitons will emerge ultimately if N is a positive integer and

$$N(N + 1) > 2h^{-9/4} > (N - 1)N, \tag{8.15a}$$

or

$$\left(\frac{N(N + 1)}{2} \right)^{-4/9} > h > \left(\frac{(N - 1)N}{2} \right)^{-4/9}. \tag{8.15b}$$

This simple result was found by Tappert and Zabusky (1971), Johnson (1973), and Ono (1972) and is in accord with the numerical solution of the Boussinesq equations by Madsen and Mei (1969) and the numerical solution of Eq. (8.5) by Johnson (1972).

It should be remarked that for still longer transition $L/\mathscr{L} = O(\varepsilon^2)$, say, an incident soliton no longer emerges as a single pulse upon reaching the shelf. Since the coefficient ν in Eq. (8.7) is slowly varying in τ (e.g., $\nu(\varepsilon\tau)$), the WKB perturbation analysis may be applied. In this way Ko and Kuehl (1978) found a low shelf trailing behind the primary pulse (see also Kaup and Newell, 1978).

The approximation (8.5), valid for wave propagation to the right, cannot account for reflection by the transition which, however, can be predicted by the more complete equations of Boussinesq (Peregrine, 1967; Madsen and Mei, 1969). In particular, Peregrine notes that weak reflection should be describable by the linearized Airy equations for variable depth, which can be handled analytically by the method of characteristics. The result is approximately a flat plateau having twice the length of the transition. If the slope of the transition is s, then the physical height of the plateau is $\frac{1}{2}s(\frac{1}{3}A/h)^{1/2}h$ where A is the amplitude of the incident soliton.

Miles (1979) points out that the invariant $\mathscr{E}$ implies energy conservation since $d\sigma \propto h^{-1/2} dx$ and $u \cong \zeta h^{-1/2}$ to the first order,

$$\mathscr{E} \propto h^{1/2} \int_{-\infty}^{\infty} \zeta^2 \, d\sigma \sim \int_{-\infty}^{\infty} \zeta^2 \, dx;$$

the last integral above is twice the total potential energy and is equal to the total energy by equipartition. The invariant $\mathcal{I}$ may be recast as

$$\mathcal{I} \propto h^{3/4} \int_{-\infty}^{\infty} \zeta \, dx$$

which has no direct physical meaning. Miles further notes that Eq. (8.5) does not conserve mass, that is

$$\mathcal{M} = \int_{-\infty}^{\infty} \zeta \, dx \neq \text{const},$$

and that $\mathcal{I}$ and $\mathcal{M}$ are both conserved only by accounting for reflection and the secondary shelf of Ko and Kuehl. This observation is consistent with the fuller solution of Peregrine (1967).

Extensions of the preceding considerations for slowly varying width and depth may be found in Miles (1979, 1980 and the references therein).

11.9 VISCOUS DAMPING OF SOLITARY WAVES

In his experiments with solitary waves, Russell (1838) also studied the effects of viscous damping. A proper theory, however, did not appear until a century later (Keulegan, 1948). Because of nonlinearity and aperiodicity of the wave, the necessary damping analysis is different in detail from that for periodic waves. Apparently unaware of Keulegan's work, Ott and Sudan (1970) studied the same problem in a more general physical context and applied a perturbation analysis similar to that in Chapter Eight. Damping of cnoidal waves has been examined more recently by Issacson (1976) and Miles (1976).

In this section we shall rederive Keulegan's results by the perturbation method[†] of Ott and Sudan.

If the length scale of the solitary wave is λ, the time scale is $\lambda(gh)^{-1/2} = \lambda/C$ with $C = (gh)^{1/2}$. The laminar boundary-layer thickness is of the order δ where $\delta \equiv (\nu\lambda/C)^{1/2}$. Since the length scale λ for solitary waves may be defined by $\lambda/h = \varepsilon^{-1/2}$ where $\varepsilon = A_0/h$, the ratio of δ to the water depth is

$$\alpha = \frac{\delta}{h} = \left(\frac{\nu}{Ch} \right)^{1/2} \varepsilon^{-1/4}. \tag{9.1}$$

In typical experiments, $O(\varepsilon) = 0.1$, $\nu = 10^{-2}$ cm s^2, $h = O(20$ cm$)$, the value of α is ~ 0.0033 which is very small. Even if ν is taken to be 1 cm s^2 in order to approximate a turbulent boundary layer, α is still just 0.033. Therefore, the inviscid approximation is expected to remain valid in most of the fluid.

[†]Details presented here were worked out by Dr. Philip L. F. Liu in 1974, at the suggestion of the author. Some errors exist in Ott and Sudan.

Let us recall the integrated equation of mass conservation

$$\frac{\partial \zeta}{\partial t} + h\frac{\partial \bar{u}}{\partial x} + \frac{\partial (\bar{u}\zeta)}{\partial x} = 0. \tag{9.2}$$

The integrated equation of momentum conservation must now include the effect of shear stress acting at the bottom

$$\frac{\partial \bar{u}}{\partial t} + \bar{u}\frac{\partial \bar{u}}{\partial x} + g\frac{\partial \zeta}{\partial x} + \frac{gh^2}{3}\frac{\partial^3 \zeta}{\partial x^3} = \frac{\tau_b}{h}. \tag{9.3}$$

For a laminar boundary layer the bottom shear stress is given by

$$\tau_b = -\nu\frac{\partial u}{\partial z}\bigg|_{z=-h}. \tag{9.4}$$

Note that u, being the local velocity, depends on z, but $\bar{u}$ does not. To help recognize the relative magnitude of each term, the following normalized variables are used:

$$\zeta' = \frac{\zeta}{A_0}, \quad x' = \frac{x}{\lambda}, \quad t' = \frac{tC}{\lambda}, \quad \bar{u}' = \bar{u}\left(\frac{A_0}{h}C\right)^{-1}$$

$$u' = u\left(\frac{A_0}{h}C\right)^{-1}, \quad z' = \frac{z+h}{\delta}. \tag{9.5}$$

The dimensionless equations are, after primes are omitted,

$$\frac{\partial \zeta}{\partial t} + \frac{\partial \bar{u}}{\partial x} + \varepsilon\frac{\partial \bar{u}\zeta}{\partial x} = 0, \tag{9.6}$$

$$\frac{\partial \bar{u}}{\partial t} + \frac{\partial \zeta}{\partial x} + \varepsilon\bar{u}\frac{\partial \bar{u}}{\partial x} + \frac{\varepsilon}{3}\frac{\partial^3 \zeta}{\partial x^3} = -\alpha\frac{\partial u}{\partial z}\bigg|_0. \tag{9.7}$$

Since $\alpha \ll \varepsilon$, we expect that viscosity may be ignored to the order $O(\varepsilon)$. The resulting equations lead to the solitary wave. Suppose that the coordinate transformation (5.36) is used, then Eqs. (9.6) and (9.7) become

$$\varepsilon\frac{\partial \zeta}{\partial \tau} - \frac{\partial \zeta}{\partial \sigma} + \frac{\partial \bar{u}}{\partial \sigma} + \varepsilon\frac{\partial}{\partial \sigma}(\bar{u}\zeta) = 0,$$

$$\varepsilon\frac{\partial \bar{u}}{\partial \tau} - \frac{\partial \bar{u}}{\partial \sigma} + \frac{\partial \bar{\zeta}}{\partial \sigma} + \varepsilon\zeta\frac{\partial \zeta}{\partial \sigma} + \frac{\varepsilon}{3}\frac{\partial^3 \zeta}{\partial \sigma^3} = -\alpha\frac{\partial u}{\partial z}\bigg|_0.$$

The sum of the two equations above gives

$$\frac{\partial}{\partial \tau}(\bar{u} + \zeta) + \frac{\partial}{\partial \sigma}(\bar{u}\zeta) + \zeta \frac{\partial \zeta}{\partial \sigma} + \frac{1}{3} \frac{\partial^3 \zeta}{\partial \sigma^3} = -\frac{\alpha}{\varepsilon} \frac{\partial u}{\partial z}\bigg|_0. \tag{9.8}$$

Let us assume that $\varepsilon^2 \ll \alpha \ll \varepsilon$; we may substitute $\bar{u} = \zeta$ into the preceding equation and obtain, with an error of $O(\varepsilon^2)$,

$$\frac{\partial \zeta}{\partial \tau} + \frac{3}{2}\zeta \frac{\partial \zeta}{\partial \sigma} + \frac{1}{6} \frac{\partial^3 \zeta}{\partial \sigma^3} = -\frac{\alpha}{2\varepsilon} \frac{\partial u}{\partial z}\bigg|_0, \tag{9.9}$$

where $\alpha/\varepsilon \ll 1$.

Without terms of $O(\alpha/\varepsilon)$, the inviscid solitary wave is, in dimensionless variables,

$$\zeta = a \operatorname{sech}^2\left[\frac{(3a)^{1/2}}{2}\left(\sigma - \frac{a}{2}\tau\right)\right]. \tag{9.10}$$

Since the bottom stress is expected to induce a slow decay, a new slow variable is introduced,

$$\tilde{\tau} = \alpha\tau \tag{9.11}$$

which characterizes the attenuation, that is, $a = a(\tilde{\tau})$. Moreover, the phase speed must also vary slowly through a, hence we introduce

$$\rho = \sigma - \frac{1}{2\alpha}\int^{\tilde{\tau}} a(\tilde{\tau}')\,d\tilde{\tau}' = \sigma - \frac{1}{2}\int^{\tau} a(\alpha\tau')\,d\tau' \tag{9.12}$$

in accordance with our earlier experience with slowly varying media. Assuming the perturbation series

$$\zeta = \zeta_0(\rho, \tilde{\tau}) + \alpha\zeta_1(\rho, \tilde{\tau}) + \cdots,$$

$$\bar{u}(x, t) = \bar{u}_0(\rho, \tilde{\tau}) + \alpha\bar{u}_1(\rho, \tilde{\tau}) + \cdots,$$

$$u(x, z, t) = u_0(\rho, z, \tilde{\tau}) + \alpha u_1(\rho, z, \tilde{\tau}) + \cdots, \tag{9.13}$$

and noting that

$$\frac{\partial}{\partial \sigma} = \frac{\partial}{\partial \rho}, \qquad \frac{\partial}{\partial \tau} = -\frac{a}{2} \frac{\partial}{\partial \rho} + \alpha \frac{\partial}{\partial \tilde{\tau}},$$

we obtain from Eq. (9.9) that

$$-\frac{a}{2} \frac{\partial \zeta_0}{\partial \rho} + \frac{3}{2}\zeta_0 \frac{\partial \zeta_0}{\partial \rho} + \frac{1}{6} \frac{\partial^3 \zeta_0}{\partial \rho^3} = 0, \tag{9.14a}$$

and

$$-\frac{a}{2}\frac{\partial \zeta_1}{\partial \rho} + \frac{3}{2}\frac{\partial}{\partial \rho}(\zeta_0 \zeta_1) + \frac{1}{6}\frac{\partial^3 \zeta_1}{\partial \rho^3} = -\frac{1}{2}\frac{\partial u_0}{\partial z}\bigg|_0 - \frac{\partial \bar{u}_0}{\partial \tilde{\tau}}. \qquad (9.14b)$$

The solution to Eq. (9.14a) is simply

$$\zeta_0 = a \operatorname{sech}^2 \frac{(3a)^{1/2}}{2}\rho. \qquad (9.15)$$

Following Ott and Sudan (1970), we introduce the operators $\mathcal{L}_0$ and $\mathcal{L}_1$ and rewrite Eqs. (9.14a) and (9.14b) as

$$\mathcal{L}_0 \zeta_0 \equiv \frac{\partial}{\partial \rho}\left[\frac{1}{6}\frac{\partial^2}{\partial \rho} + \frac{3}{4}\zeta_0 - \frac{a}{2}\right]\zeta_0 = 0, \qquad (9.16a)$$

and

$$\mathcal{L}_1 \zeta_1 = \frac{\partial}{\partial \rho}\left[\frac{1}{6}\frac{\partial^2}{\partial \rho^2} + \frac{3}{2}\zeta_0 - \frac{a}{2}\right]\zeta_1 = -\frac{1}{2}\frac{\partial u_0}{\partial z}\bigg|_0 - \frac{\partial \bar{u}_0}{\partial \tilde{\tau}}. \qquad (9.16b)$$

By straightforward partial integration, it can be shown that $\mathcal{L}_0$ and $\mathcal{L}_1$ are adjoint operators of each other, namely,

$$\int_{-\infty}^{\infty} d\rho\,(\zeta_0 \mathcal{L}_1 \zeta_1 - \zeta_1 \mathcal{L}_0 \zeta_2) = 0. \qquad (9.17)$$

Because of Eqs. (9.16a) and (9.16b), Eq. (9.17), implies a solvability condition for ζ_1

$$\int_{-\infty}^{\infty} \zeta_0 \left(-\frac{1}{2}\frac{\partial u_0}{\partial z}\bigg|_0 - \frac{\partial \bar{u}_0}{\partial \tilde{\tau}}\right) d\rho = 0 \qquad (9.18)$$

which gives a constraining equation for a. Note that up to this point the discussion is easily modified if the damping mechanism is a turbulent boundary layer.

It is now necessary to calculate the bottom stress to the leading order in terms of α. In the stationary reference frame the velocity in the boundary layer can be approximated by

$$u_0 = \bar{u}_0 + u_b,$$

where $\bar{u}_0$ is the inviscid approximation and u_b is the boundary-layer correction.

In terms of the dimensionless variables of Eq. (9.5), u_b satisfies

$$\frac{\partial u_b}{\partial t} = \frac{\partial^2 u_b}{\partial z^2} \tag{9.19}$$

to the required accuracy. In the moving reference frame of ρ, $\tilde{\tau}$, and z, the preceding equation may be written

$$-\frac{\partial u_b}{\partial \rho} = \frac{\partial^2 u_b}{\partial z^2}, \tag{9.20}$$

since $\partial/\partial t \cong -\partial/\partial\sigma \cong -\partial/\partial\rho$. The boundary conditions are

$$u_b(\rho, z, \tilde{\tau}) = -\bar{u}(\rho, \tilde{\tau}) = -\zeta_0, \qquad z = 0,$$

$$u_b \to 0, \qquad z \to \infty. \tag{9.21}$$

Since $-\rho$ plays the role of time, the "initial" condition is

$$u_b = 0, \qquad \rho \to +\infty.$$

It should be stressed that the velocities $\bar{u}_0$ and u_b are measured in the rest frame, although the space coordinate ρ refers to a moving frame. Because the disturbance $-\bar{u}_0(\rho, \tilde{\tau})$ is effectively bounded in the range $|\rho| = O(1)$, u_b is expected to vanish for $\rho \to -\infty$ also. Thus, we can take the exponential Fourier transform with respect to ρ:

$$\tilde{u}_b(k) = \int_{-\infty}^{\infty} u_b e^{-ik\rho} \, d\rho.$$

From Eq. (9.20) it follows that

$$\frac{\partial^2 \tilde{u}_b}{\partial z^2} = -ik\tilde{u}_b.$$

The solution which is bounded for $k \to \pm\infty$ is

$$u_b = -\frac{1}{2\pi} \int_{-\infty}^{\infty} dk \, e^{ik\rho} \tilde{\zeta}_0 \exp\left[\sqrt{\frac{|k|}{2}} \, z(-1 + i \operatorname{sgn} k)\right].$$

Substituting the inverse transform of $\tilde{\zeta}_0$ into the preceding formula, we obtain

$$u_b(\rho, z, \tilde{\tau}) - \frac{1}{2\pi}\int_{-\infty}^{\infty} d\rho'\, \tilde{\zeta}_0(\rho')\int_{-\infty}^{\infty} dk\, e^{ik(\rho-\rho')}e^{-\sqrt{|k|/2}\,z}$$

$$\left[\cos\left(\operatorname{sgn} k\sqrt{\frac{|k|}{2}}\, z\right) + i\sin\left(\operatorname{sgn} k\sqrt{\frac{|k|}{2}}\, z\right)\right]$$

$$-\frac{1}{\pi}\int_{-\infty}^{\infty} d\rho'\, \tilde{\zeta}_0(\rho')\int_0^{\infty} dk\, e^{-\sqrt{k/2}\,z}\left[\cos k(\rho - \rho')\cos\sqrt{\frac{k}{2}}\, z\right.$$

$$\left. - \sin k(\rho - \rho')\sin\sqrt{\frac{k}{2}}\, z\right]$$

$$= -\frac{1}{\pi}\int_{-\infty}^{\infty} d\rho'\tilde{\zeta}_0(\rho')\int_0^{\infty} dk\, e^{-\sqrt{k/2}\,z}\cos\left[k(\rho - \rho') + \sqrt{\frac{k}{2}}\, z\right].$$

$$(9.22)$$

Keulegan further transformed Eq. (9.22) to the following single integral:

$$u_b = -\frac{2}{\sqrt{\pi}}\int_0^{\infty} ds\, e^{-s^2}\zeta_0(\theta) \qquad \text{where } \theta = \frac{z^2}{4s^2} + \rho. \qquad (9.23)$$

Details are left in Appendix 11.B. After some manipulation, the shear stress on the bottom can be calculated,

$$-\frac{1}{2}\frac{\partial u_0}{\partial z}\bigg|_0 = -\frac{1}{2}\frac{\partial u_b}{\partial z}\bigg|_0 = -\sqrt{\frac{3}{\pi}}\, a^{3/2}\int_0^{\infty} ds\, \operatorname{sech}^2\left[\frac{\sqrt{3a}}{2}(\rho + s^2)\right]$$

$$\times \tanh\left[\frac{\sqrt{3a}}{2}(\rho + s^2)\right]. \qquad (9.24)$$

Since $\bar{u}_0 \cong \zeta_0$, we may differentiate Eq. (9.15) to obtain

$$-\frac{\partial \bar{u}_0}{\partial \tilde{\tau}} = -\frac{da}{d\tilde{\tau}}\operatorname{sech}^2\left(\frac{\sqrt{3a}}{2}\rho\right)\left\{1 - \frac{\sqrt{3a}}{2}\rho \tanh\left(\frac{\sqrt{3a}}{2}\rho\right)\right\}. \qquad (9.25)$$

Equations (9.24) and (9.25) may be substituted into Eq. (9.18), yielding

$$
\int_{-\infty}^{\infty} d\rho \, a \operatorname{sech}^2 \frac{\sqrt{3a}}{2} \rho \left\{ \sqrt{\frac{3}{\pi}} \, a^{3/2} \int_0^{\infty} ds \operatorname{sech}^2 \left[\frac{\sqrt{3a}}{2} (\rho + s^2) \right] \right.
$$

$$
\times \tanh \left[\frac{\sqrt{3a}}{2} (\rho + s^2) \right]
$$

$$
\left. + \left(\operatorname{sech}^2 \frac{\sqrt{3a}}{2} \rho \right) \left[1 - \frac{\sqrt{3a}}{2} \rho \tanh \frac{\sqrt{3a}}{2} \rho \right] \frac{da}{d\tilde{\tau}} \right\} = 0. \qquad (9.26)
$$

With the change of variables

$$
r = \frac{\sqrt{3a}}{2} \rho, \qquad S = (3a)^{1/4} 2^{-1/2} s,
$$

Eq. (9.26) may be transformed to

$$
\frac{da}{d\tilde{\tau}} \int_0^{\infty} dr (\operatorname{sech}^4 r)(1 - r \tanh r)
$$

$$
+ a^{5/4} \int_{-\infty}^{\infty} dr \left(\frac{2}{\pi} 3^{1/2} \right)^{1/2} \operatorname{sech}^2 r \int_0^{\infty} dS \left[\operatorname{sech}^2(r + S^2) \right] \left[\tanh(r + S^2) \right] = 0.
$$

$$
(9.27)
$$

We leave it as an exercise to show that the first single integral above is unity; the second double integral has been evaluated numerically by Keulegan and is approximately π^{-1}. Hence, from Eq. (9.27),

$$
\frac{da^{-1/4}}{d\tilde{\tau}} \cong -\frac{1}{4} \left(\frac{2 \cdot 3^{1/2}}{\pi^3} \right)^{1/2} = -0.08356
$$

which gives the law of attenuation:

$$
1 - a^{-1/4} = -0.0836 \tilde{\tau} \qquad (9.28a)
$$

or, in physical variables,

$$
\left(\frac{A_0}{A} \right)^{1/4} = 1 + 0.08356 \left(\frac{\nu \lambda}{C} \right)^{1/2} \frac{1}{\lambda} \frac{Ct}{h}. \qquad (9.28b)
$$

Using the fact that $\lambda = h(h/A_0)^{1/2}$, we have, at last,

$$A^{-1/4} = A_0^{-1/4} + 0.08356\left(\frac{\nu}{(gh)^{1/2}h^{3/2}}\right)^{1/2}\frac{Ct}{h}, \qquad (9.29)$$

where Ct is essentially the distance traveled by the solitary wave. This formula has been verified against the measurements of Russell.

Experiments with a rough bottom have been performed by Ippen and Kulin (1957), Ippen and Mitchell (1957), Özhan and Shi-igai (1977), and Naheer (1978). Theories based on an empirical formula for the bottom stress have been proposed, but they are not yet completely satisfactory when compared to experiments.

Finally, it is worth remarking that Eq. (9.18), which is the mathematical result of the Fredholm alternative, also has the physical meaning of energy conservation. Returning to physical variables, letting $x' = x - Ct$ and using $\bar{u}_0 \cong \zeta$, we may rewrite Eq. (9.18)

$$\frac{\partial}{\partial t}\rho h \int_{-\infty}^{\infty} dx'\,\bar{u}_0^2 = -\mu \int_{-\infty}^{\infty} dx'\,\bar{u}_0\frac{\partial u_0}{\partial z}\bigg|_0. \qquad (9.30)$$

The left-hand side is the rate of change of the total wave energy per unit length of crest. Consider the viscous dissipation in the boundary layer

$$-\int_{-\infty}^{\infty} dx' \int_{0}^{\infty}\mu\left(\frac{\partial u_0}{\partial z}\right)^2 dx = -\int_{-\infty}^{\infty} dx' \int_{-\infty}^{\infty}\mu\left(\frac{\partial u_b}{\partial z}\right)^2 dz.$$

The integrand may be written

$$\mu\left(\frac{\partial u_b}{\partial z}\right)^2 = \mu\frac{\partial}{\partial z}\left(u_b\frac{\partial u_b}{\partial z}\right) - \mu u_b\frac{\partial^2 u_b}{\partial z^2}. \qquad (9.31)$$

Using the physical form of Eq. (9.20)

$$-\rho C\frac{\partial u_b}{\partial x'} = \mu\frac{\partial^2 u_b}{\partial z^2},$$

we get, by integrating Eq. (9.31),

$$\int_{0}^{\infty}\mu\left(\frac{\partial u_b}{\partial z}\right)^2 dz = -\mu u_b\frac{\partial u_b}{\partial z}\bigg|_0 + \rho C\int_{0}^{\infty} dz \int_{-\infty}^{\infty} dx'\,u_b\frac{\partial u_b}{\partial x'}.$$

The last term may be integrated with respect to x' and the result is zero. Using the boundary condition at the bottom, $u_b = -\bar{u}_0$, we obtain

$$-\int_{-\infty}^{\infty} dx \int_0^{\infty} \mu \left(\frac{\partial u_b}{\partial z} \right)^2 dz = -\mu \int_{-\infty}^{\infty} dx \, \bar{u}_0 \frac{\partial u_b}{\partial z} \bigg|_0 .$$

Therefore, Eq. (9.30), or its dimensionless form (9.18), simply states that the rate of change of wave energy is due to dissipation. Equation (9.30) was first derived by Boussinesq (1878) and can be modified for other types of dissipation.

11.10   REMARKS ON MODELING LARGE-SCALE TSUNAMIS

Most of the large-scale tsunamis are caused by ruptures at the edges of great tectonic plates covering the earth. The extent of the fault of these ruptures is usually very large. For example, the Alaskan earthquake of 1964 has an approximate fault area of 100 km $\times$ 700 km, while the Chilean earthquake of 1960 has 200 km $\times$ 1000 km. While the maximum vertical displacement of the ground is only 10 m or less, the rise time at any cross section is short (< 5 s) so that a great deal of energy is suddenly imparted to the fluid.

Despite its usually low amplitude (1 m or less) in the deep ocean, tsunamis can reach large amplitudes near the coast as a result of refraction and local topography. Enormous losses of life and property have been recorded in historical accounts of tsunamis (Murty, 1977; Bolt, 1978). Clearly, efficient ways of calculating tsunami responses along the coast is of practical importance. The reliability of a forecast depends on our knowledge of the location of the epicenter, the magnitude of the earthquake, the extent of the fault area, and the type of rupture. Once estimated from seismological recordings, these data can, in principle, be used in a hydrodynamic theory to predict the propagation of water waves across the ocean and to forecast the run-up along the coast of particular concern.

Assuming that some crude estimates on the ground motion are available, one must then choose the appropriate (simple, yet accurate) equations to calculate the wave propagation. From several existing works and exchanges at a recent tsunami meeting (Hwang and Lee, 1980), a fairly clear picture has emerged for tsunamis generated at epicenters in the deep ocean. For simplicity the one-dimensional propagation will be discussed here.

For earthquakes of the Alaskan or Chilean magnitude, the sea surface displacement above the epicenter is probably no more than 1 m. For the typical values: $L = 100$ km and $h = 4$ km, the measure of nonlinearity and dispersion are

$$\varepsilon = \frac{A}{h} = 2.5 \times 10^{-4}, \qquad \mu^2 = \left(\frac{h}{L} \right)^2 = 1.6 \times 10^{-3}.$$

Thus, the long-wave approximation is suitable here. Although Boussinesq equations are the most uniformly valid, solving a simpler equation is more preferable, if appropriate. In particular, let us first focus our attention to the propagation across the ocean and assess the range of distance or time over which the simplest linear nondispersive theory is adequate.

As a rough guidance for regions not too close to shore, it is sufficient to examine the special case of constant depth and unidirectional propagation. We start with the KdV equation which includes both nonlinearity and dispersion to the leading order

$$\zeta_t + (gh)^{1/2}\left(1 + \frac{3}{2}\frac{\zeta}{h}\right)\zeta_x + \frac{h^2}{6}(gh)^{1/2}\zeta_{xxx} = 0 \qquad (10.1)$$

in physical variables, or

$$\zeta'_{t'} + \zeta'_{x'} + \frac{3}{2}\varepsilon\zeta'\zeta'_{x'} + \frac{\mu^2}{6}\zeta'_{x'x'x'} = 0 \qquad (10.2)$$

in the following dimensionless variables:

$$t = \frac{L}{(gh)^{1/2}}t', \qquad x = Lx', \qquad \zeta = A\zeta', \qquad (10.3)$$

where x and x' refer to the stationary frame. Assuming that ε and μ^2 are both small, we expand

$$\zeta' = \zeta_0 + \varepsilon\zeta_1 + \cdots.$$

The linear nondispersive equation gives the first approximation

$$\zeta_{0t'} + \zeta_{0x'} = 0. \qquad (10.4)$$

In terms of $\sigma = x' - t'$ and $\xi = x' + t'$, the solution is

$$\zeta_0 = F(\sigma), \qquad (10.5)$$

where $F = \zeta'(x', 0)$ is the initial form of ζ'. The next approximation must be

$$\left(\frac{\partial}{\partial t'} + \frac{\partial}{\partial x'}\right)\zeta_1 = \frac{\partial\zeta_1}{\partial\xi} = -\frac{3}{2}FF_\sigma - \frac{1}{6}\frac{\mu^2}{\varepsilon}F_{\sigma\sigma\sigma},$$

or

$$\varepsilon\zeta_1 = \frac{1}{2}\varepsilon\xi\left(-\frac{3}{2}FF_\sigma - \frac{1}{6}\frac{\mu^2}{\varepsilon}F_{\sigma\sigma\sigma}\right). \qquad (10.6)$$

Certainly, after a sufficiently long time or distance the perturbation expansion breaks down so that either nonlinearity or dispersion, or both, becomes important (Cole, 1968, p. 253). If the Ursell number $U_r = \varepsilon/\mu^2$ is large, nonlinearity is more important than dispersion; thus the linear and nondispersive approximation (10.4) can be used for fixed x' if $\varepsilon t' \ll 1$, that is,

$$\frac{A}{h}\frac{(gh)^{1/2}}{L}t \ll 1 \quad \text{or} \quad \left(\frac{g}{h}\right)^{1/2}t \ll \frac{L}{A}. \tag{10.7}$$

When $\varepsilon\xi$ increases to $O(1)$, Airy's nonlinear nondispersive approximation is needed. If, however, $U_r \ll 1$, then Eq. (10.4) is valid as long as $\mu^2\xi \ll 1$ or

$$\left(\frac{g}{h}\right)^{1/2}t \ll \left(\frac{L}{h}\right)^3 \tag{10.8}$$

for fixed x'. When $\mu^2\xi = O(1)$, dispersive effects must be added, but not nonlinearity. If $U_r = O(1)$, both inequalities (10.7) and (10.8) apply. When $\varepsilon t' \sim \mu^2 t' = O(1)$, the full KdV equation must be used.

To have some quantitative idea, let us take a wide fault in a deep ocean, $h = 4 \times 10^3$ m, $L = 2 \times 10^5$ m, and the maximum vertical surface displacement $A = 1$ m. Then $U_r = \varepsilon/\mu^2 = 0.625 = O(1)$ and the time for the linear nondispersive theory to break down is $t \sim (h/g)^{1/2}(L/h)^3 = 2.5 \times 10^6$ s. During this time the fastest wave has propagated the distance $x = (gh)^{1/2}t \sim 5 \times 10^5$ km, which is far greater than the typical dimension of the world oceans. Hence the linear nondispersive theory is quite adequate throughout the deep ocean, which justifies its use for transoceanic propagation as mentioned in Section 4.11.

For a fault of much smaller width, or at a short distance away from the edge of a continental shelf, the length scale L can be less. Take $h = 4$ km, $L = 2 \times 10^4$ m, and $A = 1$ m as an illustration. The time for a linear nondispersive theory to break down is $t \sim 2500$ s over which the leading wave has only propagated $x \sim 500$ km. Since $\varepsilon/\mu^2 = 0.00125$, the linear dispersive equation must be used if the distance from the fault is comparable to or greater than 500 km.

As we have discussed in Chapter Two, dispersion is important for leading waves of a transient disturbance. Let us examine how far the linear but *dispersive* theory can be applied. In a penetrating paper, Hammack and Segur (1978) showed that for the *leading wave* a more precise measure for nonlinearity involved the volume of the initial disturbance, not just A/h. We shall present their reasoning by first normalizing Eq. (10.1) with

$$\alpha = \frac{x - (gh)^{1/2}t}{h}, \qquad \tau = \frac{1}{6}\left(\frac{g}{h}\right)^{1/2}t, \qquad f = \frac{3}{2}\frac{\zeta}{h}. \tag{10.9}$$

The KdV equation then becomes

$$f_\tau + 6ff_\alpha + f_{\alpha\alpha\alpha} = 0, \tag{10.10}$$

whose linear dispersive approximation is

$$f_\tau + f_{\alpha\alpha\alpha} = 0. \tag{10.11}$$

The initial condition is assumed to be

$$f(\alpha, 0) = \phi(\alpha) \left(= \frac{\frac{3}{2}\zeta(x, 0)}{h} \right). \tag{10.12}$$

Now the dimensionless total volume for a unit length along the fault is

$$\delta = \int_{-\infty}^{\infty} \phi(\alpha) \, d\alpha = \frac{3}{2} \frac{1}{h^2} \int_{-\infty}^{\infty} \zeta(x, 0) \, dx = O\left(\frac{AL}{h^2}\right). \tag{10.13}$$

Clearly, δ is an important parameter characterizing the magnitude of the initial disturbance. Note that the Ursell number may be written

$$U_r = \frac{L\delta}{h} = \frac{AL^2}{h^3}. \tag{10.14}$$

For small δ, a perturbation solution $f = \delta f_1 + \delta^2 f_2 + \cdots$ may be sought so that

$$f_{1\tau} + f_{1\alpha\alpha\alpha} = 0 \qquad \text{with } \delta f_1(\alpha, 0) = \phi(\alpha) \tag{10.15a}$$

$$f_{2\tau} + f_{2\alpha\alpha\alpha} = -6f_1 f_{1\alpha} \qquad \text{with } f_2(\alpha, 0) = 0. \tag{10.15b}$$

The objective now is to examine when the perturbation expansion breaks down and the KdV equation is needed. The solution to Eq. (10.15a) can be found by Fourier transform

$$\delta f_1 = \frac{1}{2\pi} \int_{-\infty}^{\infty} \bar{\phi}(k) e^{i(k\alpha + k^3\tau)} \, dk, \tag{10.16}$$

where $\bar{\phi}$ is the exponential Fourier transform of $\phi(\xi)$. For large τ and fixed α, the method of stationary phase may be applied to yield

$$\delta f_1(\alpha, \tau) = \delta(3\tau)^{-1/3} \left[Ai(\chi) + C_1 \frac{U_r}{\delta} (3\tau)^{-1/3} Ai'(\chi) \right.$$

$$\left. + C_2 \left(\frac{U_r}{\delta}\right)^2 (3\tau)^{-2/3} Ai''(\chi) + O\left(\frac{U_r}{\delta}\right)^3 \frac{1}{3\tau} \right], \tag{10.17}$$

where $Ai(\chi)$ is the Airy integral with the argument

$$\chi = \frac{\alpha}{(3\tau)^{1/3}}. \tag{10.18}$$

The constants C_1 and C_2 are of order unity and depend only on the precise profile of $\phi(\alpha)$. The approximate Eq. (10.17) is valid if

$$\frac{U_r}{\delta}(3\tau)^{-1/3} \ll 1 \quad \text{or} \quad \left(\frac{g}{h}\right)^{1/2} t \gg \left(\frac{L}{h}\right)^3. \tag{10.19}$$

The second-order solution is lengthy and only the result is cited here:

$$f_2 = f_{2p} + f_{2h}, \tag{10.20}$$

where

$$f_{2p} = -\left[\int_{-\infty}^{\alpha} f_1(z, \tau) \, dz\right]^2, \tag{10.21}$$

$$f_{2h} = \left\{\int_{-\infty}^{\chi} Ai(z) \, dz + \frac{1}{2\pi}\int_{-\infty}^{\infty}\left[\tilde{\Theta}(k) - \frac{1}{ik}\right]e^{i(k\alpha + k^3\tau)} \, dk\right\}, \tag{10.22}$$

with

$$\Theta(\alpha) = \frac{1}{\delta^2}\left[\int_{-\infty}^{\alpha} \phi(\alpha) \, d\alpha\right]^2,$$

and $\tilde{\Theta}(k)$ being the Fourier transform of $\Theta(\alpha)$. f_{2p} is a particular solution satisfying the inhomogeneous governing equation, and f_{2h} is the homogeneous solution which helps satisfy the zero initial condition. For large time, we use the first terms of Eqs. (10.17) and (10.22) to approximate the perturbation series:

$$f = \delta f_1 + \delta^2 f_2 \cong \delta(3\tau)^{-1/3} Ai(\chi) + \delta^2\int_{-\infty}^{\chi} Ai(z) \, dz + \delta^2\left[\int_{-\infty}^{\chi} Ai(z) \, dz\right]^2. \tag{10.23}$$

The second-order terms do not involve powers of τ.

Evidently, for nonlinearity to remain small we must have

$$(3\tau)^{-1/3} \gg \delta \quad \text{or} \quad \left(\frac{g}{h}\right)^{1/2} t \ll \left(\frac{h^2}{AL}\right)^3 \tag{10.24}$$

otherwise the perturbation expansion breaks down and the full KdV equation is needed.

We can summarize the time range of validity for each approximation in Table 10.1. Hammack and Segur (1978) further studied theoretically and experimentally the oscillatory tails behind solitons and found them to be

Table 10.1 Choice of Approximate Equations

$U_r = \dfrac{AL^2}{h^3}$	$\tau = \left(\dfrac{g}{h}\right)^{1/2}\dfrac{t}{6}$	Approximate Equation
$\gg 1$	$\ll \dfrac{L}{A}$	Linear nondispersive
	$O\left(\dfrac{L}{A}\right)$	Nonlinear nondispersive
$\ll 1$	$\ll \left(\dfrac{L}{h}\right)^3$	Linear nondispersive
	$O\left(\dfrac{L}{h}\right)^3$	Linear dispersive
$\gtrless O(1)$	$\left(\dfrac{h^2}{AL}\right)^3 \gg \tau \gg \left(\dfrac{L}{h}\right)^3$	Linear dispersive for leading waves
	$O\left(\dfrac{h^2}{AL}\right)^3$	Nonlinear dispersive for leading waves

accurately described by the KdV equation (10.10), but not by Eq. (10.11). In reality, the finite length of the fault, and hence the two-dimensional spreading, must reduce the importance of nonlinearity in KdV or Boussinesq equations. More precise criteria should be sought by examining the two-dimensional KdV equation, as the problem warrants.

Finally, when the leading waves of a deep ocean tsunami reach a continental shelf, the depth is drastically reduced ($h \sim 100$ m, say), then for $L = 100$ km, $A = 1$ m

$$\varepsilon = \frac{A}{h} = 10^{-2}, \quad \mu^2 = \left(\frac{h}{L}\right)^2 = 10^{-6}, \quad \text{and} \quad U_r = 10^4.$$

A linear nondispersive theory is valid unless

$$O(\tau) = \frac{L}{A} = 10^5 \quad \text{or} \quad O(t) = 2 \times 10^6 \text{ s}.$$

The corresponding travel distance is $x = (gh)^{1/2}t = O(60{,}000)$ km which far exceeds the width of all continental shelves. On the other hand, if a fault of width $L = 1$ km[†] occurs near the edge of the shelf, we get for the same A, $(h/L)^2 = 10^{-2}$ and $U_r = 1$. The linear nondispersive theory remains applica-

---

[†]This is the typical crater size of a nuclear explosion. Waves generated by such explosions near the shelf edge is of interest to the deployment of submarine weapons (see the article by G. C. Wilson, *Washington Post*, p.A.3. 1980 (3/26).

ble if $\tau \ll L/A = 10^3$ or $t \ll 2 \times 10^4$ s. The corresponding distance traveled by the leading wave is $\ll 600$ km. Since the typical width of continental shelves is of the order of 200 km, KdV or Boussinesq equations may be needed as the basis of calculation.

11.11 EVOLUTION OF PERIODIC WAVES OVER CONSTANT DEPTH—HARMONIC GENERATION

It has long been known experimentally (Goda, 1967) that it is extremely difficult to generate long, simple harmonic progressive waves of finite amplitude in a shallow tank. Even when the wavemaker oscillates sinusoidally, the recorded waves at different stations along the tank differ from a sinusoidal wave markedly. In particular, within a period defined by the linearized theory, there are smaller secondary crests whose phase and size vary for different recording stations. A typical series of records are shown in Fig. 11.1 (Boczar-Karakiewicz, 1972) where the numbers in circles correspond to successive gauges spaced at equal intervals. A harmonic analysis of the wave records at various stations indicates that all the harmonics of the period ω vary periodically with respect to the distance from the wavemaker. This phenomenon suggests that nonlinearity alone can render the wave spectrum nonuniform in space or unsteady in time.

In experiments with a submerged shelf of rectangular cross section, Jolas (1960) reported that a simple harmonic incident wave sometimes produced waves of higher harmonics on the transmission side. A similar phenomenon is known in the vastly different subject of *nonlinear optics*. If a monochromatic laser beam of frequency ω and high intensity shines through a slab of quartz crystal, the transmitted waves contain frequencies ω and 2ω, that is, first and second harmonics. In the most dramatic case only second harmonics are found on the transmission side; this phenomenon is now called *second harmonic generation (SHG)*. Thus, light of one color can emerge in a different color after passing through the quartz crystal. A theoretical explanation of this optical phenomenon was given in the celebrated paper by the physicists Armstrong, Bloembergen, Ducuing and Pershan (1962) who showed the decisive role played by the mechanism of *nonlinear resonant interaction*. At the same time, very similar theories for deep-water waves were developed independently by Phillips (1960), Benney (1962), and Bretherton (1964). Since then, these ideas have brought about fundamental advances in the dynamics of surface and internal waves.

In this section we examine the effects of nonlinear interaction in shallow water with a view to understanding the observed phenomena (Mei and Ünlüata, 1971; Bryant, 1973).

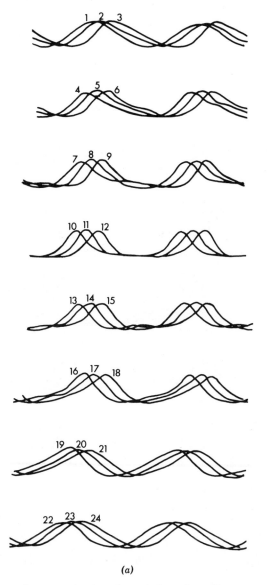

(a)

Figure 11.1 Free-surface records at various stations from the wave maker. Numbers identify the recording stations. The distance between two adjacent stations is 30 cm. Station (1) is 80 cm from the wave maker. (a) $A = 5$ cm: $H = 30$ cm, $T = 1.90$ s, $\lambda = 3.23$ m ($U_r = 0.45$); (b) $A = 5$ cm: $H = 20$ cm, $T = 2.75$ s, $\lambda = 3.86$ m ($U_r = 2.35$). (From Boczar-Karakiewicz, 1972 *Arch. Hydrotechniki*)

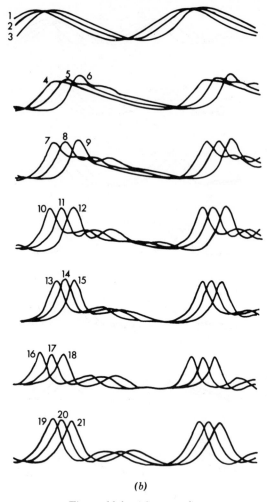

(b)

Figure 11.1   (*Continued*).

11.11.1   The Initial Development of Near-Resonant Interaction in Water of Constant Depth

Let long sinusoidal waves be generated at one end of a uniform channel of infinite length and constant depth. How do they evolve as they propagate downstream?

The KdV equation is appropriate here. For reasons explained later we choose the following form,

$$\zeta_t + \zeta_x - \frac{3}{2}\varepsilon\zeta\zeta_t - \frac{\mu^2}{6}\zeta_{ttt} = 0, \qquad (11.1)$$

which corresponds to Eq. (5.39d). Let the characteristic time scale be $2\pi/\omega$ where ω is the frequency of the fundamental harmonic, and let the characteristic horizontal length be $(gh)^{1/2}/\omega$ so that

$$\mu = \omega \left(\frac{h}{g}\right)^{1/2}.$$

At $x = 0$, the free-surface displacement is prescribed

$$\zeta(0, t) = \tfrac{1}{2}a(e^{-it} + e^{it}). \tag{11.2}$$

Let us treat ε to be small and μ^2 to be arbitrary, and first try a naive expansion in powers of ε,

$$\zeta = \zeta_0 + \varepsilon\zeta_1 + \cdots . \tag{11.3}$$

At the leading order, ζ_0 satisfies the linear dispersive equation

$$\zeta_{0t} + \zeta_{0x} - \frac{\mu^2}{6}\zeta_{0ttt} = 0. \tag{11.4}$$

The following solutions

$$e^{\pm i(Kx - \Omega t)}$$

are admissible if Ω and K are related by

$$\Omega + \frac{\mu^2}{6}\Omega^3 = K, \tag{11.5}$$

or, in dimensional form, by

$$\Omega'\left(1 + \frac{1}{6}\frac{\Omega'^2 h}{g}\right) = K'(gh)^{1/2} \quad \text{where } \Omega' = \omega\Omega, \quad K' = \frac{\omega}{(gh)^{1/2}}K. \tag{11.6}$$

In particular, if $\Omega = n$ ($\Omega' = n\omega$ in physical variables), the corresponding wavenumber is

$$K_n = n + \frac{\mu^2}{6}n^3. \tag{11.7}$$

Among the four alternative forms of the KdV equation, Eq. (5.39d) [or, Eq. (11.1)] is chosen because its implied dispersion relation (11.5) is the closest to the exact linear dispersion relation

$$\Omega = \frac{K}{\mu} \tanh K\mu, \tag{11.8}$$

even for very high Ω and K. The comparison is shown in Fig. 11.2 for $\mu = (0.05)^{1/2}$. To satisfy condition (11.2) we take

$$\zeta_0 = \tfrac{1}{2}a\{\exp[i(K_1 x - t)] + *\} = \tfrac{1}{2}a[\exp(i\phi_1) + *], \tag{11.9}$$

where

$$\phi_1 = K_1 x - t \tag{11.10}$$

and

$$K_1 = 1 + \frac{\mu^2}{6}. \tag{11.11}$$

At the next order, ζ_1 satisfies

$$\zeta_{1t} + \zeta_{1x} - \frac{\mu^2}{6}\zeta_{1ttt} = \frac{3}{2}\zeta_0\zeta_{0t}. \tag{11.12}$$

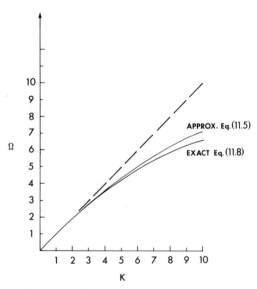

Figure 11.2 Comparison of dispersion curves for $\mu = \omega(h/g)^{1/2} = 0.05$. The straight line refers to the nondispersive limit. Ω has been normalized by the first harmonic frequency ω and K by $\omega(gh)^{-1/2}$.

The nonlinear forcing term is

$$-\tfrac{3}{8}a^2\left[ie^{2i(K_1x-t)} + *\right] = -\tfrac{3}{8}a^2\left[ie^{2i\phi_1} + *\right].$$

Because the *phase mismatch*

$$K_2 - 2K_1 = 2 + \frac{\mu^2}{6}2^3 - 2\left[1 + \frac{\mu^2}{6}\right] = \mu^2 \qquad (11.13)$$

is small, the forcing term $\exp(2i\phi_1)$ is very close to the second harmonic $\exp[i(K_2x - 2t)]$ which is a homogeneous solution of Eq. (11.12). We therefore expect the second harmonic $e^{\pm i\phi_2}$ to be nearly resonated. The explicit solution to (11.12) is

$$\zeta_1 = \frac{3a^2}{8\mu^2}\left(e^{2iK_1x} - e^{iK_2x}\right)e^{-2it} + * \qquad (11.14)$$

where a homogeneous solution has been added to satisfy the boundary condition that $\zeta_1(0) = 0$. Now $\varepsilon\zeta_1$ may be regarded as a second harmonic with a slowly and periodically varying amplitude

$$\varepsilon\zeta_1 = \frac{3}{8}\frac{\varepsilon}{\mu^2}a^2\left[\left(e^{-i\mu^2x} - 1\right)e^{i\phi_2} + *\right]$$

$$= \frac{3}{8}\frac{\varepsilon}{\mu^2}a^2\left(-2ie^{-i\mu^2x/2}e^{i\phi_2} + *\right)\sin\frac{\mu^2}{2}x. \qquad (11.15)$$

In general $\varepsilon\zeta_1$ is of the order $O(\varepsilon/\mu^2)$ and is small only if $\varepsilon \ll \mu^2$. As μ diminishes to zero, $\varepsilon\zeta_1$ becomes unbounded. By expanding Eq. (11.15) for small εx, we find

$$\varepsilon\zeta_1 = -\tfrac{3}{8}a^2\varepsilon x\left(ie^{i\phi_2} + *\right) \qquad (11.16)$$

so that the second harmonic is resonated to grow linearly as εx.

If both first and second harmonics are comparable, then their product can resonate the third harmonic $e^{\pm i\phi_3}$ also. The reason lies with the dispersion curve (Ω versus K) which is nearly straight for small μ so that $K_3 - (K_1 + K_2)$ is also small. In fact, if any two among $e^{\pm i\phi_1}$, $e^{\pm i\phi_2}$, and $e^{\pm i\phi_3}$ are present at the first order, the third can be resonated to the first order. Thus, the first three harmonics interact resonantly also. However, the corresponding phase mismatch

$$K_3 - (K_1 + K_2) \cong 3 + \frac{\mu^2}{6}3^3 - \left(1 + \frac{\mu^2}{6} + 2 + \frac{\mu^2}{6}2^3\right) = 3\mu^2 \qquad (11.17)$$

is greater than $K_2 - 2K_1$. The maximum amplitude attainable by the third harmonic can be estimated to be only one-third of the second harmonic. By an extension of the argument, still higher harmonics can be resonated by a similar nonlinear mechanism. In particular, the nth harmonic can be resonated by any of the pairs $(1, n - 1)$, $(2, n - 2)$ or $(3, n - 3)$. Since the smallest phase mismatch is

$$K_n - (K_1 + K_{n-1}) = \frac{\mu^2}{6}\left\{n^3 - \left[1^3 + (n-1)^3\right]\right\}$$

$$= \frac{\mu^2}{2} n(n-1),$$

the amplitude of the nth harmonic is roughly

$$\frac{\varepsilon}{\mu^2}\frac{2}{n(n-1)}$$

times as big as the first harmonic. The accuracy of this estimate deteriorates for large n because of the limited validity of Eq. (11.7).

The naive perturbation method as just described is clearly inadequate for $\varepsilon x = O(1)$. If the nonlinearity is so weak that only the first few harmonics are important, it is possible to use the method of multiple scales (Mei and Ünlüata, 1972). A more accurate method by Bryant (1973) is to assume a Fourier series which involves a large number of harmonics, as follows.

11.11.2 Governing Equations for Coupled Harmonics

The result of Eq. (11.16) suggests that amplitudes of the various harmonics vary in x. We therefore assume the solution

$$\zeta = \frac{1}{2}\sum_1^\infty \left[A_n(x)e^{in(x-t)} + A_n^*(x)e^{-in(x-t)}\right], \qquad (11.18)$$

where the simplest nondispersive phase function

$$\psi = x - t \qquad (11.19)$$

is used for convenience. The following alternative is also possible:

$$\zeta = \frac{1}{2}\sum_1^\infty \left[B_n(x)e^{i\phi_n} + B_n^*(x)e^{-i\phi_n}\right], \qquad (11.20)$$

with

$$\phi_n = K_n x - nt. \qquad (11.21)$$

The coefficients A_n and B_n are related by

$$A_n = B_n e^{ix(K_n - n)} = B_n e^{i(\mu^2/6)n^3 x}, \qquad n = 1, 2, 3, \ldots. \qquad (11.22)$$

Substituting Eq. (11.18) into the KdV equation (11.1), we obtain, from the linear terms,

$$\zeta_t + \zeta_x - \frac{\mu^2}{6}\zeta_{ttt} = \frac{1}{2}\sum_{n=1}^{\infty}\left[\left(\frac{dA_n}{dx} - i\frac{\mu^2}{6}n^3 A_n\right)e^{in\psi} + *\right]$$

$$= \frac{1}{2}\sum_{n=1}^{\infty}\left[\left(\frac{dB_n}{dx}\right)e^{i\phi_n} + *\right]. \qquad (11.23)$$

The nonlinear term in Eq. (11.1) can be manipulated to give

$$\frac{3}{4}\varepsilon(\zeta^2)_t = \frac{3}{16}\varepsilon\frac{\partial}{\partial t}\left\{\left[\sum_{l=1}^{\infty}\left(A_l e^{il\psi} + A_l^* e^{-il\psi}\right)\right]\left[\sum_{m=1}^{\infty}\left(A_m e^{im\psi} + A_m^* e^{-im\psi}\right)\right]\right\}$$

$$= \frac{3}{16}\varepsilon\frac{\partial}{\partial t}\left\{\sum_{l=1}^{\infty} A_l^* A_l + \sum_{n=1}^{\infty}\left\{e^{in\psi}\left[\sum_{l=1}^{\infty}{}'\alpha_l A_l A_{n-l} + 2\sum_{l=1}^{\infty} A_l^* A_{n+l}\right] + *\right\}\right\}$$

$$= \frac{3}{16}\varepsilon\sum_{n=1}^{\infty}\left\{e^{in\psi}\left[\sum_{l=1}^{\infty}{}'(-in\alpha_l)A_l A_{n-l} + (-2in)\sum_{l=1}^{\infty} A_l^* A_{n+l}\right] + *\right\}.$$

$$(11.24)$$

The inner series Σ' terminates at $\frac{1}{2}n$ for even n and $\frac{1}{2}(n-1)$ for odd n. The coefficient α_l is 2 if $l = 1, 2, 3, \ldots, \frac{1}{2}(n-1)$ and $\alpha_l = 1$ if $l = \frac{1}{2}n$. Details are given in Appendix 11.C. Substituting Eqs. (11.23) and (11.24) into Eq. (11.1), we obtain from the coefficients of $e^{in\psi}$ that

$$\left(\frac{dA_n}{dx} - i\frac{\mu^2}{6}n^3 A_n\right) = -i\varepsilon\frac{3}{8}\left\{\sum_{l=1}^{\infty}{}'(n\alpha_l)A_l A_{n-l} + \sum_{l=1}^{\infty} 2nA_l^* A_{n+1}\right\},$$

$$n = 1, 2, 3, \ldots \qquad (11.25)$$

From the coefficients of $e^{-in\psi}$, conjugate equations are obtained. Equation (11.25) is an infinite set of coupled nonlinear ordinary differential equations for all harmonics. If nonlinearity is not too severe, the importance of higher harmonics should diminish with n. We can, therefore, truncate the Fourier series after a finite number of terms (say N) and solve N equations for N unknown coefficients A_n with given initial values at $x = 0$.

If the Fourier series is truncated at $n = 1$, then

$$\frac{dA_1}{dx} - i\frac{\mu^2}{6}A_1 = 0, \tag{11.26}$$

which is possible only when nonlinearity is very weak. A correction for the second harmonic is obtained by assuming that $A_2 \ll A_1$

$$\frac{dA_2}{dx} - i\mu^2\frac{4}{3}A_2 = -i\varepsilon\frac{3}{4}A_1^2 \tag{11.27}$$

which may be written

$$e^{i4\mu^2 x/3}\frac{d}{dx}\left(e^{-i4\mu^2 x/3}A_2\right) = -i\varepsilon\frac{3}{4}A_1^2. \tag{11.28}$$

Subject to the initial conditions $A_1 = a_0$ and $A_2 = 0$ at $x = 0$, the solutions to Eqs. (11.27) and (11.28) are easily found to be

$$A_1 = a_0 e^{i(\mu^2/6)x}, \tag{11.29}$$

$$A_2 = \frac{3}{4}\frac{\varepsilon}{\mu^2}A_0^2\left(e^{i\mu^2 x/3} - e^{i4\mu^2 x/3}\right), \tag{11.30}$$

which may be shown to correspond to $\varepsilon\zeta_1$ in Eq. (11.15).

If we truncate the Fourier series after two terms, that is, disregard $A_3 = A_4 = \cdots = 0$ in the first two equations of Eq. (11.25) (Mei and Ünlüata, 1972; Bryant, 1973), we then have

$$\frac{dA_1}{dx} - i\frac{\mu^2}{6}A_1 = -i\varepsilon\frac{3}{4}A_1^*A_2, \tag{11.31}$$

$$\frac{dA_2}{dx} - i\mu^2\frac{4}{3}A_2 = -i\varepsilon\frac{3}{4}A_1^2, \tag{11.32}$$

or, by using Eq. (11.22),

$$\frac{dB_1}{dx} = -i\varepsilon\frac{3}{4}B_1^*B_2 e^{i\mu^2 x}, \tag{11.33}$$

and

$$\frac{dB_2}{dx} = -i\varepsilon\frac{3}{4}B_1^2 e^{-i\mu^2 x}. \tag{11.34}$$

The last two equations are identical to those governing two coupled harmonics in nonlinear optics and were solved by Armstrong, Bloembergen, Ducuing, and Pershan (1962) in the following manner.

11.11.3 Exact Solution of the Two-Harmonics Problem

Let

$$B_1(x) = \rho_1 e^{i\beta_1(x)} \quad \text{and} \quad B_2 = \rho_2 e^{i\beta_2(x)} \tag{11.35}$$

so that ρ_1 and ρ_2 are the amplitudes and β_1 and β_2 are the phases. Substituting Eq. (11.35) into the conjugate of Eqs. (11.33) and (11.34), we get

$$\rho_1' e^{-i\beta_1} - i\beta_1' \rho_1 e^{-i\beta_1} = iS\rho_1\rho_2 e^{i(\beta_1 - \beta_2 - \Delta x)}, \tag{11.36}$$

$$\rho_2' e^{i\beta_2} + i\beta_2' \rho_2 e^{i\beta_2} = -iS\rho_1^2 e^{i(2\beta_1 - \Delta x)}, \tag{11.37}$$

where $S = \frac{3}{4}\varepsilon$, $\Delta = \mu^2$ and primes denote derivatives with respect to x. Let us introduce

$$\theta = 2\beta_1 - \beta_2 - \Delta x. \tag{11.38}$$

Equations (11.36) and (11.37) become

$$\rho_1' - i\beta_1'\rho_1 = iS\rho_1\rho_2 e^{i\theta}, \tag{11.39}$$

$$\rho_2' + i\beta_2'\rho_2 = -iS\rho_1^2 e^{i\theta}. \tag{11.40}$$

From the real parts of the two preceding equations we get

$$\rho_1' = -S\rho_1\rho_2 \sin\theta, \tag{11.41}$$

$$\rho_2' = S\rho_1^2 \sin\theta, \tag{11.42}$$

which may be combined and integrated to give

$$\rho_1^2 + \rho_2^2 = \rho_0^2 = \rho_1^2(0) + \rho_2^2(0). \tag{11.43}$$

This relation, called the Manley-Rowe relation in electronics, asserts that a loss of energy in one harmonic must result in a gain in the other. The imaginary parts of Eqs. (11.39) and (11.40) yield

$$\beta_1' = -S\rho_2\cos\theta, \qquad \beta_2' = -\frac{S\rho_1^2}{\rho_2\cos\theta}.$$

From Eq. (11.38) it follows that

$$\theta' = 2\beta_1' - \beta_2' - \Delta = -\Delta - \left(2\rho_2 S - \frac{\rho_1^2}{\rho_2} S\right)\cos\theta \qquad (11.44)$$

which, after we eliminate S from Eq. (11.42), and use the Manley–Rowe relation (11.43), may be rewritten

$$\theta' = -\Delta + \left(-\frac{2\rho_2\rho_2'}{\rho_1^2} + \frac{\rho_2'}{\rho_2}\right)\cot\theta$$

$$= -\Delta + \left(\frac{2\rho_1'}{\rho_1} + \frac{\rho_2'}{\rho_2}\right)\cot\theta. \qquad (11.45)$$

In terms of the normalized variables

$$u = \frac{\rho_1}{\rho_0} \quad \text{and} \quad v = \frac{\rho_2}{\rho_0}, \qquad \zeta = S\rho_0 x, \qquad (11.46)$$

Eqs. (11.41), (11.42), (11.43), and (11.45) become

$$\frac{du}{d\zeta} = -uv\sin\theta, \qquad (11.47)$$

$$\frac{dv}{d\zeta} = u^2\sin\theta, \qquad (11.48)$$

$$u^2 + v^2 = 1, \qquad (11.49)$$

$$\frac{d\theta}{d\zeta} = -\delta + \left(2\frac{du/d\zeta}{u} + \frac{dv/d\zeta}{v}\right)\cot\theta. \qquad (11.50)$$

Thus, there are three unknowns u, v, and θ governed by three differential equations (11.47), (11.48), and (11.50). The dimensionless parameter for phase mismatch between the two harmonics is

$$\delta = \frac{\Delta}{S\rho_0} = \frac{\mu^3}{\frac{3}{4}\varepsilon} = \frac{4}{3}\frac{h}{a_0}(k_1 h)^2 \qquad (11.51)$$

which is essentially the inverse of the Ursell parameter.

A little manipulation of Eq. (11.50) leads to

$$\frac{d}{d\zeta}(u^2 v \cos\theta) = \delta u^2 v \sin\theta.$$

Upon combining the above result with an alternate form of Eq. (11.48)

$$\frac{\delta}{2}\frac{dv^2}{d\zeta} = \delta u^2 v \sin\theta,$$ (11.52)

we get

$$\frac{d}{d\zeta}(u^2 v \cos\theta) = \frac{\delta}{2}\frac{dv^2}{d\zeta}$$

which gives another integral

$$-\frac{\delta v^2}{2} + u^2 v \cos\theta = \text{const} = \Gamma_\delta.$$ (11.53)

The constant Γ_δ can be related to the initial data

$$\Gamma_\delta = -\frac{\delta}{2}v^2(0) + u^2(0)v(0)\cos[2\beta_1(0) - \beta_2(0)],$$ (11.54a)

$$= \Gamma_0 - \frac{\delta}{2}v^2(0),$$ (11.54b)

with

$$\Gamma_0 = u^2(0)v(0)\cos[2\beta_1(0) - \beta_2(0)]$$ (11.54c)

in terms of which Eq. (11.53) may be written

$$u^2 v \cos\theta = \Gamma_0 + \frac{\delta}{2}[v^2 - v^2(0)].$$ (11.55)

There are now two integrals for the three unknowns u, v, and θ. Only one differential equation remains to be integrated. If we eliminate u from Eq. (11.48) by using Eq. (11.49), then

$$\frac{1}{2}\frac{dv^2}{d\zeta} = v(1 - v^2)\sin\theta.$$

Squaring Eq. (11.52), we get

$$\left(\frac{1}{2}\frac{dv^2}{d\zeta}\right)^2 = u^4 v^2 \sin^2\theta = u^4 v^2(1 - \cos^2\theta)$$

$$= v^2(1 - v^2)^2 - \left\{\Gamma_0 + \frac{\delta}{2}[v^2 - v^2(0)]\right\}^2 \equiv Q(v^2)$$ (11.56)

after using Eqs. (11.49) and (11.55). The quantity $Q(v^2)$ is a cubic polynomial for v^2. Bounded solutions exist only if the cubic has three real zeros $v_a^2 < v_b^2 < v_c^2$. Letting

$$Q = \left(v_c^2 - v^2\right)\left(v_b^2 - v^2\right)\left(v^2 - v_a^2\right), \tag{11.57}$$

we get

$$\zeta = \pm \int_{v^2(0)}^{v^2(\zeta)} \frac{dv^2}{2\left[Q(v^2)\right]^{1/2}}, \tag{11.58}$$

which can be expressed in terms of elliptic integrals, as in the case of cnoidal waves. The result is

$$v^2 = v_a^2 + \left(v_b^2 - v_a^2\right)\mathrm{Sn}^2\left[\left(v_c^2 - v_a^2\right)^{1/2}(\zeta - \zeta_0), m\right], \tag{11.59}$$

with

$$m = \left(\frac{v_b^2 - v_a^2}{v_c^2 - v_a^2}\right)^{1/2} < 1. \tag{11.60}$$

The period of Sn is $4K$ with

$$K = \int_0^{\pi/2} \frac{d\theta}{\left(1 - m^2 \sin^2\theta\right)^{1/2}} \tag{11.61}$$

so that the period of Sn^2 is $2K$. In terms of the normalized coordinate ζ the dimensionless wavelength is

$$\lambda = \frac{2K}{\left(v_c^2 - v_a^2\right)^{1/2}}. \tag{11.62}$$

For the general case where both harmonics have nonzero initial values $u(0) \neq 0$ and $v(0) \neq 0$, numerical results can be found in Armstrong et al. (1962).

Let us examine the special case where $v(0) = 0$, $u(0) = 1$, then $\Gamma_0 = 0$ and

$$Q(v^2) = v^2\left\{v^4 - \left[2 + \left(\frac{\delta}{2}\right)^2\right]v^2 + 1\right\}$$

$$= v^2\left(v^2 - v_b^2\right)\left(v^2 - v_c^2\right), \tag{11.63}$$

where

$$v_b^2 = \frac{1}{v_c^2} = \left[\left(1 + \frac{\delta^2}{16} \right)^{1/2} + \frac{\delta}{4} \right]^{-2}. \tag{11.64}$$

The maximum modulation amplitude of v is v_b. From Eq. (11.60), we have

$$m = \frac{v_b}{v_c} = \frac{1}{v_c^2} = \left[\left(1 + \frac{\delta^2}{16} \right)^{1/2} + \frac{\delta}{4} \right]^{-2}. \tag{11.65}$$

As the initial amplitude a_0 increases and/or the first harmonic wavelength decreases, the Ursell parameter increases, hence δ decreases. From Eqs. (11.64) and (11.65), v_b^2 increases while m and K both increase. Thus, the maximum second harmonic amplitude increases. Since $m \cong 1 - \frac{1}{2}\delta$ and $K \cong \frac{1}{2} \ln(32/\delta)$ for small δ, (Abramowitz and Stegun, 1972, p. 591), the spatial period of modulation, which may be called the *beat* or *recurrence distance*, is, on the dimensionless scale of x [cf. Eq. (11.46c)]:

$$L \cong \frac{2K}{2S\rho_0} \cong \frac{2}{3} \frac{\delta}{\mu^2} \ln \frac{32}{\delta}$$

which decreases with δ. Thus, for smaller δ, more energy exchange between first and second harmonics takes place within a shorter distance. Recall, however, that for sufficiently small δ the third or higher harmonics may no longer be negligible.

For very small amplitudes, $\delta \gg 1$, $m \cong (2/\delta)^2 \ll 1$ and $K \cong \pi/2$. The recurrence distance is $\lambda \cong 2\pi/\delta$ in normalized terms and $2\pi h(kh)^{-3}$ in physical terms, as is also implied by Eq. (11.15). Longuet-Higgins (1977) has deduced this result more physically by considering two linearized free waves of frequencies ω and 2ω. Because of dispersion, a first-harmonic crest gains in phase relative to a second-harmonic crest. The recurrence distance is simply the distance over which the net gain is exactly one wavelength of the second harmonic.

The theory based on two interacting harmonics agrees reasonably well in both amplitude variation and recurrence distance, with the measurements of Boczar-Karakiewicz (1972), see Fig. 11.3, and of Goda (1967); comparison of recurrence distance is given in Mei and Ünlüata, (1972, Fig. 1). By extending the idea of phase gain to two cnoidal wavetrains of periods $2\pi/\omega$ and $4\pi/\omega$, and ignoring their interactions, Longuet-Higgins (1977) has also calculated the recurrence length which conforms with the experiments of Boczar-Karakiewicz. Bryant (1973) has made more accurate computations by including 11 harmonics and integrating numerically the coupled equations which are deduced from both the KdV equation and the Laplace equation with the free surface conditions approximated only to the second order in wave slope. A sample of

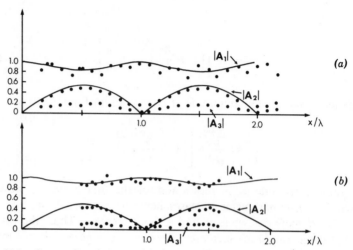

Figure 11.3 Comparison of two-harmonics theory with experiments by Boczar-Karakiewicz (1972) (comparison is due to Ü. Ünlüata, private communication). All harmonic amplitudes A_j are normalized by $|A_1(0)|$. (a) $\delta = 2.61$; (b) $\delta = 3.12$.

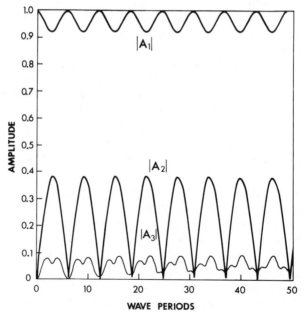

Figure 11.4 The first three harmonics using 11 coupled equations from μ-exact formulation. $\varepsilon = 0.05$; $\mu = (0.2)^{1/2}$. (From Bryant, 1973, *J. Fluid Mech*. Reproduced by permission of Cambridge University Press.)

his results is shown in Fig. 11.4; note in particular the third harmonic and its short beat length. Thus, as a periodic small-amplitude swell propagated toward a beach, second harmonics are first generated, then third and fourth..., until the depth is so small that many harmonics are present and the wave breaks, causing further small-scale fluctuations.

Qualitatively, the present results are already relevant to a submerged shelf of length L. Let a small-amplitude simple harmonic wave of frequency ω arrive from the left. Upon striking the shelf, the transmitted wave undergoes harmonic generation if the Ursell parameter on the shelf is not small. When L is half of the beat or recurrence distance, a large amount of second harmonic will enter the deep water on the transmission side of the shelf. On the other hand, when L is nearly equal to the beat length, very little second harmonic will be seen on the transmission side. This explanation is supported by the observations of Jolas (1960) for a submerged bar.

For variable depth involving a sand bar, an approximate theory has been given by Lau and Barcilon (1972). Otherwise, the numerical solution of Eq. (8.5) for slowly varying depth is straightforward. If the depth variation has the same length scale as the waves, reflection is no longer negligible and a numerical solution of Boussinesq equations is necessary (Peregrine, 1967; Madsen, Mei and Savage, 1970).

11.12 NONLINEAR RESONANCE IN A NARROW BAY

In Chapter Five the phenomenon of harbor resonance was studied on the basis of an inviscid linearized theory. For a sufficiently narrow mouth and steady forcing, the amplification was typically tenfold or larger. With some uncertainties in the friction coefficient, the effect of entrance loss was found in Chapter Six to reduce the peak amplitude. However, friction is not the only mechanism for reducing resonant amplification. Indeed, results in the preceding section suggest that nonlinearity provides an alternative mechanism in converting energy from first to higher harmonics. A quantitative assessment of nonlinear effects should be interesting and is given below for the simplest configuration of a narrow rectangular bay (Rogers and Mei, 1977).

The bay has the length L and width $2a$. The coastline is designated by the y axis and the axis of the bay coincides with the x axis. The sea depth is constant h everywhere. An incident wave arrives normally from $x \sim +\infty$.

Rogers and Mei adopted the Boussinesq equations and the following dimensionless variables:

$$t' = \omega t, \qquad (x', y') = \frac{(x, y)\omega}{(gh)^{1/2}}, \qquad z' = \frac{z}{h}$$

$$\zeta' = \frac{\zeta}{h}, \qquad \mathbf{u}' = \frac{\mathbf{u}}{(gh)^{1/2}}. \tag{12.1}$$

After primes are omitted, Eqs. (1.36) and (1.37) become

$$\zeta_t + \nabla \cdot \mathbf{u} + \nabla \cdot (\zeta \mathbf{u}) = 0, \qquad (12.2)$$

$$\mathbf{u}_t + \nabla \zeta + \tfrac{1}{2}\nabla \mathbf{u}^2 + \tfrac{1}{3}\mu^2 \nabla \zeta_{tt} = 0, \qquad (12.3)$$

where $\mu^2 = \omega^2 h/g \ll 1$. The normalized quantities $\mathbf{u}$ and ζ are of the order $O(\varepsilon)$. The error includes terms of the order $O(\varepsilon^2)$, $O(\varepsilon\mu^2)$, and $O(\mu^4)$.

Because of nonlinearity, the response contains higher harmonics

$$\left. \begin{array}{c} \zeta(x, y, t) \\ \mathbf{u}(x, y, t) \end{array} \right\} = \frac{1}{2} \sum_n \left[\begin{array}{c} \zeta_n(x, y) \\ \mathbf{u}_n(x, y) \end{array} \right] e^{-int}, \qquad n = 0, \pm 1, \pm 2,\dots \quad (12.4)$$

where ζ_{-n} and $\mathbf{u}_{-n}$ are the complex conjugates of ζ_n and $\mathbf{u}_n$, respectively. Substituting Eq. (12.4) into Eqs. (12.2) and (12.3) and separating the harmonics, we get

$$-in\zeta_n + \nabla \cdot \mathbf{u}_n + \frac{1}{2}\sum_s \nabla \cdot (\zeta_s \mathbf{u}_{n-s}) = 0, \qquad (12.5)$$

$$-in\mathbf{u}_n + \left(1 - \frac{1}{3}\mu^2 n^2\right)\nabla \zeta_n + \frac{1}{4}\sum_s \nabla (\mathbf{u}_s \cdot \mathbf{u}_{n-s}) = 0, \qquad (12.6)$$

which may be combined to give

$$(\nabla^2 + k_n^2)\zeta_n = \sum_n \left[\frac{-i}{2}n\nabla \cdot (\zeta_n \mathbf{u}_{n-s}) - \frac{1}{4}\nabla^2(\mathbf{u}_s \cdot \mathbf{u}_{n-s})\right], \qquad (12.7)$$

where

$$k_n^2 = \frac{n^2}{\left(1 - \tfrac{1}{3}\mu^2 n^2\right)}. \qquad (12.8)$$

Although there is an infinite number of terms in the Fourier series (hence an infinite set of coupled equations (12.7)), in practice, truncation after a finite number of terms is necessary. In general, the task of solving the two-dimensional coupled equations is still formidable, and the mathematical statement for the radiation condition does not appear simple. If, however, the harbor mouth is sufficiently narrow, $\delta \equiv \omega a/(gh)^{1/2} \ll 1$, the radiated waves due to the piston action at the mouth must be weak. Thus, if the nonlinear effect is unimportant in the incoming waves, it remains unimportant in the ocean. A more precise reasoning is as follows. For order estimates it suffices to consider the first harmonic which is expected to be the largest. From the linearized theory of Section 5.6.2b, it is known that the amplification factor in the harbor is proportional to $0(\delta^{-1})$, so that in the harbor $\zeta = 0(A_1/\delta)$ where A_1 is the amplitude of the incident wave. By assumption $A_1/\delta = 0(\varepsilon)$. From Eqs. (6.3) and (6.19) or Eq. (7.38), Chapter Five, the radiated wave ζ^R outside the harbor is of the order $\sim A_1 H_0^{(1)}(k_1 r)$ at resonance. Since the Hankel function is large for small r, we take the wave near the harbor entrance $r = 0(\delta)$

$$O(\zeta^R) = A_1 \ln \delta = \varepsilon \delta \ln \delta$$

and

$$\zeta^I + \zeta^R = O(\varepsilon \delta).$$

Hence the quadratic terms in the ocean are of the order $O(\varepsilon \delta \ln \delta)^2$ at least and are negligible compared to the nonlinear terms $O(\varepsilon^2)$ in the harbor. Thus, in the harbor we keep terms of $O(A_1/\delta) = O(\varepsilon)$ and $O(\varepsilon^2)$, while in the ocean we only keep the linear terms $O(\varepsilon \delta \ln \delta)$ and drop the nonlinear terms which are at most of $O(\varepsilon \delta \ln \delta)^2$.

In the ocean, the linear solution for each harmonic may be expressed by

$$\zeta_n = A_n \cos k_n x + \zeta_n^R, \tag{12.9a}$$

with

$$\zeta_n^R = \frac{k_n^2}{2n} \int_{-\delta}^{\delta} dy' U_n(y') H_0^{(1)}\left(k_n \left[x^2 + (y - y')^2 \right]^{1/2} \right). \tag{12.9b}$$

The functions U_n are just the piston velocities corresponding to the nth harmonic and are related to the surface gradient by

$$\frac{\partial \zeta_n}{\partial x} = \frac{ik_n^2}{n} U_n, \qquad |y| < \delta, \quad x = 0. \tag{12.10}$$

In the far field at a distance $O(k_1 r)$ from the entrance, the Hankel function in Eq. (12.9b) may be approximated to give

$$\zeta_n^R = \frac{k_n^2}{2n} \bar{U}_n k_n \delta H_0^{(1)}(k_n r)[1 + O(n\delta)], \qquad r = (x^2 + y^2)^{1/2} \tag{12.11}$$

where $\bar{U}_n$ is the average of U_n. Across the entrance, hence in the near field, the approximation of Eq. (12.9b) for small k_n is

$$\zeta_n^R(0, y) = \frac{k_n^2}{n} \left\{ \delta \bar{U}_n + \frac{i}{\pi} \int_{-\delta}^{\delta} dy' U_n(y') \ln \frac{\gamma k_n}{2} |y - y'| \right.$$

$$\left. + O(k_n \delta)^2 \ln k_n \delta \right\}. \tag{12.12}$$

If the average of Eq. (12.12) is taken across the entrance, then

$$\bar{\zeta}_n^R\big|_{x=0} = Z_n\bar{U}_n \equiv \bar{U}_n\frac{k_n^2\delta}{n}$$

$$\times\left\{1 + \frac{2i}{\pi}\left(\ln k_n\delta + \ln\frac{\gamma}{2}\right) + \frac{1}{4}\int\!\!\!\int_{-1}^{1}\ln|\alpha - \alpha'|\,\frac{U_n(\alpha')}{\bar{U}_n}\,d\alpha\,d\alpha'\right\}.$$

(12.13)

Taking the average of Eq. (12.10) and eliminating $\bar{U}_n$ with the help of Eq. (12.13), we get for the average of ζ_n [cf. Eq. (12.9a)]

$$-\frac{in}{k_n^2}Z_n\frac{\partial\bar{\zeta}_n}{\partial x} + A_n = \bar{\zeta}_n, \qquad x = 0, \quad |y| < \delta. \tag{12.14}$$

The quantity Z_n will be called the *entrance impedance* which depends on the profile of $U_n(y)$. Equation (12.14) will be viewed as a boundary condition which must be satisfied by the solution within the harbor. Assuming that $k_n\delta$ is small enough so that U_n may be approximated quasi-statically as in Section 2, we find the impedance explicitly:

$$Z_n = \frac{k_n}{n}k_n\delta\left(1 + \frac{2i}{\pi}\ln\frac{2\gamma k_n\delta}{\pi e}\right). \tag{12.15}$$

This approximation is good only if the higher harmonics are unimportant. That an analytical expression for Z_n is possible is due to the assumption of a straight coast and horizontal ocean bottom. For a more general coastline and ocean depth the impedance must be found numerically.

The narrowness of a harbor entrance enables us to decouple the harbor and the ocean and to study the harbor as a separate but interior boundary-value problem with $\partial\zeta_n/\partial n = 0$ on the solid boundary and Eq. (12.14) at the entrance. If the harbor shape and depth are not simple, direct numerical methods such as finite elements may be applied. For a narrow rectangular bay, the numerical work is less demanding since the problem is essentially one dimensional ($\partial\zeta/\partial y \sim \partial u/\partial y = O(\delta^2)$) everywhere except in the $O(\delta)$ neighborhood of the entrance. The justification follows from Section 4.1.1.

The one-dimensional version of Eq. (12.7) is

$$\zeta_n'' + k_n\zeta_n = -\frac{in}{2}\sum_s(\zeta_s u_{n-s})' - \frac{1}{4}\sum_s(u_s u_{n-s})'' \tag{12.16}$$

where $(\ )' = d(\ )/dx$. From Eqs. (12.5) and (12.6), $u_n' = in\zeta_n(1 + O(\varepsilon))$ and $u_n = -(i/n)\zeta_n'(1 + O(\varepsilon\mu)^2)$ for $n \neq 0$. Within the Boussinesq approximation we have

$$\sum_s (\zeta_s u_{n-s})' = \sum_s (\zeta_n u'_{n-s} + \zeta'_{n-s})$$

$$= \sum_{s \neq n} \left(i(n-s)\zeta_n \zeta_{n-s} - \frac{i}{n-s} \zeta'_s \zeta'_{n-s} \right),$$

and

$$\frac{1}{2}\sum_s (u_s u_{n-s})'' = \sum_s (u_{n-s} u''_s + u'_s u'_{n-s})$$

$$= \sum_{s \neq n} \frac{s}{n-s} \zeta'_s \zeta'_{n-s} - \sum_s (n-s)\zeta_s \zeta_{n-s}.$$

Substituting these two equations into Eq. (12.16), we obtain

$$\zeta''_n + k_n^2 \zeta_n = \frac{1}{2}\sum_s (n^2 - s^2)\zeta_s \zeta_{n-s} - \frac{1}{2}\sum_{s \neq n} \frac{n+s}{n-s} \zeta'_s \zeta'_{n-s}. \quad (12.17)$$

Since ζ_n is independent of y, it is equal to its own average. We now impose the no-flux condition at the end $x = -L$ and Eq. (12.14) at the entrance, that is,

$$\frac{\partial \zeta_n}{\partial x} = 0, \qquad x = -L, \quad (12.18)$$

$$-\frac{in}{k_n^2} Z_n \frac{\partial \zeta_n}{\partial x} + A_n = \zeta_n, \qquad x = 0. \quad (12.19)$$

The nonlinear truncated system of the boundary-value problems (12.17)–(12.19) can be solved numerically.

The zeroth harmonic corresponds to mean-sea-level changes ζ_0 or mean current u_0, which can be obtained by integrating the one-dimensional version of Eqs. (12.5) and (12.6)

$$u_0(x) = -\frac{1}{2}\sum_s \zeta_s u_{-s}, \quad (12.20)$$

$$\zeta_0(x) = -\frac{1}{4}\sum_s |u_s|^2 + \text{const.} \quad (12.21)$$

Note that Eq. (12.20) already satisfies $u_s = 0$ at $x = -l$ for all s. By properly defining the mean sea level of the ocean, $A_0 = 0$. Since $Z_0 = 0$, we have

$\zeta_0(0) = 0$ by the impedance condition (12.19), implying that

$$\zeta_0(x) = \frac{1}{4} \sum_s |u_s(0)|^2 - |u_x(x)|^2. \qquad (12.22)$$

From Eqs. (12.20) and (12.22), ζ_0 and $u_0 = O(\varepsilon^2)$, and may be omitted in the nonlinear equations for other harmonics. Spatially, the mean-sea-level change is the greatest at $x = -l$. As a function of frequency ζ_0 is resonated along with the first harmonic ζ_1.

The two-point nonlinear boundary-value problem for N harmonics can be solved by an iterative method. If the solution of the last iteration is used to linearize the nonlinear terms, then at each iterative step the resulting problem is linear and can be solved by the method of complementary functions. For details Rogers and Mei (1977) may be consulted.

Let us show some of the computed and experimental results for two bays of the same width and depth, but of different lengths, which corresponds to the first and third resonant modes for the same fundamental frequency. Now it can be inferred from the linearized theory [Eq. (6.20), Chapter Five] that the first and third resonant modes are at $L = \frac{1}{2}\pi$, and $\frac{5}{2}\pi$ for fixed k but varying L, while all peaks are of equal height. The actual resonant values of L were found experimentally to be $L = 1.227$ and 7.23. For each bay, three different incident wave amplitudes were examined by Rogers and Mei, but only two will be presented here. The dimensionless inputs are

$$L = \frac{L'\omega}{(gh)^{1/2}} = 1.227, 7.23, \qquad \delta = \frac{a\omega}{(gh)^{1/2}} = 0.169,$$

$$\mu^2 = \frac{\omega^2 h}{g} = 0.257, \qquad \varepsilon = |A_1| = 0.015, 0.04.$$

From the measurements, the magnitude and phase of the second and third harmonics A_2, A_3 were also obtained and used as inputs for computation. In Figs. 12.1 and 12.2 the numerical results are compared with the measured harmonics along the center line of the bay. Although minor discrepancies can be seen, they can be corrected after accounting for the entrance loss and for the laminar friction in the wall boundary layers of the bay. Several conclusions are apparent for the present problem:

1 For the shortest bay at quarter wavelength mode, a linear theory is sufficient if the entrance friction is accounted for.

2 As the bay becomes longer nonlinearity is more influential in generating higher harmonics; entrance and boundary-layer losses become less important. For a sufficiently long bay nonlinearity can overwhelm frictional effects.

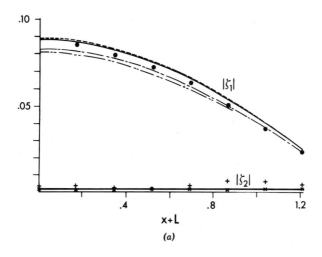

(a)

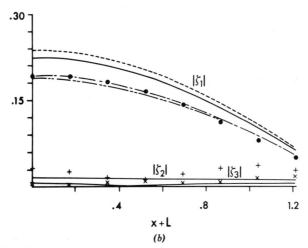

(b)

Figure 12.1 Resonance in a narrow bay. $\bigcirc$, $+$, $\times$: measured amplitudes of $|\zeta_1|$, $|\zeta_2|$, and $|\zeta_3|$. Calculated first harmonic: - - - -: linear theory; _____: inviscid nonlinear theory; — · — ·: nonlinear theory with entrance loss; — — —: nonlinear theory with entrance and boundary layer losses. (a) $L' = 1.211$ ft, $A_1 = 0.015$; (b) $L' = 1.211$ ft, $A_1 = 0.04$. (From Rogers and Mei, 1976, *J. Fluid Mech.* Reproduced by permission of Cambridge University Press.)

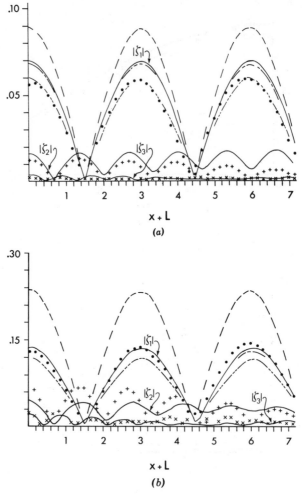

Figure 12.2 See Fig. 12.1 for caption. Calculated second and third harmonics by invisicid nonlinear theory are also shown by solid curves. (*a*) $L' = 7.136$ ft, $A_1 = 0.015$; (*b*) $L' = 7.136$ ft, $A_1 = 0.040$. (From Rogers and Mei, 1976, *J. Fluid Mech*. Reproduced by permission of Cambridge University Press.)

A numerical study of the nonlinear response in an arbitrarily shaped harbor to transient inputs has been reported by Lepelletier (1980).

APPENDIX 11.A:　EVALUATION OF CERTAIN INTEGRALS IN SECTION 11.4

The typical integral is

$$I(f) = \int_0^\infty d\zeta\, e^{-2\zeta}(2f_\zeta + f_{\zeta\zeta}). \tag{A.1}$$

By partial integration

$$I(f) = -4f(0) - f_\zeta(0) + 8\int_0^\infty d\zeta\, e^{-2\zeta}f. \tag{A.2}$$

Let $f = J_0(4\zeta^{1/2})$, then $J_0(0) = 1$, but

$$\lim_{\zeta\to 0} J_{0\zeta} = \lim_{\zeta\to 0} J_0' \frac{2}{\zeta^{1/2}} = \frac{2}{\zeta^{1/2}}\left(-\frac{4\zeta^{1/2}}{2}\right) = -4.$$

Hence,

$$I(J_0) = 8\int_0^\infty d\zeta\, e^{-2\zeta}J_0(4\zeta^{1/2}) = \frac{1}{2e^2} = 0.06767 \tag{A.3}$$

which can be found from Abramowitz and Stegun (1972).

Let $f = G(\zeta)$. From the defining equation (4.18b) we find

$$G(0) = E_2(\infty) - iE_1(\infty), \tag{A.4}$$

and

$$G_\zeta(0) = \frac{1}{\pi} - 4(E_2(\infty) - iE_1(\infty)), \tag{A.5}$$

therefore

$$-4G(0) - G_\zeta(0) = -\frac{1}{\pi}. \tag{A.6}$$

Now

$$E_G \equiv \int_0^\infty d\zeta\, e^{-2\zeta}G = -\int_0^\infty d\zeta\, e^{-2\zeta}J_0(4\zeta^{1/2})\int_0^\zeta d\xi\, e^{-2\xi}Y_0(4\xi^{1/2})$$

$$+ \int_0^\infty d\zeta\, e^{-2\zeta}Y_0(4\zeta^{1/2})\int_0^\infty d\xi\, e^{-2\xi}J_0(4\xi^{1/2})$$

$$+ (E_2(\infty) - iE_1(\infty))E_1(\infty)$$

$$= -i(E_1(\infty))^2 + \int_0^\infty d\zeta\, e^{-2\zeta}Y_0(4\zeta^{1/2})\int_0^\zeta d\xi\, e^{-2\xi}J_0(4\xi^{1/2}). \tag{A.7}$$

The last double integral above can be numerically evaluated to be 0.01431331. Upon using Eq. (A.3), we obtain

$$E_G = 0.02862 - 0.0045992i. \tag{A.8}$$

APPENDIX 11.B: REDUCTION OF AN INTEGRAL IN SECTION 11.9

Keulegan's manipulations are as follows: Equation (9.22) is first broken into two parts

$$u_b = -\frac{1}{\pi}\left\{\int_{-\infty}^{\rho}d\rho' + \int_{\rho}^{\infty}d\rho'\right\}\int_{0}^{\infty}dk[\ ] \tag{B.1}$$

with the same integrand. In the first integral, to be denoted by I_1, we let

$$k(\rho - \rho') = 2a^2 \quad \text{and} \quad \left(\frac{k}{2z}\right)^{1/2} = qa,$$

so that

$$I_1 = -\frac{4}{\pi}\int_{-\infty}^{\rho}\frac{d\rho'}{\rho - \rho'}\zeta_0(\rho')\int_{0}^{\infty}ae^{-qa}\cos(2a^2 + qa)\,da.$$

The a integral above vanishes identically [see Gradshteyn and Ryzhik, 1965, p. 499, formula 3.966(1)]. In the second integral of (B.1), we let

$$k(\rho' - \rho) = 2a^2 \quad \text{and} \quad \left(\frac{k}{2z}\right)^{1/2} = qa,$$

and use a and q to replace k and ρ' as new variables; then

$$u_b = I_2 = -\frac{4}{\pi}\int_{0}^{\infty}\frac{dq}{q}2\zeta_0\int_{0}^{\infty}ae^{-qa}\cos(2a^2 - qa)\,da.$$

The a integral here can again be found from Gradshteyn and Ryzhik [1965, 3.966(2)] as

$$\frac{\pi}{8}^{1/2}qe^{-q^2/4}.$$

With a further change from q to $2s$, Eq. (9.23) follows at once.

APPENDIX 11.C: THE SQUARE OF A FOURIER SERIES

We wish to rearrange the following product:

$$(\Sigma)^2 = \left(\sum_{l=-\infty}^{\infty}A_l e^{il\theta}\right)\left(\sum_{m=-\infty}^{\infty}A_m e^{im\theta}\right)$$

$$= \sum_{-\infty}^{\infty}\sum_{-\infty}^{\infty}A_l A_m e^{i(l+m)\theta}, \qquad A_0 = 0. \tag{C.1}$$

Let $n = l + m$, then

$$\left(\sum\right)^2 = \sum_{n=-\infty}^{\infty} e^{in\theta} \sum_{l=-\infty}^{\infty} A_l A_{n-l}. \tag{C.2}$$

Consider the l series. Using the fact that $A_0 = 0$, we break up the series as follows:

$$\sum_l \equiv \sum_{l=-\infty}^{\infty} A_l A_{n-l} = \sum_{l=-\infty}^{-1} A_l A_{n-l} + \sum_{l=1}^{n-1} A_l A_{n-l} + \sum_{l=n}^{\infty} A_l A_{n-l}. \tag{C.3}$$

In the first series we change l to $-l'$; in the last series we change $n - l$ to $-l'$. The two series can be added to give

$$2 \sum_{l=1}^{\infty} A_{-l} A_{n+l}. \tag{C.4}$$

The second series on the right-hand side of Eq. (C.3) can be rewritten depending on the parity of n, that is, whether n is odd or even. If $n = $ even $= 2p$, then

$$\sum_{l=1}^{2p-1} A_l A_{2p-l} = A_1 A_{2p-1} + A_2 A_{2p-2} + \cdots A_p A_p + \cdots A_{2p-2} A_2 + A_{2p-1} A_1.$$

Except for $A_p A_p$, the above terms may be added in pairs (e.g., the first and the last, etc.) to give

$$\sum_{l=1}^{p} \alpha_l A_l A_{2p-l} = \sum_{l=1}^{n/2} \alpha_l A_l A_{n-l}, \tag{C.5}$$

where α_l is defined by

$$\alpha_l = 2 \quad \text{if } l = 1, 2, \ldots, \frac{n}{2} - 1;$$

$$\alpha_l = 1 \quad \text{if } l = \frac{n}{2}. \tag{C.6}$$

If $n = $ odd $= 2p + 1$, $p = 0, 1, 2, \ldots$, then

$$\sum_{l=1}^{2p} A_l A_{2p+1-l} = A_1 A_{2p} + A_2 A_{2p-1} + \cdots A_p A_{p+1} + A_{p+1} A_p + \cdots A_{2p} A_1$$

$$= \sum_{l=1}^{p} 2 A_l A_{2p+1-l} = \sum_{l=1}^{(n-1)/2} \alpha_l A_l A_{n-l}. \tag{C.7}$$

The two alternatives (C.5) and (C.7) can be written as a single series, so that the second series on the right of Eq. (C.3) becomes

$$\sum_{l=1}' \alpha_l A_l A_{n-l} \tag{C.8}$$

where the upper limit of $\sum'$ is $\frac{1}{2}n$ if n is even and $\frac{1}{2}(n-1)$ if n is odd.

Finally, by combining Eqs. (C.4) and (C.8) into Eq. (C.2), we obtain

$$\left(\sum\right)^2 = \sum_{n=1}^{\infty} e^{in\theta} \left\{ \sum_{l=1}' \alpha_l A_l A_{n-l} + \sum_{l=1}^{\infty} 2 A_l^* A_{n+l} \right\} + \sum_{l=1}^{\infty} A_l A_{-l} + *.$$

$$\tag{C.9}$$

TWELVE

Some Aspects of Nonlinear Waves in Water of Intermediate or Great Depth

12.1 INTRODUCTION

For arbitrary depth, $O(kh) = 1$-∞, the effects of finite amplitude have been of long-standing interest. In older literature considerable attention was paid to the calculation of periodic progressive or standing waves. In particular, Stokes' contribution (1847) on higher-order waves of uniform amplitude set the tone of research for nearly 100 years. Much of the subsequent effort has been devoted to the maximum wave height when the depth is either constant or infinite. These classical contributions have been thoroughly summarized in Wehausen and Laitone (1960).

More recent interest in nonlinear deep-water waves owes its impetus to our need for a better understanding of the sea spectrum development. Shortly after the birth of two complementary theories of wave generation by wind, due separately to Miles (1957) and Phillips (1957), a cornerstone was laid by Phillips (1960) who introduced the mechanism of nonlinear resonant interaction in which four wavetrains slowly exchanged energy among one another if their phases were suitably matched; an era of examining nonlinear waves for their transient evolution was then launched. This resonance mechanism is now regarded as one of the most important factors in the sea spectrum formation. For an authoritative account of this vital area of oceanography the reader must refer to Phillips (1977).

Also of great interest is a special kind of resonant interaction pertaining to a wavetrain with a narrow band of frequencies and wavelengths. We are indebted to Benjamin and Feir (1967) for their decisive demonstration that

Stokes' waves are unstable to slowly modulated periodic (side-band) disturbances. The same instability mechanism was confirmed and extended by the elegant theory of Whitham (1967) and further extended by Benney and Roskes (1969). The early stage of the nonlinear evolution of a wave packet subsequent to the initial stability was pursued by Lighthill (1967), who used a pair of nonlinear equations of conservation type due to Whitham. With additional accounts for higher-order dispersive effects, Chu and Mei (1971) modified Whitham's equations and traced the evolution of a wave packet for a longer time. An alternative line of development is based on the cubic Schrödinger equation (Zakharov, 1968; Benney and Roskes, 1969), which was shown by Davey (1972) to be equivalent to the conservation equations used by Chu and Mei. Further theoretical advances received a strong boost by Zakharov and Shabat (1972) for their exact analytical solution of the cubic Schrödinger equation. Extensive efforts, including both numerical and experimental, by Yuen, Lake, and associates (see Yuen and Lake, 1980) have now unveiled many new physical secrets of Stokes' waves on deep water.

Theories for strongly nonlinear waves have also taken giant strides with the help of new computing techniques. Employing Pade's approximant, Schwartz (1974) was able to solve accurately for progressive waves of arbitrary steepness (up to the limiting height) on water of constant depth. The mathematically awesome problem of breaking has been the subject of determined attacks by Longuet-Higgins and associates. By ingeneous approximations, incisive local, global, and stability analyses, and direct numerical integration, they are gaining grounds in surmounting a defiant barrier in water wave theory, that is, nonlinearity. These pioneering contributions have already resulted in an extensive literature which is the best source of information to a serious reader (Sample papers are: Longuet-Higgins, 1975, 1978.a, b, 1979; Longuet-Higgins and Cokelet 1976, 1978; Longuet-Higgins and Fox 1977, 1978).

For engineering purposes, the presence of a variable bottom, a structure, or a coastline is of great interest. The effects of changing boundaries on nonlinear waves are, however, a formidable challenge to the researcher. Suffice it to mention that the seemingly straightforward problem of second-order forces on a large body has been a center of controversy. Handling full nonlinearity in anything other than an unbounded deep sea is still in its infancy.

The scope of the present chapter is fairly modest. On the scientific side, emphasis is limited to the slow evolution of Stokes' waves. First, general evolution equations are deduced by the method of multiple scales. Stokes' uniform wavetrain is then obtained as a special case. The initial instability due to periodic side-band disturbances is discussed next. Typical features of the nonlinear evolution of unidirectional waves are shown through numerical examples. Motivated by engineering interest, two problems of diffraction by steady Stokes' waves are also discussed. In the first problem, the body is slender and the resulting wave pattern is the object of interest. In the second, the body is a circular cylinder (hence, blunt) and the second-order force is worked out. Finally, the recent development of numerical techniques for very steep waves is sketched.

12.2 EVOLUTION EQUATIONS FOR SLOWLY MODULATED WEAKLY NONLINEAR WAVES

The objective of this section is to deduce a general set of conditions governing a slowly varying train of surface waves which are essentially sinusoidal and propagating in one direction. To this end we shall apply the method of multiple scales.[†] The following account is a mixture of Benney and Roskes (1969), Chu and Mei (1970), and Davey and Stewartson (1974).

12.2.1 Finite and Constant Depth

For simplicity, the sea bottom is taken to be horizontal at a finite depth h,

$$\frac{\partial \Phi}{\partial z} = 0, \qquad z = -h. \tag{2.1}$$

Now any analytic function $[f(x, y, z, t)]_{z=\zeta}$ may be expanded into a Taylor series about $z = 0$,

$$f(x, y, \zeta, t) = [f]_0 + \zeta \left[\frac{\partial f}{\partial z} \right]_0 + \frac{\zeta^2}{2} \left[\frac{\partial^2 f}{\partial z^2} \right]_0 + \cdots,$$

where

$$[f]_0 = f(x, y, 0, t) \cdots.$$

Let it be assumed that $\partial f/\partial z = O(kf)$ where k is the characteristic wavenumber. For small $\zeta = O(A)$ the successive terms in the expansion are progressively smaller, essentially in increasing powers of kA. The expansions for the free-surface boundary conditions [Eqs. (1.16) and (1.14), Chapter One] are

$$\left[\frac{\partial^2 \Phi}{\partial t^2} + g \frac{\partial \Phi}{\partial z} \right]_0 + \zeta \left[\frac{\partial}{\partial z} \left(\frac{\partial^2 \Phi}{\partial t^2} + g \frac{\partial \Phi}{\partial z} \right) \right]_0 + \left[\frac{\partial}{\partial t} \mathbf{u}^2 \right]_0$$

$$+ \frac{\zeta^2}{2} \left[\frac{\partial^2}{\partial z^2} \left(\frac{\partial^2 \Phi}{\partial t^2} + g \frac{\partial \Phi}{\partial z} \right) \right]_0 + \zeta \left[\frac{\partial^2}{\partial t \, \partial z} \mathbf{u}^2 \right]_0 + \frac{1}{2} [\mathbf{u} \cdot \nabla \mathbf{u}^2]_0 + \cdots = 0$$

$$\tag{2.2}$$

[†]Another formalism which is applicable to this class of problems is Whitham's method of averaged Lagrangian. For a full exposition, see Whitham (1974).

and

$$-g\zeta = \left[\frac{\partial \Phi}{\partial t}\right]_0 + \zeta\left[\frac{\partial^2 \Phi}{\partial t\,\partial z}\right]_0 + \left[\frac{u^2}{2}\right]_0 + \frac{\zeta^2}{2}\left[\frac{\partial}{\partial t}\frac{\partial^2 \Phi}{\partial z^2}\right]_0 + \zeta\left[\frac{\partial}{\partial z}\frac{u^2}{2}\right]_0 + \cdots$$

$$(2.3)$$

up to the third order $O(kA)^3$.

Let x be the direction of the carrier wave. To allow for slow modulation, we introduce the following cascade of variables:

$$x, \qquad x_1 = \varepsilon x, \qquad x_2 = \varepsilon^2 x \ldots,$$

$$y_1 = \varepsilon y, \qquad y_2 = \varepsilon^2 y \ldots, \qquad (2.4)$$

$$t, \qquad t_1 = \varepsilon t, \qquad t_2 = \varepsilon^2 t \ldots,$$

where $\varepsilon = kA \ll 1$, and assume the following perturbation expansions for the unknowns:

$$\Phi = \sum_{n=1} \varepsilon^n \phi_n, \qquad \zeta = \sum_{n=1} \varepsilon^n \zeta_n, \qquad (2.5)$$

where

$$\phi_n = \phi_n(x, x_1, x_2, \ldots; y_1, y_2, \ldots; z; t, t_1, t_2, \ldots),$$

$$\zeta_n = \zeta_n(x, x_1, x_2, \ldots; y_1, y_2, \ldots; t, t_1, t_2, \ldots). \qquad (2.6)$$

The original derivatives are first changed according to Eq. (4.3), Chapter Two, and then substituted into Laplace's equation, the perturbed free-surface conditions (2.2) and (2.3), and the bottom condition (2.1). The first three orders are collected below.

Laplace equation:

$$\left(\frac{\partial^2}{\partial x^2} + \frac{\partial^2}{\partial z^2}\right)\phi_n = F_n, \qquad n = 1, 2, 3, \qquad (2.7a)$$

where

$$F_1 = 0, \qquad (2.7b)$$

$$F_2 = -2\phi_{1_{xx_1}}, \qquad (2.7c)$$

$$F_3 = -\left[\left(\frac{\partial^2}{\partial x_1^2} + \frac{\partial^2}{\partial y_1^2}\right)\phi_1 + 2\phi_{1_{xx_2}} + 2\phi_{2_{xx_1}}\right]. \qquad (2.7d)$$

Free-surface condition (2.2):

$$\Gamma\phi_n = G_n \qquad \text{on } z = 0, \tag{2.8a}$$

where

$$\Gamma = g\frac{\partial}{\partial z} + \frac{\partial^2}{\partial t^2}, \tag{2.8b}$$

$$G_1 = 0, \tag{2.8c}$$

$$G_2 = -\left[\zeta_1\Gamma_z\phi_1 + \left(\phi_{1_x}^2 + \phi_{1_z}^2\right)_t + 2\phi_{1_{tt_1}}\right], \tag{2.8d}$$

$$G_3 = -\Big[\zeta_2\Gamma_z\phi_1 + \zeta_1\Gamma_z\phi_2 + \tfrac{1}{2}\zeta_1^2\Gamma_{zz}\phi_1 + 2\left(\phi_{1_x}\phi_{2_x} + \phi_{1_z}\phi_{2_z}\right)_t$$

$$+ \zeta_1\left(\phi_{1_x}^2 + \phi_{1_z}^2\right)_{tz} + \frac{1}{2}\left(\phi_{1_x}\frac{\partial}{\partial x} + \phi_{1_z}\frac{\partial}{\partial z}\right)\left(\phi_{1_x}^2 + \phi_{1_z}^2\right)$$

$$+ 2\phi_{2_{tt_1}} + 2\phi_{1_z}\phi_{1_{zt_1}} + 2\phi_{1_x}\phi_{1_{xt}} + 2\phi_{1_x}\phi_{1_{xt_1}}$$

$$+ 2\phi_{1_x}\phi_{1_{tx_1}} + 2\zeta_1\phi_{1_{ztt_1}} + 2\phi_{1_{tt_2}} + \phi_{1_{t_1t_1}}\Big]. \tag{2.8e}$$

Bernoulli equation on the free surface:

$$-g\zeta_n = H_n, \qquad \text{on } z = 0, \tag{2.9a}$$

where

$$H_1 = \phi_{1_t}, \tag{2.9b}$$

$$H_2 = \phi_{2_t} + \tfrac{1}{2}\left(\phi_{1_x}^2 + \phi_{1_z}^2\right) + \phi_{1_{t_1}} + \zeta_1\phi_{1_{zt}}, \tag{2.9c}$$

$$H_3 = \phi_{3_t} + \phi_{1_x}\phi_{2_x} + \phi_{1_z}\phi_{2_z} + \zeta_1\phi_{2_{zt}} + \zeta_2\phi_{1_{zt}}$$

$$+ \tfrac{1}{2}\zeta_1^2\phi_{1_{zzt}} + \tfrac{1}{2}\zeta_1\left(\phi_{1_x}^2 + \phi_{1_z}^2\right)_z + \phi_{2_{t_1}}$$

$$+ \phi_{1_x}\phi_{1_{x_1}} + \phi_{1_{t_2}} + \zeta_1\phi_{1_{zt_1}}. \tag{2.9d}$$

Condition on the bottom:

$$\frac{\partial\phi_n}{\partial z} = 0, \qquad z = -h. \tag{2.10}$$

The first-order solution is the usual linearized propagating wave. At the higher orders the nonlinear forcing terms on the free surface imply that higher harmonics must be present in the higher-order solutions. We therefore expand

$$\{\phi_n, F_n, G_n\} = \sum_{m=-n}^{n} e^{im\psi}\{\phi_{nm}, F_{nm}, G_{nm}\}, \tag{2.11a}$$

where

$$\psi = kx - \omega t \tag{2.11b}$$

is the phase. For the resulting ϕ to be real, we require that

$$\phi_{n,-m} = (\phi_{nm})^*. \tag{2.12}$$

Furthermore, the amplitudes depend on the slow variables and z

$$(\phi_{nm}, F_{nm}) = \text{functions of } (x_1, x_2, \dots, y_1, y_2, \dots, z, t_1, t_2, \dots),$$

$$G_{nm} = G_{nm}(x_1, x_2, \dots, y_1, y_2, \dots, t_1, t_2, \dots). \tag{2.13}$$

Substituting Eq. (2.11a) into the perturbation equations (2.7a), (2.8), and (2.10), we obtain at each order (n) and harmonic (m) a boundary-value problem in z:

$$\left(\frac{\partial^2}{\partial z^2} - m^2 k^2\right)\phi_{nm} = F_{nm}, \qquad -h < z < 0, \tag{2.14}$$

$$\left(g\frac{\partial}{\partial z} - m^2\omega^2\right)\phi_{nm} = G_{nm} \qquad z = 0, \tag{2.15}$$

$$\frac{\partial}{\partial z}\phi_{nm} = 0 \qquad z = -h. \tag{2.16}$$

The above boundary-value problems will be solved sequentially. For convenience the following abbreviations will be employed:

$$Q = k(z + h), \qquad q = kh,$$

$$\text{ch } q = \cosh q, \qquad \text{sh } q = \sinh q, \qquad \text{th } q = \tanh q. \tag{2.17}$$

$n = 1$: The forcing terms are

$$F_1 = G_1 = 0. \tag{2.18}$$

The solutions, which are homogeneous, are

$$m = 0: \qquad \phi_{10} = \phi_{10}(x_1, x_2, y_1, y_2, t_1, t_2, \ldots) = \phi_{10}^*, \qquad (2.19)$$

$$m = 1: \qquad \phi_{11} = -\frac{g \operatorname{ch} Q}{2\omega \operatorname{ch} q} iA, \qquad (2.20a)$$

with

$$\omega^2 = gk \operatorname{th} kh, \qquad (2.20b)$$

and $A = A(x_1, x_2, y_1, y_2, t_1, t_2, \ldots)$. The total first-order solution is

$$\phi_1 = \phi_{10} - \frac{g \operatorname{ch} Q}{2\omega \operatorname{ch} q} \left(iAe^{i\psi} + * \right), \qquad (2.21)$$

$$\zeta_1 = \tfrac{1}{2} \left(Ae^{i\psi} + * \right). \qquad (2.22)$$

Both ϕ_{10} and A are, so far, arbitrary functions of the slow variables.

$n = 2$: For $n \geq 2$, special care regarding solvability must be given to $m = 0$ (the zeroth harmonic, or the mean), and to $m = 1$ (the first harmonic), as the homogeneous problems have nontrivial solutions. For $m = 0$, the homogeneous solution is constant in z; the solvability condition for ϕ_{n0} is simply

$$\frac{1}{g} G_{n0} = \int_{-h}^{0} dz \, F_{n0}. \qquad (2.23)$$

For $m = 1$, the homogeneous solution is proportional to $\cosh k(z + h)$, and the solvability condition for ϕ_{n1} follows from Green's formula

$$\frac{1}{g} G_{n1} = \int_{-h}^{0} dz \, F_{n1} \frac{\cosh k(z + h)}{\cosh kh}. \qquad (2.24)$$

Equations (2.23) and (2.24) will lead to the so-called *evolution equations*.

For higher harmonics $|m| \geq 2$, the boundary-value problems admit no homogeneous solution. Indeed, if there were one, it would be of the form (2.20a) with (k, ω) replaced by $(mk, m\omega)$. However, mk and $m\omega$ would then be subjected to the dispersion relation (2.20b), which is not possible. Consequently, the inhomogeneous problems for $|m| \geq 2$ are always solvable.

The forcing terms for the second-order problems are

$$F_{20} = G_{20} = 0, \qquad (2.25)$$

$$F_{21} = -\frac{\omega \operatorname{ch} Q}{\operatorname{sh} q} \frac{\partial A}{\partial x_1}, \qquad F_{22} = 0, \qquad (2.26)$$

$$G_{21} = \frac{\omega^2 \operatorname{ch} q}{k \operatorname{sh} q} \frac{\partial A}{\partial t_1}, \qquad G_{22} = \frac{3i\omega^3 A^2}{4 \operatorname{sh}^2 q}. \qquad (2.27)$$

The solution for the zeroth harmonic is

$$\phi_{20} = \phi_{20}(x_1, x_2, y_1, y_2, t_1, t_2, \ldots) = \phi_{20}^*. \tag{2.28}$$

For $m = 1$, we invoke the solvability condition by substituting Eqs. (2.26) and (2.27) into Eq. (2.24), and obtain

$$\frac{\partial A}{\partial t_1} + C_g \frac{\partial A}{\partial x_1} = 0 \quad \text{with} \quad C_g = \frac{\partial \omega}{\partial k}, \tag{2.29}$$

which has been found before in Section 2.4. The solution is

$$\phi_{21} = -\frac{\omega}{2k^2 \operatorname{sh} q} (Q \operatorname{sh} Q) \frac{\partial A}{\partial x_1}. \tag{2.30}$$

We note immediately that ϕ_{21} blows up as $kh \uparrow \infty$, which is not surprising, because the limiting process violates the original premise that $O(x_1), O(x_2), \ldots > O(h)$. Therefore, we restrict the present solution to $kh = O(1)$ and deal later with the infinite depth separately.

For $m = 2$, the inhomogeneous solution is straightforward:

$$\phi_{22} = -\frac{3}{16} \frac{\omega \operatorname{ch} 2Q}{\operatorname{sh}^4 q} iA^2. \tag{2.31}$$

The total second-order solution is

$$\phi_2 = \phi_{20} - \frac{\omega}{2k^2 \operatorname{sh} q} (Q \operatorname{sh} Q) \left(\frac{\partial A}{\partial x_1} e^{i\psi} + * \right)$$

$$- \frac{3}{16} \frac{\omega \operatorname{ch} 2Q}{\operatorname{sh}^4 q} (iA^2 e^{2i\psi} + *), \tag{2.32}$$

and

$$\zeta_2 = \left\{ -\frac{1}{g} \phi_{10_{t_1}} - \frac{k}{2 \operatorname{sh} 2q} |A|^2 \right\} + \frac{1}{2\omega} \left(i\frac{\partial A}{\partial t_1} e^{i\psi} + * \right)$$

$$- \frac{q \operatorname{sh} q}{2k \operatorname{ch} q} \left(i\frac{\partial A}{\partial x_1} e^{i\psi} + * \right) + \frac{k \operatorname{ch} q(2 \operatorname{ch}^2 q + 1)}{8 \operatorname{sh}^3 q} (A^2 e^{2i\psi} + *).$$

$$\tag{2.33}$$

$n = 3$: For $m = 0$, the forcing terms are

$$F_{30} = -\left(\frac{\partial^2}{\partial x_1^2} + \frac{\partial^2}{\partial y_1^2}\right)\phi_{10},\tag{2.34}$$

$$G_{30} = \frac{\omega^3 \operatorname{ch}^2 q}{2k \operatorname{sh}^2 q}(AA^*)_{x_1} - \frac{\omega^2}{4 \operatorname{sh}^2 q}(AA^*)_{t_1} - \frac{\partial^2 \phi_{10}}{\partial t_1^2},\tag{2.35}$$

where Eq. (2.29) has been used. Condition (2.23) requires that $G_{30} = ghF_{30}$ so that

$$\frac{\partial^2 \phi_{10}}{\partial t_1^2} - gh\left(\frac{\partial^2}{\partial x_1^2} + \frac{\partial^2}{\partial y_1^2}\right)\phi_{10} = \frac{\omega^3 \operatorname{ch}^2 q}{2k \operatorname{sh}^2 q}(AA^*)_{x_1} - \frac{\omega^2}{4 \operatorname{sh}^2 q}(AA^*)_{t_1}.$$

$$\tag{2.36}$$

Physically, Eq. (2.36) describes a long wave generated by short-wave modulation. For $m = 1$, the forcing terms are

$$F_{31} = \frac{\omega}{k \operatorname{sh} q}\left[Q \operatorname{sh} Q + \frac{1}{2}\operatorname{ch} Q\right]i\frac{\partial^2 A}{\partial x_1^2}$$

$$-\frac{\omega \operatorname{ch} Q}{\operatorname{sh} q}\left(\frac{\partial A}{\partial x_2} - \frac{i}{2k}\frac{\partial^2 A}{\partial y_1^2}\right),\tag{2.37}$$

and

$$G_{31} = \frac{1}{16 \operatorname{sh}^5 q}\omega^3 k \operatorname{ch} q(\operatorname{ch} 4q + 8 - 2\operatorname{th}^2 q)i\,|A|^2 A$$

$$-\frac{\omega k}{\operatorname{sh} 2q}\left(\frac{\partial \phi_{10}}{\partial t_1} - \frac{2\omega \operatorname{ch}^2 q}{k}\frac{\partial \phi_{10}}{\partial x_1}\right)iA$$

$$+\frac{\omega \operatorname{ch} q}{2k \operatorname{sh} q}i\frac{\partial^2 A}{\partial t_1^2} - \frac{\omega^2 q}{k^2}i\frac{\partial^2 A}{\partial x_1 \partial t_1} + \frac{\omega^2 \operatorname{ch} q}{k \operatorname{sh} q}\frac{\partial A}{\partial t_2}.\tag{2.38}$$

Invoking the solvability condition (2.24), we have

$$
\frac{\partial A}{\partial t_2} + C_g\frac{\partial A}{\partial x_2} - \frac{iC_g}{2k}\frac{\partial^2 A}{\partial y_1^2} - \frac{i\omega q}{k^2 \mathrm{sh}\,2q}\mathrm{ch}^2 q\frac{\partial^2 A}{\partial x_1^2}
$$

$$
+ \frac{i}{2\omega}\frac{\partial^2 A}{\partial t_1^2} - \frac{ik^2 A}{2\omega\,\mathrm{ch}^2 q}\left(\frac{\partial\phi_{10}}{\partial t_1} - \frac{2\omega\,\mathrm{ch}^2 q}{k}\frac{\partial\phi_{10}}{\partial x_1}\right) - \frac{iq\,\mathrm{sh}\,q}{k\,\mathrm{ch}\,q}\frac{\partial^2 A}{\partial x_1\,\partial t_1}
$$

$$
+ \frac{i\omega k^2(\mathrm{ch}\,4q + 8 - 2\,\mathrm{th}^2 q)}{16\,\mathrm{sh}^4 q}\,|A|^2 A = 0. \tag{2.39}
$$

With the help of Eq. (2.29) we write

$$
\frac{\partial^2 A}{\partial t_1^2} = C_g^2\frac{\partial^2 A}{\partial x_1^2}, \qquad \frac{\partial^2 A}{\partial t\,\partial x_1} = C_g\frac{\partial^2 A}{\partial x_1^2}
$$

in Eq. (2.39). Finally, we add Eqs. (2.29) and (2.39), and consider ϕ_{10} and A as functions of x_1, y_1, and t_1 only, that is,

$$
\frac{\partial}{\partial t_1} + \varepsilon\frac{\partial}{\partial t_2} \to \frac{\partial}{\partial t_1}, \qquad \frac{\partial}{\partial x_1} + \varepsilon\frac{\partial}{\partial x_2} \to \frac{\partial}{\partial x_1},
$$

yielding

$$
\left(\frac{\partial}{\partial t_1} + C_g\frac{\partial}{\partial x_1}\right)A + i\varepsilon\left\{-\frac{1}{2}\left(\frac{\partial^2\omega}{\partial k^2}\right)\frac{\partial^2 A}{\partial x_1^2} - \frac{C_g}{2k}\frac{\partial^2 A}{\partial y_1^2}\right.
$$

$$
\left. + \frac{\omega k^2(\mathrm{ch}\,4q + 8 - 2\,\mathrm{th}^2 q)}{16\,\mathrm{sh}^4 q}\,|A|^2 A - \left(\frac{k^2}{2\omega\,\mathrm{ch}^2 q}\frac{\partial\phi_{10}}{\partial t_1} - k\frac{\partial\phi_{10}}{\partial x_1}\right)A\right\} = 0,
$$

$$
\tag{2.40}
$$

where

$$
-\frac{1}{2}\frac{\partial^2\omega}{\partial k^2} = \frac{C_g^2}{2\omega} - \frac{\omega q\,\mathrm{ch}^2 q}{k^2\,\mathrm{sh}\,2q} + \frac{q\,\mathrm{sh}\,q}{k\,\mathrm{ch}\,q}C_g > 0. \tag{2.41}
$$

Equations (2.36) and (2.40) were first obtained by Benney and Roskes. Together, they govern the coupled slow evolution of first-order amplitude A and the mean flow potential ϕ_{10}. It is significant that third-order nonlinearity affects first-order amplitude if the time and distance are ε^{-2} times greater than the wave period and length, respectively.

For two-dimensional problems $(\partial/\partial y_1 = 0)$, further simplifications are possible. If we let

$$\xi = x_1 - C_g t_1, \qquad \tau = \varepsilon t_1, \tag{2.42}$$

then

$$\frac{\partial}{\partial t_1} = \varepsilon \frac{\partial}{\partial \tau} - C_g \frac{\partial}{\partial \xi} \quad \text{and} \quad \frac{\partial}{\partial x_1} = \frac{\partial}{\partial \xi},$$

and Eq. (2.36) may be integrated once with respect to ξ, yielding

$$\frac{\partial \phi_{10}}{\partial \xi} = S(\tau) - \frac{\omega^2 \left(2\omega \, \mathrm{ch}^2 q + kC_g\right)}{4k \, \mathrm{sh}^2 q \left(gh - C_g^2\right)} |A|^2 + O(\varepsilon),$$

where $S(\tau)$ is an arbitrary function of τ. Substituting this result into Eq. (2.40), we get

$$-i\frac{\partial A}{\partial \tau} + \alpha \frac{\partial^2 A}{\partial \xi^2} + \beta |A|^2 A + \gamma A = 0, \tag{2.43a}$$

where

$$\alpha = -\tfrac{1}{2}\omega''(k) \tag{2.43b}$$

$$\beta = \frac{\omega k^2}{16 \, \mathrm{sh}^4 q}(\mathrm{ch}\, 4q + 8 - 2\,\mathrm{th}^2 q) - \frac{\omega}{2 \, \mathrm{sh}^2 2q} \frac{\left(2\omega \, \mathrm{ch}^2 q + kC_g\right)^2}{\left(gh - C_g^2\right)},$$

$$\tag{2.43c}$$

and

$$\gamma(\tau) = \frac{S(\tau)k}{2\omega \, \mathrm{ch}^2 q}\left(2\omega \, \mathrm{ch}^2 q - kC_g\right). \tag{2.43d}$$

The function $S(\tau)$ vanishes for a wavetrain beginning from rest where A and $\partial \phi_{10}/\partial \xi$ tend to zero as $\xi \to \infty$. Otherwise, the term γA may be eliminated from Eq. (2.43a) by introducing

$$A = B \exp\left(-i\int \gamma \, d\tau\right), \tag{2.44}$$

resulting in the cubic Schrödinger equation

$$-iB_\tau + \alpha B_{\xi\xi} + \beta |B|^2 B = 0 \tag{2.45}$$

which was derived for finite depth by Hashimoto and Ono (1972). A similar equation was deduced earlier for infinite depth by Zakharov (1968).

Equation (2.43) or (2.45) may be expressed in real functions by letting

$$A = a \exp\left[i \left(\int W \, d\xi - \int \gamma \, d\tau \right) \right] \tag{2.46a}$$

in Eq. (2.43), or, equivalently, by letting

$$B = a \exp\left(i \int W \, d\xi \right) \tag{2.46b}$$

in Eq. (2.45), where $W = W(\xi, \tau)$. Separating the real and imaginary parts, we get

$$\frac{\partial a^2}{\partial \tau} - 2\alpha \frac{\partial}{\partial \xi} (W a^2) = 0, \tag{2.47a}$$

$$\frac{\partial W}{\partial \tau} - \frac{\partial}{\partial \xi} \left[\alpha \left(\frac{1}{a} \frac{\partial^2 a}{\partial \xi^2} - W^2 \right) + \beta a^2 \right] = 0. \tag{2.47b}$$

These equations are in the form of conservation laws $\partial P/\partial \tau + \partial Q/\partial \xi = 0$; they were derived for deep-water waves by Chu and Mei (1970), and by Whitham (1967) without the term $(1/a)\,\partial^2 a/\partial \xi^2$. The connection between Eqs. (2.45) and (2.47) was pointed out by Davey (1972).

12.2.2 Infinite Depth

The formal limits of Eqs. (2.36) and (2.40) for $kh \to \infty$ are

$$\left(\frac{\partial^2}{\partial x_1^2} + \frac{\partial^2}{\partial y_1^2} \right) \phi_{10} = 0, \tag{2.48}$$

$$\left(\frac{\partial}{\partial t_1} + C_g \frac{\partial}{\partial x_1} \right) A + i\varepsilon \left\{ \frac{\omega}{4k^2} \left(\frac{1}{2} A_{x_1 x_1} - A_{y_1 y_1} \right) + \tfrac{1}{2}\omega k^2 |A|^2 A + k\phi_{10_{x_1}} A \right\} = 0, \tag{2.49}$$

which appear to be quite reasonable results. The kinematic or total boundary condition on the free surface requires that

$$\frac{\partial \phi_{10}}{\partial z} = 0, \qquad z = 0. \tag{2.50}$$

A possible solution for ϕ_{10} is a time-varying current

$$\frac{\partial \phi_{10}}{\partial x_1} = f(t_1), \tag{2.51}$$

where f is prescribed from upstream conditions. Equation (2.49) then governs A only. Nevertheless, the solution for ϕ_{21} is no longer meaningful as pointed out after Eq. (2.30). Inconsistency also arises at the third-order free-surface condition, (2.8a), which gives for $h \to \infty$

$$\frac{\partial^2 \phi_{10}}{\partial t_1^2} = \frac{\omega^3}{2k}(AA^*)_{x_1} \tag{2.52}$$

and renders the problem for A and ϕ_{10} overdetermined. To remedy this difficulty, Roskes (1969) reasoned that by assuming $kh \gg 1$ at the start one should allow slow modulation in the vertical direction as well. Thus, he introduced

$$z_1 = \varepsilon z, \qquad z_2 = \varepsilon^2 z \cdots. \tag{2.53}$$

However, these coordinates should only enter the long-scale quantities $\phi_{10}, \phi_{20}, \ldots$, and so on; the short-wave potential is nonzero only within a wavelength from the free surface (hence in the region $z_1, z_2, \ldots \ll 1$). Carrying out the perturbation analysis, one finds, instead of Eq. (2.36),

$$\left(\frac{\partial^2}{\partial x_1^2} + \frac{\partial^2}{\partial y_1^2} + \frac{\partial^2}{\partial z_1^2} \right) \phi_{10} = 0, \tag{2.54}$$

and instead of Eq. (2.39),

$$\left(A_{t_2} + C_g A_{x_2} \right) + i \left\{ \frac{\omega}{8k^2} \left(A_{x_1 x_1} - 2A_{y_1 y_1} \right) + \frac{1}{2} \omega k^2 |A|^2 A \right\}$$
$$+ k \left(\frac{\partial \phi_{10}}{\partial z_1} + i \frac{\partial \phi_{10}}{\partial x_1} \right) A = 0. \tag{2.55}$$

If $h = O(\varepsilon k)^{-1}$, the bottom condition can be generalized slightly so that

$$\frac{\partial \phi_{10}}{\partial n_1} = 0, \qquad z_1 = 0 \quad \text{and} \quad -h_1(x_1, y_1) \quad \text{with} \quad h_1 = O(1).$$

$$\tag{2.56}$$

Thus, the potential ϕ_{10} of the large-scale current is deflected by bottom variations, which then affect the short-wave amplitude indirectly through Eq. (2.55). Furthermore, Eq. (2.52) is replaced by

$$g\frac{\partial \phi_{20}}{\partial z_1} = -\frac{\partial^2 \phi_{10}}{\partial t_1^2} + \frac{\omega^3}{2k}(AA^*)_{x_1} \tag{2.57}$$

which provides a boundary condition for ϕ_{20} and removes the overdeterminancy on ϕ_{10} and A.

In the simplest case of constant h_1 we can take

$$\frac{\partial \phi_{10}}{\partial x_1} = \mathcal{U}(\tau) \tag{2.58}$$

which is a spatially constant current determined by the condition far upstream. Note that the magnitude of the current is $\varepsilon \, \partial \phi_{10}/\partial x = \varepsilon^2 \, \partial \phi_{10}/\partial x_1 = \varepsilon^2 \mathcal{U} = O(\varepsilon^2)$. Equation (2.55) can be further simplified by the coordinate transformation (2.42); the result is

$$\frac{\partial A}{\partial \tau} + i\left\{ \frac{\omega}{4k^2}\left(\tfrac{1}{2}A_{\xi\xi} - A_{y_1y_1}\right) + k\mathcal{U}A + \frac{\omega k^2}{2}|A|^2 A \right\} = 0 \tag{2.59}$$

The current term may be removed by $A = B\exp(-ik\int\mathcal{U}\,d\tau)$, yielding,

$$\frac{\partial B}{\partial \tau} + i\left\{ \frac{\omega}{8k^2}\left(B_{\xi\xi} - 2B_{y_1y_1}\right) + \frac{\omega k^2}{2}|B|^2 B \right\} = 0 \tag{2.60}$$

which was first derived by Zakharov (1968). In the limit of $\partial/\partial y_1 = 0$, Eq. (2.60) reduces to Eq. (2.45) with

$$\alpha = \frac{\omega}{8k^2} \quad \text{and} \quad \beta = \frac{\omega k^2}{2}. \tag{2.61}$$

12.3 UNIFORM STOKES' WAVES

We seek a solution which has no slow modulation in x and y, that is, A, $\partial \phi_{10}/\partial x_1, \partial \phi_{10}/\partial t_1,\ldots$ do not depend on $x_1, x_2,\ldots$ and $y_1, y_2,\ldots$. Equation (2.29) requires that A be independent of t_1; thus, $A = A(t_2)$. From Eq. (2.36) we get

$$\phi_{10} = \mathcal{U}x_1 - gbt_1 \tag{3.1}$$

where $\mathfrak{U}$ and b are arbitrary constants. With Eq. (3.1), Eq. (2.39) is reduced to

$$\frac{\partial A}{\partial t_2} + i\omega_2 |A|^2 A = 0, \tag{3.2}$$

where

$$\omega_2 = \frac{\omega k^2}{16\,\mathrm{sh}^4 q}(8 + \mathrm{ch}\,4q - 2\,\mathrm{th}^2 q) - \frac{1}{|A|^2}\left(k\mathfrak{U} + \frac{gk^2 b}{2\omega\,\mathrm{ch}^2 q}\right). \tag{3.3}$$

The solution is readily found to be

$$A = a_0\exp\left(-i\omega_2 a_0^2 t_2\right), \tag{3.4}$$

with a_0 real and an arbitrary phase ignored. Thus, up to $O(\varepsilon^2)$ nonlinearity changes the wave phase only. The wave profile is, to the leading order,

$$\zeta_1 = \tfrac{1}{2}\left(a_0 e^{i\tilde{\psi}} + *\right), \tag{3.5a}$$

where

$$\tilde{\psi} = (kx - \tilde{\omega}t) \tag{3.5b}$$

is the new phase, and

$$\tilde{\omega} = \omega + \varepsilon^2 \omega_2 a_0^2 \tag{3.6}$$

is the nonlinear dispersion relation. The phase speed also depends on amplitude

$$\tilde{C} = \frac{\omega}{k} + \varepsilon^2 \frac{\omega_2}{k} a_0^2. \tag{3.7}$$

The ordering parameter ε has served its purpose and may be removed.

The second-order solution is

$$\phi_2 = \phi_{20} - \frac{3}{16}\frac{\omega\,\mathrm{ch}\,2Q}{\mathrm{sh}^4 q}\left(ia_0^2 e^{2i\tilde{\psi}} + *\right), \tag{3.8}$$

$$\zeta_2 = b - \frac{k}{2\,\mathrm{sh}\,2q}a_0^2 + \frac{k\,\mathrm{ch}\,q(2\,\mathrm{ch}^2 q + 1)}{8\,\mathrm{sh}^3 q}\left(a_0^2 e^{2i\tilde{\psi}} + *\right). \tag{3.9}$$

From the solvability of ϕ_{40} it may be shown that $\partial\phi_{20}/\partial z = 0$, implying that $\phi_{20} = \phi_{20}(x_1, x_2,\ldots,y_1, y_2,\ldots)$ which does not affect the velocity field at the second order. The second harmonic in ζ_2 is always positive for $\tilde{\psi} = 0, \pi, 2\pi,\ldots$

and elevates the crests and the troughs; therefore, the total free surface is more peaked at the crests and flatter at the troughs. The second term in Eq. (3.9) represents the mean sea-level set-down which corresponds to Eq. (3.5), and has been deduced before in Section 10.5 [cf. Eq. (5.6)].

The ratio of ζ_2 to ζ_1 is of the order $ka_0/(kh)^3$ when $kh \ll 1$. Thus, Stokes' theory is valid in shallow water only if

$$\frac{ka_0}{(kh)^3} \ll 1. \tag{3.10}$$

The above ratio is precisely the Ursell parameter mentioned in the preceding chapter and is a measure of frequency dispersion versus nonlinearity.

Because of the arbitrariness of $\mathcal{U}$ and b, various expressions for the Stokes wave have been used in the literature. For example, by letting A be independent of ξ and S be a constant, we may integrate Eq. (2.43a) to obtain Eq. (3.4) with

$$\omega_2 = \beta a_0^2 + \gamma. \tag{3.11}$$

This dispersion relation can be reconciled with Eq. (3.3) if

$$\mathcal{U} = S - \frac{\omega\left(2\omega\,\mathrm{ch}^2 q + kC_g\right)}{4k\,\mathrm{sh}^2 q\left(gh - C_g^2\right)}a_0^2 \tag{3.12a}$$

and

$$b = \frac{1}{g}\mathcal{U}C_g. \tag{3.12b}$$

The preceding expressions are equivalent to those given by Davey and Stewartson (1974).

12.4   SIDE-BAND INSTABILITY OF STOKES' WAVES

The uniform Stokes waves have been applied for many years in engineering literature as a basis of computing wave forces. Massive work has been devoted to higher-order corrections in order to gain a fuller account of nonlinearity. Nevertheless it is a common experience in the laboratory that a uniform train of relatively steep waves is difficult to maintain in a long-wave tank (see, e.g., Russell and Osorio, 1957). This difficulty was first explained theoretically by the penetrating work of Benjamin and Feir (1967) who found that Stokes' waves were unstable to *side-band* disturbances, that is, disturbances whose frequencies deviated slightly from the fundamental frequency of the carrier waves.

Consider the case of finite water depth. By combining Eqs. (3.4) and (3.11) with Eq. (2.46a) we note first that $a = a_0$ and $W = 0$ for the Stokes wave. Let a' and W' be the disturbances in a and W, respectively,

$$a = a_0 + a', \qquad W = W'. \tag{4.1}$$

The linearized versions of Eqs. (2.47a) and (2.47b) are

$$\frac{\partial a'}{\partial \tau} - \alpha a_0 \frac{\partial W'}{\partial \xi} = 0, \tag{4.2}$$

$$\frac{\partial W'}{\partial \tau} + \frac{\partial}{\partial \xi}\left[\alpha \frac{a'_{\xi\xi}}{a_0} + 2\beta a_0 a' \right] = 0. \tag{4.3}$$

Since the effect of small frequency variation is to create long-scale modulations, we assume the perturbation to be slowly modulated waves:

$$a' = \bar{a}' e^{i(K\xi - \Omega\tau)},$$

$$W' = \overline{W}' e^{i(K\xi - \Omega\tau)}. \tag{4.4}$$

Equations (4.2) and (4.3) then give

$$-i\Omega\bar{a}' - \alpha a_0 iK\overline{W}' = 0, \tag{4.5}$$

$$\bar{a}'\left[\frac{\alpha}{a_0}(iK)^3 + 2\beta a_0(iK) \right] - i\Omega\overline{W}' = 0. \tag{4.6}$$

For nontrivial $\bar{a}'$ and $\overline{W}'$ the discriminant must vanish:

$$\begin{vmatrix} -i\Omega & -i\alpha a_0 K \\ iK\left(-\dfrac{\alpha K^2}{a_0} + 2\beta a_0 \right) & -i\Omega \end{vmatrix} = 0,$$

that is,

$$\Omega = \pm K\left(\alpha^2 K^2 - 2\alpha\beta a_0^2 \right)^{1/2}. \tag{4.7}$$

If $\alpha\beta < 0$, the square root is always real and the side-band disturbance is purely sinusoidal, hence is neutrally stable. If, however, $\alpha\beta > 0$, Ω can be imaginary and the side-band disturbance can increase exponentially with time, hence is unstable; this occurs when

$$K^2 < \frac{2\beta a_0^2}{\alpha}. \tag{4.8}$$

Because of Eq. (2.61), Stokes' waves on *deep water* are unstable if

$$\left|\frac{K}{k}\right| < 8^{1/2}ka_0. \tag{4.9}$$

The steeper the waves, the broader the unstable side band. The dispersion relation corresponding to Eq. (4.7) is

$$\Omega = \pm\omega\left[\frac{K^2}{8k^2}\left(\frac{K^2}{8k^2} - k^2a_0^2\right)\right]^{1/2}. \tag{4.10}$$

The unstable growth rate is zero at both $K = 0$ and $8^{1/2}k^2a_0$. The maximum growth rate occurs when

$$\left|\frac{K}{k}\right| = 2ka_0, \tag{4.11}$$

at which

$$\text{Max } \Omega = \pm i\tfrac{1}{2}\omega\left(k^2a_0^2\right). \tag{4.12}$$

The corresponding $\bar{a}'$ and $\overline{W}'$ are out of phase by $\frac{1}{2}\pi$. Generally, the growth rate Im Ω varies with the wavelength of the side-band disturbance as shown in Fig. 4.1.

A heuristic explanation of this side-band instability has been given by Lighthill (1978, p. 462). Consider a Stokes wavetrain with a slowly modulated envelope. The crests near a peak of the envelope are faster than those on either side of the peak and, therefore, tend to shorten the waves ahead and lengthen the waves behind. Now the group velocity in deep water is larger for longer waves. The rate of energy transport is lower in front and higher behind, hence accumulation occurs near the envelope peak, whose height must increase. Similarly, the trough of the envelope will tend to decrease, resulting in instability.

Controlled experiments by Feir as reported in Benjamin (1967) and by Lake and Yuen (1977) support the theory fairly well as shown in Fig. 4.2. The

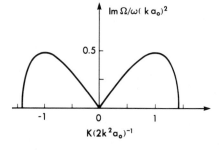

Figure 4.1 The asymptotic growth rate of the side-band amplitudes as a function of modulational wavenumber. (From Benjamin and Feir, 1967, *J. Fluid Mech.* Reproduced by permission of Cambridge University Press.)

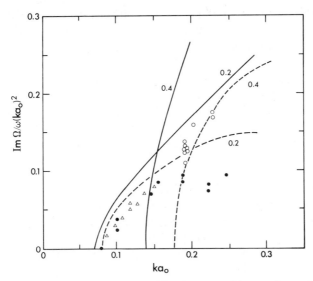

Figure 4.2 Comparison of calculated amplification rate with experimental results as a function of wave slope. Theoretical results: solid curve, by Eq. (4.10); dashed curve, Crawford et al. (1981). For $K/k = 0.2$, the Longuet-Higgins (1978b) results by exact theory cannot be distinguished from the dashed line. Experimental results: $\bigcirc$, $K/k = 0.4$ and $\bullet$, $K/k = 0.2$, by Lake et al. (1977); $\triangle$, $K/k = 0.2$ by Benjamin (1967). (From Crawford et al., 1981, *J. Fluid Mech.* Reproduced by permission of Cambridge University Press.)

discrepancy decreases for smaller wave slope or slower rate of modulation (smaller K/k).

From the expressions of α and β in Eqs. (2.43b) and (2.43c), it can be shown that the Stokes waves on finite depth are unstable to colinear side-band disturbances if $kh > 1.36$.

Benney and Roskes (1969) further studied side-band disturbances which propagated obliquely to the primary Stokes waves [cf. Eqs. (3.1) and (3.4)]. Starting from Eqs. (2.36) and (2.40) directly and letting

$$A = (a_0 + A')\exp(-i\omega_0 a_0^2 \tau),$$

$$\phi_{10} = \mathcal{U}x_1 - gbt_1 + \phi'_{10},$$

with

$$\begin{pmatrix} A' \\ \phi'_{10} \end{pmatrix} \propto \exp(i(K_1 x_1 + K_2 y_1 - \Omega t_1)),$$

they showed that in the K_1, K_2 plane there were always regions in which the Stokes waves were unstable, except for $kh = 0.38$. Sample regions of instability are plotted in Fig. 4.3. Because of the difference in the shaded area, the likelihood of instability is greater for greater water depth. In nature, modulation in any direction is possible; therefore, Stokes' waves are always susceptible to instability.

The subject of wave instability in infinitely deep water is currently being advanced at a rapid pace. For disturbances colinear with the primary waves, Longuet-Higgins (1978a, b) started from the numerically exact solution for a steady progressive wave and superimposed perturbations both shorter (superharmonics) and longer (subharmonics) than the primary wave. In particular, the subharmonic disturbance is an extension of the Benjamin-Feir wave. In the first quadrant of the plane $K/2k$ (ordinate) versus ka_0 (abscissa), the zone of instability in the first quadrant was found to be a crest-like region, instead of the whole region to the right of the straight line $K/2k = 2^{1/2}ka_0$ according to Eq. (4.9), see Fig. 4.4. Thus for sufficiently high ka_0 the waves are restabilized. In addition Longuet-Higgins (1978b) found that infinitely long perturbations could induce very rapid instability at $ka_0 \cong 0.41$. (The maximum possible steepness is $ka_0 = 0.4434$).

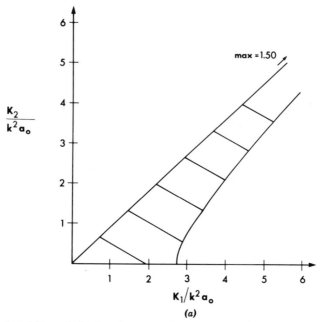

(a)

Figure 4.3   Instability of side-band waves which are inclined to the primary Stokes waves for various values of kh. Unstable regions are shaded. The maximum growth rate Im $\Omega/\omega k^2 a_0^2$ is also marked. (From Benney and Roskes, 1969, *Studies Appl. Math.* Reproduced by permission of the Editor.) (a) $kh = 2.00$; (b) $kh = 1.36$; (c) $kh = 0.7$; (d) $kh = 0.30$. At $kh = 0.38$ the instability region reduces to a straight line.

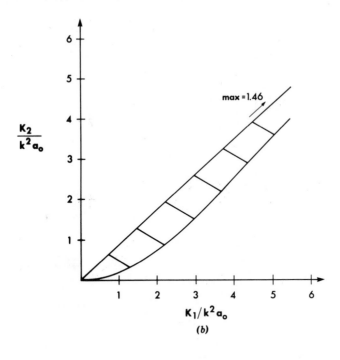

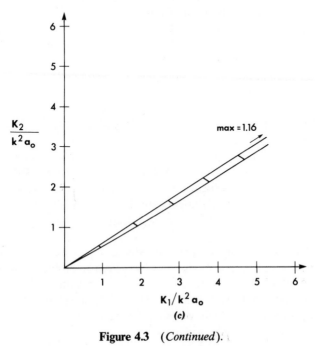

Figure 4.3 (*Continued*).

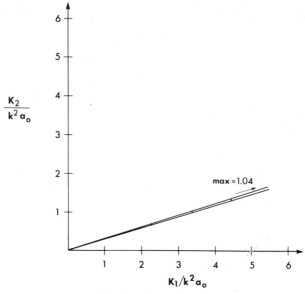

Figure 4.3   (*Continued*).

There are now two approximate theories which aim to improve the results of the cubic Schrödinger equation. One is due to Dysthe (1979) who retained the slow modulation assumption (Ω/ω, $K/k \ll 1$) but included nonlinear terms of the fourth order. His instability region was bounded only on the left by a curve which roughly coincided with the left boundary of the crest shown in Fig. 4.4. The predicted wavenumber of the most unstable side-band mode was, however, below Eq. (4.11), the exact numerical theory and the experimental values. The second approximate theory is due to Crawford et al. (1981) who applied an integro-differential equation of Zakharov (1968) which is valid to third order in wave slope but not restricted to side-band disturbances adjacent to the main waves (namely, K/k need not be small). Their instability region was also a finite crest whose left side agreed with Longuet-Higgins remarkably well, and the predicted growth rates were also closer to the measured values, as shown in Fig. 4.2. Since some of the experiments are clearly outside the realm of validity of both approximate theories, present agreement is certainly better than expected.

Instabilities due to oblique side-band disturbances have been studied by Crawford et al. (1981) using Zakharov's integro-differential equation, and by McLean et al. (1981) for very steep waves which are numerically calculated.

The subsequent evolution of initial instability must, of course, be treated as a nonlinear problem, to which we turn within the framework of the cubic Schrödinger equation (namely, small ka_0).

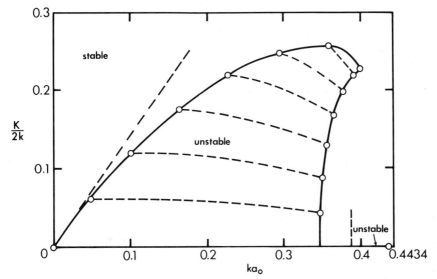

Figure 4.4 Stability diagram for numerically exact periodic nonlinear waves in deep water. The tangent at the origin corresponds to $K/2k = \sqrt{2}\,ka_0$ [cf. Eq. (4.9)]. (From Longuet-Higgins, 1978b, *Proc. R. Soc. Lond.* Reproduced by permission of the Royal Society of London.)

Exercise 4.1

Consider a progressive edge wave of the lowest mode on a plane beach of slope ε:

$$\phi = -\frac{igD}{2\Omega}\exp(-Kx_1)\exp i(Ky_1 - \Omega t_1) + * + \text{higher-order terms}$$

for $x_1 > 0$, where $x_1 = \varepsilon x$, $y_1 = \varepsilon y$, $t_1 = \varepsilon t$, and $\Omega^2 = gK$. Show by using Eq. (1.49), Chapter Eleven, and assuming $D = D(y_2, y_3, \ldots, t_2, t_3, \ldots)$ that the envelope evolves according to the following equations:

$$\left(\frac{\partial}{\partial t_2} + C_g\frac{\partial}{\partial y_2}\right)D = 0$$

and

$$i\left(\frac{\partial}{\partial t_3} + C_g\frac{\partial}{\partial y_3}\right)D - \frac{1}{2}\frac{\partial^2\Omega}{\partial K^2}\frac{\partial^2 D}{\partial y_2^2} - \frac{\Omega K^4}{4}|D|^2 D + \frac{\Omega}{4}D = 0$$

where $C_g = g/2\Omega$ is the group velocity of the edge wave along the shore and $-\partial^2\Omega/\partial K^2 = g^2/4\Omega^3$. Note that the envelope is unstable to side-band disturbances.

Exercise 4.2

Derive the instability criterion for obliquely incident side-band disturbances and confirm Fig. 4.3.

12.5 PERMANENT ENVELOPES IN DEEP WATER: NONLINEAR SOLUTIONS OF THE EVOLUTION EQUATION

Benney and Newell (1967) first showed that Eq. (2.45) admits permanent wave envelopes as solutions which are functions of $(x - Ut)$ (also see Chu and Mei, 1970; Hashimoto and Ono, 1972; Zakharov and Shabat, 1972; Scott, Chu, and McLaughlin, 1973, etc.). For illustration we consider the deep water solution by Scott, et al. Let us first nondimensionalize Eq. (2.60), with $\mathcal{U} = \partial/\partial y_1 = 0$, by letting

$$A' = \frac{A}{a_0}, \qquad \xi' = k^2 a_0 \xi, \qquad \tau' = \omega (k a_0)^2 \tau. \qquad (5.1)$$

With the primes omitted for brevity, the cubic Schrödinger equation becomes

$$-i \frac{\partial A}{\partial \tau} + \frac{1}{8} \frac{\partial^2 A}{\partial \xi^2} + \frac{1}{2} |A|^2 A = 0. \qquad (5.2)$$

We seek a solution of the form

$$A = a e^{ir(\xi - V\tau - \delta)} \qquad \text{where} \quad a = a(\xi - U\tau), \quad r, \delta = \text{const}, \qquad (5.3)$$

that is, a carrier wave with envelope a. The phase angle δ specifies the initial position. Substituting Eqs. (5.3) into (5.2) and denoting derivatives with respect to the argument by primes, we have

$$\frac{1}{8} a'' - r\left(\frac{1}{8}r + V\right)a + \frac{1}{2}a^3 + i\left(U + \frac{r}{4}\right)a' = 0. \qquad (5.4)$$

The imaginary part implies

$$r = -4U \qquad (5.5)$$

which may be used in the real part to yield

$$\frac{1}{8} a'' + 4U\left(V - \frac{U}{2}\right)a + \frac{1}{2}a^3 = 0. \qquad (5.6)$$

Multipling Eq. (5.6) by a' and integrating, we get

$$(a')^2 + 32U\left(V - \frac{U}{2}\right)a^2 + 2a^4 = C \qquad (5.7)$$

where C is a constant. Multiplying Eq. (5.7) by a^2 and defining

$$E = a^2,\tag{5.8}$$

we further get

$$(E')^2 + 64U(2V - U)E^2 + 8E^3 = 8CE,$$

which may be rewritten

$$(E')^2 = 8(E_{max} - E)(E - E_{min})E \equiv P(E).\tag{5.9}$$

Note that

$$8(E_{max} + E_{min}) = -64U(2V - U),\tag{5.10}$$

and, by definition (5.8), $E_{min} > 0$; hence, we must insist that $U(2V - U) < 0$. The cubic polynomial $P(E)$ is sketched in Fig. 5.1, and the corresponding $P(a^2)$ which is a sixth-degree polynomial is also sketched. A solution exists only if P is positive.

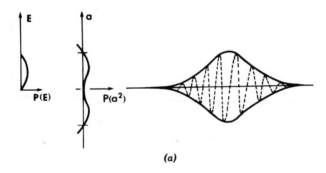

(a)

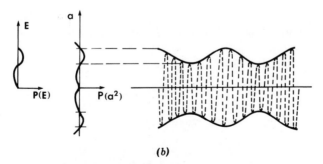

(b)

Figure 5.1   Cnoidal and solitary envelopes.

When $C = 0$, E_{min} vanishes, and Eq. (5.9) becomes

$$(E')^2 = 8E^2(E_{max} - E).$$ (5.11)

$P(E)$ has two zeroes as shown in Fig. 5.1a. Upon integration we obtain

$$E = E_{max} \operatorname{sech}^2\left[(2E_{max})^{1/2}(\xi - U\tau)\right]$$ (5.12)

or

$$a = a_{max} \operatorname{sech}\left[2^{1/2}a_{max}(\xi - U\tau)\right]$$ (5.13a)

which describes a solitary wave packet or an *envelope soliton*. Note that $a_{max} = E_{max}^{1/2}$, while U and V are related by Eq. (5.10) with $E_{min} = 0$. In particular, one may take $U = 0$, $VU = -1/16$ and $a_{max} = 1$ so that

$$a = \operatorname{sech} 2^{1/2}\xi, \qquad A = \left(\operatorname{sech} 2^{1/2}\xi\right)e^{-i(\tau/4 + \delta)}.$$ (5.13b)

More generally, $C \neq 0$ and $E_{min} > 0$ (as shown in Fig. 5.1b). The solution to Eq. (5.9) can be expressed in cosine-elliptic functions which have been discussed more fully in the shallow water theory,

$$E = E_{min} + \Delta E \operatorname{Cn}^2\{(2\Delta E)^{1/2}\gamma^{-1}(\xi - U\tau)\},$$ (5.14)

with

$$\Delta E = E_{max} - E_{min} = \text{energy amplitude},$$

$$\gamma = \left(\frac{\Delta E}{E_{max}}\right)^{1/2}.$$

The wavelength of the envelope is

$$\lambda = \frac{2^{1/2}}{(\Delta E)^{1/2}} \int_0^{\pi/2} \frac{du}{(1 - \gamma^2 \sin^2 u)^{1/2}}$$ (5.15)

(see, e. g., Chu and Mei, 1970). Again ΔE, U, and V can be specified, subject to Eq. (5.10).

For finite depth, Eq. (2.45) may be rescaled by

$$\xi = \alpha\bar{\xi} \quad \text{and} \quad \tau = \alpha\bar{\tau}$$

to give the following canonical equation:

$$-i\frac{\partial B}{\partial\bar{\tau}} + \frac{\partial^2 B}{\partial\bar{\xi}^2} + \kappa|B|^2 B = 0,$$ (5.16)

where $\kappa = \alpha\beta$. For $\kappa > 0$ (i.e., $kh > 1.36$) the extension to Eq. (5.13a) is easily verified to be

$$B = a \exp\left[-i\frac{U}{2}(\bar{\xi} - V\tau - \bar{\xi}_1)\right], \qquad (5.17)$$

where

$$a = a_{max} \operatorname{sech}\left[\left(\frac{\kappa}{2}\right)^{1/2} a_{max}(\bar{\xi} - \bar{\xi}_0 - U\bar{\tau})\right], \qquad (5.18)$$

with

$$a_{max} = \frac{2}{\kappa} U\left(\frac{U}{8} - V\right). \qquad (5.19)$$

The soliton is characterized by the free parameters a_{max} and U and by the phases $\bar{\xi}_0$ and $\bar{\xi}_1$.

For $\alpha\beta < 0$, a soliton exists which is a depression at the center and approaches a finite constant at $|\xi| \sim \infty$. The expression is

$$B = \left(\frac{2}{|\kappa|}\right)^{1/2} \frac{(\lambda - i\nu)^2 + e^{2\nu(\bar{\xi} - \bar{\xi}_0 - 2\lambda\bar{\tau})}}{1 + e^{2\nu(\bar{\xi} - \bar{\xi}_0 - 2\lambda\bar{\tau})}}, \qquad (5.20)$$

where

$$\nu = (1 - \lambda^2)^{1/2} \qquad (5.21)$$

(Zakharov and Shabat, 1973). A little manipulation shows that

$$|B| = \left(\frac{2}{|\kappa|}\right)^{1/2}\left[1 - \nu \operatorname{sech}^2\nu(\bar{\xi} - \bar{\xi}_0 - 2\lambda\bar{\tau})\right]^{1/2}. \qquad (5.22)$$

Thus, the soliton is characterized by ξ_0 and the amplitude ν. A slightly more general solution was deduced independently by Hesegawa and Tappert (1973). These solutions are known either as the *concave E-soliton* or the *envelope-hole soliton*. Note that as ν increases toward 1, λ decreases so that a deeper hole travels more slowly. In the limit of $\nu \to 1$, $\lambda \to 0$ and

$$B \to \left(\frac{2}{|\kappa|}\right)^{1/2} \tanh(\bar{\xi} - \bar{\xi}_0) \qquad (5.23)$$

which is called the *phase jump* because of the sign change from -1 at $\bar{\xi} \sim -\infty$ to $+1$ at $\bar{\xi} \sim \infty$.

Periodic permanent envelopes for both $\alpha\beta \gtrless 0$ have been given by Hashimoto and Ono (1972).

12.6 TRANSIENT EVOLUTION OF ONE-DIMENSIONAL WAVE ENVELOPE ON DEEP WATER

As the linearized instability theory must fail beyond the initial stage of exponential growth, the fuller nonlinear theory must be employed for the evolution over long periods of time. Early experiments by Feir (1967) showed that the envelope of a wave packet tended to break up into several groups strung together. Numerical solution of Eq. (2.47) by Chu and Mei (1971) confirmed this trend. Both these attempts, however, did not cover a long enough time interval to reveal the entire physical picture. A theoretical breakthrough was scored by Zakharov and Shabat (1972) who extended the technique of Gardner *et al.* (1967) and solved Eq. (5.16) exactly for both κ $(= \alpha\beta) \gtrless 0$ when the initial data $B(\xi, 0)$ vanished sufficiently fast as $|\bar{\xi}| \uparrow \infty$. Their success is based on an important observation by Lax (1968): To solve any nonlinear evolution equation

$$\frac{\partial B}{\partial \bar{\tau}} = NB \tag{6.1}$$

where N is a nonlinear operator, one must first seek two linear operators L and M, both involving only derivatives with respect to $\bar{\xi}$ and containing $B(\bar{\xi}, \bar{\tau})$ in the coefficients, such that

$$L_{\bar{\tau}} = LM - ML, \tag{6.2}$$

and

$$i\psi_{\bar{\tau}} = M\psi. \tag{6.3}$$

If the operators are found, the eigenvalues σ of the problem

$$L\psi = \sigma\psi \tag{6.4}$$

will remain constant for all time $(\bar{\tau})$; this can be verified by differentiating Eq. (6.4) with respect to $\bar{\tau}$,

$$\sigma_{\bar{\tau}}\psi = L_{\bar{\tau}}\psi + (L - \sigma)\psi_{\bar{\tau}} = LM\psi - ML\psi + (L - \sigma)\psi_{\bar{\tau}}$$

$$= (L - \sigma)(i\psi_{\bar{\tau}} - M\psi) = 0$$

where Eqs. (6.2)–(6.4) have been used. Thanks to this invariance, the initial data $B(\bar{\xi}, 0)$ alone enables one to solve Eq. (6.4) for the spectrum of σ and also $\psi(x, 0)$. For localized initial data, we expect $B(\bar{\xi}, \bar{\tau})$ to be localized in $\bar{\xi}$ too; the operator M can be simplified so that Eq. (6.3) may be easily solved for $\psi(\pm\infty, \bar{\tau})$, based on the knowledge of $\psi(\pm\infty, 0)$. Now the eigenvalues σ and the far field $\psi(\pm\infty, \bar{\tau})$ are known for all $\bar{\tau}$; the *inverse scattering theory* may be called upon to solve $B(\bar{\xi}, \bar{\tau})$ for all $\bar{\xi}$, with $\bar{\tau}$ being a parameter. The mathematical background of the last step is unfortunately very complicated, leaning heavily on the complex function theory.

For the cubic Schrödinger equation (5.16), Zakharov and Shabat found that the eigenfunction ψ is a two-component vector $\psi = (\psi_1, \psi_2)$ and that L and M are 2×2 matrices:

$$L = i \begin{bmatrix} 1 + p & 0 \\ 0 & 1 - p \end{bmatrix} \frac{\partial}{\partial \bar{\xi}} + \begin{bmatrix} 0 & B \\ B^* & 0 \end{bmatrix} \quad \text{with} \quad \kappa = \frac{2}{1 - p^2} \quad (6.5a)$$

$$M = -p \begin{bmatrix} 1 & 0 \\ 0 & 1 \end{bmatrix} \frac{\partial^2}{\partial \bar{\xi}^2} + \begin{bmatrix} \dfrac{|B|^2}{1 + p} & iB_{\bar{\xi}} \\ -iB^*_{\bar{\xi}} & \dfrac{-|B|^2}{1 - p} \end{bmatrix}. \quad (6.5b)$$

In general, the search for these operators is no trivial task, but a systematic approach for a certain class of problems has now been found by Ablowitz et al. (1974).

The theory of Zakharov and Shabat has since been elaborated for detailed implications by Satsuma and Yajima (1975)[†] and confirmed numerically and experimentally by Yuen and Lake (1975). The main physical features of these works for $\kappa > 0$ are summarized below.

1 An arbitrarily shaped envelope will eventually evolve into a finite number of solitons, plus minor oscillations which decay as $t^{-1/2}$.

2 The jth soliton is given by

$$B_j(\bar{\xi}, \bar{\tau}) = 2 \left(\frac{2}{\kappa} \right)^{1/2} b_j \left\{ \operatorname{sech} 2b_j \left(\bar{\xi} - \bar{\xi}_j - 4a_j \bar{\tau} \right) \right\}$$

$$\times \left\{ \exp \left[4i \left(a_j^2 - b_j^2 \right) \bar{\tau} - 2ia_j \bar{\xi} + i\phi_j \right] \right\}, \quad (6.6)$$

where a_j and b_j are the real and imaginary parts of the discrete eigenvalue σ_j of Eq. (6.4). The soliton has the amplitude $2(2/\kappa)^{1/2} b_j$, phase speed $4a_j$, and phases $\bar{\xi}_j$ and ϕ_j which are related to the reflection coefficient of the scattering problem (the continuous eigenvalue spectrum). In general, these complex eigenvalues σ_j may have distinct real parts; thus, the solitons may drift apart with time. A special case is the so-called N-soliton solution which corresponds to N distinct eigenvalues $\sigma_j, j = 1, \ldots, N$. If there are N solitons at $t = 0$ with the slowest one in front, then the faster solitons will overtake the slower ones at sufficiently

---

[†] I thank Dr. Harvey Segur for this reference.

large $\bar{\tau}$. The only effect of collision between a pair of solitons is a shift of phases $\bar{\xi}_i$ and ϕ_i.

3 When the initial data $B(\bar{\xi}, 0)$ is real and not antisymmetric in $\bar{\xi}$, the discrete eigenvalues are purely imaginary. In this case all solitons stay together and form *bound solitons*. As a slight extension, if the initial

$$B(\bar{\xi}, 0) = \bar{B}(\bar{\xi}) e^{iV\xi}$$

where V is a real constant and $\bar{B}(\bar{\xi})$ is real and nonantisymmetric in $\bar{\xi}$, then all solitons have the same speed $-2V$ (relative to the group velocity in the rest frame). Because each soliton oscillates at the period $4b_j^2$, the composite envelope exhibits recurrence with frequencies $4(b_i^2 - b_j^2)$. The number of recurrence frequencies is equal to the number of distinct differences between b_i^2 and b_j^2. For example, if N is the number of solitons, then for $N = 2$ there is only one recurrence frequency; for $N = 3$, there are two recurrence frequencies,..., and so on.

For the special case of

$$B(\bar{\xi}, 0) = B_0 \operatorname{sech} \frac{\bar{\xi}}{\lambda}, \qquad B_0 \text{ real}, \tag{6.7}$$

Eq. (6.4) with Eq. (6.5a) can be reduced to a hypergeometric equation for ψ; the discrete eigenvalues and the scattering coefficients for the continuous eigenvalues can be obtained explicitly (Satsuma and Yajima, 1975; Kuehl, 1976). In particular, the discrete eigenvalues are

$$\sigma_j = i\left(\left(\frac{\kappa}{2}\right)^{1/2} B_0 \bar{\lambda} - j + \frac{1}{2}\right), \qquad j = 1, 2, 3, \dots. \tag{6.8}$$

Thus, for N solitons to emerge, the initial size $B_0 \bar{\lambda}$ must be such that

$$\left(\frac{\kappa}{2}\right)^{1/2} B_0 \bar{\lambda} - N + \frac{1}{2} > 0. \tag{6.9}$$

Moreover, when

$$\left(\frac{\kappa}{2}\right)^{1/2} B_0 \bar{\lambda} = N, \qquad N = \text{positive integer}, \tag{6.10}$$

the inverse scattering problem can be solved for $B(\bar{\xi}, \bar{\tau})$ explicitly; the result only involves N bound solitons.

To apply the above criteria to Eq. (5.2) with the initial data $A(\xi, 0) = A_0 \operatorname{sech} \xi/\lambda$, the following changes are necessary:

$$B_0 \to A_0, \qquad \bar{\lambda} \to \frac{\lambda}{\alpha}, \qquad \kappa \to \alpha\beta.$$

Criteria (6.9) and (6.10) then become

$$\left(\frac{\beta}{2\alpha}\right)^{1/2} A_0\lambda > N - \frac{1}{2} \qquad \text{for } N \text{ solitons with tail} \qquad (6.11)$$

and

$$\left(\frac{\beta}{2\alpha}\right)^{1/2} A_0\lambda = N \qquad \text{for } N \text{ solitons only.} \qquad (6.12)$$

For deep water $\alpha = \frac{1}{8}$, $\beta = \frac{1}{2}$, the left side of Eqs. (6.11) and (6.12) becomes $2^{1/2}A_0\lambda$.

To confirm their analytical results, Satsuma and Yajima (1975) also computed numerically, from Eq. (5.2), a number of cases including solitons either in a bounded state or emerging with different speeds. There are now several numerical techniques (Lake, Yuan, Rungaldier, and Ferguson, 1977; Fornberg and Whitham, 1979) which are relatively straightforward to execute. They can be readily applied to, or modified for, cases yet unsolvable by the inverse scattering theory, such as semi-infinite domains, periodic boundary conditions, two space dimensions, variable coefficients for inhomogenous media, and so on. We cite below a simple implicit scheme of the Crank–Nicolsen type, which was applied by Yue (1980) to two of the three cases to be discussed. Referring to Eq. (5.2), let $\Delta\xi$ and $\Delta\tau$ be the discrete intervals on the ξ and τ axes and denote

$$A_j^n = A(j\Delta\xi, n\Delta\tau).$$

Equation (5.2) is approximated by

$$A_j^{n+1} = A_j^n - \frac{\Delta\tau}{2}\left\{\frac{i}{8(\Delta\xi)^2}\left(A_{j+1}^{n+1} - 2A_j^{n+1} + A_{j-1}^{n+1}\right)\right.$$

$$+ \frac{i}{2}|\tilde{A}_j^{n+1}|^2 A_j^{n+1} + \frac{i}{8(\Delta\xi)^2}\left(A_{j+1}^n - 2A_j^n + A_{j-1}^n\right)$$

$$\left. + \frac{i}{2}|\tilde{A}_j^n|^2 A_j^n\right\} + O(\Delta\tau^3, \Delta\xi^2). \qquad (6.13)$$

In the nonlinear terms, $\tilde{A}_j^{n+1}$ is estimated by the results at the nth time step according to an Euler scheme:

$$\tilde{A}_j^{n+1} = A_j^n - \Delta\tau\left\{\frac{i}{8(\Delta\xi)^2}\left(A_{j+1}^n - 2A_j^n + A_{j-1}^n\right)\right.$$

$$\left. + \frac{i}{2}|A_j^n|^2 A_j^n\right\} + O(\Delta\tau^2, \Delta\xi^2). \qquad (6.14)$$

As examples, numerical solutions for the following types of initial data will be discussed for physical pictures:

1 A single pulse.
2 A step envelope.
3 Periodic modulation of a uniform wavetrain.

In cases 1 and 2, the initial data are not localized, and analytical solutions are not yet possible.

12.6.1 Evolution of a Single Pulse

For convenience, we express the initial data by

$$A(\xi,0) = \text{sech}\left(\frac{2^{1/2}\xi}{\lambda}\right). \tag{6.15}$$

Recall from Eq. (5.13b) and $A(\xi,0)$ coincides with a soliton of unit height if $\lambda = 1$.

First, we take $\lambda = 2$, corresponding to an initial pulse twice as long as a soliton of equal height. It follows from Eq. (6.12) that $N = 2$ so that two and only two solitons should emerge. For this case, the explicit solution has been given by Satsuma and Yajima (1975, p. 300 after the transformation $u \to NA$, $x \to 2^{1/2}\xi/N$, and $t \to \tau/2N^2$)

$$A(\xi,\tau) = \frac{2\left[\text{ch}(3\xi/2^{1/2}) + 3e^{-i\tau/2}\text{ch}(\xi/2^{1/2})\right]}{\text{ch}\,2(2)^{1/2}\xi + 4\,\text{ch}\,2^{1/2}\xi + 3\cos\frac{1}{2}\tau}\exp\left(-\frac{i\tau}{16}\right). \tag{6.16}$$

Numerical confirmation is shown in Fig. 6.1a. From the time history of the pulse center $|A(0,\tau)|$, shown in Fig. 6.1b, the recurrence period is seen to be $\tau_0 = 12.6$, while the theoretical value is $2\pi/\frac{1}{2} = 12.566$. Typical snapshots $|A(\xi,\tau_i)|$ are given in Fig. 6.1c at quarter-period intervals to show the stages of evolution. Figure 6.1d shows the phase p defined by $A(\xi,\tau) = |A|e^{ip\pi}$ at these same instants. Note first that the phase curves for two instants separated by a recurrence period differ only by a constant. Thus, the phase variation, that is, the wavenumber, is unchanged. Also note that at the instant c_0 (or $c_1,\dots$) when the envelope has a node, the phase changes sharply, implying that the frequency distribution has a sharp trough in the front half of the group. By symmetry, there is a peak in the frequency distribution in the rear half of the group. These predictions (Chu and Mei, 1971) are in qualitative agreement with Feir's experiments. Due to the long time scales, it is difficult to check recurrence quantitatively in a laboratory unless damping is taken into account.

Let us take $\lambda = 3$ next. The corresponding N is 3 according to Eq. (6.12). The overall view is given by Fig. 6.2a. In Fig. 6.2b the history of the pulse center shows distinctly the existence of more than one recurrence period.

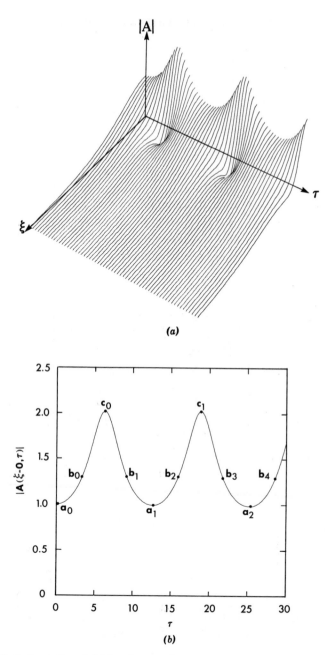

(a)

(b)

Figure 6.1 Evolution of an initial pulse envelope twice the length of the soliton of equal height $A(\xi, 0) = \mathrm{sech}(\xi/2^{1/2})$. (From Yue, 1980.) (a) The overall view for $0 < \tau < 30$, $0 < \xi < 6$. (b) Center line amplitude $|A(\xi = 0, \tau)|$ as a function of slow time τ. The recurrence period is $\tau_0 = 12.6$. The marked points a_i's and c_i's are at τ_0 apart, while b_i's are at $\frac{1}{2}\tau_0$ apart. (c) Snapshots of $|A(\xi, \tau)|$ at instants marked on (a). (d) Snapshots of the phase factor p. ($p\pi$ = phase of A) at the instants marked on (a).

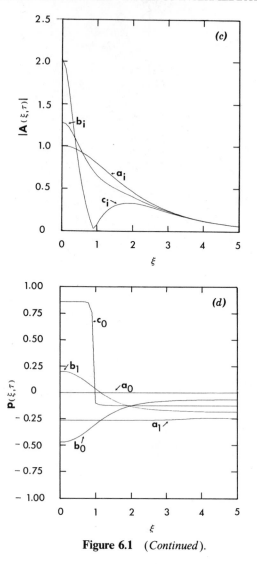

Figure 6.1   (*Continued*).

Figure 6.2*c* shows the typical profile at various stages. Again the nodes of the envelope are accompanied by rapid changes of phases which are not shown here.

Yue (1980) also computed for two initial pulses with the same carrier frequency and the same shape, but which were well separated,

$$A(\xi, 0) = \text{sech}\left[2\sqrt{2}\,(\xi - \xi_0)\right] + \text{sech}\left[2\sqrt{2}\,(\xi + \xi_0)\right].$$

In particular, each pulse was half as wide as a soliton of unit height but the

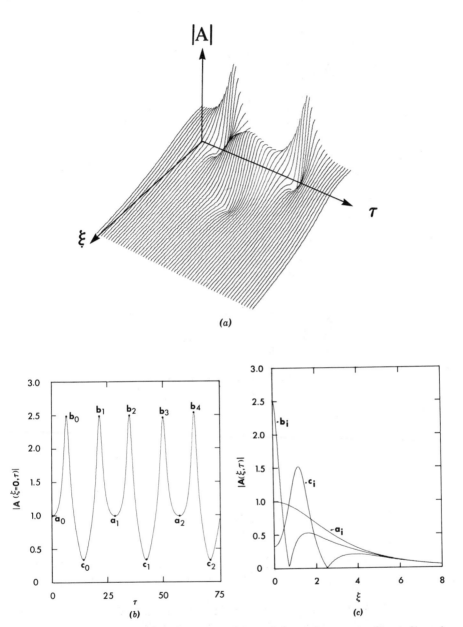

(a)

(b)   *(c)*

Figure 6.2 Evolution of an initial pulse envelope thrice as long as a soliton of equal height $A(\xi, 0) = \mathrm{sech}(2^{1/2}\xi/3)$. (From Yue, 1980.) (*a*) Overall view for $0 < \xi < 6$, $0 < \tau < 30$. (*b*) Center line amplitude $|A(\xi = 0, \tau)|$. (*c*) Snapshots of the envelope amplitude at sample instants of a recurrence period as marked in (b). (*d*) Snapshots of envelope phase at sample instants of a recurrence period.

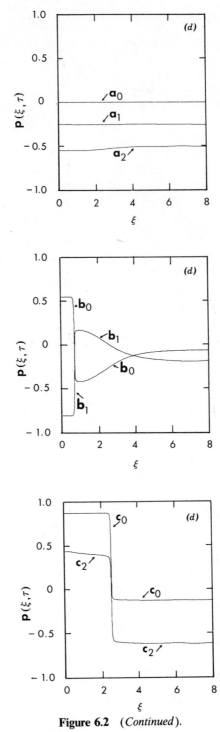

Figure 6.2 (*Continued*).

combined profile had the same area as a single soliton of unit height. His results showed that after a complex evolution the two pulses coalesced into a single soliton of height 0.4.

More generally, two initial pulses with complex amplitudes may lead to solitons traveling at different speeds (see Satsuma and Yajima for a variety of numerical examples).

12.6.2 Evolution of the Front of a Uniform Wavetrain

In Section 2.4 the front of a suddenly started sinusoidal wavetrain was found to be governed by a linear Schrödinger equation. As the envelope advanced, undulations developed behind the front in a way describable by Fresnel integrals. We now reexamine the problem by including nonlinearity. The task is to solve Eq. (5.2) with the following initial data:

$$A(\xi) = \frac{1}{2}\left[1 + \tanh\frac{\xi}{\lambda}\right]. \tag{6.17}$$

Far ahead of the front there is no disturbance,

$$A \to 0, \quad \xi \to \infty. \tag{6.18}$$

Far behind the front, a uniform Stokes wave should be approached,

$$A \sim e^{-i\tau/2}, \quad \xi \to \infty. \tag{6.19}$$

Note from the definitions of ξ and τ by Eq. (5.1), with primes omitted but implied, that a smaller $\varepsilon = ka_0$ corresponds to greater physical distance and longer time.

Figure 6.3 shows the contrast between linear and nonlinear theories for an abrupt transition $\lambda \approx 0$. The effect of finite amplitudes dramatically changes the front of the envelope, creating much stronger undulations. There are yet no appropriate experiments with which the nonlinear inviscid theory can be compared. Yue (1980) examined other values of λ but found no qualitatively new features.

12.6.3 Periodic Modulation of a Uniform Wavetrain—Evolution Beyond the Initial Stage of Instability

After the initial period of instability, the side-band disturbance grows so large that the full nonlinear effect of Eq. (5.2) becomes important. Chu and Mei (1971) calculated the nonlinear evolution for the most unstable case $K/k = 2ka_0$ by using Eqs. (2.47a) and (2.47b), but could not carry the computations beyond the first occurrence of an envelope node ($a \approx 0$). Based on Eq. (5.2) with the requirement of periodicity over one modulational period, more extensive calculations have been accomplished by Yuen and Ferguson

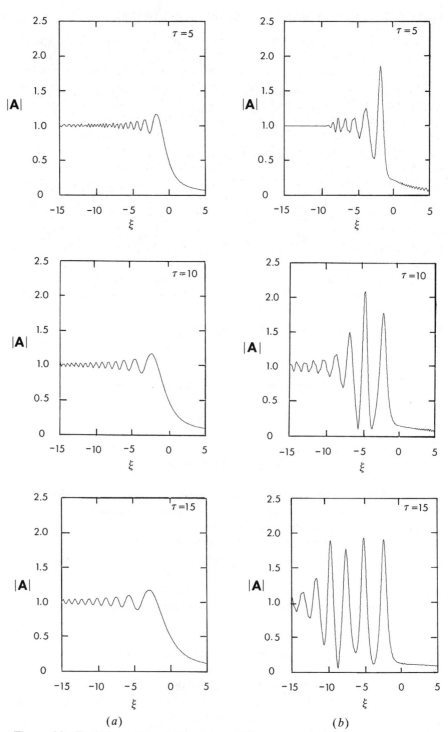

Figure 6.3 Evolution of the front of a periodic wave train. (From Yue, 1980.) (*a*) Linear theory; (*b*) nonlinear theory

(1978a, b). They found recurrence to be the dominant feature so that the envelope with a modulational wavelength evolved periodically from one crest to several and back to one again. A sample of experimental evidence is shown in Fig. 6.4. Correspondingly, energy in the lower harmonics spread to higher harmonics and then returned to the lower harmonics, and so on. The intermediate stage became progressively more complex in appearance if K/k was reduced. By varying K/k, Yuen and Ferguson found a systematic pattern for this complex variation. For $2ka_0 < K/k < 2(2)^{1/2}ka_0$, only the first harmonic in the Stokes wavetrain was unstable to the disturbance, and the evolution was the simplest. However, when K/k fell slightly below $2ka_0$, both the first and second harmonics were unstable; the evolution became more involved. Figure 6.5a shows the wavelengths and growth rates of several initial modulational disturbances, all of the form $a = a_0(1 - 0.1\cos K\xi)$ with the appropriate phase in accordance with Eqs. (4.5) and (4.6); corresponding evolutions are depicted in Fig. 6.5b.

An approximate analytical theory has been worked out also by Stiassnie and Kroszynski (1982) by ignoring the higher harmonics of the disturbance, that is, by assuming

$$B = C_0 + C_{-1}e^{-2i\pi\bar{\xi}} + C_1 e^{2i\pi\bar{\xi}}. \tag{6.20}$$

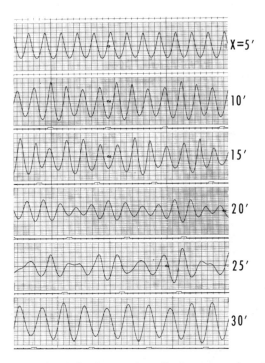

Figure 6.4 Sample records of the free surface displacement of an initially uniform wave train. Initial wave frequency is 3.6 Hz. (From Lake et al., 1977, *J. Fluid Mech.* Reproduced by permission of Cambridge University Press.)

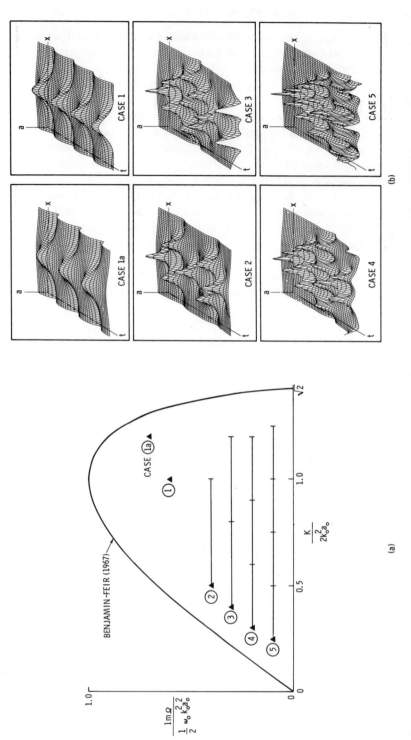

(b)

(a)

Figure 6.5 Relationship between initial conditions and long-time evolution of solutions of the nonlinear Schrödinger equation. (*a*) The perturbation wavenumber K of the various cases, and (*b*) their corresponding time evolution. The initial conditions consist of a uniform wavetrain with wavenumber k_0 and amplitude a_0 subject to a 1.0% perturbation. The numerals circled in (*a*) identify the cases in (*b*), and also correspond to the number of harmonics of the perturbation (including the primary) that lie within the unstable regime according to the stability analysis. Note that the number of unstable harmonics corresponds exactly to the number of modes that dominate the evolution. For example, Case 4 shows an evolution in which the 1st, 2nd, 3rd, 4th, 1st, and 2nd harmonics, in that order, took turns dominating the evolution as indicated by the number of peaks at various stages of evolution in the amplitude plot. (From Yuen and

where $C_0(\bar{\tau})$ corresponds to the carrier wave and $C_{\pm 1}(\bar{\tau})$ correspond to the side-band disturbances. As in the case of harmonic generation in shallow-water waves (cf. Section 11.11), we can substitute Eq. (6.20) into Eq. (5.16) to obtain two coupled nonlinear ordinary differential equations for C_0 and C_1, subject to the initial conditions $C_0(0) = 1$ and $C_{\pm 1}(0) = \delta e^{i\theta}$, C_{-1} being equal to C_1. These two equations may be solved explicitly in terms of elliptic integrals. When the coefficient $\kappa = \alpha\beta$ in Eq. (5.16) is such as to render the first harmonic unstable, $C_{\pm 1}$ oscillate at an amplitude comparable to C_0. The period of modulational recurrence so obtained agrees well with the numerically computed result by adding a large number of higher harmonics to Eq. (6.20). The approximate amplitude is, however, less satisfactory.

The regularity in the long-time evolution of Stokes' wave suggests that for waves that are not too steep, nonlinearity tends to bring coherence rather than chaos. This fact may have important ramifications on the development of the wind-wave spectrum (Lake and Yuen, 1978; Mollo-Christensen and Ramamonjiarisoa, 1978). For still steeper waves, Lake et al. (1977) observed experimentally that the carrier frequency tended to reduce with propagation distance (compare the records at 5 ft. and 30 ft. in Fig. 6.4), and that the reduction was associated with the faster growth of the lower side-band [C_{-1} in Eq. (6.20)]. By experimenting with a wide range of ka, Melville (1982) found for $ka_0 < 0.29$ and for spontaneous disturbances, that the lower side-band also grew faster than the upper side-band [C_1 in Eq. (6.20)]. In addition, breaking was prevalent for $0.21 < ka_0 < 0.29$ which partially eroded the phenomenon of recurrence. Further down the wave channel both side-bands fell to local minima while breaking ceased, and then rose again. He also found breaking to be accompanied by an amplitude reduction of the upper side-band relative to the lower side-band, thereby enhancing the frequency downshift. Beyond $ka_0 > 0.3$ three-dimensional instability dominated. It is well known that the peak of a typical wind-wave spectrum shifts to lower frequencies with increasing fetch (see Hasselman et al., 1973, Fig. 2.5). Therefore, the cited experiments suggest that this downward shift may be contributed partly by nonlinearity and not by wind alone. So far this feature has not been found from the solutions of the cubic Schrödinger equation, which predicts that opposite side-bands with equal initial amplitudes remain equal throughout cycles of rise and fall (Yuen and Ferguson, 1978a).

The important problem of nonlinear effects on the wave-spectrum development is now undergoing intensive study. Further progress in this direction deserves close attention by oceanographic engineers.

12.7   REMARKS ON VARIABLE DEPTH

For coastal interest the evolution of short waves (swell) on variable depth is of obvious importance. To avoid analytical difficulties in the general case, existing theories are all limited to slowly varying one-dimensional topography

$[h = h(x)]$. As suggested by Eq. (2.39) where x_2 and t_2 appear, bottom variation becomes as important as nonlinearity if its horizontal length scale is no greater than $O(\varepsilon^{-2})$ times the swell length, that is, $h = h(x_2)$. This class of bottom has been studied by Djordjevic and Redekopp (1978) who extended the work of Ono (1974) for slowly modulated periodic waves on very shallow water (describable by the KdV equation). On the other hand, when the horizontal scale of depth variation is only $2\pi(k\varepsilon)^{-1}$ (i.e., $h = h(x_1)$), the required analysis is dauntingly complex. Nevertheless, there are interesting phenomena in nature for which such an effort is worthwhile (Foda and Mei, 1981). We briefly comment on these works in order that the reader may pursue them further.

12.7.1 Very Mild Slope: $h = h(x_2) = h(\varepsilon^2 x)$

Djordjevic and Redekopp (1978) used the perturbation method described in Section 12.2 with the new boundary condition on the bottom:

$$\frac{\partial \Phi}{\partial z} = \varepsilon^2 \frac{dh}{dx_2} \frac{\partial \Phi}{\partial x}. \tag{7.1}$$

Instead of Eq. (2.42) they introduced

$$\xi = \frac{1}{\varepsilon} \int^X \frac{dX}{C_g(X)} - \varepsilon t, \qquad X = \varepsilon^2 x \tag{7.2}$$

[cf. Eq. (8.3), Section 11.8], and obtained for the envelope

$$-i\mu A - iA_x + \alpha' A_{\xi\xi} + \beta' |A|^2 A + \gamma' A = 0, \tag{7.3}$$

which is similar to Eq. (2.43a), where

$$\alpha' = \frac{\alpha}{C_g^3}, \qquad \beta' = \frac{\beta}{C_g}, \qquad \gamma' = \frac{\gamma}{C_g}, \qquad \gamma = \gamma(X) \tag{7.4}$$

with α, β, and γ given by Eqs. (2.43b)–(2.43d), γ being a function of X through the integration constant $S(X)$ in Eq. (2.43d). The first term in Eq. (7.3) is new and depends on the local slope

$$\mu = C_g \frac{dq}{dX} \frac{(1 - \text{th}^2 q)(1 - q\,\text{th}\,q)}{\text{th}\,q + q(1 - \text{th}^2 q)}, \qquad q = kh. \tag{7.5}$$

Since $X = O(\varepsilon)$ for $\xi = O(1)$ according to definition (7.2), Eq. (7.3) is a suitable basis for numerical solution of the initial-value problem for a given $A(\xi, X = 0)$. Introducing further approximations to Eq. (7.3), Djordjevic and Redekopp examined analytically the evolution of a soliton for both $\alpha'\beta' < 0$

and > 0. In the former case, it was necessary to limit the bottom slope still further to $O(\delta^{3/2}\varepsilon^2)$ so that the new coordinates

$$\tilde{X} = \delta^{3/2}X, \qquad \tilde{\xi} = \delta^{-1}\int^{\tilde{X}}\frac{d\tilde{X}}{C(\tilde{X})} - \delta^{1/2}\xi$$

could be introduced. A perturbation analysis was then pursued. For very shallow water $kh \ll 1$, a KdV equation with variable coefficients similar to Eq. (8.5) in Section 11.8 was obtained. Following the heuristic argument of Section 11.8, they studied the evolution of an envelope-hole soliton climbing a shelf. If the depths on two sides of the sloping regions were h_0 and h_1 with $h_0 > h_1$, fission into n envelope-hole solitons was found to occur when

$$\frac{h_1}{h_0} > \left[\frac{n(n+1)}{2}\right]^{-8/27}, \qquad h = 1, 2, 3.$$

For the case $\alpha'\beta' > 0$ everywhere, Djordjevic and Redekopp applied the method of Section 11.8 to Eq. (7.3) directly. One of their conclusions was that fission of convex solitons was possible only if the propagation was from shallower to deeper water. Numerical and experimental confirmations of these results are desirable.

The additional effects of a nonuniform current in the x direction has been studied by Turpin (1981).

12.7.2 Mild Slope: $h = h(x_1) = h(\varepsilon x)$

For a bottom slope comparable to the wave slope, Chu and Mei (1970) gave the evolution equations of short swell in conservation forms. It can be observed from their equations that second-order long waves which are natural modes trapped on a submarine ridge can be resonated by short swell through nonlinear forcing. Hints of this occurrence are already present in the theory for constant depth, as the D'Alembertian operator appears on the long scale potential ϕ_{10} in Eq. (2.36).

Munk (1949b) and Tucker (1950) were the first to record long-period *surf beats* of 1–5 min periods on natural beaches. They also found that the evolution history of the long-wave amplitude was strongly correlated to the envelope of incident short swell. Gallagher (1971) proposed that surf beats were edge waves resonated at the second order by incoming swell with a narrow-banded spectrum. The narrow bandwidth implied slow but periodic modulation in the swell envelope. If the modulation also had a spatially periodic variation along the shore in such a way that the modulation frequency and wavelength satisfied the dispersion relation of an edge wave mode (cf. Section 4.8), the trapped edge wave grew in time. Beside invoking empirical bottom friction to limit the resonant growth, Gallagher made some oversimplifications with regard to the shoaling and breaking waves. Later, King and

Smith (1978) ignored the surf zone and proposed a side-band instability mechanism; they also found it necessary to include an empirical damping term. On the other hand, more extensive field studies have been undertaken along Torrey Pines Beach, California. In particular, Huntley, Guza and Thornton (1981) established the existence of surf beats in the form of progressive edge waves. Furthermore, Guza and Thornton (1981) reported that the shore-line amplitude of the surf beat could be as high as 70% of the incident swell.

To avoid the uncertainties in bottom friction and in wave breaking, Foda and Mei (1981) developed a higher-order theory for the excitation of long waves trapped on a submerged ridge. Because of the possibility of resonance, the long waves cannot be expected to remain always at the second order but should be allowed to become comparable to the short waves; this is consistent with the field observations of Guza and Thornton. The perturbation equations can be deduced as in Section 12.2 except that the zero harmonic which represents long waves has a first-order term $\varepsilon\eta_{10}$, that is,

$$\zeta = \varepsilon\left(\eta_{10} + \eta_{11}e^{i\psi} + *\right) + \varepsilon^2\left(\eta_{20} + \eta_{21}e^{i\psi} + \eta_{22}e^{2i\psi} + *\right) + \cdots \quad (7.6)$$

where η_{10} and η_{20} depend on slow coordinates only. Because of the Bernoulli equation $g\zeta + \partial\Phi/\partial t + \cdots = 0$, the largest zeroth harmonic potential must be of $O(\varepsilon^0)$, that is,

$$\Phi = \phi_{00} + \varepsilon\left(\phi_{10} + \phi_{11}e^{i\psi} + *\right) + \varepsilon^2\left(\phi_{20} + \phi_{21}e^{i\psi} + \phi_{22}e^{2i\psi} + *\right) + \cdots.$$
$$(7.7)$$

In addition, the bottom condition now reads

$$\frac{\partial\Phi}{\partial z} = \varepsilon\frac{dh}{dx_1}\frac{\partial\Phi}{\partial x}, \qquad z = -h(x_1). \quad (7.8)$$

The extremely lengthy perturbation analysis combines all the essential steps in Sections 12.2 and 11.4 for subharmonic resonance. We only describe the main results here.

Far from the ridge $x_1 \sim -\infty$, an incident swell arrives with first-order amplitude A^0 and second-order modulation of amplitude b^0. To the second order, the swell amplitude is

$$\left\{\varepsilon A^0 + \varepsilon^2 b^0\cos Ky_1\exp\left[i\left(\frac{\Omega}{C_g^0}x_1 - \Omega t_1\right)\right]\right\}e^{i(k^0x - \omega t)} + *, \qquad x_1 \sim -\infty$$
$$(7.9)$$

where $\omega = $ const and k^0 is the swell wavenumber at $x_1 \sim -\infty$. A^0 and b^0 can be functions of $t_2 = \varepsilon^2 t$, $t_3 = \varepsilon^3 t, \ldots$, and so on.[†]

Over the ridge there is a free trapped mode of amplitude $D(t_2, t_3, \ldots)$

$$\phi_{00} = \frac{igD}{2\Omega} L_\nu(x_1) \cos Ky_1 e^{-i\Omega t_1} + * \qquad (7.10)$$

to the leading order. The modal structure $L_\nu(x_1)$ vanishes exponentially as $|x_1| \to \infty$ and ν is the modal number. Over the ridge, the first-order amplitude $|A|$ of the short swells changes according to the usual shoaling law, but the phase of A is affected by the long waves.

In the special case when the swell (A^0 and b^0) does not have any dependence on t_2 and the trapped mode is the lowest ($\nu = 0$), the evolution equation of the long-wave amplitude is

$$\frac{\partial D}{\partial t_3} + \left(\gamma_0 + A^{0^2}\gamma_1\right)D + \gamma_2 |D|^2 D = -A^0 b^0 \gamma_3, \qquad D = D(t_3), \quad (7.11)$$

where γ_0, γ_1, and γ_2 are interaction constants. γ_0 represents linear dispersive corrections to the long waves and is imaginary; γ_1 measures the nonlinear interaction between swell and the long waves, and γ_2 measures the nonlinear interaction of long waves which contain radiated first and second harmonics (in Ω). The right-hand side is the forcing term from the swell. An energy equation can be obtained by multiplying (7.11) by D^* and adding the complex conjugate of the result,

$$\frac{\partial |D|^2}{\partial t_3} + 2A^{0^2}\mathrm{Re}\,\gamma_1 |D|^2 + 2\,\mathrm{Re}\,\gamma_2 |D|^4 = -2A^0\mathrm{Re}\left(\gamma_3 b^0 D_0\right). \quad (7.12)$$

Thus, the long wave grows or decays because of energy exchange with the short waves, radiation damping ($\mathrm{Re}\,\gamma_2 > 0$ always), and forcing by resonant modulation of initial amplitude b^0. For a ridge of Gaussian profile the lowest mode of the long wave was calculated numerically and the coefficients γ_0 to γ_3 were evaluated. It was found for a range of cases that $\mathrm{Re}\,\gamma_1$ was always positive. meaning that the long wave leaked energy to first-order swell while being excited by the second-order modulation. Typical responses are plotted in Fig. 7.1. Bottom friction which can be estimated in the manner of Section 8.7.3 is seen only to hasten the attenuation but not to affect the resonance qualitatively.

Modifications for surf beats on a beach have also been made by allowing breaking of the short swell in accordance with Section 10.4. Otherwise, the long

[†] Let ε be 0.05, and the swell period be 10 s, then $t = O(10\ \mathrm{s})$, $t_1 = O(3.3\ \mathrm{min})$, $t_2 = O(1\ \mathrm{h})$, and $t_3 = O(20\ \mathrm{h})$. Thus, the scale of t_1 is in the period range of surf beats; $O(t_2)$ is typical of a storm duration.

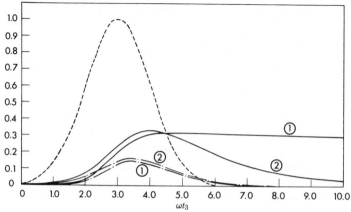

Figure 7.1 Transient response of the lowest mode trapped wave over a submerged Gaussian ridge due to a short-lived resonant modulation of the incident swell packet. Dashed curve: the Gaussian envelope $\bar{b}^0$ of the modulation; solid curve: the trapped wave amplitude D without bottom friction; dash–dot curve: the trapped wave amplitude D with bottom friction. For curves (1), the swell packet is also short-lived $\bar{A}^0 = \bar{b}^0$; for curves (2) the swell packet has constant envelope $\bar{A}^0 = $ const. (From Foda and Mei, 1981, *J. Fluid Mech.* Reproduced by permission of Cambridge University Press.)

wave was assumed not to break. Conclusions are similar to Fig. 7.1 and are omitted.

Since long waves of 1–5-min periods are of interest to the formation of giant beach cusps, oscillations of moored ships, and resonance of small harbors, there is a need for considering more realistic topography. However, the two-dimensional perturbation analysis appears so formidable that further advance may require a new analytical approach or the use of computers to perform algebraic manipulations.

12.8   DIFFRACTION OF STEADY STOKES' WAVES BY A THIN WEDGE OR A SLIGHTLY SLANTED BREAKWATER

While much attention has been paid to uninterrupted propagation of nonlinear waves in an unbounded ocean, little is known about their interaction with side walls. The latter topic is of concern to engineers and is generally much more difficult to study than the former. Even for boundaries that are formed by vertical walls spanning the entire depth, one must deal with two horizontal dimensions. In this section and the next, two problems of nonlinear diffraction are discussed when the long-time modulation is assumed to vanish, that is, when both the incident and the scattered waves are steady.

For solitary waves incident obliquely on a straight wall in shallow water, experiments have been performed by Perroud (1957) and Chen (1961) and reported by Wiegel (1964). For an angle of incidence $\theta_i > 45°$, the incident and reflected wave crests intersected at the wall with equal angles, that is, *specular reflection*. However, when the incidence angle was reduced ($20° < \theta_i < 45°$), there was a third wave crest (called the *stem*) which intersected the wall normally; the incident crest, the reflected crest, and the stem met at a point some distance away from the wall. For incidence angles less than 20°, the reflected crest disappeared, leaving only the incident crest and the stem. Because of its geometrical resemblance to the reflection of shock waves in gas dynamics (see Lighthill, 1949; Whitham, 1974), the phenomenon in shallow-water waves has also been called the *Mach stem effect* by Wiegel.

Similar experiments of oblique incidence of periodic waves in finite water depth have been performed by Nielsen (1962) and more recently by Berger and Kohlhase (1976). The kinematics of the wave crests resembled that of the solitary waves. The wave amplitude along the barrier, that is, the stem height, increased downwave for a finite distance and then leveled off gradually. At any station this amplitude increased with the angle of incidence. The width of the stem region, which generally increased with distance along the wall, was larger with smaller incidence angle and water depth. The stem width appeared to be greater for longer incident waves. Data scatter was substantial in all these experiments.

For shallow waters a theory for the Mach stems and Mach reflection of solitary waves has been proposed by Miles (1977), whose deductions have been compared with experiments by Melville (1980). Noting that the evolution equation of the present chapter is the nonlinear extension of the parabolic approximation of Section 4.10, Yue and Mei (1980) carried out a nonlinear study of grazing incidence on a thin wedge; their main findings are summarized in this section.

Recall from Section 4.10 that the parabolic approximation applies when

$$\frac{L_y}{L_x} = \varepsilon \ll 1 \quad \text{and} \quad kL_x \gg 1, \tag{8.1}$$

where L_x and L_y denote the characteristic lengths of modulation along x and y, respectively. Now for Stokes' waves the wave slope kA_0 affects the phase over the length scale $O(k^3A_0^2)^{-1}$, as Eq. (3.4) implies. Therefore, nonlinearity must be taken into account if the body length scale is such that

$$kL_x = O(kA_0)^{-2}. \tag{8.2a}$$

The parabolic approximation implies a lateral length scale such that

$$kL_y = O(kA_0) = O(\varepsilon). \tag{8.2b}$$

Equations (8.2a) and (8.2b) suggest that only

$$x_2 = \varepsilon^2 x \equiv X \quad \text{and} \quad y_1 = \varepsilon y \equiv Y \tag{8.3}$$

are the pertinent slow coordinates. If steady state is assumed, then there is no time dependence. Therefore, we can put

$$\frac{\partial}{\partial t_1} = \frac{\partial}{\partial t_2} = \frac{\partial}{\partial x_1} = 0 \tag{8.4}$$

into Eq. (2.39) and obtain again a cubic Schrödinger equation

$$2\frac{\partial A}{\partial X} - \frac{i}{k}\frac{\partial^2 A}{\partial Y^2} + iK'|A|^2 A = 0, \tag{8.5}$$

where

$$K' = k^3\Theta \equiv k^3\frac{C}{C_g}\frac{\text{ch}\,4q + 8 - 2\,\text{th}^2 q}{8\,\text{sh}^4 q} \tag{8.6}$$

is always positive. ϕ_{10} is taken to be zero.

Assume that the body is bounded by vertical walls which are symmetric about the x axis (see Fig. 8.1) and is described by

$$y = \pm y_B(x) \quad \text{or} \quad Y = \pm \varepsilon y_B \tag{8.7}$$

in normalized coordinates. Vanishing of the normal velocity on the body requires that

$$\Phi_y = \Phi_x\frac{dy_B}{dx} \quad \text{on } y = y_B,$$

or

$$\varepsilon\Phi_Y = \left[\left(\frac{\partial}{\partial x} + \varepsilon^2\frac{\partial}{\partial X}\right)\Phi\right]\frac{dy_B}{dx} \quad \text{on } Y = \varepsilon y_B. \tag{8.8}$$

A similar condition holds on $y = -y_B$. Substituting the first-order potential

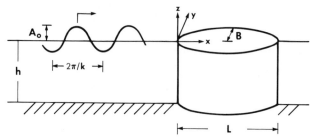

Figure 8.1 Head-sea incidence on a long and slender obstacle.

(2.21) (without ϕ_{10}) into Eq. (8.8), we find, to the leading order,

$$\frac{\partial A}{\partial Y} = ikA \frac{1}{\varepsilon} \frac{dy_B}{dx} \qquad \text{on } Y = \varepsilon y_B. \tag{8.9}$$

With considerable generality one may assume $dy_B/dx = O(\varepsilon)$ so that

$$ky_B = O(\varepsilon^{-1}). \tag{8.10}$$

For a wedge, $y_B = \varepsilon x$; Eq. (8.10) becomes simply

$$\frac{\partial A}{\partial Y} = ikA \qquad \text{on } Y = X. \tag{8.11}$$

The remaining boundary conditions are

$$A \to A_0 e^{-iK'A_0^2 X/2}, \qquad Y \to +\infty, \tag{8.12}$$

and

$$A = A_0, \qquad X = 0. \tag{8.13}$$

The reduced problem for A is of the initial-boundary-value type and can be solved numerically. In terms of the normalized variables

$$\bar{A} = \frac{A}{A_0}, \qquad \bar{X} = kX = \varepsilon^2 kx, \quad \bar{Y} = kY = \varepsilon ky, \tag{8.14}$$

Eq. (8.5) becomes

$$2i \frac{\partial \bar{A}}{\partial \bar{X}} + \frac{\partial^2 \bar{A}}{\partial \bar{Y}^2} - K|\bar{A}|^2 \bar{A} = 0. \tag{8.15}$$

The only parameter is

$$K = \left(\frac{kA_0}{\varepsilon} \right)^2 \Theta(kh), \tag{8.16}$$

where Θ is defined in Eq. (8.6) and plotted in Fig. 8.2. For fixed kh, K is a measure of nonlinearity versus the wedge angle ε.

Figure 8.3 shows the square of the envelope height (proportional to the mean setup) along the wall. The linear result ($K = 0$) oscillates in X and gradually attenuates toward 4. With increasing nonlinearity ($K = 4$), the envelope height decreases and rapidly approaches a constant. Three-dimensional views of the free surface are given in Fig. 8.4 showing that for nonlinear waves, stems are evident within a wedge next to the wall; the wedge angle

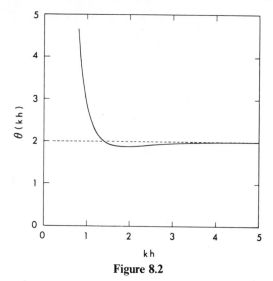

Figure 8.2

increases with nonlinearity K. Finally, in Fig. 8.5 the contour plots of the free-surface displacement show distinctly the sudden bending of wave crests within the plateau.

Outside the plateau, A^2 undulates in a way expected of diffraction. By ignoring these undulations, Yue and Mei found that a discontinuous shock approximation gave a crude estimate of the envelope height of the plateau

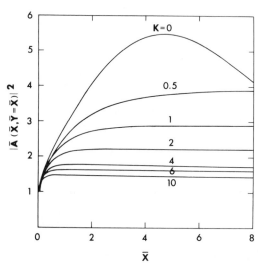

Figure 8.3 Squared-amplitude along the wall of a wedge for different values of nonlinearity K. (From Yue and Mei, 1980, *J. Fluid Mech*. Reproduced by permission of Cambridge University Press.)

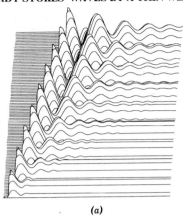

(a)

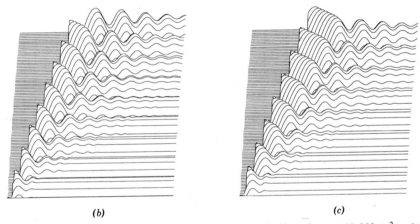

(b)   (c)

Figure 8.4 Snapshot of free surface near a wedge of half angle $\alpha = 17.55°$ ($\varepsilon^2 = 0.1$) shown in undistorted horizontal coordinates. (a) $K = 0$ (linear); (b) $K = 1$; (c) $K = 2$. Note the forward bending of crests near the wedge (shaded) for $K = 1, 2$. (From Yue and Mei, 1980, *J. Fluid Mech.* Reproduced by permission of Cambridge University Press.)

region $E_- = A_-^2$ and of the apex angle β of the plateau. The results are

$$E_- = \frac{1}{2K}\left[2K + 1 + (8K + 1)^{1/2}\right], \tag{8.17}$$

and

$$\beta = \frac{3 + (8K + 1)^{1/2}}{4}, \tag{8.18}$$

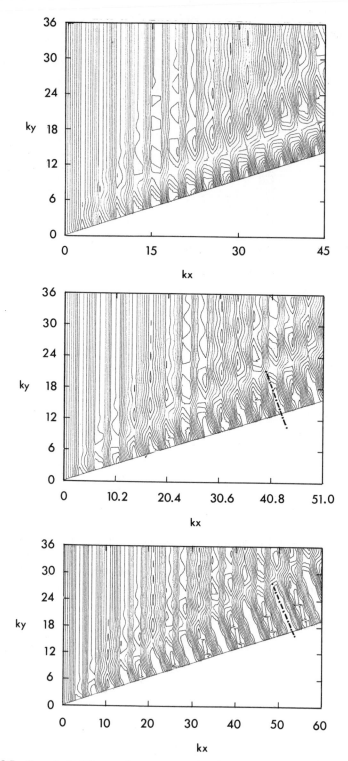

Figure 8.5 Snapshot of free-surface contours for the same three cases of Fig. 8.4. The slope of the dash–dot line is drawn according to Eq. (8.17). (From Yue and Mei, 1980, *J. Fluid Mech.* Reproduced by permission of Cambridge University Press.)

which compare well with the numerical results. The corresponding theoretical prediction on the bending angle of the stems is given by

$$\tan \delta \cong \varepsilon^3 \left[\frac{K}{2} + \frac{3 + (8K + 1)^{1/2}}{4} \right]. \tag{8.19}$$

The smallness of δ implies that the stems are nearly perpendicular to the wall, Fig. 8.5. Despite its closeness to the numerical theory, the shock approximation is a poor representation of the diffraction phenomenon outside the wedge, and hence is not elaborated here.

The present example combines the parabolic approximation and the cubic nonlinearity. Similar numerical analysis can be readily made for the nonlinear diffraction near the shadow boundary of a thin breakwater or for other problems where the direction of the incident wave is not changed significantly by diffraction (see Yue, 1980). In the next section the body is blunt, and scattering is equally important in all directions; the required analysis then becomes considerably more complicated.

12.9 SECOND-ORDER WAVE FORCES ON A FIXED BODY

Exciting forces due to waves of finite amplitude upon a large body are of vital interest in offshore engineering. So far, there is no practical numerical method for nonlinear diffraction by three-dimensional bodies. It is, therefore, natural to take a more modest step of seeking a higher-order improvement of the linearized theory. For two dimensions Lee (1968) gave a second-order theory of a semicircular barge heaving in the free surface. For the three-dimensional problem of a vertical circular cylinder, the second-order analysis turned out to be so complicated that disagreements arose among several workers (Charkrabarti, 1972; Yamaguchi and Tsuchiya, 1974; Raman and Venkatanarasalah, 1976; Issacson, 1977; Charkrabarti, 1978; Molin, 1979; Miloh, 1980; Wehausen, 1980; Hunt and Baddour, 1981). In the opinion of this author the following theory of Molin appears to be the most complete.

Let a periodic wavetrain whose first-order amplitude is A approach a body from $x \sim -\infty$, and let $\varepsilon = kA$ be the small wave slope, that is,

$$\Phi^I = \varepsilon \Phi_1^I + \varepsilon^2 \Phi_2^I + \cdots, \tag{9.1}$$

where,

$$\Phi_1^I = \text{Re}\left(\phi_1^I e^{-i\omega t} \right), \tag{9.2a}$$

$$\varepsilon \phi_1^I = -\frac{igA}{\omega} \frac{\cosh k(z + h)}{\cosh kh} e^{ikx}. \tag{9.2b}$$

Let the total potential be expanded as follows:

$$\Phi = \varepsilon\Phi_1 + \varepsilon^2\Phi_2 + \cdots, \tag{9.3a}$$

where

$$\Phi_1 = \Phi_1^I + \Phi_1^S, \qquad \Phi_2 = \Phi_2^I + \Phi_2^S, \qquad \text{and so on,} \tag{9.3b}$$

with the superscripts I and S denoting incident and scattering, respectively.

The total second-order potential is governed by

$$\nabla^2\Phi_2 = 0 \qquad\qquad \text{in } \Omega, \tag{9.4}$$

$$g\frac{\partial\Phi_2}{\partial z} + \frac{\partial^2\Phi_2}{\partial t^2} = A_2$$

$$\equiv \frac{1}{g}\frac{\partial\Phi_1}{\partial t}\frac{\partial}{\partial z}\left(g\frac{\partial\Phi_1}{\partial z} + \frac{\partial^2\Phi_1}{\partial t^2}\right) - \frac{\partial}{\partial t}(\nabla\Phi_1)^2 \quad \text{on } S_F(z = 0), \tag{9.5}$$

$$\frac{\partial\Phi_2}{\partial z} = 0 \qquad\qquad \text{on } B_0\,(z = -h), \tag{9.6}$$

and

$$\frac{\partial\Phi_2}{\partial n} = 0 \qquad\qquad \text{on } S_B. \tag{9.7}$$

The second-order part of the incident wave Φ^I satisfies Eqs. (9.4)–(9.6) and is given by Stokes' theory [cf. Eq. (3.8)],

$$\varepsilon^2\Phi_2^I = \varepsilon^2\,\text{Re}\!\left(\phi_2^I e^{-2i\omega t}\right) \tag{9.8a}$$

with

$$\varepsilon^2\phi_2^I = -\frac{3}{8}i\omega A^2\frac{\cosh 2k(z+h)}{\sinh^4 kh}e^{2ikx}. \tag{9.8b}$$

An additive constant Φ_{20} has been discarded. Now the second-order part of the scattered wave must satisfy Eqs. (9.4) and (9.6) and

$$g\frac{\partial\Phi_2^S}{\partial z} + \frac{\partial^2\Phi_2^S}{\partial t^2} = A_2 - g\frac{\partial\Phi_2^I}{\partial z} - \frac{\partial^2\Phi_2^I}{\partial t^2} \quad \text{on } S_F\,(z = 0), \tag{9.9}$$

$$\frac{\partial\Phi_2^S}{\partial n} = -\frac{\partial\Phi_2^I}{\partial n} \qquad\qquad \text{on } S_B. \tag{9.10}$$

In Eq. (9.5) the quadratic forcing term on the free surface contains zeroth and second harmonics and may be written

$$A_2 - g\frac{\partial \Phi_2^I}{\partial z} - \frac{\partial^2 \Phi_2^I}{\partial t^2} = \bar{\alpha}(x, y) + \text{Re}[\alpha(x, y)e^{-2i\omega t}]. \qquad (9.11)$$

Correspondingly, we expect Φ_2^S to contain both zeroth and second harmonics

$$\Phi_2^S = \bar{\phi}_2^S(x, y, z) + \text{Re}(\phi_2^S e^{-2i\omega t}). \qquad (9.12)$$

By the Bernoulli equation, the contribution of $\bar{\phi}_2^S$ to the pressure (hence the force on the body) is at most $O(\varepsilon^3)$ (through $\varepsilon^3 \nabla \Phi_1 \cdot \nabla \Phi_2$) and need not be considered if only the second-order forces are desired.

To analyze ϕ_2^S we must first examine the second-harmonic forcing α:

$$\alpha = -g\frac{\partial \phi_2^I}{\partial z} + 4\omega^2 \phi_2^I - \frac{i\omega}{2g}(\phi_1^I + \phi_1^S)\frac{\partial}{\partial z}\left[g\frac{\partial}{\partial z}(\phi_1^I + \phi_1^S)\right.$$

$$\left. -\omega^2(\phi_1^I + \phi_1^S)\right] + 2i\omega(\nabla \phi_1^I + \nabla \phi_1^S)^2$$

$$= \alpha_I^S + \alpha_S^S, \qquad z = 0 \qquad (9.13)$$

where α_I^S refers to all terms containing the cross products of ϕ_1^I and ϕ_1^S, and α_S^S refers to products containing ϕ_1^S only. The part containing ϕ_2^I and terms quadratic in ϕ_1^I vanishes because of the boundary condition for ϕ_2^I. By the use of Eq. (9.2b) and the asymptotic expression

$$\phi_1^S = \frac{f_1(\theta)}{r^{1/2}}\cosh k(z + h)e^{ikr}, \qquad (9.14)$$

it is clear that

$$\alpha_I^S \propto \frac{G(\theta)}{r^{1/2}}e^{ikr(1+\cos\theta)} + o(r^{-1/2}). \qquad (9.15)$$

Note that α_I^S is a radially spreading wave with the phase speed varying in θ, from $C = \omega/k$ in the forward direction to $C = \infty$ in the backward direction. Note also that for large r

$$\alpha_S^S \propto \frac{H(\theta)}{r}e^{2ikr}. \qquad (9.16)$$

Since α_I^S diminishes slowly at large r, it will be necessary to weaken the radiation condition on ϕ_2^S. Let f be a radiated second harmonic which satisfies the strong radiation condition in the form of Eq. (4.16), p. 308. Then ϕ^S can only satisfy, with f, the integral condition (6.3), p. 319; this will be checked later.

We now seek a particular solution which satisfies

$$\nabla^2\phi_2^L = 0, \qquad -h < z < 0 \tag{9.17}$$

$$g\frac{\partial\phi_2^L}{\partial z} - 4\omega^2\phi_2^L = \alpha_I^S, \qquad z = 0 \tag{9.18}$$

$$\frac{\partial\phi_2^L}{\partial z} = 0, \qquad z = -h \tag{9.19}$$

disregarding the boundary condition on the body S_B. Although the explicit solution for all r is not easy to find, the asymptotic solution for large r is not difficult as we can approximate the Laplace equation by

$$\left(\frac{\partial^2}{\partial r^2} + \frac{1}{r^2}\frac{\partial^2}{\partial \theta^2} + \frac{\partial^2}{\partial z^2}\right)\phi_2^L = o(r^{-1/2}). \tag{9.20}$$

It is easily verified that[†]

$$\phi_2^L \sim \frac{L(\theta)}{r^{1/2}}\cosh\left[k(2 + 2\cos\theta)^{1/2}(z + h)\right]e^{ikr(1+\cos\theta)} \tag{9.21}$$

meets the requirement, since

$$\left(\frac{\partial^2}{\partial r^2} + \frac{1}{r^2}\frac{\partial^2}{\partial \theta^2} + \frac{\partial^2}{\partial z^2}\right)\phi_2^L = \phi_2^L\left[-k^2(1 + \cos\theta)^2 + 2k^2(1 + \cos\theta)\right.$$

$$\left. -k^2\sin^2\theta\right] + o(r^{-1/2})$$

$$= o(r^{-1/2}). \tag{9.22}$$

To satisfy Eq. (9.18) it is only necessary to choose $L(\theta)$ such that

$$L(\theta)\left[gk(2 + 2\cos\theta)^{1/2}\sinh\left(k(2 + 2\cos\theta)^{1/2}h\right)\right.$$

$$\left. -4\omega^2\cosh\left(k(2 + 2\cos\theta)^{1/2}h\right)\right] = G(\theta). \tag{9.23}$$

Equation (9.19) is obviously satisfied.

The remainder of ϕ_2^S will be denoted by ϕ_2^F, that is,

$$\phi_2^F = \phi_2^S - \phi_2^L \tag{9.24}$$

[†]Note that ϕ_2^L is locked in phase with the forcing term α_I^S, hence the superscript L.

which must satisfy the Laplace equation, the boundary condition on the body

$$\frac{\partial \phi_2^F}{\partial n} = -\frac{\partial}{\partial n}\left[\phi_2^I + \phi_2^L\right] \qquad \text{on } S_B, \tag{9.25}$$

and the free-surface condition

$$g\frac{\partial \phi_2^F}{\partial z} - 4\omega^2 \phi_2^F = \alpha_S^S, \qquad z = 0. \tag{9.26}$$

The fictitious pressure of α_S^S behaves according to Eq. (9.16) asymptotically and is sufficiently localized to possess a two-dimensional Fourier transform. Therefore, ϕ_2^F behaves as an outgoing wave at large distances,

$$\phi_2^F \sim \frac{F(\theta)}{r^{1/2}} e^{ik_2 r} \cosh k_2(z + h), \qquad kr \gg 1, \tag{9.27}$$

where 2ω and k_2 satisfy the linear dispersion relation. For this resemblance to free waves the superscript F is used.

In summary, the asymptotic behavior of ϕ_2^S is

$$\phi_2^S = \frac{1}{r^{1/2}}\left\{ F(\theta)e^{ik_2 r}\cosh k_2(z + h) + L(\theta)e^{ikr(1 + \cos\theta)} \right.$$

$$\left. \times \cosh\left[k(2 + 2\cos\theta)^{1/2}(z + h)\right]\right\}. \tag{9.28}$$

This information rather than the complete solution for ϕ_2^S turns out to be sufficient for calculating the second-order wave forces.

Now we turn to the exciting force in the x direction

$$F_x(t) = \iint_{S_B} Pn_x \, dS, \tag{9.29}$$

where $S_B(t)$ is the instantaneous wetted surface and n_x is the x component of the unit normal pointing to the body. Expanding F_x to the second order

$$F_x = F_{0x} + \varepsilon F_{1x} + \varepsilon^2 F_{2x} + \cdots, \tag{9.30}$$

and using the Bernoulli equation, we get

$$F_{0x} = \iint\limits_{\bar{S}_B} - \rho g z n_x \, dS = 0, \tag{9.31}$$

$$F_{1x} = \iint\limits_{\bar{S}_B} - \rho \frac{\partial \Phi_1}{\partial t} n_x \, dS, \tag{9.32}$$

$$F_{2x} = \iint\limits_{\bar{S}_B} \left[-\rho \frac{\partial \Phi_2}{\partial t} - \frac{\rho}{2} (\nabla \Phi^{(1)})^2 \right] n_x \, dS$$

$$+ \iint\limits_{S_B'} \left(-\frac{\rho g z}{\varepsilon} - \rho \frac{\partial \Phi_1}{\partial t} \right) n_x \, dS, \tag{9.33}$$

where $\bar{S}_B$ is the mean wetted surface and S_B' is the fluctuating part of the wetted surface between $z = 0$ and $z = -(1/g) \partial \Phi_1 / \partial t$. F_{1x} is, of course, just the linearized exciting force. For simplicity let the body have a vertical wall near the water line. The second integral in Eq. (9.33) can be transformed into a line integral along the water line Γ

$$\iint\limits_{S_B'} \left(-\frac{\rho g z}{\varepsilon} - \rho \frac{\partial \Phi_1}{\partial t} \right) n_x \, dS = \oint_\Gamma d\Gamma \, n_x \int_0^{\zeta_1} \left(-\frac{\rho g z}{\varepsilon} - \rho \frac{\partial \Phi_1}{\partial t} \right) dz$$

$$= \oint_\Gamma \frac{\rho}{2g} \left(\frac{\partial \Phi_1}{\partial t} \right)^2 n_x \, d\Gamma.$$

Let us now separate F_{2x} into two parts,

$$F_{2x} = F_{2x}^{(1)} + F_{2x}^{(2)}, \tag{9.34a}$$

where $F_{2x}^{(1)}$ depends explicitly on Φ_1,

$$F_{2x}^{(1)} = \iint\limits_{\bar{S}_B} - \frac{\rho}{2} (\nabla \Phi_1)^2 n_x \, dS + \oint_\Gamma \frac{\rho}{2g} \left(\frac{\partial \Phi_1}{\partial t} \right)^2 n_x \, d\Gamma, \tag{9.34b}$$[†]

[†] This formula has also been derived by Lighthill (1979b).

while

$$F_{2x}^{(2)} = \iint\limits_{\bar{S}_B} -\rho \frac{\partial \Phi_2}{\partial t} n_x \, dS \qquad (9.34c)$$

depends on Φ_2.

For periodic motion the time average of $F_{2x}^{(2)}$ is zero but $F_{2x}^{(1)}$ has a nonzero mean which is just the *drift force*

$$\bar{F}_{2x} = \iint\limits_{\bar{S}_B} -\frac{\rho}{4} |\nabla \phi_1|^2 n_x \, dS + \oint_\Gamma \frac{\rho}{4g} \omega^2 |\phi_1|^2 n_x \, d\Gamma. \qquad (9.35)$$

The sinusoidal part of $F_{2x}^{(2)}$ can be further split into two terms:

$$F_{2x}^{(2)} = \left(F_{2x}^{(2)} \right)_I + \left(F_{2x}^{(2)} \right)_S, \qquad (9.36a)$$

where

$$\left(F_{2x}^{(2)} \right)_I = \text{Re} \left[2i\rho\omega \iint\limits_{\bar{S}_B} \phi_2^I n_x \, dS \, e^{-2i\omega t} \right], \qquad (9.36b)$$

$$\left(F_{2x}^{(2)} \right)_S = \text{Re} \left[2i\rho\omega \iint\limits_{\bar{S}_B} \phi_2^S n_x \, dS \, e^{-2i\omega t} \right]. \qquad (9.36c)$$

The first part, $(F_{2x}^{(2)})_I$, is straightforward. For $(F_{2x}^{(2)})_S$, Molin used Haskind's idea and introduced a fictitious radiation potential ψ for the same body swaying at the frequency 2ω, that is,

$$\bar{\Psi} = \text{Re}(\psi e^{-2i\omega t}), \qquad (9.37)$$

where

$$\nabla^2 \psi = 0 \qquad\qquad \text{in } \Omega, \qquad (9.38)$$

$$\frac{\partial \psi}{\partial n} = n_x \qquad\qquad \text{on } \bar{S}_B, \qquad (9.39)$$

$$g \frac{\partial \psi}{\partial z} - 4\omega^2 \psi = 0 \qquad\qquad \text{on } z = 0, \qquad (9.40)$$

$$\psi \sim \frac{X(\theta)}{r^{1/2}} \cosh k_2(z + h) e^{ik_2 r}, \qquad k_r r \gg 1. \qquad (9.41)$$

Now $(F_{2x}^{(2)})_S$ can be expressed as

$$\left(F_{2x}^{(2)}\right)_S = \operatorname{Re}\left[2i\omega\rho \iint\limits_{\bar{S}_B} \phi_2^S \frac{\partial\psi}{\partial n} dS\, e^{-2i\omega t}\right]. \tag{9.42}$$

Green's identity for ϕ_2^S and ψ then gives

$$\iint\limits_{\bar{S}_B} \phi_2^S \frac{\partial\psi}{\partial n} dS = \iint\limits_{\bar{S}_B} \left(\psi \frac{\partial\phi_2^S}{\partial n}\right) dS + \iint\limits_{S_F + \bar{B}_0 + S_\infty} \left[\psi \frac{\partial\phi_2^S}{\partial n} - \phi_2^S \frac{\partial\psi}{\partial n}\right] dS; \tag{9.43}$$

the integral over the sea bottom B_0 vanishes. When Eqs. (9.28) and (9.41) are used, the integral over S_∞ (cf. Eq. (6.3), p. 319) may be shown by the method of stationary phase to vanish as $r^{-1/2} e^{i\lambda r}$ where $\lambda \neq 0$. Using the boundary condition (9.39) on the left-hand side of Eq. (9.43), we get

$$\left(F_{2x}^{(2)}\right)_S = \operatorname{Re}\left\{2i\rho\omega\left[-\iint\limits_{\bar{S}_B} \frac{\partial\phi_I^{(2)}}{\partial n}\psi\, dS + \frac{1}{g}\iint\limits_{\bar{S}_B} \alpha\psi\, dS\right] e^{-2i\omega t}\right\} \tag{9.44}$$

where α is given by Eq. (9.13). Equations (9.36b) and (9.44) together give the oscillatory part of the second-order force on a body of any shape. Note that there is never a need to solve for ϕ_2 explicitly; ψ is a linear problem and α is known from the first-order solution.

For a vertical circular cylinder the linear problem is simple. Molin's numerical results which are plotted in Fig. 9.1 show that the linear theory underestimates the horizontal exciting force.

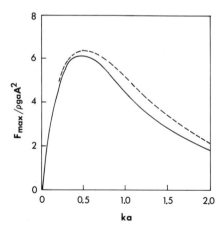

Figure 9.1 Maximum horizontal force for $h/a = 5$, $A/a = 0.19$. Solid curve: linear theory; broken curve: second-order theory. (From Molin, 1979, *Appl. Ocean Res.* Reproduced by permission of CML Publications.)

The analysis in this section illustrates the complexity of a second-order diffraction theory. The computation of the radiation problems must be at least equally involved. Together with the development of a second-order theory for the dynamical equations for a floating body, the algebra and the associated numerical task are indeed daunting. Such an effort has been reported by Papankolaou and Nowacki (1980) for the two-dimensional problem of a floating cylinder. It appears that unless the labor of the perturbation analysis itself is relieved by computer programming, the direct numerical integration of the fully nonlinear initial-value problem may be a more practical route in future studies of large-amplitude waves involving bodies (see, for example, Vinje and Brevig, 1981a, for two-dimensional bodies in a periodic array).

12.10 NUMERICAL SOLUTION FOR STEEP WAVES

As steep waves occur often in nature, techniques of direct calculation have been eagerly pursued for a long time. Historically, numerous attempts have been made to extend the Stokes expansion to higher orders with a view to obtaining the limiting profile of the highest progressive wave of permanent form. Earlier efforts of this kind were mostly limited to the first few orders and the results could not be consistently applied to near-breaking waves. Schwartz (1974) made a major breakthrough by incorporating computer-aided algebra with the so-called Padé approximant which employed rational fractions to sum series. He used the crest-to-trough height of the wave as the expansion parameter, rather than the amplitude of the first harmonic, carried the expansion to 112th order for deep water and 48th order for finite depth, and obtained accurate results for arbitrary height to near breaking. Schwartz's method has since been modified for solitary waves in shallow water by Longuet-Higgins and Fenton (1974), for deep water by Longuet-Higgins (1975), and for finite depth by Cokelet (1977), all with different expansion parameters. An alternative method for periodic waves on constant depth is based on expanding the stream function as a Fourier series in the horizontal direction, with the Fourier coefficients determined by imposing the dynamic boundary condition at a large number of surface points within a period. This method was initiated by Dean (1965) and was improved in important ways by Rienecker and Fenton (1981). These and other related studies have been reviewed by Schwartz and Fenton (1982).

From the point of view of applications, these accurate solutions of simple progressive waves provide the starting point not only for predicting wave forces on small structures, but also for calculating the slow refraction of steep waves, in a manner similar to the ray approximation of Chapter Two. Based on a set of averaged equations (Crapper, 1979; Stiassnie and Peregrine, 1979) for integral quantities which can be obtained from the numerical solution of uniform waves, Stiassnie and Peregrine (1980) calculated the shoaling of normally incident waves on a mild beach. Further studies of steep waves modulated by more general topographies are important to engineering and await future research.

For transient waves, the finite-difference technique has been developed by researchers at the Los Alamos Laboratory and elsewhere. By the so-called *marker-in-cell* method, a number of two-dimensional initial-value problems involving a free surface have been solved for a viscous fluid. The computational cost appears substantial, however, and other alternatives are desireable. See Von Kerczek (1975) for a review.

We are indebted to Longuet-Higgins and Cokelet (1976) for an efficient way to calculate for two-dimensional spatially periodic waves on deep water. This method reduces the transient two-dimensional problem to a line integral equation at each time step. The basic idea is sketched below.

Let us first introduce the normalization

$$\mathbf{x} \to \frac{\mathbf{x}}{k}, \qquad T \to t(gk)^{1/2}, \quad p \to p\rho gk^{-1}, \quad \phi \to \phi g^{1/2}k^{-3/2}$$

so that the dimensionless wavelength is 2π. A crucial step is to express the free-surface condition in Lagrangian form. Let

$$\frac{D}{Dt} = \frac{\partial}{\partial t} + \nabla\Phi \cdot \nabla, \qquad \nabla = \left(\frac{\partial}{\partial x}, \frac{\partial}{\partial y}\right).$$

We depart from earlier convention and use y to denote the vertical coordinate. The kinematic condition can be stated in two components:

$$\frac{Dx}{Dt} = \Phi_x, \qquad \frac{Dy}{Dt} = \Phi_y, \qquad \text{on } y = \zeta(x, t), \tag{10.1a}$$

or in complex variables

$$\frac{Dz}{Dt} = u + jv = w^* \qquad \text{on } y = \zeta(x, t), \tag{10.1b}$$

with $z = x + jy$ and $w(z) = u - jv$. The dynamic boundary condition is

$$\frac{\partial\Phi}{\partial t} = -P_a - y - \frac{1}{2}(\nabla\Phi)^2 \qquad \text{on } y = \zeta(x, t), \tag{10.2}$$

or

$$\frac{D\Phi}{Dt} = -P_a - y + \frac{1}{2}(\nabla\Phi)^2 = -P_a - y + \frac{1}{2}ww^*. \tag{10.3}$$

For deep water it is economical to map the fluid within a wavelength into the interior of a closed contour $C(t)$ in the complex ξ plane by

$$\xi = e^{-jz}. \tag{10.4}$$

Letting $\xi = re^{j\theta}$, we have

$$r = e^y, \qquad \text{that is, } y = \ln r \text{ and } \theta = -x. \tag{10.5}$$

All the points at $y \sim -\infty$ are mapped to the origin in the ξ plane. Now the complex potential $\chi = \Phi + j\Psi$ must be analytic inside C. Along the free surface Eq. (10.1) implies

$$
\begin{aligned}
\frac{Dr}{Dt} &= e^y \frac{Dy}{Dt} = r \frac{\partial \Phi}{\partial y} = r^2 \frac{\partial \Phi}{\partial r}, \\
\frac{D\theta}{Dt} &= -\frac{Dx}{Dt} = -\frac{\partial \Phi}{\partial x} = \frac{\partial \Phi}{\partial \theta},
\end{aligned}
\tag{10.6}
$$

while Eq. (10.3) implies

$$\frac{D\Phi}{Dt} = -P_a - \ln r + \frac{1}{2}\left\{ \left(r\frac{\partial \Phi}{\partial r} \right)^2 + \left(\frac{\partial \Phi}{\partial \theta} \right)^2 \right\}. \tag{10.7}$$

Suppose that at the initial instant $t = t_0$ the velocity potential throughout the fluid and the initial free surface $C(t_0)$ are given. We first integrate Eq. (10.6) in time to get the Lagrangian coordinates (r, θ) of all particles on the free surface after dt, and hence the new contour $C(t_0 + dt)$. Integrating Eq. (10.7) from t_0 to $t_0 + dt$, the values of Φ associated with each particle on the free surface can also be determined. The next step is to solve a Dirichlet problem for Φ everywhere within the contour $C(t_0 + dt)$, which can be handled by standard methods of integral equations as sketched in Section 7.8. In particular, the integral equation (8.5) in Section 7.8 can be applied along C in the ξ plane. Since the region within C is finite, the simple Green function $-(1/2\pi)\ln|\xi - \xi_0|$ can be used. Afterward, normal derivatives of Φ are known everywhere, including the points along C. One can then integrate Eq. (10.6) again to another time step for the new position of the free surface, and so on. The time stepping is achieved by using the Adam–Bashforth–Moulton method which is of fourth-order accuracy. For initial time stepping a Runge-Kutta scheme is used. To ensure numerical stability smoothing is necessary.

For finite but constant depth no advantage is gained by employing conformal mapping, and it is convenient to perform computations directly in the physical plane of $z = x + jy$. At a given instant t, the boundary-value problem for the velocity potential within a wavelength is slightly more involved than that for deep water. Some modification of the integral equation technique is needed and has been successfully worked out by Vinje and Brevig (1981b) as described below.

Consider the region bounded by the closed contour C which consists of the moving free surface S_F, the horizontal bottom S_B, and two vertical planes C_+ and C_- one wavelength apart (see Fig. 10.1). Since $\chi(z) = \Phi + j\Psi$ is analytic

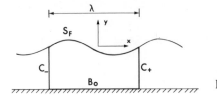

Figure 10.1

within C, Cauchy's theorem applies

$$\frac{1}{2\pi j}\oint_C \frac{\Phi + j\Psi}{z - z_0} = 0 \tag{10.8}$$

where z_0 lies outside C. Let z_0 approach a point on the boundary C. A small semicircular arc of radius ε must be introduced to leave the above integral unchanged. Along the small arc, the integral can be computed so that Eq. (10.8) becomes

$$\frac{j}{2}\{\Phi(x_0, y_0) + j\Psi(x_0, y_0)\}$$

$$+ \frac{1}{2\pi j}\fint_C \frac{dz}{z - z_0}\{\Phi(x, y, t) + j\Psi(x, y, t)\} = 0, \tag{10.9}^\dagger$$

where $\fint$ denotes the Cauchy principal value.

Separating the real and imaginary parts, we get

$$\left\{\begin{array}{c}\pi\Phi(x_0, y_0, t)\\ -\pi\Psi(x_0, y_0, t)\end{array}\right\} + \left\{\begin{array}{c}\mathrm{Re}\\ \mathrm{Im}\end{array}\right\}\fint_C \frac{dz}{z - z_0}\{\Phi + j\Psi\} = 0, \qquad \begin{array}{c}(10.10)\\ (10.11)\end{array}$$

which constitute a pair of coupled integral equations for Φ and Ψ along C for any instant t. At the initial instant $t = t_0$, the position of $C(t_0)$ is known. Let $C(t_0)$ be discretized into N nodes, with N_F nodes on S_F, N_B on B_0, and an equal number N_s of nodes on the sides C_+ and C_-. Along S_F the initial values of Φ are given at the N_F nodes, while along B the values of Ψ are given at N_B nodes. With these values in Eqs. (10.10) and (10.11), $N_F + N_B$ algebraic equations are obtained. Since Φ and Ψ must be periodic after one wavelength, one gets $2N_s$ more algebraic equations from Eqs. (10.10) and (10.11). Thus, there are N equations for N unknowns: N_F of Ψ, N_s of Φ and Ψ, and N_B of Φ. After the solution of these algebraic equations, $\chi = \Phi + i\Psi$ is known at N points on C. The complex potential within C can then be obtained by the

---

†If z_0 coincides with a sharp corner of interior angle $\alpha\pi$, the first curly brackets should be multiplied by α.

Cauchy formula

$$\chi(z, t) = \frac{1}{2\pi j} \oint_C \frac{\chi(z_0, t)}{z_0 - z} dz_0, \qquad z \text{ within } C. \qquad (10.12)$$

This result may be used to compute the right-hand side of Eqs. (10.1b) and (10.3) which can then be integrated for the Lagrangian coordinate and Φ on the free surface at $t_0 + dt$. Use of Hamming's fourth-order predictor/corrector method renders the integration stable.

In principle, the above method need not be confined to a periodic wave or to a horizontal bottom. For a strictly transient problem, a large contour may be used so that the side boundaries C_+ and C_- are in the undisturbed region, which can, of course, be very demanding on computer time.

Figures 10.2–10.4 show the sample results by Vinje and Brevig (1981b) for an initially sinusoidal wave of large slope $ka = 0.408$ on shallow and deep

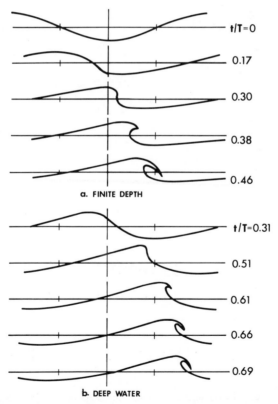

Figure 10.2 Development of plunging breaker from initial sinusoidal waves (From Vinjie and Brevig, 1981b, *Adv. Water Resources*. Reproduced by permission of CML Publications.) (*a*) Finite depth; (*b*) infinite depth.

water. The initial potentials prescribed on the free surface are those of the linear theory (hence not an exact solution). In Fig. 10.2 a tendency toward breaking is seen to develop much faster in the shallow water. Figure 10.3 gives the details of the velocity field at the instant when the crest becomes vertical at some point on the forward side. Finally, Fig. 10.4 shows the velocity fields after the crests have plunged forward.

Several variations of the numerical method just described have now been proposed. For example, in treating a transient and localized disturbance Liu (1978) has succeeded in mixing the two boundary conditions on the free surface. At any time step, the fluid domain is defined by the free surface from the previous time step, the two vertical lines far away from the disturbance, and the sea bottom. The normal derivative $\partial\Phi/\partial n$ vanishes on the vertical sides and the sea bottom. On the free surface, Liu rewrites the two boundary conditions as follows:

$$\frac{\partial\zeta}{\partial t} = \left[1 + \left(\frac{\partial\zeta}{\partial x}\right)^2 + \left(\frac{\partial\zeta}{\partial y}\right)^2\right]^{1/2}\frac{\partial\Phi}{\partial n}, \qquad z = \zeta \tag{10.13}$$

$$\frac{\partial\Phi}{\partial t} + \frac{1}{2}(\nabla\Phi)^2 + g\zeta = -\frac{P_a}{\rho} \tag{10.14}$$

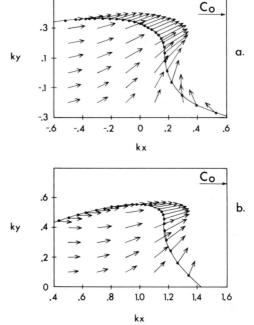

Figure 10.3 Velocity fields under plunging breakers when the wave fronts become vertical. (From Vinjie and Brevig, 1981b, *Adv. Water Resources*. Reproduced by permission of CML Publications.) (*a*) Finite depth $t/T = 0.28$; (*b*) Infinite depth $t/T = 0.49$.

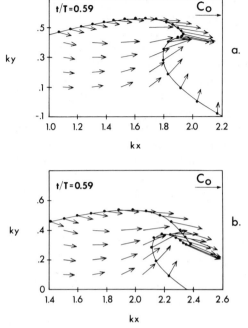

Figure 10.4 Velocity field under the deep water crest $0.1T$ and $0.15T$ after the wave front has plunged forward. (From Vinjie and Brevig, 1981b, *Adv. Water Resources.* Reproduced by permission of CML Publications.)

and then introduces relaxation parameters θ_1 and θ_2 $(0 < \theta_{1,2} < 1)$ in time differencing,

$$\zeta_j^{k+1} = \zeta_j^k + \Delta t \left\{ \left[1 + \left(\frac{\partial \zeta}{\partial x} \right)^2 + \left(\frac{\partial \zeta}{\partial y} \right)^2 \right]^{1/2} \right\}_j^k$$

$$\times \left[\theta_2 \left(\frac{\partial \Phi}{\partial n} \right)_j^{k+1} + (1 - \theta_2) \left(\frac{\partial \Phi}{\partial n} \right)_j^k \right] \quad (10.15)$$

and

$$\Phi_j^{k+1} = \Phi_j^k - \frac{\Delta t}{2} \left\{ \left[(\nabla \Phi)^2 \right]_j^k + \theta_1 \left[g\zeta_j^{k+1} + \left(\frac{P_a}{\rho} \right)_j^{k+1} \right] \right.$$

$$\left. + (1 - \theta_1) \left[g\zeta_j^k + \left(\frac{P_a}{\rho} \right) \right] \right\} \quad (10.16)$$

where k and $k + 1$ refer to the time steps and j to the spatial nodes. When ζ_j^{k+1} is eliminated from the two equations above, a linear relation between $(\partial\Phi/\partial n)_j^{k+1}$ and $(\Phi)_j^{k+1}$ is found. Now $\partial\Phi/\partial n$ on the free surface and Φ on the solid walls can be found for t_{k+1} from an integral equation of the form of Eq. (8.5) in Section 7.8 with $(1/2\pi)\ln|\mathbf{x} - \mathbf{x}_0|$ as the Green function. The free-surface displacement ζ_j^{k+1} is found from Eq. (10.15) afterward. Two initial-value problems involving radiation by local disturbances have been solved by Liu who found the method to be decidedly more efficient than the marker-and-cell method.

All these methods can be extended to three dimensions in principle, but are expensive. Further improvements and applications to practical problems can be anticipated in the near future.

THIRTEEN

Wave-Induced Stresses in a Poro-elastic Seabed

13.1 INTRODUCTION

Pipelines for transporting petroleum from an offshore terminal to the shore are usually laid directly on top of the seabed if the water depth is large, or buried in the seabed if the water depth is small. For a sufficiently shallow sea or in strong waves, the varying wave pressure can induce considerable stress and strain in the seabed, causing soil failure and pipeline breakage. Many offshore platforms are supported on piles which are drilled into the seabed, while many others of the gravity type rest directly on it. Under persistent attack by waves, not only the fatigue of the structure itself but also the dynamic stability of the seabed is of concern. The latter is also a major topic in soil dynamics and has been studied in the context of machine vibrations and earthquake engineering for some time. Now the variety of soils is unfortunately very large, ranging from solid rock, to sand, to fine clay; some soils are known to behave nonlinearly under many practical conditions. Therefore, to model the dynamic behavior of a soil realistically is an extremely difficult and as yet unfinished task. To provide some reference for design, simplified models have been used in recent theoretical studies. Indeed, a large amount of existing work on structure–foundation interaction is based on the assumption that the soil is a single-phased linearly elastic medium.

In the seabed, the fluid in the pores transmits the fluid pressure in sea water to the interior of the porous skeleton and renders the single-phase theory inadequate. In 1923, Karl Terzaghi introduced a one-dimensional theory for the consolidation of soils in which the dynamic role of the pore fluid is taken into account. This two-phase theory is widely hailed as the beginning of theoretical soil mechanics. Important generalizations have been made subsequently by Biot (1941, 1956) to three dimensions. While the fluid motion is fairly well described by Darcy's law, the solid skeleton may require different constitutive equations depending on soil type and strain magnitude. On the simplest basis of a linearly elastic soil, certain one-dimensional wave problems of interest in seismology have been investigated (Geerstma and Smit, 1961; Deresiewicz, 1960, 1961, 1962a, b). Recently, water-wave induced stresses in a porous seabed have been dealt with similarly (Yamamoto et al., 1978; Madsen, 1978b). In general, Biot's linearized equations which govern the coupled motions of the pore fluid and the solid skeleton are quite complicated. Consequently, exact and explicit solutions are hard to obtain for two- or three-dimensional engineering problems either analytically or numerically.

Granular soils have two peculiar types of behavior not accountable by Biot's original theory. There is first the static property called *dilatancy* in which volume change can be caused by applying shear stresses. Second, soils under periodic loading are known to experience a slow but steady increase of the DC component of pore pressure and the associated decrease of *effective stresses* in the soil matrix. Ultimately, the soil can be *liquefied* when the effective stress vanishes somewhere in the soil (Seed and Lee, 1966; Richart, Hall, and Woods, 1969). Recently, there have been intensive efforts to construct nonlinear constitutive equations to reproduce these two important features (see, e.g., Pande and Zienkiewicz, 1980). Many models based on different concepts and a varying number of empirical parameters have claimed comparable success for the same soil under simple test conditions in the laboratory. This perplexing state of affairs is a reflection that a sound understanding of soil physics is still lacking. Difficulties are even more severe in two or three dimensions because thorough and meaningful *in situ* experiments are not easy to perform. It is possible that fundamental progress may demand microscopic theories accounting for the intergranular dynamics, as well as new techniques of experimentation which do not interfere with the state under observation. In the meantime, the continuum approach with semi-empirical constitutive relations may remain as a pragmatic framework for a long time to come. Effective ways of treating the nonlinear Biot equations are therefore desirable. Clearly, analytical solutions based on a linear elastic theory are useful either for direct applications to small strain or for guiding the construction of efficient numerical methods for large strain.

Based on Biot's linear elastic theory, Mei and Foda (1981a, b, 1982) have developed an approximate analysis suitable for small permeability. They observed that under ordinary sea waves, the pore fluid played a rather passive role away from the top surface of the solid skeleton, namely, the *mud line*, but became an equal and separate partner of the skeleton within a *boundary layer* very near the mud line. This picture implies that a two-phase theory is essential near but *only* near the mud line, and allows many examples of practical interest to be analyzed in relatively simple ways. In addition to the immediate results, the reasoning also suggests that a two- (or multi-) phase theory is inevitable for truly nonlinear soils and that the boundary-layer approximation, which is the consequence of contrasting scales, must still be applicable for the sake of shortening computations and sharpening physical understanding.

We shall begin with an account of the basic equations of a deformable porous medium. After linearization for infinitesimal strains, boundary-layer arguments will be employed in general terms. Several examples relevant to oceanographic engineering will be discussed.

13.2 GOVERNING EQUATIONS

13.2.1 Conservation of Mass and Momentum

The most general and precise formulation of the dynamical equations for multiphase continua is called the *mixture theory*, of which Biot's theory is a special case (Atkin and Craine, 1976). Our discussion follows Prévost (1980) who amplified the classical results for soils.

Consider a mixture which is the aggregate of two phases: solid and water, each is viewed as a continuum. We define $\rho_{\tilde{\alpha}}$ as the mass of phase α within a unit volume *of the mixture*. The density of the mixture ρ is the mass of both phases contained in the unit volume:

$$\rho = \rho_{\tilde{s}} + \rho_{\tilde{w}} = \sum_{\alpha} \rho_{\tilde{\alpha}} \qquad (2.1)$$

where the subscripts s and w stand for solid and water, respectively.

The porosity* n_{α} is defined as the volume fraction of phase α in a unit volume of the mixture. If dV is the volume element of the mixture and dV_{α} the volume of phase α in dV, then

$$n_{\alpha}(\mathbf{x}, t) = \frac{dV_{\alpha}}{dV}, \qquad \alpha = s, w. \qquad (2.2)$$

Note that the material density ρ_{α}, based on the net volume of phase α, is related to $\rho_{\tilde{\alpha}}$ by

$$\rho_{\tilde{\alpha}} = n_{\alpha} \rho_{\alpha}. \qquad (2.3)^{\dagger}$$

In particular, for the solid phase ρ_s refers to the grain and $\rho_{\tilde{s}}$ to the solid skeleton.

Let us define the intrinsic Eulerian velocity $\mathbf{v}_{\alpha}(\mathbf{x}_{\alpha}, t)$ so that the net discharge of phase α through an area dA of the mixture is $q = (\rho_{\alpha}\mathbf{v}_{\alpha} \cdot \mathbf{n})\, dA_{\alpha}$,

---

*To avoid confusion, the normal vector $\mathbf{n}$ is not written in component form throughout this chapter.

†Unless specified, repetition of the species index α does not imply summation.

where dA_α is the net area occupied by phase α in dA. Assume for simplicity that the pore distribution is isotropic and statistically homogeneous over lengths much larger than the pore diameter but much smaller than any other length scales of interest. For a mixture cylinder of height dx, the volume occupied by the α phase is $dV_\alpha = dA_\alpha \, dx$ so that

$$n_\alpha = \frac{dA_\alpha}{dA} = \frac{dV_\alpha}{dV} \tag{2.4}$$

(Biot, 1956). Thus, the same discharge may be written as $q = \rho_\alpha \mathbf{v}_\alpha \cdot \mathbf{n} n_\alpha \, dA$. Note that the so-called *seepage velocity* $\mathbf{v}_{\bar\alpha}$ is defined by $\mathbf{v}_{\bar\alpha} = n_\alpha \mathbf{v}_\alpha$; consequently, $q = \rho_\alpha \mathbf{v}_{\bar\alpha} \cdot \mathbf{n} \, dA$. In summary,

$$q = \rho_\alpha \mathbf{v}_\alpha \cdot \mathbf{n} \, dA_\alpha = \rho_{\bar\alpha} \mathbf{v}_\alpha \cdot \mathbf{n} \, dA = \rho_\alpha \mathbf{v}_{\bar\alpha} \cdot \mathbf{n} \, dA. \tag{2.5}$$

Following the convention used in the mixture theory, we shall denote the material derivative following phase α by an overhead dot:

$$\dot A = \frac{\partial A}{\partial t} + \mathbf{v}_\alpha \cdot \nabla A. \tag{2.6}$$

By considering a fixed volume and assuming neither internal sources nor conversion of species, we have

$$\frac{\partial \rho_{\bar\alpha}}{\partial t} + \nabla \cdot \rho_{\bar\alpha} \mathbf{v}_\alpha = \dot\rho_{\bar\alpha} + \rho_{\bar\alpha} \nabla \cdot \mathbf{v}_\alpha = 0 \tag{2.7}$$

as the mass conservation law for phase α. Substituting Eq. (2.3) into Eq. (2.7), we have, alternatively,

$$\dot n_\alpha + n_\alpha \nabla \cdot \mathbf{v}_\alpha = \frac{-n_\alpha \dot\rho_\alpha}{\rho_\alpha}. \tag{2.8}$$

For a better physical understanding, it is useful to examine some special cases. First, if phase α is an incompressible material, we must have $\dot\rho_\alpha = 0$ so that

$$\dot n_\alpha + n_\alpha \nabla \cdot \mathbf{v}_\alpha = 0. \tag{2.9}$$

In a binary mixture of solid skeleton saturated with water the porosities satisfy

$$n_w + n_s = 1. \tag{2.10}$$

Thus, if the water is incompressible,

$$\dot n_w + n_w \nabla \cdot \mathbf{v}_w = 0, \tag{2.11}$$

but $\nabla \cdot \mathbf{v}_w$ is, in general, nonzero, in contrast with a single-phase fluid. The compressibilities of solid grain and solid skeleton must also be distinguished. If the grains are incompressible, $\dot{\rho}_s = 0$. Since

$$\dot{n}_s = \frac{\partial n_s}{\partial t} + \mathbf{v}_s \cdot \nabla n_s = -\frac{\partial n_w}{\partial t} - \mathbf{v}_s \cdot \nabla n_w$$

$$= -\frac{\partial n_w}{\partial t} - \mathbf{v}_w \cdot \nabla n_w + (\mathbf{v}_w - \mathbf{v}_s) \cdot \nabla n_w,$$

it follows from Eq. (2.8) that

$$\dot{n}_w + n_w \nabla \cdot \mathbf{v}_s = \nabla \cdot \mathbf{v}_s + (\mathbf{v}_w - \mathbf{v}_s) \cdot \nabla n_w. \tag{2.12}$$

If the solid skeleton is incompressible instead, then $\dot{\rho}_{\bar{s}} = 0$ so that

$$\nabla \cdot \mathbf{v}_s = 0 \tag{2.13}$$

from Eq. (2.7). Finally, if both the solid skeleton and grains are incompressible, then not only does Eq. (2.13) hold but $\dot{n}_s = -\dot{n}_w = 0$ upon letting $\alpha = s$ in Eq. (2.8). With $\alpha = w$ in Eq. (2.8) we then get

$$n_w \nabla \cdot \mathbf{v}_w = \frac{-\dot{\rho}_w}{\rho_w},$$

hence $\nabla \cdot \mathbf{v}_w$ is not zero unless the fluid is also incompressible.

In nature, the soil grains are about 30 times as incompressible as pure saturated water which itself is highly incompressible; therefore we shall assume $\dot{\rho}_s = 0$, that is, $\rho_s = \rho_s^0 = \text{const}$ and use Eq. (2.12). Natural water in the pores of the sea bottom usually contains tiny gas bubbles due in part to organic processes. To avoid a three-phase description for Eq. (2.8) we treat the unsaturated water as a compressible liquid; thus,

$$\dot{n}_w + n_w \nabla \cdot \mathbf{v}_w = \frac{-n_w \dot{\rho}_w}{\rho_w}. \tag{2.14}$$

The difference between Eqs. (2.12) and (2.14) is called the *storage* equation

$$n_w \nabla \cdot \mathbf{v}_w + (1 - n_w) \nabla \cdot \mathbf{v}_s + (\mathbf{v}_w - \mathbf{v}_s) \cdot \nabla n_w = \frac{-n_w \dot{\rho}_w}{\rho_w}. \tag{2.15}$$

Physically, Eq. (2.15) means that, following the fluid flow, the net fractional decrease of water density is caused jointly by the outflux of water, the dilation of solid skeleton and the change of porosity following the fluid motion relative to the solid.

Consider the surface force acting on the area element dA with unit normal $\mathbf{n}$. We define the *partial* stress vector $\mathbf{f}_{\bar{\alpha}}(\mathbf{n}, \mathbf{x}, t)$ as the force acting on phase α

per unit area of the mixture. We also define the *net partial* stress vector $\mathbf{f}_\alpha(\mathbf{n}, \mathbf{x}, t)$ as the force acting on phase α per unit net area of the α phase. It then follows that

$$\mathbf{f}_{\bar{\alpha}} = n_\alpha \mathbf{f}_\alpha. \tag{2.16}$$

In association with $\mathbf{f}_{\bar{\alpha}}$ and $\mathbf{f}_\alpha$ the partial stress tensors $\mathbf{T}_{\bar{\alpha}}$ and $\mathbf{T}_\alpha$ can be defined

$$\mathbf{f}_{\bar{\alpha}} = \mathbf{T}_{\bar{\alpha}}\mathbf{n}, \quad \mathbf{f}_\alpha = \mathbf{T}_\alpha\mathbf{n}, \quad \text{and} \quad \mathbf{T}_{\bar{\alpha}} = n_\alpha \mathbf{T}_\alpha. \tag{2.17}$$

The total stress vector $\mathbf{f}$ and the total stress tensor $\mathbf{T}$ of the mixture are

$$\mathbf{f} = \sum_\alpha \mathbf{f}_{\bar{\alpha}} \quad \text{and} \quad \mathbf{T} = \sum_\alpha \mathbf{T}_{\bar{\alpha}}. \tag{2.18}$$

A fundamental quantity in soil mechanics introduced by Terzaghi in 1923 (see Terzaghi, 1943) is the so-called *effective stress* $\mathbf{T}^e$ which is defined as the difference between the total stress in the water–solid mixture and the net partial stress in the pore fluid, that is,

$$\mathbf{T} = \mathbf{T}^e + \mathbf{T}_w. \tag{2.19}$$

The physical significance of the effective stress will be discussed later. Here we merely point out that $\mathbf{T}^e$ is neither $\mathbf{T}_{\bar{s}}$ nor $\mathbf{T}_s$. In fact, since $\mathbf{T}$ is also the sum of the partial stress tensors

$$\mathbf{T} = \mathbf{T}_{\bar{s}} + \mathbf{T}_{\bar{w}} = \mathbf{T}_{\bar{s}} + n_w \mathbf{T}_w,$$

it follows that

$$\mathbf{T}_{\bar{s}} = \mathbf{T}^e + (1 - n_w)\mathbf{T}_w. \tag{2.20}$$

In terms of the partial stress tensors we may now write the momentum equation for phase α in a unit volume of the mixture,

$$\rho_{\bar{\alpha}}\dot{\mathbf{v}}_\alpha = \nabla \cdot \mathbf{T}_{\bar{\alpha}} + \rho_{\bar{\alpha}}\mathbf{g} + \mathbf{F}_\alpha, \qquad \alpha = s, w, \tag{2.21}$$

where $\mathbf{g}$ is the body force and $\mathbf{F}_\alpha$ is the interaction force due to all other phases. Clearly, by Newton's third law,

$$\sum_\alpha \mathbf{F}_\alpha = 0 \quad \text{or} \quad \mathbf{F}_s + \mathbf{F}_w = 0 \qquad \text{for two phases.} \tag{2.22}$$

To complete the formulation which is so far exact, certain empirical assumptions for the constitutive equation and the interaction force must be added.

13.2.2 Empirical Assumptions

The empirical assumptions to be introduced are only appropriate for small-amplitude motions (Biot, 1956). First the constitutive laws: For the pore fluid we assume

$$\mathbf{T}_w = -p\mathbf{I}, \tag{2.23}$$

where $\mathbf{I}$ is the identity tensor $\mathbf{I}_{ij} = \delta_{ij}$ and p is the pore pressure. No fluid strain rates are explicitly involved as we are dealing with length scales much greater than the grain sizes. The pore pressure is assumed to be related to the fluid density by the linear equation of state

$$dp = \frac{\beta \, d\rho_w}{\bar{\rho}_w}, \tag{2.24}$$

where $\bar{\rho}_w$ is a constant reference density and β is the effective bulk modulus. To account for the likely existence of gas bubbles a simple formula discussed in Verruijt (1969) is adopted

$$\frac{1}{\beta} = \frac{1}{\bar{\beta}} + \frac{1-S}{\bar{p}}, \tag{2.25}$$

where $\bar{\beta}$ is the bulk modulus of saturated water, S is the degree of saturation which is essentially the mass ratio of the unsaturated water to the saturated water in a given volume, and $\bar{p}$ is the fluid pressure. For simplicity we assume $\bar{\beta}$, S, and $\bar{p}$ to be measured constants. It is useful to remark that a small amount of trapped air can reduce β much below $\bar{\beta}$. Taking, for example, $S = 0.99$, $\bar{p} = 1$ atm $= 10^4$ N/m^2, and $\bar{\beta} = 2 \times 10^9$ N/m^2, we get, from Eq. (2.25), $\beta = 10^6$ N/m^2.

As for the constitutive law of the solid skeleton, Terzaghi's original argument in the theory of soil consolidation is based on experimental observations of the following kind (Terzaghi, 1943). On a saturated soil layer at the bottom of a vertical tube, two alternative loadings are added at the top. If the loading is provided by raising the water level, then only the pore pressure is increased but negligible soil compression is created. If the loading is, instead, a lead shot of equal weight as the added water layer, appreciable soil compression results. Related experiments also show that the excess pore pressure does not cause shear in the soil. These observations imply that the mechanical behavior (deformation, failure, etc.) of the soil skeleton responds to the difference between the total stress $\mathbf{T}$ and the partial stress in water $\mathbf{T}_w = -p\mathbf{I}$, that is, the effective stress $\mathbf{T}^e$ as defined in Eq. (2.19). This concept is widely known as the *effective stress principle* and has been accepted as a fundamental law in soil mechanics. For small strain, Hooke's law may be assumed between the effective stress and the strain in the skeleton. Let $\mathbf{V}$ be the solid displacement which is related to $\mathbf{v}_s$ by

$$\mathbf{v}_s = \frac{\partial \mathbf{V}}{\partial t}. \tag{2.26}$$

and let the components of the effective and total stress tensors be denoted by

$$(\mathbf{T}^e)_{ij} = \sigma_{ij}, \qquad (\mathbf{T})_{ij} = T_{ij} \tag{2.27}$$

respectively. Hooke's law reads

$$\sigma_{ij} = G\left(\frac{\partial V_i}{\partial x_j} + \frac{\partial V_j}{\partial x_i} + \frac{2\nu}{1 - 2\nu}\delta_{ij}\frac{\partial V_k}{\partial x_k}\right), \tag{2.28}$$

where G is the shear modulus and ν the Poisson ratio of the skeleton. In terms of the strain components

$$\epsilon_{ij} = \frac{1}{2}\left(\frac{\partial V_i}{\partial x_j} + \frac{\partial V_j}{\partial x_i}\right),$$

Eq. (2.28) may also be written

$$\epsilon_{ij} = \frac{1}{2G}\sigma_{ij} - \frac{\nu}{2G(1 + \nu)}\sigma_{kk}\delta_{ij}. \tag{2.29}$$

Finally, the interphase force $\mathbf{F}_s = -\mathbf{F}_w$ is due, firstly, to the viscous drag caused by the relative motion. Extending the well-known empirical law of Darcy for steady flow through a rigid skeleton, the drag force on the solid skeleton in unit volume of the mixture can be written as $(n_w^2/K)(\mathbf{v}_w - \mathbf{v}_s)$, where K is the coefficient of permeability which is empirically related to n_w. In addition, the solid skeleton experiences a body force $-p\nabla n_w$; this can be seen as follows.[†] Consider the solid skeleton in the gross (mixture) volume $\mathcal{V}$ surrounded by the surface $\mathcal{S}$. The pore fluid throughout $\mathcal{V}$ exerts the body force

$$\iiint\limits_{\mathcal{V}} (1 - n_w)\nabla p \, d\mathcal{V}, \tag{2.30}$$

while the pore fluid just outside $\mathcal{S}$ exerts the surface force

$$-\iint\limits_{\mathcal{S}} (1 - n_w)p\mathbf{n} \, d\mathcal{S}.$$

[†]The interpretation is due to McTigue and Passman (1982). A more involved argument involving thermodynamics has been given by Prévost (1980).

By Gauss' theorem, the latter is transformed to the volume integral

$$\iiint\limits_{\mathscr{V}} \nabla\left[(1 - n_w)p\right] d\mathscr{V},$$

which can be added to Eq. (2.29) to give

$$- \iint\limits_{\mathscr{V}} p\nabla n_w \, d\mathscr{V}.$$

This amounts to a net body force of $-p\nabla n_w$ on the solid in a unit volume of the mixture, hence

$$\mathbf{F}_s = -\mathbf{F}_w = +\frac{n_w^2}{K}(\mathbf{v}_w - \mathbf{v}_s) - p\nabla n_w. \tag{2.31}$$

Equation (2.21) may now be written

$$n_w\rho_w\dot{\mathbf{v}}_w = -n_w\nabla p + n_w\rho_w\mathbf{g} - \frac{n_w^2}{K}(\mathbf{v}_w - \mathbf{v}_s), \tag{2.32}$$

$$(1 - n_w)\rho_s\dot{\mathbf{v}}_s = -(1 - n_w)\nabla p + \nabla \cdot \mathbf{T}^e + (1 - n_w)\rho_s\mathbf{g} + \frac{n_w^2}{K}(\mathbf{v}_w - \mathbf{v}_s). \tag{2.33}$$

In the limit of steady flow through a rigid skeleton $\partial/\partial t = \mathbf{v}_s = \nabla n_w = 0$; Eq. (2.32) reduces to the classical Darcy's law.

For small strains we shall approximate n_w by its static value n^0. We shall also assume isotropy and homogeneity:

$$n^0 = \text{const}, \qquad K = \text{const}. \tag{2.34}$$

Biot originally included in $\mathbf{F}_s$ and $\mathbf{F}_w$ a term proportional to the relative acceleration between the solid skeleton and the pore fluid. The coefficient amounts to the apparent mass and must be regarded as another empirical parameter. Several writers omit this term altogether. It will be shown later that for wave problems of ocean engineering interest the relative motion is negligible in most of the porous medium, except near the mud line where the acceleration is, however, negligible. Thus the apparent inertia is unimportant everywhere and is excluded from our discussion.

13.2.3 Static Equilibrium

At static equilibrium $\mathbf{v}_s = \mathbf{v}_w = 0$, $\partial/\partial t = 0$. The mass conservation equations (2.12) and (2.14) are identically satisfied. Using the superscript $(\ )^0$ to denote static values of density, pore pressure, and effective stress, we get from Eqs. (2.32) and (2.33)

$$0 = -\nabla p^0 + \rho_w^0 \mathbf{g} \tag{2.35}$$

and

$$0 = \nabla \cdot \mathbf{T}^{e0} - (1 - n^0)\nabla p^0 + (1 - n^0)\rho_s^0 \mathbf{g}. \tag{2.36}$$

The static pore pressure p^0 can be eliminated from Eqs. (2.35) and (2.36) to give

$$\nabla \cdot \mathbf{T}^{e0} = (1 - n^0)\left[\rho_s^0 - \rho_w^0\right]\mathbf{g} \tag{2.37}$$

which can be integrated for the static stress distribution in the solid. For slightly unsaturated pore water ($S > 0.9$, say) ρ_w^0 is nearly a constant. Thus, if $\mathbf{g}$ is gravity and the top surface of the soil is horizontal, we get

$$p^0 = \rho_w^0 g(H - z), \tag{2.38a}$$

$$\sigma_{33}^0 = (1 - n^0)(\rho_s^0 - \rho_w^0)gz, \tag{2.38b}$$

where H is the depth of the water layer above the soil. Clearly, the strain is horizontally isotropic $\varepsilon_{11} = \varepsilon_{22} = 0$ so that

$$\sigma_{11}^0 = \frac{\nu}{1 - \nu}\sigma_{33}^0. \tag{2.38c}$$

13.2.4 Linearized Equations for Water and Solid Skeleton

Let the dynamic perturbations be denoted by ( )′, that is,

$$\sigma_{ij}' = \sigma_{ij} - \sigma_{ij}^0, \qquad p' = p - p^0,$$
$$\rho_w' = \rho_w - \rho_w^0, \qquad n' = n_w - n^0. \tag{2.39}$$

The linearized version of Eq. (2.14) is

$$\frac{\partial n'}{\partial t} + n^0 \nabla \cdot \mathbf{v}_w' + \frac{n^0}{\beta}\frac{\partial p'}{\partial t} = 0 \tag{2.40}$$

after Eq. (2.24) is used. Similarly, from Eq. (2.12)

$$-\frac{\partial n'}{\partial t} + (1 - n^0)\nabla \cdot \mathbf{v}_s' = 0 \tag{2.41}$$

results. The linearized storage equation follows by adding Eqs. (2.40) and (2.41) or, directly from Eq. (2.15),

$$n^0 \nabla \cdot \mathbf{v}'_w + (1 - n^0) \nabla \cdot \mathbf{v}'_s = -\frac{n^0}{\beta} \frac{\partial p'}{\partial t}. \tag{2.42}$$

Thus, the change in pore pressure is related to the dilation rates of pore fluid and the solid skeleton.

Subtracting Eqs. (2.35) and (2.36) from Eq. (2.32) and (2.33), respectively, and linearizing, we obtain for the pore water

$$n^0 \rho_w^0 \frac{\partial \mathbf{v}'_w}{\partial t} = -n^0 \nabla p' - p^0 \nabla n' + n^0 \rho'_w \mathbf{g} - \frac{(n^0)^2}{K} (\mathbf{v}'_w - \mathbf{v}'_s), \tag{2.43}$$

and then for the solid skeleton

$$(1 - n^0) \rho_s^0 \frac{\partial \mathbf{v}'_s}{\partial t} = \nabla \cdot \mathbf{T}^{e\prime} - (1 - n^0) \nabla p' + p^0 \nabla n'$$

$$+ (1 - n^0) \rho'_s \mathbf{g} + \frac{(n^0)^2}{K} (\mathbf{v}'_s - \mathbf{v}'_w). \tag{2.44}$$

We now show that gravity can be ignored in these dynamic equations. Let P_0, ω, and L be the characteristic stress, frequency, and length, respectively. By Hooke's law the characteristic velocity scale is $O(P_0 \omega L / G)$. From Eq. (2.7), the ratio of body force to the pressure gradient is

$$O\left(\frac{\rho'_\alpha \mathbf{g}}{\nabla p'} \right) \sim \frac{(g/\omega) \rho_\alpha^0 \nabla \cdot \mathbf{v}'_\alpha}{\nabla p'} \sim \frac{\rho_\alpha^0 g L}{G} \qquad \alpha = s, w.$$

Furthermore, by using Eqs. (2.38) and (2.41) and estimating $O(H - z) \sim O(L)$, we also get

$$O\left(\frac{P_0 \nabla n'}{\nabla p'} \right) \sim \frac{\rho_s^0 g L}{G}.$$

Since the typical range of G is $10^8 - 10^{10}$ N/m^2 and $O(L) \sim 10 - 100$ m, the above ratio is in the range of $10^{-2} - 10^{-5}$ and is negligible. Therefore, the linearized momentum equations are simply

$$n^0 \rho_w^0 \frac{\partial \mathbf{v}'_w}{\partial t} = -n^0 \nabla p' - \frac{(n^0)^2}{K} (\mathbf{v}'_w - \mathbf{v}'_s), \tag{2.45}$$

and

$$(1 - n^0) \rho_s^0 \frac{\partial \mathbf{v}'_s}{\partial t} = \nabla \cdot \mathbf{T}^{e\prime} - (1 - n^0) \nabla p' + \frac{(n^0)^2}{K} (\mathbf{v}'_w - \mathbf{v}'_s). \tag{2.46}$$

Adding Eqs. (2.45) and (2.46) and invoking Hooke's law, we can show that

$$G\left(\nabla^2 \mathbf{v}_s' + \frac{1}{1-2\nu}\nabla\nabla\cdot\mathbf{v}_s'\right) - \nabla p_t' = n^0\rho_w^0\frac{\partial^2\mathbf{v}_w'}{\partial t^2} + (1-n^0)\rho_s^0\frac{\partial^2\mathbf{v}_s'}{\partial t^2}.$$

$$(2.47)$$

Equation (2.45) can be combined with the storage equation (2.42) to yield

$$K\nabla^2 p' = \nabla\cdot\mathbf{v}_s' + \frac{n^0}{\beta}\frac{\partial p'}{\partial t} - K\rho_w^0\frac{\partial}{\partial t}\nabla\cdot\mathbf{v}_w'. \qquad (2.48)$$

The coupled momentum equations (2.45) and (2.46) or, alternatively, Eqs. (2.47) and (2.48) pose analytical difficulties in general. We shall, therefore, seek an approximation.

13.3   THE BOUNDARY-LAYER APPROXIMATION

Table 3.1 lists the values of the important properties of several soils. The permeabilities K of most soils are small, suggesting intuitively that for sufficiently high frequency the fluid motion in nearly all of the interior of the soil is highly resisted by viscosity and cannot have a significant velocity relative to the solid skeleton. Near the mud line, however, drainage is much easier and relative motion must be appreciable within a thin layer. Thus, one can expect two different regions where different physical mechanisms dominate. In the *outer region*, the length scale is L which may be the wavelength or the dimensions of a large structure, while in the *boundary layer* near the mud line the transverse length scale must be much less than the tangential scale. Let us decompose all unknowns into two parts: the outer part $(\ )_{\text{out}}$ and the boundary-layer correction $(\ )_{\text{b.l.}}$, the latter being appreciable only within the boundary layer.

13.3.1   The Outer Problem

Let us introduce the following outer variables

$$\mathbf{x} = L\boldsymbol{\xi}, \qquad t = \frac{\tau}{\omega},$$

$$\begin{pmatrix}\mathbf{v}_w' \\ \mathbf{v}_s'\end{pmatrix}_{\text{out}} = \frac{P_0\omega L}{G}\begin{pmatrix}\tilde{\mathbf{u}} \\ \tilde{\mathbf{v}}\end{pmatrix} \qquad \begin{pmatrix}p' \\ \sigma_{ij}'\end{pmatrix}_{\text{out}} = P_0\begin{pmatrix}\tilde{p} \\ \tilde{\sigma}_{ij}\end{pmatrix}, \qquad (3.1)$$

where ω is the frequency and P_0 is the typical amplitude of the applied stress. The scales of $\mathbf{v}_w'$ and $\mathbf{v}_s'$ are inferred from Hooke's law. Substituting Eq. (3.1)

Table 3.1 Physical Properties of Certain Ground Materials.[a]

Material	Dolomite	Granite	Limestone	Sandstone	Sand	Dense or Firm Soil	Soil of Medium Firmness	Soft Silt or Clay
Poisson's ratio ν	0.3	0.23–0.27	0.27–0.3	0.12–0.20	0.3–0.35	0.3–0.4		0.4–0.5
Porosity n	0.035	0.13	0.13	0.18	0.25–0.3	0.25–0.3	0.3–0.5	0.15–0.3
Shear Modulus G (10^6 N/m^2)	45×10^3	30×10^3	40×10^3	10×10^3	200	120	40	15
Permeability K (m^3 s/kg)	7×10^{-13}	10^{-16}	5×10^{-14}	2×10^{-10}	$10^{-6} \times -10^{-9}$	$10^{-8}\text{--}10^{-10}$		$\leq 10^{-10}$
Shear wave velocity C_s (m/s)	3950	3230	2900		330	250	150	90
Compressional wave velocity C_p (m/s)		5790	5980	2900				
Boundary-layer thickness for $\omega = 1$ rad/s, Eq. (3.27)	0.18	0.002	0.05	1.50	10.0–1.0	7.0–0.1		0.05

[a] From Mei and Foda (1981a).

into Eq. (2.28), we get

$$\frac{\partial \tilde{\sigma}_{ij}}{\partial \tau} = \frac{\partial \tilde{v}_i}{\partial \xi_j} + \frac{\partial \tilde{v}_j}{\partial \xi_i} + \frac{2\nu}{1 - 2\nu} \delta_{ij} \frac{\partial \tilde{v}_k}{\partial \xi_k}, \tag{3.2}$$

which can be written alternatively as

$$2\frac{\partial \tilde{\varepsilon}_{ij}}{\partial \tau} = \frac{\partial \tilde{v}_i}{\partial \xi_j} + \frac{\partial \tilde{v}_j}{\partial \xi_i} = \frac{\partial \tilde{\sigma}_{ij}}{\partial \tau} - \frac{\nu}{1 + \nu} \delta_{ij} \frac{\partial \tilde{\sigma}_{kk}}{\partial \tau}. \tag{3.3}$$

Similarly, substitution into Eqs. (2.42) and (2.45)–(2.48) yields, respectively,

$$n^0 \frac{\partial}{\partial \xi_j}(\tilde{u}_j - \tilde{v}_j) + \frac{\partial \tilde{v}_j}{\partial \xi_j} = -\frac{n^0 G}{\beta} \frac{\partial \tilde{p}}{\partial \tau}, \tag{3.4}$$

$$\rho_w^0 \frac{\omega^2 L^2}{G} n^0 \frac{\partial \tilde{u}_i}{\partial \tau} = -n^0 \frac{\partial \tilde{p}}{\partial \xi_i} - \frac{(n^0 L)^2 \omega}{GK}(\tilde{u}_i - \tilde{v}_i), \tag{3.5}$$

$$\rho_s^0 \frac{\omega^2 L^2}{G}(1 - n^0)\frac{\partial \tilde{v}_i}{\partial \tau} = \frac{\partial \tilde{\sigma}_{ij}}{\partial \xi_j} - (1 - n^0)\frac{\partial \tilde{p}}{\partial \xi_i} + \frac{(n^0 L)^2 \omega}{GK}(\tilde{u}_i - \tilde{v}_i), \tag{3.6}$$

$$\nabla^2 \mathbf{v} + \frac{1}{1 - 2\nu} \nabla \nabla \cdot \tilde{\mathbf{v}} - \nabla \frac{\partial \tilde{p}}{\partial \tau} = \frac{\rho_w^0 \omega^2 L^2}{G} n^0 \frac{\partial^2 \tilde{\mathbf{u}}}{\partial \tau^2} + \frac{\rho_s^0 \omega^2 L^2}{G}(1 - n^0)\frac{\partial^2 \tilde{v}}{\partial \tau^2}, \tag{3.7}$$

and

$$\nabla^2 \tilde{p} = \frac{\omega L^2}{GK} \nabla \cdot \tilde{\mathbf{v}} + \frac{\omega L^2}{\beta K} n^0 \frac{\partial \tilde{p}}{\partial \tau} - \frac{\rho_w^0 \omega^2 L^2}{G} \frac{\partial}{\partial \tau} \nabla \cdot \tilde{\mathbf{u}}. \tag{3.8}$$

First, $\lambda = (G/\rho_s)^{1/2}\omega^{-1}$ is the shear wavelength in the solid matrix so that

$$\frac{\rho_w^0 \omega^2 L^2 n^0}{G} = O\left(\frac{L}{\lambda}\right)^2. \tag{3.9}$$

For sea waves $\omega = O(0.5 \text{ s}^{-1})$; let $\rho_s^0 = 2.5 \times 10^3 \text{ kg/m}^3$, and $G = 10^8 \text{ N/m}^2$ for dense sand, then $\lambda = O(1000 \text{ m})$. Hence, for $L \ll O(1000 \text{ m})$, inertia may be discarded[†] from Eq. (3.5) to Eq. (3.7). As an immediate consequence Eqs.

[†]If inertia is not ignored, the modified theory will lead to elastodynamics; see Mei and Foda (1981a).

(3.5) and (3.6) may be combined to give

$$\frac{\partial \tilde{\sigma}_{ij}}{\partial \xi_j} = \frac{\partial \tilde{p}}{\partial \xi_i},$$ (3.10)

and Eq. (3.7) is reduced to

$$\nabla^2 \tilde{\mathbf{v}} + \frac{1}{1 - 2\nu} \nabla \nabla \cdot \tilde{\mathbf{v}} - \nabla \frac{\partial \tilde{p}}{\partial \tau} = 0.$$ (3.11)

Next we allow $G/\beta = O(1)$ for generality and consider sufficiently small K and large L so that

$$n^{02} \omega L^2 (GK)^{-1} \gg 1,$$ (3.12)

which is easily the case, as for example, $K = O(10^{-8})$, $L \geq O(10 \text{ m})$. It follows from Eq. (3.5) or (3.6) that

$$\tilde{\mathbf{u}} \cong \tilde{\mathbf{v}},$$ (3.13)

that is, at the leading order the fluid and solid move as a single phase. Furthermore, from Eq. (3.4)

$$\nabla \cdot \mathbf{v} \cong -\frac{n^0 G}{\beta} \frac{\partial \tilde{p}}{\partial \tau}$$ (3.14)

so that the solid dilation and the change of pore pressure affect each other directly. Taking $i = j$ in Eq. (3.2) and summing over the repeated indices, we get

$$2\nabla \cdot \tilde{\mathbf{v}} = \frac{1 - 2\nu}{1 + \nu} \frac{\partial}{\partial \tau} \tilde{\sigma}_{kk},$$

which can be combined with Eq. (3.14) to give

$$\tilde{\sigma}_{kk} = -\frac{2n^0 G(1 + \nu)}{\beta(1 - 2\nu)} \tilde{p}.$$ (3.15)

As soon as the outer effective stresses are found, the outer pore pressure follows at once. With the help of Eq. (3.14), the divergence of Eq. (3.11) gives

$$\nabla^2 \tilde{p} = 0.$$ (3.16)

In view of Eq. (3.10), it is expedient to use the *total outer stress* $\tilde{\tau}_{ij}$ defined by

$$\tilde{\tau}_{ij} = \tilde{\sigma}_{ij} - \tilde{p}\delta_{ij},$$ (3.17)

so that

$$\frac{\partial \tilde{\tau}_{ij}}{\partial \xi_j} = 0. \tag{3.18}$$

Furthermore,

$$\frac{\partial \tilde{\tau}_{ij}}{\partial \tau} = \frac{\partial}{\partial t}\left(\tilde{\sigma}_{ij} - \tilde{p}\delta_{ij}\right) = \frac{\partial \tilde{v}_i}{\partial \xi_j} + \frac{\partial \tilde{v}_j}{\partial \xi_i} + \frac{\lambda_e}{G}\frac{\partial \tilde{v}_k}{\partial \xi_k}\delta_{ij}, \tag{3.19}$$

where

$$\lambda_e = G\frac{2\nu}{1-\nu} + \frac{\beta}{n}, \tag{3.20}$$

which can be called the effective Lamé constant. In terms of $\tilde{\tau}_{ij}$ and λ_e, Eqs. (3.18) and (3.19) are identical to those of elastostatics of a one-phase medium. Therefore, many known techniques of solutions can be employed. Once $\tilde{\tau}_{ij}$ is determined, the pore pressure follows from

$$\tilde{\tau}_{kk} = \tilde{\sigma}_{kk} - 3\tilde{p} = \left[-\frac{2n^0 G(1+\nu)}{\beta(1-2\nu)} - 3\right]\tilde{p}. \tag{3.21}$$

13.3.2   The Boundary-Layer Correction Near a Mud Line

For convenience we shall assume the top surface of the porous medium to be horizontal. Within a thin layer $z = O(\delta) \ll O(L)$, drainage of the pore fluid is much easier and Eq. (3.13) can no longer be true. A correction $(\ )_{b.l.}$ must be added, that is,

$$(\ ) = (\ )_{out} + (\ )_{b.l.}$$

where $(\ )_{b.l.}$ are functions of x, y, z/δ, and t. We expect $(p')_{b.l.}$ and $(\sigma_{ij}^{e'})_{b.l.}$ to be comparable to $(p')_{out}$ and $(\sigma_{ij}^{e'})_{out}$, that is, $O(P_0)$. For the strains to produce stresses of such magnitude, it is necessary that

$$G\frac{\partial}{\partial z}(v_3)_{b.l.} = O(P_0\omega), \qquad (v_3)_{b.l.} = O\left(\frac{P_0\omega\delta}{G}\right)$$

from Hooke's law. Now with inertia ignored, the curl of Eq. (2.47) gives $\nabla^2\nabla \times (\tilde{v})'_{b.l.} = 0$ which means $\partial^2\nabla \times (\tilde{v})_{b.l.}/\partial y^2 = 0$. Since $\nabla \times (\tilde{v})_{b.l.}$ vanishes outside the boundary layer, it must be identically zero throughout the boundary layer. Thus, $(v'_i/v'_3)_{b.l.} = O(\delta/L)$. We can now incorporate these order estimates in the definition of the boundary-layer variables:

$$(x, y) = L(X, Y), \qquad z = \delta Z, \qquad t = \frac{T}{\omega},$$

$$\left(p', \sigma'_{ij}\right)_{\text{b.l.}} = P_0\left(\hat{p}, \hat{\sigma}_{ij}\right)$$

$$\left(v'_{w_i}, v'_{s_i}\right)_{\text{b.l.}} = \frac{(P_0\omega L)}{G} \frac{\delta^2}{L^2}\left(\hat{u}_i, \hat{v}_i\right), \qquad i = 1, 2,$$

$$\left(v'_{w_3}, v'_{s_3}\right)_{\text{b.l.}} = \frac{P_0\omega L}{G} \frac{\delta}{L}\left(\hat{u}_3, \hat{v}_3\right).$$

(3.22)

Equation (2.47) is transformed to

$$\left(\frac{\partial^2}{\partial X^2} + \frac{\partial^2}{\partial Y^2} + \frac{1}{\varepsilon^2}\frac{\partial^2}{\partial Z^2}\right)\begin{vmatrix} \varepsilon^2\hat{v}_1 \\ \varepsilon^2\hat{v}_2 \\ \varepsilon\hat{v}_3 \end{vmatrix} +$$

$$\begin{vmatrix} \dfrac{\partial}{\partial X} \\ \dfrac{\partial}{\partial Y} \\ \dfrac{1}{\varepsilon}\dfrac{\partial}{\partial Z} \end{vmatrix}\left[\frac{1}{1-2\nu}\left(\varepsilon^2\frac{\partial\hat{v}_1}{\partial X} + \varepsilon^2\frac{\partial\hat{v}_2}{\partial Y} + \frac{\partial\hat{v}_3}{\partial Z}\right) - \frac{\partial\hat{p}}{\partial T}\right] = 0$$

where $\varepsilon = \delta/L$, and inertia has already been ignored. Equation (2.48) becomes

$$\left(\frac{\partial^2}{\partial X^2} + \frac{\partial^2}{\partial Y^2} + \frac{1}{\varepsilon^2}\frac{\partial^2}{\partial Z^2}\right)\hat{p} = \frac{\omega L^2}{GK}\left(\varepsilon^2\frac{\partial\hat{v}_1}{\partial X} + \varepsilon^2\frac{\partial\hat{v}_2}{\partial Y} + \frac{\partial\hat{v}_3}{\partial Z}\right) + \frac{\omega L^2}{\beta K}n^0\frac{\partial\hat{p}}{\partial T}.$$

With a relative error of $O(\varepsilon^2)$ the leading-order approximation is

$$\frac{\partial^2\hat{v}_i}{\partial Z^2} + \frac{1}{1-2\nu}\frac{\partial^2\hat{v}_3}{\partial X_i\partial Z} - \frac{\partial^2\hat{p}}{\partial X_i\partial T} = 0, \qquad i = 1, 2, \qquad (3.23a)$$

$$\frac{\partial^2\hat{v}_3}{\partial Z^2} + \frac{1}{1-2\nu}\frac{\partial^2\hat{v}_3}{\partial Z^2} - \frac{\partial^2\hat{p}}{\partial Z\partial T} = 0, \qquad (3.23b)$$

and

$$\frac{1}{\varepsilon^2} \frac{\partial^2 \hat{p}}{\partial Z^2} = \frac{\omega L^2}{GK} \frac{\partial \hat{v}_3}{\partial Z} + \frac{\omega L^2}{\beta K} n^0 \frac{\partial \hat{p}}{\partial T}. \tag{3.24}$$

Equation (3.23b) may be integrated

$$\frac{\partial \hat{v}_3}{\partial Z} = \frac{1 - 2\nu}{2(1 - \nu)} \frac{\partial \hat{p}}{\partial T}, \tag{3.25}$$

which may be combined with Eq. (3.24) to give

$$\frac{1}{\varepsilon^2} \frac{\partial^2 \hat{p}}{\partial Z^2} = \frac{\omega L^2}{K} \left(\frac{1}{G} \frac{1 - 2\nu}{2(1 - \nu)} + \frac{n^0}{\beta} \right) \frac{\partial \hat{p}}{\partial T}, \tag{3.26a}$$

or

$$\frac{\partial^2 \hat{p}}{\partial Z^2} = \frac{\partial \hat{p}}{\partial T} \tag{3.26b}$$

if we define

$$\delta = \left(\frac{KG}{\omega} \right)^{1/2} \left[n^0 \frac{G}{\beta} + \frac{1 - 2\nu}{2(1 - \nu)} \right]^{-1/2}. \tag{3.27}$$

Equation (3.26a) or (3.26b) is just Terzaghi's equation for one-dimensional consolidation. Once $\hat{p}$ is known, Eq. (3.25) gives the transverse component $\hat{v}_3$, while Eq. (3.23a) gives the tangential components $\hat{v}_i$.

A similar approximation of Eq. (3.5) leads to

$$\frac{\partial \hat{p}}{\partial Z} = -\frac{n\omega\delta^2}{KG} (\hat{u}_3 - \hat{v}_3). \tag{3.28}$$

With this result the solid momentum equation (3.6) gives

$$\frac{\partial \hat{\sigma}_{33}}{\partial Z} = \frac{\partial \hat{p}}{\partial Z} \quad \text{and} \quad \frac{\partial \hat{\sigma}_{i3}}{\partial Z} = 0, \quad i = 1, 2.$$

Since all correction terms vanish outside the boundary layer, we further infer

$$\hat{\sigma}_{33} = \hat{p} \quad \text{or} \quad \hat{\tau}_{33} = 0,$$

and

$$\hat{\sigma}_{i3} = \hat{\tau}_{i3} = 0,$$

that is, the total stress components τ'_{33} and τ'_{i3} are dominated by the outer part even within the boundary layer,

$$\tau'_{33} \cong P_0\tilde{\tau}_{33}, \qquad \tau'_{i3} \cong P_0\tilde{\tau}_{i3}. \tag{3.29}$$

The importance of this result will be seen later.

For a simple harmonic motion the solution of Eq. (3.26) is similar to that of the Stokes boundary layer in a viscous fluid:

$$\hat{p} = \mathscr{P}(X_i)\Gamma e^{-i\tau} \quad \text{with } \Gamma(Z) = \exp\left[(1-i)\frac{Z}{2^{1/2}}\right], \tag{3.30a}$$

where $\mathscr{P}$ is to be determined. The solid velocities are

$$\hat{v}_i = \frac{1-2\nu}{2(1-\nu)}\frac{\partial \mathscr{P}}{\partial X_i}\Gamma e^{-i\tau}, \tag{3.30b}$$

$$\hat{v}_3 = \frac{1-2\nu}{2(1-\nu)}\frac{1-i}{2^{1/2}}\mathscr{P}\Gamma e^{-i\tau}. \tag{3.30c}$$

In terms of the outer variables, we have

$$\hat{p} = \mathscr{P}(\xi_i)\Gamma e^{-i\tau}, \tag{3.31a}$$

$$\varepsilon^2\hat{v}_i = \varepsilon^2\frac{1-2\nu}{2(1-\nu)}\frac{\partial \mathscr{P}}{\partial \xi_i}\Gamma e^{-i\tau}, \tag{3.31b}$$

$$\varepsilon\hat{v}_3 = \varepsilon\frac{1-i}{2^{1/2}}\frac{1-2\nu}{2(1-\nu)}\mathscr{P}\Gamma e^{-i\tau}, \tag{3.31c}$$

with $\Gamma = \Gamma(\zeta/\varepsilon)$. The stress components are obviously dominated by $\partial\hat{v}_3/\partial\zeta$:

$$\hat{\sigma}_{ii} = \left[\frac{\nu}{1-\nu}\mathscr{P} + O(\varepsilon^2)\right]\Gamma e^{-i\tau}, \tag{3.32a}$$

$$\hat{\sigma}_{33} = \left[\mathscr{P} + O(\varepsilon^2)\right]\Gamma e^{-i\tau}, \tag{3.32b}$$

$$\hat{\sigma}_{i3} = \left[O(\varepsilon)\right]\Gamma e^{-i\tau}. \tag{3.32c}$$

The total outer stresses are

$$\hat{\tau}_{ii} = -\frac{1-2\nu}{1-\nu}\mathscr{P}\Gamma e^{-i\tau}, \tag{3.33a}$$

$$\hat{\tau}_{33} = 0, \tag{3.33b}$$

$$\hat{\tau}_{i3} = 0 \tag{3.33c}$$

for $i = 1, 2$.

13.3.3. Outline of Solving Poro-elastic Boundary-Value Problems

The results of Eq. (3.29) are particularly welcome as will now be shown in two classes of problems of practical interest.

(i) Traction prescribed on the free surface. Let the boundary values of σ'_{33}, p', and σ'_{3i} be prescribed on the mud-line

$$\tilde{p} + \hat{p} = P(\xi_j), \quad \tilde{\sigma}_{33} + \hat{\sigma}_{33} = N(\xi_j), \quad \tilde{\sigma}_{3i} + \hat{\sigma}_{3i} = T_i(\xi_j), \quad \zeta = 0 \quad (3.34)$$

with $i, j = 1, 2$. Because of Eq. (3.31) the following boundary values are known for the outer problem

$$\tilde{\tau}_{33}(\xi_j, 0) = N - P \quad \text{and} \quad \tilde{\tau}_{3i}(\xi_j, 0) = T_i(\xi_j), \quad i, j = 1, 2. \quad (3.35)$$

Together with Eq. (3.18) and the stress–strain relations, the outer problems for the total stresses are completely formulated and can be solved as ordinary elastostatic problems with traction prescribed on the mud line. Afterward, the outer pore pressure follows from Eq. (3.21), and the outer effective stresses follow from Eq. (3.17). Now the pressure boundary condition on the mud line can be invoked to determine the amplitude $P(\xi_i)$. The remaining unknowns in the boundary layer can finally be obtained from Eqs. (3.29) and (3.30).

(ii) Mixed boundary-value problems. Let the traction be prescribed only on part of the mud line S_1, and the displacements be prescribed on the rest of the mud line S_2. We now have a mixed boundary-value problem which is of obvious relevance to offshore structures. For example, we may assume S_2 to be in welded contact with the base of the rigid structure, that is,

$$v'_s \text{ given}, \quad \frac{\partial}{\partial z}(\tilde{p} + \hat{p}) = 0, \quad \text{or } z = 0. \quad (3.36)$$

The last condition follows from Eq. (2.45) by the fact that under the rigid base the normal components of fluid and solid velocities are equal and inertia is negligible. Because the boundary-layer corrections for the displacements are small to the leading order [see Eq. (3.22)], the boundary values of displacement can be approximated by the outer displacement alone. In view of the preceding subsection the outer boundary-value problem is completely specified and can be solved by any of the known methods for elastostatic problems of the mixed type [see, e.g., Mushkelishvilli (1954)]. To satisfy the condition of $\partial\tilde{p}/\partial\zeta = 0$ a boundary-layer correction can be easily added. Nevertheless, the correction $\tilde{p}$ is $O(\varepsilon)$ because of the small thickness of the boundary layer.

If both the medium properties and the external loading are uniform in the y direction, the entire problem is one of plane strain. In particular, $\epsilon_{22} = 0$ so that

$$\tilde{\sigma}_{22} = \nu(\tilde{\sigma}_{11} + \tilde{\sigma}_{33}). \quad (3.37)$$

By the definitions of linear strain components, a compatibility condition exists among them [see, e.g., Mushkelishvilli (1954)]

$$\frac{\partial^2 \epsilon_{11}}{\partial \zeta^2} + \frac{\partial^2 \epsilon_{33}}{\partial \xi^2} = 2 \frac{\partial^2 \epsilon_{13}}{\partial \xi \, \partial \zeta}. \tag{3.38}$$

Using Hooke's law in the form of Eq. (3.3) and the equilibrium conditions (3.10), (3.15), and (3.37), we may show from Eq. (3.38) that

$$\nabla^2(\tilde{\sigma}_{11} + \tilde{\sigma}_{33}) = \nabla^2(\tilde{\tau}_{11} + \tilde{\tau}_{33}) = 0. \tag{3.39}$$

Now Eq. (3.18) may be identically satisfied by the following stress function $\mathcal{F}$ of Airy:

$$\tilde{\tau}_{11} = \frac{\partial^2 \mathcal{F}}{\partial \xi^2}, \quad \tilde{\tau}_{33} = \frac{\partial^2 \mathcal{F}}{\partial \zeta^2}, \quad \tilde{\tau}_{13} = -2 \frac{\partial^2 \mathcal{F}}{\partial \xi \, \partial \zeta}. \tag{3.40}$$

It follows from Eq. (3.39) that

$$\nabla^2 \nabla^2 \mathcal{F} = 0, \tag{3.41}$$

which is called the biharmonic equation.

Note that because of Eqs. (3.37), Eqs. (3.15) and (3.21) become, respectively,

$$\tilde{\sigma}_{11} + \tilde{\sigma}_{33} = -\frac{2n^0 G}{\beta(1 - 2\nu)} \tilde{p}, \tag{3.42}$$

and

$$\tilde{\tau}_{11} + \tilde{\tau}_{33} = -2(1 + m)\tilde{p}, \tag{3.43}$$

where

$$m = \frac{n^0 G}{\beta} \frac{1}{1 - 2\nu} \quad \left(\approx \frac{G}{\beta} \text{ for } n^0 = \nu = \frac{1}{3} \right) \tag{3.44}$$

is essentially the ratio of solid elasticity to fluid elasticity.

We now turn to specific examples.

13.4 PROGRESSIVE SEA WAVES OVER A POROUS SEABED

13.4.1 Infinitely Thick Seabed

In a sea of constant depth h, let there be a plane progressive wave of amplitude A_0 and frequency ω. If the seabed is perfectly impervious, the dynamic water pressure at its top is

$$p_0 = P_0 e^{i(kx-\omega t)}, \tag{4.1a}$$

where

$$P_0 = \rho g A_0 \operatorname{sech} kh, \tag{4.1b}$$

and k is the wavenumber related to ω via the familiar dispersion relation. The small permeability suggests that the fluid pressure at the seabed can still be given by Eq. (4.1) to the leading order. A more precise estimate is as follows. The ratio of the fluid velocity in the pore to that in the sea is

$$\frac{P_0 \omega}{KG} \left(\frac{gkA_0}{\omega} \right)^{-1} \sim \left(\frac{\omega}{k} \right)^2 (C_s^2)^{-1} \sim \left(\frac{\text{sea wave speed}}{\text{shear wave speed}} \right)^2.$$

For a water depth of $h = 10$–100 m, the fastest wave speed is $(gh)^{1/2} \cong 10$–30 m/s, while $C_s \cong 100$–500 m/s. Thus, the pore water velocity is very small and the no-flux condition and Eq. (4.1), are approximately unaffected.

With k^{-1} as the outer length scale, the solution to Eqs. (3.16) and (3.39) is easily found

$$\tilde{p} = D e^{\zeta} e^{i(\xi-\tau)},$$

$$\tilde{\mathcal{F}} = (B + C\zeta) e^{i(\xi-\tau)}. \tag{4.2}$$

The corresponding total stresses are

$$\left.\begin{aligned}
\tilde{\tau}_{11} &= [B + C(2 + \zeta)] \\
\tilde{\tau}_{33} &= (-B - C\zeta) \\
\tilde{\tau}_{13} &= -i[B + C(1 + \zeta)]
\end{aligned}\right\} e^{\zeta} e^{i(\xi-\tau)}. \tag{4.3}$$

Imposing the boundary conditions on the mud line

$$\tilde{\tau}_{33} = -e^{i(\xi-\tau)}, \quad \tilde{\tau}_{13} = 0 \quad \text{on } \zeta = 0, \tag{4.4}$$

we find

$$B = 1 \quad \text{and} \quad C = -1. \tag{4.5}$$

Substituting Eq. (4.5) into Eq. (3.43), we get

$$D = -\frac{C}{1 + m} = \frac{1}{1 + m}. \tag{4.6}$$

In terms of the outer variables, the boundary-layer correction is

$$
\begin{pmatrix} \hat{p} \\ \hat{\sigma}_{11} \\ \hat{\sigma}_{33} \\ \hat{\sigma}_{13} \end{pmatrix} = \begin{pmatrix} 1 \\ \nu/(1-\nu) \\ 1 \\ 0 \end{pmatrix} \mathscr{P}\Gamma e^{i(\xi-\tau)}, \qquad \Gamma = \Gamma\left(\frac{\zeta}{\varepsilon}\right). \qquad (4.7)
$$

The pressure condition on the mud line

$$
\tilde{p} + \hat{p} = 1 \qquad \text{on } \zeta = 0
$$

determines the last unknown

$$
\mathscr{P} = \frac{m}{1+m}. \qquad (4.8)
$$

The total dynamic stresses may now be summarized:

$$
\left.
\begin{aligned}
\frac{p'}{P_0} &= \tilde{p} + \hat{p} = \left(\frac{1}{1+m}e^{\zeta} + \frac{m}{1+m}\Gamma \right) & (4.9a) \\[2mm]
\frac{\sigma'_{11}}{P_0} &= \tilde{\sigma}_{11} + \hat{\sigma}_{11} = \left[\left(-\frac{m}{1+m} - \zeta \right)e^{\zeta} + \frac{\nu}{1-\nu}\frac{m}{1+m}\Gamma \right] & (4.9b) \\[2mm]
\frac{\sigma'_{33}}{P_0} &= \tilde{\sigma}_{33} + \hat{\sigma}_{33} = \left[\left(-\frac{m}{1+m} + \zeta \right)e^{\zeta} + \frac{m}{1+m}\Gamma \right] & (4.9c) \\[2mm]
\frac{\sigma'_{13}}{P_0} &= i\zeta e^{\zeta} & (4.9d)
\end{aligned}
\right\} e^{i(\xi-\tau)}.
$$

If the pore water is perfectly saturated without air, $m \ll 1$. In the limit of $m = 0$, the boundary-layer terms disappear. For $m \gg 1$ (sandstones) $\tilde{p} \approx 0$ and the outer effective stresses are those for a dry solid. It can be shown (Foda, 1980) that Eq. (4.9) is the asymptotic limit of $\varepsilon \ll 1$ of the exact but more complicated solution by Yamamoto et al. (1978) and Madsen (1978b). We turn to a more general case before presenting some numerical results.

13.4.2   Seabed of Finite Thickness

Biot's full equations can be formally worked out in this case, but the algebra and the associated computations are complicated (Yamamoto, 1977). The boundary conditions at the mud line remain the same as in Section 13.4.1, but at the bottom of the layer ($\zeta = -H$) we assume

$$
v_{s1} = v_{s2} = 0 \qquad (4.10a)
$$

and

$$v_{w3} = 0 \quad \left(\text{thus } \frac{\partial p'}{\partial z} = 0\right). \qquad (4.10b)$$

Let us express these boundary conditions by using the relation between displacements and the Airy stress function, as derived in Appendix 13.A; the result is

$$2\tilde{V}_1 = -\frac{\partial \mathcal{F}}{\partial \xi} + \gamma \int \nabla^2 \mathcal{F}\, d\xi \qquad (4.11a)$$

$$2\tilde{V}_3 = -\frac{\partial \mathcal{F}}{\partial \zeta} + \gamma \int \nabla^2 \mathcal{F}\, d\zeta, \qquad (4.11b)$$

where

$$\gamma = \frac{\lambda_e + 2G}{2(\lambda_e + G)}. \qquad (4.11c)$$

It may be proved after straightforward algebra that the solution satisfying Eqs. (3.41) and (4.4) is

$$\mathcal{F} = \left[\cosh \zeta + \Omega_1 \sinh \zeta + \zeta(\Omega_2 \sinh \zeta - \Omega_1 \cosh \zeta)\right] e^{i(\xi - \tau)}, \qquad (4.12a)$$

with

$$\Omega_1 = \Lambda^{-1}\left[4H + 2(1 - 4\gamma)\sinh 2H\right], \qquad (4.12b)$$

$$\Omega_2 = \Lambda^{-1}\left[2 - 2(1 - 4\gamma)\cosh 2H\right], \qquad (4.12c)$$

$$\Lambda = 2(1 - 4\gamma)\cosh 2H - \left[1 + 4H^2 + (1 - 4\gamma)^2\right]. \qquad (4.12d)$$

Since

$$1 - 4\gamma = -\frac{\lambda_e + 3G}{\lambda_e + G} < 0,$$

it follows that

$$\Lambda < 0, \qquad \Omega_2 < 0.$$

We leave it as an exercise to show that

$$\tilde{p} = -\frac{1}{1+m}(\Omega_2\cosh\zeta - \Omega_1\sinh\zeta)e^{i(\xi-\tau)}, \tag{4.13a}$$

$$\tilde{\sigma}_{11} = \left[\left(1 + \Omega_2\frac{1+2m}{1+m}\right)\cosh\zeta - \Omega_1\frac{m}{1+m}\sinh\zeta\right.$$
$$\left. + \zeta(\Omega_2\sinh\zeta - \Omega_1\cosh\zeta)\right]e^{i(\xi-\tau)}, \tag{4.13b}$$

$$\tilde{\sigma}_{33} = -\left[\left(1 + \frac{\Omega_2}{1+m}\right)\cosh\zeta + \Omega_1\frac{m}{1+m}\sinh\zeta\right.$$
$$\left. + \zeta(\Omega_2\sinh\zeta - \Omega_1\cosh\zeta)\right]e^{i(\xi-\tau)}, \tag{4.13c}$$

$$\tilde{\sigma}_{13} = -i\left[(1 + \Omega_2)\sinh\zeta + \zeta(\Omega_2\cosh\zeta - \Omega_1\sinh\zeta)\right]e^{i(\xi-\tau)}. \tag{4.13d}$$

The boundary-layer correction must satisfy $\hat{p} + \tilde{p} = 1$ at $\zeta = 0$; hence

$$\mathscr{P} = 1 + \frac{\Omega_2}{1+m}. \tag{4.14}$$

The corresponding corrections $\hat{\sigma}_{ij}$ follow from Eq. (4.7) with $\mathscr{P}$ given by Eq. (4.14).

Observe that $\tilde{\sigma}_{11}$ and $\tilde{\sigma}_{33}$ are out of phase with $\tilde{\sigma}_{13}$ by $\pm\frac{1}{2}\pi$. For an infinite layer $H \to \infty$, $\Omega_1 \to 1$, $\Omega_2 \to -1$, and $A \to m/(1+m)$ so that

$$\mathscr{F} \to (1 - \zeta)e^{\zeta}e^{i(\xi-\tau)},$$

which agrees with Eqs. (4.2), (4.5), and (4.6). For perfectly saturated water, $m \ll 1$. Taking the limit of $m \to 0$, we get

$$\gamma \to \frac{1}{2}, \quad \Omega_2 \to -\frac{1 + \cosh 2H}{1 + \cosh 2H + 2H^2}, \quad \mathscr{P} \to \frac{H^2}{\cosh^2 H + H^2}. \tag{4.15}$$

In the two extremes of shallow ($H \to 0$) and thick ($H \to \infty$) layers, $\mathscr{P} \to 0$, and the boundary-layer correction is not needed at the mud line. The maximum of $\mathscr{P}$ occurs when $H = 1.2$.

At the bottom $\zeta = -H$ a minor boundary-layer correction is needed to satisfy $\partial p/\partial\zeta = 0$. We leave it as an exercise to show that the correction is

$$\hat{p} = \left[\frac{\partial\tilde{p}}{\partial\zeta}\Big|_{-H}\right]\frac{1+i}{2^{1/2}}\varepsilon\exp\left[\frac{-1+i}{2^{1/2}\varepsilon}(\zeta + H)\right]. \tag{4.16}$$

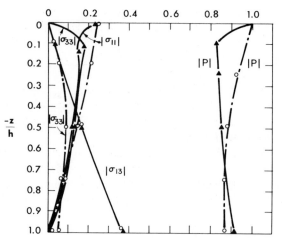

Figure 4.1 Pore pressure and effective stresses in a poro-elastic seabed under a simple progressive wave. All stresses are normalized by P_0 as defined in Eq. (4.1b). Depth of poroelastic layer = 25 m; wavelength = 325 m; wave-period 15 s, water depth = 70 m. Results are shown for: fine sand, $K = 10^{-8}$ m³ s/kg (solid curve: present theory; ▲, Yamamoto); and coarse sand, $K = 10^{-6}$ m³ s/kg (dash–dot curve: present theory, O, Yamamoto). For both sands, complete saturation is assumed ($m \cong 0$) (From Mei and Foda, 1981a, *Geophys. J. the Roy. Astron. Soc.* Reproduced by permission of Royal Astronomical Society.)

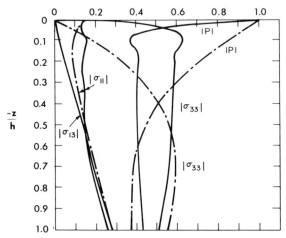

Figure 4.2 Pore pressure and effective stresses in a poro-elastic seabed under a simple progressive wave. Same data as for Fig. 4.1 except that $m = 1.0$ instead, that is, fluid is more compressible due to air entrainment. (From Mei and Foda, 1981a, *Geophys. J. the Roy. Astron. Soc.* Reproduced by permission of Royal Astronomical Society.)

In Fig. 4.1 we plot the stresses in a layer of 25 m induced by sea waves 325 m in length (period = 15 s, water depth = 70 m). These data are representative of the North Sea. The porosity is taken to be $n = \frac{1}{3}$. Two values of permeability are examined. For perfect saturation $m \cong 0$, we find $\delta = 0.98$ m for a fine sand and $\delta = 9.8$ m for a coarse sand. The correction of Eq. 4.16 has been added. The numerical but formally exact solution of Yamamoto (1977) is plotted for comparison. Note that even when the sand layer is not much thicker than the boundary layers, the approximate analytical theory is indistinguishable from the "exact" theory.

Figure 4.2 shows the stresses for a partially unsaturated sand with $m = 1$. Again for a fine sand the pores pressure and σ'_{33} approach constant values rapidly with depth. The variation is more gradual for a coarse sand. Clearly, compressibility helps distribute the loading between fluid and solid.

13.5 RESPONSE TO LOCALIZED OSCILLATING PRESSURE

13.5.1 General Solution

Consider an infinitely thick porous seabed $z < 0$. In physical variables, let the total applied stress at the mud line be

$$\tau'_{33} = -P_0 \mathbb{P}(x) e^{-i\omega t},$$

$$\tau'_{13} = 0. \tag{5.1}$$

After changing to dimensionless variables defined by Eq. (3.1), we apply to Eq. (3.41) the Fourier exponential transform with respect to ξ and obtain

$$\overline{\mathscr{F}} = \frac{\overline{\mathbb{P}}}{\lambda^2}(1 - |\lambda| \zeta) e^{|\lambda|\zeta} e^{-i\tau}, \tag{5.2}$$

where $\overline{\mathscr{F}}$ is the Fourier transform of $\mathscr{F}$ and λ is the Fourier variable. The total outer stresses are

$$\begin{pmatrix} \tilde{\tau}_{11} \\ \tilde{\tau}_{33} \\ \tilde{\tau}_{13} \end{pmatrix} = \frac{1}{2\pi} e^{-i\tau} \int_{-\infty}^{\infty} d\lambda \, e^{i\lambda\xi}(-\overline{\mathbb{P}}) e^{|\lambda|\zeta} \begin{pmatrix} 1 + |\lambda| \zeta \\ 1 - |\lambda| \zeta \\ i\lambda\zeta \end{pmatrix}. \tag{5.3}$$

Note that

$$\tilde{\tau}_{11} + \tilde{\tau}_{33} = -\frac{1}{\pi} e^{-i\tau} \int_{-\infty}^{\infty} d\lambda \, e^{i\lambda\xi} \overline{\mathbb{P}} e^{|\lambda|\zeta} \tag{5.4}$$

which can be used in Eq. (3.43) to give the outer pore pressure

$$\tilde{p} = \frac{e^{-i\tau}}{1+m} \frac{1}{2\pi} \int_{-\infty}^{\infty} d\lambda \, e^{i\lambda\xi} \overline{\mathbb{P}} e^{i\lambda|\zeta|}. \tag{5.5}$$

The outer problem is now completely solved. For the boundary-layer correction, we assume that the hydrodynamic pressure at the mud line is

$$p' = \tilde{p} + \hat{p} = Q(\xi) e^{-i\tau} \quad \text{for } \zeta = 0, |\xi| < \infty. \tag{5.6}$$

Thus, $\hat{p}$ must have the following amplitude:

$$\mathscr{P} = \left(Q - \frac{\mathbb{P}}{1+m} \right) e^{-i\tau}. \tag{5.7}$$

We now treat a specific example.

13.5.2 Disturbance Due to a Horizontal Circular Cylinder Resting on a Seabed of Infinite Thickness

Consider a long circular cylinder fixed on top of an infinitely thick seabed. When the cylinder radius a is much greater than the thickness of the boundary layer in the porous bottom, the theory of Section 13.5.1 can be applied.
    The dynamic problem can be regarded as the sum of two parts: one is due to the progressive waves over a plane bed and the other is due to the disturbance of the cylinder. The first problem has already been solved in Section 13.4.1. For simplicity we assume that the cylinder radius a is much smaller than the wavelength, so that the flow in the neighborhood is essentially the same as that due to a quasi-steady uniform current:

$$Ue^{-i\omega t} \quad \text{where } U = \frac{gkA_0}{\omega} \operatorname{sech} kh. \tag{5.8}$$

Now the flow around a circular pipe on an impervious ground has been found by Jeffreys (1929). In particular, the local velocity potential is

$$\Delta\phi(x,0) e^{-i\omega t} = \left[-Ux + U\frac{\pi a}{2} \frac{\sinh 2\pi a/x}{\sinh^2 \pi a/x} \right] e^{-i\omega t}. \tag{5.9a}$$

The corresponding pressure perturbation is

$$\Delta p(x,0) = i\rho\omega\Delta\phi. \tag{5.9b}$$

For $x/a \gg 1$, $\Delta\phi(x,0) \sim \frac{1}{3}Ux(a\pi/x)^2$, while for $x \to 0\pm$, $\phi \to \pm aU$. Thus, there is a finite jump of $2aU$ across the point of contact $x = 0, z = 0$. Since the Fourier transform of Eq. (5.9) is not easy to find, we replace it by a simpler expression which has roughly the same behavior. In dimensionless terms the applied pressure perturbation is approximated by

$$\mathbb{P}(\xi) = \frac{\Delta p}{i\pi kaP_0} = \frac{2b\xi}{\xi^2 + b^2} \quad \text{with } \xi = \frac{3x}{\pi a} \qquad (5.10)$$

where P_0 is given by Eq. (4.1b). This simulated pressure is not discontinuous, but varies sharply across the origin. By choosing $b = \frac{1}{2}$ the discrepancy from Eq. (5.9) is numerically small (see Fig. 5.1). The Fourier transform of Eq. (5.10) is easily found:

$$\overline{\mathbb{P}}(\lambda) = -2b\pi i \operatorname{sgn} \lambda e^{-b|\lambda|}. \qquad (5.11)$$

We leave it as an exercise to show that

$$\left. \begin{array}{l} \binom{\tilde{\tau}_{11}}{\tilde{\tau}_{33}} = -\dfrac{\xi}{\xi^2 + (\zeta - b)^2} \mp \dfrac{2\xi\zeta(b - \zeta)}{\left[\xi^2 + (\zeta - b)^2\right]^2} \\[4mm] \tilde{\tau}_{13} = \zeta \dfrac{\xi^2 - (\zeta - b)^2}{\left[\xi^2 + (\zeta - b)^2\right]^2} \\[4mm] \tilde{p} = \dfrac{1}{1 + m} \dfrac{\xi}{\xi^2 + (\zeta - b)^2} \end{array} \right\} 2be^{-i\tau}, \qquad (5.12)$$

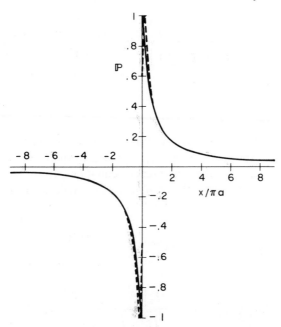

Figure 5.1 Distribution of the sea bottom pressure disturbance due to a pipe; solid curve: theoretical (Eq. (6.2)); dashed curve: the approximation [Eq. (6.3)]. (From Mei and Foda, 1981b, *Geotechnique*. Reproduced by permission of Institution of Civil Engineers.)

with $\zeta = z/\pi a$. The boundary-layer correction is

$$\hat{p} = \frac{m}{1+m} \frac{2b\xi}{\xi^2 + b^2} \exp\left(\frac{1-i}{2^{1/2}} \frac{\zeta}{\varepsilon} - i\tau\right). \tag{5.13}$$

By superposition the dynamic stresses due to the disturbance $\mathbb{P}$ are found:

$$\frac{p'}{i\pi KaP_0} = \frac{2b\xi}{1+m}\left[\frac{1}{\xi^2 + (\zeta - b)^2} + \frac{m}{\xi^2 + b^2}\Gamma\left(\frac{\zeta}{\varepsilon}\right)\right]e^{-i\tau}, \tag{5.14a}$$

$$\frac{\sigma'_{11}}{P_0} = i\pi Ka(2b)\left\{-\frac{m}{1+m}\frac{\xi}{\xi^2 + (\zeta - b)^2}\right.$$

$$\left. -\frac{2\xi\zeta(b - \zeta)}{\left[\xi^2 + (\zeta - b)^2\right]^2} + \frac{\nu}{1+\nu}\frac{\xi}{\xi^2 + b^2}\Gamma\left(\frac{\zeta}{\varepsilon}\right)\right\}e^{-i\tau}, \tag{5.14b}$$

$$\frac{\sigma'_{33}}{P_0} = i\pi Ka(2b)\left\{-\frac{m}{1+m}\frac{\xi}{\xi^2 + (\zeta - b)^2}\right.$$

$$\left. +\frac{2\xi\zeta(b - \zeta)}{\left[\xi^2 + (\zeta - b)^2\right]^2} + \frac{m}{1+m}\frac{\xi}{\xi^2 + b^2}\Gamma\left(\frac{\zeta}{\varepsilon}\right)\right\}e^{-i\tau}, \tag{5.14c}$$

$$\frac{\sigma'_{13}}{P_0} = i\pi Ka\tilde{\tau}_{13}. \tag{5.14d}$$

To get the total dynamic stresses, Eqs. (4.9a)–(4.9d) must be added to Eqs. (5.14a)–(5.14d), respectively. Attention is called to the fact that the normalizing scales are different in these two sets of equations. In the limit of $m \downarrow 0$, the boundary layer is ineffective; the combined stress field is, in the variables ξ, ζ of this section,

$$\frac{\sigma'_{11}}{P_0} = -\frac{\sigma'_{33}}{P_0} = \pi Ka\left\{-\frac{1}{3}\zeta e^{\pi Ka\zeta/3} - \frac{i4b\xi\zeta(b - \zeta)}{\left[\zeta^2 + (b - \zeta)^2\right]^2}\right\}e^{-i\tau}, \tag{5.15a}$$

$$\frac{\sigma'_{13}}{P_0} = i\pi Ka\left\{\frac{1}{3}\zeta e^{\pi Ka\zeta/3} + \frac{2b\zeta\left[\xi^2 - (\zeta - b)^2\right]}{\left[\xi^2 + (\zeta - b)^2\right]^2}\right\}e^{-i\tau}, \tag{5.15b}$$

$$\frac{p'}{P_0} = \left\{e^{\pi Ka\zeta/3} + i\pi Ka\frac{2b\xi}{\xi^2 + (\zeta - b)^2}\right\}e^{-i\tau}. \tag{5.15c}$$

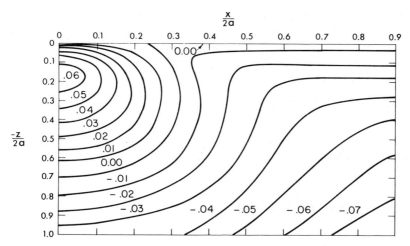

Figure 5.2 Contour lines of the dimensionless shear stress amplitude $\sigma'_{13}/\rho_w gA_0$, where A_0 is the wave amplitude and $\sigma_{13} = i\sigma'_{13}\exp(ikx - i\omega t)$. Effects of both the progressive waves and the cylinder are included. Results are independent of m. (From Mei and Foda, 1981b, *Geotechnique*. Reproduced by permission of Institution of Civil Engineers, London.)

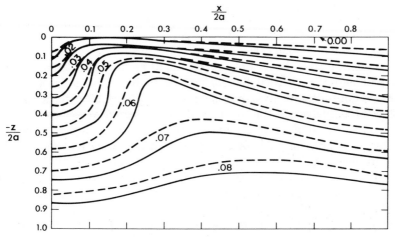

Figure 5.3 Contour lines of the dimensionless total dynamic normal stress; solid curve: $|\sigma_{11}|/\rho_w gA_0$; dashed curve: $|\sigma_{33}|/\rho_w gA_0$; for $m \cong 0$ (no air entrainment) (From Mei and Foda, 1981b, *Geotechnique*. Reproduced by permission of Institution of Civil Engineers, London.)

Note first that the dynamic disturbance due to the small cylinder is always small, $O(\pi ka)$. However, the static stress near the mud line is also small so that the former is not necessarily unimportant in failure considerations. Second, for incompressible pore fluid $m \cong 0$, the wave effect and the cylinder effect are both small but comparable near the mud line. Numerical results combining the dynamic effects of the progressive wave and of the cylinder are presented here for $n = \frac{1}{3}$. Figure 5.2 gives the effective shear stress which is independent of m. In Fig. 5.3 both σ'_{11} and σ'_{33} are shown for fully saturated water $m \cong 0$ which corresponds to Eq. (5.15). The pore pressure is dominated by the wave effect and need not be plotted here.

13.6 CONCLUDING REMARKS

The theory developed in this chapter can be applied to find the wave-induced stresses and pore pressure beneath a gravity structure. If the structure is rigid and its amplitude of motion is small compared to that of the water waves, we can use the theory of Chapter Seven to compute the wave pressure on the structure and on the sea floor. As in floating-body dynamics, the total stress field in the seabed can be decomposed into a *diffraction* problem in which the structure is held fixed, and six *radiation* problems in which the structure executes one of the six possible modes. The amplitudes of the radiation modes can be found from the dynamical equilibrium of the structure. Afterward, boundary-layer corrections should be made near the unsealed part of the mud line. By this scheme, two-phase problems in poro-elasticity are computationally no more difficult than the corresponding problems in the classical one-phase elasticity. Results so obtained should be valuable for guiding preliminary designs and for delineating the regions where soil nonlinearity is important. Mynett (1980) has studied the two-dimensional problem of a long caisson attacked by waves on one side.

For nonlinear soils the idea of superposition can no longer be applied; the computational effort needed for engineering problems is immense and the results are not easy to verify. Nevertheless, the boundary-layer approximation, which is a consequence of contrasting scales (wave period versus consolidation time), can still facilitate future efforts in nonlinear modelling of seabeds. In particular, one need only pursue a one-phase theory away from the mud line and a two-phase but *one-dimensional* theory near the mud line.

Finally, there are important problems concerning the effects of the porous seabed on ocean waves. Wave attenuation due to percolation has been treated in earlier works on the basis of rigid skeletons (see Liu, 1973, and references therein). In a deformable seabed, however, viscous dissipation is dominated by the relative motion between pore water and the soil skeleton in the boundary layer below the mud line; it is not difficult to calculate this effect by the analysis of Chapter Eight. Solid friction among grains, important in the damping of sound through marine sediments (Stoll and Bryan, 1979), may also be important here (Yamamoto, 1982). Thus, there are ample reasons for the ocean-wave researcher to look deeper into the mechanics of the sea bottom.

APPENDIX 13.A: LOVE'S RELATION BETWEEN DISPLACEMENT AND AIRY'S STRESS FUNCTION

In physical variables we write Eq. (3.19) as

$$\tau_{ij} = G\left(\frac{\partial V_i}{\partial x_j} + \frac{\partial V_j}{\partial x_i}\right) + \lambda_e \epsilon \delta_{ij}, \qquad \epsilon = \frac{\partial V_k}{\partial x_k},$$

where V_i denotes the displacement components. For plane strain

$$\epsilon = \frac{\partial V_1}{\partial x} + \frac{\partial V_3}{\partial z},$$

thus,

$$\tau_{11} + \tau_{33} = \nabla^2 \mathscr{F} = 2(\lambda_e + G)\epsilon,$$

so that

$$2G\frac{\partial V_1}{\partial x} = \tau_{11} - \lambda_e \epsilon = \frac{\partial^2 \mathscr{F}}{\partial z^2} - \frac{\lambda_e}{2(\lambda_e + G)}\nabla^2 \mathscr{F}$$

$$= -\frac{\partial^2 \mathscr{F}}{\partial x^2} + \frac{\lambda_e + 2G}{2(\lambda_e + G)}\nabla^2 \mathscr{F}.$$

Integrating with respect to x, we have

$$2GV_1 = -\frac{\partial \mathscr{F}}{\partial x} + \gamma \int \nabla^2 \mathscr{F}\, dx,$$

where

$$\gamma = \frac{\lambda_e + 2G}{2(\lambda_e + G)}.$$

Similarly,

$$2GV_2 = -\frac{\partial \mathscr{F}}{\partial y} + \gamma \int \nabla^2 \mathscr{F}\, dy.$$

REFERENCES

Several relevant publications which have not been cited in the text are also included here.

Abbot, M. B. (1979). *Computational Hydraulics*, Pitman, New York.

Ablowitz, M. J. and A. C. Newell (1973). The decay of the continuous spectrum for solutions of the Korteweg–deVries equation. *J. Math. Phys.* **14**: 1277–1284.

Ablowitz, M. J., D. J. Kaup, A. C. Newell, and H. Segur (1974). The inverse scattering transform —Fourier analysis for nonlinear problems. *Studies Appl. Math.* **LIII 4**: 249–336.

Ablowitz, M. J. and H. Segur (1981). *Solitons and the Inverse Scattering Transform*, Society Industrial and Applied Mathematics, Philadelphia.

Abramowitz, M. and I. A. Stegun (1972). *Handbook of Mathematical Functions*, Dover, New York.

Adams, N. K. (1941). *The Physics and Chemistry of Surfaces*, Oxford University Press, London.

Aranha, J. A., C. C. Mei, and D. K. P. Yue (1979). Some properties of a hybrid element method for water waves. *Int. J. Num. Methods Eng.* **14**: 1627–1641.

Armstrong, J. A., N. Bloembergen, J. Ducuing, and P. S. Pershan (1962). Interactions between light waves in a nonlinear dielectric. *Phys. Rev.* **127**: 1918–1939.

Arthur, R. S. (1946). Refraction of water waves by islands and shoals with circular bottom contours. *Trans. Am. Geophys. Union* **27**: 168–177.

Arthur, R. S. (1962). A note on the dynamics of rip currents. *J. Geophys. Res.* **67**: 2777–2779.

Atkin, R. J. and R. E. Craine (1976). Continuum theory of mixture: applications. *J. Inst. Math. Appl.* **17**: 153–207.

Bagnold, R. A. (1946). Sand movement by waves: some small scale experiments with sand of very low density. *J. Inst. Civil Eng.* **27**: 457.

Bai, K. J. and R. Yeung (1974). Numerical solutions of free-surface and flow problems. *Proc. 10th Symp. Naval Hydrodyn.* Office of Naval Research, 609–641.

Bartholomeuz, E. F. (1958). The reflection of long waves at a step. *Proc. Cambridge Philos. Soc.* **54**: 106–118.

Batchelor, G. K. (1967). *An Introduction to Fluid Dynamics*, Cambridge University Press, London.

Battjes, J. A. (1972). Set up due to irregular waves. *Proc. 13th Conf. Coastal Eng. ASCE* **2**: 1993–2004.

Battjes, J. A. (1974a). Computation of set-up long shore currents, run-up and overtopping due to wind generated waves. *Communications on Hydraulics*, Dept. of Civil Engineering, Delft University of Technology Report 74-2.

Battjes, J. A. (1974b) Surf similarity. *Proc. 14th Conf. Coastal Eng. ASCE* 466–480.

Battjes, J. A. (1975). Modeling of turbulence in the surf zone. *Proc. Symp. Modeling Techniques ASCE*. 1050–1061.

Benjamin, T. B. (1967). Instability of periodic wave trains in nonlinear dispersive systems. *Proc. R. Soc. Lond. A* **299**: 59–75.

706

Benjamin, T. B. and J. E. Feir (1967). The disintegration of wave trains on deep water. *J. Fluid Mech*. **27**: 417–430.

Benjamin, T. B. and M. J. Lighthill (1954). On cnoidal waves and bores. *Proc. R. Soc. Lond. A* **244**: 448–460.

Benney, D. J. (1962). Nonlinear gravity wave interactions. *J. Fluid Mech*. **14**: 574–584.

Benney, D. J. (1966). Long nonlinear waves in fluid flows. *J. Math. Phys*. **45**: 52–63.

Benney, D. J. and J. C. Luke (1964). On the interactions of permanent waves of finite amplitude. *J. Math. Phys*. **43**: 309–313.

Benney, D. J. and A. C. Newell (1967). The propagation of nonlinear wave envelopes. *J. Math. Phys*. **46**: 133–139.

Benney, D. J. and G. J. Roskes (1969). Wave instabilities. *Studies Appl. Math* **48**: 377–385.

Berger, U. and S. Kohlhase (1976). Mach reflection as a diffraction problem. *Proc. 15th Conf. Coastal Eng. ASCE* **1**: 796–814.

Berkhoff, J. C. W. (1972). Computation of combined refraction–diffraction. *Proc. 13th Conf. Coastal Eng. ASCE* **1**: 471–490.

Bessho, M. (1965). On the wave-free distribution in the oscillation problem of the ship. *J. Zosen Kiokai* **117**: 127–138.

Bessho, M. (1967). On the two-dimensional theory of the rolling motion of ships. *Mem. Defense Acad. Yokoyuka* **7**: 105–125.

Bigg, G. R., (1982). Diffraction and trapping of waves by cavities and slender bodies. Ph.D. thesis, Depart. of Applied Mathematics, University of Adelaide, Australia.

Biot, M. A. (1941). General theory of three-dimensional consolidation. *J. Appl. Phys*. **12**: 155–164.

Biot, M. A. (1956). Theory of propagation of elastic waves in a fluid saturated porous solid I. Low frequency range II. High frequency range. *J. Acoust. Soc. Am*. **28**: 168–191.

Boczar-Karakiewicz, B. (1972). Transformation of wave profile in shallow water—a Fourier analysis. *Arch. Hydrotechniki* **19**: 197–210.

Bohm, D. (1951). *Quantum Theory*, Prentice-Hall, Englewood Cliffs, N.J.

Bolt, B. A. (1978). *Earthquakes—A Primer*, Freeman, San Francisco.

Booij, N. (1981). Gravity waves on water with non-uniform depth and current. *Communications on Hydraulics*, Dept. of Civil Engineering, Delft University of Technology, Report No. 81-1.

Boussinesq, J. (1877). Essai sur la théorie des eaux courantes. *Mém. Prés. Acad. Sci. Paris. (Ser. 2)* **23**: 1–680.

Boussinesq, J. (1878). Complement a úne étude intitulée: "Essai sur la theorie des eaux courante" et á un memoire "Sur l'influence des frottments sur leo mouvements regulier des fluids." *J. Math. Pures Appl*. **4**: 335.

Bowen, A. J. (1969). The generation of longshore currents on a plane beach. *J. Marine Res*. **27**: 206–214.

Bowen, A. J. (1969). Rip currents, I. Theoretical investigations. *J. Geophys. Res*. **74**: 5467–5478.

Bowen, A. J. (1972). Edge waves and the littoral environment. *Proc. 13th Conf. Coastal Eng. ASCE* 1313–1320.

Bowen, A. J. and R. T. Guza (1978). Edge waves and surf beat. *J. Geophys. Res*. **83**: 1913–1920.

Bowen, A. J. and D. L. Inman (1969). Rip currents II. Laboratory and field observations. *J. Geophys. Res*. **74**: 5479–5490.

Bowen, A. J., D. L. Inman, and V. P. Simmons (1968). Wave set-down and set-up. *J. Geophys. Res*. **73**: 2569–2577.

Braddock, R. D., P. Van den Driessche, and G. W. Peady (1973). Tsunami-generation. *J. Fluid Mech*. **59**: 817–828.

Bremmer, H. (1951). The WKB approximation as the first term of a geometric optical series. *Comm. Pure Appl. Math*. **4**: 105–115.

Bretherton, F. P. (1964). Resonant interaction between waves. *J. Fluid Mech.* **20**: 457–480.

Bretherton, F. P. and C. J. R. Garrett (1969). Wave trains in inhomogeneous moving media. *Proc. R. Soc. Lond.* **A 302**: 529–554.

Brevik, I. and B. Aas (1980). Flume experiments on waves and currents, I. Rippled bed. *Coastal Eng.* **3**: 149–177.

Bryant, P. J. (1973). Periodic waves in shallow water. *J. Fluid Mech.* **59**: 625–644.

Buchwald, V. T. (1971). The diffraction of tides by a narrow channel. *J. Fluid Mech.* **46**: 501–511.

Budal, K. (1977). Theory of absorption of wave power by a system of interacting bodies. *J. Ship Res.* **21**: 248–253.

Budal, K. and J. Falnes (1975). A resonant point absorber of ocean wave power. *Nature* **256**: 478–479; **257**: 626–627.

Budal, K., J. Falnes, A. Kyllingstad, and G. Oltedal (1979). Experiments with point absorbers. *Proc. 1st Symp. Wave Energy Utilization.* Chalmers Institute of Technology, Sweden, 253–282.

Byatt-Smith, J. G. B., (1970). An exact integral equation for steady surface waves. *Proc. R. Soc. Lond.* **A 315**: 405–418.

Byatt-Smith, J. G. B. and M. S. Longuet-Higgins (1976). On the speed and profile of steep solitary waves. *Proc. R. Soc. Lond.* **A 350**: 175–189.

Carrier, G. F. (1966). Gravity waves on water of variable depth. *J. Fluid Mech.* **24**: 641–659.

Carrier, G. F. (1970). The dynamics of tsunamis. *Mathematical Problems in the Geophysical Sciences. I. Geophysical Fluid Dynamics*, American Mathematical Society, Providence, R.I., 157–181.

Carrier, G. F. and H. P. Greenspan (1957). Water waves of finite amplitude on a sloping beach. *J. Fluid Mech.* **4**: 97–109.

Carrier, G. F., M. Krook, and C. E. Pearson (1966). *Functions of a Complex Variable—Theory and Technique.* McGraw-Hill, New York,

Carrier, G. F. and R. P. Shaw (1969). *Tsunamis in the Pacific Ocean*, edited by W. M. Adams, East West Center Press, Honolulu, 377–398.

Carrier, G. F., R. P. Shaw, and M. Miyata (1971). The response of narrow mouthed harbors in a straight coastline to periodic incident waves. *J. Appl. Mech.* **38 E-2**: 335–344.

Carter, T. G., P. L. F. Liu, and C. C. Mei (1973). Mass transport by waves and offshore sand bedforms. *J. Waterways, Harbours Coastal Eng. Div. ASCE* **99**: 165–184.

Case, K. M. and W. C. Parkinson (1957). Damping of surface waves in an incompressible liquid. *J. Fluid Mech.* **2**: 172–184.

Chao, Y. Y. (1971). An asymptotic evaluation of the wave field near a smooth caustic. *J. Geophys. Res.* **76**: 7401–7408.

Chao, Y. Y. and W. J. Pierson, Jr. (1972). Experimental studies of the refraction of uniform wave trains and transient wave groups near a straight caustic. *J. Geophys. Res.* **77**: 4545–4554.

Charkrabarti, S. K. (1972). Nonlinear wave forces on vertical cylinders. *J. Hydraul. Div. ASCE* **98**: 1895–1909.

Charkrabarti, S. K. (1978). Comments on second order wave effects on large diameter vertical cylinder. *J. Ship Res.* **22**: 266–268.

Charkrabarti, S. K. and W. A. Tam (1975). Interaction of waves with a large vertical cylinder. *J. Ship Res.* **19**: 23–33.

Chen, T. G. (1961). Experimental studies on the solitary wave reflection along a straight sloped wall at oblique angle of incidence. *U.S. Beach Erosion Board Tech. Mem.* **124**.

Chen, H. S. and C. C. Mei (1974a). Oscillations and wave forces in an offshore harbor. Parsons Lab., Massachusetts Institute of Technology, Report 190.

Chen, H. S. and C. C. Mei (1974b). Oscillations and wave forces in a man-made harbor in the open sea. *Proc. 10th Symp. Naval Hydrodyn.*, Office of Naval Research, 573–594.

Chu, V. C. and C. C. Mei (1970). On slowly varying stokes waves. *J. Fluid Mech.* **41**: 873–887.

Chu, V. C. and C. C. Mei (1971). The nonlinear evolution of stokes waves in deep water. *J. Fluid Mech.* **47**: 337–352.

Cokelet, E. D., (1977). Steep gravity waves in water of arbitrary uniform depth. *Philos. Trans. R. Soc. Lond. A* **286**: 183–230.

Cole, J. D. (1968). *Perturbation Methods in Applied Mathematics*, Blaisdell, Waltham, Mass.

Collin, R. E. (1960). *Field Theory of Guided Waves*, McGraw-Hill, New York.

Courant, R. and K. O. Friedrichs (1949). *Supersonic Flow and Shock Waves*, Interscience, New York.

Courant, R. and D. Hilbert (1962). *Methods of Mathematical Physics II*. Interscience, New York.

Crapper, G. D. (1972). Nonlinear gravity waves on steady non-uniform currents. *J. Fluid Mech.* **52**: 713–724.

Crapper, G. D. (1979). Energy and momentum integrals for progressive capillary-gravity waves. *J. Fluid Mech.* **94**: 13–24.

Crawford, D. R., B. M. Lake, P. G. Saffman, and H. C. Yuen (1981). Stability of weakly nonlinear wave in two or three dimensions. *J. Fluid Mech.* **105**: 177–191.

Cummins, W. E. (1962). The inpulse response functions and ship motion. *Schiffstechnik* **9**: 101–109.

Dailey, J. W. and S. C. Stephan, Jr. (1952). The solitary wave—its celerity, profile, internal velocities and amplitude attenuation in a horizontal smooth channel, *Proc. 3rd Conf. Coastal Eng. ASCE*, 13–30.

Dalrymple, R. A. (1975). A mechanism for rip current generation on open coast. *J. Geophys. Res.* **80**: 3485–3487.

Davey, A. (1972). The propagation of a weak nonlinear wave. *J. Fluid Mech.* **53**: 769–781.

Davey, A. and K. Stewartson (1974). On three-dimensional packets of surface waves. *Proc. R.. Soc. London A* **338**: 101–110.

Davey, N. (1944). The field between equal semi-infinite rectangular electrodes on magnetic pole-pieces. *Philos. Mag.* **35**: 819–844.

Dean, R. G. (1965). Stream function representation of nonlinear ocean waves. *J. Geophy. Res.* **70**: 4561–4572.

Dean, W. R. (1945). On the reflection of surface waves by a submerged plane barrier. *Proc. Cambridge Philos. Soc.* **41**: 231–238.

De Best, A. and E. W. Bijker (1971). Scouring of a sand bed in front of a vertical breakwater. *Communications on Hydraulics*, Dept. of Civil Engineering, Delft University of Technology, Report 71-1.

Deresiewicz, H. (1960). The effect of boundaries on wave propagation in a liquid-filled porous solid. I. Reflection of plane work at a free plane boundary (non-dissipative case). *Bull. Seis. Soc. Am.* **50**: 599–607.

Deresiewicz, H. (1961). The effect of boundaries on wave propagation in a liquid-filled porous liquid. II. Love waves in a porous layer. *Bull. Seis. Soc. Am.* **51**: 51–59.

Deresiewicz, H. (1962a). The effect of boundaries on wave propagation in a liquid-filled porous solid. II. Reflection of plane waves at a full plane boundary (general case). *Bull. Seis. Soc. Am.* **52**: 595–625.

Deresiewicz, H. (1962b). The effect of boundaries on wave propagation in a liquid-filled porous solid. IV. Surface waves in a half space. *Bull. Seis. Soc. Am.* **52**: 627–638.

Dingemans, M. (1978). Refraction and diffraction of irregular waves, a literature survey. *Delft Hydraulics Laboratory Report*, W301, Part I.

Djordjevic, V. D. and L. G. Redekopp (1978). On the development of packets of surface gravity wave moving over an uneven bottom. *J. Appl. Math. Phys.* **29**: 950–962.

Donelan, M., M. S. Longuet-Higgins, and J. S. Turner (1972). Periodicity in Whitecaps. *Nature* **239**: 449–451.

Dore, B. D. (1969). The decay of oscillations of a non-homogeneous fluid within a container. *Proc. Cambridge Philos. Soc.* **65**: 301–307.

Dore, B. D. (1976). Double boundary layers in standing surface waves. *Pure Appl. Geophys.* **114**: 629–637.

Dore, B. D. (1977). On mass transport velocity due to progressive waves. *Q. J. Mech. Appl. Math.* **30**: 157–173.

Dore, B. D. (1978). Some effects of the air–water interface on gravity waves. *Geophys. Astrophs. Fluid Dynamics* **10**: 215–230.

Dysthe, K. B. (1979). Note on a modification to the nonlinear Schrödinger equation for application to deep water waves. *Proc. R. Soc. Lond. A* **369**: 105–114.

Eagleson, P. S. (1956). Properties of shoaling waves by theory and experiment. *Trans. Am. Geophys. Union* **37**: 565–572.

Eagleson, P. S. (1965). Theoretical study of longshore currents on a plane beach. Hydraulics Lab., Massachusetts Institute of Technology, Technical Report 82.

Eckart, C. (1951). Surface waves in water of variable depth. Marine Physical Lab. of Scripps Inst. Ocean. Wave Report 100–99.

Erdelyi, A. (ed) (1954). *Tables of Integral Transform. I. Bateman Manuscript Project*, McGraw-Hill, New York.

Euvrard, D., A. Jami, M. Lenoir, and D. Martin (1981). Recent progress towards an optimum coupling of finite elements and singularity distribution. *Proc. 3rd Intl. Symp. Num. Ship Hydrodyn.* Paris.

Evans, D. V. (1976). A theory for wave power absorption by oscillating bodies. *J. Fluid Mech.* **77**: 1–25.

Evans, D. V. (1978). The oscillating water column wave-energy device. *J. Inst. Math. Appl.* **22**: 423–433.

Evans, D. V. (1979). Some theoretical aspects of three-dimensional wave energy absorbers. *Proc. 1st. Symp. on Wave Energy Utilization*, Chalmers Institute of Technology, Sweden, 77–113.

Evans, D. V. (1981). Power from water waves. *Ann. Rev. Fluid Mech.* **13**: 157–187.

Evans, D. V., D. C. Jeffrey, S. H. Salter, and J. R. M. Taylor (1979). Submerged cylinder wave energy device: Theory and experiment. *Appl. Ocean Res.* **1**: 3–12.

Falnes, J. (1980). Radiation impedance matrix and optimum power absorption for interacting oscillations in surface waves. *Appl. Ocean Res.* **2**: 75–80.

Falnes, J. and K. Budal (1978). Wave power conversion by point absorbers. *Norwegian Maritime Res.* **6**: 211.

Faltinsen, O. M. and A. E. Løken (1979). Slow drift oscillations of a ship in irregular waves. *Appl. Ocean Res.* **1**: 21–31.

Feir, J. E. (1967). Some results from wave pulse experiments. Discussion of M. S. Lighthill: Some special cases treated by the Whitham theory *Proc. R. Soc. London A* **299**: 54–58.

Felsen, L. B. and N. Marcuvitz (1973). *Radiation and Scattering of Waves*, Prentice-Hall, Englewood Cliffs, N. J.

Fenton, J. (1972). A nineth-order solution for the solitary wave. *J. Fluid Mech.* **53**: 257–271.

Fenton, J. (1979). A higher-order cnoidal wave theory. *J. Fluid Mech.* **94**: 129–161.

Finkelslein, A. (1953). The initial value problem for transient water waves, Dissertation, New York University.

Foda, M. A. (1980). I. Dynamics of fluid-filled porous media. II. Excitation of surf beats in the ocean. Sc.D. Thesis, Dept. of Civil Engineering, Massachusetts Institute of Technology.

Foda, M. A. and C. C. Mei (1981). Nonlinear excitation long trapped waves by a group of short swells. *J. Fluid Mech.* **111**: 319–345.

Fornberg, B. and G. B. Whitham (1978). A numerical and theoretical study of certain nonlinear wave phenomena. *Philos. Trans. R. Soc. Lond.* **289**: 373–404.

Frank, W. (1967). Oscillation of cylinders in or below the free surface of deep fluids. Naval Ship Research and Development Center Report 2375.

Frank, W. (1967). The heaving damping coefficients of bulbous cylinders partially immersed in deep water. *J. Ship Res.* **11**: 151–153.

French, M. J. (1979). The search for low cost wave energy and the flexible bag device. *Proc. 1st. Symp. Wave Energy Utilization*, Chalmers Institute of Technology, Sweden. 364–377.

Friedrichs, K. O. (1948a). On the derivation of the shallow water theory. *Comm. Pure Appl. Math.* **1**: 81–85.

Friedrichs, K. O. (1948b) Water waves on a shallow sloping beach. *Commun. Pure Appl. Math.* **1**: 109–134.

Gallagher, B. (1971). Generation of surf beat by non-linear wave interactions. *J. Fluid Mech.* **49**: 1–20.

Galvin, C. J., Jr. (1968). Breaker type classification on three laboratory beaches. *J. Geophys Res.* **73**: 3651–3659.

Galvin, C. J. and P. S. Eaglesen (1965). Experimental study of longshore currents on a plane beach. U.S. Army Coastal Engineering Research Center Technical Memorandum 10.

Garabedian, P. R. (1964). *Partial Differential Equations*, McGraw-Hill, New York.

Gardner, C. S., J. M. Greene, M. D. Kruskal, and R. M. Miura (1967). Method for solving the Korteweg–de Vries equation. *Phys. Rev. Lett.* **19**: 1095–1096.

Gardner, C. S., J. M. Greene, M. D. Kruskal, and R. M. Miura (1974). Korteweg–de Vries equation and generalizations, VI: Methods for exact solution. *Commun. Pure Appl. Math.* **27**: 97–133.

Garrett, C. J. C. (1970). Bottomless harbours. *J. Fluid Mech.* **43**: 432–449.

Geerstma, J. and D. C. Smit (1961). Some aspects of elastic wave propagation in fluid saturated porous solids. *Geophysics* **26**: 169–181.

Gelfand, I. M. and B. M. Levitan (1955). On the determination of a differential equation from its spectral function. *Am. Math. Soc. Transl.* **1**: 253–304.

Gerwick, B. C., Jr. and E. Hognestad (1973). Concrete oil storage tank placed on North Sea floor. *Civil Eng. ASCE* **43**: 81–85.

Goda, Y. (1967). Travelling secondary wave in channels. Port and Harbour Research Institute, Ministry of Transport, Japan. Report 13: 32.

Goldstein, H. (1950). *Classical Mechanics*, Addison-Wesley, Reading, Mass.

Gradshteyn, I. S. and I. A. Ryzhik (1965). *Tables of Integrals Series and Products*, Academic, New York.

Graham, J. M. R. (1980). The forces on sharp edged cylinders in oscillatory flow at low Keulegan–Carpenter numbers. *J. Fluid Mech.* **97**: 331–346.

Grant, W. D. (1977). Bottom friction under waves in the presence of a weak current: Its relation to coastal sediment transport, Sc.D. Thesis, Dept. of Civil Engineering, Massachusetts Institute of Technology.

Grant, W. D. and O. S. Madsen (1979a). Bottom friction under waves in the presence of a weak current. Tech. Mem. ERL-MESA, National Oceanic and Atmospheric Administration.

Grant, W. D. and O. S. Madsen (1979b). Combined wave and current interaction with rough bottom. *J. Geophys. Res.* **84**: 1797–1808.

Grant, W. D. and O. S. Madsen (1982). Movable bed roughness in unsteady oscillatory flow. *J. Geophy. Res.* **87**: 469–481.

Greenspan, H. P. (1956). The generation of edge waves by moving pressure distributions. *J. Fluid Mech.* **1**: 574–590.

Greenspan, H. P. (1958). On the breaking of water waves of finite amplitude on a sloping beach. *J. Fluid Mech.* **4**: 330–334.

Greenspan, H. P. (1968). *The Theory of Rotating Fluids*, Cambridge University Press, London.

Guiney, D. C., B. J. Noye, and E. O. Tuck (1972). Transmission of waves through small apertures. *J. Fluid Mech.* **55**: 149–167.

Guza, R. T. and A. J. Bowen (1976). Finite amplitude Stokes edge waves. *J. Marine Res.* **34**: 269–293.

Guza, R. T. and D. C. Chapman (1979). Experimental study of the instabilities of waves obliquely incident on a beach. *J. Fluid Mech.* **95**: 199–208.

Guza, R. T. and R. E. Davis (1974). Excitation of edge waves by waves incident on a beach. *J. Geophys. Res.* **79**: 1285–1291.

Guza, R. T. and D. L. Inman (1975). Edge waves and beach cusps. *J. Geophys. Res.* **80**: 2997–3012.

Guza, R. T. and E. B. Thornton (1982). Swash oscillations on a beach. *J. Geophys. Res.* **87**: 483–491.

Hagen, G. E. (1975). Wave-driven generator. U.S. Pat. 4,077,213.

Hammack, J. L. and H. Segur (1978). Modelling criteria for long water waves. *J. Fluid Mech.* **84**: 359–373.

Hanaoka, T. (1959). On the reverse flow theorem concerning wave-making theory. *Proc. 9th Japan Nat'l Congr. Appl. Mech.* 223–226.

Haren, P. and C. C. Mei (1980). Rafts for absorbing wave power. *Proc. 13th Symp. Naval Hydrodyn.* The Ship Building Research Institute, Japan, 877–886.

Haren, P. and C. C. Mei (1981). Head-sea diffraction by a slender raft with application to wave-power absorption. *J. Fluid Mech.*, **104**: 505–526.

Hashimoto, H. and H. Ono (1972). Nonlinear modulation of gravity waves. *J. Phys. Soc. Japan* **33**: 805–811.

Haskind, M. D. (1944). The oscillation of a body immersed in a heavy fluid. *Prikl. Mat. Mekh.* **8**: 287–300.

Haskind, M. D. (1957). The exciting forces and wetting of ships in waves (in Russian). *Izv. Akad. Nauk SSSR Otd. Tekh. Nauk*, 7: 65–79. English version available as David Taylor Model Basin Translation No. 307.

Hasselmann, K., et al. (1973). Measurements of wind-wave growth and swell decay during the joint north sea wave project (JONSWAP). *Deutschen Hydrographischen Zeitschrit. Reihe* **A8**: 7–95.

Hayashi, T., T. Kano, and M. Shirai (1966). Hydraulic research on the closely spaced pile breakwater. *Proc. 10th Conf. Coastal Eng., Santa Barbara Specialty Conf. ASCE* 873–884.

Herbich, J. B., H. D. Murphy, and B. Van Weele (1965). Scour of flat sand beaches due to wave action in front of sea walls. *Coastal Eng., Santa Barbara Specialty Conf. ASCE.* 705–726.

Hesegawa, A. and F. D. Tappert (1973). Transmission of stationary nonlinear optical pulses in dispersive dielectric fibers. *Appl. Phys. Lett.* **23**: 142–172.

Hibberd, S. and D. H. Peregrine (1979). Surf and run-up on a beach: a uniform bore. *J. Fluid Mech.* **95**: 323–345.

Hildebrand, F. B. (1962). *Advanced Calculus for Applications*, Prentice-Hall, Englewood Cliffs, N.J.

Hill, M. N. (1962). *The Sea*, Vol. I. *Physical Oceanography*. Interscience, New York.

Ho, D. V. and R. E. Meyer (1962). Climb of a bore on a beach, Part I. Uniform beach slope. *J. Fluid Mech.* **14**: 305–318.

Holman, R. A. and A. J. Bowen (1982). Bars, bumps and holes: Models for the generation of complex beach topography. *J. Geophy. Res.* **87**: 457–468.

Horikawa, K. and H. Nishimura (1970). On the function of tsunami breakwaters. *Coastal Eng. Jap.* **13**: 103–122.

Horikawa, K. and A. Watanabe (1968). Laboratory study on oscillatory boundary layer flow. *Proc. 12th Conf. Coastal Eng.* **1**: 467–486.

Houston, J. R. (1976). Long beach harbor: numerical analysis of harbor oscillations. *U.S. Army Engineering Waterways Experiment Station, Vicksburg, MS*, Report 1, Misc. Paper H-76-20.

Houston, J. R. (1978). Interaction of tsunamis with the Hawaiian Islands calculated by a finite element numerical model. *J. Phys. Ocean.* **8**: 93–102.

Houston, J. R. (1981). Combined refraction and diffraction of short waves using the finite element method. *Appl. Ocean Res.* **3**: 163–170.

Huang, N. E. (1978). On surface drift currents in the ocean. *J. Fluid Mech.* **91**: 191–208.

Hunt, J. N. (1952). Viscous damping of waves over an inclined bed in a channel of finite width. *Houille Blanche* **7**: 836–842.

Hunt, J. N. and B. Johns (1963). Current induced by tides and gravity waves. *Tellus* **15**: 343–351.

Hunt, J. N. and R. E. Baddour (1981). The diffraction of nonlinear progressive waves by a vertical cylinder. *Q. J. Mech. Appl. Math.* **34**: 69–87.

Huntley, D. A., R. T. Guza and E. B. Thornton (1981). Field observations of surf beat. I. Progressive edge waves. *J. Geophys. Res.* **86**: 6451–6466.

Huthnance, J. M. (1981). On mass transports generated by tides and long waves. *J. Fluid Mech.* **102**: 367–387.

Hwang, L. S. and Y. K. Lee (eds.) (1980). *Tsunamis.* Proceedings of the National Science Foundation Work Shop, Tetra Tech Inc., Pasadena, Calif.

Hwang, L. S. and E. O. Tuck (1970). On the oscillation of harbours of arbitrary shape. *J. Fluid Mech.* **42**: 447–464.

Ingard, K. U. (1970). Nonlinear distortion of sound transmitted through an orifice. *J. Acoust. Soc. Am.* **48**: 32–33.

Ingard, K. U. and H. Ising (1967). Acoustic nonlinearity in an orifice. *J. Acoust. Soc. Am.* **42**: 6–17.

Inman, D. L. (1957). Wave generated ripples in nearshore sands, Beach Erosion Board U.S. Corps of Engineers Technical Memo 100.

Inman, D. L., R. J. Tait, and C. E. Nordstrom (1971). Mixing in the surf force. *J. Geophys Res.* **76**: 3493–3514.

Ippen, A. T. and Y. Goda (1963). Wave induced oscillations in harbors. The solution for a rectangular harbor connected to the open-sea. Hydrodynamics Lab., Dept. of Civil Engineering, Massachusetts Institute of Technology Report 59.

Ippen, A. T., and C. R. Kulin (1957). The effect of boundary resistance on solitary waves. *Houille Blanche* **12**: 401–408.

Ippen, A. T. and M. M. Mitchell (1957). The damping of the solitary waves from boundary shear measurement. Hydrodynamics Lab., Dept. of Civil Engineering, Massachusetts Institute of Technology Report 23.

Irribarren, C. R. and C. Nogales (1949). Protection des Ports II. *Comm. 4, 17th Int. Navig. Congr., Lisbon* 31–80.

Issacs, J. D. (1948). Discussion of "Refraction of surface waves by current" J. W. Johnson. *Trans. Am. Geophys. Union* **29**: 739–742.

Issacson, M. St. Q. (1976). The viscous damping of cnoidal waves. *J. Fluid Mech.* **75**: 449–457.

Issacson, M. St. Q. (1977). Nonlinear wave forces on large offshore structures. *G. Waterways Port Coastal and Ocean Eng. ASCE* **101**: 166–170.

Ito, Y. (1970). Head loss at tsunami breakwater opening. *Proc. 12th Conf. Coastal Eng. ASCE* 2123–2131.

James, I. D. (1974a). Nonlinear waves in the near shore region: Shoaling and set-up. *Estuary Coastal Marine Sci.* **2**: 207–234.

James, I. D. (1974b). A nonlinear theory of longshore currents. *Estuary Coastal Marine Sci.* **2**: 235–250.

Jansson, K. G., J. K. Lunde, and T. Rindby (eds.) (1979). *Proc. 1st Symp. Wave Energy Utilization*, Chalmers Institute of Technology, Sweden.

Jarlan, C. E. (1965). The application of acoustical theory to the reflective properties of coastal engineering structure. *Q. Bull. National Res. Council Canada* **1**: 23–63.

Jawson, M. A. and G. T. Symm (1977). *Integral Equation Methods in Potential Theory and Elastostatics*, Academic, New York.

Jeffreys, H. (1929). On the transport of sediments by streams. *Proc. Cambridge Philos. Soc.* **25**: 272–277.

Jeffreys, H. and B. S. Jeffreys (1953). *Methods of Mathematical Physics*, 3rd. ed., Cambridge University Press, London.

John, F. (1949). On the motions of floating bodies I. *Comm. Pure Appl. Math.* **2**: 13–57.

John, F. (1950). On the motions of floating bodies II. *Comm. Pure Appl. Math.* **3**: 45–101.

Johns, B. (1968). A boundary layer method for the determination of the viscous damping of small amplitude gravity waves. *Q. J. Mech. Appl. Math.* **21**: 93–103.

Johns, B. (1970). On the mass transport induced by oscillatory flow in a turbulent boundary layer. *J. Fluid Mech.* **43**: 177–185.

Johnson, D. W. (1919). *Shore Processes and Shoreline Development*, Hafner, New York.

Johnson, J. W. (1947). The refraction of surface waves by currents. *Trans. Am. Geophys. Union.* **28**: 867–874.

Johnson, R. S. (1972). Some numerical solutions of a variable-coefficient Korteweg–de Vries equation (with applications to soliton wave development on a shelf). *J. Fluid Mech.* **54**: 81–91.

Johnson, R. S. (1973). On the development of a solitary wave over an uneven bottom. *Proc. Cambridge Philos. Soc.* **73**: 183–203.

Jolas, P. (1960). Passage de la houle sur un seuil. *Houille Blanche* **15**: 148–152.

Jones, D. S., (1964). *The Theory of Electromagtism*, Pergamon, London.

Jonsson, I. G. (1966). Wave boundary layers and friction factors. *Proc. 10th Conf. Coastal Eng. ASCE* 127–148.

Jonsson, I. G. and O. Brink-Kjaer (1973). A comparison between two reduced wave equations for gradually varying depth. *Inst. Hydrodyn. Hydraul. Eng., Tech. Univ. Denmark Progr. Rep.* **31**: 13–18.

Jonsson, I. G. and N. A. Carlsen (1976). Experimental and theoretical investigations in an oscillatory turbulent boundary layer. *J. Hydraul. Rec.* **14**: 45–60.

Jonsson, I. G. and O. Skovgaard (1979). A mild slope equation and its application to tsunami calculations. *Marine Geodesy* **2**: 41–58.

Jonsson, I. G., O. Skovgaard, and O. Brink-Kjaer (1976). Diffraction and refraction calculations for waves incident on an island. *J. Marine Res.* **34**: 469–496.

Jonsson, I. G., O. Skovgaard, and T. S. Jacobsen (1974). Computation of longshore currents. *Proc. 14th Conf. on Coastal Eng. ASCE* 699–714.

Jonsson, I. G., O, Skovgaard, and J. O. Wang (1970). Interactions between waves and currents. *Proc. 12th Conf. Coastal Eng.* **1**: 486–501.

Kajiura, K. (1961). On the partial reflection of water waves passing over a bottom of variable depth. *Proc. Tsunami Meetings 10th Pacific Science Congress. IUGG Monograph* **24**: 206–234.

Kajiura, K. (1963). The leading wave of a tsunami. *Bull. Earthquake Res. Inst.* University of Tokyo **41**: 525–571.

Kajiura, K. (1964). On the bottom friction in an oscillatory current. *Bull. Earthquake Res. Inst. Univ. Tokyo* **42**: 147–174.

Kajiura, K. (1968). A model of the bottom boundary layer in water waves. *Bull. Earthquake Res. Inst. Univ. Tokyo* **46**. 75–123.

Kakutani, T. (1971). Effect of an uneven bottom on gravity waves. *J. Phys. Soc. Jap.* **30**: 272–276.

Kamphuis, J. W. (1975). Friction factor under oscillatory waves. *J. Waterways Harbors Coastal Eng. ASCE* **101**: 135–144.

Kaneko, A. and H. Honji (1979). Double structures of steady streaming in the oscillatory flow over a wavy wall. *J. Fluid Mech.* **93**: 727–736.

Kantorovich, L. V. and V. I. Krylov (1964). *Approximate Methods in Higher Analysis*. Noordhoff, Groningen.

Karpman, V. I. (1973). *Nonlinear Waves in Dispersive Media*, Pergamon, New York.

Kehnemuyi, M. and R. C. Nichols (1973). The Atlantic generating station. *Nuclear Eng. Institute* **18**: 477.

Keller, J. B. (1958). Surface waves on water on non-uniform depth. *J. Fluid Mech.* **4**: 607–614.

Keller, J. B. (1961). Tsunamis...Water Waves Produced by Earthquakes, edited by D. C. Cox, *Proc. Tsunami Meetings 10th Pacific Science Congress*, IUGG Monograph **24**: 154–166.

Keller, H. B., D. A. Levine and G. B. Whitham (1960). Motion of a bore over a sloping beach. *J. Fluid mech.* **7**: 302.

Keulegan, G. H. (1948). Gradual damping of solitary waves. *J. Res. Natl. Bur. Stand.* **40**: 487–498.

Keulegan, G. H. (1959). Energy dissipation in standing waves in rectangular basins. *J. Fluid Mech.* **6**: 33–50.

Keulegan, G. H. and L. H. Carpenter (1956). Forces on cylinders and plates in an oscillating fluid. National Bureau of Standards Report 4821.

King, C. A. M. (1959). *Beaches and Coasts*, Arnold, London.

King, R. and R. Smith (1978). Excitation of low frequency trapped waves. *Proc. 16th Coastal Eng. ASCE* **1**: 449–466.

Kjeldsen, S. P. and G. B. Olsen (1971). *Breaking Waves*. Film by Technical University of Denmark, Lynby, Denmark.

Ko, K. and H. H. Kuehl (1978). Korteweg–de Vries soliton in a slowly varying medium. *Phys. Rev. Lett.* **40**: 233–236.

Kober, H. (1957). *Dictionary of Conformal Representations*, Dover, New York.

Komar, P. D. (1971). Near shore circulation and the formation of giant cusps. *Geol. Soc. Am. Bull.* **82**: 2643–2650.

Komar, P. D. and M. K. Gaughan (1972). Airy wave theory and breaker height prediction. *Proc. 13th Int. Conf. Coastal Eng.* 405–418.

Korteweg, D. J. and G. de Vries (1895). On the change of form of long waves advancing in a rectangular canal and on a new type of long stationary waves. *Philos. Mag.* **39**: 422–443.

Kreisel, G. (1949). Surface Waves. *Q. Appl. Math.* **7**: 21–24.

Kuehl, H. H. (1976). Nonlinear effects on mode-converted low-hybrid waves. *Phys. Fluid.* **19**: 1972–1974.

Kyozuka Y. and K. Yoshida (1981). On wave-free floating-body in heaving oscillations. *Appl. Ocean Res.* **3**: 183–194.

Lake, B. M. and H. C. Yuen (1977). A note on some nonlinear water wave experiments and comparison of data with theory. *J. Fluid Mech.* **83**: 75–81.

Lake, B. M. and H. C. Yuen (1978). A new model for nonlinear wind waves Part I. Physical model and experimental evidence. *J. Fluid Mech.* **88**: 33–62.

Lake, B. M., H. C. Yuen, H. Rungaldier, and I. N. E. Ferguson, Jr. (1977). Nonlinear deep water waves: Theory and experiment Part II. Evolution of a continuous wave train. *J. Fluid Mech.* **83**: 49–74.

Lamb, G. L., Jr. (1980). *Elements of Soliton Theory*, Wiley–Interscience, New York.

Lamb, H. (1932). *Hydrodynamics*, Dover, New York.

Lamoure, J. and C. C. Mei (1977). Effects of horizontally two-dimensional bodies on the mass transport near the sea bottom. *J. Fluid Mech.* **83**: 415–431.

Landau, L. D. and E. M. Lifshitz (1958). *Quantum mechanics (Non-relativistic theory)*. Addison-Wesley, Reading, Mass.

Landau, L. D. and E. M. Lifshitz (1959). *Fluid Mechanics*, Pergamon, New York.

Lau, J. and A. Barcilon (1972). Harmonic generation of shallow water wave over topography. *J. Phys. ocean* **2**: 405–410.

Lau, J. and B. Travis (1973). Slow varying strokes waves and submarine longshore bars. *J. Geophy. Res.* **78**: 4489–4498.

Lautenbacher, C. C. (1970). Gravity wave refraction by islands. *J. Fluid Mech.* **41**: 655–672.

Lax, P. D. (1968). Integrals of nonlinear equations of evolution and solitary waves. *Comm. Pure Appl. Math.* **21**: 467–490.

LeBlond, P. H. and L. A. Mysak (1978). *Waves in the Ocean*, Elsevier, Amsterdam.

LeBlond, P. H. and C. L. Tang (1974). On energy coupling between waves and rip currents. *J. Geophys. Res.* **79**: 811–816.

Lebreton, J. C. and A. Marganac (1968). Calcul des mouvements d'un navire ou d'une plateforme amarrés dans la houle. *Houille Blanche* **23**: 379–390.

Lee, C. M. (1968). The second order theory of heaving cylinders in a free surface. *J. Ship Res.* **12**: 313–317.

Lee, J. J. (1971). Wave-induced oscillation in harbors of arbitrary geometry. *J. Fluid Mech.* **45**: 375–394.

Lee, J. J. and F. Raichlen (1972). Oscillations in harbor with connection basins. *J. Waterways, Harbors Coastal Eng. Div. ASCE* **98**: 311–332.

Lenoir, M. and A. Jami (1978). A variational formulation for exterior problems in linear hydrodynamics. *Comp. Methods Appl. Mech.* **16**: 341–359.

Lepelletier, T. G. (1980). Tsunamis–Harbor oscillations induced by nonlinear transient long waves. Report No. KH-R-41. Keck Laboratory, California Institute of Technology.

Lesser, M. B. and D. G. Creighton (1975). Physical acoustics and the method of matched asymptotic expansions, edited by W. P. Mason and R. N. Thurston *Phys. Acoust.* **11**: 69–149.

Li, H. (1954). Stability of oscillatory laminar flow along a wall. U.S. Army Beach Erosion Board. Tech. Memo. 47.

Lighthill, M. J. (1949). The diffraction of blast I. *Proc. R. Soc. Lond. A* **198**: 454–470.

Lighthill, M. J. (1967). Some special cases by the Whitham theory. *Proc. R. Soc. Lond. A* **299**: 28–53.

Lighthill, M. J. (1978). *Waves in Fluids*, Cambridge University Press, London.

Lighthill, M. J. (1979a). Two-dimensional analysis related to wave energy extraction by submerged resonant ducts. *J. Fluid Mech.* **91**: 253–317.

Lighthill, M. J. (1979b). Waves and hydrodynamic loading. *Proc. 2nd. Int. Conf. Behavior of Offshore Structures.* **1**: 1–40.

Lin, C. C. and A. Clark, Jr. (1959). On the theory of shallow water waves. *Tsing Hua J. of Chinese Studies, Special* **1**: 54–62.

Liu, A. K. and S. H. Davis (1977). Viscous attenuation of mean drift in water waves. *J. Fluid Mech.* **81**: 63–84.

Liu, P. L. F. (1973). Damping of water waves over porous bed. *J. Hydraul. Div. ASCE* **99**: 2263–2271.

Liu, P. L. F. (1978). Integral equation solutions to nonlinear free surface flows. *Proceedings of the Conference on Finite Elements in Water Resources.* Imperial College, London, 487–498.

Liu, P. L. F. and R. A. Dalrymple (1977). Bottom friction stresses and longshore currents due to waves with large scales of incidence. *J. Marine Res.* **36**: 357–475.

Liu, P. L. F. and G. P. Lennon (1978). Finite element modeling of near shore currents. *J. Waterways, Port, Coastal Ocean Div. ASCE* **104**: 175–189.

Liu, P. L. F., C. J. Lozano and N. Pantazarus (1979). An asymptotic theory of combined wave refraction and diffraction. *Appl. Ocean Res.* **1**: 137–146.

Liu, P. L. F. and C. C. Mei (1974). Effects of a breakwater on near-shore currents due to breaking waves. Parsons Lab., Dept. of Civil engineering, Massachusetts Institute of Technology Report 192.

Liu, P. L. F. and C. C. Mei (1976a, b). Water motion on a beach in the presence of a breakwater 1. waves 2. mean currents. *J. Geophys. Res.-Oceans Atmos.* **81**: 3079–3084; 3085–3094.

Long, R. B. (1973). Scattering of surface waves by bottom irregularities. *J. Geophys. Res.* **78**: 7861–7870.

Longuet-Higgins, M. S. (1953). Mass transport in water waves. *Philos. Trans. R. Soc.* 345: 535–581.

Longuet-Higgins, M. S. (1958). The mechanics of the boundary layer near the bottom in a progressive wave. *Proc. 6th Conf. Coastal Eng.* 184–193.

Longuet-Higgins, M. S. (1960). Mass transport in the boundary layer at a free oscillating surface. *J. Fluid Mech.* **8**: 293–305.

Longuet-Higgins, M. S. (1967). On the trapping of wave energy round islands. *J. Fluid Mech.* **29**: 781–821.

Longuet-Higgins, M. S. (1970a, b). Longshore currents generated by obliquely incident sea waves, 1 and 2. *J. Geophys. Res.* **75**: 6778–6789; 6790–6801.

Longuet-Higgins, M. S. (1970c). Steady currents induced by oscillations round islands. *J. Fluid Mech.* **42**: 701–720.

Longuet-Higgins, M. S. (1974a). Breaking waves in deep or shallow water. *Proc. 10th Symp. Naval Hydrodyn.*, Office of Naval Research, 597–605.

Longuet-Higgins, M. S. (1974b). On mass, momentum, energy and circulation of a solitary wave. *Proc. R. Soc. Lond. A* **337**: 1–13.

Longuet-Higgins, M. S. (1975). Integral properties of periodic gravity waves of finite amplitudes. *Proc. R. Soc. Lond.* **342**: 157–174.

Longuet-Higgins, M. S. (1977a). The mean forces exerted by waves on floating or submerged bodies, with applications to sand bars and wave power machines. *Proc. R. Soc. Lond. A* **352**: 463–480.

Longuet-Higgins, M. S. (1977b). On the nonlinear transformation of wave trains in shallow water. *Arch. Hydrotek.* **24**.

Longuet-Higgins, M. S. (1978a). The instability of gravity of finite amplitude in deep water I. Superharmonics. *Proc. R. Soc. Lond. A* **360**: 471–488.

Longuet-Higgins, M. S. (1978b). The instabilities of steep gravity waves of finite amplitude in deep water II. Subharmonics. *Proc. R. Soc. Lond. A* **360**: 489–505.

Longuet-Higgins, M. S. (1981). Oscillating flow over sand ripples. *J. Fluid Mech.* **107**: 1–35.

Longuet-Higgins, M. S. and E. D. Cokelet (1976). The Deformation of steep surface waves on water I. A numerical method of computation. *Proc. R. Soc. Lond. A* **350**: 1–26.

Longuet-Higgins, M. S. and E. D. Cokelet (1978) The deformation of steep surface waves on water II. Growth of normal mode instabilities. *Proc. R. Soc. Lond. A* **364**: 1–28.

Longuet-Higgins, M. S. and J. Fenton (1974). Mass, Momentum, Energy and Circulation of a Solitary Wave, II, *Proc. R. Soc. Lond. A* **340**: 471–493.

Longuet-Higgins, M. S. and M. J. H. Fox (1977). Theory of the almost highest wave: The inner solution. *J. Fluid Mech.* **80**: 721–742.

Longuet-Higgins, M. S. and M. J. H. Fox (1978). Theory of the almost highest wave II: Matching and analytic extension. *J. Fluid Mech*. **85**: 769–786.

Longuet-Higgins, M. S. and R. W. Stewart (1960). Changes in form of short gravity waves on long waves and tidal currents. *J. Fluid Mech*. **8**: 565–583.

Longuet-Higgins, M. S. and R. W. Stewart (1961). The changes in amplitudes of short gravity waves on steady non-uniform currents. *J. Fluid Mech*. **10**: 529–549.

Longuet-Higgins, M. S. and R. W. Stewart (1962). Radiation stresses and mass transport in gravity waves with applications to surf-beats. *J. Fluid Mech*. **13**: 481–504.

Longuet-Higgins, M. S. and R. W. Stewart (1964). Radiation stresses in water waves; a physical discussion with applications. *Deep-Sea Res*. **11**: 529–562.

Longuet-Higgins, M. S. and J. S. Turner (1973). A model of flow separation at a free surface. *J. Fluid Mech*. **57**: 129–148.

Longuet-Higgins, M. S. and J. S. Turner (1974). An entrainment plume model of a spilling breaker. *J. Fluid Mech*. **63**: 1–20.

Lozano, C. J. and P. L. F. Liu (1980). Refraction–diffraction model for linear surface water waves. *J. Fluid Mech*. **101**: 705–720.

Lozano, C. J. and R. E. Meyer (1976). Leakage and response of waves trapped round islands. *Phys. Fluids* **19**: 1075–1088.

Ludwig, D. (1966). Uniform asymptotic expansions at a caustic. *Comm. Pure Appl. Math*. **19**: 215–250.

Luneberg, R. K. (1964). *Mathematical Theory of Optics*, University of California Press, Los Angeles.

Madsen, O. S. (1978a). Mass transport in deep-water waves. *J. Phys. Ocean* **8**: 1009–1015.

Madsen, O. S. (1978b). Wave induced pore pressures and effective stresses in a porous sea bed. *Geotechnique* **28**: 377–393.

Madsen, O. S. and W. D. Grant (1976). Sediment transport in the coastal zone. Parsons Lab., Dept. of Civil Engineering, Massachusetts Institute of Technology Report 208.

Madsen, O. S. and W. D. Grant (1977). Quantitative description of sediment transport by waves. *Proc. 15th Conf. Coastal Eng*. 1093–1112.

Madsen, O. S. and C. C. Mei (1969). The transformation of a solitary wave over an uneven bottom. *J. Fluid Mech*. **39**: 781–791.

Madsen, O. S., C. C. Mei, and R. P. Savage (1970). The evolution of time-periodic waves of finite amplitude. *J. Fluid Mech*. **44**: 195–208.

Maeda, H., H. Tanaka, and T. Kinoshita (1980). Theoretical and experimental study of wave power absorption. *Proc. 14th Symp. Naval Hydrodyn*., The Ship Building Res. Inst. Japan 857–876.

Mallory, J. K. (1974). Abnormal waves on the south east coast of South Africa. University of Capetown Libraries, Cape Town.

Martin, D. U. and H. C. Yuen (1980). Quasi recurring energy leakage in the two space dimensional nonlinear Schrödinger equations. *Phys. Fluids* **23**: 1269–1271.

Maruo, H. (1960). The drift of a body floating on waves. *J. Ship Res*. **4**: 1–10.

Maskell, S. J. and F. Ursell (1970). The transient motion of a floating body. *J. Fluid Mech*. **44**: 203–313.

Masuda, M. (1979). Experimental full scale result of wave power machine KAIMEI. *Proc. 1st. Symp. Wave Energy Utilization*, Chalmers Institute of Technology, Sweden 349–363.

Mattioli, F. (1978). Wave-induced oscillations in harbors of variable depth. *Computers and Fluids* **6**: 161–172.

McCamy, R. C., and R. A. Fuchs (1954). Wave forces on a pile: a diffraction theory. *Tech. Memo*. No. **69**, U.S. Army Board, U.S. Army Corp. of Eng.

McCormick, M. E. (1981). *Ocean Wave Energy Conversion*, Wiley–Interscience, New York, 233 pp.

McKee, W. D. (1974). Waves on a shearing current: a uniformly valid asymptotic solution. *Proc. Cambridge Philos. Soc.* **75**: 295–301.

McKee, W. D. (1974). Waves on a shear current: a uniformly valid asymptotic solution. *Proc. Cambridge Philos. Soc.* **75**: 295–361.

McLean, J. W., Y. C. Ma, D. U. Martin, P. G. Saffman, and H. C. Yuen (1981). A new type of three-dimensional instability of finite amplitude gravity waves. *Phys. Rev. Lett.* **46**: 817–820.

McTigue, D. F. and S. L. Passman (1982). The effective stress principle and mixture theory (submitted for publication).

Mehlum, E. and J. Stamnes (1979). Power production based on focusing of ocean swells. *Proc. 1st Wave Energy Utilization*, Chalmers Institute of Technology, Sweden. 29–35.

Mei, C. C. (1966a). Radiation and scattering of transient gravity waves by vertical plates. *Q. J. Mech. Appl. Math.* **19**: 417–440.

Mei, C. C. (1966b). On the propagation of periodic water waves over beaches of small slope. Technical Note 12, Hydrodynamics Laboratory, Massachussetts Institute of Technology.

Mei, C. C. (1973). A note on the averaged momentum balance in two-dimensional water waves. *J. Marine Res.* **31**: 97–104.

Mei, C. C. (1976). Power extraction from water waves. *J. Ship Res.* **20**: 63–66.

Mei, C. C. (1978). Numerical methods in water wave diffraction and radiation. *Annual Rev. Fluid Mech.*, **10**: 393–416.

Mei, C. C. and D. Angelides (1976). Longshore currents around a conical island. *Coastal Eng.* **1**: 31–42.

Mei, C. C. and H. S. Chen (1975). Hybrid element method for water waves, *Symposium on Modeling Techniques*, 2nd Annual Symposium of the Waterways Harbors and Coastal Engineering Division American Society of Civil Engineers, Vol. 1, pp. 63–81.

Mei, C. C. and H. S. Chen (1976). A hybrid element method for steady linearized free surface flows. *Inst. J. Num. Math. Eng.* **10**: 1153–1175.

Mei, C. C. and M. A. Foda (1979). An analytical theory of resonant scattering of SH waves by thin overground structures. *Earthquake Eng. Structure Dynamics* **7**: 335–353.

Mei, C. C. and M. A. Foda (1981a). Wave induced responses in a fluid filled poro-elastic solid with a free surface—a boundary layer theory. *Geophy. J. R. Astr. Soc.* **66**: 597–637.

Mei, C. C. and M. A. Foda (1981b). Wave induced stresses around a pipe laid on a poro-elastic sea-bed, *Geotechnique.* **31**: 509–517.

Mei, C. C. and M. A. Foda (1982). Boundary layer theory of waves in a poro-elastic sea bed, *Soil Mechanics-Transient and Cyclic Loads*, G. N. Pande and O. C. Zienkiewicz (eds.) 17–35, Wiley, New York.

Mei, C. C., M. A. Foda, and P. Tong (1979). Exact and hybrid-element solutions for the vibration of a thin elastic structure seated on the sea bottom. *Appl. Ocean Res.* **1**: 79–88.

Mei, C. C. and B. LeMéhauté (1966). Note on the equations of long waves over an uneven bottom. *J. Geophys. Res.* **71**: 393–400.

Mei, C. C. and P. L. F. Liu (1973). The damping of surface gravity waves in a bounded liquid. *J. Fluid Mech.* **59**: 239–256.

Mei, C. C. and P. L. F. Liu, (1977). Effects of topography on the circulation in and near the surf zone—linear theory. *Estuary Coastal Marine Sci.* **5**: 25–37.

Mei, C. C., P. L. F. Liu and T. G. Carter (1972). Mass transport in water waves. Parsons Lab., Massachusetts Institute of Technology Report 146: 287.

Mei, C. C., P. L. F. Liu, and A. T. Ippen (1974). Quadratic head loss and scattering of long waves. *J. Waterway Harbors Coastal Eng. Div. Proc. ASCE* **100**, 217–239.

Mei, C. C. and R. P. Petroni (1973). Waves in a harbor with protruding breakwaters. *J. Waterways Harbors Coastal Eng. Proc. ASCE* **99**, 209–229.

Mei, C. C. and E. O. Tuck (1980). Forward scattering by thin bodies. *SIAM J. Appl. Math.* **39**: 178–191.

Mei, C. C. and Ü. Ünlüata (1972). Harmonic generation in shallow water waves. *Waves on Beaches*, edited by R. E. Meyer, Academic, New York, 181–202.

Mei, C. C. and Ü. Ünlüata (1978). Resonant scattering by a harbor with two coupled basins. *J. Eng. Math.* **10**: 333–353.

Melville, W. K. (1980). On the mach reflexion of a solitary wave. **98**: 285–297.

Melville, W. K. (1982). The instability and breaking of deep-water waves. *J. Fluid Mech.* **115**, 165–185.

Meyer, R. (1955). Symétrie du coefficient (complexe) de transmission de houles á travers un obstacle quelconque. *Houille Blanche* **10**: 139–140.

Meyer, R. E. (1970). Resonance of unbounded water bodies. *Mathematical Problems in Geophysical Sciences. I, Geophysical Fluid Dynamics*, American Mathematical Society, Providence, R.I. 189–227.

Meyer, R. E. (1979a). Theory of water-wave refraction. *Adv. Appl. Mech.* **19**: 53–141.

Meyer, R. E. (1979b). Surface-wave reflection by underwater ridges. *J. Physical Ocean.* **9**: 150–157.

Meyer, R. E. and A. D. Taylor (1972). Run-up on beaches. *Waves on Beaches*, edited by R. E. Meyer, Academic, New York, 357–412.

Miche, R. (1944). Mouvements ondulatoires de la mer en profondeur constante ou décroissante form limte de la houle lors de son déferlement. Application aux digues maritimes. *Ann. Pontes Chaussées* **114**: 25–78, 131–164, 270—292, 369–406.

Miche, R. (1951). Le pouvoir reflechissant des ouvrages maritime exposés á l'action de la houle. *Ann. Ponts Chaussées* **121**: 285–319.

Miles, J. W. (1957). On the generation of surface waves by shear flows. *J. Fluid Mech.* **3**: 185–204.

Miles, J. W. (1962). Transient gravity wave response to an oscillating pressure. *J. Fluid Mech.* **13**: 145–150.

Miles, J. W. (1967). Surface-wave damping in closed basins. *Proc. Soc. Lond. A* **297**: 459–475.

Miles, J. W. (1971). Resonant response of harbors: An equivalent circuit analysis. *J. Fluid. Mech.* **41**: 241–265.

Miles, J. W. (1972). Wave propagation across the continental shelf. *J. Fluid Mech.* **54**: 63–80.

Miles, J. W. (1974). Harbor seiching. *Annual Rev. Fluid Mech.* **6**: 17–35.

Miles, J. W. (1976). Damping of weakly nonlinear shallow water waves. *J. Fluid Mech.* **76**: 251–257.

Miles, J. W. (1977). Diffraction of solitary waves. *J. Appl. Math. Phy.* **28**: 889–902.

Miles, J. W. (1979). On the Korteweg-de Vries equation for a gradually varying channel. *J. Fluid Mech.* **91**: 181–190.

Miles, J. W. (1980). Solitary waves. *Annual Rev. Fluid Mech.* **12**: 11–44.

Miles, J. W. and Y. K. Lee (1975). Helmholtz resonance of harbors. *J. Fluid Mech.* **67**: 445–464.

Miles, J. W. and W. Munk (1961). Harbor paradox. *J. Waterways Harbors Div. Proc. ASCE* **87**: 111–130.

Milne-Thomson, M. N. (1967). *Theoretical Hydrodynamics*, 5th ed. MacMillan. New York, 630 pp.

Miloh, T. (1980). Irregularities in solutions of nonlinear wave diffraction problem by vertical cylinder. *J. Waterways, Port Coastal Ocean Eng. Div Proc. ASCE*: **106**: 279–284.

Minzoni, A. A., (1976). Nonlinear edge waves and shallow water theory. *J. Fluid Mech.* **79**: 369–374.

Minzoni, A. A. and G. B. Whitham (1977). On the excitation of edge wave on beaches. *J. Fluid Mech.* **79**: 273–287.

Miura, R. M. (1974). The Korteweg–de Vries equation: a model equation for nonlinear dispersive waves *Nonlinear waves*, edited by S. Leibovich and A. R. Seabass, Cornell University Press, Ithaca, New York, 212–234.

Miura, R. M. (1976). The Korteweg–de Vries equation—a survey of results. *SIAM Rev.* **18**: 412–459.

Molin, B. (1979). Second order diffraction loads upon three-dimensional bodies. *Appl. Ocean Res.* **1**: 197–202.

Mollo-Christensen, E. and A. Ramamonjiariosa (1978). Modeling the presence of wave groups in a random wave field. *J. Geophys. Res.* **83**: 4117–4122.

Momoi, T. Tsunami in the vicinity of a wave origin (I-IV) *Bull. Earthquake Res. Inst. Univ. Tokyo* I: (1964a). **42**: 133–146; II: (1964b). **42**: 369–381; III: (1965). **43**: 53–93; IV: (1965b). **43**: 755–772.

Monin, A. S. and A. M. Yaglom (1971). *Statistical Fluid Mechanics*. M.I.T. Press, Cambridge, 769 pp.

Moore, D. (1970). The mass transport velocity induced by the free oscillation of a single frequency. *Geophys. Fluid Dynamics* **1**: 237–247.

Moraes, C.deC. (1970). Experiments of wave reflection on impermeable slopes. *Proc. 12th Conf. on Coastal Eng. ASCE* 509–521.

Morse, P. M. and H. Feshbach (1953). *Methods of Theoretical Physics I and II*, McGraw-Hill, New York.

Morse, P. M. and K. U. Ingard. (1968). *Theoretical Acoustics*, McGraw-Hill, New York.

Munk, W. H. (1949a). The solitary wave and its application to surf problems. *N.Y. Acad. Sci.* **1**: 376–424.

Munk, W. H. (1949b). Surf beats. *Trans. Am. Geophys. Union* **30**: 849–854.

Munk, W. H., F. E. Snodgrass, and G. F. Carrier (1956). Edge waves on the continental shelf. *Science* **123**: 127–132.

Munk, W. H. and M. Wimbush (1969). A rule of thumb for wave breaking over sloping beaches. *Oceano.* **9**: 56–59.

Murty, T. S. (1977). *Seismic Sea Waves/Tsunamis*, Bulletin 198, Fisheries and Environment, Ottawa, Canada.

Mushkelishvilli, N. I. (1954). *Some Basic Problems of the Mathematical Theory of Elasticity*, Noordhoff, Leyden, The Netherlands.

Mynett, A. E. (1980) Dynamic Stresses and Pore Pressures in a Poro-Elastic Foundation Beneath a Rigid Structure, Sc.D. Thesis, Civil Eng., Massachusetts Institute of Technology.

Mynett, A. E., D. D. Serman, and C. C. Mei (1979). Characteristics of Salter's cam for extracting energy from ocean waves. *Appl. Ocean Res.* **1**: 13–20.

Naheer, E. (1978). The damping of solitary waves. *Int. J. Hydraul. Res.* **16**: 235–249.

Nayfeh, A. H. (1973). *Perturbation Methods*. Wiley, New York.

Newman, J. N. (1960). The exciting forces on fixed bodies in waves. *J. Ship Res.* **6**: 10–17.

Newman, J. N. (1965). Propagation of water waves past long two-dimensional obstacles. *J. Fluid Mech.* **23**, 23–29.

Newman, J. N. (1967). The drift force and moment on ships in waves. *J. Ship. Res.* **11**: 51–60.

Newman, J. N. (1974). Second-order, slowly-varying forces on vessels in irregular waves. *Proc. Int. Symp. on the Dynamics of Marine Vehicles and Structures in Waves*, Univ. College, London.

Newman, J. N. (1975). Interaction of waves with two dimensional obstacles: a relation between the radiation and scattering problems. *J. Fluid Mech.* **71**: 273–282.

Newman, J. N. (1976). The interaction of stationary vessels with regular waves. *Proc. 11th Symp. Naval Hydrodyn.*, Office of Naval Research 491–502.

Newman, J. N. (1979). Absorption of wave energy by elongated bodies. *Appl. Ocean Res.* **1**: 189–196.

Nielsen, A. H. (1962). Diffraction of periodic waves along a vertical breakwater for small angles of incidence. University of California–Berkeley IER Technical Report HEL 1–2.

Noda, E. K. (1972). Rip currents. *Proc. 13th Conf. Coastal Eng.* 653–668.

Noda, E. K. (1974). Wave-induced nearshore circulation. *J. Geophys. Res.* **79**: 4097–4106.

Noda, H. (1968). A study on mass transport in boundary layers in standing waves. *Proc. 11th Conf. Coastal Eng. ASCE* 227–235.

Ogawa, K. and K. Yoshida (1959). A practical method for a determination of long gravitational waves. *Records Oceangraphic Works Jap* **5**: 38–50.

Ogilvie, T. F. (1960). Propagation of waves over an obstacle in water of finite depth, University of California–Berkeley, Inst. Eng. Res. Report 82–14.

Ogilvie, T. F. (1963). First and second order forces on a cylinder submerged under the free surface. *J. Fluid Mech.* **16**: 451–472.

Ogilvie, T. F. (1964). Recent progress toward the understanding and prediction of ship motions. *Proc. 5th Symp. Naval Hydrodyn.* Office of Naval Research 3–97.

Ono, H. (1972). Wave propagation in an inhomogeneous anharmonic lattice, *J. Phys. Soc. Jap.* **32**, 332–336.

Ono, H. (1974). Nonlinear wave modulation in inhomogeneous media. *J. Phys. Soc. Japan* **37**: 1668–1672.

Ostendorf, D. W. and O. S. Madsen (1979). An analysis of longshore current and associated sediment transport in the surf zone, Parsons Laboratory, Dept. Civil Engineering, Massachusetts Institute of Technology Report 241.

Ostrovsky, L. A. and E. N. Pelinovskiy (1970). Wave transformation on the surface of a fluid of variable depth. *Atmos. Oceanic Phys.* **6**: 552–555.

Ott, E. and R. N. Sudan (1970). Damping of solitary waves. *Phys. Fluids* **13**: 1432.

Özhan, E. and H. Shi-igai (1977). On the development of solitary waves on a horizontal bed with friction. *Coastal Eng.* **1**: 167–184.

Özsoy, E. (1977). Dissipation and wave scattering by narrow openings. Coastal and Oceang. Engineering Lab. University of Florida UFL/COEL/TR-037.

Pande, G. N. and O. C. Zienkiewicz (eds). (1982). *Soils Mechanics: Cyclic and Transient loading* Wiley, New York.

Papanikolaou A. and H. Nowacki (1981). Second-order theory of oscillating cylinders in a regular steep wave. *Proc. 13th Symp. on Naval Hydrodyn.*, Office of Naval Research. 303–333.

Papoulis, A. (1968). *Systems and Transforms with Applications in Optics*, McGraw-Hill, New York.

Peregrine, D. H. (1967). Long waves on a beach. *J. Fluid Mech.* **27**: 815–827.

Peregrine, D. H. (1976). Interaction of water waves and currents. *Adv. Appl. Mech.* **16**: 10–117.

Peregrine, D. H. and R. Smith (1975). Stationary gravity waves on non-uniform free streams. *Math. Proc. Cambridge Philos. Soc.*, **77**: 415–438.

Perroud, P. H. (1957). The solitary wave reflection along a straight vertical wall at oblique incidence. University of California–Berkeley IRE Technical Report 99–3.

Phillips, O. M. (1957). On the generation of waves by turbulent wind. *J. Fluid Mech.* **2**: 415–417.

Phillips, O. M. (1960). On the dynamics of unsteady gravity waves of finite amplitude Part I. *J. Fluid Mech.* **9**: 193–217.

Phillips, O. M. (1977). *Dynamics of the Upper Ocean*, 2nd ed. Cambridge University Press, London.

Pierson, W. J., Jr. (1962). Perturbation analysis of the Navier-Stokes equations in Lagrangian form with selected linear solutions. *J. Geophys. Res.* **67**: 3151–3160.

Pleass, C. M. (1978). The use of wave powered system for desalinization–a new opportunity. *Proc. Symp. Wave Tidal Energy. 1*, Paper D1-1, BHRA.

Pocinki, L. S. (1950). The application of conformal transformations to ocean wave refraction problems. *Trans. Am. Geophys. Union* **31**: 856–860.

Prévost, J. H. (1980). Mechanics of continuous porous media. *J. Eng. Sci.* **18**: 787–800.

Price, W. G. and R. E. D. Bishop, (1972). *Probabilistic Theory of Ship Dynamics*, Wiley, New York.

Putnam, J. A. (1949). Loss of wave energy due to percolation in a permeable sea bottom. *Trans. Am. Geophys. Union* **30**: 349–356.

Putnam, J. A., W. H. Munk and M. A. Traylor (1949). The prediction of longshore currents. *Trans. Am. Geophys Union* **30**: 337–345.

Radder, A. C. (1979). On the parabolic equation method for water wave propagation. *J. Fluid Mech.* **95**: 159–176.

Raman, H., G. V. Prabhakara Rao, and P. Venkatanarasalah (1975). Diffraction of nonlinear waves by a circular cylinder. *Acta Mechanica* **23**: 145–158.

Raman, H. and P. Venkatanarasalah (1976). Forces due to nonlinear waves on vertical cylinders. *J. Waterways Harbor, Coastal Engineering Div. ASCE* **102**: 301–316.

Rayleigh, L. (1883). On the circulation of air observed in Kundt's tubes, and some allied acoustical problems. *Philos, Trans. R. Soc. London* **175**: 1–21.

Rayleigh, L. and J. W. Strutt (1897). On the passage of waves through aperatures in plane screens, and allied problems. *Philos. Mag.* **43**: 259–272.

Riabounchinsky, D. (1932). Sur L'anologue hydraulique des mouvements dun fluid compressible. Institut de France Academic des Sciences. *Comptes Rendus.* **195**: 998.

Richart, F. E., Jr., J. R. Hall, Jr. and R. D. Woods (1969). *Vibration of Soils and Foundations*, Prentice-Hall, Englewood Cliffs, N.J.

Richey, E. P. and C. K. Sollitt (1969). Attenuation of deep water waves by a porous-walled breakwater. Dept. of Civil Engineering University of Washington, Seattle, Technical Report 25.

Rienecker, M. M. and J. D. Fenton (1981). A Fourier approximation for steady water waves. *J. Fluid Mech.* **104**: 119–137.

Riley, N. (1967). Oscillatory viscous flows: review and extension. *J. Inst. Math. Appl.* **3**: 419–434.

Risser, J. F. (1976). Transient Response in Harbors, Master of Science Thesis. Dept. of Civil Engineering, Massachusetts Institute of Technology.

Rockliff, N. (1978). Finite amplitude effects in free and forced edge waves. *Math.-Proc. Cambridge Philos. Soc.* **83**: 463–479.

Rogers, S. R. and C. C. Mei (1977). Nonlinear resonant excitation of a long and narrow bay. *J. Fluid Mech.* **88**: 161–180.

Roseau, M. (1952). Contribution a la theorie des ondes liquides de gravite en profondeur variable. *Pub. Sci. Tech. Due Ministére de l'Air*, 275.

Roseau, M. (1976). *Asymptotic Wave Theory*, North-Holland, Amsterdam,

Roskes, G. (1969). Wave Envelopes and Nonlinear Waves, Ph.D. Thesis, Dept. of Mathematics, Massachusetts Institute of Technology.

Russell, J. S. (1838). Report of the Committee on Waves. *Rep. Meet. Br Assoc. Adv. Sci. 7th.* Liverpool, 1837, John Murray, London, 417–496.

Russell, J. S. (1845). Report on waves. *Rep. Meet. Brit. Assoc. Adv. Sci. 14th York*, 1844, John Murray, London, 311–390.

Russell, R. C. H. and J. D. C. Osorio (1957). An experiment investion of drift profiles in a dosed channel. *Proc. 6th Conf. Coastal Eng. ASCE* 171–193.

Salter, S. (1974). Wave power, *Nature* **249**: 720–724.

Salter, S. H. (1979). Recent progress on ducks. *Proc. 1st Symp. Wave Energy Utilization*, Chalmers Institute of Technology, Sweden. 36–76.

Sarpkaya, T. and M. St. Q. Issacson (1981). *Mechanics of Wave Forces on Offshore Structures*, Van Nostrand Reinold, New York.

Satsuma, J. and N. Yajima (1974). Initial value problems of One-dimensional self-modulation of nonlinear waves in dispersive media. *Suppl. Progr. Theor. Phys.* **55**, 284–306.

Sauvage de Saint Marc, M. G., and M. G. Vincent (1955). Transport littoral formation de fleches et de tombolos. *Proc. of 5th Conf. Coastal Eng. ASCE* 296–328.

Schlichting, H. (1968). *Boundary Layer Theory*, 6th ed, McGraw-Hill, New York.

Schonfeld, J. C. (1972) Propagation of Two Dimensional Short Waves. *Delft University of Technology Manuscript* (in Dutch).

Schwartz, L. W. (1974). Computer extension and analytic continuation of Stokes' expansion for gravity waves. *J. Fluid Mech.* **62**: 553–578.

Schwartz, L. W. and J. D., Fenton, (1982). Strongly nonlinear waves. *Ann. Rev. Fluid Mech.* **14**: 39–60.

Scott, A. C., F. Y. F. Chu, and D. W. McLaughlin (1973). The soliton: a new concept in applied science. *Proc. IEEE* **61**: 1443–1483.

Seed, H. B. and K. L. Lee (1966). Liquefaction of saturated sands during cyclic loading. *J. Soil Mech. Found. Div. ASCE* **92**: 105–134.

Seed, H. B. (1968). The Fourth Terzaghi Lecture: Landsides during earthquakes due to liquefaction. *J. Soil Mech. Found. Div. ASCE* **94**: 1053–1122.

Segur, H. (1973). The Korteweg–de Vries equation and water waves. Solution of the equation: Part I. *J. Fluid Mech.* **59**: 721–736.

Serman, D. D. (1978). Theory of Salter's wave energy device in random sea. Master of Science Thesis, Dept. of Civil Engineering, Massachusetts Institute of Technology.

Seto, H. and Y. Yamamoto (1975). Finite element analysis of surface wave problems by a method of surperposition. *Proc. Inst. Int. Conf. Num. Ship Hydrod.* David Taylor Naval Ship Res. and Devel. Center, 49–70.

Shen, M. C. and R. E. Meyer (1963a). Climb of a bore on a beach-Part II. Nonuniform beach slope. *J. Fluid Mech.* **16**: Part I, 108–112.

Shen, M. C. and R. E. Meyer (1963b). Climb of a bore on a beach-Part III. Run-up. *J. Fluid Mech.* **16**: Part I, 113–125.

Shepard, F. P. (1963). *Submarine Geology*, 2nd ed., Harper and Row, New York.

Shepard, F. P. and H. R. Wanless (1973). *Our Changing Coastlines*, McGraw-Hill, New York.

Shimano, T. M. Hom-ma, and K. Horikawa (1958). Effect of a jetty on nearshore currents. *Coastal Eng. Japan* **1**: 45–58.

Simon, M. J. (1981). Wave energy extractions by submerged cylindrical resonant duct. *J. Fluid Mech.* **104**: 159–181.

Sitenko, A. G. (1971). *Lectures in Scattering Theory*, Pergamon, New York.

Skovgaard, O., I. G. Jonsson, and J. A. Bertelsen (1976). Computation of wave heights due to refraction and friction. *J. of Waterways Harbors and Coastal Eng. Div. ASCE.* **102**: 100–105.

Sleath, J. F. A. (1970). Velocity measurements close to the bed in a wave tank. *J. Fluid Mech.* **42**: Part I, III–123.

Smith, J. D. and S. R. McLean (1977). Spatially averaged flow over a wavy surface. *J. Geophy. Res.* **82**: 1732–1746.

Smith, R. (1976). Giant waves. *J. Fluid Mech.* **77**: 417–431.

Smith, R., and T. Sprinks (1975). Scattering of surface waves by a conical island. *J. Fluid Mech.* **72**: 373–384.

Sneddon, I. N. (1951), *Fourier Transforms*, McGraw-Hill, New York.

Snodgrass, F. E., W. H. Munk, and G. R. Miller (1962). Long period waves over California's continental borderland, part I. background spectra. *J. Marine Res.* **20**: 3–30.

Sonu, C. J. (1972). Field observation of nearshore circulation and meandering currents. *J. Geophys. Res.* **77**: 3232–3247.

Srokosz, M. A. (1979). Some theoretical aspects of wave power absorption. Ph.D. Thesis, Dept. of Appl. Math., University of Bristol.

Srokosz, M. A. (1980). Some relations for bodies in a canal, with an application for wave power absorption, *J. Fluid Mech.* **99**: 145–162.

Stiassnie, M. and U. I. Kroszynski (1982). Long-time evolution of an unstable water-wave Train. *J. Fluid Mech.* **116**: 201–225.

Stiassnie, M. and D. H. Peregrine, (1979). On averaged equations for finite amplitude water waves. *J. Fluid Mech.* **94**: 401–407.

Stiassnie, M. and D. H. Peregrine, (1980). Shoaling of finite-amplitude surface waves on water of slowly-varying depth. *J. Fluid Mech.* **97**: 783–805.

Stoker, J. J. (1948). The formation of breakers and bores. *Comm. Pure and Appl. Math.* **1**: 1–87.

Stoker, J. J. (1956). On radiation conditions. *Comm. Pure Appl. Math.* **9**: 577–595.

Stoker, J. J. (1957). *Water Waves*, Interscience, New York.

Stokes, G. G. (1847). On the theory of oscillatory waves. *Trans. Cambridge Philos. Soc.* **8**: 441–455. Reprinted in *Math. Phys. Papers* Cambridge University Press, London, Vol. 1, 314–326.

Stoll, R. D. and G. M. Bryan (1970). Wave attenuation in saturated sediments. *J. Acoust. Soc. Am.* **47**: 1440–1447.

Stuart, J. T. (1966). Double boundary layers in oscillatory viscous flow. *J. Fluid Mech.* **24**: 673–687.

Su, C. L. (1973). Asymptotic solutions of resonances in harbors with connected basins. *J. Waterways Harbor Coastal Eng. Div. ASCE* 375–391.

Susbielles, G. and Ch. Bratu, (1981). *Vagues et Quvrages Pétroliers en Mer* Editions Technip 27, Paris.

Svensen, I. A. (1967). The wave equation for gravity waves in water of gradually varying depth. Technical University Denmark, Coastal Engineering Lab. Progress Report 15.

Tam, C. K. W. (1973). Rip currents. *J. Geophys. Res.* **78**: 1937–1943.

Tappert, F. D., (1977). The Parabolic Approximation, *Wave Propagation and Underwater Acoustics*, 224–287, edited by J. B. Keller and J. S. Papadakis, Springer-Verlag, Berlin.

Tappert, F. D. and N. J. Zabusky (1971). Gradient-induced fission of solitons. *Phys. Rev. Lett.* **27**: 1774–1776.

Terrett, F. L., F. D. C. Osorio and G. H. Lean (1968). Model studies of a perforated breakwater, *Proc. 11th Conf Coastal Eng. ASCE* 1104–1109.

Terzaghi, K. (1954). *Theoretical Soil Mechanics*, Wiley, New York,

Thomas, J. R. (1981). The absorption of wave energy by a three-dimensional submerged duct. *J. Fluid Mech.* **104**: 189–215.

Thornton, E. B. (1970). Variation of longshore current across the surf zone. *Proc. 12th Coastal Eng. Conf. ASCE* 291–308.

Tong, P. P., T. H. H. Pian, and S. J. Lasry (1973). A hybrid-element approach to crack problems in plane elasticity. *Int. J. Num. Methods Eng.* **7**: 297–308.

Tong, P. and J. N. Rossettos (1976). *Finite Element Method*, M.I.T. Press, Cambridge, Mass.

Tuck, E. O. (1971). Transmission of water waves through small apertures. *J. Fluid Mech.* **49**: 65–73.

Tuck, E. O. (1975). Matching problems involving flow through small holes. *Adv. Appl. Mech.* **15**: 90–158.

Tuck, E. O. (1976). Some classical water-wave problems in variable depth, *Waves on Water of Variable Depth*, edited by D. G. Provis and R. Radak, Lecture Notes in Physics no 64, Springer-Verlag, New York.

Tuck, E. O. and L. S. Hwang (1972). Long wave generation on a sloping beach. *J. Fluid Mech.* **51**: 449–461.

Tucker, M. J. (1950). Surf beats: Sea waves of 1 to 5 min. period. *Proc. R. Soc. Lond. A* 202: 565–573.

Turpin, F. M. (1981). Interaction between waves and current over a variable depth. Master's Thesis, Dept. of Civil Engineering, Massachusetts Institute of Technology.

Ünlüata, Ü. and C. C. Mei (1970). Mass transport in water waves. *J. Geophys. Res.* 7611–7618.

Ünlüata, Ü. and C. C. Mei (1973). Long wave excitation in harbors-an analytic study. Dept. of Civil Engineering, Massachusetts Institute of Technology Technical Report 171.

Ünlüata, Ü. and C. C. Mei (1975). Effects of entrance loss on harbor oscillations. *J. Waterways Harbors and Coastal Eng. Div. ASCE* **101**: 161–180.

Ünlüata, Ü. and C. C. Mei (1976). Resonant scattering by a harbor with two coupled basins, *J. Eng. Math.* **10**: 333–353.

Ursell, F. (1947). The effect of a vertical barrier on surface waves in deep water. *Proc. Cambridge Philos. Soc.* **47**: 374–82.

Ursell, F. (1948). On the waves due to the rolling of a ship. *Q. J. Mech. Appl. Math.* **1**: 246–52.

Ursell, F. (1952). Edge waves on a sloping beach. *Proc. R. Soc. Lond. A* **214**: 79–97.

Ursell, F. (1953). The long wave paradox in the theory of gravity waves. *Proc. Cambridge Phil. Soc.* **49**: 685–694.

Ursell, F. (1964). The decay of the free motion of a floating body. *J. Fluid Mech.* **19**: 305–319.

Ursell, F. (1981). Irregular frequencies and the motion of floating bodies. *J. Fluid Mech.* **105**: 143–156.

U.S. Army Coastal Engineering Research Center (1975). *Shore Protection Manual*, Vols. I–III, U.S. Government Printing Office, Washington, D.C.

Van Dorn, W. G. (1966). Boundary dissipation of oscillatory waves. *J. Fluid Mech.* **24**: 769–779.

Van Dorn, W. G. (1968). Tsunamis. *Contemp. Phys.* **9**: 145–164.

Verruijt, A. (1969). Elastic storage of aquifers. *Flow through Porous Media*, edited by R. J. M. DeWiest, Academic, New York.

Vinje, T. and P. Brevig (1981a). Nonlinear ship motions. *Proc. 3rd Int. Symp. Num. Ship Hydrodyn. Paris.*

Vinje, T. and P. Brevig (1981b). Numerical simulation of breaking waves. *Adv. Water Res* **4**: 77–82.

Vitale, P. (1979). Sand bed friction factors for oscillatory flows. *J. Waterways, Port Coastal Ocean Div. ASCE* **105**: 229–245.

Von Kerczek, C. (1975). Numerical solution of naval free surface hydrodynamics problems. *Proc. 1st Int. Conf. Num. Ship Hydrodyn.*, edited by J. W. Schot and N. Salvesen, David Taylor Naval Ship Research and Development Center, 1–47.

Wait, J. (1962). *Electromagnetic Waves in Stratified Media*, Pergamon, New York.

Wang, C. Y. (1968). On high-frequency oscillatory viscous flows. *J. Fluid Mech.* **32**: 55–68.

Watson, G. N. (1958). *A Treatise on the Theory of Bessel Functions*, Cambridge University Press, London.

Weggel, J. R. (1972). Maximum breaker height. *J. Waterways, Harbours Coastal Eng. Div. ASCE.* **98**: 529–548.

Wehausen, J. V. (1967). Initial value problem for the motion in an undulating sea for a body with a fixed equilibrium position. *J. Eng. Math.* **1**: 1–19.

Wehausen, J. V. (1971). The motion of floating bodies. *Ann. Rev. Fluid Mech.* **3**: 237–268.

Wehausen, J. V. (1980). Perturbation methods in diffraction. *J. Waterways Port Coastal Ocean Eng. Div.* **2**: 290–291.

Wehausen, J. V., and E. V. Laitone (1960). *Handbuch der Physik*, edited by W. Flügge, vol. 9, pp. 446–778. Springer-Verlag, Berlin.

West, B. J. (1981). *Deep Water Gravity Waves*. Springer-Verlag, Berlin.

Whitham, G. B. (1962). Mass, momentum and energy flux in water waves. *J. Fluid Mech.* **12**: 135–147.

Whitham, G. B. (1965). Nonlinear dispersive waves. *Proc. R. Soc. Lond. A* **283**: 238–261.

Whitham, G. B. (1967). Nonlinear dispersion of water waves. *J. Fluid Mech.* **27**: 399–412.

Whitham, G. B. (1974). *Linear and Nonlinear Waves*, Wiley–Interscience, New York.

Whitham, G. B., (1976). Nonlinear effects in edge waves. *J. Fluid Mech.* **74**: 353–368.

Wiegel, R. L. (1960). A presentation of cnoidal wave theory for practical application. *J. Fluid Mech.* **7**: 273–286.

Wiegel, R. L. (1964). *Oceanographical Engineering*, Prentice-Hall, Englewood Cliffs, N.J.

Wilson, B. W., and A. Tørum (1968). The tsunami of the Alaskan earthquake 1964: engineering evaluation. Technical Memo. No. 25 Coastal Engineering Research Center, U.S. Army Corps. of Engineers.

Wirt, L. and T. Higgins (1979). DAM-ATOLL: Ocean Wave Energy Extraction, *Marine Technology Society Meeting*, New Orleans, La., Oct. 11–12.

Wooley, M., and J. Platts (1975). Energy on the crest of a wave. *New Scientist* **66**: 241–243.

Wu, T. Y. (1957). Water waves generated by the translatory and oscillatory surface disturbance. California Institute of Technology, Engineering Division Report No. 85-3.

Yamaguchi, M., and Y. Tsuchiya. (1974). Non-linear effect of waves on wave pressure and wave force on a large cylindrical pile. *Proc. Civil. Eng. Japan* **229**: 41–53 (in Japanese).

Yamamoto, T. (1977). Wave induced instability in seabeds. *Proc. ASCE Specialty Conf.: Coastal Sediments*. Charleston, South Carolina.

Yamamoto, T. (1982). Nonlinear mechanics of ocean wave interaction with sediment beds. *Appl. Ocean Res.* **4**: 99–106.

Yamamoto, T., H. L. Koning, H. Sellmeigher, and E. V. Hijum (1978). On the response of poro-elastic bed to water waves. *J. Fluid. Mech.* **87**: 193–206.

Yeung, R. W. (1975). A hybrid integral equation method for time-harmonic free surface flow. *Proc. 1st Int. Conf. Num. Ship Hydrodyn.* David Taylor Naval Ship Research and Development Center 581–607.

Yeung, R. W. (1982). Numerical methods in free-surface flows. *Ann. Rev. Fluid Mech.* **14**: 395–442.

Yue, D. K. P. (1980). Numerical theory of Stokes wave diffraction at grazing incidence, Sc.D. Thesis, Department of Civil Engineering, Massachusetts Institute of Technology.

Yue, D. K. P., H. S. Chen, and C. C. Mei (1976). Water wave forces on three-dimensional bodies by a hybrid element method. Parsons Laboratory, Dept. of Civil Engineering, Massachusetts Institute of Technology Technical Report 215.

Yue, D. K. P., H. S. Chen, and C. C. Mei (1978). A hybrid element method for diffraction of water waves by three-dimensional bodies. *Int. J. Num. Methods Engineering* **12**: 245–266

Yue, D. K. P., and C. C. Mei (1980). Forward diffraction of Stokes waves by a thin wedge. *J. Fluid Mech.* **99**: 33–52.

Yuen, H. C., and W. E. Ferguson, Jr. (1978a) Relationship between Benjamin-Feir instability and recurrence in the nonlinear Schrödinger equation. *Phys. Fluids* **21**: 1275–1278.

Yuen, H. C., and W. E. Ferguson, Jr. (1978b) Fermi–Pasta–Ulam recurrence in the two space dimensional nonlinear Schödinger equation. *Phys. Fluids* **21**: 2116–18.

Yuen, H. C., and B. M. Lake (1975). Nonlinear deep water waves: Theory and experiment. *Phys. Fluids* **18**: 956–960.

Yuen, H. C., and B. M. Lake (1980). Instabilities of waves on deep water. *Ann. Rev. Fluid Mech.* **12**: 303–334.

Zabusky, N. J. (1968). Solitons and bound states of the time independent Schrödinger equation. *Phys. Rev.* **168**: 124–128.

Zabusky, N. J., and C. J. Galvin (1971). Shallow-water waves, the Korteweg–deVries equation and solitons, *J. Fluid Mech.* **47**: 811–824.

Zabusky, N. J., and M. D. Kruskal (1965). Interaction of solitons in a collisionless plasma and the recurrence of initial states. *Phy. Rev. Lett.* **15**: 240.

Zakharov, V. E. (1968). Stability of periodic waves of finite amplitude on the surface of a deep fluid. *J. Appl. Mech. Tech. Phys.* **2**: 190–194.

Zakharov, V. E., and A. B. Shabat (1972). Exact theory of two-dimensional shelf-focusing and one-dimensional self modulation of waves in nonlinear media. *Sov. Phys. JETP* **34**: 62–69.

Zakharov, V. E. and A. B. Shabat (1973). Interaction between solitons in a stable medium. *Sov. Phys. JETP* **37**: 823–828.

Zenkovich, V. (1967). *Processes of Coastal Development*, Oliver and Boyd, London.

Zienkiewicz, O. C. (1971). *Finite Element Method in Engineering Science*, McGraw-Hill. New York.

Author Index

Aas, B., 106–108, 708
Abbot, M. B., 517, 706
Ablowitz, M. J., 554, 557, 559, 633, 706
Abramowitz, M., 155, 189, 591, 601, 706
Adams, N. K., 391, 706
Angelides, D., 496, 498, 500, 719
Aranha, J. A., 332, 336, 344, 706
Armstrong, J. A., 578, 587, 590, 706
Arthur, R. S., 74, 84, 502, 706
Atkin, R. J., 675, 706

Baddour, R. E., 657, 713
Bagnold, R. A., 434, 706
Bai, K. J., 170, 335, 706
Barcilon, A., 593, 716
Bartholomeuz, E. F., 120, 124, 706
Batchelor, G. K., 3, 367, 706
Battjes, J. A., 467–470, 484, 502, 706
Benjamin, T. B., 542, 605, 620, 622, 623, 706, 707
Benney, D. J., 505, 550, 553, 578, 606, 607, 614, 623, 624, 707
Berger, U., 651, 707
Berkhoff, J. C. W., 87, 707
Bertelsen, J. A., 86, 724
Bessho, M., 320, 327, 707
Bigg, G. R., 162, 167, 707
Bijker, E. W., 434, 709
Biot, M. A., 674, 679, 707
Bishop, R. E. D., viii, 723
Bloembergen, N., 578, 587, 706
Boczar-Karakiewicz, B., 578, 579, 591, 592, 707
Bohm, D., 130, 141, 707
Bolt, B. A., 572, 707
Booij, N., 108, 168, 707

Boussinesq, J., 504, 572, 707
Bowen, A. J., 156, 438, 453, 469, 470, 472–475, 484, 502, 533, 539, 707, 712
Braddock, R. D., 707
Bratu, C., 330, 725
Bremmer, H., 136, 707
Bretherton, F. P., 90, 97, 98, 578, 708
Brevig, P., 665, 667, 669–671, 726
Brevik, I., 106–108, 708
Brink-Kjaer, O., 87, 714
Bryan, G. M., 704, 725
Bryant, P. J., 578, 584, 586, 591, 592, 708
Buchwald, V. T., 194, 708
Budal, K., 346, 351, 357–359, 708, 710
Byatt-Smith, J. G. B., 542, 543, 708

Carlsen, N. A., 413, 714
Carpenter, L. H., 254, 284, 715
Carrier, G. F., 26, 156, 190, 217, 238, 240, 241, 244–246, 251, 468, 523, 527, 528, 530, 708, 721
Carter, T. G., 419, 426, 428, 429, 431, 433, 434, 442, 708, 719
Case, K. M., 404, 708
Chao, Y. Y., 74, 708
Chapman, D. C., 712
Charkrabarti, S. K., 316, 317, 657, 708
Chen, H. S., 170, 235, 236, 238, 335, 336, 708, 719, 727
Chen, T. G., 650, 708
Chu, F. Y. F., 628, 724
Chu, V. C., 606, 607, 616, 630, 632, 636, 641, 647, 708, 709
Clark, A., Jr., 504, 716
Cokelet, E. D., 606, 665, 666, 709, 717
Cole, J. D., 50, 574, 709

729

Collin, R. E., 249, 280, 709
Courant, R., 513, 517, 709
Craine, R. E., 675, 706
Crapper, G. D., 107, 665, 709
Crawford, D. R., 623, 626, 709
Creighton, D. G., 716
Cummins, W. E., 372, 709

Dailey, J. W., 542, 709
Dalrymple, R. A., 480, 482, 483, 502, 709,
   717
Davey, A., 606, 607, 616, 709
Davey, N., 217, 709
Davis, R. E., 533, 712
Davis, S. H., 450, 533, 716
Dean, R. G., 665, 709
De Best, A., 434, 709
Deresiewicz, H., 674, 709
de Vries, G., 545, 715
Dingemans, M., 98, 709
Djordjevic, V. D., 646, 647, 709
Donelan, M., 58, 709
Dore, B. D., 384, 395, 406, 410, 429, 442,
   443, 449, 450, 710
Ducing, J., 578, 587, 706
Dysthe, K. B., 626, 710

Eagleson, P. S., 71, 452, 478, 710, 711
Eckart, C., 154, 710
Erdelyi, A., 44, 710
Euvrard, D., 710
Evans, D. V., 346, 349, 358, 364, 710

Falnes, J., 346, 357, 358, 708, 710
Faltinsen, O. M., 710
Feir, J. E., 605, 620, 622, 636, 707, 710
Felsen, L. B., 140, 710
Fenton, J., 542, 549, 665, 710, 717, 723, 724
Ferguson, I. N. E., Jr., 641, 645, 715, 727
Feshbach, H., 222, 721
Finkelstein, A., 22, 710
Foda, M. A., 170, 205, 646, 648, 650, 674,
   685, 686, 698, 701, 703, 710, 719
Fornberg, B., 559, 711
Fox, M. J. H., 606, 717, 718
Frank, W., 320, 342, 711
French, M. J., 345, 347, 711
Friedrichs, K. O., 111, 513, 517, 533, 709,
   711
Fuchs, R. A., 315, 718

Gallagher, B., 647, 711
Galvin, C. J., 468, 469, 478, 711, 728
Garabedian, P. R., 52, 711

Gardner, C. S., 554, 555, 557, 559, 632, 711
Garrett, C. J. R., 90, 97, 98, 208, 329, 708,
   711
Gaughan, M. K., 470, 715
Geerstma, J., 674, 711
Gelfand, I. M., 557, 558, 711
Gerwick, B. C., Jr., 253, 711
Goda, Y., 236, 469, 470, 578, 591, 711
Goldstein, H., 287, 711
Gradshteyn, I. S., 602, 711
Graham, J. M. R., 254, 711
Grant, W. D., 412–415, 417, 480, 711, 718
Greene, J. M., 554, 711
Greenspan, H. P., 156, 395, 468, 523, 527,
   528, 530, 708, 711
Guiney, D. C., 217, 712
Guza, R. T., 156, 533, 539, 648, 707, 712,
   713

Hagen, G. E., 346, 347, 712
Hall, J. R., Jr., 674, 723
Hammack, J. L., 574, 576, 712
Hanaoka, T., 326, 712
Haren, P., 162, 363, 712
Hashimoto, H., 616, 628, 631, 712
Haskind, M. D., 300, 326, 712
Hasselman, K., 645, 712
Hayashi, T., 253, 254, 258, 262–264, 712
Herbich, J. B., 434, 712
Hesegawa, A., 631, 712
Hibberd, S., 530, 712
Higgens, T., 346, 727
Hijum, E. V., 727
Hilbert, D., 513, 709
Hildebrand, F. B., 65, 712
Hill, M. N., 2, 712
Ho, D. V., 530, 712
Hognestad, E., 253, 711
Holman, R. A., 438, 712
Homma, M., 495, 724
Honji, H., 434, 715
Horikawa, K., 268, 413, 414, 495, 713, 724
Houston, J. R., 179–181, 238, 713
Huang, N. E., 450, 713
Hunt, J. N., 406, 419, 424, 426, 450, 657,
   713
Huntley, D. A., 648, 713
Huthnance, J. M., 450, 713
Hwang, L. S., 234, 530, 531, 572, 713, 725

Ingard, K. U., 259, 713, 721
Inman, D. L., 434, 472, 475, 485, 502, 533,
   707, 712, 713

Ippen, A. T., 236, 254, 263, 275, 571, 713, 719
Irribarren, C. R., 467, 713
Ising, H., 713
Issacs, J. D., 89, 713
Issacson, M. St. Q., viii, 254, 283, 330, 564, 657, 713, 723
Ito, Y., 268, 275, 714

Jacobson, T. S., 476, 479, 484, 714
James, I. D., 502, 714
Jami, A., 335, 344, 710, 716
Jannson, K. G., 345, 714
Jarlan, C. E., 253, 714
Jawson, M. A., 341, 714
Jeffrey, D. C., 345, 710
Jeffreys, B. S., 29, 714
Jeffreys, H., 29, 700, 714
John, F., 284, 296, 342, 379, 382, 714
Johns, B., 395, 419, 420, 424, 426, 450, 713, 714
Johnson, D. W., 451, 714
Johnson, J. W., 89, 714
Johnson, R. S., 560, 561, 563, 714
Jolas, P., 578, 593, 714
Jones, D. S., 164, 714
Jonsson, I. G., 86, 87, 89, 106, 108, 411, 413–415, 476, 479, 481, 484, 485, 714, 724

Kajiura, K., 30, 35, 36, 45, 46, 120, 135, 136, 411–414, 715
Kakutani, T., 561, 715
Kamphuis, J. W., 415, 715
Kaneko, A., 434, 715
Kano, T., 253, 254, 712
Kantorovich, L. V., 210, 715
Karpman, V. I., 554, 715
Kaup, D. J., 563, 706
Kehnemuyi, M., 238, 715
Keller, H. B., 527, 530
Keller, J. B., 60, 715
Keulegan, G. H., 254, 284, 406, 564, 569, 570, 602, 715
King, C. A. M., 451, 715
King, R., 647, 715
Kinoshita, T., 366, 718
Kjeldsen, S. P., 467, 715
Ko, K., 563, 564, 715
Kober, H., 715
Kohlhase, S., 651, 707
Komar, P. P., 470, 502, 715
Koning, H. L., 727
Korteweg, D. J., 545, 715

Kreisel, G., 306, 322, 323, 715
Krook, M., 26, 241, 708
Kroszynski, U. I., 643, 725
Kruskal, M. D., 554, 711, 728
Krylov, V. I., 210, 715
Kuehl, H. H., 563, 564, 634, 715
Kulin, C. R., 571, 713
Kyllingstad, A., 708
Kyozuka, Y., 320, 715

Laitone, E. V., viii, 284, 553, 605, 726
Lake, B. M., 606, 622, 633, 643, 645, 709, 715, 727
Lamb, G. L., Jr., 554, 557, 716
Lamb, H., 342, 716
Lamoure, J., 435, 437–439, 450, 716
Landau, L. D., 18, 559, 716
Lasry, S. J., 169, 725
Lau, J., 432, 593, 716
Lautenbacher, C. C., 234, 716
Lax, P. D., 559, 632, 716
Lean, G. H., 254, 725
LeBlond, P. H., viii, 2, 502, 716
Lebreton, J. C., 716
Lee, C. M., 268, 716
Lee, J. J., 217, 219, 231, 234, 236, 268, 275, 716
Lee, K. L., 674, 724
Lee, Y. K., 572, 713, 720
LeMehauté, B., 511, 719
Lennon, G. P., 503, 717
Lenoir, M., 335, 344, 710, 716
Lepelletier, T. G., 553, 600, 716
Lesser, M. B., 716
Levine, D. A., 530, 715
Levitan, B. M., 557, 558, 711
Li, H., 411, 716
Lifschitz, E. M., 18, 559, 716
Lighthill, M. J., 346, 542, 606, 622, 651, 662, 707, 716
Lin, C. C., 504, 716
Liu, A. K., 450, 482, 716
Liu, P. L. F., 168, 254, 263, 275, 391, 395, 419, 426, 429, 431, 433, 434, 442, 480, 483, 486, 491, 492, 494, 495, 502, 503, 564, 670, 672, 704, 708, 716, 718, 719
Løken, A. E., 371, 710
Long, R. B., 417, 717
Longuet-Higgins, M. S., viii, 58, 90, 98, 140, 158, 159, 161, 162, 365, 417, 419, 420, 426, 427, 442, 443, 449, 450, 453, 462, 465, 472, 474–479, 482–485, 542, 543, 591, 606, 623, 624, 626, 627, 665, 666, 708, 709, 717

Lozano, C. J., 87, 168, 491, 717, 718
Ludwig, D., 74, 86, 718
Luke, J. C., 550, 553, 707
Lunde, J. K., 345, 714
Luneberg, R. K., 63, 66, 74, 718

Ma, Y. C., 719
McCamy, R. C., 315, 718
McCormick, M. E., 345, 719
McKee, W. D., 104, 719
McLaughlin, D. W., 628, 724
McLean, J. W., 626, 719, 724
McLean, S. R., 417
McTigue, D. F., 680, 719
Madsen, O. S., 412–415, 417, 450, 480, 503,
    511, 560, 563, 593, 674, 695, 711, 718,
    722
Maeda, H., 366, 718
Mallory, J. K., 104, 718
Marcuvitz, N., 140, 710
Marganac, A., 716
Martin, D., 710, 719
Martin, D. U., 718
Maruo, H., 325, 366, 718
Maskell, S. J., 378, 718
Masuda, M., 347, 718
Mattioli, F., 234, 718
Mehlum, E., 347, 719
Melville, W. K., 645, 651, 720
Meyer, R., 323, 720
Meyer, R. E., 59, 87, 140, 158, 530, 712,
    718, 720, 724
Miche, R., 469, 529, 530, 720
Miles, J. W., 55, 144, 184, 202, 205, 208,
    217, 225, 268, 563, 564, 605, 651, 720
Miller, G. R., 141, 724
Milne-Thomson, M. N., 124, 720
Miloh, T., 657, 720
Minzoni, A. A., 156, 533, 539, 720
Mitchell, M. M., 571, 713
Miura, R. M., 554, 711, 721
Miyata, M., 217, 708
Molin, B., 657, 663, 664, 721
Mollo-Christensen, E., 645, 721
Momoi, T., 35, 721
Monin, A. S., 443, 721
Moore, D., 450, 721
Moraes, C. deC., 468, 721
Morse, P. M., 222, 259, 721
Munk, W. H., 141, 156, 184, 202, 205, 208,
    225, 452, 467, 470, 647, 720, 721, 723,
    724
Murphy, H. D., 434, 712

Murty, T. S., 572, 721
Mushkelishvilli, N. I., 692, 693, 721
Mynett, A. E., 336, 337, 350, 352, 704, 721
Mysak, L. A., viii, 2, 716

Naheer, E., 571, 721
Nayfeh, A. H., 50, 74, 721
Newell, A. C., 563, 706, 707
Newman, J. N., viii, 318, 323, 326, 327,
    329, 349, 351, 353, 360, 364, 366, 371,
    721, 722
Nichols, R. C., 238, 715
Nielsen, A. H., 651, 722
Nishimura, H., 268, 713
Noda, E. K., 502, 722
Noda, H., 434, 502, 722
Nogales, C., 467, 713
Nordstrom, C. E., 485, 713
Nowacki, H., 665, 722
Noye, B. J., 217, 712

Ogawa, K., 136, 722
Ogilvie, T. F., 120, 346, 372, 376, 722
Olsen, G. B., 467, 715
Oltedal, G., 708
Ono, H., 560, 561, 563, 616, 628, 631, 646,
    712, 722
Osorio, J. D. C., 254, 620, 723, 725
Ostendorf, D. W., 503, 722
Ostrovsky, L. A., 563, 722
Ott, E., 564, 567, 722
Özhan, E., 572, 722
Özsoy, E., 263, 264, 722

Pande, G. N., 674, 722
Pantazarus, N., 491, 717
Papanikolaou, A., 665, 722
Papoulis, A., 41, 722
Parkinson, W. C., 404, 708
Passman, S. L., 680, 719
Peady, G. W., 707
Pearson, C. E., 26, 241, 708
Pelinovskii, E. M., 563, 722
Peregrine, D. H., 59, 104, 107, 505, 512,
    530, 563, 564, 593, 665, 712, 722, 725
Perroud, P. H., 650, 722
Pershan, P. S., 578, 587, 706
Petroni, R. P., 221, 227, 228, 720
Phillips, O. M., viii, 90, 411, 453, 454, 458,
    578, 605, 722
Pian, T. H. H., 169, 725
Pierson, W. J., Jr., 74, 444, 708, 722
Platts, J., 346, 727

Pleass, C. M., 345, 722
Pocinki, L. S., 68, 83–85, 723
Prabhakara Rao, G. V., 723
Prévost, J. H., 675, 680, 723
Price, W. G., viii, 723
Putnam, J. A., 452, 723

Radder, A.C., 167, 723
Raichlen, F., 231, 234, 716
Raman, H., 657, 723
Ramamonjiariosa, A., 645, 721
Rayleigh, Lord, 419, 542, 723
Redekopp, L.G., 646, 647, 709
Riabouchinsky, D., 512, 723
Richart, F.E., Jr., 674, 723
Richey, E.P., 253, 723
Rienecher, M.M., 665, 723
Riley, N., 439, 723
Rindby, T., 345, 714
Risser, J.F., 240, 246, 247, 251, 723
Rockliff, N., 533, 539, 723
Rogers, S.R., 206, 553, 593, 598–600, 723
Roseau, M., 148, 723
Roskes, G.J., 606, 607, 614, 617, 623, 624, 707, 723
Rossettos, J.N., 168, 725
Rungaldier, H., 715
Russell, J.S., 541, 564, 571, 723
Russell, R.C.H., 620, 723
Ryzhik, I.A., 602, 711

Saffman, P.G., 709, 719
Salter, S.H., 346, 351, 710, 723
Sarpkaya, T., viii, 254, 283, 330, 723
Satsuma, J., 633–635, 724
Sauvage de Saint Marc, M.G., 486, 724
Savage, R.P., 560, 593, 718
Schlichting, H., 283, 724
Schonfeld, J.C., 87, 724
Schwartz, L.W., 469, 665
Scott, A.C., 554, 628, 724
Seed, H.B., 674, 724
Segur, H., 554, 557, 558, 574, 576, 706, 712, 724
Sellmeigher, H., 727
Serman, D.D., 296, 336, 372, 721, 724
Seto, H., 335, 724
Shabat, A.B., 559, 606, 628, 631–633, 728
Shaw, R.P., 217, 245, 708
Shen, M.C., 530, 724
Shepard, F.P., 32, 451, 724
Shi-igai, H., 571, 722
Shimano, T., 495, 724

Shirai, M., 253, 254, 712
Simmons, V.P., 472, 475, 707
Simon, M.J., 346, 351, 724
Sitenko, A.G., 148, 724
Skovgaard, O., 86, 89, 106, 107, 476, 479, 484, 714, 724
Sleath, J.F.A., 724
Smit, D.C., 674, 711
Smith, J.D., 417, 724
Smith, R., 87, 104, 648, 715, 722, 724
Sneddon, I.N., 37, 724
Snodgrass, F.E., 140, 156, 721, 724
Sollitt, C.K., 253, 723
Sonu, C.J., 502, 724
Sprinks, T., 87, 724
Srokosz, M.A., 358, 359, 725
Stamnes, J., 347, 719
Stegun, I.A., 155, 189, 591, 601, 706
Stephan, S.C., Jr., 542, 709
Stewart, R.W., 90, 98, 453, 462, 465, 472, 718
Stewartson, K., 607, 709
Stiassnie, M., 643, 665, 725
Stoker, J.J., viii, 26, 53, 376, 513, 517, 521, 523, 530, 725
Stokes, G.G., 156, 504, 605, 725
Stoll, R.D., 704, 725
Strutt, J.W., 723
Stuart, J.T., 439, 725
Su, C.L., 725
Sudan, R.N., 564, 567, 722
Susbielles, G., 330, 725
Svensen, I.A., 725
Symm, G.T., 341, 714

Tait, R.J., 485, 713
Tam, C.K.W., 502, 725
Tam, W.A., 316, 317, 708
Tanaka, H., 366, 718
Tang, C.C., 502, 716
Tappert, F.D., 162, 560, 563, 631, 712, 725
Taylor, A.D., 530, 720
Taylor, J.R.M., 710
Terrett, F.L., 254, 258, 725
Terzaghi, K., 673, 678, 679, 690, 725
Thomas, J. R., 351, 725
Thornton, E. B., 453, 474, 485, 648, 712, 713, 725
Tong, P., 168, 170, 719, 725
Travis, B., 432, 716
Traylor, M. A., 452, 723
Tsuchiya, Y., 657, 727

Tuck, E. O., 120, 124, 127, 162, 167, 217, 234, 530, 531, 712, 713, 720, 725
Tucker, M.J. 647, 726
Turner, J.S., 58, 709, 718
Turpin, F.M., 647, 726
Tørum, A., 727

Ünlüata, Ü., 203, 208, 217, 220, 231, 268, 277, 278, 443, 578, 584, 586, 591, 720, 726
Ursell, F., 156, 346, 378, 504, 718, 726

Van den Driessche, P., 707
Van Dorn, W.G., 726
Van Weele, B., 434, 712
Venkatanarasalah, P., 657, 723
Verruijt, A., 679, 726
Vincent, M.G., 486, 724
Vinje, T., 665, 667, 669–671, 726
Vitale, P., 417, 726
Von Kerczek, C., 666, 726

Wait, J., 136, 726
Wang, C.Y., 439, 726
Wang, J.O., 106, 108, 714
Wanless, H.R., 451, 724
Watanabe, A., 413, 414, 713
Watson, G.N., 39, 726
Weggel, J.R., 726
Wehausen, J.V., viii, 284, 318, 372, 376, 553, 605, 657, 726

West, B.J., viii, 727
Whitman, G.B., 29, 90, 98, 156, 530, 533, 539, 554, 559, 606, 607, 616, 651, 711, 715, 720, 727
Wiegel, R.L., 547, 548, 650, 651, 727
Wilson, B.W., 727
Wimbush, M., 467, 721
Wirt, L., 346, 727
Woods, R.D., 674, 723
Wooley, M., 346, 727
Wu, T.Y., 55, 727

Yaglom, A.M., 443, 721
Yajima, N., 633–635, 724
Yamaguchi, M., 657, 727
Yamamoto, T., 335, 674, 695, 698, 699, 705, 724, 727
Yeung, R.W., 170, 335, 344, 706, 727
Yoshida, K., 136, 320, 715, 722
Yue, D.K.P., 162, 170, 332, 335, 336, 344, 637, 639, 641, 642, 651, 654–657, 706, 727
Yuen, H.C., 606, 622, 633, 641, 645, 709, 715, 718, 719, 727

Zabusky, N.J., 554, 559, 560, 563, 725, 728
Zakharov, V.E., 559, 606, 616, 618, 626, 628, 631–633, 728
Zenkovich, V., 434, 451, 495, 496, 501, 502, 728
Zienkiewicz, O.C., 168, 674, 722, 728

Subject Index

Absorber, 345
  beam-sea, 345, 347
  head-sea, 345, 347, 360
  omnidirectional, 345, 346, 355
  point, 345
Absorption (capture) width, 353, 358, 363, 364
Added mass, 302, 315, 349, 376
  matrix, 303
Addition theorem, 39
Airy:
  equation, 31, 91, 509, 511, 530, 533
  function, 31, 73
  integral, 575
  stress function, 693, 696
  theory, 513–530
Amplification factor, 161, 203, 213, 216, 228, 232, 594
Amplitude spectrum, 47–49, 70
Angle of incidence, 67, 129, 480, 499, 650
Averaged Equations, 453
  of mass conservation, 454
  of momentum conservation, 455
Averaged Lagrangian, 98, 607

Bandwidth, 180, 335, 647
Bay, 199, 202
Beach cusps, 650
Benjamin-Feir instability, see Side-band instability)
Bernoulli equation, 4, 8, 13, 22, 112, 121, 256, 288, 368, 508, 609, 648, 659, 662
Bessel function, 37, 382
Bessho-Newman relations, 327, 329
Biharmonic equation, 693
Biot's theory of poroelasticity, 674, 675–684

Bottom friction, 114, 475, 476, 649
Boundary layer, 8, 9, 55, 384, 386, 396, 402, 403, 407, 421, 436, 446–450, 571
  meniscus, 392
  oscillatory, 387
  poro-elastic, 674, 684–692, 695, 699, 700, 702
  thickness, 386, 404, 564, 685
  turbulent, 384, 411–417
Bound states, 140, 555, 556
Boussinesq equations, 510–512, 540, 549, 550, 560, 573, 577, 578, 593, 597
Boussinesq's theory, 510
Breaker:
  collapsing, 468
  line, 71, 452, 472
  plunging, 468, 669
  spilling, 469
  zone, 452
Breaking, 71, 111, 184, 451, 453, 464, 467, 470, 496, 518, 525, 526, 530, 563, 647
  empirical criterion of, 467
  of standing waves, 467
Breakwater, 183, 193, 215, 221, 233, 236, 268, 437, 463
  hydraulic, 107
  offshore, 485, 486, 492
  perforated, 253, 254, 269
  pile, 263, 269
  protruding, 220–231
  slotted, 254
Bulk modulus, 679

Calculus of variations, 65
Canal, 199, 205
Capillary wave, 2, 18, 35

Capture width, *see* Absorption width
Carrier wave, 628, 645
Cauchy-Poisson problem, 20, 35, 374
Cauchy's theorem, integral, residue, 24, 48, 251
Caustic, 70, 72–74, 79, 102
Center of buoyancy, 292
Center of mass, 285, 287, 291, 298
Center of rotation, 291
Characteristics, 514–520, 522, 523
Circular depth contours, 74–86
Circular cylinder, 312
Cnoidal wave, 502, 543–549, 552
Conformal mapping, 122, 195
Conservation:
　of crests, 63
　of wave action, 62, 97
Consolidation, 690, 704
Constant state, 517
Constraining force, 288
Constraining torque, 291, 295
Creeping flow theory, 442, 443–450
Critical radius, 77
Current:
　induced by breaking, 451–503
　induced by viscosity, 419–450
　longshore, 451, 452, 471
　refraction by, 89–108
　rip, 495, 503
　shear, 100–103
　uniform, 99
　varying, 104

Damping, 113, 114, 276, 384
　effect of air on, 406–411
　of progressive waves, 400, 406
　rate, 385, 388, 394, 395, 410, 416
　of solitry waves, 564–572
　of standing waves, 415
Darcy's law, 674, 680
Depth-averaged equations, 454–458
Depth discontinuity, 116, 130, 140, 159
Diffraction, 59, 86, 108, 193, 282, 300, 318, 321, 326, 342, 487, 489, 490, 657
　and refraction, 86
　of Stokes waves, 650–665
Discharge coefficient, 256, 258
Dispersion, 12, 511, 591
Dispersion relation, 12, 304, 305, 404, 511, 582, 647, 661, 694
　for deep water waves, 12
　for nonlinear waves, 619
　for shallow water waves, 12

Dissipation, 390, 392, 393, 405, 484, 572
Doppler's effect, 99
Drag coefficient, 315, 316
Drift force, 284, 365, 371, 663
Drift moment, 366, 370
Dynamic boundary condition, 5, 93, 285, 296, 385, 408, 409

Eddy viscosity, 9, 413, 443, 449, 485
Edge wave, 154, 156, 532, 538, 539
　subharmonic resonance of, 532–539
Effective stress, 674, 678, 680, 682, 687
　principle, 679
Eigenfrequency, 140, 187
Eigenvalue, 187, 189, 304, 533, 555, 556, 559, 633
Eikonal equation, 62, 64, 66
Elliptic integral, 267, 476, 480, 544, 546, 590
Energy, 16, 49, 100, 119
　conservation, 153, 323, 387, 563
　flux, 16, 17, 49, 63, 136
Envelope, 17, 47, 58, 627, 628, 633, 639
　Gaussian, 47, 239
　soliton, 630
Envelope-hole soliton, 631, 647
Equipartion, 17, 564
　theorem, 416–418
Equivalent linearization, 261, 263, 412
Euler's constant, 279
Euler-Lagrange equation, 65, 66, 172, 332
Eulerian drift (or streaming), 419, 424, 426
Evanescent mode, 88, 306
Evolution equation, 607, 611, 628
Excess momentum flux, 458
Exciting force, 302, 356, 376, 378, 661, 662

Far field, 120, 194, 199, 255, 257, 435
Fermat's principle, 63, 65, 66
Finite element method, 168, 174, 330
Fission of solitons, 559, 560
Floating body, 282–296, 344
　angular momentum of, 290–296
　linearized equations for, 296–300
　linear momentum of, 287–290
　transient motion of, 371–378
Flow separation, 253, 283, 317
Fourier series, 223, 266, 586, 594, 602
Fourier transform, 22, 23, 36, 47, 375, 379, 554, 568, 661, 699, 700
Fredholm alternative, 52, 398, 537, 568
Frequency, 6, 57, 61, 110
　absolute, 95, 96
　intrinsic, 95, 97

Frequency response, 300, 375
Fresnel integrals, 57, 206, 641
Friction coefficient (factor), 261, 414, 475,
    479–483
Functional, 65, 178, 236, 331
Fundamental harmonic, 276–278, 581
Fundamental mode, 188

Gas-dynamic analogy, 512
Gauss theorem, 63, 171, 331, 367, 398, 681
Gelfand-Levitan equation, 557, 558
Generalized displacement, 287
Generalized force, 293–302
Generalized normal, 287, 293, 301
Geometrical optics approximation, 29, 89, 487
Green's formula (theorem), 52, 61, 81, 97,
    112, 154, 172, 308, 309, 331, 340, 342,
    398, 611
Green's function, 209, 211, 308, 309, 331,
    335, 340, 342, 344, 379–383
Green's law, 71, 563, 667, 672
Group velocity, 2, 14–18, 27–29, 48, 52, 58,
    95, 100
  absolute, 96
  intrinsic, 100

Hagen-Cockerell raft, 360
Hankel function, 155, 160, 206, 221, 595
Hankel transform, 37, 38
Harbor, 183, 184
  circular, 217, 219
  of complex geometry, 234
  with coupled basins, 231–234
  entrance, 183, 205, 209
  with protruding breakwater, 220–231
  rectangular, 206–218
Harbor oscillations, 183–252, 268
  nonlinear, 593
Harbor paradox, 205, 215, 278
Harbor response to transient input, 238–247
Harmonic generation, 578–593
Harmonics, 578, 583, 590, 595, 610, 611,
    624, 643
Haskind-Hanaoka relation (theorem), 326,
    328, 349, 354, 356, 378
Head loss, 254, 272, 273
Head-sea incidence, 162
Helmholtz equation, 111, 164–165, 166, 172,
    187–189, 194, 222, 226, 233, 248, 312,
    343, 464
Helmholtz mode, 207, 216, 217, 228, 231,
    232, 245, 275, 276, 278

Hooke's law, 679, 680, 684
Hybrid element method, 169–181, 234, 238,
    284, 330, 335

Impedance, 190, 363, 596
Impulse, 372
Impulsive pressure, 22
Incidence, 70
  glancing, 70, 79
  normal, 71, 116, 193
  oblique, 100, 127, 495
Induced streaming, 422, 442
Inertia coefficient, 315, 316
Inner approximation (solution), 165, 166, 208,
    529
Inner expansion, 196, 199, 202, 209, 211
Integral equation, 223, 225, 270, 340, 558,
    667, 668
Interference, 133, 134, 357
Internal gravity wave, 1
Interpolating function, 168, 173
Inverse scattering theory, 554, 632
Irregular frequency, 341–344

Jacobi symbol, 39, 209
Jordan's Lemma, 24, 380
Jost functions, 149, 150
Joukowski transformation, 195

Karman constant, 411, 484
Kinematic boundary condition, 5, 94, 285,
    385, 408, 409, 454, 455
Kinetic energy, 16
Keulegan-Carpenter number, 254, 284
Kockin's H-function, 311, 312
Korteweg-deVries (KdV) equation, 549, 551,
    554, 560, 561, 573–578, 580, 581, 585,
    646, 647

Lagrangian, 98, 108
Lagrangian coordinates, 445, 667, 669
Lagrangian drift, 419, 424, 426, 427
Lagrangian equations of motion, 443
Laguere polynomial, 155
Laplace equation, 36, 87, 195, 331, 372, 488,
    506, 511, 608, 660, 661
Laplace transform, 21, 23, 39, 50, 381
Leading wave, 30–34, 44, 574
Lee shore, 84, 501
Leibniz rule, 62, 454, 459
Longshore current, 452, 471–485
Long wave, 12, 109–182, 183, 253

Mach stem, 651
Mass transport, 419–450
  in long crested wave, 427
  under partially standing wave, 427
  in progressive wave train, 427
  near small body, 434–439
Mean sea level, 475, 492, 494, 598
Mean-square response, 272
Mechanical energy, 387
Meniscus, 391
Method:
  characteristics, 513–517
  matched asymptotics, 120, 194, 203, 209,
    231, 235, 259
  multiple scales, 20, 50, 397, 535, 550, 606
  stationary phase, 25, 33, 41–43, 324, 370,
    664
  steepest descent, 241
Mild-slope equation, 59, 86–89
Mixture theory, 675
Moment of inertia, 293
Mud line, 674, 694, 697

Natural boundary condition, 170, 172
Navier-Stokes equations, 2, 384–420, 439,
  443, 445
Near field, 120, 122, 124, 195, 255, 595
Nodal line, 188, 189
Non-Helmholtz mode, 212, 217, 275
Nonlinear diffraction, 60
Nonlinear evolution:
  in shallow water, 554–559, 578–593
  of Stokes waves, 632–645
Nonlinear resonance in a bay, 593–600
Numerical method:
  by finite differences, 517
  by finite elements, 234, 330
  by integral equation, 234
  for steep waves, 665–672

Oblique incidence, 100, 127
Optical theorem, 325, 370
Optimum efficiency, 349, 351, 356, 362
Outer approximation (solution), 165, 166, 202,
  208, 528, 684
Outer expansion, 196, 197, 200

Parabolic approximation, 162, 163, 166, 486,
  651
Parallel depth contours, 66
Partial stress tensor, 677, 678
Partial wave, 221, 356
Partial wave expansion, 41, 160, 181, 182

Permanent envelope, 628–631
Permanent long wave, 540
Permeability, 680, 685
Perturbation analysis, 395, 399, 409, 564
Phase lines, 10, 62, 63
Phase mis-match, 583
Plane strain, 692
Poisson's ratio, 680, 685
Pore pressure, 679, 682, 687
Poro-elasticity, 673–705
Porosity, 675, 685
Potential energy, 17, 418
Power absorption devices, 344–364
Power take-off, 345, 363
Pressure working, 390, 393
Principal-valued integral, 251, 382
Propagating mode, 306, 308
Pumping mode, 207, 216

Quality factor, 192
Quarter-wave length mode, 205

Radiation, 300, 318, 319, 326, 342
  condition, 53, 113, 114, 117, 163, 172,
    185, 194, 307, 319, 321, 379, 536, 594,
    659
  damping, 184, 185, 190, 192, 204, 214,
    243, 273, 304, 320, 349, 649
  damping matrix, 303
  potential, 301, 355
  stress, 97, 98, 453, 461, 462, 464–466,
    475, 477, 493
Ray, 63, 75, 83, 452, 497, 498, 501
  approximation (theory), 59–108
  channel, 63, 82
  geometry, 66–70, 74
  separation factor, 64, 81, 83
Recurrence, 591, 593, 636, 639
Refraction, 59–108, 129, 497, 665
Reflection, 135, 139, 140, 431
  coefficient, 119, 128, 130, 132, 259, 323,
    326, 340, 345, 555, 633
Resonance:
  on beach, 649
  in narrow bay, 202–205, 593–600
  in rectangular harbor, 206–219
  subharmonic, 532–539
Resonant:
  interaction, 578, 605
  mode, 213, 229
  peak, 203, 204, 214, 272, 274
  spectrum, 212
Response curve, 203

Restoring force, 296, 302, 374
  matrix, 302
Reynolds number, 216, 258, 275, 459
Reynolds stress, 423, 437, 441, 462, 473, 484
Ridge, 69, 80, 81, 131, 141, 649
Riemann invariant, 514, 516, 519, 522
Ripples, 413, 417, 434
Roughness height, 414
Run-up, 52, 527

Saddle point, 241
Salter's cam (or duck), 337, 346, 347–351
Saturation, 679
Scattering, 109, 199, 201, 253, 282
  by circular cylinder, 312–317
  by circular sill, 159
  by depth discontinuity, 116, 127, 132
  matrix, 118, 146, 147, 150
  by ridge, 130
Schwarz-Christoffel transformation, 125, 200, 217
Schrodinger equation, 55, 148, 489, 641
  cubic, 606, 616, 626, 633, 652
Seabed, 673, 693, 695
Secondary boundary layer, 443
Second order wave force, 661, 664
Sediment, 430, 437, 451
  sorting, 432
  transport, 426
Seepage velocity, 676
Set-down, 472
Set-up, 475
Shadow boundary, 487, 488, 495
Shear current, 100–104
Shear modulus, 680, 685
Shear wave, 686
Shoal, 78
Shoaling, 647, 662
  zone, 452, 463, 471, 477, 491, 492, 501
Side-band disturbances, 606, 621, 623, 626, 628, 641, 645
Side-band instability, 620–628
Simple wave, 518
Slowly varying current, 86–108
Slowly varying depth, 59–86, 135–140, 645–650
Slow modulation, 607
Snell's law, 66, 76, 473
Solitary wave, 541–543, 548, 564, 566, 650
Soliton, 554, 558, 561, 562, 576, 633, 637, 638
Solvability condition, 52, 61, 112, 537, 567, 611, 612, 614, 619

Standing waves, 119, 203
  in circular basin, 188, 189, 401
  nonlinear, 524, 550–553
  in rectangular basin, 187, 188
Steepest descent method, 241
Steep waves, 665–672
Stiffness matrix, 174
Stokes boundary layer, 407, 428, 439, 441, 442
Stokes drift, 419, 425
Stokes problem, 398
Stokes waves, 469, 606, 618, 623, 624, 641, 643, 645, 651
  instability, 620
Stresses in seabed:
  due to progressive wave, 693, 695, 697
  near cylinder, 700, 702
Strouhl number, 263, 275, 384
Sturm-Liouville equation, 113, 148, 161, 304
Subharmonic resonance, 532–539
Superelement, 169, 170, 179, 234, 331–333
Surf beat, 647
Surf parameter, 468, 526, 530
Surf zone, 452, 463, 474, 484, 501, 502, 648
Swell, 1, 9, 533, 645, 646, 648, 649

Terzaghi's equation, 690
Time-average, 389
Tombolo effect, 486, 496
Total stress, 680, 687, 691, 694
Traction, 692
Transient propagation 20–58
Transient response:
  of floating body, 371–378
  in harbor, 238–247
Transient waves, 20–58, 140–146
Transmission coefficient, 119, 128, 130, 132, 150, 259, 260, 263, 267, 322, 323, 325, 326, 340, 345, 555
Trapped modes, 144, 146, 150, 649
Tsunami, 2, 32, 39, 44, 180, 181, 183, 245, 268
  modelling of large, 572–578
Turbulence in surf zone, 484
Turbulent boundary layer, 459
Turbulent fluctuations, 461
Turning points, 72, 147
Two-phase theory, 674–681

Ursell parameter, 504, 552, 574, 575, 588, 591, 593, 620

Variational approximation, 225
Variational formulation, 331
Variational principle, 168, 170, 225, 250
Viscosity, 9, 419, 443, 458
Vortex shedding, 283, 316
Vorticity, 3, 439–442

Wave:
  action, 62, 97, 102
  packet, 47, 239, 243, 246

power, 344
slope, 7, 458, 591, 651, 657
trapping, 69, 70, 80, 109, 157
Wavelength, 5–7, 9, 31, 109, 187, 283, 545
Wavenumber, 10, 27, 61, 128, 212, 216, 607
Wavetrain, 27, 50, 55, 184, 344, 452, 605, 636
Weak formulation, 332, 336
WKB approximation, 59–63, 71, 90, 94, 135–159, 528, 561
Wronskian, 150, 205, 313